Sugarcane Biorefinery, Technology and Perspectives

Sugarcane Biorefinery, Technology and Perspectives

Edited by

FERNANDO SANTOS
Rio Grande do Sul State University,
Porto Alegre, Brazil

SARITA CÂNDIDA RABELO
College of Agricultural Sciences,
São Paulo State University (UNESP/Botucatu Campus),
Botucatu, Brazil

MARIO DE MATOS
Senior Associate, Schaffer/Africaglobal

PAULO EICHLER
Pontifical Catholic University of Rio Grande do Sul,
Porto Alegre, Brazil

uergs Universidade Estadual do Rio Grande do Sul

BIOREFINERY Sustainable Solutions

ELSEVIER

ACADEMIC PRESS
An imprint of Elsevier

Academic Press is an imprint of Elsevier
125 London Wall, London EC2Y 5AS, United Kingdom
525 B Street, Suite 1650, San Diego, CA 92101, United States
50 Hampshire Street, 5th Floor, Cambridge, MA 02139, United States
The Boulevard, Langford Lane, Kidlington, Oxford OX5 1GB, United Kingdom

British Library Cataloguing-in-Publication Data
A catalogue record for this book is available from the British Library

Library of Congress Cataloging-in-Publication Data
A catalog record for this book is available from the Library of Congress

ISBN: 978-0-12-814236-3

For Information on all Academic Press publications
visit our website at https://www.elsevier.com/books-and-journals

Publisher: Charlotte Cockle
Acquisition Editor: Nancy Maragioglio
Editorial Project Manager: Devlin Person
Production Project Manager: Nilesh Kumar Shah
Cover Designer: Mark Rogers

Typeset by MPS Limited, Chennai, India

Contents

List of contributors ix
Preface xiii

1. Sugarcane world scenario **1**
Mario de Matos, Fernando Santos and Paulo Eichler

Introduction 1
Historical itinerary of sugarcane 3
History of the main products of sugarcane 7
World scenario of sugarcane and its main products 8
Socioeconomic and environmental benefits of using ethanol from sugarcane 16
Perspectives 17
References 19

2. By-products of the sugarcane industry **21**
Fernando Santos, Paulo Eichler, Grazielle Machado, Jaqueline De Mattia and Guilherme De Souza

Introduction 21
Sugar and ethanol production process 22
Bagasse and straw 26
Molasses 29
Vinasse 30
Filter cake 31
Yeasts 33
Other by-products of the sugarcane sector 34
Chemical potential of sugarcane residues 36
Future perspectives for sugarcane biorefinery 42
References 45
Further reading 47

3. Sugarcane: biorefinery, technology, and perspectives **49**
Saulo Philipe Sebastião Guerra, Marcelo Scantamburlo Denadai, André Luiz Merthan Saad, Emanuel Rangel Spadim and Marcius Xavier Rodrigues da Costa

Introduction 49
Sugarcane straw bales ash content 53

Water content in sugarcane straw bales 58
References 64
Further reading 64

4. Sugarcane: food production, energy, and environment 67
Evaristo E. de Miranda and Marcelo F. Fonseca

Introduction 67
Sugarcane: the main agricultural product of Brazil 68
Sugarcane: food and energy production 70
Sugarcane in São Paulo State 75
Sugarcane and the environment 79
Conclusion 85
References 86
Further reading 87

5. Sustainability of sugarcane for energy purposes 89
Carla Silva, Rui Pacheco, Danilo Arcentales and Fernando Santos

Introduction 89
Flex-fuel vehicle technologies and well-to-wheels 97
Conclusions 100
Acknowledgments 100
References 100

6. Energy cane breeding 103
Geraldo Veríssimo de Souza Barbosa, João Messias dos Santos, Carlos Assis Diniz, Danilo Eduardo Cursi and Hermann Paulo Hoffmann

Introduction 103
Sugarcane 104
Energy cane 108
References 115

7. Energy from sugarcane 117
Luís Augusto Barbosa Cortez, Ricardo Baldassin, Jr. and Eduardo de Almeida

Sugarcane: main Brazilian energy crop 118
Residues energy from sugar/ethanol mills 122
Main conversion routes for sugarcane bioenergy 130
Energy cane: an opportunity to improve agricultural productivity 135
The future markets of sugarcane bioenergy 135

Final comments 136
References 137
Further reading 139

8. **Chemical and energy potential of sugarcane** **141**
Sarita Cândida Rabelo, Lívia Beatriz Brenelli de Paiva, Thaynara Coradini Pin, Luisa Fernanda Rios Pinto, Laura Plazas Tovar and Pedro Yoritomo Souza Nakasu

Introduction 141
Energy potential of sugarcane: technology platforms of energy production based on energy form 142
Chemical potential of sugarcane 148
Conclusion 159
References 159

9. **Sugarcane for renewable chemistry** **165**
Sílvio Vaz Jr.

Introduction 165
Sugarcane as a raw material for renewable chemistry 168
Perspectives for sugarcane bioproducts 170
Challenges involved 175
Conclusion 176
References 177

10. **Use of sugarcane vinasse to biogas, bioenergy, and biofertilizer production** **179**
Anderson Carlos Marafon, Karina Ribeiro Salomon, Eduardo Lucena Cavalcante Amorim and Fernanda Santana Peiter

Sugarcane crop and bioethanol production in Brazil 179
Characterization and chemical composition of sugarcane vinasse 180
Alternatives for the use of sugarcane vinasse 182
Perspectives of the vinasse sugarcane use in Brazil 191
References 193
Further reading 194

11. **Production of second-generation ethanol from sugarcane** **195**
Fernando Santos, Paulo Eichler, José Humberto de Queiroz and Fernando Gomes

Structural organization of lignocellulosic biomass 197

Recalcitrance of lignocellulosic biomass 200
Production of second-generation ethanol from sugarcane 202
Pretreatments 204
Types of pretreatments 206
Hydrolysis 212
Fermentation 217
Inclusion of second-generation ethanol production annex to standard distillery 221
References 223
Further reading 227

12. Biopolymers of sugarcane 229

Thalita Mendonça de Resende and Marcelo Moreira da Costa

Introduction 229
Sugarcane 231
Biopolymer 232
Sugarcane bagasse 233
Cellulose 236
Perspectives and challenges for the production of biopolymer from sugarcane 249
References 251
Further reading 254

13. Bioelectricity of sugarcane: a case study from Brazil and perspectives 255

Zilmar José de Souza

Introduction 256
What is bioelectricity? 256
The electric matrix and power installed by bioelectricity 257
São Paulo: responsible for 44% of bioelectricity in 2018 261
Productivity in the generation of sugarcane bioelectricity 263
One point of attention: the decontracting of bioelectricity in the coming years 269
The potential of bioelectricity yet to enjoy 272
Expansion of biomass is far from indicated in the 10-year plan in Brazil 272
The potential of bioelectricity and the RenovaBio 273
An agenda for bioelectricity 275
References 278
Further reading 279

Index *281*

List of contributors

Eduardo Lucena Cavalcante Amorim
Technology Center, Federal University of Alagoas, Maceió, Alagoas

Danilo Arcentales
Dom Luiz Institute (IDL), Faculty of Sciences, University of Lisbon, Lisbon, Portugal; Faculty of Engineering in Earth Sciences, Higher Polytechnic School of Coastline (ESPOL), Guayaquil, Ecuador

Luís Augusto Barbosa Cortez
Interdisciplinary Center for Energy Planning-NIPE, University of Campinas-UNICAMP, Campinas, Brazil

Danilo Eduardo Cursi
Sugarcane Breeding Program, RIDESA/UFSCar, Araras, Brazil

Marcelo Moreira da Costa
Federal University of Viçosa, Viçosa, Brazil

Marcius Xavier Rodrigues da Costa
São Paulo State Scientific Police, São Paulo, Brazil

Eduardo de Almeida
School of Chemical Engineering-FEQ, University of Campinas-UNICAMP, Campinas, Brazil

Mario de Matos
Senior Associate, Schaffer/Africaglobal

Jaqueline De Mattia
Rio Grande do Sul State University, Porto Alegre, Brazil

Evaristo E. de Miranda
Brazilian Agricultural Research Corporation (Embrapa), Brasilia, Brazil

Marcelo Scantamburlo Denadai
Botucatu College of Technology, Botucatu, Brazil

Lívia Beatriz Brenelli de Paiva
Interdisciplinary Center of Energy Planning, University of Campinas-UNICAMP, Campinas, Brazil

José Humberto de Queiroz
Department of Biochemistry and Molecular Biology, Federal University of Viçosa, Viçosa, Brazil

Thalita Mendonça de Resende
Adfert Additives Industry and Trade, Uberlandia, Brazil

Guilherme De Souza
State Secretariat of Environment and Infrastructure, Porto Alegre, Brazil

Zilmar José de Souza
FGV, Rio de Janeiro, Brazil, Sugarcane Industry Union (UNICA), São Paulo, Brazil

Geraldo Veríssimo de Souza Barbosa
Federal University of Alagoas, Maceió, Brazil; Sugarcane Breeding Program, RIDESA/UFAL, Campinas, Brazil

Carlos Assis Diniz
Sugarcane Breeding Program, RIDESA/UFAL, Campinas, Brazil

João Messias dos Santos
Federal University of Alagoas, Maceió, Brazil; Sugarcane Breeding Program, RIDESA/UFAL, Campinas, Brazil

Paulo Eichler
Pontifical Catholic University of Rio Grande do Sul, Porto Alegre, Brazil

Marcelo F. Fonseca
Brazilian Agricultural Research Corporation (Embrapa), Brasilia, Brazil

Fernando Gomes
Federal Rural University of Rio de Janeiro, Rio de Janeiro, Brazil

Saulo Philipe Sebastião Guerra
Paulista State University Julio de Mesquita Filho, São Paulo, Brazil

Hermann Paulo Hoffmann
Federal University of São Carlos, São Carlos, Brazil; Sugarcane Breeding Program, RIDESA/UFSCar, Araras, Brazil

Ricardo Baldassin Jr.
Agronomic Institute – IAC, Campinas, Brazil

Grazielle Machado
Pontifical Catholic University of Rio Grande do Sul, Porto Alegre, Brazil

Anderson Carlos Marafon
Embrapa Coastal Boards, Rio Largo Research and Development Unit, Rio Largo, Brazil

Pedro Yoritomo Souza Nakasu
Faculty of Chemical Engineering, University of Campinas-UNICAMP, Campinas, Brazil

Rui Pacheco
Institute of Bioengineering and Biosciences, Higher Technical Institute, University of Lisbon, Lisbon, Portugal

Fernanda Santana Peiter
São Carlos School of Engineering, University of São Paulo, São Carlos, Brazil

Thaynara Coradini Pin
Faculty of Chemical Engineering, University of Campinas-UNICAMP, Campinas, Brazil

Luisa Fernanda Rios Pinto
Faculty of Chemical Engineering, University of Campinas-UNICAMP, Campinas, Brazil

Sarita Cândida Rabelo
College of Agricultural Sciences, São Paulo State University (UNESP/Botucatu Campus), Botucatu, Brazil

André Luiz Merthan Saad
Paulista State University Julio de Mesquita Filho, São Paulo, Brazil

Karina Ribeiro Salomon
Technology Center, Federal University of Alagoas, Maceió, Brazil

Fernando Santos
Rio Grande do Sul State University, Porto Alegre, Brazil

Carla Silva
Dom Luiz Institute (IDL), Faculty of Sciences, University of Lisbon, Lisbon, Portugal

Emanuel Rangel Spadim
Paulista University, São Paulo, Brazil

Laura Plazas Tovar
Institute of Environmental, Chemical and Pharmaceutical Sciences, Department of Chemical Engineering, Federal University of São Paulo (UNIFESP/Diadema Campus), Diadema, Brazil

Sílvio Vaz Jr.
Brazilian Agricultural Research Corporation – National Research Center for Agroenergy (Embrapa Agroenergy), Brasilia, Brazil

Preface

The concept of sustainability, much sought in recent years, is linked to the partial and gradual replacement of exhaustible sources by renewable energy sources. This sustainability path fits the concept of biorefinery, which aims to convert biomass into biofuels as well as bioenergy and bioproducts of commercial interest, in order to find a solution that combines economic viability with environmentally friendly production.

In the context of the most cultivated agroenergy biomass in the world, sugarcane is one of the most ecological and sustainable, being fundamental for the economy. In addition, it is important for the reduction of carbon dioxide emissions, quality of life in urban centers, and global improvement of the environment. Though its origin is unknown, many believe that sugarcane was first cultivated in New Guinea, where it was considered a wild and ornamental plant.

Sugarcane consists of the stalk—formed by several culms and made up of fibers (cellulose, hemicelluloses, and lignin), sugars (sucrose, fructose, and glucose), and leaves—composed of green leaves and dried leaves/sheaths, also called straw. With regard to sugarcane mass composition, the three main components (straw, bagasse, and juice) are practically in the same proportions and together represent 43% of the plant mass content, and the remaining 57% are represented by water. In energy terms, the same proportionality is verified among the sugarcane constituents, giving about 7400 MJ of energy content.

Straw and bagasse are mostly used for a number of agronomic benefits, for example, weed control, moisture retention, animal feed, and improving physical, chemical, and biological properties of soil. However, due to the significant amount that is generated, its composition and, consequently, the chemical and energetic potential of the vegetable fiber, it is possible to use part of these materials in a more noble use, both chemically and energetically, without any damage to agronomic benefits through processes involving the different biorefinery platforms—chemical route, biochemical route, and thermochemical route.

The biorefinery concept is gaining more prominence with great investment potential and has been identified as one of the most promising routes for the creation of new industries. This vision of technological development will have great relevance and importance regarding the evolution of the world agricultural and industrial sectors, directly reflecting on the generation of jobs and income in different segments of society.

Thinking about this optimistic scenario, which involves the integral and sustainable use of sugarcane through the different biorefinery platforms, we decided to publish the book *Sugarcane Biorefinery, Technology, and Perspectives*. The book is divided into 13 chapters and is prepared by some of the leading authorities on each of the topics addressed. The work is an important contribution to the dissemination of knowledge and the consolidation of sugarcane biorefinery in the world. The future of biorefinery has begun!

Good reading!

Fernando Santos

CHAPTER 1

Sugarcane world scenario

Mario de Matos, Fernando Santos and Paulo Eichler

Contents

Introduction 1
Historical itinerary of sugarcane 3
History of the main products of sugarcane 7
World scenario of sugarcane and its main products 8
Approximate values for the 2016/17 harvest 11
Approximate values for the 2017/18 harvest 12
The sugar and ethanol sectors in Africa 12
The sugar, ethanol, and energy sectors in Brazil 14
Socioeconomic and environmental benefits of using ethanol from sugarcane 16
Perspectives 17
References 19

Introduction

Sugarcane, now known worldwide for its high productivity, participation in high technology processes, high-quality raw material, and, especially for the potential of sugar and ethanol production, has undergone several modifications throughout history. It is a semiperennial plant of the grass family, originating from hot to tropical temperate regions of Asia, especially India. Its main characteristic is the production of sugars (mainly sucrose, glucose, and fructose), concentrating on its culm. The aerial part of the plant consists of stalk, green leaves, and dry leaves. The upper part of the plant has a higher humidity and the lower part has a lower humidity, with dry (or dead) leaves (Fig. 1.1). The sugarcane presents a C4 photosynthetic cycle, with leaves in the form of spears, sprouting in stalks, and abundant tillering in the initial phase of development (Santos et al., 2013). The plant has approximately 57% of water in its mass composition, the remainder being divided between straw, bagasse, and sugar.

This culture has been shown, throughout history, as an important product, widely marketed, and of great interest during several periods

Sugarcane Biorefinery, Technology and Perspectives
DOI: https://doi.org/10.1016/B978-0-12-814236-3.00001-9

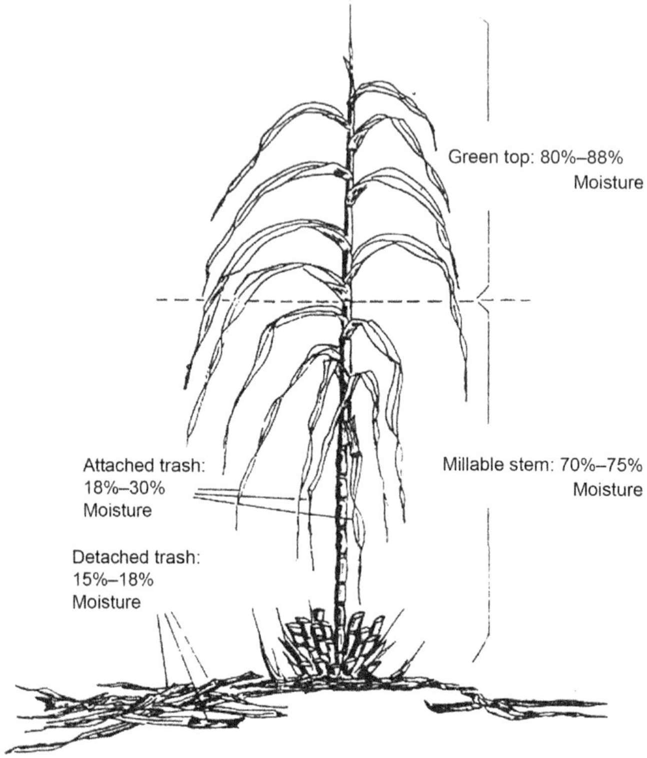

Figure 1.1 Sugarcane plant illustration. *Alexander AG (1985) The energy cane alternative. Amsterdam: Elsevier Science Publishers B.V., 509 pp.*

of history. Since sugar cane was used for sugar production it was responsible for the creation of new commercial routes in the old Mediterranean until its establishment in the Americas, when it is used for ethanol production.

In this chapter, we show the history of sugarcane, from its probable origin in Southeast Asia to the present day where sugarcane is present on all continents; being one of the most important crops in the world, generating hundreds of thousands of direct and indirect jobs. Due to the importance of its main products (sugar, ethanol, and energy), sugarcane has become a crucial source of income and development for several tropical countries like Brazil. Thus it is important to show the development of these countries with the use of sugarcane, so that is possible to know how impacting it is for humanity.

Historical itinerary of sugarcane

For a few years, India and Papua New Guinea disputed the origin of this plant as a native of these countries. The fact that some of the spontaneous species of *Saccharum spontaneum* and *Saccharum robostun* still being found in Papua New Guinea ended up giving reason for choosing this country that is now officially considered their homeland. The domestication of this species would give *Saccharum barberi* widespread in India and *Saccharum sinensis* implanted in China after many years. Thus Indians would have been the first to extract the sugarcane juice to produce raw sugar around 500 BCE. At that time, Emperor Darius, upon arriving in India, observed that there were "plants that produced honey without the need of bees" (Santos et al., 2018).

The name *Saccharum* attributed to sugarcane has its origin in the word of Transkrit, Karkara of India that was later called Sakkar or Sukkar by the Arabs after having introduced sugarcane from India to the Middle East through Persia.

A few centuries later, after being brought from Persia to Europe by Alexander the Great, the Romans would call it much later *Saccharum* name with which it would be adopted and later attributed by Linnaeus in 1753 for its botanical classification that lasts until our days.

However, sugarcane has been widely cultivated, in addition to Egypt, in Spain (where there is still the oldest known sugarcane mill in the world) which uses stone grinders, and also introduced and cultivated in other parts of the Mediterranean, such as Venice, from which later to transit in the 15th and 6th centuries for the Madeira Islands and the Canary Islands. Notoriously, from the Canary Islands it was introduced in the New World to Hispaniola Island (Dominican Republic) in CE 1493 and to Brazil in CE 1532.

In the mid-14th century, there are records that the production of sugarcane in the Mediterranean region (Crete, Greece, North Africa, and Cyprus) was related to the use of slave labor. Like the other agricultural productive means, the economy of the region was based on the slave work as productive force. At this time, the regions around the Mediterranean Sea had large trades and were considered developed for the time, influencing European countries like Portugal and Italy. Later, these methods of production and management of sugarcane were transferred to Brazil in the colonial period (Schwartz, 1988).

After the takeover of Constantinople, there was a monopoly on the production and sale of sugar. To end the monopoly, the Portuguese crown decides, as an alternative, to stimulate the production of sugarcane in its colonies of the American continent. Favorable tropical climate, fertile soil, and abundance of water pointed to Brazil as a land of favorable, even ideal, conditions for the cultivation of *Saccharum officinarum*, which originally came from the South Pacific and India (Le Couter and Burreson, 2006). Brazil would then affirm at the beginning as the place of preference for excellence of this culture that quickly expanded and became the main export wealth of the, then, Portuguese Empire, constituting a monopoly that would only be competed with the Dutch in the 17th century after its introduction in the Caribbean. Fig. 1.2 illustrates the passage of sugar culture around the world, starting from its origins in New Guinea, through India, the Mediterranean until its arrival in Brazil, around CE 1500.

In addition to the strong trade competition in the Mediterranean region, Portugal saw the need to effectively occupy the colonies of the Americas, which necessarily required the development of productive activities to justify the investments of the Portuguese crown. In this way, sugarcane production, as in many other countries, has been especially rooted in the history of Brazilian territory and in its economy since the 16th century. Sugarcane was a valuable Portuguese settlement that

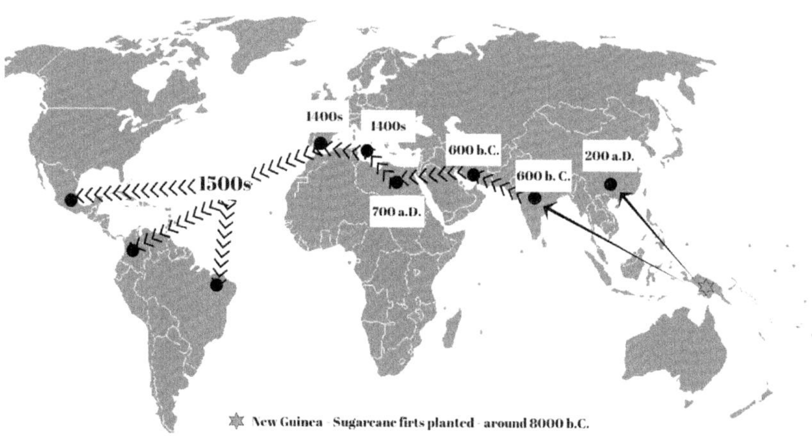

Figure 1.2 The path of sugarcane culture: from its beginning to its arrival in America.

enabled the mercantilist colonizing project of Portugal. Unfortunately, in addition to the devastation of forests, the colonial sugar economy also relied on the production model adopted by the Europeans in the new world. This model was based on the tripod landowner, that is, the landscapes dominated by sugarcane plantations were based on large land usage, monoculture, and slave labor (Prado, 2006). In this way, however good, the trade involved in sugarcane and its products brought social impacts in Brazil that are still felt today.

In Brazil, sugar production was established between CE 1530 and 1540, with the formation of small mills, of the "trapiche" type, moved by oxen and horses or hydraulic force (Fig. 1.3). Initially, as there was not sufficient slave labor, indigenous workers were used, who were the native population. As the mills grew, there was a need to increase the labor force, with the transition to slave labor. In this way, thousands of Africans were taken to Brazil to work in the mills. The first merchants,

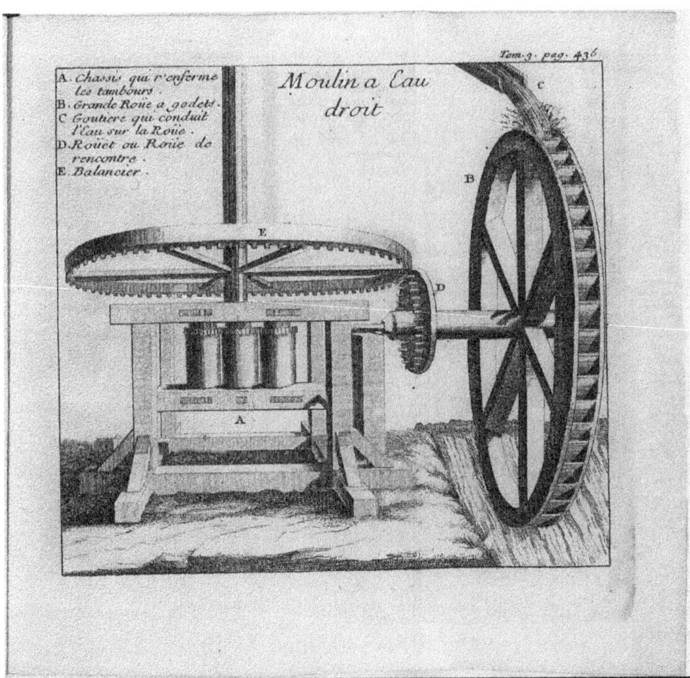

Figure 1.3 Hydropowered mill described by Jean Baptiste on his voyage in India between 1693 and 1706. *Labat, J.B., 1742. Nouveau voyage aux isles de l'Amerique. Paris.*

members of Fernando de Noronha's group, during trips and stops on Madeira Island, Azores, and São Tomé, negotiated the first sugarcane plant cultivated in Brazil. In order to implant the sugar culture in Brazil, the land was divided into lots and donated to nobles, according to the wishes of the Portuguese crown. In 1516 the Portuguese monarch D. Manuel I, the *Venturoso*, decreed that everyone from Portugal to Brazil should bring instruction and equipment for large-scale sugarcane production. To this end, large amounts of money were traded on loans from Portugal and Holland to the land donors who were to produce the cane (Novinsky et al., 2016).

Cane growing successively passed from the Spanish Island to Cuba in 1760 as soon as this island is occupied by England where it also knows a great expansion favored by the demand of the sugar of the Americas in Europe. With the revolution in Haiti in 1795, a few years later, and destruction of cane plantations in Spanish territories, the French installed the culture in Louisiana in the late 17th century, thus initiating the American sugar industry.

The growth of sugarcane culture in the New World and in other continents has also seen different advances in the area of breeding after the creation of the noble variety POJ 2878 in Indonesia, which for many years has been the genetic reference in the hybridization crosses of many breeders from different research centers and varietal selection from several countries. Also going through the creation of many varieties of the Coimbatore Research Institute in India in 1912. The introduction of new varieties from the Coimbatore Research Institute also contribute to the growth of the sugar industry worldwide namely by using fuzz from Coimbatore to selected the well known NCo varieties from Mount Edgecombe in South Africa.

The main contribution to the enormous success of the growth and expansion of this culture goes mainly to the genetic valorization and introduction of irrigation techniques, where information about the first *Java* productions in 1840 from 2000 kg of cane per hectare is increased to 10 tons in 1910 and 20 tons in 1940.

Also, in the early 20th century, with the increased production of internal combustion vehicles, fuel ethanol began to become extremely interesting. As a result, Brazil, which already had large sugarcane production, created incentives for the production of ethanol from the crop, which helped solidify sugarcane cultivation.

History of the main products of sugarcane

Both sugar and cane culture have a very similar history. Sugarcane extraction takes place approximately 8000 BCE in Southeast Asia. Approximately 100–500 BCE in India, a manufacturing process is discovered that drives the use and commercialization of sugar: the transformation of sugarcane juice into sugar crystals. Thus it is possible to commercialize the product over great distances, because it improves its conservation and reduces the volume of transport and also the way of transport. With the progress of commercialization, China is soon also interested in the practice of sugar production, where reports (between CE 600 and 700) describe Chinese visits to India to learn the processes of sugar production. By the 7th century, it was reported that one of the earliest documented plantations of sugarcane in China, brought by the help of Buddhist monks in special missions of Chinese emperors (Sen, 2003).

For some time in the Middle East, the use of cane sugar for medicinal purposes (around the 1st century) was also known, between Greeks and Romans. With increasing consumption of sugar by European countries via Mediterranean routes, their price and importance increased; leading countries such as Cyprus, Crete, Zanzibar, and Greece to also have sugarcane production to export sugar, around the 10th century. At the time of the Crusades, there were also reports from the "Holy Land" of caravans carrying what was called "sweet salt."

In Brazil, sugar was the first product to be exported in large quantities. Its great demand begins in the colonial period in CE 1520, having great importance for the development of the country. During the decades of 1530–40, the production of sugar was characterized by small sugar mills (Fig. 1.3). At the beginning of the first mills, the main product was what we now call brown sugar, a less-refined sugar that was the main product of sugarcane at the time. Despite being the largest Brazilian economy in colonial times, this industry is not exhausted during this period, and its effects can be seen throughout the 17th, 18th, 19th, and 20th centuries, entering the tropical territory of the continents of America and Africa. In the mid-19th century, due to the increase of sugar beet in the international market, the sugarcane crop suffered a financial crisis. In addition to competition, sugarcane farming was extremely dependent on slave labor and, due to the extinction of African traffic and the emergence of emancipationist laws, at the end of the 19th century, the main colonial crops were in crisis. Table 1.1 shows the occupation of slaves at the end of the

Table 1.1 Occupation of slaves in Brazil between 1870 and 1887.

Occupation	Frequency	Percentual
Craftsman	22	3.9
Domestic	26	4.7
Carriage driver	22	3.9
Nurse	4	0.8
Foreman	2	0.4
Worker	8	1.4
Herdsman	4	0.8
Sea worker	10	1.8
Agriculture	459	82.3
Total	557	100

Source: Adapted from Filho, W.F., 2006. Crossroads of Freedom (in Portuguese: Encruzilhadas da Liberdade). Unicamp Publishing, São Paulo, Brazil.

19th century in Brazil. It is noted that more than 80% of the slaves had agriculture as occupation, mainly the plantations of sugarcane.

The fermentation of sugar for alcohol production dates back more than 9000 years, by the Chinese, who used the technique to make alcoholic beverages. The technique of fermentation was passed over the years, being used in several drinks in history as beer, wine, and "sake." By the 9th century, an Arab chemist, Al-Kindi, described the distillation of wine. From there, the distillation is used in several other drinks, with reports from the 12th century until the 14th century. Only in the late 18th century, a German-Russian chemist, Johann Tobias, obtained pure ethanol from chemical synthesis. In the following years, with advances in the chemical area, it became possible to use alcohol on a larger scale, where in 1840 in the United States, ethanol was used in lanterns. Then, in the early 1990s, ethanol was used in fuel-injected engines, primarily in early Ford models such as the Ford Model T, which could use both gasoline and ethanol. Then Brazil would follow the same US model, using cars with ethanol combustion engines, where the fuel was produced from sugarcane. Alcohol development and incentive programs, such as the pro-alcohol program in Brazil, made possible the great growth in sugarcane cultivation, as well as an increase in ethanol production, making the country one of the largest producers of ethanol to date (Santos et al., 2018).

World scenario of sugarcane and its main products

From a historical appreciation of the past of culture, we now turn to a comparative exercise of world sugarcane productions in the fourth

Table 1.2 Sugarcane production, in thousands of tons of sugarcane (TS), harvested area, in hectares (Ha), and sugarcane productivity, in tons per hectare (TH), between 1961 and 2014 for the four largest producers in the world.

Year	1961	1974	2000	2014
		(TS) Tons of sugarcane ($\times 10^3$)		
Brazil	59,377	95,624	3,27,705	7,36,109
India	1,10,001	1,40,805	2,99,325	3,52,142
China	12,416	25,578	69,299	1,26,153
Thailand	2000	13,339	54,052	1,03,697
World	4,47,977	6,48,516	12,558,88	18,842,46
		(Ha) Hectares harvested		
Brazil	13,666,40	20,566,91	48,459,90	104,196,78
India	24,13,000	27,52,100	42,19,700	50,12,000
China	2,92,600	5,96,800	11,885,89	17,683,88
Thailand	63,000	2,58,609	8,93,353	13,53,025
World	89,118,79	11,932,313	19,397,902	27,124,723
		(TH) Tons of sugarcane per hectares		
Brazil	43.4	46.5	67.6	70.6
India	45.6	51.1	70.9	70.2
China	42.4	42.8	58.3	58.3
Thailand	31.7	51.6	60.5	76.6

Source: Adapted from FAOSTAT, 2019. Food and Agriculture Organization of the United Nations. FAOSTAT Statistics Database. FAO, Rome.

countries responsible for 70% of the world's productions (Brazil, India, China, and Thailand), where Brazil accounts for about 30% of production (Table 1.2 and Fig. 1.4). These data have simply informative, economical, and quantitative value, since different ecological systems of production, crop water regimes, extensivism (LISA), intensivism (HISA), etc., are compared. In Fig. 1.5, it is possible to observe the world production of sugarcane from the 1960s until last recorded. The steady increase in production is evident, as sugarcane products are increasingly inserted into the world economy (Table 1.3).

These figures indicate a 21.3% global consumption increase forecast in the indicated decade (Fig. 1.3). If we consider that 80% of sugar is produced from sugarcane, we can say that if this proportion remains linear, there will be a verified surplus of 36 million tons of sugar. This corresponds to 316 million tons of sugarcane, which at an average of 9 tons of sugarcane per ton of sugar and an average production yield of 70 t/ha of sugarcane,

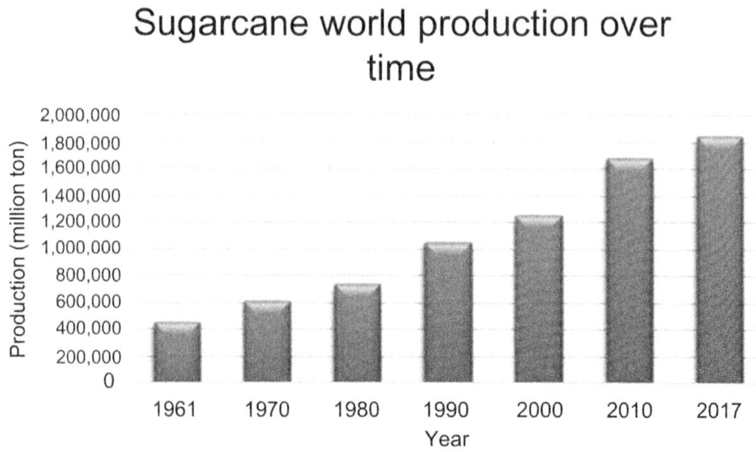

Figure 1.4 Evolution of world production for the four major sugarcane-producing countries, in thousands of tons of sugarcane, from 1961 to 2014. *Adapted from FAOSTAT, 2019. Food and Agriculture Organization of the United Nations. FAOSTAT Statistics Database. FAO, Rome.*

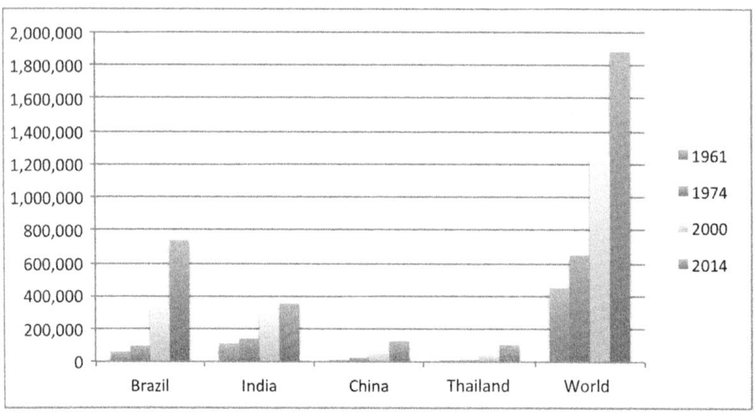

Figure 1.5 World sugarcane production through time. *Adapted from FAOSTAT, 2019. Food and Agriculture Organization of the United Nations. FAOSTAT Statistics Database. FAO, Rome.*

will represent an increase of 4.5 million hectares above of the current sugarcane area in Brazil. This provides for a plantation of 450,000 ha per year to be planted between 2015 and 2025. With 667,000 ha planted between 2013/15 and 2017 according to the same statistics, we will have 3.5 million hectares to be planted between 2017 and 2025, with Brazil already accounting for more than a third of world sugar production.

Table 1.3 Estimates of sugar production increases (in million tons) from 2015 to 2025.

Year	Thousand tons	% Increase
2015	169	
2016	173	2.4
2017	175	1.2
2018	178	1.7
2019	182	2.2
2020	186	2.2
2021	190	2.2
2022	194	2.1
2023	197	1.5
2024	201	2.0
2025	205	2.0

Source: Adapted from OECD/FAO, 2018. OECD-FAO Agricultural Outlook 2018–2027. OECD Publishing Paris/FAO, Rome.

Approximate values for the 2016/17 harvest

The volume of the crop in Brazil (2016/17) was significantly lower than in the previous year, although the percentage of sugarcane destined for sugar production was higher in this harvest—46.4% in 2016/17 compared to 40% of the previous harvest.

In India, sugar production reached 12.9 million tonnes at the end of January 2017, 10% below the previous harvest.

Sugar production in China in the 2016/17 harvest was estimated to be about 9.9 million tonnes, or 13.8% above production in the previous year.

In Thailand, the 2016/17 crop began 11 days later than the previous year. Official Thai statistics revealed on February 8, 2017, that 46.82 million tonnes had already been produced, a figure significantly lower than the 2015/16 crop which at the same time announced a production of 5.43 million tonnes. It is believed that this fall in production was mainly due to the worst drought in recent years, as seen in the previous year.

In sub-Sahelian Africa, the estimation for this harvest was 7.58 million tons, representing an increase of 223,000 tons due in particular:

- the positive progress noted in this harvest in the Grand Expansion Plan of Ethiopia and
- to the gradual improvement of production verified in Southern Africa, where the climate was more stable after 2 years of drought.

However, a deficit of 2.94 million tonnes of sugar due to fundamental problems still exists due to:

- tighter sugarcane production bases and persistent endemic global sugar deficits;
- indices of critical levels of stocks for consumption;
- lack of production in India and import possibilities;
- unpredictable USD exchange rate fluctuations;
- stronger equity markets; and
- higher crude oil prices.

Approximate values for the 2017/18 harvest

According to data published by the USDA for the 2017/18 crop worldwide, assured growth in the production and sale of world sugar was expected to increase by a further 9 million tons to reach a total of 180 million tons due mainly to the increase of sugarcane production in Brazil, China, European Union, and Thailand.

Despite this increase in production, world reserves will fall by 2%, or 8 million tonnes. Production in Brazil was expected to grow by about 500,000 tons to reach a total of 39.7 million tons, due to a more favorable climate and a 1% increase in planted area. Exports will also increase by 500,000 tonnes to reach 28.5 million tonnes. Thus there will be an increase in sugar reserves and consumption will remain stable.

Production in China grew by 1 million tonnes to a total of 10.5 million tonnes. The country's sugar reserves will continue to be reduced. An investigation into imports is currently under way and may have an impact on price policy.

In India, there was a trend expected in the recovery of 18% in production, being 25.8 million tons, due mainly to the increase in crop areas and higher yields of the crop.

In the case of Thailand, an increase of 1.2 million tonnes is expected to reach a total of 11.2 million tonnes, due mainly to improved agricultural productivity following a 2-year drought recovery. Thailand's exports could increase from 400,000 tonnes to a total of 8.4 million tonnes.

The sugar and ethanol sectors in Africa

The African continent deserves particular attention given its great sugar-energy growth in recent decades. The link between sugarcane production and the use of its molasses by-products for the production of

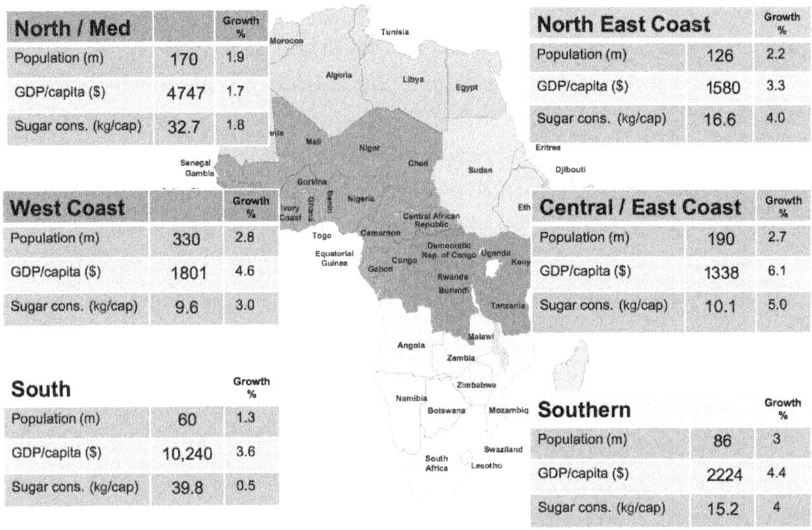

Figure 1.6 Sugar consumption in kg/capita in the different economic regions of North and sub-Saharan Africa. *SCC, 2019. Soft Commodity Consulting. UK. Available from: <www.softcommodityconsulting.com>.*

first-generation ethanol and the bagasse for cogeneration of electricity are current practices of sugarcane producers in sub-Saharan African countries where prices of imported fossil fuels and energy (mostly thermal) are very high. This also includes countries producing and exporting crude oil such as Nigeria. In general, prices in Africa are twice (or more) than those practiced in the United States (Fig. 1.6).

The transport sector is practically almost entirely dependent on petroleum products, with few existing support and protection policies being in place to promote the production and use of alternative biofuels. There is a worrying prospect of a nearly double global requirement in 2040 with projected consumption of crude oil at about 2.2 million barrels per day, 55% of which is for gasoline and 40% for diesel.

Although ethanol production in Africa is small, some countries already have legislation on the use of ethanol and gasoline, actively promoting green energy in the transport sector. While Brazil produces an average of 7000 L of ethanol with one hectare of 75 tonnes (Santos et al., 2015) in sub-Saharan Africa, this figure drops to 4000 L where cane is produced in 37 countries on large plantations of irrigated or pluvial regime and where 10%–20% of the cane is complementary produced in small family agricultural areas of 0.5–10 ha.

South Africa produces about 30% of ethanol followed by Malawi, Uganda, Sudan, Tanzania, Sierra Leone, Mauritius, Kenya, Swaziland, Zambia, and Zimbabwe. In most of these countries with some exceptions like Sierra Leone (with the Addax Project of 20,000 ha of cane only for ethanol), ethanol is essentially obtained from the fermentation of molasses.

The production of ethanol directly through the conversion of cane cultivated for this purpose not only constitutes a greater added value to the product, but also promotes the growth of the agricultural sector, also developing the rural sector as a whole, creating new jobs and significantly reducing imports of fossil fuels and the exchange rates resulting therefrom. The same fuel is also used in rural and domestic electrification, transport and agricultural vehicle feeding, etc.

The generalization and development of this energy in Africa will stabilize the supply of energy in the different countries that practice it, diversifying options and reducing the costs of importing fossil fuels.

According to the World Bank, sub-Saharan Africa is projected to grow to 2.6% by 2017. The countries that are most likely to grow are, according to the same source, Ethiopia with +8.3%, Tanzania with +7.2%, and Côte d'Ivoire with +6.8%.

This represents a positive counterpart to the sharp stagnation of up to 1.3% in 2016, and is expected to strengthen somewhat in 2018. This growth is a consequence and reflects the recovery of global commodity prices and improvements in domestic conditions. Most of the growth will also come from Angola and Nigeria—the largest exporters of crude oil. However, investment is expected to recover very gradually, further reflecting tight liquidity conditions on the oil exporters mentioned above and low investor confidence in South Africa.

The sugar, ethanol, and energy sectors in Brazil

The estimated production of sugarcane in Brazil for the 2018/19 crop is 615.8 million tons, showing a reduction of 2.8% in relation to the previous crop. The area to be harvested is also showing a drop compared to the previous harvest, around 1.1%, with a total of 8.63 million hectares. The last recorded harvest also showed a drop on sugar production, destining most of the production for ethanol industry. It is estimated that sugar production is expected to reach 31.73 million tons, a reduction of approximately 16% of the previous harvest. Meanwhile, ethanol production is up 18.6% over the previous crop, producing a total of 32.31 billion liters of ethanol. Of this total, most ethanol is hydrous ethanol (66.8%) and the remainder is treated as anhydrous ethanol (Conab, 2018).

It is interesting to note that in Brazil, unlike other countries, the sugar and alcohol agroindustry operates in a positive and sustainable environment. The products produced, such as ethanol, have the premise of reducing the use of fossil fuels, helping to mitigate the environmental problems associated with the burning of fossil fuels. Also, most sugarcane residues are used in the industry for energy recovery and, consequently, further reducing the environmental impacts of that crop. This makes Brazil one of the largest and best cases of biorefinery, using almost fully sugarcane biomass for the generation of products with high added value.

Another peculiarity of this culture in Brazil is the participation in the national energy matrix. In 2017 the share of renewable sources in the energy matrix is among the highest in the world, 42.9%, being a worldwide example in renewable sources. Still, sugarcane biomass represented 17% of Brazil's domestic energy supply in the same year, being the largest percentage of renewable energies (Fig. 1.7; EPE, 2019). These numbers are significant and have shown the world the possibility of significantly reducing the consumption of fossil fuels, replacing them with biofuels, reducing dependence on oil and greenhouse gas (GHG) emissions. Currently, ethanol from sugarcane is the best fuel capable of meeting the growing demand for renewable energy in the world, and is also low cost and less polluting. It is estimated that by replacing gasoline with ethanol,

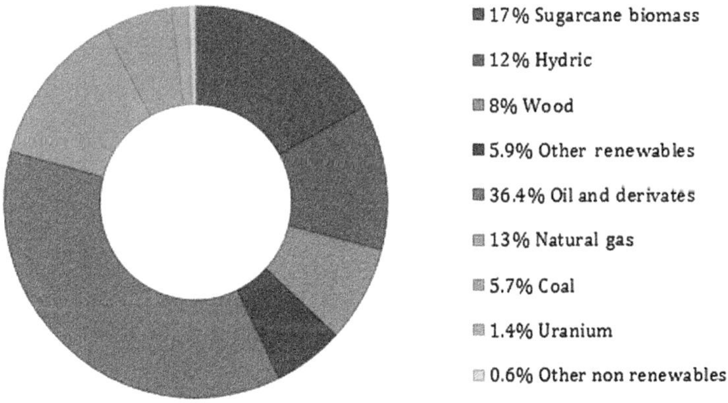

Figure 1.7 Internal energy supply in Brazil. *EPE, 2019. Energetic Research Company (in Portuguese: Empresa de Pesquisa Energética). National Energetic Balance, Base Year 2018. Rio de Janeiro, Brazil.*

a 60% reduction in emissions is possible, thus helping to mitigate global warming, since the burning of fossil fuels accounts for approximately 82% of GHG emissions (Santos et al., 2018).

Socioeconomic and environmental benefits of using ethanol from sugarcane

Although much of the literature agrees with the improvements in the use of sugarcane for ethanol production, the subject is still very controversial, with several points still in question such as: net balance of energy generated/spent, competition between food and energy, use of water (which is high), and the direct and indirect impacts of the exploitation of land for the production of biomass for energy.

As for the net balance, Brazil has shown a positive case, and the expenditure of fossil fuel used to produce ethanol fuel is low compared to US corn ethanol.

On the question "food versus energy," it is difficult to reach a consensus. Several factors come into question, where they may be related to both the type of land use (if it was used for food cultivation previously) and also with the indirect factors that are linked to the final application of the product. Therefore such a discussion will not be put on the table for now.

As for the use of water, it is notorious that this is a point to be improved in the current cane crop. It is estimated that while American ethanol production requires 3–4 L of water per liter of ethanol produced, Brazilian production demands 15 times more, showing the need to develop techniques for better water use. There are water recycling initiatives, which can reduce current demand to 12 L of water per liter of ethanol. Also, another solution is the use of subsurface drippers, which can reduce water expenditure and, together with solutions such as oxygation, can increase production by increasing the oxygenation of the roots of the plant (Bhattarai et al., 2005).

About emissions, in the life cycle of ethanol, CO_2 emissions are 80% lower than gasoline emissions (0.6 kg of CO_2 per liter of gasoline) and GHGs emitted during the production of sugarcane are 70% lower than gasoline. Still, there is modern carbon capture technology, which could further increase the potential for reducing GHG emissions in ethanol production (Santos et al., 2013).

Ripoli et al. (2000) estimated that in Brazil alone, about 12.5 million people could be supplied with electricity using sugarcane residues, which at the time was more than the population of Portugal, demonstrating that the energy was not properly utilized at the time. Even today, as there is no production of surplus electricity, the population still does not benefit from this possible distributed generation. With the increase of decentralized production, a better distribution of energy could occur, which, if surplus, could benefit a large part of the population.

Perspectives

Concerning the process of converting sugarcane into ethanol, it is expected in the long term that the overall conversion efficiency, as well as the reduction of energy consumption and process costs, will be increased. This can be possible with improved technology and greater process innovation. Currently, it is estimated that about 70% of the cost of production of ethanol refers to the production of sugarcane. In this context, one of the greatest challenges is the development of new plant cultivars, obtaining greater adaptability to specific regions, increasing their productivity and, consequently, reducing their production costs. Also noteworthy is the importance of technological advances in the agricultural sector, with improved harvesting processes, due to precision farming, more products can be harvested and separated more quickly and more easily. It is also expected to advance in this sector of harvest to also include the separation of sugarcane straw, which can generate greater yields of energy or ethanol production with this agricultural residue. Also, with the advent of second-generation ethanol production, it is expected that new technologies may enable the use of waste for its production, potentially doubling the production of cane ethanol (Santos et al., 2018).

For the production of sugar, the raw material, that is, sugarcane, also ends up being a limiting factor. Several factors influence sugar production as plant characteristics, stage of maturation, management practices, edaphoclimatic conditions, and genetic variety. Also, the treatment of sugarcane juice is another limiting factor, because it is the process that aims maximum recovery of sucrose, reflecting the quality and final quantity of the product (Santos et al., 2013).

Analyzing the history of sugarcane, its great benefits and development obtained by the countries that use and cultivate the plant, we can see that the sugarcane industry is of great importance worldwide. Especially in countries such as Brazil, India, and China, the sugar and ethanol industry has a strong presence in the economy, contributing to several products in both the domestic and the foreign markets. This is mainly due to the adaptability of the plant to the tropical climates, obtaining high agricultural and energy productivity with low consumption of fossil energy used in the ethanol production process.

Even with the progress of the sugarcane crop, there are still points where, through research and technological innovation, the productive and economic potential of sugarcane and its products can be increased. Some points that are highlighted apply to the agricultural and industrial model. For improvement in the agricultural sector the points to improve are:

- productivity of sugarcane (through bioengineering and basic research on photosynthesis);
- integration of sugarcane with the environment (study of the impact of climate changes on cultivation);
- nutrition of sugarcane and fertilizers (new fertilizers and method of delivery of nutrients);
- control of pests and defenses (biological control agents, impact of agrochemicals and use of pheromones);
- use of residues in production (use of vinasse as fertilizer);
- harvesting (recovery of straw for biorefinery); and
- management and monitoring (precision farming).

In new industrial model, the points that can be emphasized for improvement are:

- logistics and transportation of sugarcane (integrated logistics for sugarcane, straw, bagasse, ethanol, and vinasse, more efficiently);
- reduction of water use in processes;
- reduction of the use of electric energy in the processes (to promote surplus generation, with sale to grid);
- valuation of sugarcane by-products (trade in other biorefinery waste products);
- second-generation ethanol (improving pretreatment and conversion processes); and
- control and monitoring of the industrial process (automation).

Finally, with the perception of future convergence in substitution of fossil fuels for products that are renewable, and also in view of the worldwide trend of the use of sustainable processes, the sugarcane crop is a great example for this transition of technologies. Also, with a growing market for fuels and higher energy demand, there is room for improvement in current processes, making feasible practices such as the production of second-generation ethanol or other biomass-to-liquids fuels. Also, the full use of biomass, even more so in the case of sugarcane, where the residues are already notoriously beneficial, could increase the potential gain with the crop.

References

Alexander, A.G., 1985. The Energy Cane Alternative. Elsevier Science Publishers B.V, Amsterdam, 509 pp.

Bhattarai, S.J., et al., 2005. Oxygation unlocks yield potentials of crops in oxygen-limited soil environments. Adv. Agron. 88:, 313−377.

CONAB, 2018. (National Supply Company of Brazil). Sugarcane Crop Traking 2017/2018 (in Portuguese: Acompanhamento de Safra brasileira de cana 2017/2018). Brasilia, Brazil, v. 4, n. 4.

EPE, 2019. Energetic Research Company (in Portuguese: Empresa de Pesquisa Energética). National Energetic Balance, Base Year 2018. Rio de Janeiro, Brazil.

FAOSTAT, 2019. *Food and Agriculture Organization of the United Nations*. FAOSTAT Statistics Database. FAO, Rome.

Filho, W.F., 2006. *Crossroads of Freedom* (in Portuguese: Encruzilhadas da Liberdade). Unicamp Publishing, São Paulo, Brazil.

Labat, J.B., 1742. Nouveau voyage aux isles de l'Amerique. Pére Labat, Paris.

Le Couter, P., Burreson, J., 2006. Buttons of Napoleon: 17 molecules that changed history (in Portuguese: Os botões de Napoleão: as 17 moléculas que mudaram a história). Trad. M.L.X.A. Rio de Janeiro, Brasil.

Novinsky, A., et al., 2016. *The Jews That Built Brazil* (in Portuguese: Os Judeus que construíram o Brasil). Planeta Publishing, São Paulo, Brazil.

OECD/FAO, 2018. OECD-FAO Agricultural Outlook 2018−2027. OECD Publishing, Paris/FAO, Rome.

Prado, J.C., 2006. *Economic History of Brazil* (in Portuguese: História econômica do Brasil). Brasiliense, São Paulo.

Santos, F., Colodette, J., Queiroz, J.H., 2013. *Bioenergy and Biorefinery—Sugarcane and Forestry Species* (in Portuguese: Bioenergia e Biorrefinaria—Cana-de-Açúcar eEspécies Florestais), first ed. Editora UFV, Viçosa, MG.

Santos, F., Borem, A., Caldas, C., 2015. Sugarcane Agricultural Production, Bioenergy and Ethanol. Academic Press, San Diego, CA.

Santos, F.S., Fernandes, O.W., Caldas, C., Eichler, P., 2018. *Production Technology: Sugarcane and Cachaça* (in Portuguese: Tecnologia de Produção Cana-de-Açúcar e Cachaça), first ed. Mecenas Publishing, Brazil.

SCC, 2019. Soft Commodity Consulting. UK. Available from: <www.softcommodity-consulting.com>.

Sen, T., 2003. Buddhism, Diplomacy, and Trade: The Realignment of Sino-Indian Relations, 600−1400. Asian Interactions and Comparisons, a joint publication of the University of Hawaii Press and the Association for Asian Studies, Manoa, pp. 38−40.

Schwartz, S.B., 1988. In: Motta, T.L.T. (Ed.), *Internal Secrets: Mills and Slaves in Colonial Society* (in Portuguese: Segredos Internos: Engenhos e escravos na sociedade colonial). Companhia de Letras, São Paulo.

Ripoli, T.C.C., et al., 2000. Energy potential of sugarcane biomass in Brazil. Scientia Agricola 57 (4), 677−681.

CHAPTER 2

By-products of the sugarcane industry

Fernando Santos, Paulo Eichler, Grazielle Machado, Jaqueline De Mattia and Guilherme De Souza

Contents

Introduction	21
Sugar and ethanol production process	22
Bagasse and straw	26
Molasses	29
Vinasse	30
Filter cake	31
Yeasts	33
Other by-products of the sugarcane sector	34
Chemical potential of sugarcane residues	36
Future perspectives for sugarcane biorefinery	42
References	45
Further Reading	47

Introduction

Brazil is the world's largest producer of sugarcane and sugar, ranking first in the production of ethanol from sugarcane (FAOSTAT, 2019). The forecast for the 2018/19 harvest indicates that approximately 640 million tons of sugarcane are going to be processed, generating approximately 35 million tons of sugar and 30 billion liters of ethanol, in an area estimated at 8.7 million hectares (CONAB, 2019).

The units that process sugarcane are classified in (1) sugar mills, (2) sugar mills with ethanol distilleries, and (3) autonomous distilleries (ethanol producers). There are currently 378 sugar, ethanol, and mixed production units registered in the Sugarcane and Agroenergy Department of the Ministry of Livestock and Supply (MAPA, 2019). The best units produce on average about 7000 L of ethanol and about 11 tons of sugar per hectare of processed cane.

Sugarcane Biorefinery, Technology and Perspectives
DOI: https://doi.org/10.1016/B978-0-12-814236-3.00002-0

In the last decades, there has been a great technological advance in the sugar-energy sector, both in relation to the productive chain and industrial chain, associated with management improvements. These technological advances are enabling the production units to commercially exploit, in addition to sugar and ethanol, other products (by-products) of high added value. The sugarcane industry then becomes a precursor to future "biorefinery" facilities, at relatively low cost, for the high availability and mix of approximately one-third of sucrose with two-thirds of lignocellulosic (bagasse and straw) biomass.

Sugar and ethanol production process

The main products of the sugar and ethanol sector are sugar, which supplies the food market, anhydrous ethanol, used as an additional fuel for gasoline, and hydrous ethanol, which serves flex vehicles, as well as a small market for nonenergy uses. In the sugar and ethanol production process, the initial steps are similar (Fig. 2.1).

Once transported to the production unit, the sugarcane is usually washed and sent to the broth preparation and extraction system, through a set of four to seven milling units. The broth is extracted and the sugars separated from the fiber (bagasse), which, in turn, goes to the energy plant of the producing unit. In order to increase the yield of the extraction, the chopped and shredded cane passes through successive washes with hot water, extracting their sugars and, at the end, passes through a drying roller, from which leaves the bagasse to be used in the boilers. Produced in the mill or diffuser, the sugarcane juice can then be used to produce sugar or ethanol.

For the production of sugar, the broth goes through a series of treatment steps, which include physical actions (sieving, heating, flashing) and chemical (reactions promoted by the addition of chemicals, polymers, etc.) with the aim of eliminating nonsugars, colloids, turbidity, and color, as well as favoring sedimentation as much as possible (Cavalcante, 2012). In this stage, the maximum sucrose recovery occurs and good quality of the final product is obtained.

After the chemical treatment, the broth is heated to eliminate microorganisms by sterilization, complete the chemical reactions with the alkalizing agent, coagulate and flocculate the insoluble impurities and remove the gases. Then the broth is taken to the decanter or clarifier, where the flocculated impurities, also called sludge, are separated. The

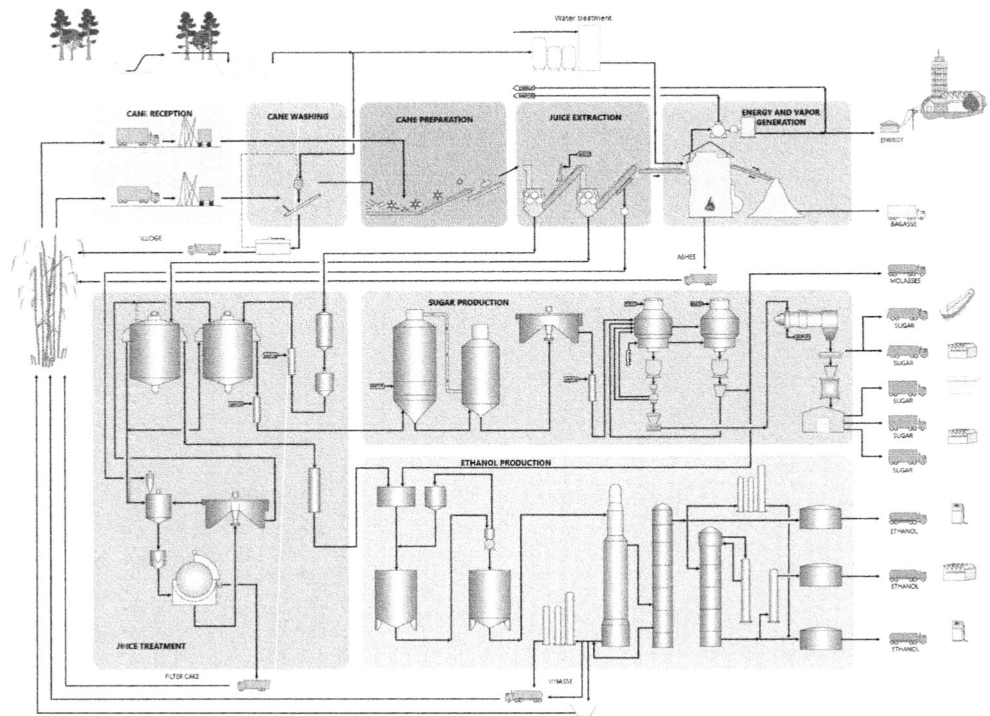

Figure 2.1 Simplified flowchart of a sugar and ethanol production unit.
Modified from Santos, F. et al., 2013. Bioenergia e Biorrefinaria — Cana-de-Açúcar e Espécies Florestais, first ed. Viçosa, MG: Editora UFV.

clarified broth exits from the upper part of the trays, which are already free of most impurities. In the decanters, only the physical separation between the broth and the impurities (sludge) occurs. The removed sludge is then added bagasse and filtered to recover the sucrose still contained therein, while the residual filter cake is used in the sugarcane crop as fertilizer.

The treated broth is then concentrated in multieffect evaporators and cookers for the crystallization of the sucrose. In this process, not all the available sucrose in the cane is crystallized and the residual solution rich in sugar (molasses) may return more than once to the process in order to recover more sugar. The final broth, called molasses, which does not return to the sugar production process, still contains some sucrose and a high content of reducing sugars (glucose and fructose), and can be used as raw material for the production of ethanol through fermentation.

Finally, the sugar is routed to the drying stage, which is based the reduction of its moisture by the simultaneous cooling, to humidity levels that allow its storage for more or less long periods, without presenting significant changes of its characteristics, that is, preserving quality for consumption as a food product.

The ethanol production process is based on the fermentation of both cane juice directly as well as mixtures of broth and molasses. This is the most used process in Brazil, the United States and, in general, in other countries. This complementarity of sugar and ethanol production reflects the synergy existing in the Brazilian production system.

As already described, the first stages of ethanol production are similar to the sugar production process. Once treated, the cane juice is evaporated to adjust its sugar concentration and eventually mixed with the molasses, giving rise to the wort. Next, the wort goes to the fermentation units, where it is added to the yeasts (*Saccharomyces cerevisiae*) and fermented for an average period of 6–10 hours, giving rise to the wine (fermented mixture, with a concentration of 7%–10% of alcohol). Higher times can indicate contaminations, low viability of yeasts, low concentration of yeasts in the fermentative medium, excess sugars in the wort, among other factors. Those problems need to be corrected with the utmost urgency, since it works with fermenters with hundreds of cubic meters of volumetric capacity (Vasconcelos, 2012). The most used fermentation process in the distilleries of Brazil is the Melle–Boinot, whose main characteristic is the recovery of the yeasts of the wine through its centrifugation (Fig. 2.2).

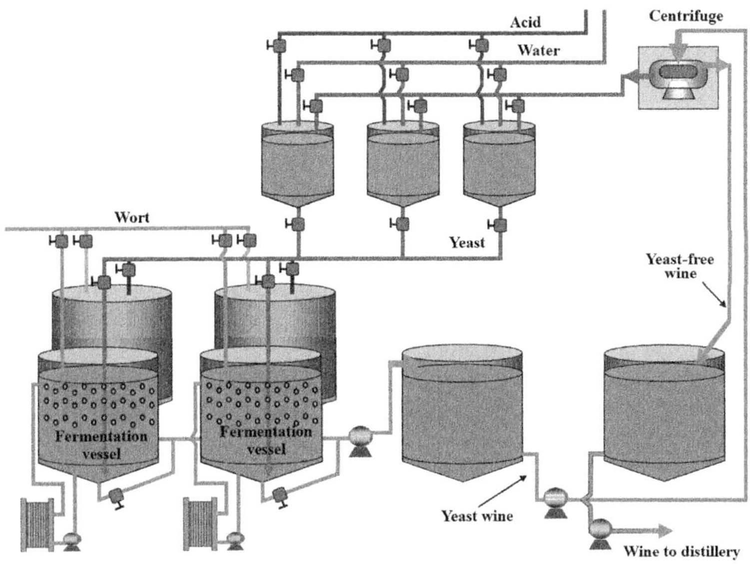

Figure 2.2 Melle–Boinot ethanolic fermentation process.
Modified from Vasconcelos, J.N., 2012. Fermentação etanólica. In: F.A. Santos, et al. (Eds.), Cana-de-açúcar: Bioenergia, Açúcar e Etanol – Tecnologias e Perspectivas, second ed. Viçosa, Brazil.

Thus after fermentation, the yeasts are recovered and treated for further use, while the wine is taken to the distillation columns. This process presents economic advantages due to the reuse of yeasts in the fermentation process.

Distillation is a physical operation aimed at separating components from a mixture, according to the relative volatility of the components. The ethanol is initially recovered in the hydrated form with approximately 92.8–94.7 INPM (m/m), corresponding to about 6% water by weight, leaving the vinasse as residue, usually in the proportion of 10–13 L per liter of absolute ethanol produced. In this process, other liquid fractions are also separated, giving rise to the second alcohols and "fusel" oil, or "Finkel." The hydrated ethanol may be stocked as the final product or may be shipped to the dewatering column. But since it is an azeotropic mixture, its components cannot be separated by simple distillation. The most commonly used technology is dehydration by the addition of the cyclohexane, forming a ternary azeotropic mixture with a lower boiling point than that of anhydrous ethanol. In that process, in the dehydration column, the cyclohexane is added to the top, and the anhydrous ethanol is withdrawn at the bottom, with approximately 99.7 INPM or 0.4% water by weight. The ternary mixture withdrawn from the top is then condensed and decanted, while the water-rich part is sent to the cyclohexane recovery column (Santos, 2013).

Ethanol dehydration can still be done by adsorption with molecular sieves or by extractive distillation with monoethyleneglycol (MEG), which stand out due to the lower energy consumption and also the higher costs. Due to the growing demands of the foreign market, several ethanol-producing plants are opting for molecular sieves, since with them it is possible to produce anhydrous ethanol free of contaminants. It is expected a great spread of its use in the coming years, in the quest to meet the demanding standards of the international market.

It is important to point out that this traditional model of sugar and ethanol production is already giving way to a "new model" capable of producing, in addition to sugar and ethanol, new value-added products through biorefinery.

Bagasse and straw

The use of by-products from the sugar-energy sector, which is now generated in a significant quantity, should increase considerably in the coming years with the expansion of sugarcane plantations and the installation of new production units, transforming them from cost sources to sources of revenue (Fig. 2.3).

Bagasse is the result of sugarcane juice extraction processes. As it is a by-product, it has practically no cost of production or transportation, so it is highly valued, mainly for being substitute of fossil fuel and wood in the generation of steam and electric energy, allowing the energy self-sufficiency of the producing units and, in some, the commercialization of surplus electricity. Also, bagasse is used in the production of cellulosic ethanol and furfural. Furfural is used as a solvent for the refining of lubricating oils, wood resins, and vegetable oils, as well as furfuryl alcohol as raw material for furanic polymers, anticorrosives, polymers of urea, modified formaldehyde, fragrances, and solvent of resins and dyes.

Sugarcane straw, left on the soil surface after mechanical harvesting, consists basically of green leaves, dried leaves, stalks, and plant tips. In the coming years, there will be a great availability of sugarcane straw, and two attitudes, not necessarily excluding each other, may be taken: (1) leaving the straw in the field (tillage) or (2) recovering it as raw material for the production of fuels or chemicals.

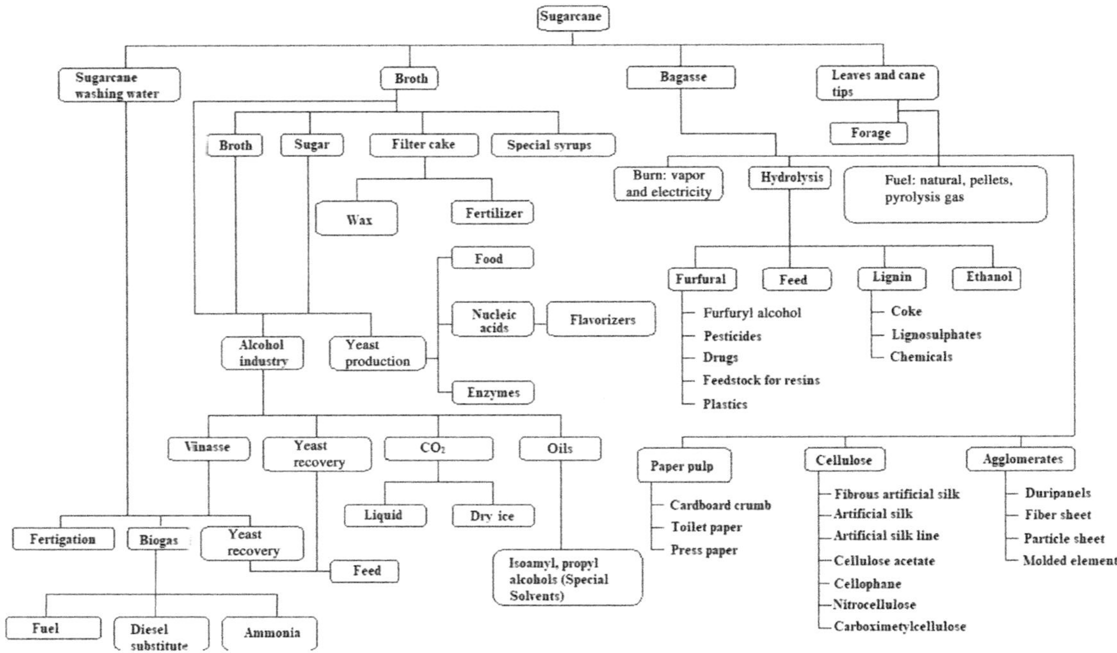

Figure 2.3 Sugar-energy sector by-products and potential products.

Agronomically, it is interesting to keep the straw in the field, as it contributes to the improvement of the physical, chemical, and biological properties of the soil, to the control of erosion and weeds, to retention of humidity, in addition to increasing the soil microbiota. However, the permanence of the straw in the soil can cause some problems, such as delay in sprouting, immobilization of mineral nutrients, mainly nitrogen, difficulties in the operation of agricultural machines and implements, as well as a higher incidence of pests and diseases in subsequent cultivation.

Therefore aiming at the utilization of straw for the production of fuels and chemical products, without affecting its benefits in the soil, Hassuani et al. (2005) reached the following conclusions:

1. The straw should be removed under these conditions:
   a. in inhabited areas and, or, close to highways, due to the risk of accidental or criminal fires;
   b. in areas subject to electrical storms (top relief);
   c. before soil preparation, in areas with soil pest infestation, when it is necessary to eliminate crop residues;
   d. in very wet winter regions, especially in poor drainage soils, which affect sprouting.
2. The straw can be removed by technical–economic analysis in these cases:
   a. where the cover of the straw in the soil makes it difficult to sprout the cane;
   b. in areas with high rates of pest that are favored by thatch;
   c. in places where reduced soil preparation cannot be employed due to infestation by plants not controlled by straw or soil pest whose control depends on the planting of the soil.
3. The straw may be partially removed under these conditions:
   a. during or after the harvest of the cane, it is recommended to leave about 7.5 tons/ha of straw on the ground, evenly distributed; and when the straw is approximately 60 cm above the planting lines of cultivars that show low sprouting.

It can therefore be stated that the amount of sugarcane straw left on the soil surface has a number of advantages and disadvantages, depending on the agronomic conditions involved and taking into account, in particular, costs in the process of collecting and transporting the straw to the agribusiness. Braunbeck and Cortez (2005), when discussing aspects of straw recovery, concluded that, on average, 70% of straw should be recovered.

Unlike sugarcane bagasse, which is already in the industry, the costs for straw recovery are considered high. In energy terms, straw represents approximately one-third of the energy potential of sugarcane, which is currently wasted or used less favorably (Buckeridge et al., 2010; Santos et al., 2012). With the technical and economic viability of the production of cellulosic ethanol and the search for energy and environmental sustainability of the production chain, given the demands of the international market, the recovery and the use of sugarcane straw should occupy a prominent place in the sugar-energy sector.

Molasses

In the process of producing sugar, there is the production of a by-product called residual molasses, which, depending on the region, is considered as poor molasses, final molasses, or simply molasses. It is a dense, viscous liquid, of dark brown color, rich in sugars, containing small percentage of water. Its density varies from 1.4 to 1.5 g/mL and is produced at the rate of 40 kg/tons of cane. The yield in ethanol is 280–320 L/tons.

There are several factors that interfere with the composition of molasses, for example, cultivar, age, health status, maturation cycle, cultivation system, fertilization, and sugarcane treatments, as well as climatic conditions, sugar production, cane harvesting, and storage time.

This by-product is widely used for the production of ethanol in the distilleries annexed to the mills, but can also be used for animal feed or for the cultivation of fungi and bacteria in other fermentation processes aimed at the production of chemicals and pharmaceuticals, as well as biological yeast. In East Asian countries, molasses is widely used in the fermentation process for the production of monosodium glutamate, acids (citric, formic), and amino acids (lysine).

Molasses powder is an organoleptic, energetic, and flavoring supplement for animal feed, a direct input for cattle farmers and manufacturers of feed and mineral salts. This product takes advantage in relation to molasses in natura for several reasons, mainly due to the greater ease in the formulation of animal feed and also in the operations of transport, handling, and storage, given its plastic packaging. The molasses in natura presents greater difficulty in transportation, requiring care and storage in special tanks. In addition, the molasses in natura presents greater possibility of fermentation. Molasses powder is also used in steelmaking to coat forms for casting steel.

Vinasse

Vinasse is the final residue from the distillation of the fermentation wine to obtain ethanol. It has brown coloration and darkens as it is oxidized by exposure to air. It is the most important and most worrying liquid effluent of the agro-energy industry. For each liter of ethanol, an average of 10–15 L of vinasse is produced. The problem caused by this quantity of vinasse produced is associated with the toxicity of ethanol to the yeasts used, forcing a reduction in the alcohol content during the final phase of the fermentation process. Thus for the same volume of wine, there is a reduction in the volume of ethanol produced and an increase in the quantity of vinasse generated.

In the mid-1970s, vinasse presented high BOD—biochemical oxygen demand (between 30,000 and 40,000 mg/L), low pH (4–5), unconverted sugars, unfermented carbohydrates, dead yeasts, and high content of mineral salts, which caused great environmental damage because it was thrown directly into the rivers. With the emergence of fertigation and new technologies for the treatment of vinasse, these problems were practically solved, using vinasse to be used as fertilizer in sugarcane plantations and thus contributing to the increase of agricultural productivity and, consequently, reducing the costs with chemical fertilizers. However, if applied in excess, it can have several negative effects, for example, on the quality of the sugarcane delivered in the agribusiness, interfering in the process of clarifying the broth, especially when destined to the production of sugar (Santos, 2013).

The fertigation system should be started with the observation of two essential items: the soil properties and the physical–chemical characteristics of the vinasse. The soil–vinasse application rate in the soil according to the potassium concentration (Rocha et al., 2012) is determined from the physical–chemical characteristics of the soil and vinasse. When applied in adequate doses, the vinasse causes improvements in the chemical and physical properties of the soil.

The addition of vinasse, together with the incorporation of organic matter, which has been increasingly practiced in sugarcane plantations, can improve the physical conditions of the soil and promote greater mobilization of nutrients, also due to the greater solubility provided by the residue (Da Silva et al., 2007). In addition to fertigation, vinasse can also be harnessed to:

- animal feed—by concentrating vinasse;
- production of unicellular proteins—through aerobic fermentation;
- biogas production—by means of anaerobic fermentation.

Filter cake

It can be defined as the by-product that is eliminated in the process of decanting the sugarcane juice during the treatment step for the production of sugar and/or ethanol. It can be obtained through three types of processes: rotary vacuum filter, filter press, and diffuser separation. Using average data from the producing units, the production of the filter cake in quantitative terms, as a function of the generation process, can be observed in Table 2.1.

The chemical composition of the filter cake depends on several factors: soil type, cultivars, type of harvest, degree of extraction of the juice, lime dosage, and other products used in the clarification, methods of filtration employed, among others. The in natura cake contains approximately 75% water; its average composition is presented in Table 2.2.

In the vast majority of the sugar and ethanol production units, the filter cake is used as fertilizer in the sugarcane fields. This is the most widespread use of this by-product, mainly due to the large amount of nitrogen, phosphorus, calcium, and organic matter that the pie provides

Table 2.1 Filter cake production in industrial process.

Broth extraction system	Generation process	Cake production (kg/tons)
		Processed sugarcane
Mill	"Oliver" filter	28 to 35
Mill	Press filter	18 to 22
Diffusion	Separator	5 to 6

Table 2.2 Typical composition of sugarcane filter cake (% dry weight).

Components	Content (%)
Wax, lipids, and resins	4–14
Fiber	15–30
Sugars	1–15
Proteins	5–15
Total ashes	9–20
SiO_2	4–10
CaO	1–4
P_2O_5	1–3
MgO	0.5–1.5

to the soil. Vitti et al. (2012) recommend the use of filter cake according to the following conditions and methods of application:

1. Method of application:
   a. in total area—at the time of reform or expansion of the cane field and, in some cases, also in a ratoon;
   b. localized—mainly in the planting groove, and can be applied in strip in the areas of ratoon plants, in the line of sugarcane.
2. Process for use:
   a. in natura—used mainly in the winter cane planting (in the south-–central region of Brazil) in the furrow irrigation, aiming at the supply of nutrients, together with the water and the beneficial thermal effect of the cake. However, because the humidity is high, in order to reach adequate amounts of nutrients, it is necessary to apply high dosages;
   b. "conditioned"—that is, the filter cake is used, which has undergone a drying process, together with the physical conditioning to improve the application characteristics. In this sense, as the concentration of nutrients occurred and improved the flow by applicators, it becomes a very interesting alternative.
   c. "enriched"—that is, in addition to undergoing the conditioning process, the filter cake undergoes an aerobic composting process, being mixed with other raw materials, aiming to raise the nutrient concentration standard with a product of lower humidity, expanding the areas of application, since the product allows to be transported in greater distances due to the smaller dosages.

As in the case of vinasse, the controlled use of the filter cake is recognized as a good practice in sugarcane cultivation from an environmental and productive point of view, because it allows the total recycling of by-products of the sugarcane industry, increased soil fertility, water abstraction for irrigation, reduction of the use of chemical fertilizers, and costs.

Another utility of the filter cake is in obtaining wax. During the agroindustrial process, only 40% of the lipid material present in the cane is dispersed in the broth, with the remainder in the bagasse. Of the amount present in the broth, about 95% is concentrated in the filter cake (Laguna et al., 1996). This percentage meets the quality criteria required for its industrial use, presenting great importance for the food, pharmaceutical, chemical, cosmetic, cleaning, and polishing sectors.

Yeasts

They are heterotrophic, unicellular, aclorophylated microorganisms, and may present anaerobic or aerobic metabolism. They are very widespread in nature, soil, powder, and fruits in general, and can be carried by wind and insects.

The composition of the yeasts (Table 2.3) may vary depending on a number of factors, such as the substrate used, the yeast species, the fermentation method, cell age, and drying conditions (Desmonts, 1966). In addition to high protein content, yeast products are rich in B vitamins (B1, B2, B6, pantothenic acid, niacin, folic acid, and biotin), in minerals, in macro- and microelements, particularly selenium and dietary fiber.

Among the yeasts, the most studied for the production of ethanol is *S. cerevisiae*, which draws attention to its nutritional composition and, for this reason, is widely used in the production of several other products. Another interesting feature is that it ferments sugars to ethanol with high yields even in aerobiose. *S. cerevisiae*, when the sugar concentration is high, preferentially directs pyruvate to ethanol production, although the energy yield is lower than could be achieved from the citric acid cycle and oxidative phosphorylation. This phenomenon is called the Crabtree effect and it is postulated that it offers an advantage in ecological niches, because the ethanol produced in these conditions exert an inhibitory effect on competing microorganisms.

The yeasts present great industrial application for their potential to metabolize hexoses, pentoses, organic acids, hydrocarbons, and their ability to produce alcohols to heterologous proteins. Other applications of

Table 2.3 Dry yeast composition (*Saccharomyces cerevisiae*).

Components	Yeast (*S. cerevisiae*)
Dry weight (%)	86.71
Proteins (%)	26.95
Minerals (%)	4.92
Calcium (%)	0.57
Phosphorus (%)	0.52
Magnesium (%)	0.17
Manganese (%)	26
Ferro (mg/kg)	518
Zinc (mg/kg)	173
Potassium (mg/kg)	2

yeast are in the commercial production of reasonable amounts of alcohol; dehydrogenase; hexokinase; lactate dehydrogenase; glucose-6-phosphate dehydrogenase; coenzyme A; nucleosides of diphosphopyridine; and adenine mono, di, and triphosphate of guanine, cytidine, and uridine (Almazán, 1997). Yeasts could also be used to improve growth and yield rates in cattle, pigs, and poultry as a consequence of their beneficial action on the intestinal flora.

The improvement of the ethanol quality requires the selection of yeasts, which must present the following characteristics: high fermentation speed, dominance and permanence during the harvest, good fermentation capacity, high conversion of sugars into ethanol, small production of glycerol, low formation tolerance to high concentrations of substrate and ethanol, resistance to acidity and high temperatures, genetic stability, flocculence (when eliminating centrifuges), good fermentative efficiency (high yield in ethanol), high productivity and high speeds cell growth, ethanol production, and substrate consumption (Vasconcelos, 2012).

Therefore it is observed that yeasts have high capacity to produce new products and derivatives, which, if used wisely, could contribute significantly in several areas, such as health, food, and industry.

Other by-products of the sugarcane sector

It should be noted that the sugarcane agro-industry, which traditionally originates a diversity of products (sugar and ethanol) and by-products (bagasse, straw, molasses, vinasse, filter cake, and yeasts), has been diversifying and incorporating new technologies of production of "new products" obtained from the by-products of sugarcane (Table 2.4).

Carbon dioxide (CO_2) produced in units of ethanolic fermentation is generally released into the atmosphere, but can be purified, deodorized, liquefied, and stored under pressure for other purposes, such as the production of soft drinks and dry ice, sodium bicarbonate, and use in the treatment of effluents. It is of very high purity (approximately 99.9%) and of biological origin. In industry, about 1 kg of CO_2 is generated per liter of ethanol produced. The producing units have been structured for the beneficiation of carbon dioxide. The main applications are carbonation of beverages, inertization of atmosphere for welding, casting, fire extinguishers, refrigeration, aerosol propellant, tertiary recovery of oil wells, and transport of solids by pipelines (Rabelo et al., 2012).

Table 2.4 Recovery of sugar-energy agro-industry products.

By-products	Categories/ technologies	"New products"
Molasses	Biotechnological	• Citric acid • Amino acids: lysine • Pesticides • Nitrogen fixator • Silage Innoculum
MolassesBagasseVinasse	Chemicals	• Industrial inputs (technical dextran, calcium gluconate, mannitol, sorbitol, and biodegradable surfactants) • Furfural (xylose liquor, furfuryl alcohol, furan–epoxy compounds, wood preservative, cast resin) • Plastics (PHB and PHB/hl, PHA mcl/PHB hpe) • Inputs for the pulp and paper industry • Concentrated vinasse
MolassesBagasse	Veterinary products	• Antidiarrheal preparation • Iron–dextran complex • Probiotic
Bagasse	Alimentos	• Yeast, fructose, and glucose derivatives • Fructooligosaccharides • Inverted syrups via enzymatic route • Edible mushrooms of the species *Pleurotus ostreatus*
Bagasse	Biological	• Fertilizer compound
Bagasse	Structural	• Agglomerates of bagasse/ cement • MDF agglomerates

PHA, polyhydroxyalkanoate.

Biodegradable or bioplastic is a special type of polymer biosynthesized by bacteria from sugars and other forms of carbon, with characteristics very close to that of synthetic polymers, obtained from petroleum. The main difference is that it decomposes easily in nature. While polyethylene terephthalate packages, called PET and used mainly for soft drinks, should take more than 200 years to degrade and traditional plastics, more than a 100 years, biodegradable plastic resins decompose in around 1 year, depending on the advantage of being produced from renewable sources, such as starch, sugars, or fatty acids. Some Brazilian companies are investing in R&D for the production of bioplastics on a large scale.

Citric acid has been produced for decades in Brazil by the fermentation process, in which the fungus *Aspergillus niger* is used in molasses substrate dissolved in water. Citric acid is an ingredient used for food preservation as well as flavoring. It also serves for cleaning industrial equipment and manufacturing detergents and other hygiene and cleaning products.

Among the amino acids that can be produced through the fermentation of sugars, lysine stands out, whose market is growing every day, due to its application in several industrial segments (meat, food, pharmaceutical, etc.).

The use of the by-products of the agro-energy industry associated to the development of new technologies opens perspectives for the production of numerous products with higher added value, besides the great environmental contribution.

Chemical potential of sugarcane residues

The use of sugarcane and its residues as feedstock in conversion, treatment, and processing to obtain high value-added products through sustainable processes is the objective of a biorefinery. In biorefineries, biomass conversion processes may be integrated to allow the production of a wide range of biofuels, bioenergy, biomaterials, and biochemicals. Some examples of high value-added products, which can be produced from sugarcane and its residues are shown in Fig. 2.4. There are four main lignocellulosic biomass conversion processes, involving different technologies: biochemical processes, such as fermentation and enzymatic conversion; thermochemical processes, such as pyrolysis and gasification; chemical processes, such as acid hydrolysis and transesterification; and physical processes, such as distillation (Fitzpatrick et al., 2010; Alvim et al., 2014). Fig. 2.5 shows main products, which can be obtained through each biorefinery platform.

Biochemical conversion platform in biorefineries includes fermentation processes to produce ethanol and other chemicals as alcohols and organic acids; anaerobic digestion to produce biogas and biofertilizers; enzymatic

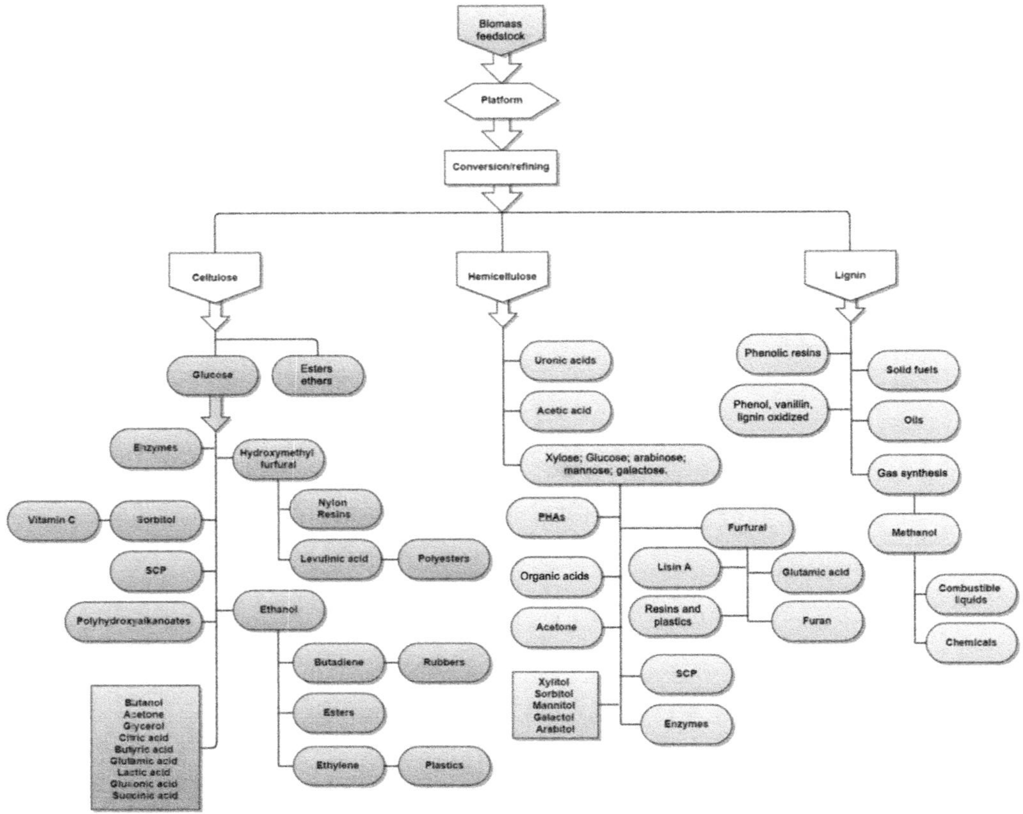

Figure 2.4 Some high value-added products obtained from lignocellulosic feedstock.
PHA, polyhydroxyalkanoate; *SCP*, single cell protein. *Boneberg, B.S., et al., 2016. Biorefinery of lignocellulosic biopolymers. Rev. Eletrônica Científica Da UERGS 2 (1), 79–100.*

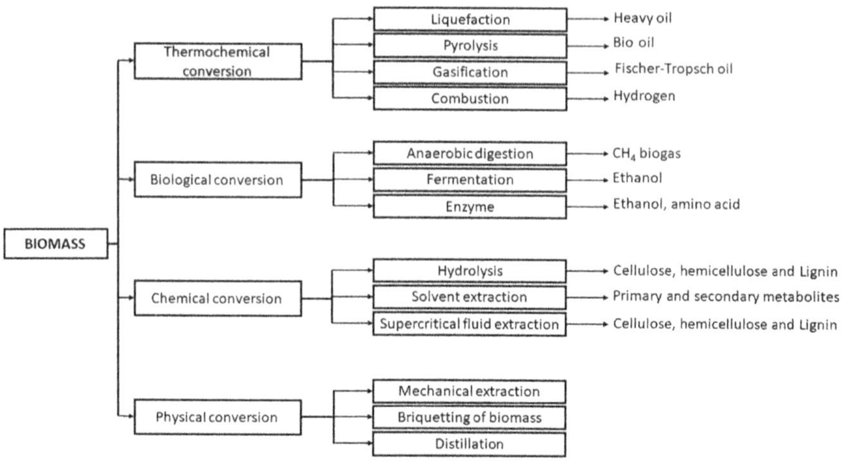

Figure 2.5 Lignocellulosic biomass conversion processes.
FT, Fischer–Tropsch.
From Rodrigues, J.A.R., 2011. Do engenho à biorrefinaria. A usina de açúcar como empreendimento industrial para a geração de produtos bioquímicos e biocombustíveis. Quím. Nova 34 (7).

processes as enzymatic hydrolysis to release structural sugars present in lignocellulosic biomass (Santos et al., 2013; Rodrigues, 2011).

Sugar release from lignocellulosic recalcitrant structure requires a pretreatment step, which causes disturbance in its structure exposing cellulose and hemicelluloses to enzymatic attack. Enzymatic hydrolysis of these polysaccharides releases monosaccharides which can be further submitted to fermentation processes. It is possible to accomplish hydrolysis and fermentation processes separately or combine these two processes in a simultaneous saccharification and fermentation process. Besides that, once polysaccharides present in biomass contain both hexoses and pentoses, it is possible to ferment these two types of sugars separately or use a microorganism capable of converting both hexoses and pentoses to bioproducts (Vaz Junior, 2011).

The viability of biomass conversion to high value-added products is directly linked to efficiency and velocity of conversion cellulose and hemicelluloses to monosaccharides. Polysaccharides hydrolysis can be acid or enzymatic, and this last one is less toxic to yeasts. Enzymatic hydrolysis shows yields higher than 0.85 g of glucose/g of cellulose, under mild temperatures, around 40°C–45°C in atmospheric pressure. However, long process duration (48–72 hours), catalytic deactivation through enzymatic activity inhibition, and high cost of enzymes are process-limiting factors (Lora et al., 2008).

Among the bioproducts that can be obtained through biochemical processes of conversion of sugarcane and its residues, it is possible to highlight biofuels such as ethanol, butanol, and 2,3-butanodiol produced through fermentation of sugars derived from polysaccharides present in lignocellulosic biomass. Some enzymes, including α-amylases, cellulases, and xylanases, can be produced through simultaneous saccharification and fermentation processes using sugars from biomass as substrate. Some organic acids, such as itaconic, succinic, lactic, and citric acids can be produced by applying glucose derived from cellulose as carbon source in fermentation processes (Sindhu et al., 2016).

Glucose obtained from cellulose can also be used as substrate to produce biopolymers, like polyhydroxyalkanoates (PHAs) and polylactic acid. Fig. 2.6 shows main routes to produce different types of biopolymers from lignocellulosic biomass. Other biopolymers that can be produced from lignocellulosic biomass are xanthan gum and dextran. Xanthan gum is an exopolysaccharide that can be obtained by fermentation of sugars obtained from biomass, widely used in food, pharmaceutical, cosmetics, paint, textile, agricultural products, and petroleum industries (Pu et al., 2018). Dextran, also an exopolysaccharide, is produced through fermentation applying lacto microorganisms, such as *Leuconostoc mesenteroides*, *Lactobacillus brevis*, *Streptococcus mutants*, and *Weissela confusa*. This biopolymer is applied in the food, pharmaceutical, and medical industries (Rosca et al., 2018).

Some fungi, as example basidiomycete *Trametes hirsuta*, produce efficient lignolytic enzimes (laccases), which allow them to produce ethanol through biochemical conversion processes from lignocellulosic residues without requiring pretreatment and hydrolysis steps prior to fermentation. Studies in the literature point that this fungus has huge conversion potential, not only of glucose, but also of xylose, present in large amount in sugarcane residues. Using this type of fungus in ethanol production is interesting, once it uses both pentoses and hexoses and it is capable of degrading lignocellulosic residues, not requiring enzymatic hydrolysis step, which has a high cost (Okamoto et al., 2011).

Another bioproduct that can be obtained from both glucose and xylose through biochemical conversion process is biohydrogen. Lai et al. (2014) studied biohydrogen production from sugarcane bagasse pretreated with diluted sulfuric acid and showed that bagasse is a suitable substrate to produce biohydrogen due to large amounts of glucose and xylose, along with small amounts of inhibitors.

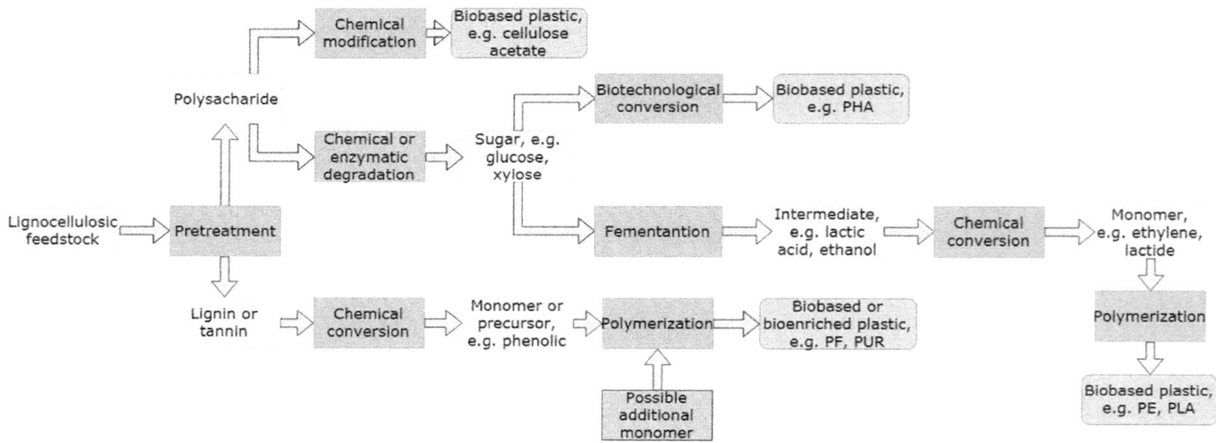

Figure 2.6 The main routes for production of biopolymers from lignocellulosic feedstock.
PHA, polyhydroxyalkanoate; *PLA*, poly-lactic acid; *PUR*, poly-urethanes; *PE*, poly-ethylene; *PF*, phenol-formaldehyde.
Adapted from Brodin, M., et al., 2017. Lignocellulosics as sustainable resources for production of bioplastics—A review. J. Clean. Prod. 162, 646–664.

Anaerobic digestion process is a biochemical conversion process in which biogas is the main product, composed mostly of methane and carbon dioxide, which can be purified to biomethane and used as biofuel. Several microorganism's groups are involved in anaerobic digestion. In first step, called hydrolytic phase, anaerobic and facultative bacteria secrete enzymes that degrade high molecular weight organic compounds to simple compounds. In acidogenic phase, simple compounds are cleaved to hydrogen, carbon dioxide, alcohols, and volatile organic acids. In the third step, organic acids are converted to acetic acid by acetogenic bacteria. Methanogenic bacteria produce methane from carbon dioxide, hydrogen, and acetic acid produced in previous steps. Solid and liquid residues from anaerobic digestion can be used as biofertilizers (Holm-Nielsen et al., 2009; Neves et al., 2009).

Chemical conversion processes are exclusively based on chemical reactions. Among the chemical processes in biorefineries, it is possible to highlight transesterification, hydro-processing, chemical cracking, and Fischer–Tropsch (FT) synthesis. Many of these chemical processes are used in petrochemical industry and have been adapted to be used in biorefineries. One of the chemical processes largely applied is acid hydrolysis, which allows production o levulinic acid from hexoses and furfural from pentoses (Santos et al., 2013; Rodrigues, 2011).

Furfural is an important chemical that can be produced from sugarcane and its residues through chemical conversion processes. Xylans, one of the types of hemicelluloses present in abundance in lignocellulosic biomass, are mainly composed of pentoses and generally are the largest constituent of hemicelluloses in grasses, including sugarcane residues. Furfural is a versatile chemical that can be further converted to many other chemicals including furan and furfuryl alcohol, and it is largely utilized in several applications in petrochemical, plastic, pharmaceutical, and agrochemical industries (Machado et al., 2016). Some examples of chemical derived from furfural are shown in Fig. 2.7.

Xylose, present in hemicelluloses, can be applied in xylitol production through chemical conversion processes as well. Xylitol is a sweetening agent that can be prepared through batch hydrogenation of xylose, by applying high pressures to solubilize hydrogen and high temperatures to increase hydrogenation rate (Sousa-Aguiar et al., 2014).

Hydro-processing uses hydrogen and a catalyst to produce long-chain hydrocarbons similar to naphtha, which after purification results in products like gasoline and kerosene. Catalytic cracking can be applied in

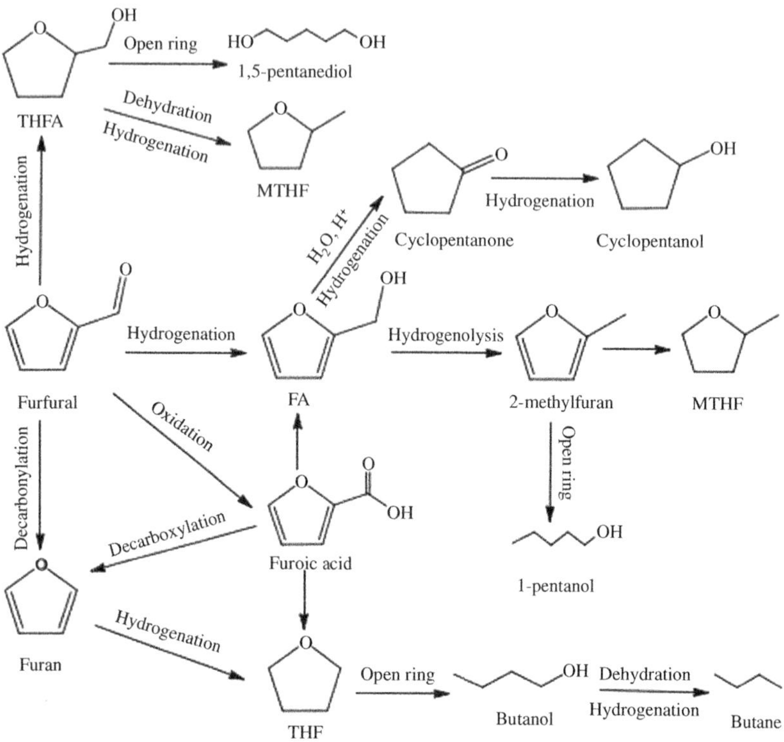

Figure 2.7 Conversion of furfural into several high value-added chemicals.
THFA, tetrahydrofurfuryl alcohol; *MTHF*, methyltetrahydrofuran; *FA*, furfuryl alcohol; *THF*, tetrahydrofuran.
From Yan, K., et al., 2014. Production, properties and catalytic hydrogenation of furfural to fuel additives and value-added chemicals. Renew. Sustain. Energy Rev. 38, 663–676.

bio-oil obtained through thermochemical processes, which in the presence of a catalyst results in biofuels and other chemical compounds. FT synthesis consists of biofuels production from syngas, composed of H_2 and CO, derived from thermochemical processes in biomass (gasification). Products obtained through this synthesis are linear chain aliphatic hydrocarbons and in small amounts branched hydrocarbons, unsaturated hydrocarbons, and primary alcohols (Almeida, 2008).

Future perspectives for sugarcane biorefinery

Although the energy-cane concept is recent, the idea of biomass production for energy is very well known, especially for the use of wood crops such as Eucalyptus. The sugarcane culture as an energy source is more interesting for its high productivity and possibility to use as direct energy source or as feedstock for biofuels such as ethanol. As the use of land for high productivity energetic crops is recent, it has risen some concerns. The main one is the competition of land with food crops. In Brazil, despite sugarcane being considered a food crop, it is the major source of the renewable energy matrix, representing 17% of total energy supply source. Such is the success of sugarcane in Brazil, as energy source, that in 2018 represented more than double of the wood and vegetable coal (8%) energy supply [Empresa de Pesquisa Energética (EPE), 2018). Though prices of sugarcane bagasse and straw are defined by food chain production, being sugar or ethanol, not representing commercial value despite cane residues generate so much energy. However, with vision of sugarcane high potential for energy production, mitigation of environmental problems and capacity to generate income, significant changes are expected to happen in next years.

Recent biotechnology progress allowed the development of new sugarcane varieties, called energy cane, which does not focus on producing greater amounts of juice (for production of sugar and ethanol), but rather enhances production of possible bioenergy from the plant. Not only productivity has been increased with new varieties but also it has been possible to produce the plant with specific characteristics such as amount of sugar, fiber content, and total biomass content. Some varieties are shown in Table 2.5, comparing between those characteristics of different varieties of sugarcane. The use of energy cane can have some advantages such as renewable energy generation; mitigation of greenhouse gas effect; high conversion of atmospheric carbon into biomass; multiple renewable sources of energy; higher resistance to biotic and abiotic stress; offers less competition with food (comparing to trees); cultivation management is already well established (harvest, transport, and irrigation); and has less problems with harvesting time, which regular

Table 2.5 Comparison between commercial hybrid to an energy cane clone.

Characteristics	Commercial (RB72454)	Energy cane (Clone)
Pol % cane	14.60	8.74
Fiber % cane	12.05	19.80
Stalks t/ha	148	155
Fiber t/ha	17.08	30.63

Source: Adapted from Santos, F., et al., 2015. Sugarcane Agricultural Production, Bioenergy and Ethanol. Academic Press/Elsevier, San Diego, CA.

sugarcane needs specific time to produce more sugars, in energy cane there are no such problems (Santos et al., 2015).

Integral utilization of sugarcane residues for energy production, although highly appealing for environmental and financial reasons, is still not economically attractive. Firstly, due to low market value, bagasse and straw are less attractive than other biomass and, as they have low density, transport is particularly costly. Since bagasse is generated inside processing factory, is easier to use it for energy (thermal or electrical), but to utilize cane straw would be costly because there is the need to collect it from the fields. Also, because leaves collection would be from the ground, along with straw, other pollutants such as sand and dirt would be collected as well, increasing the cost of preprocessing this residue for energy purposes. Therefore one of the major challenges of integral use of sugarcane residues is around solutions for economically feasible uses of cane straw: its transportation, preprocessing, and choosing the best biorefinery exploitation of this residue. In Fig. 2.8, it is possible to see some products and

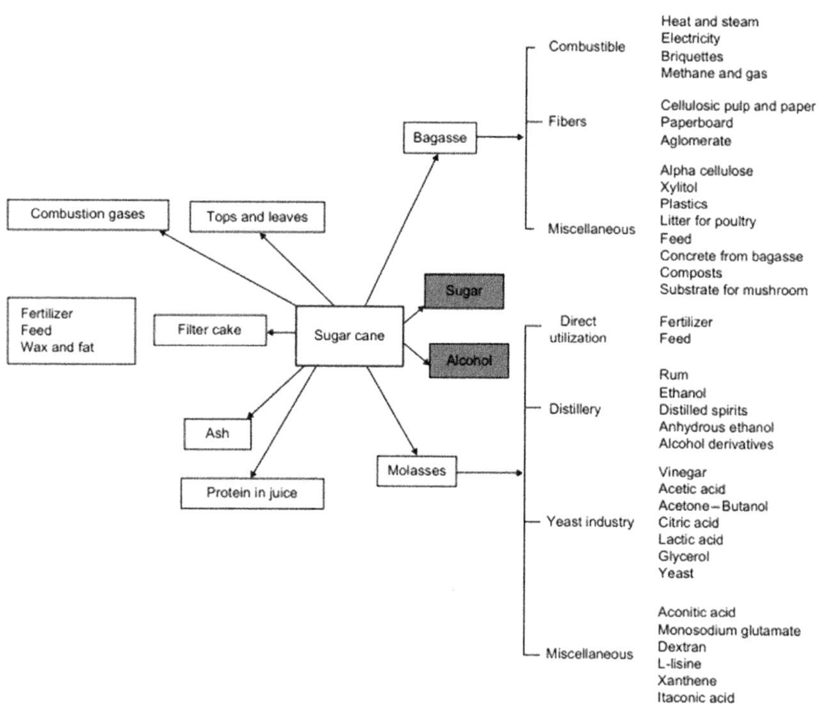

Figure 2.8 Possible products in sugarcane industry.
From Santos, F., et al., 2015. Sugarcane Agricultural Production, Bioenergy and Ethanol. Academic Press/Elsevier, San Diego, CA.

by-products of sugarcane industry. Despite sugar and ethanol being considered main products of this culture, other products such as electricity, heat, plastics, and other obtained from cane residues are becoming increasingly more interesting as technology allows their economic feasibility.

In an overview of Brazil's sugarcane case, it is possible to confirm that this culture has an incredible potential for chemical and thermochemical processes. Overall, studies showed positive advances in the use of cane residues for biorefinery processes, resulting in a wide range of possible value-added products. It is important to note that because different countries have different edaphoclimatic conditions, it is possible to have different results not only in biomass chemical composition but also on subsequent biorefinery products. Although, several countries are reporting good expectations for sugarcane production and its many possible biorefinery products (Hiloidhari et al., 2018).

One of the greatest biorefinery bottleneck is the economic feasibility of processes. Because many technologies are involved in most final products, costs of pretreatment of biomass, especially deconstruction, and further downprocessing of biomass, economic feasibility is still an issue. Therefore with future improvements on processes for second-generation ethanol production, especially for biomass pretreatments, it is expected that integral use of sugarcane and its residues become financially interesting.

References

Almazán, O., 1997. Applied biotechnology: the state of the art. In: Proceedings of the XXII Congress of the ISSCT, pp. 23–28, Cartagena-Colombia.

Almeida, M.B.B., 2008. Bio-óleo a partir da pirólise rápida, térmica ou catalítica, da palha da cana-de-açúcar e seu co-processamento com gasóleo em craqueamento catalítico (Master Thesis), Rio de Janeiro—Brazil.

Alvim, J.C., et al., 2014. Biorrefinarias: conceitos, classificação, matérias primas e produtos. J. Bioen. Food Sci 1 (3), 61–77.

Boneberg, B.S., et al., 2016. Biorefinery of lignocellulosic biopolymers. Rev. Eletrôn. Cient. UERGS 2 (1), 79–100.

Braunbeck, O.A., Cortez, L.A.B., 2005. O cultivo da cana-de-açúcar e o uso dos resíduos. In: Rosillo-Calle, F., Bajay, S.V., Rothman, H. (Eds.), Uso da biomassa para a produção de energia na indústria brasileira. Unicamp, Campinas, pp. 215–246.

Brodin, M., et al., 2017. Lignocellulosics as sustainable resources for production of bioplastics – a review. J. Clean. Prod. 162, 646–664.

Buckeridge, M.S., Goldman, G.H., 2010. Routes to Cellulosic Ethanol, 1 ed. Springer.

Cavalcante, C.S. Processo de produção do açúcar. In: Santos F. A., et al., 2012. Cana-de-açúcar: Bioenergia, Açúcar e Etanol – Tecnologias e Perspectivas. 2 ed., Viçosa, Brazil.

Cavalcante, C.S., 2012. Processo de produção do açúcar. In: Santos, F.A., Borém, A., Caldas, C. (Eds.), Cana-de-Açúcar: Bioenergia, Açúcar e Etanol – Tecnologias e Perspectivas, second ed, Revisada e ampliada., Viçosa, MG, pp. 400–450.
CONAB. Companhia Nacional de Abastecimento. Acompanhamento de safra Brasileira: Cana-de-Açúcar, segundo levantamento, Agosto/2019. Brasília, CONAB, 2019.
Da Silva, M.A.S., Griebeler, N.P., Borges, L.C., 2007. Uso de vinhaça e impactos nas propriedades do solo e lençol freático. Rev. Brasil. Eng. Agríc. Amb. 11 (n. 1), 108–114.
Desmonts, R., 1966. Tecnologia da produção dos fermentos secos de destilaria. Boletim Informativo da APM. Piracicaba 8 (2), 1.
EPE, Empresa de Pesquisa Energética, 2018. Balanço Energético Nacional 2018: Ano base 2017. EPE, Ministerio de Minas e Enegia, rio de Janeiro.
FAOSTAT. Food and Agriculture Organization of the United Nations, Statistical Data. Available at: fao.org/faostat. (accessed 09.03.2019.).
Fitzpatrick, M., et al., 2010. A biorefinery processing perspective: treatment of lignocellulosic materials for the production of value-added products. Biores. Technol. 101, 8915–8922.
Hassuani, S.J., Verde Leal, M.R.L., Macedo, I.C., 2005. Biomass power generation: sugarcane bagasse and trash. Project BRA/96/G31 PNUD – CTC. Unipress Disc Records do Brasil. Série Caminhos para Sustentabilidade, Piracicaba.
Hiloidhari, M., et al., 2018. Bioelectricity from sugarcane bagasse co-generation in India – an assessment of resource potential, policies and market mobilization opportunity for the case of Uttar Pradesh. J. Clean. Prod .
Holm-Nielsen, J.B., et al., 2009. The future of anaerobic digestion and biogas utilization. Biores. Technol. 100, 5478–5484.
Laguna, A.G. et al., 1996. Policonasol, una mezcla de alcoholes alifáticos primarios superiores para el tratamento de complicaciones atereoescleróticas tales como la hiperagregabilidad plaqueteria, loa acidentes isquêmicos, tromboses e incluso su efectividad contra úlceras gástricas quimicamente inducidas y su processo de obtención de la caña. Cuban patente CU 22229A1.
Lai, Z., et al., 2014. Optimization of key factors affecting hydrogen production from sugarcane bagasse by a thermophilic anaerobic pure culture. Biotechnol. Biofuels 7 (n. 119), 11 p.
Lora, E.E.S., et al., 2008. Rotas termoquímica e bioquímica para biocombustíveis: estado-da-arte, oportunidades e desafios para o Brasil. 7 Congresso Internacional Sobre Geração Distribuída e Energia No Meio Rural. UNIFOR, Fortaleza, pp. 1–10.
Machado, G., et al., 2016. Literature review on furfural production from lignocellulosic biomass. Nat. Res. 7 (n. 3), 115–129.
MAPA. Ministério da Agricultura, Pecuária e Abastecimento. Sistema de Acompanhamento da Produção Canavieira – Cadastro de instituições. Brasília, 2019. Available at: agricultura.gov.br. (accessed 09.03.2019.).
Neves, L.C.M., et al., 2009. Biogas production: new trends for alternative energy sources in rural and urban zones. Chem. Eng. Technol. 32 (n. 8), 1147–1153.
Okamoto, K., et al., 2011. Direct ethanol production from starch, wheat bran and rice straw by the white rot fungus *Trametes hirsuta*. Enzyme Microb. Technol. 48, 273–277.
Pu, W., et al., 2018. A comprehensive review of polysaccharide biopolymers for enhanced oil recovery (EOR) from flask to field. J. Ind. Eng. Chem. 61, 1–11.
Rabelo, S.C., Costa, A.C., Rossel, C.E.V., 2012. Aproveitamento de resíduos industriais. In: Santos, F.A., Borém, A., Caldas, C. (Eds.), Cana-de-Açúcar: Bioenergia, Açúcar e Etanol – Tecnologias e Perspectivas, second ed., Revisada e ampliada., Viçosa, MG, pp. 515–536.
Rocha, M.H., et al., 2012. Resíduos da produção de biocombustíveis: vinhaça e glicerina. In: Lora, E. E. S., et al. Biocombustíveis. V. 1 and 2. Rio de Janeiro, p. 691-809.

Rodrigues, J.A.R., 2011. Do engenho à biorrefinaria. A usina de açúcar como empreendimento industrial para a geração de produtos bioquímicos e biocombustíveis. Quím. Nova 34 (n. 7).
Rosca, I., et al., 2018. Biosynthesis of dextran by *Weissella confusa* and its in vitro functional characteristics. Int. J. Biol. Macromol. 107, 1765–1772.
Santos, F., et al., 2013. Bioenergia e Biorrefinaria – Cana-de-Açúcar e Espécies Florestais, first ed. Editora UFV, Viçosa, MG.
Santos, F., et al., 2015. Sugarcane Agricultural Production, Bioenergy and Ethanol. Academic Press/Elsevier, San Diego.
Santos, F.A., Queiroz, J.H., Colodette, J.L., Fernandes, S.A., Guimarães, V.M., Rezende, S.T., 2012. Potencial da palha de cana-de-açúcar para produção de etanol. Rev. Quím. Nova XY:1-7.
Sindhu, R., et al., 2016. Bioconversion of sugarcane crop residue for value added products—an overview. Renew. Energy 98, 203–215.
Sousa-Aguiar, E.F., et al., 2014. Some important catalytic challenges in the bioethanol integrated biorefinery. Catal. Today 234, 13–23.
Vasconcelos, J.N. Fermentação etanólica. In: Santos F. A., et al., 2012. Cana-de-açúcar: Bioenergia, Açúcar e Etanol – Tecnologias e Perspectivas. 2 ed., Viçosa, Brazil.
Vaz Junior, S. (Ed.), 2011. Biorrefinarias: cenários e perspectivas. Embrapa Agroenergia, Brasília, DF.
Vitti, G.C., Luz, P.H.C. ,Altran, W.S., 2012.Nutrição e adubação. In: Santos, F.A., Borém, A., Caldas, C. (Eds.), Cana-de-Açúcar: Bioenergia, Açúcar e Etanol – Tecnologias e Perspectivas, second ed., Revisada e ampliada. – Viçosa, MG, pp. 73–117.
Yan, K., et al., 2014. Production, properties and catalytic hydrogenation of furfural to fuel additives and value-added chemicals. Renew. Sustain. Energy Rev. 38, 663–676.

Further Reading

BNDES & CGEE, 2008. Banco Nacional para o Desenvolvimento Social e Econômico; Centro de Gestão e Estudos Estratégicos. Bioetanol de cana-de-açúcar: Energia para o desenvolvimento sustentável. BNDES & CGEE, Rio de Janeiro, 316 p.
Camhi, J.D., 1979. Tratamento do vinhoto, subproduto da destilação de álcool. Br. Açucareiros 94 (n. 1), 18–23.
CONAB, 2018. Companhia Nacional de Abastecimento. Acompanhamento de safra Brasileira: Cana-de-Açúcar, segundo levantamento, September 2018. CONAB, MAPA, Brasília.
Cortez, L.A.B., 2010. Bioetanol de Cana-de-Açúcar: P&D para Produtividade Sustentabilidade. Edgard Blucher Ltda, São Paulo, pp. 761–772.
Cortez, L., Magalhães, P., Happ, J., 1992. Principais subprodutos da agroindústria canavieira e sua valorização. Rev. Br. Ener. 2 (n. 2), 111–146.
Cortez, L.A.B., Rossel, C.E.V., Jordan, R.A., Leal, M.R.L.V., Lora, E.E.E., 2010. Necessidades de P&D na área industrial em vinhaça. In: Cortez, L.A.B. (Ed.), Em Bioetanol de Cana-de-Açúcar: P&D para Produtividade Sustentabilidade. Edgard Blücher Ltda, São Paulo, pp. 619–636.
Felipe, M.G.A., 2010. A qualidade da matéria-prima na produção de etanol de cana-de-açúcar. In: Cortez, L.A.B. (Ed.), Bioetanol de Cana-de-Açúcar: P&D para Produtividade Sustentabilidade. Edgard Blücher Ltda, São Paulo, pp. 553–559.
Franco, T.T.; Garzón, C.S.L., 2010. Novas possibilidades de negócios do setor sucroenergético: alcoolquímica e biorrefinaria. In: Gómez, E.O., Souza, R.T.G., Rocha, G.J.M., Almeida, E., Cortez, L.A.B. A palha de cana-de-açúcar como

matéria-prima para processos de segunda geração. In: Cortez, L.A. B. (Ed.). Bioetanol de Cana-de-Açúcar: P&D para Produtividade Sustentabilidade. Edgard Blücher Ltda, São Paulo, pp. 637–659.

Leal, M.R.L.V., 2010. Cana energia. In: Cortez, L.A.B. (Ed.), Bioetanol de Cana-de-Açúcar: P&D para Produtividade Sustentabilidade. Edgard Blücher Ltda., São Paulo, pp. 751–760.

Matsuoka, S., Bressiani, J., Maccheroni, W., Fouto, I., 2012. Bioenergia da cana. In: Santos, F.A., Borém, A., Caldas, C. (Eds.). Cana-de-Açúcar: Bioenergia, Açúcar e Etanol – Tecnologias e Perspectivas, second ed., Revisada e ampliada., Viçosa, MG, pp. 547–577.

Moura, A.G., Castillo, E.F., Palacio, J.C.E., Reno, M.L.G., Lora, E.E.S., Venturini, O.J., et al., 2012. Biocombustíveis de primeira geração – bioetanol pela rota convencionale 2. In: Lora, E.E.S., Venturini, O.J. (Eds.), Biocombustíveis, v. 1. Interciência Ltda, Rio de Janeiro, pp. 359–409.

Mutton, M.A., Rosseto, R., Mutton, M.J.R., 2010. Utilização agrícola da vinhaça. In: Cortez, L.A.B. (Ed.), Bioetanol de Cana-de-Açúcar: P&D para Produtividade Sustentabilidade. Edgard Blücher Ltda, São Paulo, pp. 423–440.

Olivério, J.L.; Hilst, A.G.P., 2005. DHR-Dedini Hidrólise Rápida – Revolutionary Process for Producing Alcohol from Sugar Cane Bagasse. In: International Society Of Sugar Cane Technologists Congress, 25, Guatemala, Janeiro/Fevereiro.

Paturau, J.M., 1969. By-products of the Cane Sugar Industry. Elsevier Publishing Company, New York, 274 p.

Penatti, C.P., Araújo, J.V., Donzelli, J.L., Souza, S.A.V., Forti, J.A., Ribeiro, R., 2005. Vinasse: a liquid fertilizer In: International Society of Sugar Cane Technologists (ISSCT), Guatemala, 26, Proceeding, pp. 403–411.

Ripoli, T.C.C., Ripoli, M.L.C., 2004. Biomassa de cana-de-açúcar: colheita, energia e transporte. Piracicaba. 302 p.

CHAPTER 3

Sugarcane: biorefinery, technology, and perspectives

Saulo Philipe Sebastião Guerra, Marcelo Scantamburlo Denadai, André Luiz Merthan Saad, Emanuel Rangel Spadim and Marcius Xavier Rodrigues da Costa

Contents

Introduction	49
Sugarcane straw bales ash content	53
Water content in sugarcane straw bales	58
References	64
Further reading	64

Introduction

The sugarcane in Brazil presents an economical importance since the colonial period from the 16th century, being that the first seedlings came from the Portuguese and were cultivated in the northeast region, being an agricultural landmark for the production of sugar and industrial in the country. The Brazilian sugar was exported mainly to the European countries, fertile soil, where large tracts of land and favorable climate favored a cultural productivity.

The cultivation and productive system of the colonial period for our present have undergone great transformations. One of them is related to labor, which has become more qualified and technified due to the increasingly competitive market. Mechanization and precision agriculture in sugarcane were fundamental for the standardization of planting and harvesting operations, efficiently improving resource utilization and lowering the operational costs of soil preparation, crop management, harvesting, and transportation.

The behavior of society related to environmental issues favored the development of new research and technologies in the sector, the use of biomass sugarcane and ethanol as renewable energy sources leverage the

Sugarcane Biorefinery, Technology and Perspectives
DOI: https://doi.org/10.1016/B978-0-12-814236-3.00003-2

Brazilian sugarcane industry. Sugarcane straw that was previously burned in the field became a valuable material, with the potential to be used in energy cogeneration and or second-generation ethanol (SGE).

The production of sugarcane according to data from the National Company of Supplying (Conab) and Ministry of Agriculture Livestock and Food Supply (Mapa) of the crop 2016/2017 harvest was 665.6 million tons, and the estimated harvest 2017/2018 is 647.63 million tons. This result is due to the rainfall distribution, which affected the production of some producing states such as Goias and Minas Gerais, ending up in losses of productivity. Sugar production is forecast at 38.9 million tons for this crop, being higher than the 2016/2017 harvest of 38.7 million tons, pushed for better market conditions. Ethanol production is expected to be up to 27.8 billion liters, lower than the 2016/2017 harvest of 30.5 billion liters.

Sao Paulo State, according to data from the Brazilian Agriculture Directory (Agrianual) in 2016 was the largest producer with 400,789 thousand tons, in an area of 4,983,930 ha, representing 52% of the production area. Goias State is in the second place with 70,433 thousand tons, and an area of 907,214 ha, 10.4% of Brazilian production. Minas Gerais State, third position, with 69,436 thousand tons, an area of 903,387 ha (10.1%), fourth Mato Grosso do Sul State with 52,200 thousand tons, in an area of 729,112 ha (7%), followed by Alagoas with 386,207 ha. In the other producing states, the areas are smaller, with representations below 3%.

The average productivity of Brazilian sugarcane is currently 76.9 tons/ha, an increase of 9.1% over the previous harvest of 70.4 tons/ha, according to data from the National Company of Supplying (Conab) in 2016. This increase is related to proper soil management, new crop varieties, and the correct and adequate use of fertilizers and agricultural practices.

The demand for sugarcane production has been consolidated by the use of ethanol as sources of renewable fuel since 1975 with the National Alcohol Program (Proálcool). This program was created with the purpose of diversifying the Brazilian energetic matrix, facing the world oil crisis in that period. Currently, Brazil has one of the largest fleets of biofuel cars, consisting of almost 57% of Flex cars (gasoline and ethanol), 42.8 million vehicles circulating according to data from the National Association of Automobile Manufacturers (Anfavea) and the National Association of Components for Automotive Vehicles (National Union of the Component Industry for Motor Vehicles, 2016).

With the advancement of the energy potential of sugarcane, the use of biomass from bagasse and straw for the generation of energy and SGE is extremely important for the country's energy sector. According to Ripolli (1999), countries such as Brazil have immense agricultural areas under favorable climatic conditions and cannot afford biomass as an integral part of their energy matrix. The author stands out the importance of sugarcane as the crop that has the best conditions to increase the biomass share, due to the extensions of area, better adaptation conditions to transform bagasse and straw from the harvest, in electric energy.

Data from the National Energy Balance, provided by the Ministry of Mines and Energy (National Energy Balance, 2016), present that 41.2% of the country's availability of domestic energy is renewable, with 16.9% coming from sugarcane biomass (biomass, wood, charcoal, ethanol, and other primary renewable sources) are the industries and the transport sector, which represent about 65% of the country's energy consumption.

Knowing the positive aspects of the sugar-energy sector, one of the negative consequences is related to socio-environmental aspects, which are always pointed out by the media. Because it is a profitable or profitable activity than food production, many family farmers lease their lands to the ethanol and sugar mills, causing changes in farm work and the availability of food supplies in some Brazilian regions. In contrast, Unicamp's rector professor José Tadeu Jorge pointed out in 2008 (Levy, 2008) that the expansion of sugarcane plantations to meet the growing demand for biofuels does not threaten food production, but the situation may change if Brazil does not adopt an efficient agricultural policy over the long term. According to the dean, the expansion areas of sugarcane are advancing in areas of old pastures or degraded, which do not compete with grain production, which are records in production and export in the country. For him, the only case in which biofuel production affects food availability concerns the United States if it targets its corn production for ethanol production.

Brazil has the largest agricultural area in the world, between 80 and 100 million hectares, therefore, enormous potential for the occupation and growth of energy production and also food (Embrapa, 2006). A factor that is also evident in relation to environmental issues, justifying the investments of the sugar-energy sector in relation to the adequate management of Brazilian production (LIMA et al., 2014) and (Gilio, 2015) reveals a lack of studies that assess the effects of the recent changes in the sector, such as the prohibition of burning sugarcane, changes caused by mechanization, alterations related to tenure and land leases, regions producers, and incentive policies.

The sustainability of the sugarcane industry lies in the changing behavior and culture of society, facing environmental issues of preservation and global warming. The use of sugarcane biomass represents a value above the energy produced by hydroelectric plants 11.3% and firewood 8.2%, for example, according to the National Energy Balance (Ben, 2016). If we compare the benefits of biomass to fossil fuels and petroleum products, which currently account for 58.8% of the country's energy matrix, we can, in addition to environmental factors, such as the use of renewable and clean energy, greenhouse effect, mention the high competitiveness in relation to the cost and the generation of jobs from the agriculture to terminals of distribution of energy.

Sugarcane in addition to the production of sugar and ethanol, bagasse, and straw is also used for the production of energy in times of seasonality, such as the winter period in the absence of rainfall and the production of SGE. SGE produced from biomass, according to data from the Raízen company, in 2017, increases production by 50% without increasing the area of cultivation, having a production of biofuel even during the off-season of sugarcane.

The technological advance also occurs not only in the production within the plants but also in energy sugarcane plantations. According to data from the site Sugar gave rise (2017), the productivity of this variety is between 85% and 121% in relation to traditional sugarcane, and the average productivity of excellent sugarcane grows from 100 to 220 tons/ha. Although the energy sugarcane has a lower total recoverable sugars (TRS), it doubles due to the amount produced per hectare. In relation to biomass, the amount of bagasse produced per hectare increases from 25 to 96.2 tons/ha. This characteristic of energy sugarcane is at the crosses directed to produce higher content of fiber and rhizome to take advantage of the better nutrient conditions of the soil.

The use of biomass in addition to technological and sustainable advances leads to a new concept in relation to integrated systems of raw materials, processes, products, waste, such as biorefineries. Biorefineries are plants that integrate biomass conversion processes into biofuels, chemical inputs, oils, materials, food, feed, and energy according to the Sugarcane, sugar and alcohol producers' council of Sao Paulo State (2006).

The biorefineries aims to optimize the use of resources and minimize the impacts of production, integrating several sectors or areas of conversion; biochemical; chemical; thermochemical; and microbiological industries, focusing on the use of biomass so that chains of values similar to those of derived from petroleum, but with less impact on the environment. Developed and developing countries, such as Brazil, are mobilizing large resources and public and private efforts aimed at the utilization of biomass, adding value to productive chains.

The products developed from coproducts and residues, according to Embrapa, are those with the highest added value of productive chains, examples of biorefinery in the petrochemical, pharmaceutical, automotive, civil construction, and agribusiness industries. In biorefineries, there are products of great volume and low added value, such as biofuels, as well as products of small volumes, but of high added value, such as chemical specialties and additives.

The possibilities of growth and development of the sugar-energy sector and biorefineries in the use of biomass are enormous for the future. Brazil, as a pioneer in the use of biofuels, can boost the generation of jobs, income, industrialization, and the conquest of foreign markets for export.

Sugarcane straw bales ash content

In Brazil, several processes have been used to collect sugarcane straw in sugarcane plantations, among which:

- Bulk pickup: At mechanical harvesting of sugarcane, the harvester primary and secondary extractors separate the billets, which are taken to the mill, from the straw and other residues that are left on the ground and can go to quantities of up to 32 tons of dry mass (tDM). In this kind of picking up process, the biomass is bundled by a tractor with rakes, and posteriorly collected and chopped by a self-propelled straw chopper that it in trucks that take the biomass to the mill. Usually, this process is used in short distances due to the low density of the material.
- Integral harvest: in this picking up process, the sugarcane harvester operates with the primary and secondary extractors off, not separating the straw from the billets. The straw is loaded with the billets and taken to the mill for further separation. The advantage of this process is because it does not need any other machine to collect the remaining straw in the field, but a disadvantage is the reduction of the sugarcane mass transportation due to the reduction of load density.

Figure 3.1 Bailing of large prismatic sugarcane straw bales.

- Baling: this is the Brazilians producers mostly used process. Among the balers used for biomass energy purposes, the large cylindrical and prismatic bales (Fig. 3.1) stand out. This is the process that demands the largest quantity of machines, due to the need of straw bundling in the field, baling, picking up the bales, temporary storaging, loading in large trucks, and finally arriving at the mill. This process is good for logistics because it does not interfere with the dynamics of sugarcane harvesting and transport, and also because it allows the biomass transportation at greater distances due to the load density increasing. This load density increase can reduce up to 15 times the initial volume of the straw, making bales of up to 220 kg.DM/m^3.

Every shown process has its advantages and disadvantages, and each company chooses its one according to its characteristics. Most of the straw suppliers have been working with the baling, but in fields too close to the mills, in order of up to 15 km away, it has been used the bulk pickup. In spite of the low biomass density in this process, for short distances, the costs get bigger in the baling due to the quantity of machines and its fixed costs.

Nowadays, most of the collected material is used for electricity generation by sugar and ethanol mills, but some companies have also researched the feasibility of producing SGE. A possible novelty to the Brazilian market is the production of pulp and paper from sugarcane straw, which a company recently started commercial trials in the Sao Paulo State.

Brazil, as mentioned, has representative areas of sugarcane production and can supply a large amount of biomass, but some problems have limited its use and in many cases have increased its cost of use. The two major factors related to these problems are ash content and water content of straw.

The ash content is the less important, as it can be controlled by the producer in some way. In some studies, it was observed that the ash content is directly related to water content during baling, and can be predicted and managed.

With the exception of the integral harvesting process, in which the straw is taken to the mill with the sugarcane, the straw-collecting processes use rakes to bundle the biomass for both balers and self-propelled straw chopper collecting. There are several models of rakes currently in the market, with different working capacities and efficiency. This efficiency refers to the ash content that will be baled and transported to the mill. Basically, the rakes bundles biomass for making tracks, and in this process, it also drags soil particles. The greater the wetness of the straw, the easier the retention of these soil particles in the biomass.

This is a seemingly simple solution, and biomass should be collected in the lowest possible water content, but several factors have influence in this result, such as climate, precipitation, latitude, season, amount of field remaining straw, and vigor of sugarcane. After harvesting the sugarcane, in addition to the straw that is already dry along the stems, there are also green leaves from the top of the plant. Most balers cannot operate under these water content conditions, and there is an interest in leaving this material in the field to reduce water and chlorine content.

Depending on the region, there is a seven to nine months period of sugarcane harvesting. During these months, the temperature and precipitation oscillate affecting the time necessary for drying the straw in the field. On the rainy months, the drying time is longer, and the larger the amount of straw on the soil, the longer it will take for the biomass to lose water.

With all these factors acting together, a short period is available for picking the straw, because with favorable conditions in a few days, sugarcane sprouts, making it impossible to collect the biomass. It should also be emphasized that the wetness of the straw changes along the day, being higher in the morning and lower in the afternoon. In this context, the producer collects the biomass with great variations of water content, being possible to pack from 8% up to 15% in the beginning and the end of the

Figure 3.2 Dry sugarcane straw cleaning process in a mill.

day, respectively. Observing the whole harvest, water content of up to 40% at baling will be found.

As shown, the higher the water content, the greater the ash content. The ash content depends on several factors as well as the water content, even during one day. This fact is an aggravating factor for the management of a straw-processing unit, because it will lead to higher costs due to the high wear of the boilers, waste disposal, and also the cleaning of the straw (Fig. 3.2).

The Brazilian mills consider as an acceptable ash content the scale of 3%–5% (mass/mass ratio), but, depending on the conditions, some bales shown up to 32% ash content in our studies (Fig. 3.3).

It is very important to know the biomass ash content for the business decision making, as it makes possible to control the high ash loads incoming, but as the analysis demands a time, the bale is just unloaded, not mattering how it is. In this context, a research was conducted in two industrial plants at São Paulo state countryside into studying the correlation between water content and ash content.

The straw was obtained in bales and sampled by probes right after the entrance weight scales of the industrial plants, as the same equipment's and procedures used in the sampling methodology of Consecana (2006). It collected 22% of the loads, which arrived in 2 years of bale supply at these units.

Figure 3.3 Visible ash content in sugarcane straw bales sampling.

The water content was found by ovendrying at 105°C up to obtaining constant weight, defined by the following equation:

$$WC = \left(1 - \frac{DW}{WW}\right) \times 100 \tag{3.1}$$

where WC is the water content observed (%), DW is the dry straw weight, and WW is the wet straw weigh.

The sugarcane straw ash content was obtained by the complete organic constituents burning in a muffle until calcination (800°C), calculated by the equation subsequently:

$$AC = \left(\frac{FM \times 100}{IM}\right) - 1.38 \tag{3.2}$$

where AC is the calculated ash content (%), IM the initial straw mass before calcination, FM the final straw mass after calcination, and 1.38 corresponds to the ash content of the straw (white).

In all, 4983 samples were taken during the research period, resulting in the linear equation below with high adjustment ($R^2 = 0.9497$):

$$AC = (0.3094 \times MC) - 0.4 \tag{3.3}$$

where AC is the estimated ash content (%) and MC is the water content (%).

The equation obtained offers an alternative methodology for having faster ash content, which contributes to the faster and more accurate decision making for producers and mills.

Figure 3.4 Flat (left) and perforated (right) balers prechamber panels.

In order to obtain a better quality of baled biomass, in addition to the proposed equation, adaptations were also made in the balers, which were submitted to field trials for validation. Among the various mechanisms and baler parts is the baling prechamber, which predates the compaction chamber. Inside the prechamber, the straw is drawn by a metallic flat surface. In the evaluated baler, this part was composed of two panels that were replaced by panels with several perforations of approximately 10 mm in diameter (Fig. 3.4).

Both research treatments were performed under the same terrain, baler, operator, and schedule conditions, in order to reduce the interference of external factors. By installing the perforated panels to the baler, it was observed a reduction of up to 10% of ash content in the bales when compared to the ones baled with the flat panels.

Water content in sugarcane straw bales

As mentioned previously, another factor that has caused great problems in the use of biomass is water content, causing problems in different aspects. transporting the water limits the quantity of useful load per truck, increasing the biomass costs. A study was carried out to verify the increase in the cost due to the increase in the water content of the bales. From the same distance and cost of transport with a truck, it was simulated what would be the costs of the range of 10%–65% water content. To calculate the cost increase, the value of 10% of water content was considered as reference (Table 3.1).

From the calculated data, it was possible to obtain a linear equation ($R^2 = 0.9329$) to estimate the increase in transport costs as a function of water content at the moment of loading of the bales (Fig. 3.5).

$$CI = TC \times [(2.6782 \times MC) - 43.667] \qquad (3.4)$$

where CI is the increase of transportation cost (US$/t), TC is the transport base cost (US$/t), and MC is the bale water content (%).

From the equation, producers and entrepreneurs can more accurately estimate the cost of transporting the straw, depending on the water content of the straw during harvesting. For this, the decision maker must know the cost of shipping at US$/t (based on 10% of water content) in the function of distance. Generally, this base cost is known within power plant, but is not detailed in the function of water content, and the use of the equation can be plausible.

In addition to the impact of the transportation cost caused by water content, another problem is related to the storage of the bales in the mills. Regardless of the way bales are used, there is a certain level of storage, which can be classified as short (up to 1 week), medium (1 week to 2 months), and long term (over 2 months).

Short-term storage usually occurs in bale producers, who after baling require a few hours or days to deliver the bales to their final destination, usually the power plants. In this logistics, the bales stay dispersed in the field or in small piles of up to six bales high until they are taken to the power plant, being subject to the occurrence of rains. Medium-term storage is usually associated with the companies that process the bales. When these companies operate in the same period as the sugarcane harvest, a small stock is formed at the power plant aiming at not stopping the industry due to a momentary interruption of production and transportation of bales. For example, in southeastern Brazil, after a 20 mm rainfall it may take up to a week for the straw to dry and the baling operation to resume.

In the long-term storage, we are referring to the companies designed to operate througout the year, whether for the production of electricity, ethanol, or cellulose, independent of the sugarcane harvest. For this, every day more amount of bales is produced than the daily consumed by the industry, aiming to supply the off-season months of the sugarcane. In this

Table 3.1 Increased transport costs of sugarcane straw in bales due to the increase of water content.

Maximum load (tons)	Water content of bales (%)	Bales weight (tonsDM)	Water weight (tons)	Freight cost (US$/load)	Cost (US $/tonsDM)	Cost increase (%)
74.0	10	66.6	7.4	200.00	1.50	0.0
74.0	15	62.9	11.1	200.00	1.59	5.9
74.0	20	59.2	14.8	200.00	1.69	12.5
74.0	25	55.5	18.5	200.00	1.80	20.0
74.0	30	51.8	22.2	200.00	1.93	28.6
74.0	35	48.1	25.9	200.00	2.08	38.5
74.0	40	44.4	29.6	200.00	2.25	50.0
74.0	45	40.7	33.3	200.00	2.46	63.6
74.0	50	37.0	37.0	200.00	2.70	80.0
74.0	55	33.3	40.7	200.00	3.00	100.0
74.0	60	29.6	44.4	200.00	3.38	125.0
74.0	65	25.9	48.1	200.00	3.86	157.1

Figure 3.5 Loading of bales in trucks adapted for sugarcane straw.

Figure 3.6 Covered patio for storing bales of sugarcane straw in the medium term.

sense, straw bales can be stored for up to a year in the industry yard, which can reach more than 50 hectares. Due to the large extensions of these patios, it is very expensive to construct roofs to protect the biomass from rainfall (Fig. 3.6). In sugarcane-producing regions, it can rain from 1000 to 1500 mm annually, all of this water being dumped on the bales.

All this water inside the bales, whether being before the baling or after the storage, together with billets and other constituents of the straw, lead to the fermentation of the biomass. In our preliminary studies, we obtained temperatures up to 70°C inside small piles of bales in the first weeks of storage. This increase in temperature due to water content may be related to the fires occurred in the bales storage of two large Brazilian companies.

In this context, some preliminary work was carried out to monitor water content inside the bales after rainfall. In bales dispersed in the field, which were packed with 10% water content, after 24 hours of a 15 mm rain, the water content inside the bales varied between 10% and 65% (short-term storage). From these works, experiments were carried out with bale piles in medium-term storage, being monitored up to 70 days (Fig. 3.7).

Nine piles of 27 bales each were field mounted in different treatments, and 294 water content points were sampled in each pile (Fig. 3.8). Each treatment consisted in the variation between type of straw (chopped or not), density, and amount of rain accumulated in the bales. The chopped straw and not chopped is related to the existence or not of a chopper coupled to the baler. The lowest and highest density used in the experiment were 147 and 193 $kg.MS/m^3$.

A total of 2646 water content points were sampled in the nine different piles, all of which were analyzed via geostatistics from ArcMac 10.3 software. All treatments presented a medium to high spatial dependence, with 69% being the lowest value of the Spatial Dependency Index (SDI) found, and considering this condition, geostatistical maps of water content distribution were made for the nine piles (Fig. 3.9).

The type of bale interfered in the distribution of water content in the sugarcane piles stored without cover. For medium storage periods, bales

Figure 3.7 Field-mounted bale piles to monitor water content after rainfall.

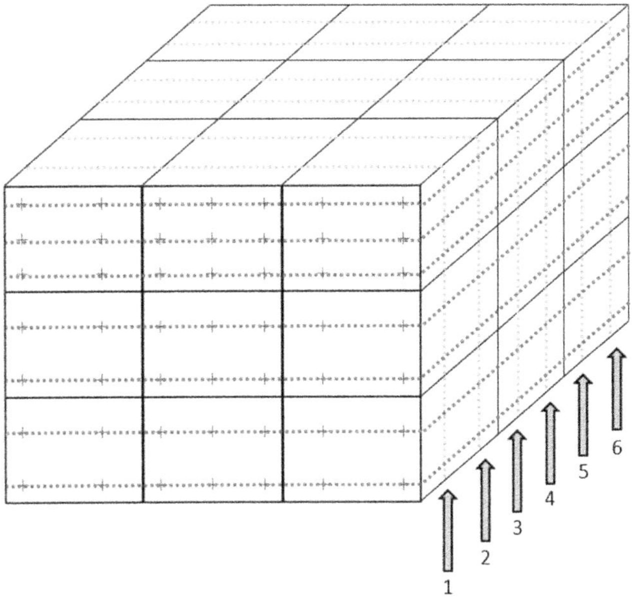

Figure 3.8 Sketch of the water content sampling points in the sugarcane bale piles assembled in the experiment.

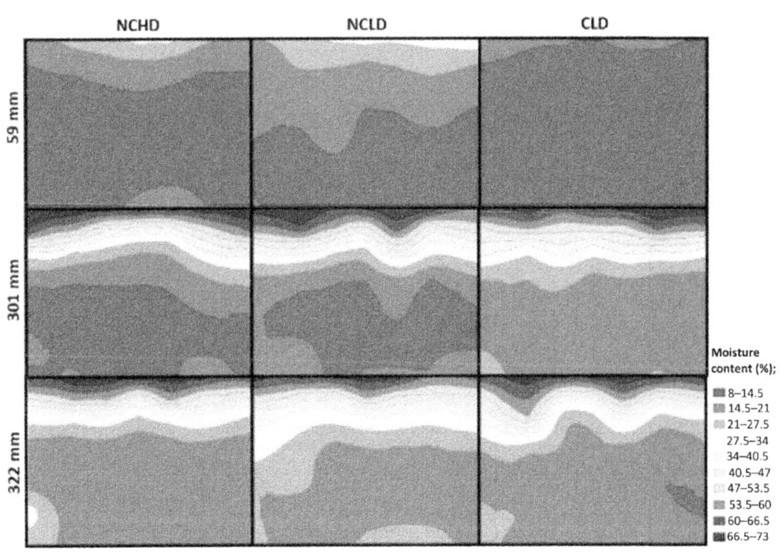

Figure 3.9 Geostatistical maps of the water content distribution in nine different treatments of straw type, density, and amount of accumulated rainfall. *CLD*, Chopped and low density; *NCHD*, not chopped and high density; *NCLD*, not chopped and low density.

of chopped straw are more recommended, but in longer periods, bales of nonchopped straw are preferable due to the lower retention of water content. In the longer storage period evaluated (greater accumulation of rainfall), chopped piles presented potential for natural drying in the field after rainwater retention.

All these dynamics presented in this chapter, mainly involving water content and mineral impurity, must be carefully analyzed by producers and entrepreneurs, as they are highly complex. The equations and results presented in this chapter aim to exploit this abundant biomass in the sugarcane-producing areas with greater efficiency, lower costs, increasing the profitability of the productive systems, and also reducing the risks. More research is being carried out, especially on long-term storage, as well as the routes of use of the straw for the farms and industries to be increasingly efficient in the generation of alternative energy.

References

Gilio, L 2015. Analysis of the socioeconomic impacts of the expansion of the sugarcane industry. Ph.D thesis, Luiz de Queiroz College ofAgriculture.

Lima, N.C., Silva de Souza, G.H., 2014, 'Rural and agroindustrial organizations', vol. 16, no. 4, pp 532–544. Available from: <http://www.redalyc.org/pdf/878/87832899010.pdf>. [15 July 2017].

Manzatto, C.V., Freitas Junior, E., Peres, J.R.R., 2002. Agricultural Use of Brazilian Soils. Embrapa, Rio de Janeiro.

National Energy Balance, 2016. Available from: <http://www.cbdb.org.br/informe/img/63socios7.pdf>. [25 July 2017].

National Union of the Component Industry for Motor Vehicles, 2016. Available from: <http://www.sindipecas.org.br/area-atuacao/?co = s&a = frota-circulante>. [30 June 2017].

NovaCana.com. *Sugarcane gave rise to 17.5% of all energy in Brazil in 2016*, 2017. Available from: <https://www.novacana.com/n/cana/meio-ambiente/cana-de-acucar-origem-17-energia-brasil-2016-140717>. [23 June 2017].

Ripoli, T.C.C., Molina Junior, W.F., Ripoli, M.L.C., 1999. Bagasse and straw in electricity production'. Global climate change and brazilian agriculture conference. EMBRAPA, Campinas, p. 37.

Sugarcane, sugar and alcohol producers' council of Sao Paulo State, 2006. Instruction manual. CONSECANA, Piracicaba.

Further reading

Brazilian Agricultural Research Corporation n.d., *Biorefineries*, EMBRAPA, Brasília. Available from: <http://www.infoteca.cnptia.embrapa.br/bitstream/doc/908142/1/biorrefinariamodificadoweb.pdf>. [25 June 2017].

Levy, C., 2008. The balance recipe. Unicamp Journal. . Available from: <https://www.unicamp.br/unicamp/unicamp_hoje/jornalPDF/ju402pag03.pdf>. [22 June 2017].

Manzatto, C.V., Freitas Junior, E., Peres, J.R.R., 2002. Agricultural Use of Brazilian Soils. Embrapa, Rio de Janeiro.

National Association of Manufacturers of Automotive Vehicles, 2016. Yearbook of the Brazilian Automotive Industry. ANFAVEA, Brasília. Available from: <http://www.automotivebusiness.com.br/abinteligencia/pdf/Anfavea_anuario2016.pdf>. [30 June 2017].

National Confederation of Agriculture and Livestock of Brazil, 2016. The potential of Brazil. CNA, Brasília. Available from: <https://ainfo.cnptia.embrapa.br/digital/bitstream/item/19030/1/Producao-agricola-mundial.pdf>. [28 July 2017].

National Energy Balance, 2016. Available from: <http://www.cbdb.org.br/informe/img/63socios7.pdf>. [25 July 2017].

National Energy Efficiency Plan, 2016. Available from: <http://www.ons.org.br/download/avaliacao_condicao/operacao_energetica/RE-3-0101-2016_PEN2016_SumarioExecutivo.pdf >. [30 June 2017].

National Petroleum Agency n.d., *Biofuels*. Available from: <http://www.anp.gov.br/biocombustíveis>. [28 July 2017].

National Supply Company, 2017. Follow-up of the Brazilian sugarcane crop. CONAB. Available from: <http://www.conab.gov.br/OlalaCMS/uploads/arquivos/16_12_27_16_30_01_boletim_cana_portugues_-3o_lev_-_16-17.pdf>. [22 July 2017].

National Union of the Component Industry for Motor Vehicles, 2016. Available from: <http://www.sindipecas.org.br/area-atuacao/?co = s&a = frota-circulante>. [30 June 2017].

NovaCana.com. *Sugarcane gave rise to 17.5% of all energy in Brazil in 2016*, 2017. Available from: <https://www.novacana.com/n/cana/meio-ambiente/cana-de-acucar-origem-17-energia-brasil-2016-140717>. [23 June 2017].

Raízen n.d., *Second generation ethanol*. Available from: <http://www.raizen.com.br/energia-do-futuro-tecnologia-em-energia-renovavel/etanol-de-segunda-geracao>. [28 July 2017].

Ripoli, T.C.C., Molina Junior, W.F., Ripoli, M.L.C., 2000. Energy potential of sugar cane biomass in Brazil. Scientia Agricola 57 (4), 677–681. Available from: <http://www.scielo.br/scielo.php?pid = S0103-90162000000400013&script = sci_abstract&tlng = pt>. [25 June 2017].

Yearbook of Brazilian Agriculture Agrianual 2016. Available from: <http://www.agrianual.com.br>. [30 June 2017].

CHAPTER 4

Sugarcane: food production, energy, and environment

Evaristo E. de Miranda and Marcelo F. Fonseca

Contents

Introduction 67
Sugarcane: the main agricultural product of Brazil 68
Sugarcane: food and energy production 70
  Food 70
  Agroenergy 72
Sugarcane in São Paulo State 75
Sugarcane and the environment 79
Conclusion 85
References 86
Further reading 87

Introduction

Sugarcane is one of the main and most environmental friendly crop of Brazilian agriculture. In addition to generating electricity through the bagasse, sugarcane production keeps Brazil as the largest producer and exporter of sugar and assures the running of a huge fleet of ethanol and anhydrous ethanol-fueled vehicles. In the 2015/16 crop production, 10.8 million hectares of sugarcane were planted. Production reached 666.8 million tons of sugarcane, processed into 33.8 million tons of sugar and 30.2 thousand m^3 of ethanol (ÚNICA, 2017). The proportion of sugar/ethanol (food/energy) production varies from year to year following the international prices of each commodity always causing great economic and territorial outcome, particularly for the southeast region and São Paulo State. The sugarcane production chain includes many sustainable management practices, outlined by a model whose territorial management aligns production and environmental preservation. Besides the good agronomic practices in rural properties, the maintenance of native vegetation

Sugarcane Biorefinery, Technology and Perspectives
DOI: https://doi.org/10.1016/B978-0-12-814236-3.00004-4

is present in sugarcane production areas. Numerical and cartographic data (to be published) on sugarcane production and the environment were obtained from the information found on the Rural Environmental Registry (CAR, in Portuguese), which is a management tool established by the Forest Code of 2012, and are presented in this paper. In the state of São Paulo, they depict the size and the territorial distribution of areas dedicated to preservation of native vegetation in rural properties dedicated to sugarcane production in the state of São Paulo.

Sugarcane: the main agricultural product of Brazil

Population, longevity, income, and food consumption continuously grow across the globe. World population is increasing and people are living longer. In response, world food production and consumption grow every year. The current world population is estimated at 7.6 billion (UN, 2017). The challenge is to feed an estimated additional of over 2 billion people the next four decades. The Brazilian animal and vegetal food production are sufficient to meet the basic food needs of its population and other one billion people. The country is one of the world's leading producers and/or exporters of sugar, soybeans, beef, poultry, coffee, citrus, corn, and other products. Brazilian agriculture is also a major producer of agrofuels (solid, liquid, gaseous, and energy fuels), in addition to vegetable and animal fibers.

Brazil's trade surplus remains positive thanks to the agribusiness, whose exports totaled more than US$ 84.9 billion in 2016 and represent almost half of total exports (AGROSTAT, 2017), having sugar as the main product. The gross agricultural production value was R$ 531 billion in 2016 and the five main products were: livestock (R$ 144 billion, excluding milk and eggs), soybeans (R$ 116 billion), sugarcane (R$ 53 billion), corn (R$ 42 billion), and milk (R$ 28 billion). Agricultural participation in the total gross domestic product was 23% (CNA, 2016). The country has significant participation in the international market, leading both the production and export of a dozen of agricultural products, as shown in Table 4.1.

Sugarcane was first introduced in Brazil by Martim Affonso de Souza in the beginning of 16th century, at São Vicente captaincy (presently the coast of Sao Paulo State). The first sugar mill constructed there promoted the territorial expansion toward the northeast of Brazil. After a few decades, Portugal assumed the monopoly of sugar production (Miranda,

Table 4.1 Brazil: products, production, export, and international trade.

Main agricultural products	World ranking		Participation in international trade (exports, %)
	Production/metric tons	Export/metric tons	
Sugar	1°/38,700,117	1°/28,933,000	48
Coffee	1°/3,060,000	1°/1,918,000	25
Orange juice	1°/1,092,900	1°/1,080,939	76
Soybeans	2°/113,013,400	1°/63,577,000	43
Chicken meat	2°/14,223,900	1°/4,307,000	38
Beef	2°/8,458,600	2°/1,349,000	19
Corn	3°/92,832,600	2°/21,842,000	22
Soy oil	4°/7,800,000	2°/1,475,000	13
Soybean meal	4°/30,800,000	2°/15,244,000	23
Cotton	5°/2,236,000	4°/805,000	8
Pig meat	4°/3,609,000	4°/720,000	10

Source: Data from CONAB - Companhia Nacional de Abastecimento (National supply company). Brazilian grain Harvest Monitoring: 2017. Brasilia. Available at: <http://www.conab.gov.br>. Accessed on: 03 Oct. 2019; USDA, United States Department of Agriculture, 2017. Disponível em: <https://www.usda.gov/> (17 jul. 2017).

2008). At that time, sugarcane production, crop postharvest outflow, and export of sugarcane products were boosted by good adaptation of crop to local soil and climatic conditions and the strategic location of sugar mills to export. These conditions made sugarcane to become the country's major commodity for a long period.

Sugarcane production chain faced considerable transformation since then, but its socioeconomic relevance remains the same. In 2014 113 countries had some sugarcane production in their territory (FAOSTAT, 2017). More than 10 million hectares was planted in Brazil, with more than 700 million tons of sugarcane harvested, yielding an average of 71 tons/ha. Table 4.2 presents the 10 top-ranking countries in sugarcane production.

Table 4.2 shows the impressive participation of Brazil (44%) in the total area harvested and quantity of sugarcane produced. The economic repercussions are very significant, given the diversity of industrialized products derived from the production of sugarcane. According to a study conducted by Neves and Trombin (2014), the value generated by the sugar-energy sector in the 2013/14 crop was over 43 billion dollars and represented almost 2% of the gross domestic product (GDP) of 2013. This expanded view can be observed in Table 4.3 by means of the sector-specific GDP given by the final products of the entire sugarcane production chain.

Table 4.2 Main sugarcane-producing countries in the world.

Position	Country	Harvested area (ha)	Produced quantity (tons)	Productivity (tons/ha)
1	Brazil	10,419,678	736,108,487	71
2	India	5,012,000	352,142,000	70
3	China	3,528,838	251,764,769	71
4	Thailand	1,353,025	103,697,005	77
5	Pakistan	1,140,492	62,826,458	55
6	Mexico	761,834	56,672,829	74
7	Colombia	401,132	36,508,450	91
8	Australia	375,216	30,517,650	81
9	Indonesia	472,676	28,600,000	61
10	United States	351,470	27,600,190	79
	Total	23,816,361	1,686,437,838	Average = 73

Source: Data from FAOSTAT, Food and Agriculture Organization of the United Nations, 2014. Available at: <http://www.fao.org/faostat/en/#home>.

Table 4.3 Estimated gross domestic product of the sugarcane sector in 2013 (US$ million).

Sugarcane by-product	Local market	Export market	Total
Hydrous ethanol	12,861	591	13,452
Anhydrous ethanol	8890	1076	9966
Nonenergy ethanol	655	–	655
Sugar	6927	11,110	18,037
Bioelectricity	894	–	894
Bioplastic	90	210	300
Yeast and additive	21	34	55
Carbon credit	–	0.27	0.27
Total	30,338	13,021	43,359

Source: Data from Neves, M.F., Trombin, V.G, 2014. A Dimensão do Setor Sucroenergético. Mapeamento e Quantificação da Safra 2013/14. Ribeirão Preto: Markestrat, Fundace, FEA-RP/USP.

Sugarcane: food and energy production

Food

Sugar

Sugar is, at the same time, a staple food and a commodity produced in many parts of the world. Sugarcane is responsible for 85%–87% of world sugar production; the rest of sugar is produced from sugar beet. The production involves agricultural and industrial processes that are very labor and capital intensive. The world production was 171 million tons in

2016, but consumption is even higher: 174 million tons, requiring the use of world stocks and pushing for an increase in production. The increase in demand is due to population growth and buying capacity in several world regions and increase in consumption of processed foods and sugar-based sweeteners such as sucralose. The highest rise per capita in sugar consumption is presently in Asia, where income and the rural exodus are growing rapidly. The largest sugar consumers in the world are also the largest producers: Brazil, India, China, Thailand, Pakistan, Mexico, and Colombia. Brazilian production is equivalent to the sum of the next six countries in the ranking. Brazilian consumption per capita is 52 kg/year, while the world average is 22 kg/year (CONAB, 2015).

Sugarcane liquor (Cachaça)

The sugarcane liquor—the sugarcane brandy produced exclusively in Brazil—is the second most consumed alcoholic beverage in the country, surpassed only by beer. It is the preferred distillate of Brazilians, whose consumption is almost five times greater than that of whiskey and vodka. The Brazilian sugarcane liquor market has 4200 trademarks registered in Ministry of Livestock Agriculture and Supply (MAPA) and generates about 400,000 direct and indirect jobs. Cane spirit may contain 38%–54% ethanol by volume at 20°C. The alcoholic content should be maintained below 48% to be considered as sugarcane liquor, either for industrial or artisanal products. The sugarcane liquor can be stored in barrels of different types of wood, for a year or more (aged sugarcane liquor). In this case, the sugarcane is harvested without previous burning and processed on the same day and the fermentation is done with natural yeast for 24 hours.

More than 40,000 small farmers are involved with sugarcane liquor production in Brazil. There are crops that are cultivated only for this purpose and the beverage integrates a set of cultural traditions. The production capacity is estimated at 1.2 billion liters/year, but the official national production is around 800 million liters/year; the state of São Paulo accounts for most of the industrial production (44% of the national total) and accounts for half of the exports. Minas Gerais State leads artisanal production, with 300 million liters/year (IBRAC, 2017).

Exports—US$ 13.9 million for 8.3 million liters in 2016—exporting to 60 countries, particularly in Europe, where Germany is the largest consumer (AGROSTAT, 2016). Certified sugarcane liquor holds the National Conformity Mark, following the standards established by the

Ministry of Agriculture. The certification system encourages improvements in the quality of the beverage and in the production process; indicates compliance with health, safety, environmental and social responsibility requirements; adds value to brand names; opens up new markets; informs and protects the consumer; and facilitate purchasing decisions. In recent years, the national market for sugarcane liquor has grown around 40% due to the incentive created by various programs to promote the beverage, besides the upgrade of status, being consumed by a consumer that is more demanding and with more buying capacity (COOCACHAÇA, 2015).

Agroenergy

Thanks to agriculture, Brazil has one of the planet cleanest energy production system. Renewable energy generated by biomass alone accounts for more than 30% of the Brazilian energy production. It represents, at least, 75 million tons of oil equivalent (TEP) per year in agroenergy (BEN, 2017). In other words, the solar energy is transformed into chemical energy by plants through photosynthesis and stored in stems and leaves or in animal fat, which are fed by plants. This chemical energy can be converted into liquid fuel (biodiesel, ethanol), solid (firewood, coal), and/or gaseous (biogas).

Agroenergy efficiency happens with tropical crops where production cycles are long. This allows plants to accumulate more chemical energy. In temperate countries, the climate limits the photosynthetic activity to a maximum of 120 days per year, as is the case of short cycle crops—wheat, oats, maize, and oilseeds. It requires one-third or even a quarter of the time used by sugarcane, energetic forests, and other tropical long-cycle crops (cassava, oil palm, pasture), whose photosynthesis is practically constant throughout the year. In addition, long-cycle crops occupy the land for several years until they are renewed, while short cycle crops need to be planted annually with higher fossil fuel consumption in land preparation, erosion problems, etc.

Brazilian agroenergy system is one of the most profitable and sustainable in the planet, due to the use of innovative technologies of mechanization, fertilization, pest and disease control, waste recycling, among others.

Ethanol and anhydrous alcohol

The bi-fuel car is an example of success: from 1.98 million small vehicles licensed in 2016 in Brazil, 88% were flex-fuel (MME, 2017). In 2015 41.2% of Brazilian energy came from renewable sources (BEN, 2017), while the world average was 13.5% and that of the Organization for Economic Cooperation and Development countries was only 9.4% (IEA, 2016). In other words, more than 90% of the energy used in developed countries is of fossil origin (oil, gas, and coal), high CO_2 and other pollutant gas emitter, or come from nuclear power plants.

These are the sources that best contribute to the renewable part of the Brazilian energy matrix: sugarcane (ethanol and cogeneration of electricity), with 16.9%; hydroelectric power plants, with 11.3%; energy forests (firewood and coal), with 8.2%; vegetable oils and cattle fat in biodiesel, energy exploration of agricultural residues and other renewable sources, such as wind and solar, adding up to 4.7%. The bioelectricity produced by agriculture alone (all biomass sources, with a large share from sugarcane) has an installed capacity of 14,619 MW (SINGLE, 2017), more than the total of solar, wind, and nuclear sources together (12,237 MW), and more than natural gas (13,705 MW).

In 2010 Brazil reached a record of sugarcane production of more than 700 million tons. In 2016 even with the economic crisis, 667 million tons were produced and transformed into 33.8 million tons of sugar and 30.2 billion liters of ethanol and anhydrous ethanol (ÚNICA, 2017). This production filled the tanks of 24.9 million vehicles powered by alcohol or biofuels (58.4% of the total fleet). In addition, the anhydrous alcohol was mixed with gasoline for fueling another 13.5 million gasoline-powered vehicles, in proportions ranging from 18% to 25%.

The increasing consumption of ethanol and anhydrous ethanol creates a permanent need for expanding production. If 50% of the Brazilian vehicles use ethanol in 2020, the production of sugarcane needed will be 1.2 billion tons. One and half billion tons will be needed if 80% of vehicles use ethanol. That means raising sugarcane production by 150%. However, innovative second-generation ethanol extraction technologies, using bagasse, can positively change this picture. Ethanol and anhydrous ethanol are presently efficient reducers of CO_2 emissions in the Brazilian fuel matrix by totally or partially replacing the use of gasoline. There is also no emission of nitrogen oxide or sulfur and the emission of particulates and other pollutants is less. This guarantees the improvement of the air quality in the Brazilian large cities, with proven benefit for the health of the population and the environment.

Bioenergy

Each tonne of sugarcane generates, in average, 250 kg of bagasse and 200 kg of straw and tips. In the past, these by-products were burned in boilers with low energy efficiency. Today, they are valuable by-products. Processing equipment was improved and now they produce enough steam to spin turbines and generate electricity, bioelectricity. Steam production doubled from 200 to over 400 tonnetonss/h and the generation capacity of the turbines from 30 to 60 MW or more. The plants moved from self-sufficiency in energy (during harvest time) to the generation of surpluses, furnishing to the National Integrated Energy System (SIN).

The available capacity for energy cogeneration of sugarcane plants connected to the energy network reached 11,086 MW (UNICA, 2017) and could reach 22,000 MW by 2020, surpassing the capacity of Itaipu hydroelectric plant (14,000 MW). One tonne of bagasse can generate more than 300 kWh for the energy network. A ton of straw can generate 500 kWh. The average consumption of a Brazilian household in 2015 was 166 kWh (BEN, 2017). One hectare of cane can supply with bioelectricity around eight households for 1 year, and provide fuel for their cars.

There is still the possibility of giving a different use to the vinasse, presently being used for ferti-irrigation, previously considered an undesirable and polluting waste. Some sugar-processing plants are using vinasse through bio digestion, to provide power to boilers, having methane to power generator or a turbine and generate energy. The capacity to generate energy depends essentially on the quality of the vinasse and the quantity of organic material in the residue. The production of 1 L of ethanol generates 10–14 L of vinasse. To obtain 30 billion liters of ethanol and anhydrous ethanol, the processing plants produce between 300 and 420 billion liters of vinasse. That is, the bioelectricity generation potential of this waste varies between 1600 and 2100 MW per year. Moreover, the new production of bioenergy from bagasse and sugarcane straw does not cause energy losses or additional emissions of greenhouse gases. It also does not require long-distance transmission—as it happens with Amazonian hydroelectric plants—given the proximity and capillarity of the energy distribution network and consumption centers.

Currently, bioelectricity is present in 20 Brazilian states, with São Paulo accounting for 65% of the country's total. Bioelectricity from sugarcane is available between April and November in the Center–South, whose sugarcane harvest takes about 210 days. In the North and Northeast, the harvest is from September to February and takes about 160 days. Although available for only 6–8 months, bioelectricity is added to the network at the time of the dry season, when rivers are at the lowest level and there is a sharp drop in hydroelectric production. It is exactly during this period that fossil fuel-fired thermoelectric plants (coal, diesel, or gas) need to supply energy to the network. Limiting the use of these fossil fuels in the electric energy network by replacing it with bioelectricity, Brazil will be entitled to carbon credits, and a better predictability of energy available.

Sugarcane in São Paulo State

About 90% of the area planted with sugarcane in Brazil is located in the Center-South of Brazil, mainly in the State of São Paulo, responsible for more than half of the total, surpassing the 5.5 million hectares (IBGE, 2017).[1] The value of sugarcane production generated in São Paulo is higher than all other states together. With the introduction of the National Alcohol Program (PROÁLCOOL) in the 1970s, sugarcane established a new territorial dynamics in São Paulo, expanding its area, not only in traditional sugarcane regions as Piracicaba and Ribeirão Preto, but also in the northwest and west of Sao Paulo, such as Assis, Araçatuba and São José do Rio Preto (Miranda, 2010). Table 4.4 shows the ranking of the 20 main Brazilian municipalities that produce sugarcane; 9 of them are in the state of São Paulo.

Looking at the top 100 sugarcane-producing municipalities, the presence of São Paulo's participation increases even more: there are 64 municipalities with a total of 2,304,571 ha of sugarcane area and 174,577,355 tons produced (IBGE, 2017). The territorial distribution of sugarcane production in the state of São Paulo is presented in Fig. 4.1.

With all this significant numerical and spatial importance, the size of the sugar and alcohol sector involving the breadth of products and by-products generated from the production of sugarcane, can be seen from the perspective of compilations in relation to the others agricultural products. Table 4.5 shows the significance of sugarcane and its agro-industrial products as compared to other agricultural products, in the State of Sao Paulo (the 20 main crops with more than 10,000 ha cultivated in 2014).

Table 4.6 compares the territorial dimension of sugarcane in relation to forestry activity of its main crops (eucalyptus, pinus, and rubber tree).

[1] The planted area data presented by different research institutions show slight variations due to the use of different methodologies.

Table 4.4 Sugarcane area (ha) and quantity produced (ton) in the main Brazilian producing municipalities.

Ranking	Municipality	State	Area (ha)	Production (tons)
1	Morro Agudo	SP	95,000	7,600,000
2	Quirinópolis	GO	74,396	6,758,505
3	Uberaba	MG	73,720	6,266,200
4	Rio Brilhante	MS	72,389	6,384,095
5	Nova Alvorada do Sul	MS	71,339	5,380,819
6	Barretos	SP	65,500	5,240,000
7	Frutal	MG	61,972	5,001,040
8	Guaíra	SP	60,000	5,100,000
9	Jaboticabal	SP	57,550	4,604,000
10	Paraguaçu Paulista	SP	54,161	3,859,025
11	Santa Vitória	MG	53,500	3,477,500
12	Mineiros	GO	52,000	2,808,000
13	Rancharia	SP	51,639	4,131,153
14	Piracicaba	SP	49,000	3,185,000
15	Goiatuba	GO	47,900	3,822,428
16	Barra do Bugres	MT	47,706	3,653,516
17	Ituverava	SP	47,500	3,800,000
18	Costa Rica	MS	47,174	3,721,661
19	Guararapes	SP	46,561	3,259,270
20	Denise	MT	45,623	2,793,040
	Brazil		10,161,622	748,636,167

Source: Data from IBGE, Instituto Brasileiro de Geografia e Estatística, 2017. Produção Agrícola Municipal. Disponível em: <https://sidra.ibge.gov.br/pesquisa/pam/tabelas> (04 jul. 2017).

The area occupied by forests is 18% of sugarcane area in the State of São Paulo. On the other hand, what happens with pastures is the opposite (Table 4.7). Pasture occupies 20% more than the total area of sugarcane in São Paulo.

Comparing the last two agricultural censuses, carried out by the IBGE in 1995 and 2006, natural pastures showed a tendency to increase, the opposite of planted pastures, which tended to decrease. The average pasture area per grazing animal (ha) also declined in the last decades, from 2.3 ha in 1950 to 1.1 ha for the year of the last national agricultural census (2006).

In degraded pasture areas and also in pasture areas that are capable to support more animals, sugarcane cultivation can invariably find suitable places for its expansion, integrating cattle raising with agriculture in a consortium. Figs. 4.2 and 4.3 represent, respectively, the areas of natural and cultivated

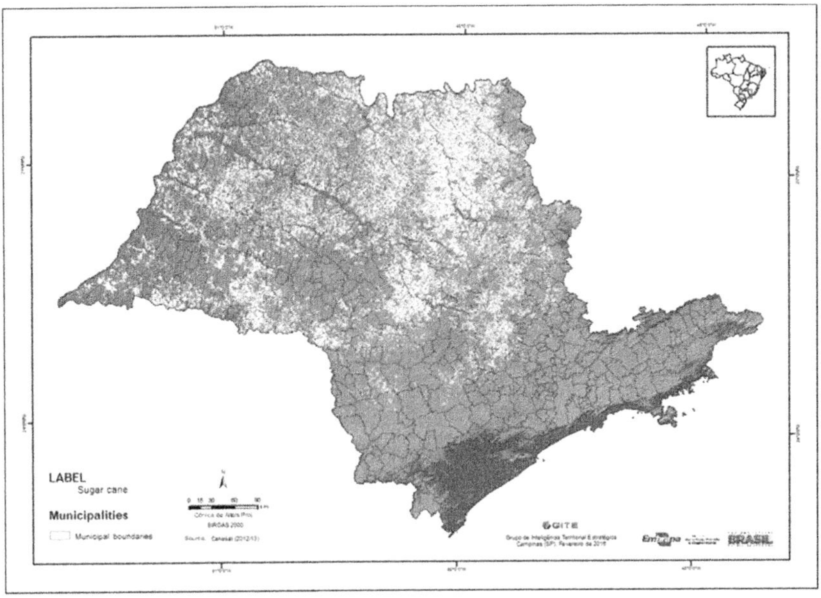

Figure 4.1 Spatial distribution of sugarcane on a digital topographic model in the State of São Paulo.

pasture in the State of São Paulo, made by Rural Development Office (RDO).[2]

Lastly, Table 4.8 presents the value of main 20 agricultural products in the State of São Paulo, showing the importance of sugarcane.

The production value of sugarcane in São Paulo represents 47% of the 20 main agricultural products of São Paulo. It represents more than three times the value of the second most "profitable" agricultural product, the cattle beef. The state's more than 5.5 million hectares of sugarcane are present in 484 municipalities (75% of the total) and in 194 of these municipalities (Fig. 4.4) there are more than 10,000 ha of sugarcane in their territory.

In these sugarcane-producing municipalities, there are many sugar, alcohol, and electric energy production plants, besides its immense production infrastructure, which is of fundamental importance when analyzed from the social and economic point of view. Despite the crisis that the sugarcane industry has experienced in recent years (NOVACANA, 2017;

[2] There are 40 RDOs distributed in the various regions of the State of São Paulo. These territorial units are used as the basis for collecting data from the São Paulo agriculture.

Table 4.5 Sugarcane cultivated area in the State of São Paulo as compared to the area occupied by other main agricultural products.

ID	Product	Area (ha)	% Relative	% Accumulated
1	Sugarcane	5,570,732.05	66.92	66.92
2	Corn	804,752.73	9.67	76.59
3	Soy	709,371.75	8.52	85.11
4	Orange	418,510.06	5.03	90.14
5	Coffee	203,472.78	2.44	92.59
6	Bean	112,281.44	1.35	93.93
7	Peanut	99,251.10	1.19	95.13
8	Wheat	75,623.00	0.91	96.04
9	Banana	54,531.20	0.66	96.69
10	Manioc	49,186.82	0.59	97.28
11	Potato	27,635.88	0.33	97.61
12	Sorghum	20,316.40	0.24	97.86
13	Lemon	15,227.97	0.18	98.04
14	Rice	13,986.60	0.17	98.21
15	Grape	13,078.00	0.16	98.37
16	Cotton	11,650.80	0.14	98.51
17	Tomato	11,389.17	0.14	98.64
18	Triticale	11,040.00	0.13	98.77
19	Mango	10,873.69	0.13	98.91
20	Lettuce	10,774.50	0.13	99.03
	Total	8,243,686	–	–

Source: Data from Institute of Agricultural Economics (IAE), 2014. Grouped crops. Areas in production in 2014.

Table 4.6 Sugarcane and the three main forestry products in São Paulo.

Product	Area (ha)	% Relative	% Accumulated
Sugarcane	5,570,732.05	100.00	–
Eucalyptus	842,680.80	85.24	85.24
Pinus	91,336.30	9.24	94.47
Rubber tree	54,622.47	5.53	100.00
Total forestry	988,639.57	100.00	–

Source: Data from EA, Instituto de Economia Agricola (Institute of Agricultural Economics), 2014. Database: São Paulo Production Statistics. Available at: <http://www.iea.sp.gov.br/out/bancodedados.html> (18 Apr. 2015).

ÚNICA, 2017), a survey carried out in a study showed the current situation of all these plants in the country (CTBE, 2017). It was found 376 active production plants, with 53% of them located in the Southeast region, 42%

Table 4.7 Sugarcane and pastures in the State of São Paulo.

Product	Area (ha)	% Relative	% Accumulated
Sugarcane	5,570,732.05	100.00	–
Cultivated pasture	5,749,570.35	85.49	85.49
Natural pasture	948,683.27	14.11	99.59
Pasture for seed production	27,510.60	0.41	100.00
Total pasture	6,725,764.22	100.00	–

Source: Data from Institute of Agricultural Economics (IAE), 2014.

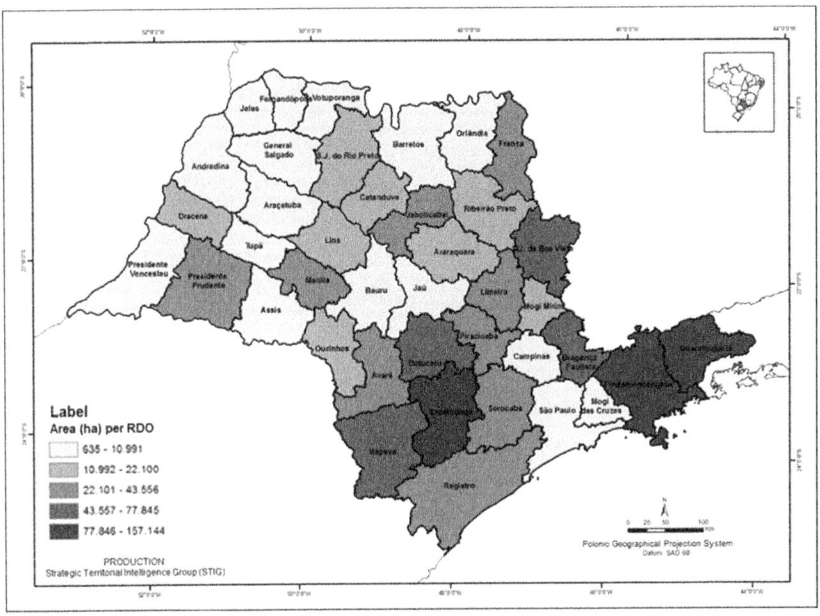

Figure 4.2 Natural pasture area in each RDO of the State of São Paulo.

in the state of São Paulo. The geocoded location of the plants in the state of São Paulo, distributed by biome, can be observed in Fig. 4.5. The study identified, for the year 2014, 203 sugarcane processing plants in the state, with 168 active plants and 35 inactive plants (EMBRAPA/GITE, 2014).

Sugarcane and the environment

Many ecological studies of Brazilian agriculture look exclusively at environmental impact of production systems, without considering the areas in

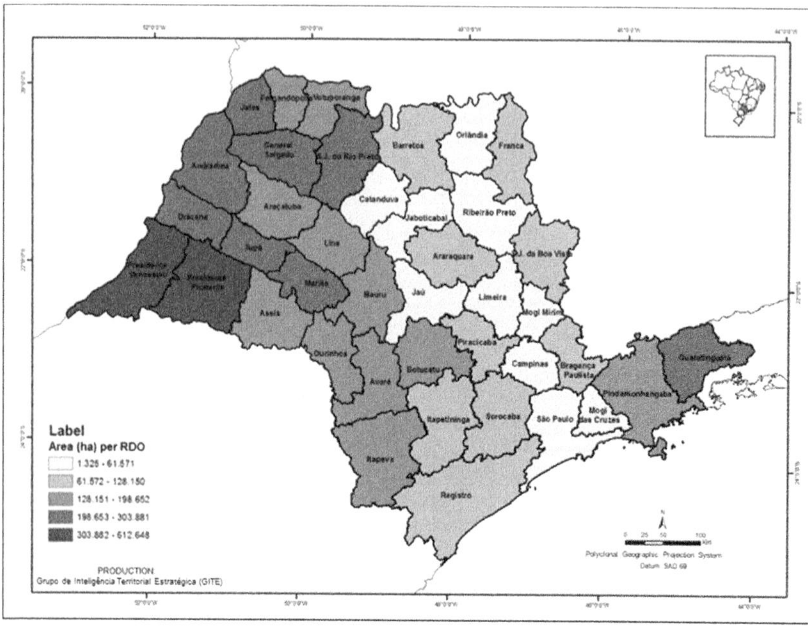

Figure 4.3 Pasture area cultivated in each RDO of the State of São Paulo.

rural properties that are not explored and maintained with native vegetation by farmers, interacting with neighboring agricultural areas. The environmental impact in exploited and preserved areas is permanent and dynamic and can be positive or negative. The understanding of environmental processes in agriculture cannot disregard this perspective of the presence of exploited and preserved areas in rural properties. This more comprehensive and complete perspective of the rural property makes it possible to understand what are of interest of the Brazilian agribusiness, its territorial dimension and its economic, agronomic and technological organization.

This kind of analysis, however, is insufficient if limited to samples of rural properties. Both the preserved and permanent areas acquire another dimension when considered in the scale of landscape, river basins, region, biome, country, and even the planet. Multiscale surveys of the territorial dimension of agriculture, from local to global, represent a major scientific challenge. In ecology, as in physics, scale creates the phenomenon (Forman and Godron, 1986).

The national policy for sugarcane seeks to expand on a sustainable basis, with economic, environmental and social criteria. The Agroecological

Table 4.8 Production value of the main 20 agricultural products in the State of São Paulo.

ID	Product	Unit price	Production	Unit	Value of production (R$)
1	Sugarcane	61.34	408,989,376	tons	25,087,408,345
2	Beef	120.72	65,007,069	15 kg	7,847,653,370
3	Chicken meat	2.33	1,582,876,976	kg	3,688,103,354
4	Orange for industry	8.99	221,330,614	Box of 40, 8 kg	1,989,762,076
5	Chicken's egg	59.09	32,705,730	Box 30 dz	1,932,581,611
6	Milk	1.05	1,674,787,100	Litre	1,758,526,455
7	Coffee benefited	382.46	4,419,180	Bag 60 kg	1,690,159,273
8	Soy	60.81	27,124,000	Bag 60 kg	1,649,410,452
9	Corn	23.59	60,325,133	Bag 60 kg	1,423,069,885
10	Table orange	13.63	73,776,885	Box 40–8 kg	1,005,578,684
11	Potato	57.14	15,905,510	Bag 50 kg	908,840,819
12	Banana	19.22	46,901,167	Box 21 kg	901,440,021
13	Table tomato	32.31	23,751,734	25 kg	767,418,526
14	Pig meat	70.92	7,839,402	15 kg	555,970,418
15	Tangerine	27.33	15,280,706	Box 26 kg	417,621,530
16	Table grape	2.74	147,349,530	kg	403,737,712
17	Lemon	19.23	20,169,130	Box 27 kg	387,852,370
18	Eraser	2.19	166,727,714	kg	365,133,693
19	Bean	102.47	3,528,300	Bag 60 kg	361,544,891
20	Peanuts in shell	29.71	10,800,676	Bag 25 kg	320,888,072
	Total	–	–	–	53.5 billion

Source: Data from Institute of Agricultural Economics (IAE), 2014.

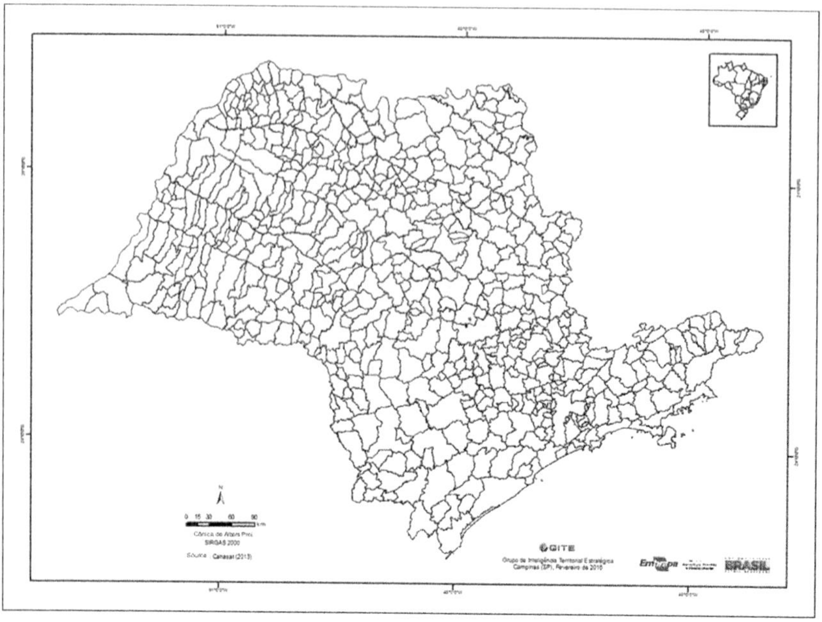

Figure 4.4 Municipalities with more than 10,000 ha of sugarcane in the state of São Paulo.

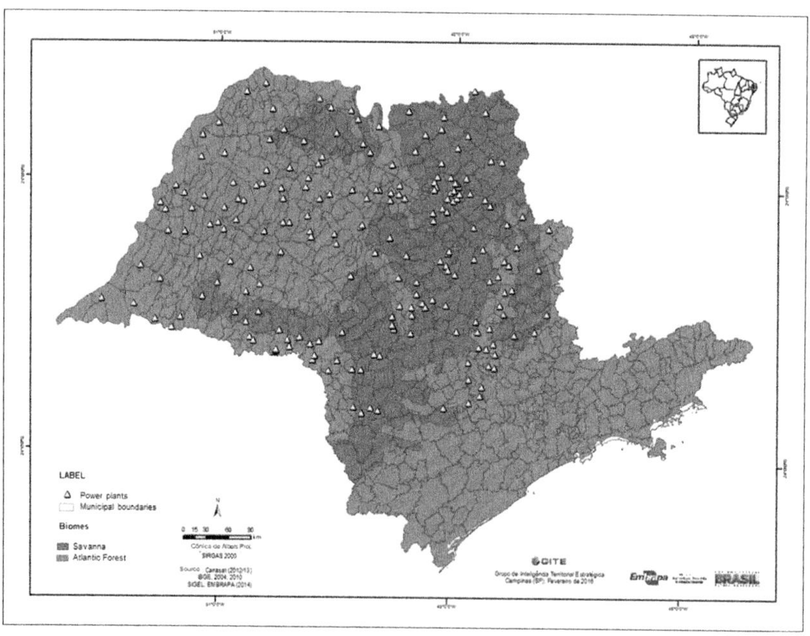

Figure 4.5 Location of sugar and alcohol production plants in biomes at the state of São Paulo.

Zoning of sugarcane regulates the planting in the country and considers the environment and the economic aptitude of regions. Expansion of cultivation in the Amazon, Pantanal, Upper Paraguay Basin and in areas of native vegetation is not allowed. From the technological standpoint, conservationist practices of soil and water and recycling of agricultural and industrial waste prevail. Mechanization replaces burnings at harvest, requiring flat areas, which contributes to the restoration of permanent preservation areas (PPA).

A trifold of sugarcane characteristics makes it an unbeatable producer of sugar, biofuel, and electricity in a sustainable way. They are the tropical climate in production areas, the intrinsic and genetically enhanced characteristics of the plant and the modern production technologies available for the production chain.

In tropical areas, sugarcane "generates" energy 11 months of the year. Once harvested, it sprouts rapidly, since the roots are already established in the soil. Thanks to new varieties, irrigation and soil conservation techniques, the same sugarcane area can be harvested from 6 to 8 years without replanting and in some cases even more than that. Together with soybeans and corn, sugarcane is one of the crops that most incorporate new production technologies, with many varieties adapted to different climatic situations and soils. Large-scale integrated management of pests and diseases, such as airplane fungi application and insects for biological control of pests, grown in large quantities on agricultural properties.

While soybean production accounts for 52% of the demand for agricultural pesticides in Brazil, sugarcane requires only 10% (SINDIVEG, 2016). The use of precision agriculture is increasing, as is the recycling of waste and effluents originating from sugarcane processing. Vinasse, filter cake, and boiler ash return to the soil with its nutrients and no environmental impacts, thanks to proper collecting, transportation, and distribution techniques.

The mechanized harvest reduces the use of fire, year after year. In São Paulo, 91.3% of the sugarcane harvest is already mechanized and without burning (SMA/CPLA, 2016). And it will be 100% by 2020, avoiding smoke and soot and increasing carbon credits, bioelectricity, saving inputs and better yields. In the Center–South region, productivity surpasses 100 tons/ha, making possible the country to meet the national demand for agroenergy, even with sugarcane occupying only 1.2% of the national territory and 13% of the cultivated agricultural area.

After the creation and implementation of the Rural Environmental Registry (CAR, in Portuguese), a management tool established in Chapter VI, Article 29 of related legislation, within the scope of the National Environmental Information System (SINIMA), established by the new Forest Code (Law 12.651/2012), a unique opportunity has been created to emphasize the importance given to the environment by farmers across the country. Preliminary studies with data provided by the Brazilian Forestry Service (BFS, 2016) indicate that in Brazil, more than 176 million hectares of vegetation are preserved inside of these properties (EMBRAPA/GITE, 2017).

In São Paulo State, over 302,000 rural properties were registered in the CAR, by the end of 2016, covering more than 17 million hectares. The area reserved for preservation by the farmers reached 3.6 million hectares. This means that around 20% of the total area of rural properties are preserved by farmers. It is understood as preserved area the ones classified as remnant of native vegetation, areas of permanent preservation, areas of restricted use and, if present, legal reserve area.

What is the participation of sugarcane farms in the amount of this preserved area in São Paulo? Preliminary calculations involving the analysis

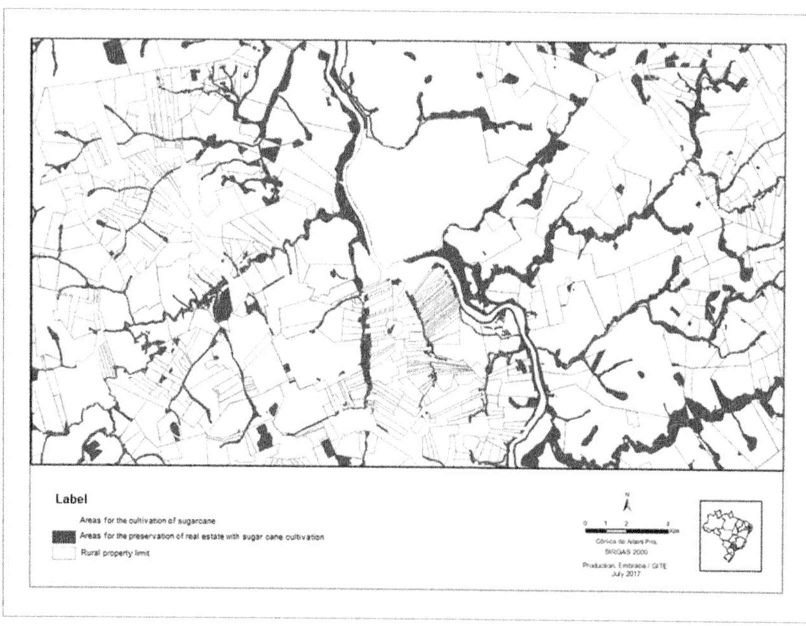

Figure 4.6 Areas of vegetation preserved in rural properties in the midst of sugarcane cultivation, connected and associated to the hydrographic network.

and geocoded crossing of sugarcane areas (CANASAT, 2016) and rural properties registered in CAR (SICAR, 2017) identify 97,000 registered rural properties (32% of the total of the state) planting sugarcane in their farms (about 5 million hectares).

These rural properties in São Paulo dedicate about 1.7 million hectares to the preservation of native vegetation in the form of PPAs, legal reserves, forest remnants, aquatic ecosystems, etc. It represents 49.6% of the entire preserved areas in registered rural properties throughout the state. In other words, almost half of the area dedicated to preservation of native vegetation in the rural properties of São Paulo is in 97,000 sugarcane farms.

Fig. 4.6 shows a cartographic representation of a region in the northeast–central part of São Paulo, where the preserved vegetation is seen bordering sugarcane plantations, with a spatial distribution that connects almost all the remnants along the hydrographic network.

Conclusion

The sugarcane farmers linked to the agro-industrial complex provide relevant service to the Brazilian urban–rural society, contributing to the development of the food, energy and environment segments. Some of the contributions of sugarcane agro-industry are production of sugar, fuel ethanol that improves air quality in large cities, bioelectricity, yeasts as animal feed, cogeneration of electric energy, and the maintenance of areas for environmental preservation. Future perspectives indicate advances in second-generation ethanol and genetic improvement of species adaptable to different climates and soils, with the participation of several of research institutions (public and private), universities, etc.

The State of São Paulo, with more than 5.5 million hectares planted and with an annual production value of more than 25 billion reals (almost half of the total of the 20 main agricultural products), is synonymous of development, generating dividends and taxes, direct and indirect jobs, moving the local and regional economic circuit, through the trade of fertilizers, agrichemicals, combines, tractors, implements, seedlings, equipment, services, etc.

In recent years, the massive mechanization of harvesting has not only reduced the use of fires, which is expected to be completely extinguished by 2020, representing gains in air quality and consequently human health.

It also opened possibilities for improving yields and protecting water sources, soil, and all the environmental diversity of farms. The care of the environment occurs in both the exploited and preserved areas in rural properties: preliminary data point to about 1.7 million hectares dedicated to the preservation of native vegetation in sugarcane farms, only in the state of São Paulo, according to a geocoded analysis of the data of the Rural Environmental Registry of 2016 (EMBRAPA, 2017).

References

AGROSTAT, Estatísticas de Comercio Exterior do Agronegócio Brasileiro (Brazilian Agribusiness Foreign Trade Statistics), 2016. Available at: < http://indicadores.agricultura.gov.br/agrostat/index.htm>.

AGROSTAT, 2017. Estatísticas do Comércio Exterior do Agronegócio Brasileiro. Disponível em: <http://sistemasweb.agricultura.gov.br/pages/AGROSTAT.html> (08 jun. 2017).

BEN, Balanço Energético Nacional, 2017. Empresa de Pesquisa Energética - EPE. Rio de Janeiro/RJ. Disponível em: <https://ben.epe.gov.br> (14 jun. 2017).

BFS, Brazilian Forestry Service (Serviço Florestal Brasileiro), 2016. Available at <http://www.florestal.gov.br/> Accessed: 01 Jan 2017. CANASAT. Mapa da colheita (Mapa da colheita), 2016. Available at: < http://www.dsr.inpe.br/laf/canasat/colheita.html>. Accessed: 28 abr. 2017.

CNA, Confederação da Agricultura e Pecuária do Brasil, 2016. Boletins. Disponível em: <http://www.cnabrasil.org.br/central-comunicacao/boletins> (10 dez. 2016).

CONAB, Companhia Nacional de Abastecimento, 2015. Dados de estimativas da safra. Disponível em: <http://www.conab.gov.br> (18 set. 2015).

COOCACHAÇA, 2015. Cooperativa de Produção e Promoção da Cachaça de Minas. Disponível em: <http://www.sebraemercados.com.br/numeros-da-cachaca-no-brasil> (27 out. 2015).

CTBE, 2017. Laboratório Nacional de Ciência e Tecnologia do Bioetanol. Mapeamento das Unidades de Produção de Açúcar, Álcool e Energia Elétrica no Brasil. Núcleo de Agricultura de Precisão. Centro Nacional de Pesquisa em Energia e Materiais. Boletim CTBE/CNPEM. Campinas/SP. 6 p.

EMBRAPA, Empresa Brasileira de Pesquisa Agropecuária, 2017. Agricultura e preservação ambiental: Uma primeira análise do Cadastro Ambiental Rural. Disponível em: <https://www.cnpm.embrapa.br/projetos/car/index.html> (17 jul. 2017).

EMBRAPA, Empresa Brasileira de Pesquisa Agropecuária. Grupo de Inteligência Territorial Estratégica (GITE), 2017. Disponível em: <https://www.embrapa.br/gite> (21 jun. 2017).

EMBRAPA/GITE, Empresa Brasileira de Pesquisa Agropecuária (Brazilian Agricultural Research Company)/Grupo de Inteligência Territorial Estratégica (Strategic Territorial Intelligence Group), 2014. Available at < https://www.embrapa.br/gite/> Accessed 01 Jan 2015.

FAOSTAT, Food and Agriculture Organization of the United Nations, 2017. Disponível em: <http://www.fao.org/faostat/en/#home> (04 jul. 2017).

Forman, R.T.T., Godron, M., 1986. 620 p Landscape Ecology. John Wiley & Sons, New York.

IBGE, Instituto Brasileiro de Geografia e Estatística, 2017. Produção Agrícola Municipal. Disponível em: <https://sidra.ibge.gov.br/pesquisa/pam/tabelas> (04 jul. 2017).

IBRAC, Instituto Brasileiro da Cachaça, 2017. Disponível em: <http://www.ibrac.net> (25 jan. 2017).
IEA, International Energy Agency, 2016. Disponível em: <https://www.iea.org> (16 dez. 2016).
Miranda, J.R., 2008. História da Cana-de-Açúcar, vol. 1. Komedi, Campinas, 168 p.
Miranda, E.E.de., 2010. Questões Ambientais (Impactos locais e globais) e energéticos: a expansão da cana-de-açúcar e a ocupação das terras no estado de São Paulo. In: Cortez, L.A.B. (Ed.), Bioetanol de cana-de-açúcar: P&D para produtividade e sustentabilidade. Blucher, Fapesp, 992 p.
MME, Ministério de Minas e Energia, 2017. Disponível em: <http://www.mme.gov.br> (27 jun. 2017).
Neves, M.F., Trombin, V.G., 2014. A Dimensão do Setor Sucroenergético. Mapeamento e Quantificação da Safra 2013/14. Ribeirão Preto: Markestrat, Fundace, FEA-RP/USP.
NOVACANA Portal, 2017. Crise afeta investimento de usinas em canaviais e reduz safra na região de Araraquara (SP). Disponível em: <https://www.novacana.com/n/cana/safra/crise-investimento-usinas-canaviais-safra-araraquara-sp-160517> (06 jul. 2017).
SICAR, Sistema Nacional de Cadastro Ambiental Rural, 2017. Cadastro Ambiental Rural (CAR). Disponível em: <http://www.car.gov.br> (25 de jan. 2017).
SINDIVEG, Sindicato Nacional da Indústria de Produtos para Defesa Vegetal, 2016. Disponível em: <http://sindiveg.org.br> (16 dez. 2016).
SMA/CPLA, Secretaria de Meio Ambiente do Estado de São Paulo. Coordenadoria de Planejamento Ambiental, 2016. Disponível em: <http://www.ambiente.sp.gov.br/cpla> (16 dez. 2016).
UN, United Nations, 2017. Available at www.un.org.
ÚNICA, União da Indústria de cana-de-açúcar, 2017. Disponível em: <http://www.unicadata.com.br> (08 jun. 2017).
USDA, United States Department of Agriculture, 2017. Disponível em: <https://www.usda.gov/> (17 jul. 2017).

Further reading

BRASIL, 2017. Novo Código Florestal (Lei Nº 12.651, de 25 de maio de 2012). Disponível em: <http://www.planalto.gov.br/ccivil_03/_ato2011-2014/2012/lei/L12651.htm> (14 jun. 2017).
CATI, Coordenadoria de Assistência Técnica Integral, 2016. Principais atividades agrícolas do Estado de São Paulo. Disponível em: <http://www.cati.sp.gov.br> (18 dez. 2016).
DSR/OBT/INPE, Divisão de Sensoriamento Remoto, Coordenação Geral de Observação da Terra, Instituto Nacional de Pesquisas Espaciais, 2014. Banco de Dados Geomorfométricos do Brasil (TOPODATA). Disponível em: <http://www.dsr.inpe.br/topodata> (10 nov. 2014).
IEA, Instituto de Economia Agrícola, 2015. Banco de dados: Estatísticas da Produção Paulista. Disponível em: <http://www.iea.sp.gov.br/out/bancodedados.html> (18 abr. 2015).
INPE, Projeto Canasat, Adami, M., Mello, M.P., Aguiar, D.A., Rudorff, B.F.T., Souza, A.F., 2012. A web platform development to perform thematic accuracy assessment of sugarcane mapping in South-Central Brazil. Remote Sens. 4, 3201–3214. Disponível em: <http://www.mdpi.com/2072-4292/4/10/3201>.

INPE, Projeto Canasat, Rudorff, B.F.T., Aguiar, D.A., Silva, W.F., Sugawara, L.M., Adami, M., Moreira, M.A., 2010. Studies on the rapid expansion of sugarcane for ethanol production in São Paulo State (Brazil) using Landsat Data. Remote Sens. 2, 1057–1076. Disponível em: <http://www.mdpi.com/2072-4292/2/4/1057>.

ONU, Organização das Nações Unidas, 2017. Perspectivas da População Mundial: revisão de 2017. Disponível em: <https://esa.un.org/unpd/wpp/> (07 jul. 2017).

SFB, Serviço Florestal Brasileiro, 2017. Números do Cadastro Ambiental Rural. Download dos dados do CAR. Disponível em: <http://www.car.gov.br/publico/imoveis/index> (17 jan. 2017).

CHAPTER 5

Sustainability of sugarcane for energy purposes

Carla Silva, Rui Pacheco, Danilo Arcentales and Fernando Santos

Contents

Introduction 89
Flex-fuel vehicle technologies and well-to-wheels 97
Conclusions 100
Acknowledgments 100
References 100

Introduction

The pursuit of fossil-free systems with minimum waste disposal covering food, energy, materials, and chemicals production is a major driver for the development of biorefineries where all these products could be produced simultaneously. The idea to have a dedicated unit to produce exclusively biofuels should be avoided, or its benefits over biorefinery should be proven. To select the most feasible biorefinery configuration, economic as well as environmental aspects must be studied. The former relates to a technoeconomic assessment (TEA), the latter is usually tackled in a life cycle assessment (LCA) framework.

Both TEA and LCA studies are designed to address a specific question and contain different assumptions, data sources, and uncertainties. Therefore, it is not surprising that the results vary widely across the studies and care must be taken in making direct comparisons between them. For example, a different geographical location/different country will have different labor, insurance of equipment and materials costs, different electricity and product prices, different currencies, different electricity generation mixes, different transportation/distribution distances, and even different agriculture productivities induced by different climate/weather conditions. Additionally, different studies may refer to different chronologies (different reference dollars or euros) and different Intergovernmental Panel on

Sugarcane Biorefinery, Technology and Perspectives
DOI: https://doi.org/10.1016/B978-0-12-814236-3.00005-6

Climate Change (IPCC) assessment report (AR) CO_2eq values. Table 5.1 includes the 100-year time horizon global warming potentials (GWP) relative to CO_2. The AR5 values are the most recent, but AR2, AR3, and AR4 report values are also listed because they are sometimes used for inventory and reporting purposes.

Historically, the culture of sugarcane has been considered as the main economic activity since the Brazilian colonial period, when slave labor was still used. Since then, with the rise in sugar consumption, the economic importance of the crop has been further increased. At this time, sugarcane mills in Brazil produced only sugar. The Pró-Álcool program (1975) induced the increase in distilleries and sugarcane mills producing both sugar and ethanol in an integrated fashion. More recently, electricity production is also a by-product of the sugarcane mills, and therefore the actual sugarcane mill is a biorefinery, producing food, electricity, and biofuel. Fig. 5.1 shows the sugarcane production historical evolution, the ethanol production, and the electricity sold to the grid. The number of registered flex-fuel cars is also represented to show its link with ethanol availability and sugarcane industry growth.

From 2005 to 2014, co-generated electric energy dispatch increased from 1.1 to 19.4 TWh, and biomass became the third electric energy generation source in Brazil, after hydro and fossil sources (Vasconcelos and Carpio, 2017). Further details on the world scenario of sugarcane and its main products are discussed in Chapter 1, Sugarcane world scenario, of this book.

In terms of sustainable assessment of the sugarcane mills, for example, TEA and LCA could both be used to find out the merit of a sugarcane biorefinery in comparison with the alternatives to produce ethanol and other coproducts.

Table 5.1 Differences in global warming potentials (GWP) for 100-year time horizon, source Intergovernmental Panel on Climate Change (IPCC) assessment reports (ARs).

	Designation	Chemical formula	AR2-1996	AR3-2001	AR4-2007	AR5-2013
Common among carbon footprint	Carbon dioxide	CO_2	1	1	1	1
	Methane	CH_4	21	23	25	28
	Nitrous oxide	N_2O	310	296	298	265

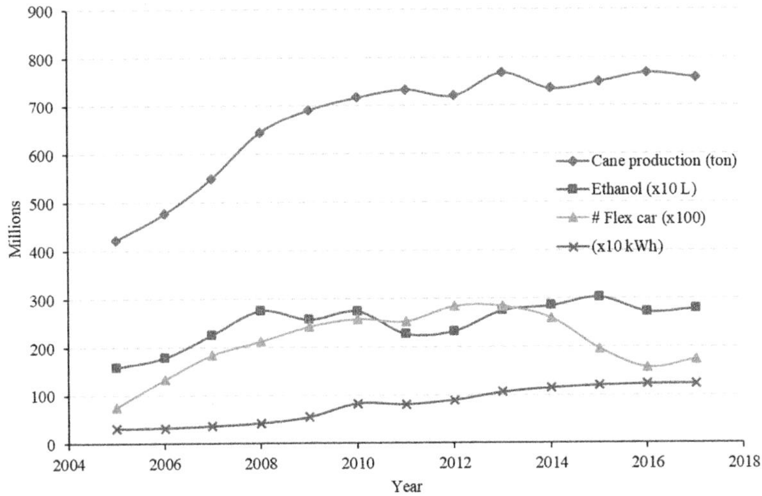

Figure 5.1 Historical trends showing the increase in cane production, ethanol production, number of flex-fuel vehicles registered, and surplus electricity from bagasse (ANFAVEA, 2018; FAO, 2019; EPE, 2019; CONAB (National Supply Company of Brazil), 2018).

The following studies were reviewed from literature to have the inventory values of Table 5.2 mainly from Brazil Center–South region (Macedo et al., 2008; Vasconcelos and Carpio, 2017; Seabra et al., 2011; Souza et al., 2012). The quantification of wastes is stressed out because they can be further processed in the sugarcane biorefinery. Regarding straw and tips, studies recommend the use of about 50% of the total collected to maintain the field, control erosion, and to keep moisture, temperature, and soil fertility. The remaining 50% is used for burning in boilers, producing electric energy. In addition, it is used to produce cellulosic ethanol, also known as second-generation (2G) ethanol. The sugarcane wastes from harvesting (straw) and juice extraction (bagasse) are produced in large quantities, about 280 million tons of bagasse and straw per year, and they are likely to increase in the near future as new industrial plants are implemented. Currently, sugarcane residues are mostly burned for the production of heat and electricity at the sugar mill. However, they could also be used as feedstocks for the production of other high-value products in the context of the lignocellulosic biorefinery (del Río et al., 2015). This would be a 2G biorefinery integrated with the first generation (1G), whose schematic is represented in Fig. 5.2.

Table 5.2 Input–output inventory for 1G biorefinery from different sources in the literature. The uptake of CO_2 from sugarcane culture adopted is 653 $kgCO_2$ per tonne of sugarcane (Cavalett et al., 2012).

Stage	Inputs		Outputs	
Agriculture (Souza et al., 2012)	Nitrogen (N)	58.7 kg/ha/year	Sugarcane	89 ton/ha/year
	Phosphate (P_2O_5)	20.63 kg/ha/year		
	Potassium (K_2O)	30.39 kg/ha/year		
	Limestone ($CaCO_3$)	400 kg/ha/year		
	Herbicide	5.00 kg/ha/year		
	Insecticide	2.00 kg/ha/year		
	Cropping practices	33.49 L_{diesel}/ha/year		
	Harvesting	49.09 L_{diesel}/ha/year		
	Water	140 m^3/ha/year		
Agriculture (Seabra et al., 2011)	274 L_{diesel}/ha/year		Sugarcane	86.7 ton/ha/year
	N	777 g/t_{cane}	Cane trash yield	140 kg_{dry}/t_{cane}
	P_2O_5	249 g/t_{cane}		
	K_2O	980 g/t_{cane}		
	$CaCO_3$	5183 g/t_{cane}		
	Herbicides	44 g/t_{cane}		
	Insecticides	3 g/t_{cane}		
Industrial (Souza et al., 2012)	Lime	1066.52 $g/ton_{sugarcane}$	Bagasse	0.27 ton/ton_{cane}
	Phosphoric acid	23.28 $g/ton_{sugarcane}$	Filtercake	24 kg/ton_{cane}
	H_2SO_4	6.31 $g/L_{ethanol}$	Stillage	14 $L/L_{ethanol}$
	Lubricants	13.93 $g/ton_{sugarcane}$	Ethanol	85 $L/ton_{sugarcane}$
Industrial (Seabra et al., 2011)	Lubricants	10.3 g/t_{cane}	Stillage	11 L/L
	Sulfur	156 g/bag	Filtercake	31 kg/t_{cane}
	Lime	880 g/t_{cane}	Boiler cogeneration	
	Sulfuric acid fermentation	7.4 g/L	Ash	2 kg/t_{cane}
			Soot	12 kg/t_{cane}
			Bagasse	0.264 ton/ton_{cane}
			Ethanol	86.7 L/ton_{cane}
			Electricity surplus	10.7 kWh/ton_{cane}

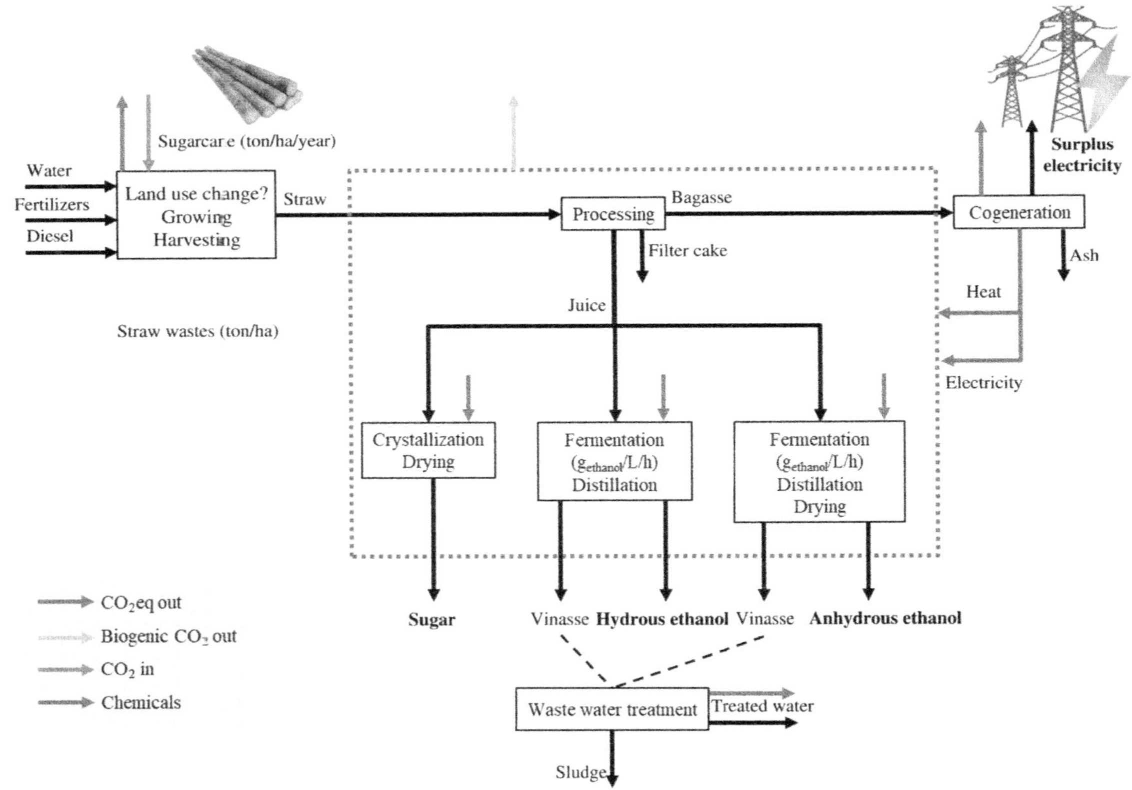

Figure 5.2 Process material, emissions, and energy flows for a sugarcane traditional 1G biorefinery mill (can be used for sugarcane, energy cane, and sweet sorghum) (Souza et al., 2015). Partition of 40% sugar, 39% hydrous ethanol, and 21% anhydrous ethanol is shown in Seabra et al. (2011).

For year-round operation, the sugar mill could adopt other feedstocks, besides sugarcane. Energy cane and sweet sorghum have different harvest times, but can be processed for bioethanol using the same equipment (Kim and Day, 2011).

The CO_2 intensity of electricity in Brazil is low due to the high percentage of renewables, c.73 gCO_2eq/kWh carbon intensity (Gnansounou et al., 2015); electricity used in the processes, 14 kWh/ton cane (Macedo et al., 2008) is nowadays mainly suppressed by bagasse cogeneration and surplus electricity is sold to the grid, being a product of the biorefinery and no longer an input.

The use of land use change (LUC), when the natural landscape is transformed by human activity, and biogenic emissions of CO_2 as a result of sugar fermentation processes, can also be taken into account when analyzing production processes (Paixão et al., 2018).The values of fossil energy consumption and CO_2eq emissions with no LUC, no biogenic emissions from fermentation and credits to surplus electricity are from the literature: for the reference case (Seabra et al., 2011), 2008 data for sugar production, fossil energy use 721 kJ/kg_{sugar} and 234 g CO_2eq/kg_{sugar}, respectively. For the ethanol life cycle, these values are 80 kJ/MJ and 21.3 g CO_2eq/MJ. Other study for the Center–South Region of Brazil claims 202–238 $MJ/MJ_{ethanol}$ and 24 g CO_2eq/MJ (Cavalett et al., 2013). Considering both agricultural and industrial phases, the total emissions of hydrous and anhydrous ethanol production for 2005/2006 were evaluated as 417 and 436 g CO_2eq/L, respectively (Macedo et al., 2008), or 19.8–20.7 g CO_2eq/MJ (density 0.785 kg/L, lower heating value 26.9 MJ/kg (Heywood, 1988)), respectively.

Other possible biorefineries could be projected to further convert the process residues to 2G ethanol or even 3G ethanol. Those several virtual biorefineries are usually simulated by using the AspenPlus software to get all virtual materials and energy flows (Bonomi et al., 2011). These virtual inventories are then used in the LCA and TEA analysis. For example, comparison of three biorefineries in terms of the internal return rate (IRR), a TEA metric (Corrêa do Lago et al., 2012). Fifty percent of the straw waste (crop residues) was used in the soil and the other 50% used in the cogeneration unit to produce extra electricity to the grid (biorefinery #1, Table 5.3), and to produce ethanol from cellulose (biorefinery #2, Table 5.3) and from cellulose and hemicellulose (biorefinery #3, Table 5.3). More specifically, 50% of sugarcane juice was used for sugar production and the other part of the juice, together with molasses from

Table 5.3 Technoeconomic assessment (TEA) analysis, based on internal return rate (IRR) metric, of virtual biorefinery from sugarcane (Corrêa do Lago et al., 2012).

	Biorefinery #1 optimized 1G	Biorefinery # 2 Integrated optimized 1G and 2G	Biorefinery #3 Future integrated optimized 1G and 2G
Ethanol yield L/ ton_{cane}	89.3	110.7	131.5
Electricity surplus to the grid kWh/ton_{cane}	185.8	92.8	72.7
US$2009/L	0.288	0.317	0.253
IRR per year	16.9%	12.2%	18.4%

sugar processing, was used to produce ethanol. Furthermore, 50% of the sugarcane crop residues were hydrolyzed together with surplus of bagasse (amount that was not used to attend the process heat and electricity demand) to produce 2G ethanol, and crop residues were used to attend energy supply of the process. No LCA global warming impact assessment was made, and only the value of CO_2 uptake 178 kg of carbon per 1000 kg of sugarcane stalks during the plant-growing phase is mentioned.

Gnansounou et al. (2015) used TEA and LCA metrics to select the best integrated sugarcane-based biorefineries. They were integrated biorefineries producing 1G and 2G ethanol, sugar, molasses (for animal feed), and electricity in the context of Brazil. The metrics prospective economic performance (PEP) (for TEA) and climate change impact (CO_2eq) are used in a reference system that produces gasoline (allowing a vehicle to move the same km), sugar, animal feed and electricity, and in the integrated biorefineries. They concluded that integrated 1G2G biorefinery reduced 56%–59% the CO_2eq emissions in comparison with the reference system and even got lower emissions than a dedicated 1G mill. The TEA metric showed that the 1G mill had a higher economic performance, followed by the 1G2G with four main products: ethanol, sugar, electricity, and animal feed. This example compares a basket of products with other system producing the same products, and, so, is not only a well-to-wheel study of the liquid fuel ethanol.

Another example found in the literature considers sugarcane ethanol and palm oil biodiesel joint production in a biorefinery (joint sugarcane ethanol biodiesel—JSEB) whose products are ethanol, glycerin, kernel oil and cake, and electricity to the Brazilian grid. In this case, no sugar was produced (Souza et al., 2012). The biodiesel internally produced was used to suppress the diesel fuel needs in managing soils, harvesting, and transportation. Life cycle CO_2eq emissions of ethanol, based on the traditional, system were 20.2 g/MJ, without LUC emissions. In comparison, CO_2eq emissions of ethanol produced by the JSEB were 15.6 g/MJ, a 23% reduction in life cycle emissions when compared to the traditional system. This example shows a cradle-to-gate comparison of biorefinery systems.

1G2G3G ethanol biorefinery integration systems, for example, exploring microalgae to consume the CO_2 released in subprocesses and used as raw material for ethanol production is not yet fully explored (Teixeira et al., 2016) and could be a topic of future research. The only study found regarding algae use to collect the biogenic CO_2 from fermentation and produce biodiesel to use in all agriculture operations was in an ethanol distillery (Souza et al., 2015). The authors showed a potential reduction of 10%–50% of CO_2eq emissions in comparison to a traditional Brazilian sugarcane ethanol distillery.

Despite this intense research in Brazil, other American countries are likely to follow the Brazilian example. For instance, the major sources of biomass in Ecuador are agricultural residues, animal manure, woody biomass, industrial effluents, and municipal solid wastes (Zafar & Decker, 2009). The Ecuadorian's agricultural production benefits from its geographic location, and favorable weather conditions having an average 12-hour day luminosity (Purdy and Vega, 2012).

Biofuels production in Ecuador is poor comparing to Brazil. Ecuadorian sugarcane production is mostly concentrated in the coastal region, closely to Guayaquil (MAGAP) and it has been like that since the 1990s decade. For the year 2001, Ecuador counted with a sugarcane-cultivated area of 69,085 ha. After 4 years, the total cultivated area increased to 135,000 ha (75,000 ha for sugar) whereof solely 10,000 ha were dedicated for ethanol production. For the same year, the Ecuadorian Ministry for Agriculture, Livestock, Aquaculture and Fisheries (MAGAP, acronym in Spanish) estimated that Ecuador had 675,932 ha of suitable land for sugarcane production. A year later, by 2006, the total sugarcane-cultivated area increased to 147,270 ha. This year it was determined that almost 60% of the national sugarcane crops belonged to individual farmers and the remaining 40% belonged to Valdez and San Carlos sugar companies (Morales Enríquez, 2010). In the year 2009, Ecuador had a daily capacity to produce 136,000 L of ethanol, resulting in an estimated production of 50 million liters per year (Domínguez, 2012). The use of ethanol on fuels started in Guayaquil and Duran by 2010, where the Ecuadorian government launched a pilot program denominated Ecopaís fuel (E-5), a blended gasoline type extra of 85 octanes with 5% ethanol processed through sugarcane production. Initially, the government aimed to increase the blend of gasoline with ethanol from 5% to 15% by 2016. For this reason, the government started promoting sugarcane production in small and large scale. After the introduction of Ecopaís, it was possible to observe a decrease in emissions of CO, NO_x, and HC (Paredes, 2015). In comparison with the previous year (2011), Ecuador increased by 10% the sugarcane-cultivated area, having an installed capacity to distill around 185,000 L of alcohol per day. From that amount of liters, uniquely 20,000 L per day were used for the production of ethanol (Purdy and Vega, 2012). Therefore, alcohol production was around 20.5 million liters in 2012, where 39% of the total amount was used for biofuels production (Grupo Spurrier, 2013). The installed capacity by 2012 was of 36 million liters of ethanol. However, in order to reach the target of having a blended gasoline with 15% (400 million liters of ethanol per year), it is a requirement to sow larger extensions of sugarcane and build new distilleries.

Flex-fuel vehicle technologies and well-to-wheels

When the comparison is the liquid fuel ethanol production system with gasoline refining to use in a flex-fuel vehicle, a WTW framework is used. For example, in Europe, the carbon footprint (as CO_2eq) of ethanol from sugarcane produced in Brazil does not account for LUC, nor biogenic CO_2 emissions. Table 5.4 shows the Well-to-tank (WTT) part of energy and CO_2 balance. From here, the "Primary energy input (1 + MJexpended)/MJ_{final})" is 3.09 MJ/MJ_{EtOH} and "total CO_2 emission 24.8 gCO_{2eq}/MJ_{EtOH}." The overall WTW using E5 to E85 blends are 26–37 MJ_{fossil}/100 km and 36–53 gCO_2eq/km, a reduction of more than 60% of the CO_2 equivalent emissions (Edwards et al., 2013).

With the expanded greenhouse gases, regulated emissions, and energy use in transportation model (GREET), Wang et al. (2008) examined the WTW energy use and CO_2eq emissions of sugarcane-derived ethanol produced in Brazil and used to fuel light-duty vehicles in the United States. The sugarcane-to-ethanol pathway evaluated in the GREET model comprises fertilizer production, sugarcane farming, sugarcane transportation, and sugarcane ethanol production in Brazil; ethanol

Table 5.4 Brazilian sugarcane-to-ethanol for use in Europe. Excess bagasse used for electricity production. SCET1 pathway from Edwards et al. (2013) with inventory data mostly from Macedo et al. (2008). Intergovernmental Panel on Climate Change (IPCC) 100 values from AR4.

Brazil product to European context		Energy expended	GHG emissions g $CO_2eq/MJ_{ethanol}$			
Standard steps	**Actual steps**	**$MJ/MJ_{ethanol}$**	**Total**	**as CO_2**	**as CH_4**	**as N_2O**
Production and conditioning at source	Sugarcane cultivation + local transport	0.09	17.9	6.54	3.33	7.99
Transformation at source	Ethanol production	1.88	− 1.4	− 1.73	− 0.06	0.36
	Of which credit for electricity from surplus bagasse	− 0.06	− 2.5	− 2.37	− 0.14	− 0.03
Transportation to market	Ethanol long distance transport	0.09	6.7	6.62	0.03	0.02
Transformation near market	*NA*					
Conditioning and distribution	Distribution	0.02	1.1	1.11	0.01	0.02
	Dispensing at retail site	0.01	0.5	0.48	0.03	0.00
Total WTT		2.09	24.8			

WTT, well-to-tank.

transportation to United States ports and then to US refueling stations; and ethanol use in E85 vehicles. Results for sugarcane ethanol were compared with those for petroleum gasoline. Their analysis showed that sugarcane ethanol can reduce CO_2eq by 78%. This value was recently updated to 40%–62% (Wang et al., 2015).

Both European and United States advantages in using ethanol from sugarcane produced in Brazilian biorefineries were explored in conventional internal combustion engine vehicles, using 85% anhydrous ethanol blend with gasoline, or hydrous ethanol 5%–10% blend with gasoline. As seen in Fig. 5.1, the flex-fuel technology could be another way to boost ethanol consumption and stimulate more efficient sugarcane from 1G2G3G biorefineries. With the future electrification of the road vehicles the plug-in hybrid flex-fuel technology could be the best choice in terms of fuel flexibility (ethanol–gasoline–electricity), decarbonization, and use of endogenous resources (Silva & Arcentales, 2019).

Looking to the WTT part, for the same industrial emission burden, the higher the ethanol productivity, the better. Depending on the feedstock and fermentation conditions (e.g., temperature, agitation speed, and initial sugar concentration), ethanol productivity values can range from 0.13 to 6.55 $g_{ethanol}/L_{bioreactor\ medium}/h$. To increase ethanol production in industrial scenarios, engineered yeast strains can be used. Those may have an enhanced tolerance to ethanol concentration in the medium, pH fluctuations, temperature, osmotic stress, toxic inhibitors, and/or have a capacity to perform alcoholic fermentation of xylose (a pentose) additionally to glucose (a hexose) (Mohd Azhar et al., 2017).

An industrial value of this productivity, which actually occurs in sugarcane mills, is not publicly known. If we try to estimate it to 1G mills: sugarcane juice density varies between 1044.5 and 1189.5 kg/m^3 (Astolfi-Filho et al., 2010) and from the Table 5.2, inventory for 1G mills, bagasse, and filtercake amount ~290 $kg/1000\ kg_{cane}$. This means an availability of ~215 L_{juice}/ton_{cane}. If we take the bioreactor medium as 215 L and assume 60 hour fermentation, this would mean a 1G productivity of ~5 $g_{ethanol}/L_{bioreactor\ medium}/h$. These are very rough claims and of course should be taken with caution. So, if this ethanol productivity was doubled by means of engineered yeast and because results of the CO_2eq are expressed as per the production of ethanol, the doubled production would mean half the emission burden. The yeast capability to process pentose will boost 2G ethanol production.

Conclusions

Integrated 1G2G biorefineries, processing food (sugar), ethanol (biofuel), heat, and electricity from bagasse (bioenergy for self-consumption) and surplus electricity to the Brazilian grid seems to be favorable from the point of view of technoeconomic metrics and the global warming metric. Typically, integrated biorefineries show reductions in CO_2eq of 23%–60%. The engineering yeast to further increase ethanol productivity for the same conditions will improve the WTT values used to compare with conventional gasoline vehicles because the metric is expressed as $MJ/MJ_{ethanol}$ and $gCO_2eq/MJ_{ethanol}$. For 1G mills, the latter is typically on the order of 20 g/MJ, without land use issues and disregarding biogenic emissions.

Acknowledgments

Publication supported by FCT—project UID/GEO/50019/2019—Instituto Dom Luiz.

Rui Pacheco is pleased to acknowledge the Fundação para a Ciência e Tecnologia (FCT) for the provision of the scholarship SFRH/BD/102763/2014.

References

ANFAVEA, 2018. Brazilian Automotive Industry Yearbook 2018.

Astolfi-Filho, Z., Minim, L.A., Telis-Romero, J., Minim, V.P.R., Telis, V.R.N., 2010. Thermophysical properties of industrial sugar cane juices for the production of bioethanol. J. Chem. Eng. Data 55 (3), 1200–1203. Available from: http://doi.org/10.1021/je900590t.

Bonomi, A., Mariano, A.P., Jesus, C.D.F., de, Franco, H.C.J., Cunha, M.P., Dias, M.O., et al., 2011. The Virtual Sugarcane Biorefinery (VSB). Brazilian Bioethanol Science and Technology Laboratory (CTBE).

Cavalett, O., Chagas, M.F., Erguy, N.R., Sugawara, E.T., Cardoso, Terezinha F., Bonomi, A., 2012. *Sugarcane Life Cycle Inventory*. Campinas, São Paulo. Available from: http://lnbr.cnpem.br/wp-content/uploads/2016/08/MeT-082012.pdf.

Cavalett, O., Chagas, M.F., Seabra, J.E.A., Bonomi, A., 2013. Comparative LCA of ethanol versus gasoline in Brazil using different LCIA methods. Int. J. Life Cycle Assess. 18 (3), 647–658. Available from: https://doi.org/10.1007/s11367-012-0465-0.

CONAB (National Supply Company of Brazil), 2018. Sugarcane Crop Tracking 2017/2018. Brazilia. Available from: https://www.conab.gov.br/info-agro/safras/cana.

Corrêa do Lago, A., Bonomi, A., Cavalett, O., Pereira da Cunha, M., Pinheiro Lima, M. A., 2012. Sugarcane as a carbon source: the Brazilian case. Biomass Bioenergy. Available from: http://doi.org/10.1016/j.biombioe.2012.09.007.

del Río, J.C., Marques, G., Lino, A.G., Lima, C.F., Colodette, J.L., Gutiérrez, A., 2015. Lipophilic phytochemicals from sugarcane bagasse and straw. Industr. Crops Prod. Available from: http://doi.org/10.1016/j.indcrop.2015.09.064.

Domínguez, J.M., 2012. Are Biofuels Socially Accepted in Guayaquil? In *International Association of Agricultural Economists (IAAE) Triennial Conference 18–24 August 2012*. Foz do Iguaçu, Brazil.

Edwards, R., Larive, J.-F., Rickeard, D., Weindorf, W., 2013. Well-to-wheels analysis of future automotive fuels and powertrains in the European context WELL-TO-TANK (WTT) Report. Version 4. Joint Research Center of the EU (JRC), Ispra, Italy. Available from: http://doi.org/10.2790/95629.

EPE, 2019. National Energetic Balance, Base Year 2018. Rio de Janeiro.

FAO, 2019. FAOSTAT Statistics Database. Rome.

Gnansounou, E., Vaskan, P., Pachón, E.R., 2015. Comparative techno-economic assessment and LCA of selected integrated sugarcane-based biorefineries. Bioresour. Technol. Available from: http://doi.org/10.1016/j.biortech.2015.07.072.

Grupo Spurrier, 2013. Biocombustibles PROECUADOR.

Heywood, J.B., 1988. Internal Combustion Engine Fundamentals. McGrawHill Series in Mechanical Engineering (21).

Kim, M., Day, D.F., 2011. Composition of sugar cane, energy cane, and sweet sorghum suitable for ethanol production at Louisiana sugar mills. J. Industr. Microbiol. Biotechnol. Available from: http://doi.org/10.1007/s10295-010-0812-8.

Macedo, I.C., Seabra, J.E., Silva, J.Ea.R., 2008. Green house gases emissions in the production and use of ethanol from sugarcane in Brazil: the 2005/2006 averages and a prediction for 2020. Biomass Bioenergy 32 (7), 582–595. Available from: http://doi.org/10.1016/j.biombioe.2007.12.006.

Mohd Azhar, S.H., Abdulla, R., Jambo, S.A., Marbawi, H., Gansau, J.A., Mohd Faik, A. A., Rodrigues, K.F., 2017. Yeasts in sustainable bioethanol production: A review. Biochemistry and Biophysics Reports. Available from: https://doi.org/10.1016/j.bbrep.2017.03.003.

Morales Enríquez, J.D. (2010). LA INDUSTRIA DE LOS BIOCOMBUSTIBLES EN ECUADOR: el estado de situación, 137.

Paixão, S.M., Alves, L., Pacheco, R., Silva, C.M., 2018. Evaluation of Jerusalem artichoke as a sustainable energy crop to bioethanol: energy and CO_2eq emissions modeling for an industrial scenario. Energy. Available from: https://doi.org/10.1016/j.energy.2018.02.145.

Paredes, V., 2015. Estado del desarrollo de biocombustibles en el Ecuador. Guayaquil.

Purdy, E., Vega, H., 2012. Ecuador biofuels annual Ecuador biofuel industry, bioethanol, biodiesel, production, forecast 2012. *Quito*. Available from: https://gain.fas.usda.gov/Recent GAIN Publications/Biofuels Annual_Quito_Ecuador_6-28-2012.pdf.

Seabra, J.E.A., Macedo, I.C., Chum, H.L., Faroni, C.E., Sarto, C.A., 2011. Life cycle assessment of Brazilian sugarcane products: GHG emissions and energy use. Biofuels Bioprod. Biorefining. Available from: http://doi.org/10.1002/bbb.289.

Silva, C.M., Arcentales, D., 2019. Exploring the introduction of plug-in hybrid flex fuel buses in Ecuador. In *Special session-Smart Mobility and Energy Transitions, Energies*. Available from: https://doi.org/10.3390/en12122244.

Souza, S.P., de Ávila, M.T., Pacca, S., 2012. Life cycle assessment of sugarcane ethanol and palm oil biodiesel joint production. Biomass Bioenergy. Available from: https://doi.org/10.1016/j.biombioe.2012.04.018.

Souza, S.P., Gopal, A.R., Seabra, J.E.A., 2015. Life cycle assessment of biofuels from an integrated Brazilian algae-sugarcane biorefinery. Energy 81, 373–381. Available from: https://doi.org/10.1016/j.energy.2014.12.050.

Teixeira, A.C.R., Sodré, J.R., Guarieiro, L.L.N., Vieira, E.D., de Medeiros, F.F., Alves, C.T., 2016. A review on second and third generation bioethanol production. Available from: https://doi.org/10.4271/2016-36-0515.

Vasconcelos, P.S., Carpio, L.G.T., 2017. Bagasse, straws, tips and vinasse: from sugarcane waste to a clean and renewable bioenergy source. Int. J. Adv. Soc. Sci. Human 5.

Wang, M., Han, J., Dunn, J.B., Cai, H., Elgowainy, A., 2015. Well-to-wheels energy use and greenhouse gas emissions of ethanol from corn, sugarcane and cellulosic biomass for US use. Efficiency and Sustainability in Biofuel Production: Environmental and Land-Use Research. Available from: http://doi.org/10.1088/1748-9326/7/4/045905.

Wang, M.Q., Wu, M., Huo, H., Liu, J.H., 2008. Life-cycle energy use and greenhouse gas emission implications of Brazilian sugarcane ethanol simulated with the GREET model. Int. Sugar J. Available from: https://doi.org/10.1109/ICBME.2015.7404112.

Zafar, S., Decker, C.S., 2009. Development of biomass energy systems in Ecuador What Is Biomass? (May). Retrieved from: https://businessdocbox.com/Green_Solutions/70288416-Development-of-biomass-energy-systems-in-ecuador.html.

CHAPTER 6

Energy cane breeding

Geraldo Veríssimo de Souza Barbosa, João Messias dos Santos, Carlos Assis Diniz, Danilo Eduardo Cursi and Hermann Paulo Hoffmann

Contents

Introduction 103
Sugarcane 104
- Taxonomy and genetics 104
- Sugarcane breeding 105
- Germplasm bank and sugarcane hybridization 105
- Selection of superior individuals 106
- Sugar yield gain 107
- Potential yield of sugarcane 108

Energy cane 108
- Strategies to select energy cane 110
- Hybridization 111
- Selection 113
- Advantages and challenges of energy cane 114

References 115

Introduction

Sugarcane (*Saccharum* spp.) is grown in all tropical and subtropical regions of the world, covering about 130 countries. It is a plant with high photosynthetic metabolism—C4 plant, being very efficient in the conversion of radiant energy to chemical energy. Harvesting the stalks and extracting the juice (composed of water and sugars—sucrose, glucose, and fructose), there is a surplus of bagasse (composed of fibers—cellulose, hemicellulose, and lignin). Among the main products derived from this raw material, some may be cited, such as sugar, ethanol, electricity, molasses, rum, bagasse, yeast, filter cake, vinasse, among others. The energy cane can be understood as the one that provides greater lignocellulosic biomass, with the full utilization of the cane (stalks, leaves, and straw), and thus raise the ethanol production through hydrolysis of the lignocellulosic

Sugarcane Biorefinery, Technology and Perspectives
DOI: https://doi.org/10.1016/B978-0-12-814236-3.00006-8

biomass, increase electric energy generation, besides optimizing the production of other fiber derivatives—biopolymers, briquettes, pellets, Medium Density Fiberboard (MDF) agglomerates, among others. This chapter aims to discuss the advances achieved with the genetic improvement of sugarcane, as well as to present strategies for selecting energy cane cultivars, with a view to increasing sugar and renewable energy production by traditional industries.

Sugarcane

Taxonomy and genetics

The sugarcane belongs to the family Poaceae, tribe Andropogenae, and to the genus *Saccharum*. It is a perennial grass, of great size, that reproduces sexually, preferably by allogamy. As an adult, the plant has a height of 2–6 m and can be harvested in several cycles without the need for new planting. The main parts of the plant are stalks, leaves, and roots. The stalk of a mature plant contains on average 70% water and 30% dry matter (sugars, salts, and fibers), components of variable proportion depending on the species or hybrid and the crop condition. The stalk varies in form and color and is formed by nodes and internodes. Propagation is asexual by planting a portion of the stalk, which contains buds present on the nodes, which sprout and then early formed tillers give rise to thicker and heavier stalks forming the clumps. The flowers are hermaphrodite and the fruit is a true seed. Genetic breeding occurs through the germination of its true seeds, which is a strategy used by breeding programs to obtain variability and selection of new clones (Henry, 2010; James, 2004).

Roach and Daniels (1987) report that in the *Saccharum* genus there are six species, originating in Southeast Asia and New Guinea, with the following characterization: *Saccharum officinarum*—tall plants, thick stalks, high sugar content, and low fiber content; known as "noble cane"; *Saccharum spontaneum*—plants of short stalks, thin, fibrous, low sugar content, vigorous, and abundant tillering plants and well-developed root system; *Saccharum robustum*—very tall plants, thick, fibrous, and low sugar content; *Saccharum barberi*—low- to medium-sized stalks, thin, fibrous, and low sugar content; *Saccharum sinense*—high stalks, thin, fibrous and regular sugar content, abundant, and strong roots; *Saccharum edule*—plants characterized by inflorescences used for human consumption.

Piperidis et al. (2010) and D'Hont et al. (1998) discuss the cytogenetics of sugarcane: it is a polyploid plant, that is, it contains more than two copies of each of its chromosomes. There are several levels of ploidy: The *S. officinarum* species has in each cell $2n = 80 = 8x$ ($x = 10$ chromosomes); in the *S. spontaneum* species, the subtypes $2n = 40$, 64, 80, 96, 112, and $128 = 5x$, $8x$, $10x$, $12x$, $14x$, and $16x$ ($x = 8$ chromosomes) are more common. The modern cultivars are the result of interspecific genetic crosses carried out in the first half of the 20th century between *S. officinarum* and *S. spontaneum* and have between 100 and 130 chromosomes, 80% of which are derived from *S. officinarum*, 10% from *S. spontaneum*, and 10% from recombinants of the two species. Among the studied crops, sugarcane is the one that has the most complex genome, and the complexity of this genome is greater when it is considered the frequent aneuploidy in the present cultivars, because the somatic cells have an abnormal number of chromosomes (increase or decrease).

Sugarcane breeding

In the sugarcane breeding, the necessary conditions are: (1) to have a Germplasm Bank with wide genetic diversity among accesses of species of the genus *Saccharum*, related species and hybrids of several research centers in the world; (2) creation of genetic variation through hybridization; and (3) selection of superior individuals. The main sugar-producing countries of the world have genetic breeding programs, which aim to obtain superior clones in sugar yield (Ming et al., 2006).

Germplasm bank and sugarcane hybridization

Through hybridization and successive selections, two components are fundamental: stalk mass and sugar content by stalk mass (Jackson, 2005). After more than a century of the first hybrids development, obtaining cultivars with high yields in sugar remains a great challenge, given the complexity of the cane genome.

The main sugarcane breeding programs in the world differ in the combination of parents used in hybridizations, from basic species of the *Saccharum* complex, hybrids imported from other sites to superior clones selected in the breeding program itself (Burnquist et al., 2000).

The sugarcane flowering process is quite sensitive to environmental conditions. This applies to floral induction, inflorescence emergence, and pollen fertility. Among the factors that cause differentiation of the apical meristem in the floral primordium, the relative maturity of the plant development, the photoperiod (11 hours 30 minutes to 12 hours of dark period), the temperature (night minimums above 18°C associated with diurnal maximums below 32°C), soil moisture and fertility can be highlighted. However, there are some artificial techniques to modify the flowering period of each genotype, both anticipating and retarding, and thus enabling the synchronization and achievement of the desired crossing combination. Both "darkroom" and "illumination field" are artifacts used by breeding programs to contribute to this purpose. Hybridization stations located at latitudes close to 10 degrees are the ones that have the most profuse natural flowering and the production of true seeds in greater quantity and better quality. In regions of higher latitudes, in temperate climate, it is necessary to use artificial chambers for temperature and photoperiod control, and thus to promote the flowering induction. The sugarcane flowering is seasonal, that is, at a given latitude, it occurs always in the same period of the year, and with high consistency for each clone; in the southern hemisphere occurs from March (autumn equinox), going through June, as the latitude increases, the ratio of 2 days for each degree; in the northern hemisphere, the model is exactly that of an image in the mirror relative to the south, starting in September and extending through December (Moore and Berding, 2014; Ethirajan, 1987).

Selection of superior individuals

Traditionally, breeding programs distribute thousands of seedlings (heterozygous individuals) per year from true seeds resulting from hybridizations of previously selected parents. In the first selection stage, some programs adopt the visual mass selection method, in which individuals are selected in plant cane or first ratoon crop based on their phenotypic values—this is considered a less efficient criterion; other programs adopt the plant cane selection of the best families (with higher genotypic values) and the process follows in the first ratoon crop with the selection of the best individuals within the best families—it is considered the most efficient criterion. In the subsequent stages of the two criteria (selection of individuals or families), clonal selections are adopted for another two or three stages, followed by experimentation in several growing environments, until obtaining and releasing cultivars with higher sugar yields. According to the growing environment, the characteristics of disease tolerance, early maturing, nonlodging, nonflowering, tolerance to water deficit and tolerance to low temperatures are also considered in the selection of superior clones

(Ming et al., 2006; Kimbeng and Cox, 2003). It takes around 10–12 years to develop a new cultivar with higher sugar yield potential than the commercial standards.

Historically, sugarcane varieties are replaced mainly because of its susceptibility to some of the major diseases. Despite the effort to develop superior clones and cane cultivars, the occurrence of a new disease causes a great impact to the breeding program. The sugarcane diseases with major economic importance are: Sugarcane Mosaic Virus, Fiji Disease Virus; Ratoon Stunting Disease (*Leifsonia xyli* subsp. *Xyli*), Leaf Scald (*Xanthomonas albilineans*), Smut (*Sporisorium scitamineum*), Brown rust (*Puccinia melanocephala*); and Orange Rust (*Puccinia kuehnii*) (Rutherford, 2014; Cheavegatti-Gianotto et al., 2011; Matsuoka et al., 2009).

Sugar yield gain

In addition to the sugarcane management practices, the genetic improvement had a significant contribution to the sugar yield gain in the world sugarcane sector, resulting from the substitution rate of cultivars (Ming et al., 2006). Breeding programs around the world report significant gains in sugar yield, but often it is confused with the negative and positive effects of agronomic and environmental factors (Heinz, 1987). According to Berding and Skinner (1987), this gain was in the order of 1.1% per year in the Hawaii sugar industry from 1908 to 1984; Simmonds (1979) estimated a gain of 0.64% per year in Barbados, which was higher in biomass than in sugar content; in Australia, between 1948 and 1975, a gain of 1.9% per year was obtained where half of which was attributed to genetic improvement (Berding et al., 2004); in Florida/USA, from 1968 to 2002, the gain was in the order of 1.6% per year (Edmé et al., 2005); in Brazil, from 1970 to 2011, there was a gain of 4% per year, half of which was due to the continuous substitution of cultivars (Barbosa et al., 2012).

Lingle et al. (2010) report that there is an apparent limit on the accumulation of sugar from sugarcane in Louisiana/USA. In turn, Edmé et al. (2005) did not show this limit in Florida/USA. Moore (2005) states that in a century there has been great growth of commercial sugar yield in Australia, but it is possible that it has already reached the maximum. Jackson (2005) argues that in the last four decades, the sugar yield increases in Australia were due to cultivars with more stalk mass, without improving the sugar content. This author further claims that theoretically the selection of sugarcane clones to increase the sugar content is more effective than the selection of clones with highest stalk mass. It is important to point out that the results of sugar yield gains, having a higher contribution of stalk mass and lower sugar content, contradict the magnitudes of heritability estimates for the two traits, which according to Skinner et al. (1987), the sugar content (estimated by the total soluble solids content—Brix) has medium to high magnitude, with values of 0.65 for individual selection and 0.90 for family selection, while the trait stalk mass has a low magnitude, with values below 0.17 for individual selection and 0.90 for family selection.

Potential yield of sugarcane

Waclawovsky et al. (2010) indicate that the annual potential yield of cane stalks is in the order of 381 t/ha. In Brazil (Agrovale Mill, Bahia state, Lat. 09°26′18″S, Lon. 40°30′18″W), under full irrigation regime, in the 1-year cycle, RB92579 (cultivar of RIDESA/UFAL) presented commercial sugar yield of 212 t/ha of stalks; and in experimental condition, the yield of this cultivar was 299 t/ha (Busato Farm, Bahia state, Lat. 13°14′51″S, Lon. 43° 24′53″W). It is noteworthy that in the last four decades there has been a significant gain in sugar yield in all the sugarcane regions of the world with the adoption of modern cultivars, contributing to a closer approximation of the maximum theoretical yield level. It should be noted that in the main sugarcane-producing countries the average cane yield is less than 25% of this maximum theoretical yield, ranging from 70 to 80 t/ha, with sugar contents between 13% and 15% and fiber between 12% and 14%. It should also be noted that the commercial sugar yield of the cultivar RB92579 was 29.5 t/ha, associated with 31.5 t/ha of fiber. Inman-Bamber (2014) reported maximum commercial sugar yield ranging from 22 to 31 t/ha in the tropics and 17–22 t/ha in the subtropics, values lower than that obtained in Brazil with the RB92579 cultivar.

Energy cane

In the traditional sugarcane industries, which are focused on sugar and ethanol production, cane fiber levels had to be in a narrow range with a minimum content to generate electricity that would allow sugar and ethanol production to be processed and a maximum that would not cause a reduction in the industrial efficiency to obtain these products. Therefore breeders used to adopt selection strategies in order to obtain cultivars with high sugar content, without considering fiber content. With the advancement of new technologies and especially the greater appeal of renewable energy sources, sugarcane fiber is becoming a major component of this raw material which has motivated sugarcane breeding programs to develop the energy cane.

Alexander, in the 1980s, was the pioneer in the conceptualization of the use of energy cane in sugarcane mills in Puerto Rico. Expert in sugarcane physiology, he defined energy cane as a plant to be harvested integrally, using all the lignocellulosic biomass for energy purposes (stalks, leaves, straw, and bagasse). Energy cane presents different characteristics from the sugarcane, such as larger stalks population per area (between 130,000 and 210,000 stalks per hectare), plants with stalks of greater stature, thinner and more fibrous diameter, narrow leaves and abundant and deep root system, with two to three times greater biomass yield (Alexander, 1985).

In the last decade, researchers gave greater emphasis to energy cane, conceptualizing it and comparing it with sugarcane due to the combination of sugar, fiber, and water contents.

Tew and Cobill (2008) presented three distinct types of cane:

1. Traditional sugarcane—cane variety containing about 75% water, 12% fiber, and 13% sugar. This raw material provides juice for sugar and ethanol production, and fiber for electricity production. To obtain a traditional sugarcane cultivar, breeding programs focus on the selection of genotypes with high sugar contents.
2. Type I energy cane—cane variety selected to maximize sugar and fiber yield. This type of cane, which was conceptualized by Alexander in 1985, has lower water content (65%), fiber ranging from 13% to 17% (an increase of up to 40% in relation to the traditional sugarcane variety), and small reduction or maintenance of the sugar content. Thus this raw material, in addition to providing juice for sugar and ethanol production, contributes with more fiber for the greater production of electricity, lignocellulosic ethanol, as well as other derivatives of economic value for the sugarcane industry;
3. Type II energy cane—cane variety selected to maximize fiber yield (fiber content above 30%), with insignificant sugar content and lower water content (60%). This raw material is not of interest to the current sugarcane industry, being required by other agro-industry sectors that need biomass for the energy generation.

In Mauritius, Santchurn et al. (2014) also classified different types of cane varieties with high biomass, fiber, and biomass yield.

Based on these energy cane conceptualizations, breeding programs have adopted different criteria and strategies for the development of new cultivars, which have high fiber content and high biomass yield.

In the development of the modern commercial sugarcane varieties, derived from interspecific hybridization performed in the early 20th century between *S. officinarum* and *S. spontaneum* it used to be common to obtain segregant populations for the following traits: number of stalks per area, height and stalk diameter, sugar and fiber content. In these populations, clones with high yield of biomass were found but used to be discarded because of low sugar yield. However, from the 1960s onwards, the concept of energy cane began to be developed with the use of wild species of the genus *Saccharum*, in particular *S. spontaneum*, to obtain clones with greater biomass yield potential, becoming an integral part of some breeding programs, such as Australia (Berding and Roach, 1987), Puerto Rico (Alexander, 1985), Barbados, India, Taiwan, Louisiana, and Hawaii (Ming et al., 2006).

Strategies to select energy cane

Selection strategies for the development of different types of energy cane should be clearly defined at the beginning of any breeding program. Most institutions have established their strategies based on the relative proportion of sugar and fiber that is desired. The types of cane conceptualized by Alexander (1985), Tew and Cobill (2008), and Santchurn et al. (2014), can serve two distinct productive segments:

1. the current sugarcane industries aiming to optimize the production of sugar, ethanol, and electricity. This energy cane is characterized by an increase of up to 40% in the amount of fiber and does not significantly alter the sugar content in relation to the cultivars currently grown, resulting in increased biomass yield;
2. companies producing biomass for other energy purposes: steam production, heat, thermal energy, electricity, solid fuels (briquettes and pellets), among others. This type of cane is characterized by having a fiber content above 30% and insignificant sugar content (below 5%). When the production of cellulosic ethanol is consolidated, this cane profile can be used on a commercial scale by biorefineries, as well as other companies to obtain products of economic interest derived from the fiber.

In order to develop these energy cane cultivars that support these market segments, it is necessary to have significant genetic diversity, especially for sugar and fiber contents in the species of the *Saccharum* complex.

Currently, in the main cane-producing countries, there is germplasm collection with modern hybrids and accessions of the genus *Saccharum* (*S. officinarum, S. barberi, S. sinense, S. robustum, S. edule, and S. spontaneum*), *Erianthus*, and *Miscanthus*. The main collections are maintained by the following institutions: Sugarcane Breeding Institute (SBI), Coimbatore–India; United States Department of Agriculture (USDA), Miami–Florida–United States of America; Flowering and Crossing Station of Serra do Ouro/Inter-university Network for the Development of Sugarcane Industry (RIDESA), Murici–Alagoas–Brazil; Sugarcane Technology Center (CTC), Camamu–Bahia–Brazil; Instituto Agronômico de Campinas (IAC), Uruçuca–Bahia–Brazil; Sugar Research Australia (SRA), Gordonvale–Queensland–Australia; South African Sugarcane Research Institute (SASRI), KwaZulu–Natal–South Africa, Mauritius Sugar Industry Research Institute (MSIRI), Moka Road–Réduit–Mauricius; West Indies Central Sugarcane Breeding Station (WICSCBS), Groves–St. George–Barbados. These breeding programs have great experience in obtaining sugarcane and energy cane cultivars. Fig. 6.1 illustrates the Flowering and Crossing Station of Serra do Ouro to obtain RB (Republic of Brazil) cultivars of RIDESA (Lat. 9° 13′S, Lon. 35°50′W, Alt. 515m).

In order to obtain energy cane cultivars, between 6 and 8 years, most breeding programs have adopted a similar scheme: it maintains a germplasm collection with high genetic diversity, performs hybridizations, produces high amount of seedlings, selects superior individuals according to the desired type, and obtains the energy cane cultivars for grower's adoption (Fig. 6.2). However, each breeding program uses criteria according to its interests and particularities of the environments of its region.

Hybridization

Genetic crosses are performed between elite sugarcane hybrids and accessions of *S. spontaneum*, *S. robustum*, *Erianthus*, and *Miscanthus*, which are generally used as male parent (pollen supplier). In the parental selection to be used in the hybridizations, each breeding program adopts criteria according to their experiences.

During the releasing pollen process, egg fertilization, formation, and maturation of caryopsis, some techniques are used for the maintenance of stalk metabolism, such as marcotting and floral preservatives (the nutrient acid solution developed in Hawaii is the most used one). Despite the successful

Figure 6.1 Flowering and crossing station of Serra do Ouro (Murici, Alagoas, Brazil, Lat. 9°13′S, Lon. 35°50′W, Alt. 515m), managed by the Sugarcane Breeding Program of the Federal University of Alagoas to obtain RB cultivars of RIDESA. (A) Aerial view; (B) abundant flowering field; (C) hybridization; (D) panicle with caryopsis.

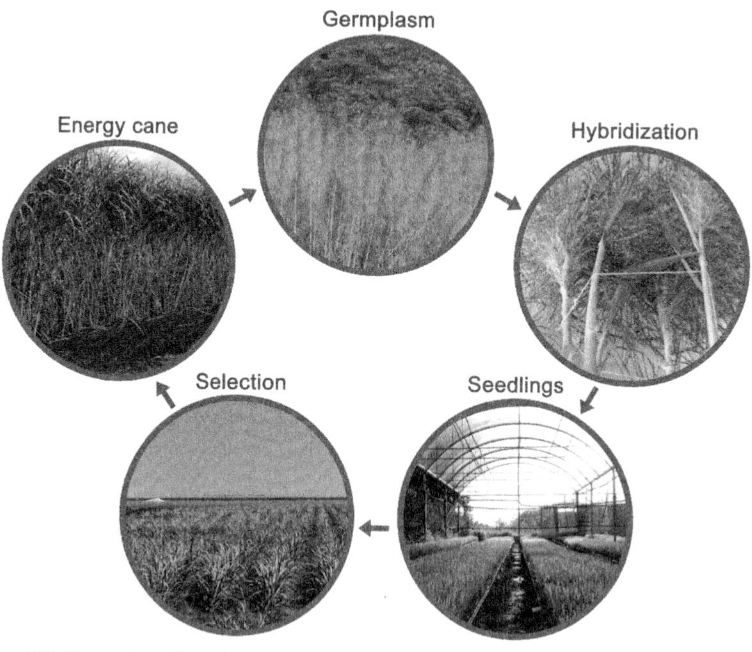

Figure 6.2 Procedure for obtaining energy cane cultivars.

use of the Hawaiian solution in the process of sugarcane hybridization, the same has not been observed for the hybridization of energy cane in accessions of *S. spontaneum*. To overcome this problem, an alternative is to use the Hawaiian solution only for the sugarcane hybrids and remaining the *S. spontaneum* plants in its original pots or directly in the field.

Selection

In the mass selection of energy cane, clones should preferably present the following characteristics: clumps with intense tillering and erect growth habit, stalks with high stature, without pithiness, high unit weight, internodes with mean diameter, absence of aerial rooting, narrow, erect and bulky leaves, absence of flowering, good plant health (absence of pests and diseases), and fiber and sugar contents according to the desired segment for the commercial production of the biomass (Fig. 6.3). In addition, the clone must be vigorous and resistant to mechanical damage.

The main advances achieved for biomass yield in the selection of energy cane cultivars have been obtained in F1 generation. In addition, such energy cane hybrids produce much more without evidence of yield decline over the cycles; in sugarcane hybrids, the opposite has been observed (Legendre and Burner, 1995). On the other hand, it is possible to recover the characteristic of high sugar content and others of interest, through backcrossing with energy cane hybrids selected in F1. Selected energy cane clones generally have abundant and deep root system, with a greater amount of dry matter than the sugarcane genotypes (Fig. 6.4).

Figure 6.3 Energy cane clone selected by RIDESA/UFAL.

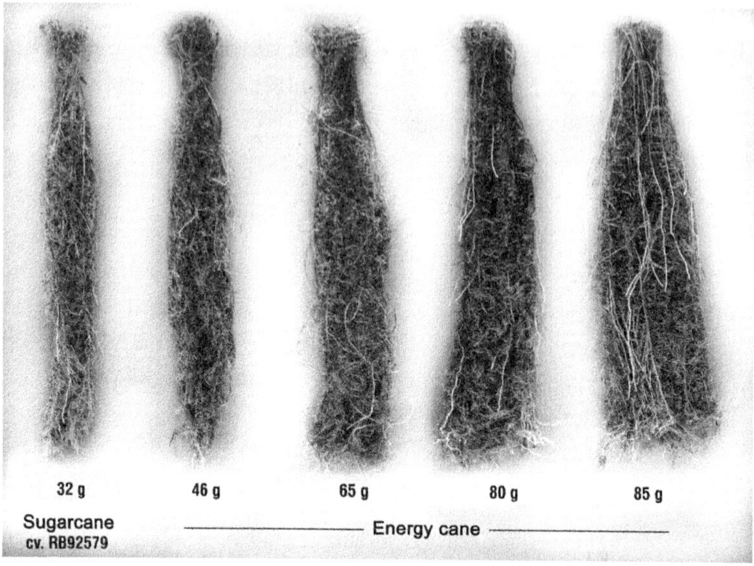

Figure 6.4 Dry mass (g) of sugarcane roots (cv. RB92579) and of RB clones of energy cane.

Although energy cane clones may contain lower sugar content, sugar yield may be higher than sugarcane clones, because stalk yield of energy cane is much higher than sugarcane (Alexander, 1985).

According to Ming et al. (2006), the genetic gain obtained with the selection strategies adopted for energy cane may be higher, because the fiber and sugar values are more flexible, resulting in a greater number of clones with yield potential for the breeders, and thereby, the selection process may be more effective since such restrictive criteria are not adopted as the selection of sugarcane cultivars.

Advantages and challenges of energy cane

Energy cane has the following advantages in relation to sugarcane: (1) multiplication rate between two and three times higher; (2) less susceptibility to pests; (3) greater water and nutrients use efficiency; (4) greater resistance to damage trampling, with consequent increase of yields and longevity; (5) can be grown in marginal areas without competing with food production; (6) integral harvesting system without burning, reducing environmental impact and generating greater income opportunity for the growers. Many of these advantages contribute to a lower energy cane production cost.

There are three major problems in the selection of energy cane clones: high incidence of smut disease (*S. scitamineum*), high flowering level, and low stalk mass. Thus the great challenge of the genetic improvement for obtaining and consolidating commercial energy cane cultivars is to use more effective strategies to overcome these three major problems.

For the adoption of energy cane cultivars by the agro-industry sector, there are still many technological bottlenecks, such as (1) development of mechanized harvesters for high productivity and fiber content cultivars and (2) in the industrial processing, to improve the efficiency of mills for the juice extraction. In addition, energy cane cultivars with low sucrose content and greater reducing sugar content (glucose and fructose) make sugar crystallization difficult.

Finally, the adoption of energy cane by growers will depend on the economic values of the final products of the raw material processing (sugar, ethanol, electricity, biochemicals, solid fuels, among others).

References

Alexander, A.G., 1985. The Energy Cane Alternative. Elsevier, Amsterdam.

Barbosa, M.H.P., Resende, M.D.V., Dias, L.A.S., Barbosa, G.V.S., Oliveira, R.A., Peternelli, L.A., et al., 2012. Genetic improvement of sugar cane for bioenergy: the Brazilian experience in network research with RIDESA. Crop Breed. Appl. Biotechnol. 12, 87–98.

Berding, N., Roach, B.T., 1987. Germplasm collection, maintenance, and use. In: Heinz, D.J. (Ed.), Sugarcane Improvement Through Breeding. Elsevier, Amsterdam, pp. 143–210.

Berding, N., Skinner, J.C., 1987. Traditional breeding methods. COPERSUCAR International Sugarcane Breeding Workshop. COPERSUCAR, São Paulo, pp. 269–320.

Berding, N., Hogarth, M., Cox, M., 2004. Plant improvement of sugarcane. In: James, G. L. (Ed.), Sugarcane, second ed. Blackwell Science, Victoria, pp. 20–53.

Burnquist, W.L., Redshaw, K. Gilmour, R.F. 2000. Evaluating sugarcane R&D performance: evaluation of three breeding programs. In: Proceedings of the Twenty-Seven International Society of Sugar Cane Technologists, pp. 1–15.

Cheavegatti-Gianotto, A., Abreu, H.M.C., Arruda, P., Bespalhok Filho, J.C., Burnquist, W.L., Creste, S., et al., 2011. Sugarcane (*Saccharum X officinarum*): a reference study for the regulation of genetically modified cultivars in Brazil. Trop. Plant Biol. 4, 62–89.

D'Hont, A., Ison, D., Alix, K., Roux, C., Glaszmann, J.C., 1998. Determination of basic chromosome numbers in the genus *Saccharum* by physical mapping of ribosomal RNA gene. Genome 41, 221–225.

Edmé, S.J., Miller, J.D., Glaz, B., Tai, P.Y.P., Comstock, J.C., 2005. Genetic contribution to yield gains in the Florida sugarcane industry across 33 years. Crop Sci. 45, 92–97.

Ethirajan, A.S., 1987. Sugarcane hybridization techniques. COPERSUCAR International Sugarcane Breeding Workshop. COPERSUCAR, São Paulo, pp. 129–138.

Heinz, D.J., 1987. Sugarcane improvement: current productivity and future opportunites. COPERSUCAR International Sugarcane Breeding Workshop. COPERSUCAR, São Paulo, pp. 55–70.
Henry, R.J., 2010. Basic information on the sugarcane plant. In: Henry, R.J., Kole, C. (Eds.), Genetics, Genomics and Breeding of Crop Plants. Science Publishers, New Hampshire, pp. 1–7.
Inman-Bamber, G., 2014. Sugarcane yields and yield-limiting processes. In: Moore, P., Botha, F.C. (Eds.), Sugarcane Physiology, Biochemistry & Functional Biology. Wiley, New Jersey, pp. 579–600.
Jackson, P.A., 2005. Breeding for improved sugar content in sugarcane. Field Crops Res. 92, 277–290.
James, G.L., 2004. An introduction to sugarcane. In: James, G.L. (Ed.), Sugarcane, second ed. Blackwell Science, Victoria, pp. 1–19.
Kimbeng, C.A., Cox, M.C., 2003. Early generation selection of sugarcane families and clones in Australia: a review. J. Am. Soc. Sugarcane Technol. 23, 20–39.
Legendre, B.L., Burner, D.M., 1995. Biomass production of sugarcane cultivars and early-generation hybrids. Biomass Bioenergy 8, 55–61.
Lingle, S.E., Johnson, R.M., Tew, T.L., Viator, R.P., 2010. Changes in juice quality and sugarcane yield with recurrent selection for sucrose. Field Crops Res. 118, 152–157.
Matsuoka, S., Ferro, J., Arruda, P., 2009. The Brazilian experience of sugarcane ethanol industry. In Vitro Cell. Develop. Biol. Plant 45, 372–381.
Ming, R., Moore, P.H., Wu, K.-K., D'Hont, A., Glaszmann, J.C., Tew, T.L., et al., 2006. Sugarcane improvement through breeding and biotechnology. In: Janick, J. (Ed.), Plant Breeding Reviews, vol. 27. John Wiley & Sons Inc, New Jersey, pp. 15–118.
Moore, P.H., 2005. Integration of sucrose accumulation processes across hierarchical scales: towards developing an understanding of the gene-to-crop continuum. Field Crops Res 92, 119–135.
Moore, P., Berding, N., 2014. Flowering. In: Moore, P., Botha, F.C. (Eds.), Sugarcane Physiology, Biochemistry & Functional Biology. Wiley, New Jersey, pp. 379–410.
Piperidis, N., Piperidis, G., D'Hont, A., 2010. Molecular cytogenetics. In: Henry, R.J., Kole, C. (Eds.), Genetics, Genomics and Breeding of Crop Plants. Science Publishers, New Hampshire, pp. 9–18.
Roach, B.T., Daniels, J.A., 1987. Review of the origin and improvement of sugarcane. COPERSUCAR International Sugarcane Breeding Workshop. COPERSUCAR, São Paulo, pp. 1–31.
Rutherford, R.S., 2014. Mechanisms of resistance to pest and pathogens in sugarcane and related crop species. In: Moore, P., Botha, F.C. (Eds.), *Sugarcane Physiology, Biochemistry & Functional Biology*. Wiley, New Jersey, pp. 435–482.
Santchurn, D., Ramdoyal, K., Badaloo, M.G.H., Labuschagne, M.T., 2014. From sugar industry to cane industry: evaluation and simultaneous selection of different types of high biomass canes. Biomass Bioenergy 61, 82–92.
Simmonds, N.W., 1979. The impact of plant breeding on sugarcane yields in Barbados. Trop. Agricul. 56 (4), 289–300.
Skinner, J.C., Hogarth, D.M., Wu, K.K., 1987. Selection methods, criteria, and indices. In: Heinz, D.J. (Ed.), Sugarcane Improvement Through Breeding. Elsevier, Amsterdam, pp. 409–453.
Tew, T.L., Cobill, R.M., 2008. Genetic improvement of sugarcane (*Saccharum* spp.) as an energy crop. In: Vermerris, W. (Ed.), Genetic Improvement of Bioenergy Crops. Springer, New York, pp. 249–272.
Waclawovsky, A.J., Sato, P.M., Lembke, C.G., Moore, P.H., Souza, G.M., 2010. Sugarcane for bioenergy production: an assessment of yield and regulation of sucrose content. Plant Biotechnol. J. 8, 263–276.

CHAPTER 7

Energy from sugarcane

Luís Augusto Barbosa Cortez, Ricardo Baldassin Jr. and Eduardo de Almeida

Contents

Sugarcane: main Brazilian energy crop 118
Residues energy from sugar/ethanol mills 122
  Bagasse 122
  Straw 127
  Vinasse 128
Main conversion routes for sugarcane bioenergy 130
  Production of electricity 130
  Hydrolysis 131
  Fast pyrolysis 132
Energy cane: an opportunity to improve agricultural productivity 135
The future markets of sugarcane bioenergy 135
Final comments 136
References 137
Further reading 139

The concept "bioenergy" is relatively recent and has been the subject of controversy and discussion, given its possible competition with food production. However, due to the constant crises in oil supply with correspondent price instability and, mainly, the harmful effects on the environment, the development of bioenergy crops for energy shown as a viable, sustainable, and rational alternative to supply part of the current demand for oil and its derivatives, mainly in the transport sector. In this context, this chapter aims to present the energy characteristics of sugarcane and the current and potential uses of the byproducts generated in the production of bioethanol and surplus bioelectricity. With a long-term vision, the potentially applicable conversion routes such as hydrolysis, fast pyrolysis, and "sugarcane bioenergy" are introduced.

Sugarcane Biorefinery, Technology and Perspectives
DOI: https://doi.org/10.1016/B978-0-12-814236-3.00007-X

Sugarcane: main Brazilian energy crop

Sugarcane is a plant originally from Oceania (New Guinea) and Asia (mainly India and China) and was brought by the Portuguese to Brazil during the colonial period, specifically for the production of sugar. Sugarcane presents C4 photosynthetic cycle, is characterized as perennial grass, erect (measuring between 2 and 5 m) (BNDES and CGEE, 2008). Brazil is the leading country in sugarcane production, both for energy and sugar production. Most of it is planted relatively near the Tropic of Capricorn since the plant yields better with hot humid summers and cooler drier winters.

Currently only the aerial part of the plant is considered as useful, being basically composed by several stems constituted by fibers (cellulose, hemicellulose, and lignin) and sugars (sucrose, fructose, and lactose) and contain more than 80% of the plant water—and the leaves—composed of green leaves and dry leaves, commonly called straw.

The commonly called "trash" is a residue originated in the agricultural phase of sugar and alcohol production. It consists of green and dry leaves and some stems (located on top of the plant) left in the soil after the mechanical harvesting of them. Also, fragments of stalks and other plant remains on the surface of the terrain, which aggregate to the terrain process, are considered part of the straw or trash (Braunbeck et al., 2008). The bagasse—residue of the industrial phase—is the term used for the residual mass fraction obtained after the milling process.

As far as the sugarcane composition is concerned, the three main components (straw, bagasse, and sugar) present practically the same proportions and together account for 42.5% of the mass content of the plant; the remaining 57.5% is represented by water. In energy terms, the same proportionality is verified and gives the sugarcane an energy content of the order of 7400 MJ per tonne of sugarcane (Table 7.1).

Straw and bagasse are composed mainly of carbon and oxygen, which together account for approximately 90% of the dry mass of the components, and are present in the same proportions, that is, straw and bagasse are products with similar chemical compositions (except in the case of nitrogen, but in both cases the values are <1%). Regarding the physical properties, there are significant differences in terms of volatiles, fixed carbon, ash, and moisture (Table 7.2). The differences in values between the components and between the same component are acceptable, as they are strongly influenced by the soil and climate conditions (climate, soil type,

Table 7.1 Mass and energy composition of sugarcane straw, bagasse, and sugar (dry basis).

Components	Mass		Energy	
	(kg/t)	(%)	(MJ)	(%)
Straw	140	32.9	2500	33.8
Bagasse	135	31.8	2400	32.4
Sugar	150	35.3	2500	33.8
Total	425	100	7400	100

Source: Data from Leal, M.R.L.V., 2007. The potential of sugarcane as an energy source. In: International society of Sugarcane Technologists Congress, vol. 36. Anais, Durban, South Africa, 29 July, vol. 26, pp. 23–24; Leal, M.R.L.V., 2010a. Technological evolution of sugarcane processing for ethanol and electric power generation. In: Cortez, L.A.B. (Ed.), Sugarcane Bioethanol—R&D for Productivity and Sustainability. Blucher, São Paulo, pp. 561–576; Leal, M.R.L.V., 2010b. Energy cane. In: Cortez, L.A.B. (Ed.), Sugarcane Bioethanol—R&D for Productivity and Sustainability, pp. 751–760.

Table 7.2 Ultimate and proximate analysis of sugarcane components.

Components	Ultimate analysis[a]					Proximate analysis[a]				
	C	H	N	O	S	FC	VM	Ash	H_2O	HHV[b]
Dry leaves[c]	46.2	6.2	0.5	43.0	0.1	11.6	84.5	3.9	13.5	17.4
Green leaves[c]	45.7	6.2	1.0	42.8	0.1	15.7	80.6	3.7	67.7	17.4
Tops[c]	43.9	6.1	0.8	44.0	0.1	16.4	79.3	4.3	82.3	16.4
Bagasse[c]	44.6	5.8	0.6	44.5	0.1	18.0	79.9	2.2	50.2	18.1
Straw[d]	41.6	5.8	0.45	52.1	0.08	6.9	81.6	11.7	9.9	17.7

[a]% mass, dry basis.
[b]High heating value (MJ/kg).
[c]*Source*: Hassuani, S.J., 2005. Biomass power generation: sugar cane, bagasse and trash. Piracicaba: PNUD-CTC, Série Caminhos para a Sustentabilidade, 216 pp. The samples of the components were obtained in *in natura*, straw and top samples were collected in the cane field and the bagasse obtained after the milling process of the stalks at the mill.
[d]*Source*: Pelàez Samaniego, M.R., 2007. Uso de biocombustível da pirólise rápida da palha de cana em um motor de ciclo Otto. Dissertação (Mestrado) Faculdade de Engenharia Mecânica, FEM, Unicamp, Campinas, 100 pp. A mixture of green and dry leaves.

fertilizers, sugarcane variety, cut number, precipitation, among others), the harvesting time, as well as the way the samples were handled and conditioned until the analysis.

Regarding the physical properties of the residues, the minerals are present in different proportions in each of the four components, especially the potassium oxide (K_2O) in the green leaves and the tops; calcium oxide (CaO) in the dry leaves; and iron oxide (Fe_2O_3) and aluminum oxide

Table 7.3 Mineral analysis of sugarcane components (g/kg).

Minerals	Dry leaves	Green leaves	Tops	Bagasse
P_2O_5	0.5	2.0	2.5	0.5
K_2O	2.7	13.3	29.5	1.7
CaO	4.7	3.9	2.6	0.7
MgO	2.1	2.2	2.5	0.5
Fe_2O_3	0.9	0.5	0.2	2.3
Al_2O_3	3.5	1.4	0.5	2.3
CuO	< 0.00006	< 0.00006	< 0.00006	–
ZnO	0.009	0.015	0.035	–
MnO_2	0.169	0.120	0.155	0.062
Na_2O	0.123	0.128	0.119	0.045

Source: Data from Hassuani, S.J., 2005. Biomass power generation: sugar cane, bagasse and trash. Piracicaba: PNUD-CTC, Série Caminhos para a Sustentabilidade, 216 pp.

Table 7.4 Chemical analysis of *in natura* sugarcane components (cellulose, hemicellulose, and lignin).

Components	Dry leaves[a]	Green leaves[a]	Trash[a]	Bagasse[b]
Cellulose	44.5 ± 0.5	40.5 + 0.8	40.1 + 0.4	46.86 + 0.22
Hemicellulose	30.4 ± 0.3	30.8 + 0.8	30.7 + 0.2	27.50 + 0.14
Lignin	12.3 + 0.2	22.8 + 0.2	22.9 + 0.2	26.27 + 0.71
Ash	7.5 + 0.3	2.1 + 0.2	2.2 + 0.2	1.61 + 0.02
Extractives	3.7 + 0.1	2.5 + 0.1	3 + 0.3	–
Total	98.4 + 0.3	98.4 + 0.4	98.9 + 0.3	102.24 + 0.44

[a]Olivares Gómez et al. (2010).
[b]Gouveia et al. (2009).

(Al_2O_3) in the bagasse. Other minerals, such as phosphorus, magnesium, copper, zinc, manganese, and sodium, are also found, however, in lower proportions (Table 7.3).

The lignocellulosic material of sugarcane (fibers), composed of cellulose, hemicellulose, and lignin, represents 11%–16% of the dry matter (Rossell, 2006) in different concentrations in the straw and in the bagasse (Table 7.4).

The juice produced after cane crushing is a solution composed basically of water and sugars (mainly sucrose), which together can add up to 99% of the juice. There are also other components in low concentration, such as organic solids (amino acids, fats, waxes, dyes, acids, etc.) and inorganic solids (SiO_2, K_2O, P_2O_5, CaO, MgO, Na_2O, Fe_2O_3, SO_3, and Cl), which together represent the remaining 1% (Table 7.5).

Table 7.5 Composition of sugarcane juice.

Rossell (2006)[a]		Araújo (1982) and Vasconcelos (2012)[b]	
Sucrose	70–91	Sucrose	16
Glucose	2–4	Glucose	0.3
Fructose	2–4	Fructose	0.1
Total of sugars	75–93	Total of sugars	18
Salts	3–5	Water	81.0
Inorganic acids	1.5–4.5	Ash	0.4
Organic acids	1.0–3.0	Sulfuric acid	0.5
Carboxilic acids	1.1–3.0		
Amino acids	0.5–2.5	Brix	19.5
Proteins	0.5–0.6	pH	5.5
Starch	0.001–0.05	–	–
Gums	0.3–0.6	–	–
Waxes and fats	0.05–0.15	–	–
Colorants and other	3.0–5.0	–	–

[a]% mass, dry basis.
[b]% mass, wet basis.

The cultivation of sugarcane for bioenergy deserves special mention, given the good vegetative performance of the plant. It is observed that one tonne of sugarcane (stem + bagasse) is equivalent to 0.17 tonnes of oil. Therefore considering an average yield of 80 tonnes of stalks plus 20 tonnes of straw per hectare (~42.5 tonnes of dry matter), field production can reach up to 730 GJ per hectare, or 17.4 tonnes of oil. This means that taking into account the 2017/2018 harvest of 635.51[1] million tonnes of sugarcane (MAPA, 2018), the gross energy production in Brazilian sugarcane plantations may have reached approximately 50 million tonnes of oil equivalent (toe) (MME, 2018). When compared to US bioenergy production from maize, it can be seen that in Brazil, sugarcane presents higher energy productivity per planted area (Table 7.6).

However, despite the success of the Brazilian sugar and ethanol production program and the use of primary energy from sugarcane in the form of fuel ethanol and electricity, only 26% of the energy present in the cane is useful (Fig. 7.1). This is due to inefficient processes or nonuse of bagasse and straw in the production of excess electric energy or ethanol. The impact is evident when the primary energy of sugarcane is analyzed

[1] This figure does not consider the sugarcane straw since historically, the sugarcane was harvested after burning.

Table 7.6 Comparison between US corn and Brazilian sugarcane energy productivity.

Indicators	US corn			Brazilian sugarcane		
	Grains	Cob	Trash	Sugar	Bagasse	Trash
Agricultural productivity (t/ha) (dry basis)	10.6[a]	2.0[a]	6.9[a]	14.5	14	14
		19.5			42.5	
HHV (MJ/kg)	18.8[b]	17.7[b]	15.8[b]	15.9	17.9	17.9
Energy productivity (GJ/ha)	199	35	109	231	251	251
		344			732	

[a]Pordesimo et al. (2004).
[b]Patzek (2004).

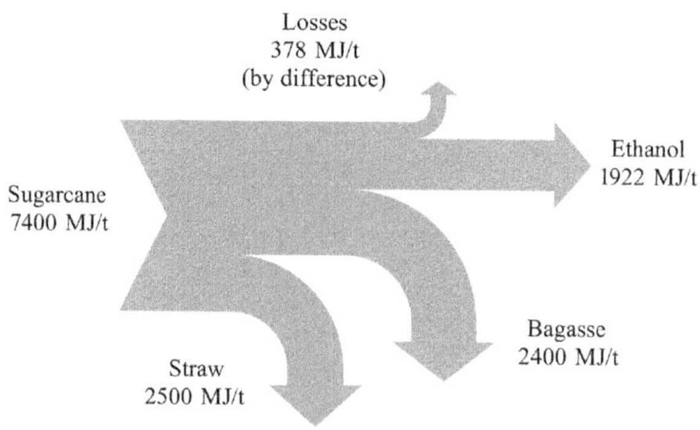

Figure 7.1 Sankey diagram: primary energy conversion of sugarcane in Brazil.

in the Brazilian energy matrix, which was only 16.5% in 2017 (considering only molasses, juice, and bagasse) (MME, 2018).

Residues energy from sugar/ethanol mills

Bagasse

After the extraction of the sugarcane juice, four fractions were obtained: fibrous material (45%), water (51%–49%), nonsoluble solids (2%–3%), and soluble solids and extractives (2%–3%) (Triana et al., 1990).

Morphologically, the outer fibers[2] (50%)—rich in fibrovascular bundles and other fibrous forms that guarantee resistance and hardness to the

[2] Fibers present in the stem skin.

vegetal tissue—abundant quantities of parenchymal cells (30%)—with porous walls are present in the constitution of the bagasse as a function of the sucrose reserve and the transport of nutrients in small distances—and segments of vessels (15%)—with the function of conducting nutrients and distinct from the parenchymal cells by their great length. In addition, the epidermal cells (5%), which give rise to the lining tissue and are characterized by irregular borders, are present in lesser amounts (Sanjuàn et al., 2001; Triana et al., 1990).

In terms of dimensions, the sugarcane bagasse particles differ according to the sugarcane juice extraction systems used by mills. The most common processes are called "milling" and "diffusion." The currently most used process is the milling process, in which the juice is separated by the pressure of cylindrical rollers on the layer of the defibered sugarcane. In the diffusion process, the separation takes place through the washing of sucrose absorbed in the defibered sugarcane layer (MAPA, 2018). Fig. 7.2 presents the characteristics dimensions of sugarcane fractions obtained by milling and diffusion processes.

According to the morphology or particle sizes the sugarcane bagasse can be classified in three different fractions with peculiar characteristics: rind, fiber, and pith (CGEE, 2009). The rind fraction is the largest one when compared to the others, presenting as roughly rectangular blades. The fiber consists of fibrovascular bundles and other fibrous forms, guaranteeing resistance to the sugarcane stems, also presenting parenchyma

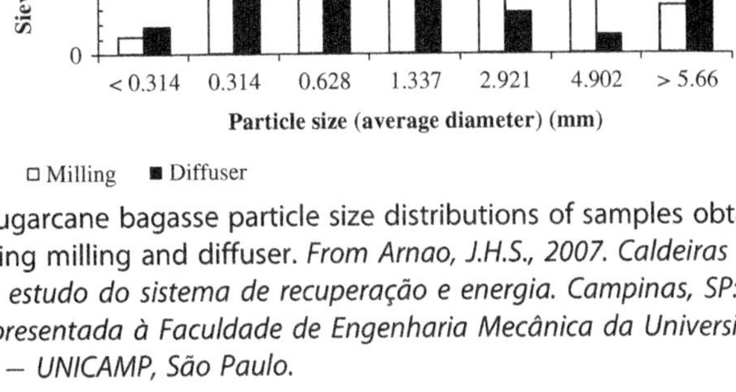

Figure 7.2 Sugarcane bagasse particle size distributions of samples obtained by juice extraction using milling and diffuser. *From Arnao, J.H.S., 2007. Caldeiras aquatubulares de bagaço – estudo do sistema de recuperação e energia. Campinas, SP: [s.n.]. Tese de doutorado apresentada à Faculdade de Engenharia Mecânica da Universidade Estadual de Campinas – UNICAMP, São Paulo.*

cells. Geometrically, the fiber particles may be represented by an elliptical base prism. The pith fraction predominantly presents parenchymal vegetable tissue, is formed by spongy particles with a flat shape and, in dimensional terms, has the form of powder (CGEE, 2009). Fig. 7.3 shows the three different bagasse fractions and Table 7.7 shows the chemical properties associated with them.

About 92% of all the generated bagasse at the mill is used for the heating process into own mill, remaining 8% of surplus bagasse (Macedo, 2005). Some researchers analyzed some scenarios assuming a process of enzymatic hydrolysis acid-diluted, concluding that if surplus bagasse was used as feedstock for second-generation ethanol, it can increase the productivity by up to 2200 L of ethanol per hectare (Leite et al., 2009).

Bagasse is a very important raw material for the expansion of ethanol production, but the technology for conversion of cellulose present in the bagasse into fermentable sugar by hydrolysis process still has a series of challenges.

One of these key challenges is to achieve a homogeneous material with high surface contact in order to allow increases in yields of hydrolysis, either acid, or enzymatic. Focusing on enzymatic hydrolysis, it seems to be the best alternative to disrupt bagasse sugar chains to obtain by subsequent processes bioproducts as ethanol, xylitol, microbial protein, bioplastics, paper, etc.

However, enzymatic hydrolysis still presents other important bottlenecks that need to be overcome, such as the high costs. The high cost is mainly affected by high enzymes prices and some steps of the process, for example, physical pretreatments that decrease the bagasse particle size and allow better contact between cellulose and enzymes but with high energy consumption because the bagasse should be submitted to milling, grinding, and chipping or a combination of these.

In order to propose alternatives to the physical pretreatment, consuming less energy than other ways of obtaining particles of different sizes it was analyzed the performance of a device, called pneumatic classifier equipment, developed at the School of Agricultural Engineering (Feagri/UNICAMP). Based on its operation principle on fluid dynamic theories, the pneumatic classifier equipment makes the separation of bagasse in an air column by dragging the small particles to the top and collecting the coarse particles at the bottom (Almeida, 2012).

The enzymatic hydrolysis of the small particles separated by the equipment, around 37% of the bagasse total mass at ambient equilibrium

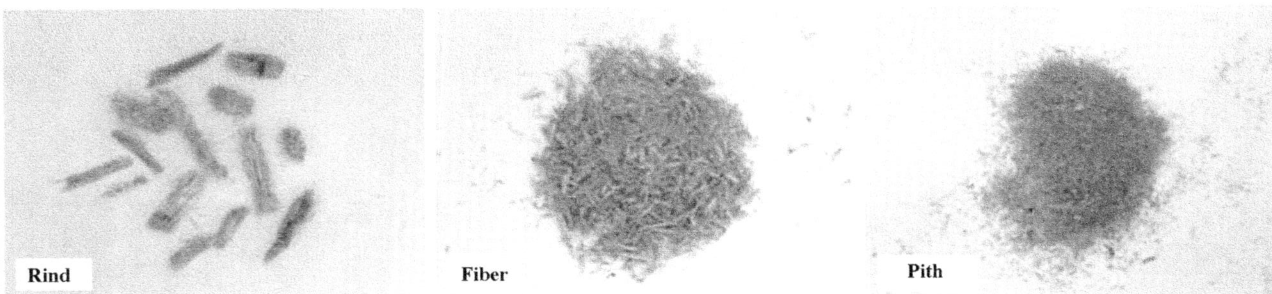

Figure 7.3 Sugarcane bagasse fractions: rind, fiber, and pith. *Adapted from Almeida, E., 2012. Estudo da separação pneumática de frações de bagaço de cana e sua influência na hidrólise enzimática. Dissertação (Mestrado) — Faculdade de Engenharia Química, Universidade Estadual de Campinas, São Paulo.*

Table 7.7 Sugarcane bagasse chemical composition (% mass, dry basis).

Component	Whole bagasse	Fiber	Pith
Cellulose	46.6	47.0	41.2
Hemicellulose	25.2	25.1	26.0
Lignin	20.7	19.5	21.7
Organosoluble	2–3	—	—
Water-soluble	2–3	—	—
Ashes	2.6	1.4	5.4
Water content	47–52	—	—

Source: Adapted from CGEE (2009) and Triana et al., 1990. Atlas del Bagazo de la Caña de Azucar, Cuba-9/GEPLACEA/PNUD/ICIDCA, México.

moisture content, demonstrates that it was the possible conversion of cellulose to glucose of 64% after 72 hours without any extra pretreatment (Almeida, 2012).

This conversion result proved to be as so good as results reported by many studies in the literature using different pretreatments for this biomass. Thus the pneumatic classifier equipment allowed the continuous separation of sugarcane bagasse particles obtaining more homogenous fractions of material presenting in their constitution some cells with characteristics that provide better accessibility of cellulose by the enzymes.

Nowadays theoretical studies of sugarcane bagasse particles pneumatic separation have been conducted aiming to evaluate the particles' separation with a desired range of sizes. Simulation results have shown that there is great accuracy between experimental and virtual tests. Thus the simulations of bagasse particle separations have been shown to be a powerful tool for the development of new equipment that allow the feasible separation of the bagasse particles as well as extending the models to improve studies of others several processes that use the fluidization of sugarcane bagasse as drying, pyrolysis, and gasification (Almeida et al., 2016).

Bagasse is vital for the sugar mills operations, for steam and electricity generation. The bagasse energy content (high heating value) is approximately 18–20 MJ/kg and, according to Leal (2010a), generate around 585 kWh per tonne of sugarcane (Fig. 7.4).

Facing the 2017/2018 harvest (MAPA, 2018), the total bagasse generation in the sugarcane plantations may have reached 175 million tonnes—considering that bagasse represents around 27.5% of the sugarcane stalks (mass basis). In some sugarcane mills, mainly those using less-efficient

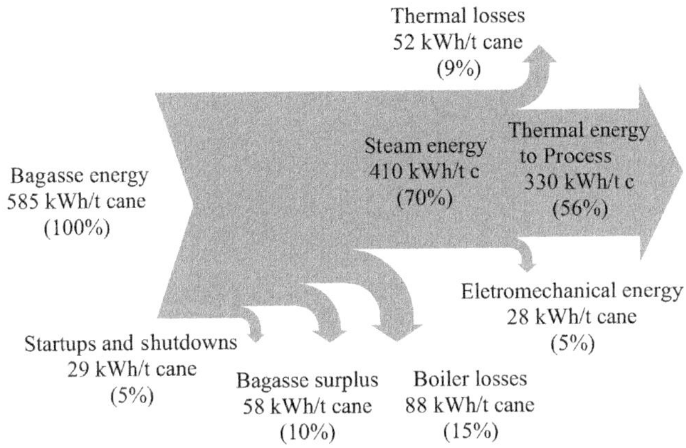

Figure 7.4 Sankey diagram: bagasse energy conversion in a typical Brazilian sugarcane mill. *From Leal, M.R.L.V., 2010a. Technological evolution of sugarcane processing for ethanol and electric power generation. In: Cortez, L.A.B. Sugarcane Bioethanol—R&D for Productivity and Sustainability. São Paulo, Blucher, pp. 561–576; Leal, M.R.L.V., 2010b. Energy cane. In: Cortez, L.A.B. Sugarcane Bioethanol—R&D for Productivity and Sustainability, pp. 751–760.*

cogeneration systems, the bagasse surplus is a reality, reflecting the low attractiveness of surplus electricity production.

Straw

Despite the high energetic potential of the straw, its use is not yet fully developed. As nonburning practice is being implemented in Brazil, straw is becoming available. Technical difficulties associated with recovery in the field, transportation to the mill, cleaning, and costs are the main obstacles. As more and more mills are cogenerating electricity in Brazil, more efforts have been made in overcoming the technical barriers.

Besides the energy use, straw is also being a dispute for soil protection. According to Rossetto et al. (2010), several studies show the benefits of maintaining straw in sugarcane plantations, in order to minimize soil degradation. The authors report that the degradation can be physical, chemical, and biological and that, in this scenario, the maintenance of the straw cover can contribute to each of these different processes. The authors point to erosion reduction (removal of thinner soil particles by rain and wind), maintenance of soil moisture (reduction of water evaporation), and nutrient replenishment as the main benefits of maintaining straw in the soil. However, it is still a controversial and unanswered issue to determine the minimum and ideal quantity of litter to be left in the field since it depends on each situation.

As straw and bagasse have similar chemical composition, it is expected that the energy use of straw follows the same trends of bagasse, such as electricity and enzymatic hydrolysis in order to increase the productivity of ethanol in the mills. In this scenario, the straw is indicated as a possible complement energy source of the bagasse.

When the production of straw in the Brazilian sugarcane plantations was evaluated, it was estimated that in the 2017/2018 harvest, the total produced may have reached approximately 140 million tonnes—considering the productivity of 14.4 tonnes per hectare (dry basis) (Hassuani, 2005) and about 9–10 million hectares of planted area. In the 2017/2018 harvest, mechanized harvesting accounted for more than 90% of all sugarcane harvested in the state of São Paulo, where more than 60% of sugarcane is planted in Brazil.

Vinasse

Vinasse is the liquid residue from the distillation of the product of the alcoholic fermentation of cane juice, molasses, or the molasses–juice mixture.

Due to the suspension of organic and mineral solids, the vinasse presents high chemical oxygen demand and oxygen biochemistry, besides presenting low pH and high corrosive power, being therefore considered as a high polluting potential residue.

Produced in the order of 10–15 L per liter of ethanol, the main chemical constituents of the vinasse are potassium and sulfur; however, significant amounts of phosphorus, nitrogen, calcium, magnesium, and some micronutrients are found in it.

According to Pinto (1999), the composition of vinasse depends on a number of factors, such as:

- nature and composition of the raw material used (e.g., must from molasses obtained from the manufacture of sugar, if the molasses is diluted in water or juice of the last mills, or if the must is from pure juice);
- nature and composition of wines;
- fermentation system employed;
- additives used in fermentation, chemicals, acids, antibiotics, and nutrients based on nitrogen and phosphorus;

- type of apparatus used in distillation;
- yeast breed used;
- water quality used; and
- the system of work and influence of operators.

In Table 7.8, the characteristics of the vinasse, resulting from molasses, juice, and the molasses–juice mixture are presented.

According to Cortez et al. (2010a), due to its polluting potential, it is extremely important to conduct studies aimed at reducing the volume of vinasse, as well as to develop forms and processes for its correct treatment.

Among the possible methods for treating vinasse, the following stand out:

- *Biodigestion*: Production of biogas (methane gas) and biofertilizers through fermentation or aerobic or anaerobic digestion.
- *Concentration*: Reduction of the amount of water from the vinasse in order to facilitate its handling, for later use as fertilizer, or transformation into protein for feed or even for combustion or incineration in boilers.

Table 7.8 Characteristics of sugarcane vinasse.

Parameter	Molasses	Juice	Mixed
pH	4.2–5.0	3.7–4.6	4.4–4.6
Temperature (°C)	80–100	80–100	80–100
BOD (mg/L O_2)	25,000	6000–16,500	19,800
BOD (mg/L O_2)	65,000	15,000–33,000	45,000
Total solids (mg/L)	81,500	23,700	52,700
Volatile Solids (mg/L)	60,000	20,000	40,000
Fixed solids (mg/L)	21,500	3700	12,700
Nitrogen (mg/L N)	450–1,600	150–700	480–710
Phosphor (mg/L P2O5)	100–290	10–210	9–200
Potassium (mg/L K_2O)	3740 7830	1200–2100	3340–4600
Calcium (mg/L C)	450–5180	130–1540	1330–4570
Magnesium (mg/L MgO)	420–1520	200–490	580–700
Sulfate (mg/L SO4)	6400	600–760	3700–3730
Carbon (mg/L C)	11,200–22,900	5700–13,400	8700–12,100
C/N relation	16–16.27	19.7–21.07	16.4–16.43
Organic matter (mg/L)	63,400	19,500	38,000
Reducing substances (mg/L)	9500	7900	8300

Source: Data from Pinto, C.P. Tecnologia da digestão anaeróbia da vinhaça e desenvolvimento sustentável. 1999. Dissertação (Mestrado) – Universidade Estadual de Campinas, Campinas, SP, 1999.

Another technical solution for vinasse disposal may be combustion or incineration. Vinasse may be used as fuel in boilers for the production of steam and/or electric energy or simply incineration in a process similar to the paper industry with black liquor.

Nowadays vinasse is used by the mills, almost entirely, in the fertirrigation of sugarcane, aiming at the replacement of part of the nutrients to the soil, thus reducing the demand for fertilizers and correctives and, consequently, the expenses with them. However, this process is carried out without previous treatment of the vinasse.

One of the main ways to use vinasse for energy purposes is by biodigestion. In this process, the vinasse is subjected to the action of microorganisms in reactors, generating as one of the products—methane gas, which can be used as fuel and subsequently burned in boilers or internal combustion engines. However, the main bottleneck of this route is the high cost of biogas produced, making it unviable economically compared to other fuels (mainly diesel), although vinasse biodigestion is recently gaining popularity in Brazil.

In the case of vinasse as fuel in boilers, Cortez et al. (2010a) report experiences and results of this technique; however, the authors note the need to increase the solid concentration to values greater than 50%, as the reduced solids concentration reduces the flame adiabatic temperature. However, for Cortez et al. (2010a), despite past efforts in the development and studies of vinasse combustion, as well as the importance of the theme, all projects were abandoned. But the authors report that the German company SAAKE has been developing a fluid burner with characteristics similar to that of vinasse and intends to commercialize this technology in Brazil, where it has already been conducting vinasse combustion tests.

Main conversion routes for sugarcane bioenergy

Production of electricity

In the Brazilian sugarcane mills, electric energy is generated in Rankine topping cycles, with heat supply in the boilers from the burning of sugarcane bagasse. The typical (old) configuration uses a boiler with steam generation at 22 bar, superheated steam temperature between 280°C and 320°C, steam consumption of the order of 500 kg per tonne of cane and backpressure turbines.

According to Ensinas et al. (2010), the generation of surplus electric power is only possible if more modern systems are used. Currently the new plants already employ a more modern cogeneration system, with boilers operating at 60–65 bar and 480°C–520°C and backpressure turbines (Ensinas et al., 2010; Leal, 2010a). In this scenario, Walter and Nogueira (2008) and Walter et al. (2005) show that the generation of exportable electricity can reach 40.5 kWh per tonne of sugarcane.

However, Ensinas et al. (2010) and Leal (2010a) state that improvements in the process and technologies used should be seen in the coming years, which will increase the efficiency of cogeneration systems and reduce the demand for steam in the process. More modern cogeneration systems should operate as 100–120 bar per 520°C boilers and condensation/extraction or backpressure/condensation turbines, which the authors say are already available in the country.

According to the Brazilian Ministry of Mines and Energy, the contribution of sugarcane bioelectricity to the Brazilian electricity matrix was 8.2% (35 TWh) in 2017, where about 40% was internally consumed (sugar–ethanol production) and 60% was sold to the electricity grid (MME, 2018).

In the long term, biomass gasification integrated with gas turbine systems (BIG/GT technology) is considered one of the most promising technologies for the generation of surplus electric power in the plants. However, some technical–economic bottlenecks still need to be overcome for its viability and commercialization.

Hydrolysis

It is recognized until today as the route with the greatest potential to promote the increase of ethanol production (without expanding the agricultural frontier), as well as to increase the conversion efficiency of the primary energy of sugarcane.

Hydrolysis is the most common method of converting cellulose present in biomass (such as sugarcane bagasse) into monosaccharide components—especially glucose—which, among many utilities, can be produced from ethanol by a fermentation process.

Currently the two main routes investigated are acid hydrolysis and enzymatic hydrolysis. In the case of the acid, the production of glucose is essentially from the degradation of cellulose in an acid medium (diluted or concentrated), with sulfuric or hydrochloric acid being generally employed. In the case of the enzyme, the reducing sugars are produced by enzymes (usually groups of enzymes), especially endoglucanase, exoglucanase, and β-glucosidase, and auxiliary enzymes, such as glucuronidase, xylanase, gluconase, among others, which add to the hemicellulose attack (Olsson et al., 2005; Sánches and Cardona, 2008).

Enzymatic hydrolysis is a relatively slow process (may take days) when compared to acid hydrolysis (occurs in minutes, hours). In addition, the production costs of enzymes are high. However, this route is considered to be the most advantageous route and is considered the most promising route for the production of second-generation ethanol.

The principal disadvantages of the acid route are:

- degradation of glucose in hydroxymethylfurfural, xylose in furfural;
- the production of other components toxic to the yeast that inhibit the subsequent fermentation activity; and
- the high cost associated with the great acid demands of the process performed in concentrated acid medium.

In the case of the enzymatic route, the drawbacks are associated with the high costs of producing enzymes and over a long process time.

According to Leite et al. (2009), considering a more intense use of sugarcane energy, in which 50% of the straw would be recovered to substitute part of the bagasse in the cogeneration cycles (thus making possible a 50% bagasse surplus), by hydrolysis. Then, it would be possible to increase ethanol production by up to 4000 L per hectare, that is, total production could reach up to 10,000 L of ethanol per hectare, leading to an increase of over 66% in ethanol production.

Currently efforts have been made to make this route feasible, technically and economically, although important technical problems still need to be solved. The most important hydrolysis projects in Brazil are the Raízen Project and the Granbio Project. Raízen claims to be producing 2G ethanol and the Granbio Project faces important engineering difficulties. Other projects are in progress around the world. An appraisal of hydrolysis projects is given by Lynd (2018).

At the research level on sugarcane hydrolysis, important contributions are given by the Brazilian Bioethanol National Laboratory (CTBE), The University of Campinas (UNICAMP), and the University of São Paulo.

Fast pyrolysis

It is pointed out as the main alternative for lignocellulosic biomass pretreatment, mainly for converting solid biomass coproducts (or residues) into liquid. The main characteristic is the energy compaction, where it is possible to increase the energy density by more than 1500% (Baldassin et al., 2011). This characteristic, associated with the viability of the decentralized processing of biomass in small plants, and easily handling and storage, can be a solution to overcome the high cost of biomass transportation, usually by trucks. However, despite the existence of several studies on this technology and its potential impacts and benefits (USLU et al., 2008; Badger e Fransham, 2006, Rogers and Brammer, 2009; Trippe et al., 2010; Pootakham e KUMAR, 2010a,b; Kurian et al., 2013; Kersten e Garcia-Perez, 2013), some technical–economic bottlenecks still need to be overcome for its viability and commercialization at a large scale (Meier et al., 2013).

The liquid produced in the fast pyrolysis is an emulsion of oxygenated compounds and water. It may be dark brown (in most cases), reddish brown (after steam filtration and charcoal removal), or dark green (presence of high concentrations of hydrogen). The oxygenated emulsion is composed of two distinct and nonsoluble phases, defined as a light aqueous fraction and a heavy nonaqueous fraction. The light fraction (also known as pyrollenous acid, acid extract, or pyrolytic acid) is composed of low molecular weight compounds and has a viscosity similar to water (at 25°C), high translucency, orange coloration, high acidity, strong odor, and high water content. The heavy fraction (commonly called bio-oil, or pyroligneous tar, wood oil, or pyrolysis oil) is composed of high molecular weight compounds and has high viscosity, high opacity, dark color, strong odor, and lower presence of water (when compared to the aqueous fraction) (Yang et al., 2014; Hossain e Davies, 2013).

The bio-oil (heavy fraction) is the most interesting, and its chemical composition is similar to the biomass used for its production; however, the properties (viscosity, water content, organic compounds, and others) is strongly influenced by the pyrolysis technology and the process conditions (Meier et al., 2013). The main applications of bio-oils are for heating or steam production (direct combustion in boilers, including cofiring), and electricity production (turbines and internal combustion engines), or as raw material for advanced biofuels or chemicals (upgrading, gasification, chemical extraction, fractionation) (Bridgwater, 2012).

Baldassin Jr. (2015) carried out an investigation of fast pyrolysis of sugarcane, and the results demonstrated that it was possible to convert the whole sugarcane plant (fiber and sugar) into bio-oil. The whole plant processing can be an alternative for reducing the transportation costs of the

Table 7.9 Ultimate and proximate analysis of sugarcane bio-oils.

Composition	Whole sugarcane	Straw	Bagasse
Carbon	48	52.9	52.2
Oxygen[a]	45.1	38.7	40.4
Hydrogen[a]	5.7	6.55	6.42
Nitrogen[a]	0.7	1.3	0.7
Sulfur[a]	0.1	0.1	0.1
Ash[a]	0.4	0.5	0.2
Water content[b]	14	10	12
HHV (MJ/kg)[c]	20.29	22.75	20.8
LHV (MJ/kg)[c]	19.08	21.36	19.44

[a]% mass, dry basis.
[b]% mass, wet basis.
[c]Dry basis.
Source: Data from Baldassin Jr., R., 2015. Pirólise rápida de cana-de-açúcar integral em reator piloto de leito fluidizado. 165f. Dissertação (Doutorado) – Faculdade de Engenharia Agrícola, Universidade Estadual de Campinas, São Paulo.

sugarcane, mainly taking into account some specific applications (e.g., for advanced biofuels and chemicals productions). The results also demonstrated that the boil-oil from whole sugarcane was similar to bio-oils from bagasse and straw (Table 7.9).

According to Meier et al. (2013), around 50 localities have been identified as developing pyrolysis activities, with much of the work being done experimentally and academically in research centers, but with some important commercial initiatives. The most relevant research, demonstration, and commercial applications have been developed in the United Kingdom, United States, Canada, Germany, Finland, and the Netherlands. The authors point out participation and collaboration of large companies in various pyrolysis works worldwide, especially: Honeywell company (UOP), Chevron Technology Ventures, Fibria Celulose, Felda Global Ventures, and Metso. The authors state that there is not a single fast pyrolysis plant in the world operating for commercial purposes since it is considered that the EMPYRO project is a demonstration project.

In Brazil, some research activities in fast pyrolysis technology have been developed (e.g., UNICAMP, Federal University of Minas Gerais, and others), and a commercial project is in progress led by Fibria Celulose SA, in partnership with Ensyn.

Table 7.10 Comparison of three sugarcane varieties.

Characteristics		Typical Brazilian	Mauritius	Barbados
Pol % cane	(%)	14.5	19.9	12
Fiber % cane	(%)	13.5	17.5	26
Straw % cane	(%)	14	NI	25
Productivity	(t/ha/year)	68.7	NI	100
Total fiber	(t/ha/year)	19.3	NI	51
Primary energy	(GJ/ha/year)	520	NI	1100
	(MJ/t)	7400	NI	11,200

NI, Not informed.

Energy cane: an opportunity to improve agricultural productivity

The term "energy cane" is a recent one, and it means the development of new sugarcane varieties aimed at increasing total sugarcane biomass instead of sugar only. The main idea is to increase the energy conversion efficiency of primary sugarcane energy, agricultural productivity, or total bioenergy per hectare.

Some initiatives in Mauritius and Barbados include the development of new sugarcane or energy cane varieties. Table 7.10 shows some data referring to these new varieties of sugarcane and Brazilian sugarcane.

In Brazil, efforts to develop energy cane are made by VIGNIS. The company is owned by the founders of CanaVialis (bought and closed Monsanto) and headquartered in the city of Campinas, state of São Paulo. The company's proposal is to act on the exclusive genetic improvement for energy cane, aiming at the production of high productivity varieties adapted to the different Brazilian biomes, as well as developing customized cultivars for the possible future demands of the mills (Vignis, 2012).

The future markets of sugarcane bioenergy

The traditional markets for sugarcane bioenergy are bioethanol as gasoline substitute and bioelectricity, as commented before.

According to the International Energy Agency, the future of bioethanol for transportation is considered to be important in the strategy to reduce greenhouse gas (GHG) emissions. This is well discussed by Leite et al. (2009) who also discuss its participation in the transportation sector even with the introduction of electric vehicles.

Bioethanol from sugarcane can have low GHG emissions and therefore continue to be competitive with solar or wind electricity. However, other efforts should be made in different areas (agriculture, conversion, and final use) to extend bioethanol long-run lifetime.

Production of electric energy may be a more durable market for two reasons. First, the technologies for power electricity production are more mature and second, the complementarity of sugarcane bioenergy with other renewables, such as hydro, solar, and wind.

In the medium and long terms, biofuels from aviation or biojet fuels also can be an important market, although the "drop-in" requirement establishes a difficult barrier to be overcome.

At last, sugarcane biofuels can also be used for maritime transportation, an area which does not seem to have many competitors. Sugarcane or wood can be used to produce pyrolysis oil which can be blended with maritime fuel oil mitigating GHG emissions.

In any circumstance, it is very important to produce sustainable feedstock and special attention should always be given to minimize inputs and negative externalities.

Final comments

The development of biomass crops for energy purposes, although not recent, is still controversial, but it is an irreversible process as the advantages over the exploitation and use of fossil fuels are significant. Only through biomass, we will be able to produce significant volumes of liquid fuels for the transportation sector (in the short term), thus reducing dependence on oil.

In the case of Brazil, the adoption of sugarcane as the main bioenergy crop has shown to be ideal, given the plant's adequacy for Brazilian climate and soil and reasonable agricultural and energy productivity. It also presents a low consumption of fossil energy in the production process of ethanol, a good potential for expansion of supply, and good competitive cost as fuel in Brazil.

Besides the traditional transportation and electricity markets in which sugarcane already plays an important role in Brazil, there are other interesting possibilities, such as the aviation and maritime sectors. In both cases, there are significant challenges to be overcome.

References

Almeida, E., 2012. Estudo da separação pneumática de frações de bagaço de cana e sua influência na hidrólise enzimática. Dissertação (Mestrado) – Faculdade de Engenharia Química, Universidade Estadual de Campinas, São Paulo.

Almeida, E., Spogis, N., Silva, M.A., 2016. CFD DEM coupling validation for pneumatic separation of sugarcane bagasse. In: Aiche Annual Meeting, vol. 16. Anais, San Francisco, CA, November 13–18.

Araújo, J.A., 1982. de. Obtenção de etanol por fermentação alcoólica. Rio Largo, IAA-PLANALSUCAR (COONE), 83 pp.

Arnao, J.H.S., 2007. Caldeiras aquatubulares de bagaço – estudo do sistema de recuperação e energia. Campinas, SP: [s.n.]. Tese de doutorado apresentada à Faculdade de Engenharia Mecânica da Universidade Estadual de Campinas – Unicamp, São Paulo.

Badger, P.C., Fransham, P., 2006. Use of mobile fast pyrolysis plants to densify biomass and reduce biomass handling costs—a preliminary assessment. Biomass Bioenergy 30, 321–325.

Baldassin Jr. R., 2015. Pirólise rápida de cana-de-açúcar integral em reator piloto de leito fluidizado. 165f. Dissertação (Doutorado) – Faculdade de Engenharia Agrícola, Universidade Estadual de Campinas, São Paulo.

Baldassin Jr. R., et al., 2011. Energetic compaction of sugarcane trash by fast pyrolysis. In: BBEST – Brazilian Bioenergy Science and Technology Conference, vol. 1, Campos de Jordão.

BNDES, CGEE, 2008. Banco Nacional para o Desenvolvimento Social e Econômico; Centro de Gestão e Estudos Estratégicos. Bioetanol de cana-de-açúcar: Energia para o desenvolvimento sustentável. BNDES e CGEE, Rio de Janeiro, 316 pp.

Braunbeck, O.A., Magalhães, P.S.G., Garcia, M.O., 2008. Colheita e recuperação de biomassa. In: Cortez, L.A.B., Lora, E.E.S., Olivares Gómez, E. (Eds.), Biomassa para energia. Editora da Unicamp, Campinas, 734 pp.

Bridgwater, A.V., 2012. Review of fast pyrolysis of biomass and product upgrading. Biomass Bioenerg 38, 68–94.

Cortez, L.A.B., et al., 2010a. R&D needs in the industrial production of vinasse. In: Cortez, L.A.B. Sugarcane Bioethanol—R&D for Productivity and Sustainability. Blucher, São Paulo, pp. 619–636.

CGEE, Centro de Gestão e Estudos Estratégicos (Management and Strategic Studies Center), 2009. Available at: <https://www.cgee.org.br>.

CTBE. Laboratório Nacional de Ciência e Tecnologia do Bietanol. Disponível em: <www.bioetanol.org.br>.

Ensinas, A.V., Arnao, J.H.S., Nebra, S.A., 2010. Increasing energetic efficiency in sugar, ethanol, and electricity producing plants In: Cortez, L.A.B. Sugarcane Bioethanol—R&D for Productivity and Sustainability. Blucher, São Paulo, pp. 583–600.

Gouveia, E.R., Trajano do nascimento, R., Souto-Maior, A.M., 2009. Validação de metodologia para caracterização química de bagaço de cana-de-açúcar. Revista Química Nova 32 (6), 1500–1503.

Hassuani, S.J., 2005. Biomass Power Generation: Sugar Cane, Bagasse and Trash. PNUD-CTC, Série Caminhos para a Sustentabilidade, Piracicaba, 216 pp.

Hossain, A.K., Davies, P.A., 2013. Pyrolysis liquids and gases as alternative fuels in internal combustion engines – a review. Renew. Sust. Energ. Rev. 21, 165–189.

Kersten, S., Garcia-Perez, M., 2013. Recent developments in fast pyrolysis of ligno-cellulosic materials. Curr. Opin. Biotechnol. 24, 414–420.

Kurian, J.K., Nair, G.R., Hussain, A., Raghavan, G.S.V., 2013. Feedstocks, logistics and pre-treatment processes for sustainable lignocellulosic biorefineries: a comprehensive review. Renew. Sust. Energ. Rev. 25, 205–219.

Leal, M.R.L.V., 2007.The potential of sugarcane as an energy source. In: International Society of Sugarcane Technologists Congress, vol. 36, Durban, Anais, South Africa, 29 July, vol. 26, pp. 23–24.

Leal, M.R.L.V., 2010a. Technological evolution of sugarcane processing for ethanol and electric power generation. In: Cortez, L.A.B. Sugarcane Bioethanol—R&D for Productivity and Sustainability. Blucher, São Paulo, pp. 561–576.

Leal, M.R.L.V., 2010b. Energy cane. In: Cortez, L.A.B. Sugarcane Bioethanol—R&D for Productivity and Sustainability, pp. 751–760.

Leite, R.C.C., Leal, M.R.L.V., Cortez, L.A.B., Griffin, W.M., Scandiffio, M.I.G., 2009. Can Brazil replace 5% of the 2025 gasoline world demand with ethanol? Energy 34, 655–661.

Lynd, L.R., 2018. 2G 2.0 In: Cortez, L.A.B., Leal, M.R.L.V., Nogueira, L.A.H. Sugarcane Bioenergy for Sustainable Development – Expanding Production in Latin America and Africa. Routledge, London, 417 pp.

Macedo, I.C., 2005. A Energia da Cana-de-Açúcar - Doze estudos sobre a agroindústria da cana-de-açúcar no Brasil e a sua sustentabilidade. SP: UNICA – União da Agroindústria Canavieira do Estado de São Paulo, São Paulo, 237 pp.

MAPA, Ministério da Agricultura, Pecuária e Abastecimento (Ministry of Agriculture, Livestock and Supply), 2018. Brasília, 2018. Available at: <agricultura.gov.br> Accessed at 01/01/2019.

Meier, D., van de beld, B., Bridgwater, A.V., Elliott, D.C., Oasmaa, A., Preto, F., 2013. State-of-the-art of fast pyrolysis in IEA bioenergy member countries. Renew. Sust. Energ. Rev. 20, 619–641.

MME, Ministerio de Minas e Energia (Ministry of Mines and Energy), 2018. Available at: <http://www.mme.gov.br/>, Brazil.

Olivares Gómez, E., et al., 2010. Sugarcane trash as feedstock for second generation processes. In: Cortez, L.A.B. Sugarcane Bioethanol—R&D for Productivity and Sustainability. Blucher, São Paulo, pp. 637–659.

Olsson, L., et al., 2005. Bioethanol production from lignocellulosic material. In: Dumitriu, S. (Ed.), Polysaccharides—Structural Diversity and Functional Versatility., second ed. Marcel Dekker, New York, pp. 957–993.

Patzek, T.W., 2004. Thermodynamics of the corn-ethanol biofuel cycle. Crit. Rev. Plant Sci. 23, i. 6, pp. 519–567.

Pelàez Samaniego, M.R., 2007. Uso de biocombustível da pirólise rápida da palha de cana em um motor de ciclo Otto. Dissertação (Mestrado) – Faculdade de Engenharia Mecânica, FEM. Unicamp, Campinas, 100 pp.

Pinto, C.P., 1999. Tecnologia da digestão anaeróbia da vinhaça e desenvolvimento sustentável. Dissertação (Mestrado) – Universidade Estadual de Campinas, Campinas, SP.

Pootakham, T., Kumar, A., 2010a. Bio-oil transport by pipeline: a techno-economic assessment. Bioresour. Technol. 101, 7137–7143.

Pootakham, T., Kumar, A., 2010b. A comparison of pipeline versus truck transport of bio-oil. Bioresour. Technol. 101, 414–421.

Pordesimo, L.O., Edens, W.C., Sokhansanj, S., 2004. Distribution of aboveground in corn stover. Biomass Bioenerg. 36, 337–343.

Rogers, J.G., Brammer, J.G., 2009. Analysis of transport costs for energy crops for use in biomass pyrolysis plant networks. Biomass Bioenerg. 33, 1367–1375.

Rossell, C.E.V., 2006. Produção de etanol de cana-de-açúcar: qualidade da matéria prima. In: Workshop—Produção de etanol, diretrizes de políticas públicas para a pesquisa científica e tecnológica em bioenergia no estado de são paulo, vol. 1, Lorena, 10 de novembro de. Disponível em: <http://www.inovacao.unicamp.br/etanol/report/docs-Iworkshop_cana.php> (acesso em: 03 de agosto de 2012.).

Rossetto, R., et al., 2010. Fertility maintenance and soil recovery in sugarcane crops. In: Cortez, L.A.B. Sugarcane Bioethanol—R&D for Productivity and Sustainability. Blucher, São Paulo, pp. 381−404.

Sánches, O.J., Cardona, C.A., 2008. Trends in biotechnological production of fuel ethanol from different feedstocks. Bioresour. Tecnhnol. 99, 5270−5295.

Sanjuán, R., et al., 2001. Morphological and chemical composition of pitch and fiber from Mexican sugarcane bagasse. Eur. J. Wood Wood Prod. 59, 447−450.

Triana, O., et al., 1990. Atlas del Bagazo de la Caña de Azucar, Cuba-9/GEPLACEA/PNUD/ICIDCA, México.

Trippe, F., Fröhling, M., Schultmann, F., Stahl, R., Henrich, E., 2010. Techno-economic analysis of fast pyrolysis as a process step within biomass-to-liquid fuel production. Waste Biomass Valor. 1, 415−430.

Uslu, A., Faaij, A.P.C., Bergman, P.C.A., 2008. Pre-treatment technologies, and their effect on international bioenergy supply chain logistics. Techno-economic evaluation of torrefaction, fast pyrolysis and pelletisation. Energy 33, 1206−1223.

Vasconcelos, J.N. Fermentação etanólica. In: Santos F. A., et al. Cana-de-açúcar: Bioenergia, Açúcar e Etanol – Tecnologias e Perspectivas. 2 ed., Viçosa, Brazil, 2012.

Vignis, 2012. Disponível em <http://www.vignis.com.br> (acesso em 30 de julho de.)

Walter, A.C., et al., 2005 Levantamento do Potencial Nacional de Produção de Eletricidade nos segmentos sucroalcooleiro, madeireiro e em usinas de beneficiamento de arroz. Relatório final para o PNUD/MME, Unicamp, Campinas, julho.

Walter, A.C., Nogueira, L.A.H., 2008. Sistemas de produção de eletricidade a partir da biomassa. In: Cortez, L.A.B., Lora, E.E.S., Olivares Gómez, E. (Eds.), Biomassa para energia. Editora da Unicamp, Campinas, pp. 573−607.

Yang, S.I., Wu, M.S., Wu, C.Y., 2014. Application of biomass fast pyrolysis part I: pyrolysis characteristics and products. Energy 66, 162−171.

Further reading

Conab. <https://www.novacana.com/n/cana/safra/conab-levantamento-safra-2018-19-cana-de-acucar-210818>.

Conab, 2008. Companhia Nacional de Abastecimento. Perfil do setor do açúcar e do álcool no Brasil. Conab, Ministério da Agricultura, Pecuária e Abastecimento, Brasília, 76 pp.

Cortez, L.A.B., Brossard Perez, L.E., 1997. Experiences on Vinasse disposal – part III: combustion of Vinasse #6 fuel oil emulsions. Braz. J. Chem. Eng. 14 (1), 9−18.

RAO, P.S. 2012. Genetic potential of sugarcane germplasm for higher biomass production to generate energy. In: workshop produção de etanol, diretrizes de políticas públicas para a pesquisa científica e tecnológica em bioenergia no estado de são paulo, 8., Painel 2: Energy Cane: the Barbados experience, Campinas, 5 de outubro de 2007. Disponível em <http://www.apta.sp.gov.br/cana/ver_work.php?work_id = 51> (acesso em 03 de agosto de 2012.)

CHAPTER 8

Chemical and energy potential of sugarcane

Sarita Cândida Rabelo, Lívia Beatriz Brenelli de Paiva, Thaynara Coradini Pin, Luisa Fernanda Rios Pinto, Laura Plazas Tovar and Pedro Yoritomo Souza Nakasu

Contents

Introduction 141
Energy potential of sugarcane: technology platforms of energy production based on energy form 142
Routes for energy production 143
Energy storage applications of sugarcane-derived materials 147
Chemical potential of sugarcane 148
Technology platforms of chemical production 149
Microbial metabolic pathway and sugarcane-derived products 153
Advanced bio-based polymers and derived materials 155
Conclusion 159
References 159

Introduction

The circular economy needs innovative designs and process technologies to provide materials, chemicals, facilities (i.e., energy) across the three-sector economic model: extraction of raw materials, manufacturing, and services. Bioenergy and renewable products consumption has grown significantly, contributing to the reduction in greenhouse gas (GHG) emissions, creation of new jobs, increased tax revenue for the countries, and improved family income (Renewable Fuels Association, 2016).

Several driving forces are responsible for the growing global interest in a bio-based economy: establishing blending mandates for biofuels and stringent regulations to ensure a sustainable, renewable energy production which has been increasing in developed and developing countries since 2006. Moreover, expanding biofuels beyond their traditional usage in the

Sugarcane Biorefinery, Technology and Perspectives
DOI: https://doi.org/10.1016/B978-0-12-814236-3.00008-1

road transport sector such as in aviation, electricity generation, and even maritime transport (United Nations Conference on Trade and Development, 2016). Another one is advancing technologies to improve human life, which has generated novel opportunities to create advanced bio-based chemicals and derived polymers and materials (Moraes et al., 2015). Political and private strategies in many countries matured from a shallow biofuels concept to more in-depth forms of bioenergy, and also aiming to increase biomass utilization for the production of biofuels and biomaterials.

In addition, most commercially available organic chemicals and polymers are derived from fossil feedstocks. When thinking of gradually reducing utilization of fossil sources, it is important to design chemical pathways to efficiently replace petrochemicals. Sugarcane is considered an important commercial crop worldwide, and, Brazil is its largest producer, followed by India, China, and Thailand (FAO, 2018). Sugarcane's primary function entails sugar and alcohol production. However, there is a wide range of new applications for the extracted sugars, lignocellulosic materials (sugarcane bagasse and straw), and processing by-products (vinasse, filter cake) in biofuels, materials, chemicals, and facilities (Fig. 8.1).

Energy potential of sugarcane: technology platforms of energy production based on energy form

Current discussions in bioenergy consider the inherent crop potential. Previous studies have primarily focused on selecting high-efficiency crops

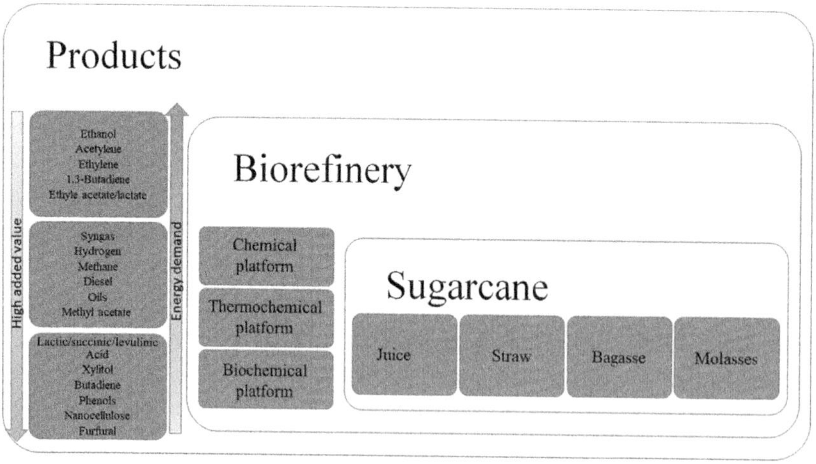

Figure 8.1 A biorefinery based on sugarcane.

in terms of productivity and energy supply. Nowadays, the focus has shifted to developing highly sustainable biorefineries. In case of sugarcane, recent studies described the potential for energy production platforms to integrate all the fraction from the cane crop.

Routes for energy production

A wide array of sustainable energy production routes have been investigated by many researchers worldwide. The energy potential in biomass can be explored chemically, thermochemically, or biochemically, or even by integration of such paths. However, there are still some gaps that might require further development. In the following section, the main processes for sugarcane biomass utilization will be addressed.

1G and 2G ethanol production

1G ethanol production technology from sugarcane is already consolidated in the world. However, there are still opportunities for investment in research, development, and innovation-seeking improvements in conversion processes. Factors such as outdated or inefficient equipment in mills and nonintegral use of sugarcane — considering the straw remaining in the field — often require improvements, and that may pave a rapid and cost-efficient route to 2G ethanol with relatively low capital expenses retrofits (Menandro et al., 2017; Dias et al., 2012).

1G ethanol is produced from sugarcane juice or molasses (or a mixture thereof) depending on the processing plant. Sugar mills can be classified into three types of installation: sugar mills exclusively producing sugar, those producing both sugar and bioethanol or annexed distilleries, and the ones producing only bioethanol autonomous distilleries. However, in addition to sugar and alcohol, the sugar–alcohol industry also cogenerates electrical energy to its basic production mix. In Brazil, annexed plants are more common due to their production flexibility that floats depending on the market demands, which is part of the reason for the success of ethanol production in the country (Moraes et al., 2015).

While great potential remains untapped for the 1G ethanol production, more attention has been now given to 2G ethanol and biomaterials especially in the European Union, the United States, China, and Brazil (Milanez et al., 2015). 2G ethanol appears as a promising alternative to increase biofuels' production. It is also in line with worldwide goals — and ratified by the Paris Agreement — to expand the renewable energy matrix.

However, 2G ethanol production has faced major scientific and technological bottlenecks. Adjustments in facilities and changes in technological routes have been required to meet the expectations and deadlines for biofuels development in the short and medium term (United Nations Conference on Trade and Development, 2016).

Bioelectricity from sugarcane bagasse and straw

Transformation of bagasse into heat and bioelectricity in the mills has been long history in Brazil and Africa in the so-called cogeneration process (United Nations Conference on Trade and Development, 2016). During sugarcane harvesting, most part of the straw (tops and leaves) are left in the fields, which can also be burnt in boilers in the cogeneration systems (Menandro et al., 2017; Khatiwada et al., 2012).

At first, cogeneration was aimed for the energetic autonomy of sugarcane plants, then, by the introduction of new high-pressure boilers, it has also become an interesting source of bioenergy technologies (Carpio and Simone de Souza, 2017). In countries like Brazil and Nepal, bioelectricity from bagasse and straw is generated during sugarcane's harvesting season, which also happens to be the dry season in such countries, therefore complementing hydroelectrical power generation (Khatiwada et al., 2012).

Thus sugarcane biomasses have a great potential to generate surplus electricity, and many countries, such as Brazil (Dantas et al., 2013; Dias et al., 2012), India (Hiloidhari et al., 2018; Khatiwada et al., 2012), Nepal (Khatiwada et al., 2012), and Mauritius (To et al., 2018), have started to integrate it in the national grid. On the other hand, the use of traditional, unprocessed biomass tends to decline in the next decades, leading to an increase of the fuel pellets, which already play important roles in the bioenergy market. This growing industrial demand for pellets and briquette coupled with sustainability issues has encouraged many companies to produce fuel pellets from nonwoody biomass. Pellets and briquettes of sugarcane bagasse and straw have similar characteristics in terms of NO, CO, and SO_2 emissions and conversion efficiency when compared with wood products (Cardozo et al., 2014). However, there are some disadvantages such as their high ash and moisture contents, which could pose problems to the gasifier (Erlich et al., 2005) and demand high energy consumption during the drying process respectively. Then new challenges and opportunities for the commercialization of fuel pellets from different feedstocks have been created for current technologies.

Biogas from by-products

Disposal of large amounts of energy-valued waste is a main challenge in the waste management of the Brazilian sugarcane industry. As a sustainable solution, the biogas process can be a suitable technology for dealing with sugarcane waste products and for providing valuable commodities such as biogas and fertilizers. Additionally, proper treatment of bagasse, straw, and residues (filter cake and vinasse) may avoid greenhouse gas emissions, air pollution, and environmental contamination of soil and water (Moraes et al., 2015). The digestate from the biogas process can be used as fertilizer on agriculture fields, reducing mineral fertilizers consumption and, consequently, GHG emissions.

In addition, biogas can be produced from process streams (pentose liquor) or even other by-products such as enzymatic hydrolysis residues and vinasse from the 2G ethanol production (Rabelo et al., 2011). On the other hand, different properties from these by-products such as lower nutrient levels, may require alternative solutions for their use. Suitable legislations are also needed to stimulate adequate utilization of such residues (Souza et al., 2015).

Thermochemical route

The circular economy of sugarcane is guaranteed by the thermochemical route to valorize waste streams that could provide important products based on a superior life-cycle assessment.

The thermochemical conversion of biomass materials involves chemical changes when heat is supplied to the feedstock. This process offers a few advantages from the conventional biochemical mechanism, once it profits from the scaleup processing of feedstocks in a more compact and drier way (Nicodème et al., 2018). There are currently three different thermochemical processes available for the sugar industry by-products, as can be illustrated in Fig. 8.2.

There is a vast potential for thermochemical processes once several products such as chemicals, biofuels, electricity, and heat can be obtained from it. One way to evaluate the potential of the biomass as a feedstock is to compare their enthalpy energy in terms of higher heating values (HHVs), which is an important characteristic and factor to be considered for applications in terms of usable energy (Conag et al., 2018). Sugarcane bagasse has a HHV above 17–20 MJ/kg (Yan et al., 2017; Ahmed and Gupta, 2012), making it suitable to be burnt or converted into other

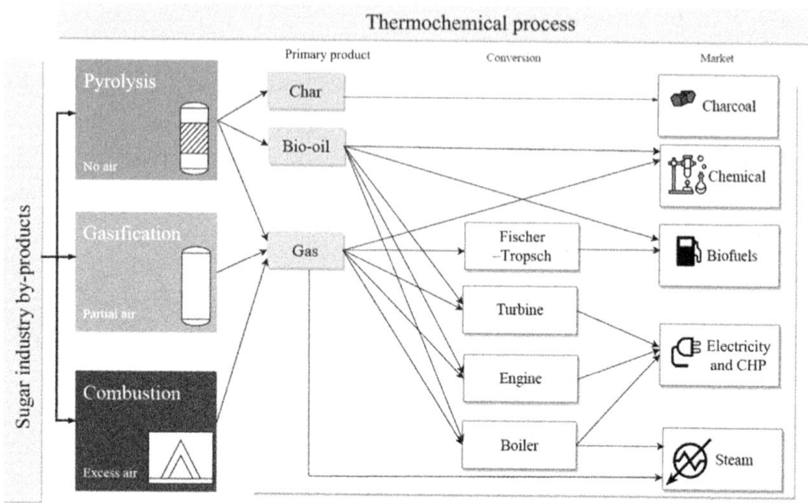

Figure 8.2 Thermochemical route from sugarcane by-products and market.

value-added products. Hereafter, the different processing routes are detailed.

Pyrolysis

Pyrolysis is the thermal decomposition of organic materials in the absence of oxygen or partial presence of oxygen. It occurs at high temperatures, generally between 300°C and 800°C (Anderson et al., 2008). Pyrolysis of sugarcane by-products produces volatile products (composed mainly of CH_4, H_2, CO, CO_2), and also carbon-rich nonvolatile residues, char and bio-oil, with a vast array of applications. Nowadays, pyrolysis has the lowest process cost to convert biomass into fuels or chemicals (Kasmuri et al., 2019). After many years of study and developing technologies for the process, the pyrolysis method is finally growing as a commercial application focused on the production of bio-oil and biochar.

Gasification

Gasification is the thermal conversion of carbonaceous materials into syngas or fuel gas. Syngas is one of the products with most applications including the production of methanol, hydrogen, or its conversion into synthetic fuel (Haro et al., 2013). In this regard, the Fischer–Tropsch is a catalytic process that is actively used to transform syngas into biofuels as liquid hydrocarbons (Hu et al., 2012). Typical operating temperature and pressure for this process is between 200°C and 500°C and 25 and 60 bar, respectively. On the other hand, methanol production requires temperatures and pressure reaction between 220°C and 275°C and 100 bar, respectively. Methanol can be used as a transportation fuel or for the production of synthetic gasoline and olefins. In order to produce gasoline, methanol needs to be vaporized at temperatures around 300°C–320°C; the vapors form dimethyl ether at around 400°C–420°C. In contrast, lower temperatures (375°C–400°C) are required to produce olefins. According to Nicodème et al. (2018), methanol can be converted into olefins (56.4%), gasoline (35.7%), and paraffins (6.5%). The gasification process is the most used worldwide, and according to Solarte-Toro et al. (2018), there are more than 250 gasification facilities in operation.

Combustion

The most common thermal conversion route includes direct combustion to combine heat and power from sugarcane bagasse for power generation. It basically generates CO_2 and water as products, but there can also be few amounts of NO_x, SO_x, and other impurities associated with the used fuel (Shukla and Kumar, 2017). Nowadays these technologies are mainly consolidated in industries, and the most used are Stirling engines and Rankine steam power cycles. However, the production is limited to hot gases and electricity generation.

When thinking of carbon neutral processes, the most attractive thermal conversions of sugarcane bagasse are combustion and gasification, because such processes are the best options in terms of technical–economic and life-cycle assessment (Centeno-González et al., 2017). Moreover, a rigorous selection of the energy vectors for the process must be performed considering energy efficiency and economics.

Energy storage applications of sugarcane-derived materials

When it is important to talk about ways to transition to a bio-based economy and to harness green energy from biomass, it is also paramount to make sure how to store energy for portable electronic devices [lithium ion batteries (LIBs)] and pave the way for more efficient electric cars. Biomass itself presents features that could replace oil-derived carbon materials — or even inorganic materials — in energy store applications such as low cost, availability, and renewability (Gao et al., 2017). Hence, sugarcane's main fractions such as cellulose and lignin could be used as binders in battery electrodes.

Binders are made up of nonelectrochemical-active polymers, but they can provide mechanical support to electrode-active materials (Nirmale et al., 2017; Li et al., 2011). Binders are mixed in a slurry with the electrochemical-active material together with other chemicals by addition of a solvent. Slurries should be ideally homogeneous at room temperature to ensure high efficiency.

Therefore native cellulose can be used as a binder, but due to its insolubility in water, a solvent like an ionic liquid is required to solubilize into the slurry. For instance, the aprotic ionic liquid 1-ethyl-3-methyl imidazolium acetate was able to used, recycled, and coated on a typical battery current collector (Jeong et al., 2012). Cellulose can also be derivatized into water-soluble materials such as carboxymethyl cellulose and microfibrillated cellulose nanoparticles (Jabbour et al., 2010; Nyström et al., 2010). Although not much explored, lignin can also be used as a binder. There are reports on the use of lignin as a binder in $LiFePO_4$ and graphite electrodes (Lu et al., 2016), and a study of durable silicon nanoparticles anode whose binder and conductive materials were replaced by lignin (Chen et al., 2016).

Transformation of sugarcane into biomass-derived carbon materials (BDCMs) that could be used in electrodes is also possible. In order to improve BDCMs' performance, activation processes are required to increase porosity, wettability, and electrical conductivity (Gao et al., 2017). The activation of BDCMs can be performed either physically or chemically by thermal processing with or without chemical-activating agents, although a combination of both processes is also possible (Sevilla and Mokaya, 2014). The main types of BDCMs' applications entail developing rechargeable batteries such as LIBs, lithium–sulfur batteries, or sodium-ion batteries (Gao et al., 2017).

Chemical potential of sugarcane

Several innovative technologies are being studied for conversion of sugarcane into commodity chemicals such as resins, inks, additives, and solvents; and major chemicals, including plastics, cosmetics, and pharmaceuticals, among others. The development of technological innovations adopted by these sectors led to a significant increase in sugarcane exploration.

Sugarcane fractionation is considered the most important step for efficient biomass utilization and for more economically reliable processes in a biorefinery. However, it is considered one of the most expensive steps in biorefinery and to be more attractive, it must be cost-effective and produce high-value products of interest.

Lignocellulose fractionation into its major components — cellulose, hemicelluloses, and lignin — is crucial to develop economically viable sugarcane biorefineries (Fig. 8.3). Depending on the process, lignocellulose fractionation can convert cellulose and hemicelluloses directly into monomers or oligomeric sugars, promote components' solubilization, and lignin modification.

One ideal scenario has cellulose and lignin being recovered separately in solid streams, while hemicelluloses and sugar degradation products — such as furfural and 5-hydroxymethyl-furfural — can be recovered in a water-soluble fraction. Cellulose-rich streams can be hydrolyzed to hexose (C6) sugars (e.g., bioethanol production) or treated to obtain crystalline cellulose for further conversions platform (e.g., nanocellulose material). Hemicelluloses can be potentially utilized in pentose (C5) fermentation or recovered to be converted into added-value products such as furfural and levulinic acid. Lignin-rich streams are useful as a polymeric coproduct of high value and acetic acid (released from the hemicellulose backbone) can be sold separately (Mussatto and Dragone, 2016; Sannigrahi and Ragauskas, 2013).

Technology platforms of chemical production

Sugarcane bagasse is naturally recalcitrant due to the combination of complex macromolecules such as lignin, hemicelluloses, and cellulose. Fractionation of bagasse is necessary to separate the macropolymers; biomass fractionation can be obtained by performing a pretreatment. Once fractionation occurs, depolymerization and chemical upgrading allow the production of building blocks that would form the base of the chemical platform. In a petrochemical refinery, the primary output mainly consists of several key building blocks such as methanol, ethylene, and propylene; these building blocks are converted into polymers, plastics, and several fine chemicals (IEA Bioenergy, 2010). Similarly a biorefinery could also present a small number of key building blocks, but with oxygen-rich molecules with multiple functional groups.

Chemicals from carbohydrates

In 2004 a team of scientists in the United States screened several key building blocks that could be produced from carbohydrates via chemical

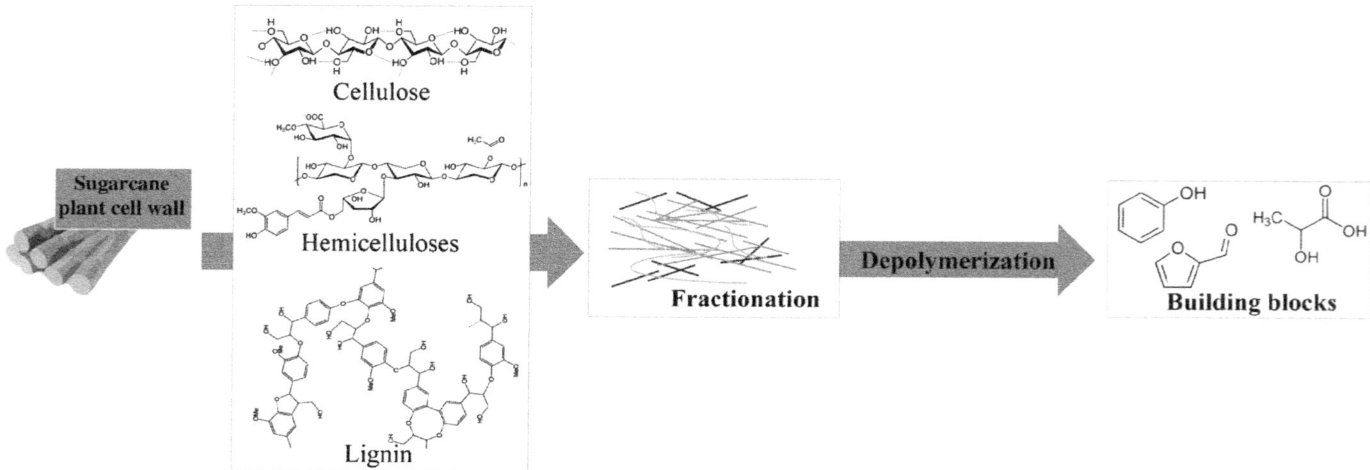

Figure 8.3 Structure and main components of sugarcane.

or biological conversions (Werpy and Petersen, 2004). The report selected 12 chemicals from the initial list, which was then upgraded in 2010 based on technological developments (Bozell et al., 2000). The compound classes consist of three main functional groups — alcohols, organic acids, and furanics (Fig. 8.4) — that could be produced via biochemicals and/or chemical pathways.

Furanics can be produced through homogeneous or heterogeneous acid-catalyzed dehydration of hexose and pentose sugars. They can be converted to diesel range biofuels through condensation reactions, reduced by hydrogenation to yield several solvents and fuel additives (Huber and Dumesic, 2006). Other furanics include agricultural chemicals such as herbicides, perfumes, and flavoring agents like furan and nitrofuran, plastic resins, synthetic fibers, and dyes (Rackemann et al., 2016; Dias et al., 2010). The main representatives of the alcohols class are ethanol and polyols (sorbitol, xylitol, and glycerol). Despite being considered a "super commodity" due to its high-volume production as a biofuel, ethanol can also be a feedstock for the production of chemicals. Ethanol

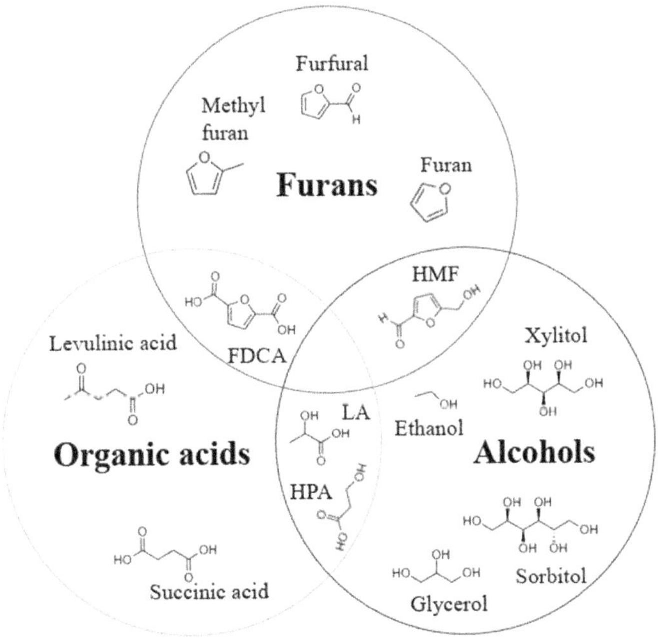

Figure 8.4 Top chemicals from carbohydrates and their multifunctionalities. *FDCA*, 2,5-Furandicarboxylic acid; *HMF*, 5-hydroxymethylfurfural; *HPA*, 3-hydroxy-propanoic acid; *LA*, lactic acid.

catalytic dehydration leads to ethylene which is a C2 building block to produce polymers such as high-density polyethylene, low-density polyethylene, polyvinyl chloride, and polyethylene terephthalate. Ethanol's complete oxidation produces acetic acid, which is a precursor of polyvinyl acetate and cellulose acetate. Polyols such as xylitol and sorbitol can be synthesized via catalytic hydrogenation of sugars but can also be produced via biochemical pathways, although the chemical route is more technically matured. They are used as sweeteners; in the case of sorbitol, it can also be a raw material for the production of polyurethanes, surfactants, and ascorbic acid. Glycerol is one of the most versatile chemicals as a building block; a wide range of chemicals may be produced from it including ethylene glycol, propylene glycol, acetol, and epichlorohydrin.

Organic acids — such as lactic, succinic, and levulinic — are also an important part of the biorefinery and their organic functions allow interesting diverse derivatization. Lactic acid is a C3 building block that has been used in the food and beverage industries as a preservative and pH-adjusting agent. It is also used in the pharmaceutical and chemical industries as a solvent and as starting material in the production of lactate esters (IEA Bioenergy, 2010). Succinic acid is a C4 building block industrially produced by the hydrogenation of maleic acid or anhydride. Succinic acid can be converted into 1,4-butanediol which can then be converted into tetrahydrofuran or into polymers like polybutylene terephthalate and polybutylene succinate. Levulinic acid is a C6 building block that can be chemically converted into pyrrolidones, lactones, and levulinate esters. Levulinate esters are used in the food industry as solvents and plasticizers. γ-Valerolactone (GVL) can be produced from levulinic acid by catalytic hydrogenation. It is a C5 cyclic ester with applications in the production of perfumes, food additives, and green solvents (Climent et al., 2014).

Chemicals from lignin

As emerging fractionation technologies allow efficient carbohydrate and lignin separation, it is also important to design chemical pathways to generate platform chemicals from lignin. Depolymerization methods generate a range of monomeric compounds that usually require a chemical upgrading or derivatization before commercialization. Additionally, the complexity of the product mixture obtained in the depolymerization often offers problems with purification. Funneling is an alternative upgrading strategy to ease such problems, and it was nicely reviewed in a study by Schutyser et al. (2018). The main idea is to convert a broad and heterogeneous mixture into a small pool of central platform chemicals/intermediates, which in turn can be transformed into desired products. Some approaches entail reductive depolymerization to yield low molecular weight alkanes, cycloalkanes, and phenolics; oxidative depolymerization to produce ketones, phenolics, and carboxylic acids. Other approaches include acid/base catalyzed, solvolytic, and thermal depolymerization.

Microbial metabolic pathway and sugarcane-derived products

Currently, both industry and society face significant challenges regarding the use of natural resources and the production of bio-based products with increasing restrictive political regulations. Thus, replacing petroleum-based products by their bio-based counterparts is a key factor for increasing global market share. Microbial fermentation of sugars from sugarcane as well as their crop and agro-processing wastes has been explored for the production of several products such as organic acids (lactic, polymalic, succinic, itaconic, gluconic, etc.) and higher alcohols (butanol and xylitol) (Fig. 8.5).

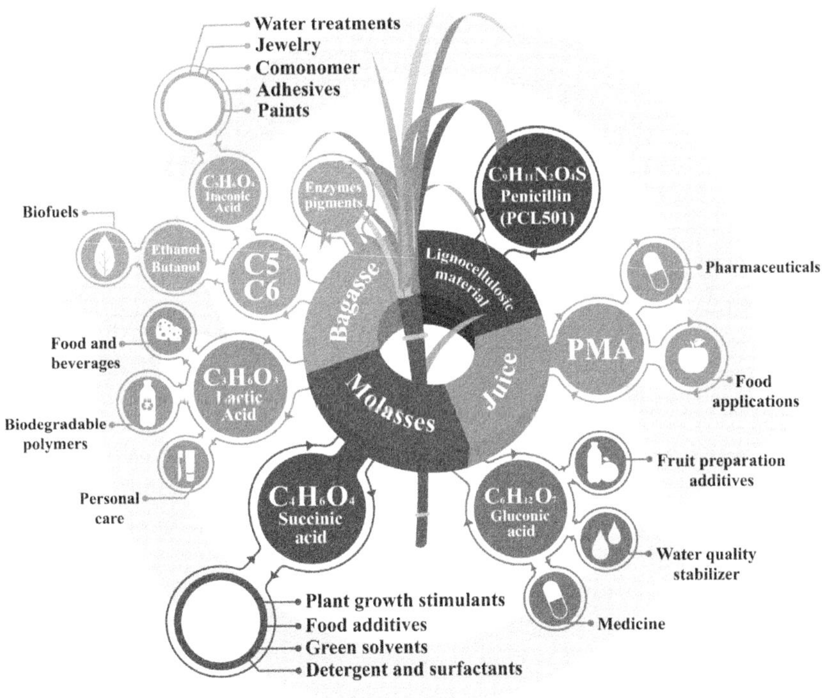

Figure 8.5 Bioeconomic and potential sugarcane-derived products from microbial metabolic pathway.

Recently the organic acid market was valued at US$16,837 million, expecting to reach US$29,197 million by 2023 (Organic Acid Market, 2017). This remarkable growth is due to the renewability characteristic of such products and as an approach to reduce GHGs and depletion of fossil fuels.

Among the organic acids, lactic acid has the highest growth rate in the world market. Lactic acid is mainly used to produce polylactic acid, which has a significant demand from various industries as pharmaceutical precursors, packaging, food additives, polyester, etc. (Oliveira et al., 2018a). Microorganisms such as lactic bacteria and fungi have been widely used to produce lactic acid, such as *Lactobacillus plantarum*, using sugarcane bagasse hemicellulosic hydrolysate (Oliveira et al., 2018b) and molasses (Komesu et al., 2017).

On the other hand, polymalic, succinic, itaconic, and gluconic acids should also drive the demand for organic acids globally in the near future. There is a high potential of molasses and sugarcane juice for the production of polymalic acid using the yeast *Aureobasidium pullulans*, with several interesting properties for their pharmaceutical use and food production (Feng et al., 2018; Wei et al., 2017).

Molasses with enhanced characteristics were fermented with *Actinobacillus succinogenes* 130Z to produce succinic acid (Cao et al., 2018) which has a number of emerging applications, such as plasticizers, solvents, detergent, and lubricants.

Currently, the large potential of itaconic acid involves a market of US $80.8 million and can reach up to US$102.3 million by 2022 (Market Research Report, 2018). The demand for sustainable manufacturing fostered the use of sugarcane bagasse for itaconic acid production under the integrated biorefinery concept (Nieder-Heitmann et al., 2018).

On the other hand, bioconversion of glucose in sugarcane juice by *Gluconobacter oxydans* (Pal et al., 2018) can be achieved for the production of gluconic acid, which constitutes one of those important organic acids, with a current market demand of US$51.6 billion.

The circular bioeconomy approach concept implies increasingly integrated industries, technologies, and processes. Although ethanol is a biofuel with a well-established fermentative route, there is a growing interest in butanol due to its better fuel properties. For example, a strategy can be established for a company competition based on the flexibility of ethanol–butanol production between the eucalyptus and sugarcane industrial sectors (Assumpção et al., 2018), where a plant could process both eucalyptus cellulose and the sugarcane hemicellulosic hydrolysate depending on the season.

Alternatively, establishing the circular economy creates more diverse market outlets for microbial metabolic-based products. Thus in addition to organic acids and higher alcohols, other products such as natural pigments, enzymes, and antibiotics are also interesting alternatives for incorporation in biorefineries, and some common studies are found in the literature:

- An attractive production route for a red pigment with high thermal stability was proposed, using sugarcane bagasse hydrolysate as fermentation medium for *Monascus ruber* Tieghem IOC 2225 (Hilares et al., 2018).
- Production of enzymes, such as cellulases, xylanases, pectinases, and laccases, during the value-adding process of agro-industrial sugarcane waste (bagasse as a single source of carbon) was evaluated using six strains of filamentous fungi (Ferreira et al., 2018).
- Production of penicillin from sugarcane lignocellulosic materials by *Penicillium chrysogenum* (PCL501) is not only profitable and sustainable but is also an environmentally friendly alternative approach to the usual production of penicillin (Onyegeme-Okerenta et al., 2018).

In this context, it is evident that some features of microbial metabolism still require a careful selection. For instance, improving microorganism selection and stress-resistance may allow higher production yields of the desired product and constitutes a key area of technological research and development, allowing process cost reduction and enabling large-scale production, among other aspects.

Advanced bio-based polymers and derived materials

Bioenergy feedstock producing countries are no exceptions to the trend of replacing petro-based materials in the next years, and Brazil sugarcane is a potential raw material (Khoo, 2015). Research efforts to design novel and tailored bio-based polymers and materials out of sugarcane are essential to allow their multiple use with environmental and economical positive effects. Bio-based polymers and materials from renewable resources have become a reality, and their current progress and future trends will be revisited in the light of the sugarcane industry.

Nanomaterials

Nanostructured materials can be obtained from cellulose, hemicelluloses, and lignin present in the cell walls of the sugarcane plant, and because of that they inherit an ability to assemble into highly hierarchical and multidimensional nanostructures (Martin-Martinez et al., 2018). Furthermore, controlling the nonspecific forces that assemble these materials in the gas–liquid–solid interfaces may produce dispersions, hydrogels, foams, and emulsions enabling effective processing into one-dimensional or multidimensional structures (Mohammadinejad et al., 2016). The resulting systems can, in turn, be applied in advanced functional materials in different segments.

Nanomaterials derived from cellulose exhibit high surface areas with powerful outsider interactions. As a result, nanocellulose possesses unique properties of high strength and stiffness combined with low weight. Additionally, modification techniques (e.g., acetylation, esterification, and etherification polymerization) can be used to introduce charged or hydrophobic moieties, tailoring the nanocellulose surface (Shak et al., 2018; Thomas et al., 2018).

Nanocellulose sourced from biomass and plants can be classified into cellulose nanocrystals (also referred to as nanocrystalline cellulose or cellulose nanowhiskers) and cellulose nanofibrils (also referred to as nanofibrillated cellulose) which contain both amorphous and crystalline cellulose domains (Abitbol et al., 2016). Nanocellulose can be used as reinforcements in nanocomposites based on polymers, metals, metal oxides, and carbon whenever strength, flexibility, and specific nanostructuration are required.

Nanocellulose-based polymers and materials have a wide spectrum of applications with the potential to outperform the conventional petroleum-based ones when combined with other matrices and systems (Thomas et al., 2018). The most relevant applications include

- As enforcement in nanocarbon composites for enhanced mechanical strength, electrical conductivity, and biocompatibility, especially for carbon quantum dot, nanotube, and graphene composites.
- As enforcement in organic composites, such as poly(styrene-*co*-butyl acrylate), polyurethane, chitosan, and epoxy resins, increasing the strength and biocompatibility of the resulting matrix.
- As enforcement in foams and aerogels and acting as stabilizers in Pickering emulsions systems.
- Combination with inorganic nanoparticles (e.g., metals and oxides) to improve their thermal, mechanical, and optical properties.
- As templates — in the form of nanocrystals — to obtain mesoporous and chiral plasmonic films, especially mesoporous metal oxides.

Some applications of nanocellulose and nanocellulose-based polymers and materials are illustrated in Fig. 8.6.

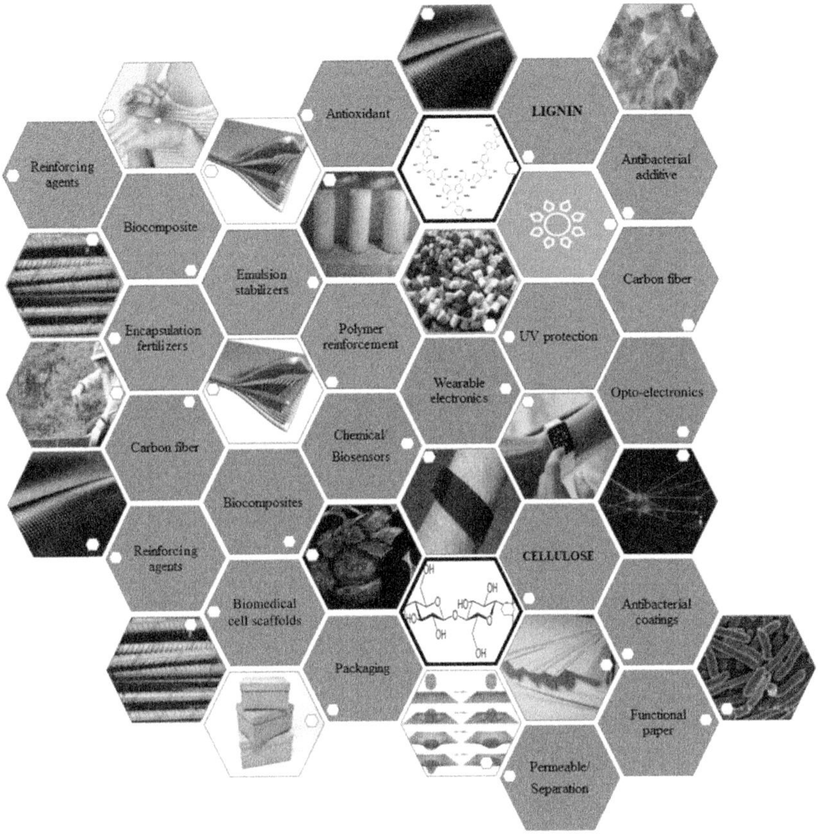

Figure 8.6 Most recent applications of cellulose, nanocellulose, nanocellulose-based polymers and materials, lignin, and lignin nanoparticles.

Attractiveness for biomedical applications is focused toward new patents. Developing chemical sensors for health monitoring; hydrogels, threads, and scaffolds for tissue engineering based on nanocelluloses opens an optimistic scenario for medical devices innovations with potential for commercialization (Dumanli, 2017).

According to the Global Market Insights, Inc., the nanocellulose market size is expected to surpass US$1 billion by 2024, driven by the sustainability concern. Paper processing, coatings, gas, and food are pointed as the major application segments in the global nanocellulose market (Herald Keeper, 2018).

Lignins can also be converted into functional nanoparticles. The main source of lignins available for use on a large scale is produced as downstream during fractionation or pretreatment of lignocellulosic biomass. Although lignins are considered as versatile materials with unique properties, industrial applications are still hampered by their heterogeneity and poor solubility in various organic solvents and water (Ayyachamy et al., 2013). Therefore lignin nanoparticles and other nanostructured materials have gained interest during the last years since they exhibit considerably distinct properties from the original lignins and are nontoxic and ecologically friendly (Ago et al., 2017). In addition, it is an opportunity to integrate them into biomass conversion processes in biorefineries with value-added bioproducts.

Lignin nanoparticles can be obtained by different approaches according to the tailored structures desired. Variations in solvent, pH, crosslink, and polymerization are the most common methods found in the literature (Beisl et al., 2017). Ice segregation, mechanical treatment, aerosol processing, and carbon dioxide antisolvent also successfully produce lignins nanoparticles. In particular, aerosol technologies employing continuous atomized reactors have a great potential to produce size-controlled lignin nanoparticles in a single step with interesting physicochemical properties (Assis et al., 2018).

Nevertheless, research on the technoeconomical feasibility of lignin extraction processes from sugarcane bagasse such as using high concentrated solutions, minimizing solvent consumption, and maximizing solvent recovery, are crucial for lignin nanoparticles commercialization.

Carbon fiber composites

Carbon fibers are mainly produced from polyacrylonitrile, which is considered as an expensive fossil-based polymer and obtained via wet spinning. Carbon fiber-reinforced composite materials (epoxy or polymers) are extensively used in aeronautics, sports equipment, and luxury cars despite their high price and intricate manufacturing (Peebles, 2018; Frank et al., 2012). Thus developing efficient methods for obtaining carbon fibers from renewable sources such as cellulose and lignin can leverage the platform market of advanced biopolymers and derived composites.

The major limitations hampering large-scale production of cellulose-based carbon fibers are the low yields in the pyrolysis method. On the other hand, fibers impregnated with a catalyst can reach high conversion levels. The drawbacks related to the lignin-based fibers are due to their thermoplastic nature: long stabilization time, nonestablished melt spinning, and limited mechanical properties of the resulting composite. Moreover, lignin properties and quality can vary depending on the source and isolation method. Substantial efforts must be undertaken to obtain high-performance lignins and cellulose-derived carbon fibers to decrease their production cost and increase high-end applications.

Conclusion

Considering all those perspectives and advances, it appears to evident the potential of sugarcane as a strong market supplier bioplatform, providing chemicals, biopolymers, biofuels, and materials in the light of a circular bioeconomy. Thus it is possible to reshape sugarcane's economic role, where the so-called waste becomes a raw material for the manufacture of new products via chemical, biochemical, and thermochemical conversion routes of lignocellulosic materials whose chemical structure is as diverse and functional as fossil matrices. However, technological advances nowadays are taking place at a faster pace than ever and establishing them has become a reality with the help of political subsidies and policies. The global transition to a bio-based economy needs to happen simultaneously in all key areas of development or our society, that is, transportation, food, and energy.

References

Abitbol, T., Rivkin, A., Cao, Y., Nevo, Y., Abraham, E., Ben-Shalom, T., et al., 2016. Nanocellulose, a tiny fiber with huge applications. Curr. Opin. Biotechnol. 39, 76–88.

Ago, M., Tardy, B.L., Wang, L., Guo, J., Khakalo, A., Rojas, O.J., 2017. Supramolecular assemblies of lignin into nano- and microparticles. MRS Bull. 42 (5), 371–378.

Ahmed, I.I., Gupta, A.K., 2012. Sugarcane bagasse gasification: global reaction mechanism of syngas evolution. Appl. Energy 91 (1), 75–81.

Anderson, C.J., Pas, S.J., Arora, G., Kentish, S.E., Hill, A.J., Sandler, S.I., et al., 2008. Effect of pyrolysis temperature and operating temperature on the performance of nanoporous carbon membranes. J. Membr. Sci. 322 (1), 19–27.

Assis, C.A., Greca, L.G., Ago, M., Balakshin, M.Y., Jameel, H., Gonzalez, R., et al., 2018. Techno-economic assessment, scalability, and applications of aerosol lignin micro- and nanoparticles. ACS Sustain. Chem. Eng. 6 (9), 11853–11868.

Assumpção, Dd.C., Rivera, E.A.C., Tovar, L.P., Ezeji, T.C., Filho, R.M., Mariano, A.P., 2018. Resolving mismatches in the flexible production of ethanol and butanol from eucalyptus wood with vacuum fermentation. Bioproc. Biosyst. Eng. 41 (11), 1651–1663.

Ayyachamy, M., Cliffe, F.E., Coyne, J.M., Collier, J., Tuohy, M.G., 2013. Lignin: untapped biopolymers in biomass conversion technologies. Biomass Convers. Bior. 3 (3), 255–269.

Beisl, S., Miltner, A., Friedl, A., 2017. Lignin from micro- to nanosize: production methods. Int. J. Mol. Sci. 18 (6), 1244.

Bozell, J.J., Moens, L., Elliott, D.C., Wang, Y., Neuenscwander, G.G., Fitzpatrick, S.W., et al., 2000. Production of levulinic acid and use as a platform chemical for derived products. Resour., Conserv. Recy. 28 (3), 227–239.

Cao, W., Wang, Y., Luo, J., Yin, J., Xing, J., Wan, Y., 2018. Succinic acid biosynthesis from cane molasses under low pH by *Actinobacillus succinogenes* immobilized in luffa sponge matrices. Bioresour. Technol. 268, 45–51.

Cardozo, E., Erlich, C., Alejo, L., Fransson, T.H., 2014. Combustion of agricultural residues: an experimental study for small-scale applications. Fuel 115, 778–787.

Carpio, L.G.T., Simone de Souza, F., 2017. Optimal allocation of sugarcane bagasse for producing bioelectricity and second generation ethanol in Brazil: scenarios of cost reductions. Renew. Energy 111, 771–780.

Centeno-González, F.O., Silva Lora, E.E., Villa Nova, H.F., Mendes Neto, L.J., Martínez Reyes, A.M., Ratner, A., et al., 2017. CFD modeling of combustion of sugarcane bagasse in an industrial boiler. Fuel 193, 31–38.

Chen, T., Zhang, Q., Pan, J., Xu, J., Liu, Y., Al-Shroofy, M., et al., 2016. Low-temperature treated lignin as both binder and conductive additive for silicon nanoparticle composite electrodes in lithium-ion batteries. ACS Appl. Mater. Interf. 8 (47), 32341–32348.

Climent, M.J., Corma, A., Iborra, S., 2014. Conversion of biomass platform molecules into fuel additives and liquid hydrocarbon fuels. Green Chem. 16 (2), 516–547.

Conag, A.T., Villahermosa, J.E.R., Cabatingan, L.K., Go, A.W., 2018. Predictive HHV model for raw and torrefied sugarcane residues. Waste Biomass Valori. 10 (7), 1929–1943. Available from: https://doi.org/10.1007/s12649-018-0204-2.

Dantas, G.A., Legey, L.F.L., Mazzone, A., 2013. Energy from sugarcane bagasse in Brazil: an assessment of the productivity and cost of different technological routes. Renew. Sust. Energy Rev. 21, 356–364.

Dias, A.S., Lima, S.P.M., Valente, A.A., 2010. Furfural and furfural-based industrial chemicals. In: Pignataro, B. (Ed.), Chemistry and Molecular Sciences: Advances in Synthetic Chemistry. John Wiley & Sons, Ltd, Federal Republic of Germany, pp. 165–186.

Dias, M.O.S., Junqueira, T.L., Jesus, C.D.F., Rossell, C.E.V., Maciel Filho, R., Bonomi, A., 2012. Improving second generation ethanol production through optimization of first generation production process from sugarcane. Energy 43 (1), 246–252.

Dumanli, A.G., 2017. Nanocellulose and its composites for biomedical applications. Curr. Med. Chem. 24 (5), 512–528.

Erlich, C., Öhman, M., Björnbom, E., Fransson, T.H., 2005. Thermochemical characteristics of sugar cane bagasse pellets. Fuel 84 (5), 569–575.

FAO, 2018. World food and agriculture - statistical pocketbook. Rome. <http://www.wipo.int/amc/en/mediation/rules>. (accessed 17.12.18.)

Feng, J., Yang, J., Yang, W., Chen, J., Jiang, M., Zou, X., 2018. Metabolome- and genome-scale model analyses for engineering of *Aureobasidium pullulans* to enhance polymalic acid and malic acid production from sugarcane molasses. Biotechnol. Biofuels 11 (1), 94.

Ferreira, F.L., Dall'Antonia, C.B., Shiga, E.A., Alvim, L.J., Bom Pessoni, R.A., 2018. Sugarcane bagasse as a source of carbon for enzyme production by filamentous fungi. Hoehnea 45 (1), 134–142.

Frank, E., Hermanutz, F., Buchmeiser, M.R., 2012. Carbon fibers: precursors, manufacturing, and properties. Macromol. Mater. Eng. 297 (6), 493–501.

Gao, Y.-P., Zhai, Z.-B., Huang, K.-J., Zhang, Y.-Y., 2017. Energy storage applications of biomass-derived carbon materials: batteries and supercapacitors. New J. Chem. 41 (20), 11456–11470.

Haro, P., Ollero, P., Villanueva Perales, Á.L., Vidal-Barrero, F., 2013. Potential routes for thermochemical biorefineries. Biofuel. Bioprod. Biorefin. 7 (5), 551–572.

Herald, K., 2018. At 33.9% growth nanocellulose market will cross $1bn by 2024. Available from: <http://heraldkeeper.com/market/33-9-growth-nanocellulose-market-will-cross-1bn-2024-121927.html> (accessed 19.01.19.)

Hilares, R.T., de Souza, R.A., Marcelino, P.F., da Silva, S.S., Dragone, G., Mussatto, S.I., et al., 2018. Sugarcane bagasse hydrolysate as a potential feedstock for red pigment production by *Monascus ruber*. Food Chem. 245, 786–791.

Hiloidhari, M., Araújo, K., Kumari, S., Baruah, D.C., Ramachandra, T.V., Kataki, R., et al., 2018. Bioelectricity from sugarcane bagasse co-generation in India—an assessment of resource potential, policies and market mobilization opportunities for the case of Uttar Pradesh. J. Clean. Prod. 182, 1012–1023.

Hu, J., Yu, F., Lu, Y., 2012. Application of Fischer—Tropsch synthesis in biomass to liquid conversion. Catalysts 2 (2), 303.

Huber, G.W., Dumesic, J.A., 2006. An overview of aqueous-phase catalytic processes for production of hydrogen and alkanes in a biorefinery. Catal. Today 111 (1), 119–132.

IEA Bioenergy, 2010. Bio-based chemicals - value added products from biorefineries, Task 42 Biorefinery. Available from: <https://www.ieabioenergy.com/publications/bio-based-chemicals-value-added-products-from-biorefineries/> (accessed 18.01.19.)

Jabbour, L., Gerbaldi, C., Chaussy, D., Zeno, E., Bodoardo, S., Beneventi, D., 2010. Microfibrillated cellulose—graphite nanocomposites for highly flexible paper-like Li-ion battery electrodes. J. Mater. Chem. 20 (35), 7344–7347.

Jeong, S.S., Böckenfeld, N., Balducci, A., Winter, M., Passerini, S., 2012. Natural cellulose as binder for lithium battery electrodes. J. Power Sources 199, 331–335.

Kasmuri, N.H., Kamarudin, S.K., Abdullah, S.R.S., Hasan, H.A., Som, A.M., 2019. Integrated advanced nonlinear neural network-simulink control system for production of bio-methanol from sugar cane bagasse via pyrolysis. Energy 168, 261–272.

Khatiwada, D., Seabra, J., Silveira, S., Walter, A., 2012. Power generation from sugarcane biomass — a complementary option to hydroelectricity in Nepal and Brazil. Energy 48 (1), 241–254.

Khoo, H.H., 2015. Review of bio-conversion pathways of lignocellulose-to-ethanol: sustainability assessment based on land footprint projections. Renew. Sust. Energy Rev. 46, 100–119.

Komesu, A., Wolf Maciel, M.R., Maciel Filho, R., 2017. Central composite experimental design applied to evaluate the lactic acid concentration by short path evaporation. Chem. Eng. Process 117, 89–94.

Li, Y., Zhang, W.-B., Hsieh, I.F., Zhang, G., Cao, Y., Li, X., et al., 2011. Breaking symmetry toward nonspherical janus particles based on polyhedral oligomeric silsesquioxanes: molecular design, "click" synthesis, and hierarchical structure. J. Am. Chem. Soc. 133 (28), 10712–10715.

Lu, H., Cornell, A., Alvarado, F., Behm, M., Leijonmarck, S., Li, J., et al., 2016. Lignin as a binder material for eco-friendly Li-ion batteries. Materials 9 (3), 127.

Market Research Report, 2018. Itaconic acid market by derivative (styrene butadiene, methyl methacrylate, polyitaconic acid), application (SBR latex, synthetic latex, chillant dispersant agent, superabsorbent polymer), and region - global forecast to 2022. Available from: <https://www.marketsandmarkets.com/Market-Reports/itaconic-acid-market-35695455.html> (accessecd 10.01.19.)

Martin-Martinez, F.J., Jin, K., López Barreiro, D., Buehler, M.J., 2018. The rise of hierarchical nanostructured materials from renewable sources: learning from nature. ACS Nano 12 (8), 7425–7433.

Menandro, L.M.S., Cantarella, H., Franco, H.C.J., Kölln, O.T., Pimenta, M.T.B., Sanches, G.M., et al., 2017. Comprehensive assessment of sugarcane straw:

implications for biomass and bioenergy production. Biofuels Bioprod. Biorefin. 11 (3), 488–504.

Milanez, A.Y., Nyko, D., Valente, M.S., Sousa, L.C., Bonomi, A.M.F.L.J., Jesus, C.D.Fd, et al., 2015. De promessa a realidade: como o etanol celulósico pode revolucionar a indústria da cana-de-açúcar: uma avaliação do potencial competitivo e sugestões de política pública. BNDES Setorial 41, 237–294.

Mohammadinejad, R., Karimi, S., Iravani, S., Varma, R.S., 2016. Plant-derived nanostructures: types and applications. Green Chem. 18 (1), 20–52.

Moraes, B.S., Zaiat, M., Bonomi, A., 2015. Anaerobic digestion of vinasse from sugarcane ethanol production in Brazil: challenges and perspectives. Renew. Sust. Energy Rev. 44, 888–903.

Mussatto, S.I., Dragone, G.M., 2016. Chapter 1 - Biomass pretreatment, biorefineries, and potential products for a bioeconomy development. In: Mussatto, S.I. (Ed.), Biomass Fractionation Technologies for a Lignocellulosic Feedstock Based Biorefinery. Elsevier, Amsterdam, pp. 1–22.

Nicodème, T., Berchem, T., Jacquet, N., Richel, A., 2018. Thermochemical conversion of sugar industry by-products to biofuels. Renew. Sust. Energy Rev. 88, 151–159.

Nieder-Heitmann, M., Haigh, K.F., Görgens, J.F., 2018. Process design and economic analysis of a biorefinery co-producing itaconic acid and electricity from sugarcane bagasse and trash lignocelluloses. Bioresour. Technol. 262, 159–168.

Nirmale, T.C., Kale, B.B., Varma, A.J., 2017. A review on cellulose and lignin based binders and electrodes: small steps towards a sustainable lithium ion battery. Int. J. Biol. Macromol. 103, 1032–1043.

Nyström, G., Mihranyan, A., Razaq, A., Lindström, T., Nyholm, L., Strømme, M., 2010. A nanocellulose polypyrrole composite based on microfibrillated cellulose from wood. J. Phys. Chem. B 114 (12), 4178–4182.

Oliveira, R.A., Komesu, A., Vaz Rossell, C.E., Maciel Filho, R., 2018a. Challenges and opportunities in lactic acid bioprocess design—from economic to production aspects. Biochem. Eng. J. 133, 219–239.

Oliveira, R.A., Vaz Rossell, C.E., Venus, J., Cândida Rabelo, S., Maciel Filho, R., 2018b. Detoxification of sugarcane-derived hemicellulosic hydrolysate using a lactic acid producing strain. J. Biotechnol. 278, 56–63.

Onyegeme-Okerenta, B.M., Okochi, V.I., Chinedu, S.N., 2018. Penicillin production by *Penicillium chrysogenum* PCL 501: effect of UV induced mutation. Internet J. Microbiol. 12 (1), 1–7.

Organic Acid Market, 2017. Organic acids market by type (acetic acid, citric acid, formic acid, lactic acid, itaconic acid, succinic acid, gluconic acid, ascorbic acid, fumaric acid, and propionic acid), source (biomass, molasses, starch, chemical synthesis, agro-industrial residue), end-user (food & beverage, animal feed, chemicals & industrial, pharmaceuticals, personal care, agriculture) - global opportunity analysis and industry forecast, 2017-2023. Available from: <https://www.alliedmarketresearch.com/organic-acids-market> (accessed 10.01.19.)

Pal, P., Kumar, R., Banerjee, S., 2018. Purification and concentration of gluconic acid from an integrated fermentation and membrane process using response surface optimized conditions. Front. Chem. Sci. Eng. Available from: https://doi.org/10.1007/s11705-018-1721-z.

Peebles, L.H., 2018. Carbon Fibers: Formation, Structure, and Properties. CRC Press, Boca Raton, FL.

Rabelo, S.C., Carrere, H., Maciel Filho, R., Costa, A.C., 2011. Production of bioethanol, methane and heat from sugarcane bagasse in a biorefinery concept. Bioresour. Technol. 102 (17), 7887–7895.

Rackemann, D.W., Zhang, Z., Doherty, W.O.S., 2016. Chapter 8 - Conversion of sugarcane carbohydrates into platform chemicals. In: O'Hara, I.M., Mundree, S.G. (Eds.), Sugarcane-Based Biofuels and Bioproducts. John Wiley & Sons, Ltd, New Jersey, pp. 207–236.
Renewable Fuels Association, 2016. Fueling a high octane future. Available from: <https://ethanolrfa.org/wp-content/uploads/2016/02/RFA_2016_full_final.pdf> (accessed 14.12.18.).
Sannigrahi, P., Ragauskas, A.J., 2013. Chapter 10 - Fundamentals of biomass pretreatment by fractionation. In: Wyman, C.E. (Ed.), Aqueous Pretreatment of Plant Biomass for Biological and Chemical Conversion to Fuels and Chemicals. John Wiley & Sons, Ltd, Chichester, pp. 201–222.
Schutyser, W., Renders, T., Van den Bosch, S., Koelewijn, S.F., Beckham, G.T., Sels, B. F., 2018. Chemicals from lignin: an interplay of lignocellulose fractionation, depolymerisation, and upgrading. Chem. Soc. Rev. 47 (3), 852–908.
Sevilla, M., Mokaya, R., 2014. Energy storage applications of activated carbons: supercapacitors and hydrogen storage. Energy Environ. Sci. 7 (4), 1250–1280.
Shak, K.P.Y., Pang, Y.L., Mah, S.K., 2018. Nanocellulose: recent advances and its prospects in environmental remediation. Beilstein J. Nanotech. 9, 2479–2498.
Shukla, A., Kumar, S.Y., 2017. A comparative study of sugarcane bagasse gasification and direct combustion. Int. J. Appl. Eng. Res. 12 (24), 14739–14745.
Solarte-Toro, J.C., Chacón-Pérez, Y., Cardona-Alzate, C.A., 2018. Evaluation of biogas and syngas as energy vectors for heat and power generation using lignocellulosic biomass as raw material. Electron. J. Biotechnol. 33, 52–62.
Souza, G.M., Victoria, R.L., Joly, C.A., Verdade, L., M., 2015. Bioenergy & Sustainability: Bridging the Gaps. Scientific Committee on Problems of the Environment (SCOPE), São Paulo.
Thomas, B., Raj, M.C., Athira, B.K., Rubiyah, M.H., Joy, J., Moores, A., et al., 2018. Nanocellulose, a versatile green platform: from biosources to materials and their applications. Chem. Rev. 118 (24), 11575–11625.
To, L.S., Seebaluck, V., Leach, M., 2018. Future energy transitions for bagasse cogeneration: lessons from multi-level and policy innovations in Mauritius. Energy Res. Soc. Sci. 35, 68–77.
United Nations Conference on Trade and Development, 2016. Second generation biofuel markets: state of play, trade and developing coutry perspectives. Available from: <https://unctad.org/en/PublicationsLibrary/ditcted2015d8_en.pdf> (accessed 14.12.18.).
Wei, P., Cheng, C., Lin, M., Zhou, Y., Yang, S.-T., 2017. Production of poly(malic acid) from sugarcane juice in fermentation by *Aureobasidium pullulans*: kinetics and process economics. Bioresour. Technol. 224, 581–589.
Werpy, T., Petersen, G., 2004. Top value added chemicals from biomass: volume I -- Results of screening for potential candidates from sugars and synthesis gas - DOE/GO-102004-1992; TRN: US200427%%671. U.S. Department of Energy Office of Scientific and Technical Information (accessed 01.08.04.).
Yan, X., Ma, J., Wang, W., Zhao, Y., Zhou, J., 2017. The effect of different catalysts and process parameters on the chemical content of bio-oils from hydrothermal liquefaction of sugarcane bagasse. BioResources 13 (1), 997–1018.

CHAPTER 9

Sugarcane for renewable chemistry

Sílvio Vaz Jr.

Contents

Introduction 165
Sugarcane as a raw material for renewable chemistry 168
Perspectives for sugarcane bioproducts 170
Chemical routes 173
Biochemical routes 174
Thermochemical routes 175
Challenges involved 175
Conclusion 176
References 177

Introduction

In countries with a large production of sugarcane, as Brazil, efforts have been made in order to promote the economic potential of the plant biomass, in accordance with the use of renewable sources to develop a sustainable chemistry, by means its use in biorefineries (Brazilian Center for Management and Strategic Studies, 2010; United States Department of Energy, 2004). It is important to mention that the use of residual biomass is crucial to enable the production of cheap feedstock for biorefineries. Bioeconomy and renewable chemistry are closely related as a proposal for an innovative design to obtain good results for the usage of sugarcane in the chemical industry.

The development of chemicals from biomass to replace petrochemicals has boosted the renewable chemistry approach, where renewable sources of raw materials are preferred instead of oil. Nevertheless, this approach remains, in its majority, in the R&D stage but with renewable chemicals already exploited in some segments—although petrochemical prices are the baseline for commodities (Strrathof and Bampouli, 2017)—showing

Sugarcane Biorefinery, Technology and Perspectives
DOI: https://doi.org/10.1016/B978-0-12-814236-3.00009-3

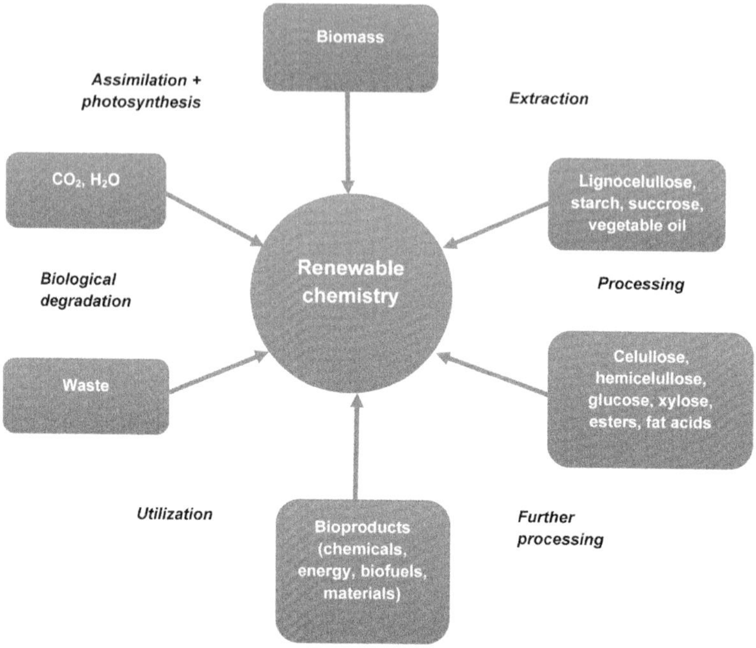

Figure 9.1 The renewable chemistry approach. *Adapted from Metzger, J.O., Eissen, M., 2004. Concepts on the contribution of chemistry to a sustainable development. Renewable raw materials, CR Chim. 7, 6–7, 569–581.*

the economic potential in the present and near future. Fig. 9.1 shows the renewable chemistry approach.

Bioeconomy, in a simple way, "(. . .) comprises those parts of the economy that use renewable biological resources from land and sea—such as crops, forests, fish, animals and micro-organisms—to produce food, materials and energy" (European Investment Bank, 2017). Here, we can consider chemicals as a strong component of this concept because they can be obtained from a diversity of plant biomass: lignocellulose (e.g., woods, woody-biomass), oleaginous (e.g., soybean, cotton), starch (e.g., corn, potato), and saccharose (e.g., sorghum, sugarcane). The use of agricultural biomass stands out as the first source of supply. Fig. 9.2 shows the bioeconomy concept in a circular mode, what is suitable for agroindustrial biomass as raw material for chemical industry.

Green chemistry emerges in the 1990s as a new philosophy in academia and industry to break old paradigms of chemistry such as large waste generation and intensive use of petrochemicals through a holistic view of processes in laboratories and industries (Anastas and Kirchhoff, 2002). This

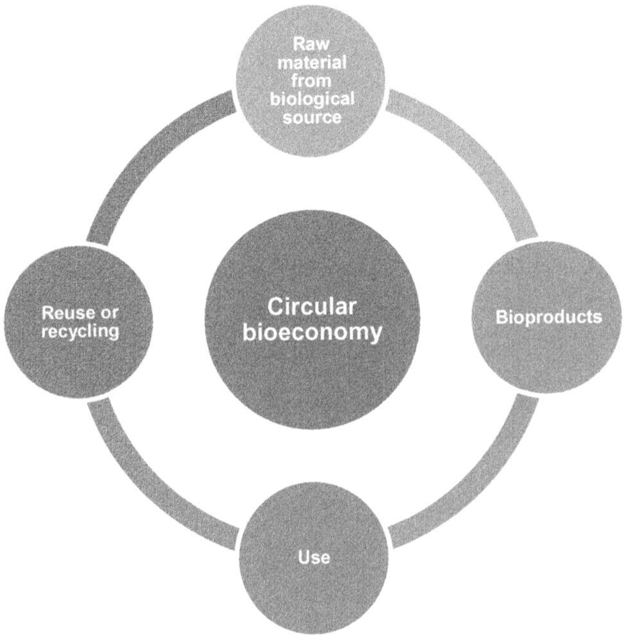

Figure 9.2 Components for a circular bioeconomy based on agroindustrial biomass.

approach, described in 12 principles, proposes to consider, among other aspects, the reduction of waste generation, the atomic and energy economy, and the use of renewable raw materials (Anastas and Warner, 1998). In the case of plant biomass, the seventh principle—*use of renewable raw materials*—stands out as a great strategic opportunity for segments related to several areas of chemical industry worldwide (e.g., pharma, polymers, resins, agrochemicals, cosmetics, among others).

Renewable raw materials are an extremely strategic asset for large biomass-producer countries, such as Brazil, United States, Canada, Finland, Germany, China, India, among others (European Investment Bank, 2017). These raw materials—the agroindustrial biomass, as sugarcane—are an abundant and cheap feedstock for the transformation processes of chemistry or the conversion processes applied to biomass, which are biocatalytic, chemocatalytic, fermentative, and thermochemical.

Chemicals are the bioproducts with the highest potential to add value into a generic biomass chain, given the importance of the conventional chemical industry and the fine chemical industry in different sectors of the economy, being possible to highlight compounds that can be used as building blocks, synthetic intermediates, polymers, specialties, among others; such idea can be greatly explored by biorefineries (Kamm et al., 2006). On the other hand, the need to develop technologies to obtain these products presents considerable bottlenecks to be overcome, related to technical, scientific, and market issues. Thus the use of biomass through chemistry opens new business opportunities and wealth generation for a large number of countries, as well as promoting a less negative impact on the environment and the sustainable chemistry.

Especially for the Brazilian case, sugarcane-based agroindustry has a great potential to share wealth based on a solid industrial leadership in this sector, when allied to technical and scientific development.

Sustainability criteria, comprising environmental, social, and economic impacts, are especially relevant for the substitution of oil by biomass in chemical industry, as we can see in Fig. 9.3.

Sugarcane as a raw material for renewable chemistry

Table 9.1 lists the major worldwide sugarcane (*Saccharum* spp.) producers. As expected, these producers are established in the tropical regions, except the United States. Then, we have a forecast of availability in these countries of 1.2 billion tonnes of sugarcane to be processed by biorefineries.

The use of sugarcane as raw material for renewable chemistry is due to its potential to supply the following compounds and materials:

- Sugars: sucrose (broth), glucose (derived from cellulose and sucrose inversion), and xylose (derived from hemicellulose).
- Ethanol: from sucrose fermentation.
- Natural polymers or macromolecules: cellulose, hemicellulose, and lignin, which are present in the bagasse and straw (lignocellulosic biomass); it can be converted into hexose, pentose, and phenol sources, respectively.
- Vinasse rich in organic matter.
- Other compounds such as higher alcohols and carbon dioxide (CO_2) in a high purity content for food and chemical applications.

The average mass composition of the sugarcane plant is presented in Table 9.2.

Sucrose, which is a disaccharide formated by fructose and glucose, is used for ethanol production by fermentation and for commercial sugar production, through separation and crystallization. However, sucrochemistry searches to obtain other molecules of higher value from sucrose, glucose, and xylose—constituents of cellulose and hemicellulose, respectively.

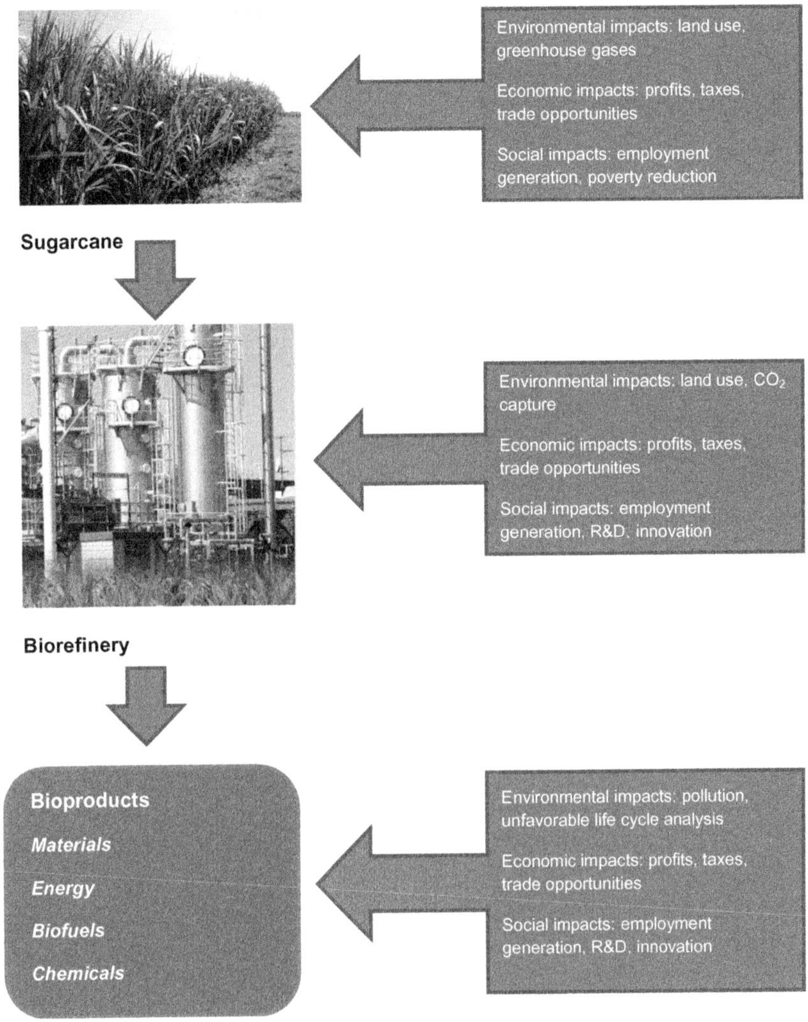

Figure 9.3 Examples of sustainability criteria applied to the sugarcane use by renewable chemistry. Impacts could be measured as positive or negative.

Table 9.1 Worldwide sugarcane producers in the year of 2013, according to Food and Agriculture Organization of the United Nations (2017)

Country	Production (k tonnes)
Brazil	739,267
India	341,200
Mexico	61,182
Colombia	34,876
Indonesia	33,700
United States	27,906

Table 9.2 Mass composition of the sugarcane plant, according to Dinardo-Miranda et al. (2008).

Component	Average (% w/w)
Fiber	8–14
Fructose	0.2–1
Glucose	0.2–1
Sucrose	14–24
Various organic compounds	0.8–1.8
Various inorganic compounds	0.2–0.7
Water	75–82

Bagasse is frequently used for animal feeding and production of bioelectricity by cogeneration, as in Brazil; then, the sugarcane mills are self-sufficient in the use of electricity (Vaz Jr., 2014a). The use of bagasse and straw for second-generation (2G) ethanol production is a topic that has a large amount of published literature; however, the commercial production is starting, with barriers to be overcome, such as reducing the cost of development of enzymes and yeasts that ferment the pentoses from hemicellulose (Sarkar et al., 2012). Vinasse has been used both in generating biogas and in soil fertigation; however, a frequent monitoring of its application is required due to the high content of ions and organic matter, which can alter the physicochemical properties of the soil, with subsequent leaching of these ions (NO_3^-, PO_4^{3-}, K^+, etc.) promoting groundwater pollution (Da Silva et al., 2007).

Table 9.3 shows the main sugarcane components and derivatives with economic potential for renewable chemistry. Ethanol stands out as the main product because it can be obtained from three of the four sugarcane components (sucrose, bagasse, and straw). Furthermore, it can be used to produce other derivatives for chemical industry, as "green" plastics.

Perspectives for sugarcane bioproducts

Table 9.4 presents a description of target products with high added value that can be obtained from sugarcane by means of different processes. In this table, only building blocks and end-use products can be observed. They were defined according to the characteristics of Brazilian biomass, data from national and international scientific literature, and demand from national chemical industry and related industries. Building blocks are molecules from which a number of other chemical products of economic

Table 9.3 Sugarcane components and derivatives with relevance for renewable chemistry.

Raw material	Main composition	Usages
Sucrose	Glucose and fructose	Commercial sugar Ethanol Renewable chemicals
Bagasse	Lignin, cellulose, hemicellulose, inorganics, and water	Animal food Bioelectricity Renewable chemicals Ethanol 2G Fibers and polymers
Straw	Lignin, cellulose, hemicellulose, inorganics, and water	Ethanol 2G Soil recovery
Vinasse (aqueous effluent)	Solubilized organic matter, inorganic solids, solubilized inorganic salts, and water	Biogas Biofertilizer

interest are originated. Synthesis intermediates are those compounds used in small quantities as fine chemical reagents—production of drugs, pesticide, cosmetics, among others. Chemical specialties must be considered, which are restricted to certain purposes of use.

We can highlight that these bioproducts have not reached the stage of industrial maturity in their majority.

The current technological scenario, taken from the information shown in Tables 9.2 and 9.3, suggests a production of a wide range of products in a model similar to an oil refinery. Indeed, the American Chemical Society published in 2014 a scenario evaluation for renewable chemicals considering building block molecules for biobased polymers (Bomgardner, 2014), adding information to Table 9.4. The following compounds were highlighted:

- 3-hydroxypropionic acid from bacterial fermentation of glucose from sucrose—likely in 10 years for commercial step;
- sorbitol from hydrogenation of glucose from sucrose—likely in 10 years; and
- levulinic acid from acid-catalyzed dehydration of sugars (hexose or pentose)—maybe likely in 10 years.

These three compounds can be produced from sugarcane.

Another revision of 2015 conducted by Biofuels Digest appointed some renewable chemicals with high market potential, which can be obtained from sugarcane components (Biofuels Digest, 2015):

Table 9.4 Bioproducts from sugarcane identified as high potential for renewable chemistry, according to Vaz Jr. (2017).

Target compound	Precursor	Route	Status
2,5-Furanodicarboxylic acid	Glucose from cellulose	Organic synthesis Fermentation	In development: improvement of catalysts, biocatalysts, microorganisms, yield, others
Succinic acid	Xylose from hemicellulose	Fermentation	In development: improvement of microorganisms and yields
Antioxidants	Lignin	Catalytic cracking	In development: improvement of catalysts, yield, others
Derivatives from cellulose (acids, esters, nitrates, ethers, etc.)	Cellulose	Organic synthesis	Established industrial processes
2G ethanol	Glucose from cellulose Xylose from hemicellulose	Fermentation	In production but with improvement of enzymes, microorganisms, yields, and cost reduction
Phenols	Lignin	Catalytic cracking	In development: improvement of catalysts, yield, others
Furfural	Xylose from hemicellulose	Organic synthesis	Established industrial processes: still needing to improve catalysts, yields, others
Gas of synthesis ($CO + H_2$)	Lignocellulosic biomass	Gasification	Established industrial processes based on petroleum; still needing to improve yields for biomass
5-Hydroxymethylfurfural	Cellulose	Organic synthesis	Established industrial processes: still needing to improve catalysts, yields, others
Sulfonated lignins	Lignin	Organic synthesis	Established industrial processes
Xylitol	Xylose from hemicellulose	Organic synthesis	In development: improvement of catalysts, yield, others

- 1,4-diacids: succinic, fumaric, and malic—obtained from xylose and glucose;
- 2,5-furan dicarboxylic acid—obtained from glucose;
- 3-hydroxypropionic acid—obtained from glucose; and
- glucaric acid—obtained from glucose.

We expected that those products will be more environmentally friendly than petroleum-based molecules, contributing to the sustainability of raw materials (sugarcane) and their conversion processes in biorefineries. The main processes to be applied in a sugarcane biorefinery are described in the following sections.

A biorefinery refers to the technologies and processes used for the conversions of biomass into five types of products presented: energy, chemical commodities or bulk chemicals, biofuels, materials, and fine chemicals. The technologies are compiled into routes or processes, which are divided into biochemical (fermentation and enzymatic catalysis), chemical (majority, chemocatalytic), and thermochemical (combustion, gasification, and pyrolysis).

Chemical routes

Chemical routes are those based exclusively on chemical reactions (forming or breaking). Organic synthesis is very relevant for the biomass conversion. Most of the times, a biomass component is extracted, purified, and, subsequently, used as starting material in a synthetic route. Often catalysts are used to increase the yield of product of interest and to decrease the reaction time—the chemical processes are chemocatalytic. Thus it is possible to notice that several aspects of green chemistry, such as the use of renewable feedstock, catalysts, and the reduction of waste production (Anastas and Warner, 1998).

In the case of utilization of cellulose and hemicellulose from lignocellulosic waste (sugarcane bagasse and straw), first these polymers and their constituent sugars must be obtained, especially glucose (hexose) and xylose (pentose), respectively, to generate products of industrial interest, such as ethanol 2G (Kamm et al., 2006). For lignin, the breaking of its molecular structure is initially attempted to release mainly phenolic compounds which may be tested, for example, as monomers for various routes of preparation. Obtaining building blocks and synthesis intermediates is the usual approach used by R&D projects to add greater value on bioenergetic chains in a biorefinery (Bozell and Petersen, 2010; United States

Department of Energy, 2007, 2004). Building blocks, such as furfural and xylitol (derived from xylose constituent of hemicellulose) and 5-hydroxymethylfurfural (derived from glucose), among others, can add great value to carbohydrates, which can be extended to derivatives of lignin (Vaz Jr., 2014a,b; Bomgardner, 2014; Bozell and Petersen, 2010; Kamm et al., 2006), as reported in Table 9.3.

It is worth commenting on the development and use of catalysts for these processes, given their importance for improving yields and selectivities (enantioselectivity, regioselectivity, and stereoselectivity). Examples are metals (soluble and insoluble salts and complexes) have been applied to heterogeneous catalysis (Ni, Pd/C, Ru/C, Co-Mo, Ni-Mo, Ru/Al_2O_3, etc.) to reduce lignins (Bozell and Petersen, 2010; Zakzeski et al., 2010); and metal complexes of V, Mn, Co, Pd, Fe, Re, and Cu acting as homogeneous and heterogeneous catalysts for cellulose oxidation, among other reactions (Collinson and Thielemans, 2010).

Biochemical routes

Biochemical routes considered for sugarcane transformation are fermentation, for the production of first generation ethanol (1G) and other chemicals, such as alcohols and other organic acids and anaerobic digestion, for production of biogas and biofertilizer (mineralized fraction). Enzymatic catalysis contributes to increase the rate of metabolic reactions involved in biochemical conversion routes.

The biochemical processes have major operational similarity to the chemical processes in respect to the stages of compositional analysis and characterization of the feedstock, pretreatment (whenever necessary), structural identification, and study of industrial potential. However, the main feature of these processes relates to the use of microorganisms (fungi, bacteria, and yeast), which have the biochemical mechanisms that allow the synthesis of organic chemicals, as ethanol. For example, production of ethanol (1G and 2G) by the yeast *Saccharomyces cerevisiae* by means fermentation of sucrose, also produces CO_2 (Sarkar et al., 2012)—a carbon dioxide with a high purity degree and heat for industrial purposes.

In anaerobic fermentation of the organic matter present in the vinasse, it produces, mainly, methane and CO_2 (biogas), as well as mineral residue rich in inorganic salts of nitrogen, sulfur, and carbon (biofertilizer) (Salomon and Lora, 2009). In this case, bacterial consortia is used, formed by *Acinetobacter, Arthrobacter, Bacillus, Brevbacterium*, among others (Cutright, 2002).

The bacteria *Lactobacillus delbrueckii* can be used for the production of lactic acid via fermentation of glucose and *Anaerobiospirillum succiniciproducens* for the production of succinic acid through fermentation of pentoses and hexoses (Bozell and Petersen, 2010). However, despite the high potential for production of a large amount of chemicals, the slow rate of reaction (kinetic of conversion) and the difficulty to separate final products (downstream step), in some cases, can limit the use of bioprocesses in sugarcane-biorefineries for other products than ethanol and some organic acids.

On the other hand, enzymes such as cellulase, β-glucosidase, and xylanase are widely used in the hydrolysis of cellulose and hemicellulose to release glucose and xylose, respectively (Sarkar et al., 2012).

Thermochemical routes

As for chemical and biochemical routes, steps of compositional analysis and characterization of feedstocks, the structural identification and study of industrial potential are common. Nevertheless, the main characteristic of these processes is the use of thermal energy that leads to combustion, gasification, or pyrolysis. Through the combustion, it generates thermal energy (heat) (Nussbaumer, 2003); through the gasification it generates synthesis gas (syngas), which is mainly composed of carbon monoxide and hydrogen to be used in organic synthesis of various molecules for chemical industry (Akay and Jordan, 2011), by the Fischer–Tropsch reaction (Gökalp and Lebas, 2004). Fast pyrolysis produces bio-oil (rich in polar and nonpolar organic compounds) and biochar (a soil fertilizer). Cogeneration is a combined thermal process, in which electric energy is produced. The biomass combustion generates heat, which heats the water, generating steam, moving the turbines for the production of bioelectricity—it can either be used by the biorefinery itself or be sold to the electric grid.

As seen in Table 9.4, gasification is a process that can add value to the biomass waste. Syngas and fast pyrolysis can produce renewable chemicals to substitute several petrochemicals.

Challenges involved

Many of the bioproducts shown in this chapter are subject to extensive literature production, mainly in the areas of organic chemistry, biotechnology, and chemical engineering. However, in some cases, there are no processes established for the industrial production of them, which points to the existence of scientific, technical, and economic challenges, separate or concurrent. It is important to consider, however, that the worldwide market of chemicals involves values around USD100 billion per year, of which about 3% are bioproducts or biomass derivatives, The estimate is that the total market share reaches 25% by the year 2025 (Vijayendran, 2010). These figures give an idea about the possibilities and the risks involved. For specialties and fine chemicals, the current participation of renewable—about 25% in both segments—can reach a market share of 50%, while for polymers, the current 10% can reach 20%, also in 2025 (Biotechnology Industry Organization, 2017).

The technical challenges involve the technological development or improvement that allow the scale-up of processes developed in the laboratory, as separation methods, process optimization, energy efficiency, among others—a clear example for these observations is the 2G ethanol.

Failure to overcome this kind of challenges can derail the production of certain bioproducts, as a value-added molecule, which can show great market potential in its R&D stage. Therefore a well-planned R&D stage must have a proper technological support, in order to make the laboratory scale achieve an industrial scale.

One of the main economic challenges concerns attracting and allocating funds in R&D&I projects and, subsequently, in demonstration of technology projects. Rise and the possible decline of said "green" chemicals needs to be considered in the budget planning of development projects or for renewable compounds production, since previous international scenarios related to the chemical industry draw attention to market external factors (Biotechnology Industry Organization, 2017; Bomgardner, 2014). Nowadays, shale gas could be a threat for renewable chemicals based on its lower cost.

Nevertheless, the existing demand for ethanol 1G and the beginning of ethanol 2G production will provide a favorable economic environment for the sugarcane in chemical industry.

Conclusion

This chapter attempts to show the great economic potential of sugarcane for renewable chemistry. Building blocks and synthesis intermediates are good examples of potential bioproducts in a medium and large horizon, integrated to ethanol, sugar, and bioenergy production. Energy and biofuel are very important products in the sugarcane biorefinery, in which conceptual model is the petrochemical refinery.

When evaluating products mentioned as potential in other regions of the world, it is noted that they do not always reflect local needs, justifying the use of own information obtained in studies conducted in order to better adapt for the national scenarios, technical and scientific planning, and investment resources.

Chemical, biochemical, and thermochemical processes are fundamental to exploit the full economic potential of the sugarcane. However, it is still required strong actions to overcome several challenges, comprising mainly pretreatments, synthesis routes, catalysts, and microorganisms. However, is very important to establish an integrated sugarcane biorefinery to achieve the real economic potential for the renewable chemistry.

Scientific, technical, and economic challenges must be overcome by the government and private sector. It will allow sugarcane to turn into a viable alternative to the oil replacement for some products (e.g., organic acids, polymers, alcohols, syngas, among others) in chemical industry, causing positive environmental, social, and economic impacts as a model of bioeconomy based on renewable biomass.

References

Anastas, P.T., Kirchhoff, M.M., 2002. Origins, current status, and future challenges of green chemistry. Accounts Chem. Res. 35 (9), 686–694.

Anastas, P.T., Warner, J.C., 1998. Green Chemistry: Theory and Practice. Oxford University Press, New York.

Akay, G., Jordan, C.A., 2011. Gasification of fuel cane bagasse in a downdraft gasifier: influence of lignocellulosic composition and fuel particle size on syngas composition and yield. Energ. Fuel. 25 (5), 2274–2283.

Biofuels Digest, 2015. The DOE's 12 top biobased molecules – what became of them? <http://www.biofuelsdigest.com/bdigest/2015/04/30/the-does-12-top-biobased-molecules-what-became-of-them/> (accessed 03.08.17.).

Biotechnology Industry Organization, 2017. Biobased chemicals and products: a new driver for green jobs. <http://www.bio.org/articles/biobased-chemicals-and-products-new-driver-green-jobs> (accessed 03.08.17.).

Bomgardner, M.M., 2014. Biobased polymers. Chem. Eng. News 92 (43), 10–14.

Bozell, J.J., Petersen, G.R., 2010. Technology development for the production of biobased products from biorefinery carbohydrates - the US Department of Energy's Top 10 revisited. Green Chem. 12 (4), 539–554.

Brazilian Center for Management and Strategic Studies, 2010. Química verde no Brasil: 2010 – 2030. CGEE, Brasília.

Collinson, S.R., Thielemans, W., 2010. New materials focusing on starch, cellulose and lignin. Coordin. Chem. Rev. 254 (15-16), 1854–1870.

Cutright, T.J., 2002. In: Ghassemi, A. (Ed.), 'Biotechnology principles' in *Handbook of Pollution and Waste Minimization*. Marcel Dekker, New York, pp. 189–232.

Da Silva, M.A.S., Griebeler, N.P., Borges, L.C., 2007. Uso de vinhaça e impactos nas propriedades do solo e lençol freático. Revista Brasileira de Engenharia Agrícola e Ambiental 11 (1), 108–114.

Dinardo-Miranda, L.L., Vasconcelos, A.C.M., Landell, M.G.A. (Eds.), 2008. Cana-de-açúcar. Instituto Agronômico de Campinas, Campinas.

European Investment Bank, 2017. Study on Access-to-Finance Conditions for Investments in Bio-based Industries and the Blue Economy. European Commission, Luxembourg.

Food and Agriculture Organization of the United Nations, 2017. *Faostat*. <http://faostat3.fao.org/home/E> (accessed 03.08.17.).

Gökalp, I., Lebas, E., 2004. Alternative fuels for industrial gas turbines (AFTUR). Appl. Therm. Eng. 24 (11-12), 1655–1663.

Kamm, B., Gruber, P.R., Kamm, M. (Eds.), 2006. Biorefineries: industrial processes and products: status quo and future directions. Wiley-VCH, Weinheim.

Metzger, J.O., Eissen, M., 2004. Concepts on the contribution of chemistry to a sustainable development. Renewable raw materials. CR Chim. 7 (6-7), 569–581.

Nussbaumer, T., 2003. Combustion and co-combustion of biomass: fundamentals, technologies, and primary measures for emission reduction. Energ. Fuel. 17 (6), 1510–1521.

Salomon, K.R., Lora, E.E.S., 2009. Estimate of the electric energy generating potential for different sources of biogas in Brazil. Biomass Bioenerg. 33 (9), 1101–1107.

Sarkar, N., Ghosh, S.K., Bannerjee, S., Aikat, K., 2012. Bioethanol production from agricultural wastes: an overview. Renew. Energy 37 (1), 19–27.

Straathof, A.J.J., Bampouli, A., 2017. Potential of commodity chemicals to become bio-based according to maximum yields and petrochemical prices. Biofuel. Bioprod. Bior. 11 (5), 798–810. Available from: https://doi.org/10.1002/bbb.1786.

United States Department of Energy, 2007. Top Value Added Chemicals From Biomass: Results of Screening for Potential Candidates From Biorefinery Lignin. US-DOE, Springfield.

United States Department of Energy, 2004. Top Value Added Chemicals From Biomass: Results of Screening for Potential Candidates From Sugars and Synthesis Gas. US-DOE, Springfield.

Vaz Jr, S., 2014a. Perspectives for the Brazilian residual biomass in renewable chemistry. Pure Appl. Chem. 86 (5), 833–842.

Vaz Jr, S., 2014b. A renewable chemistry linked to the Brazilian biofuel production. Chem. Biol. Technol. Agric. 1, 13.

Vaz Jr, S., 2017. Sugarcane-biorefinery. Advances in Biochemical Engineering/ Biotechnology. Springer, Berlin.

Vijayendran, B.J., 2010. Bio products from bio refineries – trends, challenges and opportunities. J. Bus. Chem. 7 (3), 109–115. pp.

Zakzeski, J., Bruijnincx, P.C.A., Jongerius, A.L., Weckhuysen, B.M., 2010. The catalytic valorization of lignin for the production of renewable chemicals. Chem. Rev. 110 (6), 3552–3599.

CHAPTER 10

Use of sugarcane vinasse to biogas, bioenergy, and biofertilizer production

Anderson Carlos Marafon, Karina Ribeiro Salomon, Eduardo Lucena Cavalcante Amorim and Fernanda Santana Peiter

Contents

Sugarcane crop and bioethanol production in Brazil 179
Characterization and chemical composition of sugarcane vinasse 180
Alternatives for the use of sugarcane vinasse 182
Fertirrigation in natura 182
Anaerobic biodigestion 184
Vinasse concentration 189
Other uses 191
Perspectives of the vinasse sugarcane use in Brazil 191
References 193
Further Reading 194

Sugarcane crop and bioethanol production in Brazil

Alcohol as well as sugar and electricity from cogeneration systems, is one of the main products of the sugarcane industry in Brazil. The annual production of ethanol in Brazil is about 30 million m^3. Considering that to each liter of ethanol produced other 12 L of vinasse is generated, about 360 billion of liters of this by-product is produced annually in the country (Mauad et al., 2017).

The sugarcane gets to the mill where it is washed and ground for the extraction of the juice. This juice contains sucrose, glucose, yeast, nitrogenous matter, etc. Afterward, the juice is submitted for clarification, concentration, and centrifugation for the attainment of the commercial sugar and syrup. After going through vacuum cooking process for the attainment of a lower quality sugar, it is transformed into a final syrup, which is also called poor syrup or molasses, which is sent to the fermentation tanks.

Sugarcane Biorefinery, Technology and Perspectives
DOI: https://doi.org/10.1016/B978-0-12-814236-3.00010-X

Many distilleries use the sugarcane juice directly after milling as raw material for ethanol production. After the fermentation, the resulting liquid is called wine, which also goes through a centrifugation process to recover the ferment (yeast) that will be reused in the fermentation process. Then this wine is sent to the distillation columns for the separation of alcohol, which in turn produces a residue called vinasse (Salomon and Lora, 2009).

Developing biorefineries aimed at resource recovery is a growing trend to devise alternative systems for obtaining energy and materials in a more sustainable way. Within this concept is the use of sugarcane vinasse industries to recover water, nutrients, and energy as products of interest (Peiter, 2018). Nonetheless, the characteristics that make vinasse a rich material to supply the needs of plants are the same that make it an effluent with a high pollutant load. There are still doubts about the environmental safety of fertigation practice in soil, groundwater, and atmosphere. From an economic perspective, there are problems with the high cost of large volumes of this liquid even in sugarcane growing areas (Rabelo et al., 2015).

Sugarcane vinasse is one of the most polluting residues produced by Brazilian ethanol industries, mainly because of its harmful effects on the environment, such as high organic matter load and acidity. In Brazil, almost all of the produced vinasse was used as fertilizer and in the process of irrigating the sugarcane itself. The problem is that this practice harms the environment and wastes the potential for better uses of vinasse, such as, for example, to generate electricity, transforming vinasse into biogas.

Since the 1980s electricity generated from bagasse in cogeneration systems is one of the by-products of sugarcane industries, an alternative source of energy from a material previously considered waste. Nevertheless, using vinasse could improve the energy, economic, and environmental potential of sugarcane biorefineries. The most common methods of using the vinasse include (1) anaerobic digestion as an energy pathway and to produce a supernatant rich in nutrients; and (2) concentration processes, such as evaporators and filter membranes, to recycle water and produce a material with a high content of nutrients and organic matter (Peiter, 2018).

Characterization and chemical composition of sugarcane vinasse

Sugarcane vinasse is the main by-product from the sugar–ethanol industry. Sugarcane vinasse, also named distillery water, wastewater, or stillage, is an aqueous solution of organic solids and minerals, besides residual contents of sugars, ethanol, and volatile compounds with a high molecular weight (Rocha et al., 2012).

Sugarcane vinasse is a dark brown liquid, with an acid nature, that leaves the ethanol distillation plant at a temperature of approximately 107°C. Its odor goes from astringent to nauseant; its quality is related to the sugar residual content, which in turn, causes a putrefying process as soon as the vinasse is discharged, releasing fetid gases that make the surrounding environment unbearable (Salomon and Lora, 2009).

Vinasse is an acidic compost (pH: 3.5–5), dark brown slurry, with high organic content. In general, this effluent presents dark color and consists of basically water (93%) and organic solids and minerals (7%), of which 75% are organic and biodegradable compounds and the other 25% are minerals (Laime et al., 2011). Due to the significant levels of nutrients, mainly potassium, wastewater is used as a soil amendment and fertilizer in the sugarcane plantations. This liquid is transported to these areas using trucks, canals, or pipes. It can be applied directly to the soil or can be spread by spraying using sprayers or sprinklers (Rabelo et al., 2015).

Vinasse is considered very polluting due to the presence of high organic load that causes the proliferation of microorganisms that deplete the dissolved oxygen in the water, causing damages to the availability of drinking water, besides the very low pH. The vinasse usually exhibits high pollutant content mainly characterized by its low pH, high corrosion ability, and large organic matter content, in the distillation process of the must (fermented broth). Its polluting power can be 100 times as powerful as domestic sewage, having an elevated biochemical oxygen demand (BOD; Laime et al., 2011).

Vinasse is an ethanol waste by-product, an organic material rich in potassium (K), nitrogen (N), calcium (Ca), and magnesium (Mg). The chemical composition of vinasse depends on the characteristics of the soil, the variety of sugarcane, the period of the harvest and the industrial process used for the production of ethanol (Salomon and Lora, 2009). It is important to emphasize the high K concentration in relation to the other nutrients in this residue. In addition to these nutrients, vinasse contains organic compounds (organic acids, alcohols, glycerol) that are converted into methane by anaerobic bacteria (Soares et al., 2014).

Ethanol can be manufactured by using sugarcane juice, molasses (a by-product in the cooking stage), or a mixture of both as input. Fermentation of these components generates wine, which is processed

Table 10.1 Physicochemical characteristics of sugarcane vinasse (Cortez et al., 1998).

Characteristics	Molasses	Juice	Mixed
pH	4.2–5	3.7–4.3	4.4–4.6
Temperature (°C)	80–100	80–100	80–100
BOD (g/L O_2)	25	6–16	19.8
COD (g/L)	65	15–33	45
Total solids (g/L)	81.5	23.7	52.7
Volatile matter (g/L)	60	20	40
Fixed matter (g/L)	21.5	3.7	12.7
Nitrogen (g/L; N)	0.45–1.6	0.15–0. 7	0.48–0.71
Phosphor (g/L; P_2O_5)	0.1–0.29	0.1–0.21	0.09–0.2
Potassium (g/L; K_2O)	3.74–7.83	1.2–2.1	3.34–4.
Calcium (g/L; CaO)	0.45–5.18	0.13–1.54	1.33–4.57
Magnesium (g/L; MgO)	0.42–1.52	0.2–0.49	0.58–0.7
Sulfate (g/L; SO_4)	6.4	0.6–0.76	3.70–3.73
Carbon (g/L; C)	11.2–22.9	5.7–13.4	8.7–12.1
C/N relation	16–16.27	19.7–21	16.4–16.43
Organic matter (g/L)	63.4	19.5	38
Reduced substances (g/L)	9.5	7.9	8.3

BOD, Biochemical oxygen demand; *COD*, chemical oxygen demand.

into distillation columns. The latter process results in the generation of ethanol as the main product and vinasse as wastewater.

The main characteristics of the vinasse produced by the process to attain alcohol out of molasses, juice, and their mixture are presented in Table 10.1.

Alternatives for the use of sugarcane vinasse

Fertirrigation in natura

The biodigestion process of vinasse reduces its organic load but maintains its fertilizing power. The organic matter present in the vinasse is degraded into simpler and easily available compounds, making the nutrients partially solubilized. The reduction of the C/N ratio promoted by biodigestion favors the application of biomass of vinasse digested as biofertilizer (Cortez et al., 2007).

Vinasse presents in its constitution a series of elements that make it different in relation to the other fertilizers and soil conditioners. It presents practically all the elements that can be part of a chemical recovery of the soils, not only in the surface but also in subsurface. Being fluid, it

Figure 10.1 Vinasse fertirrigation in a sugarcane cultivation area. *Photo: Saulo Coelho.*

penetrates the soil and proceeds to recompose not only chemical, but also physical and biological conditions.

The sugarcane vinasse resulting from the ethanol distillation is deposited on the decantation tanks before be conducted through canals or transported with tank trucks to the cultivation areas to be applied on the soil (Fig. 10.1).

In Brazil, the vinasse is directly applied to the soil as a fertilizer. But for this practice to be carried out, is necessary to do an analysis of the characteristics of the soil, to define the appropriate amount can be used. The main difficulties related to the final disposal of the vinasse are usually the high BOD rates whose values range from 30 to 40 g/L and the low pH, which varies between 4 and 5 because of the contained organic acids. Among the alternatives for the use of sugarcane vinasse, the fertirrigation in natura is the most commonly used, as it requires a low initial investment (tubes, pumps, trucks, and decantation tanks), low maintenance cost, fast application, does not require complex technologies, and increases crop yield (Christofoletti et al., 2013).

Exaggerated vinasse use as a fertilizer can cause environmental damage, such as groundwater contamination with potassium (impairing the absorption of calcium and other elements), soil salinization (by the additional intake of sodium and chlorine), leaching of metals and sulfates, release of bad smell, and greenhouse gas emissions such as nitrous oxide, which is about 300 times more polluting than carbon dioxide (CO_2)

(Soares et al., 2014). For these reasons, environmental agencies set the maximum quantity of vinasse per hectare. The amount of vinasse based on the K^+ saturation on the cation-exchange capacity to a maximum of 5% and in the requirement and capacity of extraction and export of this nutrient by the crop (Cetesb, 2006).

Anaerobic biodigestion

Anaerobic biodigestion is a highly efficient wastewater treatment method that could potentially be used to treat sugarcane vinasse. This process consists of the biodegradation of the organic load of vinasse to produce biogas and biodigested vinasse (Cortez et al., 2007). Therefore anaerobic biodigestion is an alternative of great economic as well as environmental interest in the treatment of vinasse, as the biogas produced, once purified, has calorific value similar to that of natural gas, with the advantage of being a renewable and easily available fuel (Szymanski et al., 2010).

Anaerobic biodigestion may be considered the primary alternative for managing vinasse in sugarcane biorefineries. The technique presents important advantages over fertirrigation, including a reduction in the polluting organic load of vinasse, the potential recovery of bioenergy from biogas, and the potential for enhancing the profitability of biorefineries through the generation of surplus electricity, based on the burning of biogas in prime movers (Fuess et al., 2018; Moraes et al., 2015).

In Brazil, the pioneer project of the sugarcane vinasse biodigestion was used in 1981 by the Paisa Agroindustrial Distillery, located at municipality of Penedo, Alagoas State. This test was carried out in an Upflow Anaerobic Sludge Blanket (UASB) reactor in a pilot plant with a capacity of 11 m^3 of biogas, producing 13.1 L of biogas (65% methane, i.e., CH_4) per liter of vinasse. After this, two more biodigesters of 24 m^3 and one in an industrial scale with capacity for 500 m^3 were installed (Rocha et al., 2012).

The alternative of anaerobic biodigestion of the organic load of vinasse has increasingly being used in the ethanol industry. Usually, the applied technologies to the anaerobic treatment of sugarcane vinasse are the UASB reactors, conventional digesters, and covered ponds. All of them are characterized by a low application rate, since a very large volume of reactor (or pond) is required for each m^3 of stillage to be treated. A modern reactor with internal circulation was developed. This system presents higher efficiency in the anaerobic biodigestion process than the UASB reactor. Independent of the reactor model, the anaerobic digestion needs to contemplate on two main steps: (1) equalization, with the vinasse recirculation in the tank to complete substrate homogenization; and the (2) conditioning, which includes pH adjustment and the mineral nutrients supply (mainly nitrogen: N and phosphorus: P) (Rocha, 2009).

If the startup of the bioreactor and the maintenance of active biomass are adequate, suitable biogas can be generated. In anaerobic digestion processes, biological degradation of organic components is achieved with no requirement of molecular oxygen. Most of the carbon atoms originating in the waste material are reduced to CH_4.

Biodigestion processes facilitate the mobilization of nutrients from the organic matter to the liquid phase. Thus N is converted into ammonium, and organic P is hydrolyzed to soluble P. Temperature control is fundamental for maintaining optimal bacterial growth and conversion processes in anaerobic microbial systems. The optimum growth temperature of anaerobic microorganisms is 35°C or higher.

According to Moraes et al. (2015), anaerobic digestion consists of a set of complex and sequential metabolic processes that occur in the absence of molecular oxygen and depend on the activity of at least three distinct groups of microorganisms to promote the stable and self-regulating fermentation of organic matter, resulting mainly into methane and CO_2 gases (Speece, 1996; Leitão et al., 2006; O'Flaherty et al., 2006; Madsen et al., 2011). These groups of microorganisms include acidogenic (or fermentative) bacteria, acetogenic (or syntrophic) bacteria, and methanogenic archaea (Mosey, 1982; Guiot et al., 1992; Wirth et al., 2012). In the presence of sulfate, sulfite, or thiosulfate, there is also activity from sulfate-reducing bacteria, responsible for the reduction of oxidized sulfur compounds to sulfide dissolved in the effluent ($HS^-/S^{-2}/H_2S$) and to hydrogen sulfide (H_2S) in the biogas (O'Flaherty et al., 2006). Fig. 10.2 illustrates the scheme of the anaerobic digestion of complex organic matter and identifies the respective groups of microorganisms involved in each step.

The removal of nutrients in anaerobic biodigestion systems is negligible, which means that the fertilizing potential of vinasse is maintained in the biodigested effluents (Cortez et al., 1998; Moraes et al., 2015; Salomon et al., 2011) (Table 10.2).

Biogas composition

The recovery of bioenergy through biodigestion was considered for the hydrogen (H_2) biogas and CH_4-rich biogas streams obtained from the bioconversion of vinasse during the acidogenic and methanogenic (single-phase or combined) steps, respectively (Table 10.3).

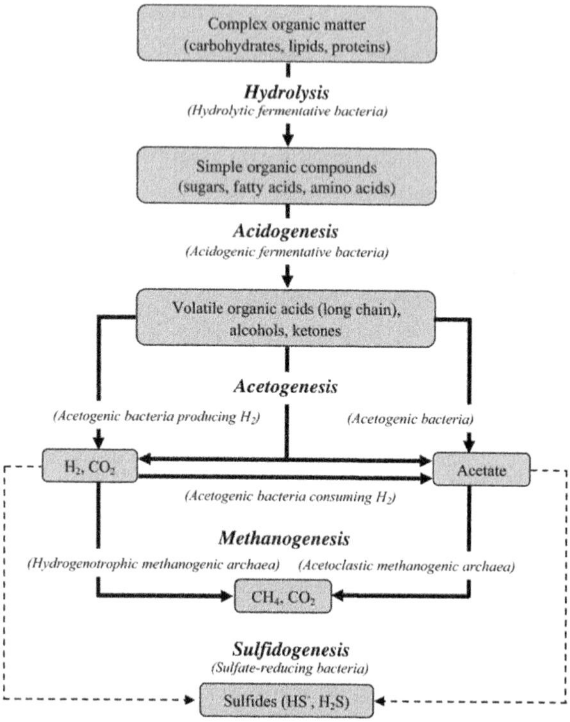

Figure 10.2 Scheme of the anaerobic digestion of complex organic matter, depicting the steps and microbial populations involved (Moraes et al., 2015).

Table 10.2 Physicochemical characterists of sugarcane vinasse submitted to biodigestion.

Characteristics	Vinasse (before biodigestion)	Vinasse (after biodigestion)
pH	4	6.9
COD (g/L)	29	9
N total (g/L N)	0.55	0.6
N ammoniacal (g/L)	0.04	0.22
Potassium (g/L K_2O)	1.4	1.4
Phosphor (g/L P_2O_5)	0.017	0.032
Sulfate (g/L)	0.45	0.032

The biodigestion reduces the vinasse chemical oxygen demand (COD) from about 30 to 75 kg/m^3. The COD concentration is a measure of the amount of oxygen required to chemically oxidize all organic compounds

Table 10.3 Composition of vinasse sugarcane derivated biogas (Salomon and Lora, 2009).

Component	Percentage (%)
Methane	40–75
Carbon dioxide	25–40
Nitrogen	0.5–2.5
Oxygen	0.1–1
Hydrogen sulfide	0.1–0.5
Ammonia	0.1–0.5
Carbon monoxide	0–0.1
Hydrogen	1–3

to water and CO_2. The amount of biogas produced from 1 m^3 of vinasse varies between 7 and 15 Nm^3 of biogas. Considering a vinasse with COD concentration of 30 kg/m^3, the volume of reactor required to treat each 1 m^3 of vinasse can vary from 3^3 to 15 m^3.

Some factors such as pH and the nutritional needs of microorganisms can influence the biodigestion process. Therefore during the biodigestion, the pH is corrected by the addition of alkaline substances, such as 50% sodium hydroxide. Some nutrients, such as nitrogen and phosphorus, can also be provided based on the physicochemical characterization of vinasse (Rocha, 2009).

Anaerobic biodigestion of vinasse results in the formation of two products: biogas and biodigested vinasse. Biogas can include many applications such as heating, combined heat, and power generation (cogeneration), transportation fuel (after being upgraded to biomethane), or upgraded to natural gas quality for a wide range of uses. The use of vinasse to produce biogas allows the thermal energy generated by the biogas combustion to be used for the concentration of the vinasse or the biogas, originating from the concentrated biodigested vinasse, which maintains the fertilizer characteristics of the vinasse. The biodigested vinasse can be used as a liquid fertilizer to be applied directly or previously concentrated application on the soil. Both concentrated vinasse in natura and the biodigested vinasse can be used to produce solid fertilizers after composting, crushing, mixing, granulating, and packaging (Szymanski et al., 2010).

The biodigested vinasse is later used as a fertilizer. Although it presents a reduced organic load, it maintains its original properties as a fertilizer. On the other hand, biogas is mainly used to produce energy due to its high methane content. In the sugar–ethanol industry, biogas can be used to operate gas turbines combined to an electric generator; substitute part of the fuels used in the agroindustry during the harvesting time; or used in boilers to generate vapors and to mill sugarcane (Cortez et al., 2007).

After the biodigestion process with ethanol production, stillage is still suitable for use in fertirrigation, since its nutrients (nitrogen, phosphorus, and potassium) are not removed during the biogas generation process. The biodigestion process promotes the degradation of organic matter into simpler and easily available compounds for the biological activity of the soil, with partially solubilized nutrients. As a result of the production of CH_4 and CO_2, there is a reduction in the C/N ratio, which favors its application as a biofertilizer, because the resulting material is more easily assimilated by the soil biological activity (Cortez et al., 1998; Szymanski et al., 2010).

Biogas desulfurization

Anaerobic biodigestion avoids most of the problems related to its application in the soil. Vinasse normally contains a high amount of sulfate due to the use of sulfuric acid in the production process (Moraes et al., 2015). During biodigestion, sulfate is converted to sulfide in the anaerobic reactor, resulting in a significant amount of H_2S in the biogas. H_2S is extremely corrosive and needs to be taken out of the biogas before it can be used in a boiler or engine. For the removal of H_2S from the biogas, it is necessary to install a desulfurizer, which consists basically of a gas scrubber and a bioreactor. The H_2S-containing biogas is washed in the washer which is filled with carrier media. For washing, an alkaline solution is sprayed onto this media at the top of the washer, coming into contact with the biogas entering from below. The H_2S then passes from the gas phase to the liquid phase, and the treated biogas leaves the washer by gravity. The sulfide-containing alkaline solution flows through the scrubber into the bioreactor where it contacts specific bacteria which, in the presence of oxygen, convert the sulfide to elemental sulfur. The regenerated lavage water from the bioreactor is then conducted back to the washer to remove more H_2S from the biogas. The sulfur formed in the bioreactor is discharged through a sedimentation tank, leaving the system as a concentrated high purity sulfur slurry, which can be used as fertilizer or fungicide. Thus from the vinasse, a biogas without H_2S is generated, from which it is possible to remove CO_2 and compress the resulting biomethane, using it as fuel (Peiter, 2018).

Table 10.4 Characteristics of vinasse in natura and concentrated at 35 and 65 Brix.

Characteristics	4 Brix (in natura)	35 Brix	65 Brix
pH	4.4–4.6	4.6–5.0	4.6–5.0
Temperature (°C)	80–100	50–60	50–60
BOD (g/L O_2)	20	173.2	321
COD (g/L)	45	393.7	731.2
Total solids (g/L)	52.7	461.1	856.3
Soluble solids (g/L)	40	350	650
Insoluble solids (g/L)	12.7	111.1	206.3
Nitrogen (g/L N)	0.48–0.71	4.2–6.2	7.8–11.5
Phosphor (g/L P_2O_5)	0.1–0.2	0.1–1.75	0.14–3.25
Potassium (g/L K_2O)	3.34–4.6	29.2–40.2	54.2–74.7
Calcium (g/L CaO)	1.33–4,57	11.6–39.9	21.6–74.2
Magnesium (g/L MgO)	0.58–0.7	5.1–6.1	9.4–11.3
Sulfate (g/L SO_4)	3.7–3.73	0.6–0.76	3.70–3.73
Carbon (g/L C)	11.2–22.9	32.3–32.6	60.1–60.6
Vinasse/ethanol relation	12	14	0.74
Organic matter (g/L)	63.4	19.5	38
Reduced substances (g/L)	9.5	7.9	8.3

BOD, Biochemical oxygen demand; *COD*, chemical oxygen demand.

Vinasse concentration

Besides environmental issues, high transportation costs to bring vinasse to crop areas and excess of liquid applied to the soil are some of the major problems in fertigation practice. In this case, the concentration of vinasse stands out as a solution, decreasing its volume and recovering water for use in different applications. Table 10.4 presents the characteristics of in natura and concentrated vinasse at 35 and 65 Brix. The Brix index (1 g-solute/100 g-solution) is an indicator commonly used in the sugarcane sector to represent the concentration of solids contained in a solution (Peiter et al., 2019).

Evaporation

The concentration of sugarcane vinasse by evaporation is an alternative for the use of this residue, since fertigation cannot always dispose off total volume of vinasse produced. The product obtained in this process is used in the production of livestock feed and to improve the quality of vinasse as a fertilizer. It can also be burned in special boilers generating energy or decreasing the water use in the facility, and the condensate removed by evaporation can be treated and reused by the factory (Christofoletti et al., 2013).

Evaporation is a technology commonly used in sugarcane mills to concentrate juice to produce sugar and can be extended to vinasse as it is well established in the sector. However, the obstacle of evaporation technology is the high requirement of steam, as it is one of the operation units that consume more energy in sugar and alcohol plants. These systems bring significant additional thermal dissipation to distilleries. To reduce this depletion, several strategies are used, such as implementing multiple-effect evaporators (Madaeni and Zereshki, 2010; Pina et al., 2017).

In a multiple-effect evaporator, the evaporators are assembled in sequence, where the evaporated vapor from the first effect is the energy source for the second effect and so on. The evaporated water from the last effect passes through a condenser, finishing the process. The amount of energy saved is defined by the ratio between the total evaporated water and the steam provided for the first effect. The multiple-effect configurations can also save refrigerated water in the condenser because this equipment operates only for condensing the vapor generated in the last effect (Carvalho and Silva, 2011; Cortes-Rodríguez et al., 2018).

The concentration of vinasse in natura by evaporation requires a stainless steel evaporator due the presence of high H_2S content. The requirement of large areas, vast energy consumed, and fouling are the primary technical problems related to evaporator systems.

Concentration of biodigested vinasse

The abovementioned drawbacks in evaporation for vinasse concentration have motivated the search for new methods of concentration, such as membrane filtration. Some studies have addressed the use of membrane processes, such as microfiltration and reverse osmosis, for that purpose (Amaral et al., 2016; Madaeni and Zereshki, 2010) since this technology involves less energy and smaller footprint than evaporators do. However, membrane performance can be affected by fouling formation when significant levels of solids and organic matter are present in the liquid feed, as in the case of vinasse (Peiter et al., 2019).

Given the hindrances in concentrating in natura vinasse, the incorporation of an anaerobic bioreactor could be favorable as a pretreatment step for the membrane filtration process. The scheme in Fig. 10.3 embodies a reasonable configuration for a vinasse biorefinery composed of an anaerobic reactor and a reverse osmosis unit. Since vinasse leaves the distillation at high temperatures, a storage tank for cooling it should be employed to bring the liquid to a temperature within the recommended range for

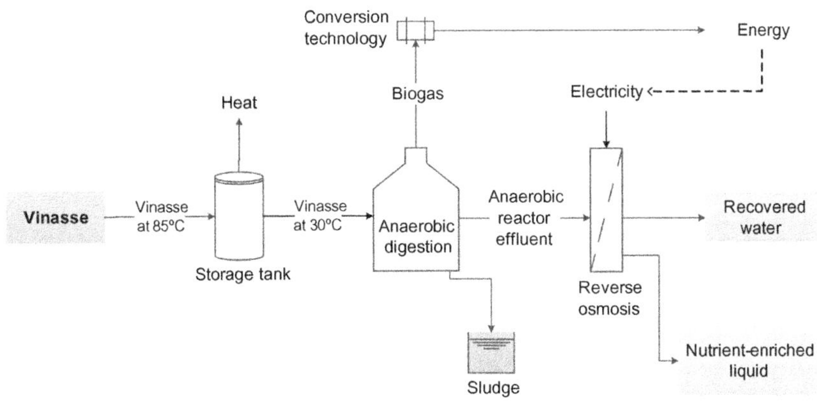

Figure 10.3 Scheme of a vinasse biorefinery.

anaerobic digestion. Even the evaporators could constitute a concentration step after the anaerobic digestion process. The method applied will depend on the characteristics of the enterprise and the region in which the biorefinery will be implemented. In some cases, as in developed countries, the evaporators are more attractive with well-known technology, and reverse osmosis, despite being a promising technology, is still very expensive (Peiter et al., 2019).

Other uses

The concentrated vinasse also can be used to direct combustion in boilers, but needs purification to eliminate the H_2S, with exception to the biodigested vinasse.

The aerobic fermentation of vinasse with yeasts provides a compound rich in proteins, amino acids, and vitamins, which can serve as a promising alternative raw material for animal feed production. Besides the *Saccharomyces cerevisiae*, other yeast species may be used (such as *Torula utilis*, *Candida utilis*, *C. solani*, *C. tropicalis*, *C. javanica*, *C. brumpti*, and *C. macedoniensis*). The yield obtained is about 840 g of final product for every 100 L of vinasse (Rocha et al., 2012).

Perspectives of the vinasse sugarcane use in Brazil

Increasing exploitation of energy and materials has motivated the search for alternative sources to avoid scarcity of existing natural resources. There are social and political incentives for companies to adapt their scope to fit this premise, reducing environmental impacts arising from their processes. In this sense, the concept of biorefinery to recover resources from wastewater has been gaining increasing attention. If applied to vinasse treatment, a biorefinery can add more value to products from the sugar and alcohol industry (Gupta and Verma, 2015; Osaki and Seleghim, 2017).

The retrieval of energy and the production of diverse products, including the use of vinasse and other wastes, would be an application of the currently important concepts of biorefinery and sustainability for the Brazilian ethanol industry (Moraes et al., 2015).

The use of concentrated vinasse and its anaerobic digestion are excellent ways of handling vinasse to mitigate the environmental and logistic problems, besides generating new economic opportunities to the mills. Concentration decrease in vinasse volume by using technologies such as evaporation, membrane filtration, and anaerobic digestion reduces the organic matter load and produces methane gas that can be used as fuel.

The concentrated vinasse can be used as a fertilizer. Nowadays the high cost of chemical fertilizers and the threats to environment have become important impetus to study the recycling of large quantities of organic residues produced as by-products of alcohol agroindustries. Fortunately it is recognized that large quantities of vinasse are low cost materials that can be processed into organic fertilizers and used as soil improvers through composting, fermentation, and a series of granulating processes.

In addition to producing heat and electricity, the use of biogas in the plant's boilers can reduce the volume of bagasse consumed by the mills, releasing it for other uses, such as sales or application in the production of cellulosic or second-generation (2G) ethanol.

Considering that biorefineries operate for about 200 days a year, pausing in the interharvest period, there are no studies examining the successful restart of the thermophilic treatment plant, the role of the microbial community in the anaerobic digestion of vinasse, and persistence of these communities within the reactors during the period in which the plant remains stationary (Ferraz Júnior et al., 2016).

Although the literature has discussed the sugarcane vinasse biodigestion at length, there is a lack of studies that analyze the different possible technological arrangements for its treatment. This analysis is important to help industry decision-makers in choosing the configuration to be implemented. Therefore an assessment of these technologies can provide a comprehensive indication of opportunities and advantages in terms of implementing an efficient system designed to reclaim resources from vinasse.

References

Amaral, M.C.S., Andrade, L.H., Neta, L.S.F., Magalhaes, N.C., Santos, F.S., Mota, G.E., et al., 2016. Microfiltration of vinasse: sustainable strategy to improve its nutritive potential. Water Sci. Technol. 73, 1434–1441.

Carvalho, T.C., Silva, C.L., 2011. Reduction of vinasse volume by the evaporation process. In: 21st Brazilian Congress of Mechanical Engineering, Natal, Brazil.

CETESB, 2006. Norma técnica P 4.231. Vinhaça: critérios e procedimentos para aplicação no solo agrícola. São Paulo, CETESB, 12 p.

Christofoletti, C.A., Escher, J.P., Correia, J.E., Marinho, J.F.V., Fontanetti, C.S., 2013. Sugarcane vinasse environmental implications of its use. Waste Manage. 33, 2752–2761.

Cortes-Rodríguez, E.F., Fukushima, N.A., Palacios-Bereche, R., Ensinas, A.V., Nebra, S. A., 2018. Vinasse concentration and juice evaporation system integrated to the conventional ethanol production process from sugarcane – heat integration and impacts in cogeneration system. Renew. Energ. 115, 474–488.

Cortez, L.A.B., Freire, W.J., Rosillo-Calle, F., 1998. Biodigestion of vinasse in Brazil. Int. Sugar J. 100 (1196), 403–413.

Cortez, L.A.B., Silva, A., de Lucas junior, J., Jordan, R.A., de Castro, L.R., 2007. Biodigestão de Efluentes. In: Cortez, L.A.B., Lora, E.S. (Eds.), Biomassa para Energia. Editora da UNICAMP, Campinas, pp. 493–529.

Ferraz Júnior, A.D.N., Koyama, M.H., Araújo JR, M.M., Zaiat, M., 2016. Thermophilic anaerobic digestion of raw sugarcane vinasse. Renew. Energ. 89, 245–252.

Fuess, L.T., Klein, B.C., Chagas, M.F., Rezende, M.C.A.F., Garcia, M.L., Bonomi, A., et al., 2018. Diversifying the technological strategies for recovering bioenergy from the two-phase anaerobic digestion of sugarcane vinasse: an integrated technoeconomic and environmental approach. Renew. Energ. 122, 674–687.

Guiot, S., Pauss, A., Costerton, J., 1992. A structured model of the anaerobic granule consortium. Water Sci. Technol. 25, 1–10.

Gupta, A., Verma, J.P., 2015. Sustainable bio-ethanol production from agro-residues: a review. Renew. Sust. Energ. Rev. 41, 550–567.

Laime, M.O., Fernandes, P.D., Oliveira, D.C.S., Freire, E.A., 2011. Possibilidades tecnológicas para a destinação da vinhaça: uma revisão. Rev. Tróp. - Ciênc. Agr. Biol. 5 (3), 17.

Leitão, R.C., Van haandel, A.C., Zeeman, G., Lettinga, G., 2006. The effects of operational and environmental variations on anaerobic wastewater treatment systems: a review. Bioresour. Technol. 97, 1105–1118.

Madsen, M., Holm-Nielsen, J.B., Esbensen, K.H., 2011. Monitoring of anaerobic digestion processes: a review perspective. Renew. Sust. Energ. Rev. 15, 3141–3155.

Madaeni, S.S., Zereshki, S., 2010. Energy consumption for sugar manufacturing. Part I: evaporation versus reverse osmosis. Energ. Convers. Manage. 51 (6), 1270–1276.

Mauad, F.F., Ferreira, L.C., Trindade, T.C.G., 2017. Energia renovável no Brasil: análise das principais fontes energéticas renováveis brasileiras. EESC/USP, São Carlos, 349 p.

Moraes, B.S., Zaiat, M., Bonomi, A., 2015. Anaerobic digestion of vinasse from sugarcane ethanol production in Brazil: challenges and perspectives. Renew. Sust. Energy Rev 44, 888–903.

Mosey, F., 1982. New developments in the anaerobic treatment of industrial wastes. Water Pollut. Control 81, 540–552.

O'flaherty, V., Collins, G., Mahony, T., 2006. The microbiology and biochemistry of anaerobic bioreactors with relevance to domestic sewage treatment. Rev. Environ. Sci. Biotechnol. 5, 39–55.

Osaki, M.R., Seleghim, P., 2017. Bioethanol and power from integrated second generation biomass: a Monte Carlo simulation. Energ. Convers. Manage. 141, 274–284.

Peiter, F.S., 2018. Análise comparativa de tecnologias de concentração: design de biorrefinarias recuperação de recursos da vinhaça. Tese (Doutorado em Engenharia Hidráulica e Saneamento) – Escola de Engenharia de São Carlos, Universidade de São Paulo, São Carlos, 160 f.
Peiter, F.S., Hankins, N.P., Pires, E.C., 2019. Evaluation of concentration technologies in the design of biorefineries for the recovery of resources from vinasse. Water Res. 157, 483–497.
Pina, E.A., Palacios-Bereche, R., Chavez-Rodriguez, M.F., Ensinas, A.V., Modesto, M., Nebra, S.A., 2017. Reduction of process steam demand and water-usage through heat integration in sugar and ethanol production from sugarcane–evaluation of different plant configurations. Energy 138, 1263–1280.
Rabelo, S.C., Costa, A.C., da, Rossel, C.E.V., 2015. Industrial waste recovery. In: Santos, F., Borém, A., Caldas, C. (Eds.), Sugarcane: Agricultural Production, Bioenergy, and Ethanol. Elsevier Inc, pp. 365–381.
Rocha, M.H., 2009. Uso da análise do ciclo de vida para a comparação do desempenho ambiental de quatro alternativas para tratamento da vinhaça. Dissertação. Universidade Federal de Itajubá - Instituto de engenharia mecânica, Itajubá, 263 p.
Rocha, M.H., Neto, A.E., Salomon, K.R., Lora, E.E., Venturini, O.S., del Olmo, O.S., et al., 2012. Resíduos da produção de biocombustíveis: Vinhaça e glicerina. In: Lora, E.E., Venturini, O.S. (Eds.), Biocombustíveis. InterCiência, Rio de Janeiro, pp. 692–809.
Salomon, K.R., Lora, E.E.S., 2009. Estimate of the electric energy generating potential for different sources of biogas in Brazil. Biomass Bioenerg. 33, 1101–1107.
Salomon, K.R., Lora, E.E.S., Rocha, M.H., del Olmo, O.A., 2011. Cost calculations for biogas from vinasse biodigestion and its energy utilization. Sugar Ind 136, 217–223.
Soares, M.R., Casagrande, J.C., Nicoloso, R.S., 2014. Uso da vinhaça da cana-de-açúcar como fertilizante: eficiência agronômica e impactos ambientais. In: first ed Palhares, J. C.P., Gebler, L. (Eds.), Gestão Ambiental na Agropecuária, 2. Embrapa, Brasília, pp. 145–198.
Speece, R.E., 1996. Anaerobic Biotechnology for Industrial Wastewaters. Archae Press, Nashville, 416 p.
Szymanski, M.S.E., Balbinot, R., Schirmer, W.N., 2010. Biodigestão anaeróbica da vinhaça: aproveitamento energético do biogás e obtenção de créditos de carbono – estudo de caso. Sem.: Ciên. Agr. 31, 901–912.
Wirth, R., Kovacs, E., Maroti, G., Bagi, Z., Rakhely, G., Kovacs, K., 2012. Characterization of a biogas-producing microbial community by short-read next generation DNA sequencing. Biotechnol. Biofuels 5, 41–49.

Further Reading

Hankins, N.P., Singh, R., 2016. Emerging Membrane Technology for Sustainable Water Treatment. Elsevier, 480 p.
Lin, H., Peng, W., Zhang, M., Chen, J., Hong, H., Zhang, Y., 2013. A review on anaerobic membrane bioreactors: applications, membrane fouling and future perspectives. Desalination 314, 169–188.

CHAPTER 11

Production of second-generation ethanol from sugarcane

Fernando Santos, Paulo Eichler, José Humberto de Queiroz and Fernando Gomes

Contents

Structural organization of lignocellulosic biomass 197
Recalcitrance of lignocellulosic biomass 200
Production of second-generation ethanol from sugarcane 202
Pretreatments 204
Types of pretreatments 206
  Physical pretreatments 206
  Chemical pretreatments 206
  Biological pretreatments 208
  Combined pretreatments 208
Hydrolysis 212
  Acid hydrolysis 213
  Enzymatic hydrolysis 214
Fermentation 217
  Fermentative microorganisms 217
  Ways of conducting the fermentative process 218
  Presaccharification 220
  Simultaneous saccharification and cofermentation 220
  Consolidated bioprocess 221
Inclusion of second-generation ethanol production annex to standard distillery 221
References 223
Further reading 227

With the impetus to replace fossil fuels (oil, coal, and gas), at least partially, interest in alternative sources of energy, including biomass, solar, wind, hydroelectric, geothermal, etc., has arisen in recent years. Among these alternative sources, biomass has been extensively studied, especially for its potential to produce solid fuels, liquids, electricity, biogas, bio-oil, and various chemical products. For this reason, most developed or

Sugarcane Biorefinery, Technology and Perspectives
DOI: https://doi.org/10.1016/B978-0-12-814236-3.00011-1

developing countries, such as Brazil, are mobilizing large investments so that biomass has a significant role in the energy matrix, especially in obtaining biofuels that are renewable and sustainable.

As a result of the photosynthesis reaction, the biomass depends essentially on solar energy, water, and carbon dioxide (CO_2). Under these conditions, tropical and subtropical regions are favored because of the greater availability of solar energy, arable land, and water (Matsuoka et al., 2012). These conditions are perfectly met by Brazil, which still has the advantage of having 27% of potential arable land in the world, but uses only about 8% of its land for cultivation (Embrapa, 2019; Conway, 2003).

Biomass can be obtained from nonwoody vegetables, woody vegetables, organic residues, as well as from biofluids (Fig. 11.1) (Cortez et al., 2008). The products obtained depend on both the biomass used and the processing technology.

Among the main sources of biomass are forests, especially eucalyptus, sugarcane, and elephant grass, plants that develop well in the tropics and subtropics because they have high photosynthetic efficiency for their C4 photosynthetic metabolism. The high biomass accumulation rate is due to the intense photosynthetic activity and the long vegetative period (Santos et al., 2012).

In the context of the production of liquid fuels, several countries now consider the use of biomass for the production of fuel ethanol, a promising alternative capable of leveraging economic development and, above all, contributing significantly to the reduction of greenhouse gases (GHGs). GHG emissions from burning biomass fuels are 75% lower when

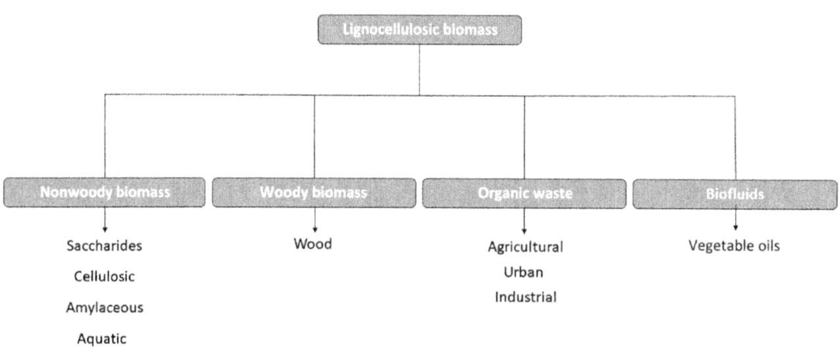

Figure 11.1 Sources of biomass. *Adapted from Cortez, L.A.B., Lora, E.E.S., Olivares Gómez, E. 2008. Biomassa para energia. Editora da Unicamp, Campinas, SP, 733 pp.*

compared to gasoline emissions, while all CO_2 produced is reabsorbed by biomass in subsequent plantations (IPCC, 2007).

For this growth to continue and improve the use of biomass efficiency, it is necessary that besides the technological advance, management models are developed that guarantee the technical, economic, and environmental sustainability of this energy source (Cortez et al., 2008).

Structural organization of lignocellulosic biomass

The main chemical components of lignocellulosic biomass are cellulose, hemicelluloses, and lignin, as well as small amounts of ash and extractives. The concentration of each of these components depends on the type of biomass, tissue type, plant age, and growth conditions (Table 11.1).

Cellulose is the most important natural polymer in the world and one of the main constituents of the cell wall of plants (Fig. 11.2). It provides strength and stiffness, preventing the cell from swelling, with consequent rupture of the plasma membrane, when osmotic conditions favor the excessive entry of water into the cell. It is a linear (nonbranched) polymer, and its structure can be classified into three organizational levels. The first is defined by the sequence of β-D-glucopyranosidic residues linked by

Table 11.1 Chemical composition of different biomass sources.

Biomass	Cellulose (%)	Hemicelluloses (%)	Lignin (%)
Sugarcane straw	40–44	30–32	22–25
Sugarcane bagasse	32–48	19–24	23–32
Hardwood	43–47	25–35	16–24
Softwood	40–44	25–29	25–31
Corn straw	35	25	35
Corn cob	45	35	15
Cotton	95	2	0.3
Wheat straw	30	50	15
Sisal	73.1	14.2	11
Rice straw	43.3	26.4	16.3
Corn forage	38–40	28	7–21
Coconut fiber	36–43	0.15–0.25	41–45
Banana fiber	60–65	6–8	5–10
Barley straw	31–45	27–38	14–19

Data from Gómez, E.O., Souza, R.T.G., Rocha, G.J.M., Almeida, E., Cortez, L.A.B., 2010. A palha de cana-de-açúcar como matéria-prima para processos de segunda geração. In: Cortez, L.A.B. (Ed.). Bioetanol de cana-de-açúcar: P&D para produtividade sustentabilidade. Edgard Blücher Ltda, São Paulo, pp. 637–659.

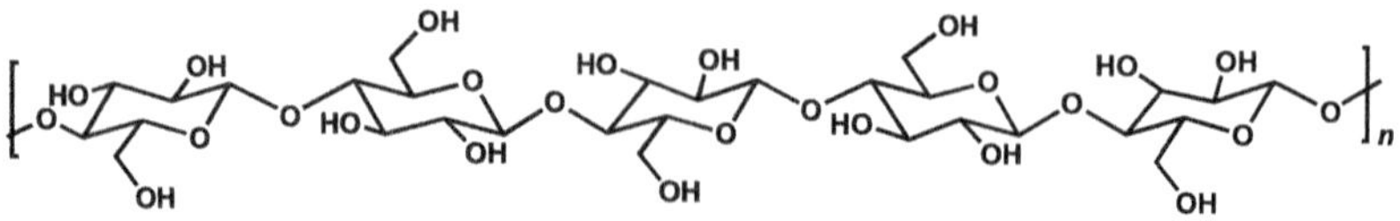

Figure 11.2 Schematic representation of the cellulose molecule. *From Santos, F.A., Queiroz, J.H., Colodette, J.L., Fernandes, S.A., Guimarães, V.M., Rezende, S.T., 2012. Potencial da palha de cana-de-açúcar para produção de etanol. Química Nova, 35 (5), 1004–1010.*

covalent bonds, forming the homopolymer of anhydroglucose with glycosidic β-(1→4) bonds of general formula $(C_6H_{10}O_5)_n$. The second level describes the molecular conformation, that is, the spatial organization of the repetitive units, and is characterized by the distances of the bonds and their angles, as well as the intramolecular hydrogen bonds. The third level defines the association of the molecules, forming aggregates with a certain crystalline structure. These aggregates confer high tensile strength, rendering the cellulose insoluble in water and in a large number of other solvents.

Hemicelluloses are complex heteropolysaccharides, with smaller chains than those of cellulose, composed of D-glucose, D-galactose, D-mannose, D-xylose, L-arabinose, D-glucuronic acid, and 4-*O*-methyl-glucuronic acid. Its chemical composition varies according to the plants, the type of tissue, the stage, the growing conditions, and the storage sites. Both xylans and mannan are found in conifers and hardwoods, but the type and proportion differ significantly between them. In the conifers, the main constituents are glucomannan (10%–15%), arabinoglucouronoxilanes (7%–15%), and galactoglucomananas (5%–8%), while on the hardwoods are glycoronoxilanes (15%–35%) and glucomananas (2%–5%). In grass and cereals, arabinoxylans have been found as dominant hemicelluloses of the cell wall of the supporting tissues.

The structure of the hemicelluloses presents branches that interact easily with the cellulose, giving stability and flexibility to the aggregate. Compared with cellulose, they are more susceptible to acid hydrolysis as they offer greater accessibility to the mineral acids commonly used as catalysts. This reactivity is usually attributed to the amorphous character of these polysaccharides.

Lignin, after cellulose, is the most abundant macromolecule among biomasses (Fig. 11.3). It is an amorphous heteropolymer consisting of three different phenylpropane units: *p*-coumaril alcohol, coferyl alcohol, and synapyl alcohol. The lignin structure is not homogeneous and there

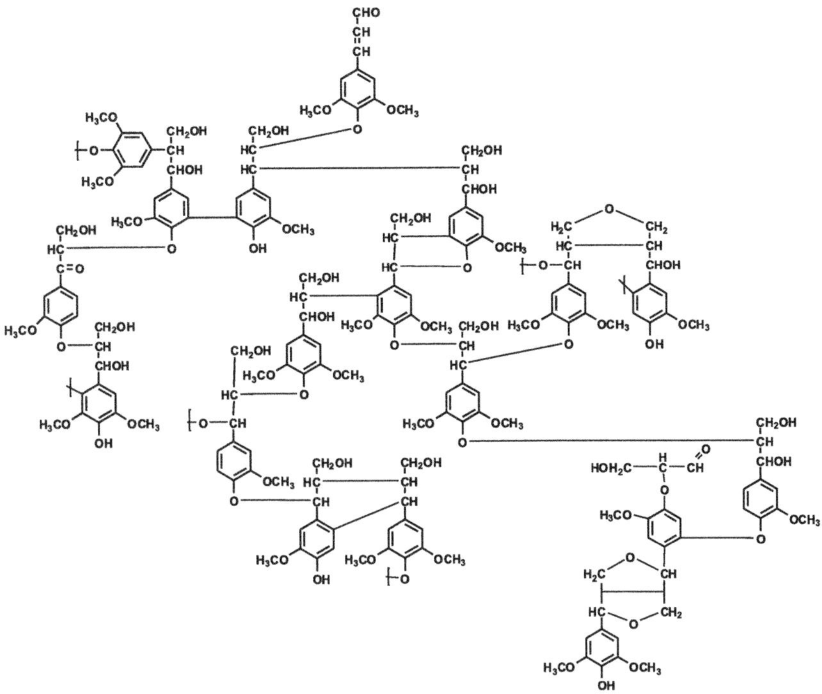

Figure 11.3 Schematic representation of eucalyptus lignin. *From Santos, F.A., Queiroz, J.H., Colodette, J.L., Fernandes, S.A., Guimarães, V.M., Rezende, S.T., 2012. Potencial da palha de cana-de-açúcar para produção de etanol. Química Nova 35 (5), 1004–1010.*

are amorphous regions and globular structures. In addition to lignin content varying between conifers, hardwoods, and grasses, the distribution of H (*p*-hydroxyphenyl), G (guaiacyl), and S (syringyl) units also vary. In hardwoods, lignins consist of units of G and S, as well as some traces of H, while conifers are mainly composed of G units. However, grasses incorporate larger amounts of H. In the enzymatic hydrolysis process, lignin acts as a physical barrier to enzymes that may be irreversibly captured by lignin and hence influence the amount of enzyme required for the hydrolysis as well as hinder recovery of the enzyme after hydrolysis.

There are different types of binding between complexes: lignin, phenolic compounds, and carbohydrates, such as (1) ester bond, (2) ether bond, (3) hydroxycinnamic ester acid, (4) hydroxycinnamic ether, (5) "bridge" ferulic acid, (6) "bridge" dihydrodifferulic ester, and (7) "bridge" di-ester dihydrodiferulic ether. Understanding the distribution of lignin and carbohydrates in the cell wall can make the fractionation and conversion processes more efficient.

The difference in the chemical composition of the lignocellulosic biomass refers not only to the amount of noncellulosic polysaccharides, but also to their type and also to the abundance of proteins and phenolic compounds.

The extractives are organic compounds that are deposited on the lume of cells and in other voids. They have low molecular weight and are soluble in water, neutral organic solvents, or volatilized. They can be extracted using polar and nonpolar solvents. The extractives are often responsible for certain characteristics of the plant, such as color, smell, natural resistance to rot, taste, and abrasive properties. In extractives, terpenoids and steroids, fats, waxes and phenolic substances, including stilbenes, lignans, tannins, and flavonoids can be found. In addition to extractables are also found in the cell wall inorganic compounds (ashes), proteins, and pectic substances. The sum of these components varies in each species of lignocellulosic material and represents approximately 5%–20% of all material.

Recalcitrance of lignocellulosic biomass

Lignocellulosic biomass is the largest source of natural carbohydrates in the world. The difficulty in converting lignocellulosic biomass into liquid fuel is attributed to its recalcitrance, since they are made up of cellulose fibers wrapped in an amorphous matrix of hemicelluloses and lignin. This amorphous matrix acts as a natural barrier to attack from microorganisms and/or enzymes, making these materials structurally rigid and unreactive.

As with all plant cells, the cell wall of biomass is composed mainly of cellulose, hemicelluloses, and lignin, as already mentioned (Fig. 11.4). Cellulose reactivity is determined by the functional groups present in its structure and by the physicochemical interactions between the macromolecules (cellulose, hemicelluloses, and lignin).

The intramolecular hydrogen bonds between the hydroxyls provides cellulose resistance. The intermolecular interactions between hydroxyls are responsible for the formation of the vegetal fiber, in which six to eight molecules of cellulose line up parallel to form a microfibril, eliminating the water molecules and making the fiber extremely long and resistant (Fig. 11.5). On the surface of the microfibrils, the hemicelluloses that cover the cellulose are adhered, forming the so-called cellulose–hemicellulose domain of the cell wall (Buckeridge et al., 2010).

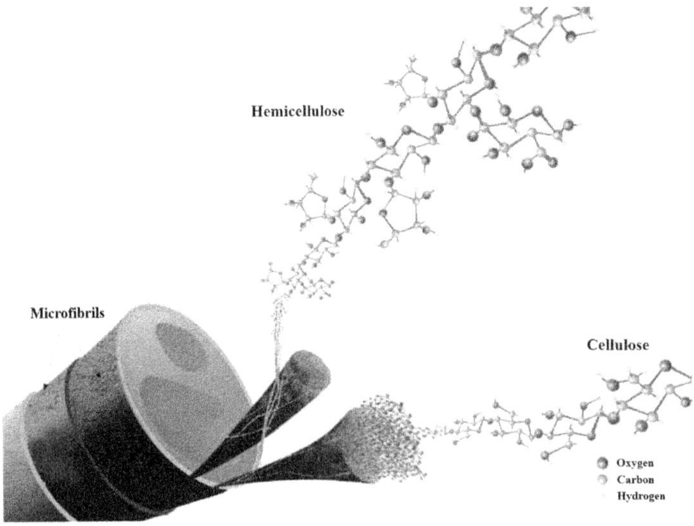

Figure 11.4 Recalcitrant structure of lignocellulosic biomass. *Adapted from Buckeridge, M.S., Santos, W.D., Souza, A.P., 2010. As rotas para o etanol celulósico no Brasil. In: Cortez, L.A.B. (Ed.). Bioetanol de cana-de-açúcar: P&D para Produtividade Sustentabilidade. Edgard Blücher Ltda., São Paulo, pp. 365–380.*

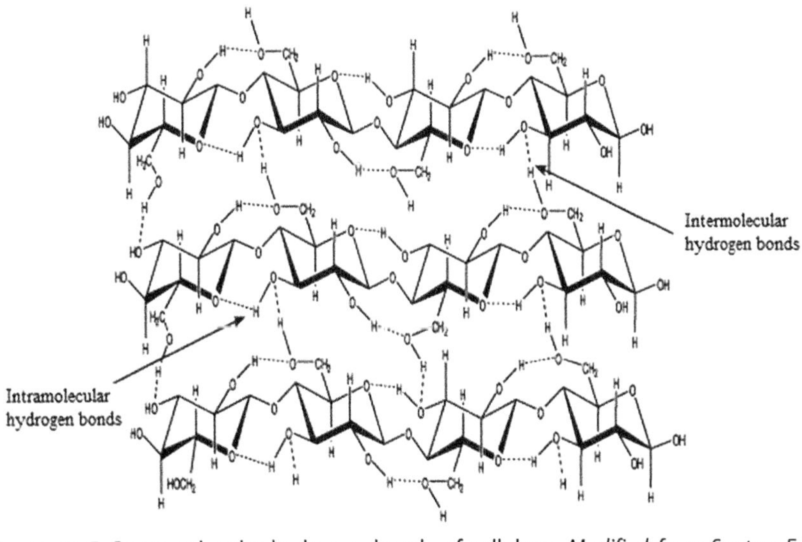

Figure 11.5 Supramolecular hydrogen bonds of cellulose. *Modified from Santos, F.A., Queiroz, J.H., Colodette, J.L., Fernandes, S.A., Guimarães, V.M., Rezende, S.T., 2012. Potencial da palha de cana-de-açúcar para produção de etanol. Química Nova 35 (5), 1004–1010.*

The hemicelluloses prevent cellulose molecules of parallel fibers from collapsing among themselves, but also allow weak interaction between one fiber and another, forming a network. The cellulose–hemicelluloses domain is immersed in a domain formed by pectins, that is, highly branched sugars, which determine porosity of the cell wall (Buckeridge et al., 2008).

Inter- and intramolecular hydrogen bonds are responsible for the maintenance of the crystalline regions and make the cellulose highly resistant to acid, alkaline, or enzymatic hydrolysis. The formation of hydrogen bonds is the main factor influencing the physical properties of cellulose, such as solubility, reactivity, and crystallinity. The interaction between the hydroxyl groups via hydrogen bonds between the chains limits the access of agents to the functional groups of the chain, which hinders both their solubilization and reactions.

The presence of hemicelluloses and lignin involving the cellulose chains also makes difficult the access of reagents and catalysts, being necessary treatment for the removal of these components, facilitating the access of the enzymatic complex to the cellulose. Several authors have attributed the recalcitrance of the lignocellulosic biomass, mainly to the crystallinity, the degree of polymerization, and the low accessibility of the cellulolytic enzymes to the substrate (Buckeridge et al., 2010; Chang and Holtzapple, 2000).

Production of second-generation ethanol from sugarcane

The energy production based on lignocellulosic biomass is an important route that has been studied worldwide. The interest in obtaining the second-generation ethanol has been growing a lot, as a consequence of cellulose being the substance with the highest concentration in biomass and presenting high efficiency for ethanol production when submitted to hydrolysis reactions.

In the process of obtaining second-generation ethanol, the objective is to deconstruct the cell wall structures, to obtain the polysaccharides as a source of fermentable sugars. However, it has already been mentioned how complex the recalcitrant structure of lignocellulosic biomass is.

Considering the great availability of biomass produced by the agroenergy industry, bagasse, and straw stand out in the production of second-generation ethanol. Bagasse is a promising material for ethanol production as it is available at the plant as it is a "by-product" of the first-generation ethanol production process, and the best cogeneration technologies allow the generation of larger quantities of energy from excess bagasse. It is also important to consider that, with the gradual end of the burning of sugar-cane, there will be a significant increase in the quantity and availability of straw in the field.

The average productivity of sugarcane per hectare in the world is 70 tonnes per hectare. And for each tonne of processed cane, about 140 kg of straw and 140 kg of bagasse are generated on a dry basis, that is, about 10 tonnes of straw and 10 tonnes of bagasse per hectare. Considering that both bagasse and straw have 40% cellulose and that all glucose will be converted to ethanol, the full use of sugarcane (culm, straw, and bagasse) can significantly increase ethanol production per hectare, passing off the current 7000 L to approximately 13,000 L without the need to expand the cultivated area, in addition to being able to reduce production costs (Fig. 11.6) (Santos et al., 2012). In energy terms, bagasse and straw

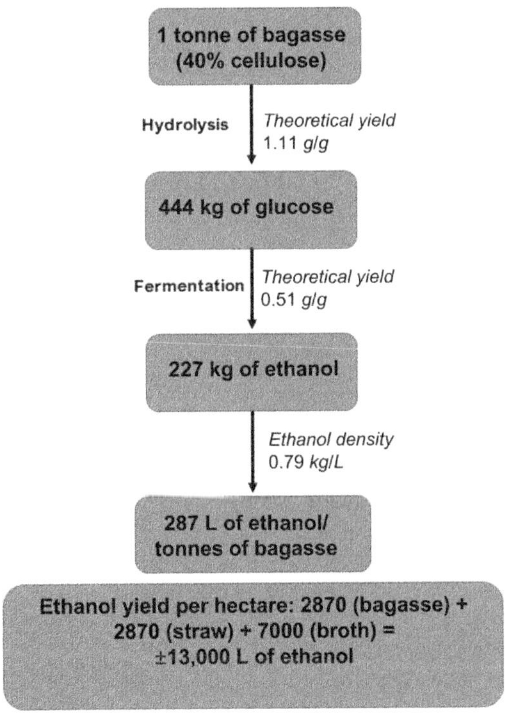

Figure 11.6 Theoretical yield of second-generation ethanol production per hectare of processed cane. *Adapted from Santos, F.A., Queiroz, J.H., Colodette, J.L., Fernandes, S.A., Guimarães, V.M., Rezende, S.T., 2012. Potencial da palha de cana-de-açúcar para produção de etanol. Química Nova 35 (5), 1004–1010.*

represent approximately two-thirds of the potential energy of sugarcane (Magalhães and Braunbeck, 2010; Rabelo et al., 2012).

Nowadays technologies to obtain second-generation ethanol involve (1) pretreatment, which aims to alter or remove lignin and hemicelluloses, increasing the surface area, and decreasing the degree of polymerization and crystallinity of the cellulose; (2) the hydrolysis step, to obtain the sugars from the cellulose and hemicelluloses; and (3) the conversion of these sugars into ethanol via fermentation and distillation (Fig. 11.7) (Santos et al., 2012).

Pretreatments

Pretreatment consists of one of the most relevant operational steps in terms of direct cost, as well as significantly influencing the costs of previous and subsequent steps. Basically pretreatment is required to modify the biomass structure in order to make the cellulose more accessible to the enzymes that convert the carbohydrate polymers into fermentable sugars. Efficient pretreatment can substantially reduce the amount of enzymes needed in the cellulose saccharification process and hence the production costs. According to Galbe and Zacchi (2010), an effective pretreatment must obey a series of desirable characteristics:

- low investment and operational cost for pretreatment;
- minimum waste production;
- rapid reactions and/or noncorrosive chemical reagents to minimize reactor cost;
- high product returns on subsequent operations with minimal costs;
- low enzyme loading adequate to achieve digestibility of greater than 90% of the pretreated cellulose; and
- facilitate the recovery of lignin and other constituents for conversion into high added value coproducts and purification steps.

In general terms the pretreatment of sugarcane (bagasse and straw), aiming at the production of second-generation ethanol, must present a cellulosic pulp with high accessibility and fiber reactivity to acidic or enzymatic hydrolytic agents, assure adequate recovery of pentoses and limit the formation of inhibitory compounds to yeasts and enzymes, associated with the low cost involved in the pretreatment process. Fig. 11.8 illustrates the effect of pretreatment on biomass.

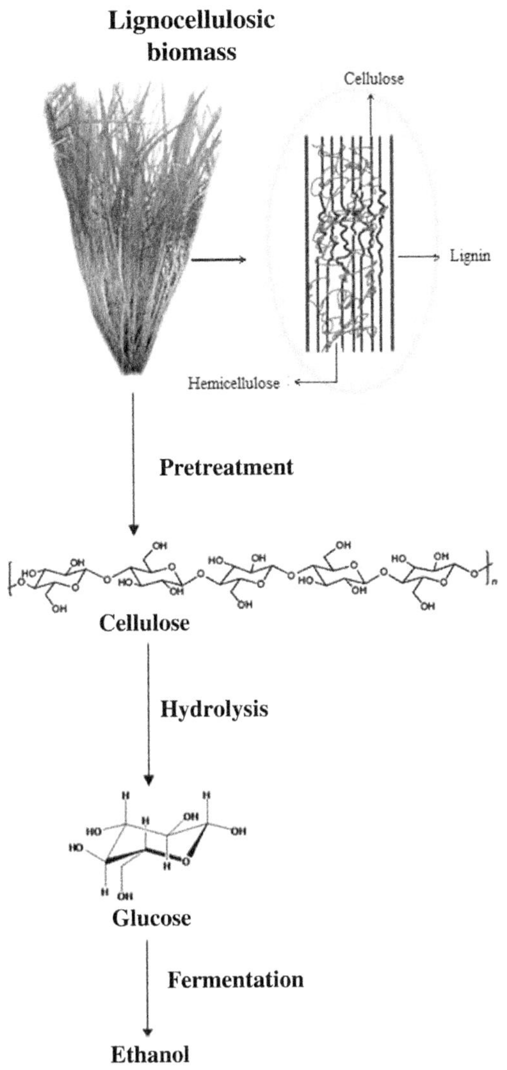

Figure 11.7 Schematic representation of second-generation ethanol production. *Modified from Santos, F.A., Queiroz, J.H., Colodette, J.L., Fernandes, S.A., Guimarães, V. M., Rezende, S.T., 2012. Potencial da palha de cana-de-açúcar para produção de etanol. Química Nova 35 (5), 1004–1010.*

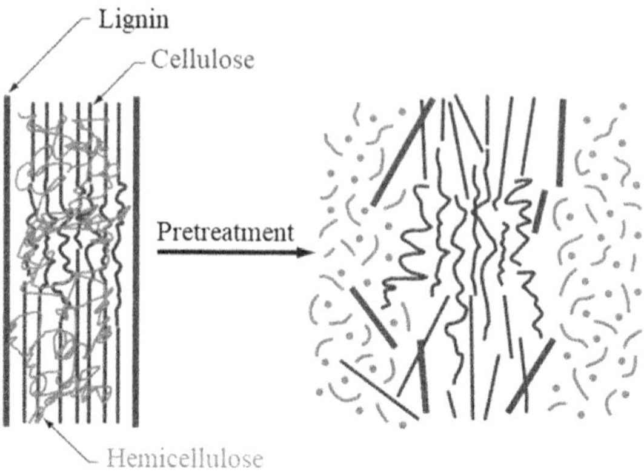

Figure 11.8 Effect of pretreatment on lignocellulosic biomass. *Adapted from Genomics. Department of Energy Genome Programs. Disponível em: <www.genomics.energy.gov> (accessed 10.12.).*

Types of pretreatments

Many methods of pretreatment have been suggested over the last few decades, and each has its advantages and disadvantages. They can be basically divided into the following categories: physical, chemical, biological, or a combination of these methods.

Physical pretreatments

They are usually divided into two categories: mechanical and nonmechanical. In mechanical (grinding and milling), impact forces reduce particle size and crystallinity and increase specific surface and bulk density. In nonmechanical pretreatments, the cellulosic substrate is subjected to the action of external agents in order to cause changes in the structures of the original material. This pretreatment has the disadvantage of inefficiency in increasing biomass digestibility and the consumption of large amounts of energy in the process. Physical pretreatment combined with heating and adding chemicals can be an interesting option.

Chemical pretreatments

They are those in which chemical agents such as acids, bases, and organic solvents are used. They are intended to increase the surface of the substrate by swelling the fibers and the modification or removal of hemicelluloses and/or lignin in order to make the cellulose more accessible to the enzymes.

Acid pretreatment consists of placing the biomass in direct contact with inorganic, diluted, or concentrated acids to solubilize the hemicelluloses in order to obtain more accessible cellulose. The most common diluted acid pretreatment consists of immersing the material in an acid solution of approximately 4% (w/w) and then heating to temperatures in the range of 140°C–200°C for a period of several minutes to 1 hour. Different reactions occur during pretreatments with acids, and one of them is the hydrolysis of hemicelluloses, especially xylans and glycans. Depending on the pretreatment conditions, there may be formation of furfural and hydroxymethylfurfural by the degradation of pentoses and hexoses, respectively, which may negatively influence the fermentation step (Palmqvist et al., 2000). An advantage of this pretreatment is the solubilization of hemicelluloses, increasing the accessibility of the cellulose to the enzymes, and a higher yield compared to the basic hydrolysis. On the other hand, there is a risk of corrosion of the process equipment, as well as the formation of volatile carbon degradation products, which in many cases slow the conversion to ethanol. However, volatile products can be converted to methane. The condensation and precipitation of solubilized lignin components is also an undesired reaction as it decreases digestibility. Pretreatments with concentrated acid for the production of ethanol are not attractive processes, due to the great risk of production of inhibitory components by the degradation of carbohydrates. In a study of optimum pretreatment conditions that used dilute sulfuric acid (H_2SO_4), the best condition was found at 155°C, using H_2SO_4 at 1.5% in 25 minutes of reaction. This same study used these conditions showing an economic feasibility study of a theoretical ethanol production plant reaching production values of 187.5 L of ethanol per tonne of raw biomass of sugarcane straw, one of the values found in the literature (Mesa et al., 2017).

In alkaline pretreatment, dilute alkaline solutions are generally used under mild conditions compared to acidic systems. The main effect of this pretreatment is the removal of lignin. Soaking the material in an alkaline solution, such as sodium, potassium, or ammonium hydroxide, followed by heating, leads to the decrease in the crystallinity degree of polymerization of cellulose. It also causes breakdown of lignin–carbohydrate bonds, as well as disruption of the lignin structure. In some cases, it can be conducted at room temperature, but this requires high reaction times of the order of hours, days, or weeks. Unlike acid pretreatments, a limitation occurs because some bases can be converted into salts unrecoverable or incorporated as salts in the biomass through the pretreatment reactions

(Chang and Holtzapple, 2000). It is also worth noting that this method is more effective in agricultural residues and grasses than in woody materials, as these materials generally contain less lignin. For woody materials, the alkali concentration must be increased considerably, so that the process is similar to kraft pulping (Galbe and Zacchi, 2010). A dilute alkaline pretreatment is normally used when it is known that there may be a large increase in the yield of ethanol production. Using an alkaline pretreatment with 3% NaOH at 60°C for 9 hours, it was possible to obtain the highest glucose recovery (20.6%) and a high hydrolysis efficiency (72%) using a high cellulose grass (42.7%) (Siripong, 2018).

Ionic liquids have also been recently studied as a pretreatment option for lignocellulosic biomass to increase saccharification yield and ethanol production. In general, ionic liquids are salts that, during their liquid state (usually at high temperatures and pressures), have the ability to break recalcitrant biomass bonds due to their strong ion–electrostatic interaction (Zhang et al., 2017). Although the use of ionic liquids in biomass has been reported since 2002, only in recent years there has been more interest in the technique due to the development of new ionic liquids and a better understanding of the pretreatment processes using this technique.

Biological pretreatments

They involve the use of microorganisms (fungi and bacteria) or enzymes (laccases) to delignify and reduce the degree of polymerization of cellulose and hemicelluloses. The advantage of these pretreatments is the low energy requirement. Although effective, clean, and without production of undesirable metabolites, the process is rather slow to be applied industrially. Another disadvantage is that these microorganisms are relatively not selective in biomass; thus besides degrading hemicelluloses and lignin, they degrade cellulose (Sun and Cheng, 2002). Some microorganisms, mainly filamentous fungi, have been studied in solid-state fermentation as a pretreatment of lignocellulosic biomass and even as precursors for the production of enzymes such as cellulases for the enzymatic hydrolysis processes (Yu et al., 2016).

Combined pretreatments

This category includes combinations of physical and chemical pretreatments, such as steam pretreatment with the addition of a catalyst (acid or alkaline). A typical pretreatment in this category is the ammonia fiber explosion (AFEX), which is also alkaline, increasing the reactivity of the cellulose fraction due to the swelling of it, combined with hydrolysis of hemicelluloses and fiber disintegration. The biomass is treated with liquid ammonia for a period of 10–60 minutes at moderate temperatures (100°C) and high pressures (30 bar). Up to 2 kg of ammonia is used per kilogram of dry biomass. One of the disadvantages of this pretreatment is the low solubilization of hemicelluloses and lignin (Fuentes, 2009). Like other alkaline pretreatment methods, AFEX is more efficient in agricultural waste and has no efficiency with wood due to its higher lignin content. Therefore in this case, ammonia recycled percolation pretreatment is more suitable for short-fiber wood and agricultural residues, since it is less efficient for long-fiber wood (Galbe and Zacchi, 2010). Another example of combined treatment is the use of continuous double screws together with some solvent (hot water or glycerol), without the need for strong chemical reagents or the need for high temperatures and pressures. Also, this pretreatment was efficient in the use in continuous systems, good for large-scale use, and with the potential to reduce environmental costs and impacts (Moro et al., 2017). Also, recently there has been interest in the use of other technologies to aid in the disintegration of biomass such as microwaves (Bezerra and Ragauskas, 2016; Chen et al., 2011), ionic liquids (Bian et al., 2014), and various combinations thereof (Moghaddam et al., 2014; Bahrani et al, 2015) to increase the yield of the pretreatment with the intention of improving the final product (ethanol).

The association between one or more pretreatments and the structure and composition of the biomass will determine the success of the pretreatment chosen as a facilitator of the hydrolysis process. Table 11.2 presents the main characteristics that differentiate some pretreatments. The pretreatments with diluted acid, calcium hydroxide, biological, steam explosion, and AFEX basically remove hemicelluloses, with minimal changes in lignin structure. However, the organosolve pretreatment removes lignin and hemicelluloses.

Knowing how to evaluate each pretreatment method for the advantages and disadvantages in the case of each biomass is essential for the viability of second-generation ethanol production. For this, economic evaluations are focused not on the determination of the cost of ethanol, but on the comparison between the technological options available.

For bagasse and cane straw, some methods have proven to be adequate, but all still need to be optimized to reduce high-cost expectations. In the agroindustry, sugarcane is subjected to a physical pretreatment at

Table 11.2 Effect of different pretreatments on biomass.

Pretreatment		Chemical characteristics			Advantages	Disadvantages
		Cellulose	Hemicellulose	Lignin		
Physical	Ball mill	Intensive decrease in crystallinity	Does not remove	Does not remove	Reduces crystallinity	High energy consumption
Chemical	Diluted acid	Little depolimerization	80%–100% removal	Little removal	Easy conditions, high production of xylose	Difficult acid recovery, corrosive and costly
	Sodium hydroxide	Significant swelling	Considerable solubility	Considerable solubilization, >50%	Effective removal of esters	Expensive reagent
	ARP Percolation with ammonia	Less than 5% depolymerization	~50% solubility	~70% solubilization	Effective delignification	Expensive reagent
	Calcium hydroxide	Low depolymerization	Significant solubilization	Partial solubilization (~40%)	Effective lignin and acetyl removal, low cost	Less effectiveness due to poor solubility of lime
	Ozonolysis	No depolymerization observed	Small solubilization	Solubilization above 70%	Effective delignification under mild conditions	Expensive, need for more ozone
	Organosolve	Significant swelling	Significant, almost complete	Significant, can be almost complete	High production of xylose,	Recovery of solvent is expensive

					effective delignification	
	Ionic liquids (ILs)	Swelling depending on IL	Dissolve depending on IL	Dissolve lignin depending on IL	Reduction of cellulose crystallinity and lignin recovery	High cost and some ILs are toxic
Biological	Biological	20%–30% depolymerization	Over 80% solubilization	~40% delignification	Low energy requirement, effective delignification	Loss of cellulose, low hydrolysis rate
Combined	Steam explosion	Little depolymerization	80%–100% removal	Little removal, but structure change occurs	Energy efficient, no cost of recycling	Degradation of xylan as an inhibitory product
	Ammonia fiber explosion	Decrease in the degree of crystallinity	Over 60% solubility	10%–20% solubilization	Less xylan loss, no formation of inhibitors	Ammonia recovery, not effective for high concentration of lignin
	Dilute acid and Microwave	Swelling occurs and also loss of crystallinity	Almost complete removal	No change in lignin structure	Effective removal of lignin and hemicelluloses	Requires a lot of energy and special equipment

Source: Adapted from Santos, F.A., Queiroz, J.H., Colodette, J.L., Fernandes, S.A., Guimarães, V.M., Rezende, S.T., 2012. Potencial da palha de cana-de-açúcar para produção de etanol. Química Nova 35 (5), 1004–1010.

the time of its production, that is, after the sugarcane milling, and becomes a heterogeneous set of particles with sizes varying between 1 and 25 mm, with an average of 20 mm. Owing to the reduced granulometry, the bagasse does not require grinding prior to the physical–chemical pretreatment, which represents an advantage in terms of raw material preparation cost. However, low density and low bagasse compaction pose a problem in terms of reactor feed operation, as well as the difficulty in conducting pretreatment with solids loads greater than 50% (Bonomi, 2010).

Bagasse has a high absorption capacity of liquids as well as reduced "hardness." In addition, the high humidity of the bagasse from the mills (45%–50%) facilitates the impregnation of this biomass with acidic and alkaline solutions. This aspect is of fundamental importance with respect to the efficiency of acid–hydrolytic pretreatment, which requires adequate concentrations of hydroxyl ions (H_3O^+), formed from water and dissociated acid. The lack of water in the biomass would result in lower formation of hydroxyl ion, as well as reduced availability of the transport fluid inside the biomass, and consequent loss of efficiency of the hydrolytic capacity. Therefore in this case, the challenge is to determine the optimum quantity of water in the bagasse, in order to guarantee effective impregnation of the biomass, while obtaining a solids load in the reactor, mainly in steam explosion processes. Autocatalytic steam pretreatments, or in the presence of catalysts, hot water and dilute sulfuric acid are more promising methods for sugarcane (Bonomi, 2010).

There are several pretreatment projects in pilot scale, in demonstration phase, and already in commercial use aiming at the production of second-generation ethanol. However, there is no "best pretreatment," that is, the most efficient for all types of lignocellulosic biomass. Each type of biomass requires a particular pretreatment method to minimize substrate degradation, maximize sugar yield and, at the same time, reduce consumption of chemical inputs and energy.

Hydrolysis

After the pretreatment step, cellulose hydrolysis is required to obtain fermentable sugars. The hydrolysis yield reaches 20% in processes where no pretreatment occurs and exceeds 90% when the process is preceded by pretreatment (Hamelinck et al., 2005). Hydrolysis is indicated as the route with the greatest potential to promote the increase of ethanol production (without expanding the agricultural frontier) as well as increase the conversion efficiency of sugarcane's primary energy. In this step, the cellulose is converted into glucose, which can be catalyzed by acids (diluted or concentrated) or by enzymes.

Acid hydrolysis

The processes of obtaining second-generation ethanol from biomass using acid catalysts allow, under appropriate pressure and temperature conditions, solubilization of hemicelluloses and cellulose, without significantly altering lignin. In acid hydrolysis process, they may be catalyzed with dilute or concentrated acids.

In concentrated acid hydrolysis, the hemicelluloses and cellulose present in the lignocellulosic biomass are hydrolyzed using aqueous solutions of strong mineral acids such as sulfuric, hydrochloric, nitric, or phosphoric acids at a moderate temperature. In this category, biomass should be dried to obtain moisture content below 10% in order to avoid acid dilution in the hydrolysis stage. In this process, the hemicellulose macromolecule is primarily hydrolyzed over cellulose, and due to this preference, monosaccharides released from hemicellulose are exposed to the reaction medium for a long time, causing their degradation and loss of these sugars. It is relatively a slow process requiring high energy demand for the recovery of the acid used for economic and environmental reasons.

The dilute acid process is the oldest technology in converting biomass into second-generation ethanol. This process involves two hydrolysis reactions. In the former, the hemicellulose fraction is converted into sugars and other coproducts using diluted acid and steam. The process temperature is in the range 100ºC−190ºC, pressure of 15 psi, and time between 10 and 30 minutes. The acid commonly used in hydrolysis is 2% (mL/mL) diluted sulfuric acid, in acidic solution ratio (using 1:30 g/mL solids ratio), with an average yield of 0.76 g/g. The conversion of sugars not only reduces yields, but furfural and other coproducts may inhibit the fermentation step (Tizazu and Moholkar, 2018). In order to remove the inhibitory substances produced in the processes, second-stage sugars are treated with calcium hydroxide and small amounts of sodium sulfate may need to be added before fermentation. On average, yields are 89% for mannose, 82% for galactose, but only 50% for glucose (Hamelinck et al., 2005).

In industrial processes, acid hydrolysis has been performed with H_2SO_4. The very timid commercialization of ethanol produced from the

Table 11.3 Hydrolysis conditions of dilute sulfuric acid.

Temperature (ºC)	Time (minutes)	Acid (%)	Maximum sugar (%)
170	30	2.2	88
140	30	1.2	94
121	60	1.2	70
180	15	0.5	76
120	60	1.0	75
140	40	1.0	86

Source: Adapted from Singh, J., Suhag, M., Dhaka, A., 2015. Augmented digestion of lignocellulose by steam explosion, acid and alkaline pretreatment methods: a review. Carbohydr. Polym. 117, 624–631.

acid hydrolysis of sugarcane bagasse is related to technical and operational difficulties that result in a high cost of the final product. Part of this cost is due to the fact that for hydrolysis to occur efficiently it is necessary to heat the polysaccharide in the acid solution. The ideal temperature for hemicelluloses' breakage is between 100°C and 120°C and the ideal H_2SO_4 concentration is around 3% (Buckeridge et al., 2010). In the specific case of sugarcane, this cost is minimized due to the fact that part of the bagasse is burned to feed the boilers and produce the electricity consumed in the process. Because the recovery of concentrated acid is very expensive in industry, hydrolysis with dilute acids is more often done, but also cost less to include a wide range of biomass such as hardwood, softwood, herbaceous, agricultural waste, and even urban solid waste. Table 11.3 shows the different yields of sugars with varying hydrolysis conditions using dilute H_2SO_4 (most common) (Singh et al., 2015). The values were taken from the literature where different biomasses were used. Thus it is understood that for each biomass there should be an optimal condition mainly due to the particularities of the recalcitrant interactions of each biomass.

Enzymatic hydrolysis

In this case, the hydrolysis is catalyzed by cellulases. It is an enzymatic complex that shows a high degree of synergism represented by endoglucanases, exoglucanases, and β-glycosides. Endoglucanases randomly hydrolyze the inner regions of the amorphous structure of the cellulose, releasing minor oligosaccharides. These enzymes are responsible for the rapid solubilization of the cellulose fiber. The products generated by the endoglucanases thus become substrates for the exoglucanases, which cleave the oligosaccharides, releasing mainly cellobiose. The main representatives of exoglucanases are cellobiohydrolases, which can act on both the reducing and nonreducing ends of cellulose. β-Glycosides prevent the accumulation of cellobiose, breaking it into glucose. However, as the concentrations of these monomers increase in the medium, the activities of the cellulases are inhibited. Cellobiose is a stronger inhibitor than glucose, and it is interesting that it is rapidly hydrolyzed to glucose by β-glucosidases (Sun and Cheng, 2002).

The enzyme may be produced in a separate reactor in the ethanol-producing unit or purchased from a supplier. In the context of the sugar-energy sector, the production of in-house enzymes (using the ethanol-producing unit), using part of the pretreated bagasse (about 30%–40%) as a substrate, emerges as a potentially interesting technological alternative. Among the advantages are the lack of transportation costs, the possibility of using diluted formulations, a lower cost of purification and concentration, a lower complexity of product conservation, and the possibility of using hemicellulosic hydrolysates in the cultivation of microorganisms (Bonomi, 2010).

Enzymatic hydrolysis has some advantages compared to the acid, for example, greater biocatalyst specificity, lower rates of glucose degradation, moderate temperature and pressure conditions, and lower separation costs due to the lower formation of inhibitory coproducts. However, it is still necessary to advance the development of enzymes to make feasible the implantation of enzymatic hydrolysis processes in commercial scale (Perdices et al., 2012).

Generally the enzymatic hydrolysis process presents advantages associated with obtaining yields greater than 0.85 g of glucose/g of cellulose, under temperatures of 40ºC–50ºC and atmospheric pressure. However, operational aspects related to the long process duration (48–72 hours), catalytic deactivation by inhibition of enzymatic activity, as well as high cost of enzymes, have led to uncertainties regarding the economic viability of the enzymatic hydrolysis process in the context of ethanol production from the lignocellulosic biomass (Bonomi, 2010).

Several factors can affect the enzymatic hydrolysis of lignocellulosic biomass. They refer to:

Substrate: Substrate concentration (one of the most important factors affecting production and the initial rate of enzymatic hydrolysis of cellulose); structural characteristics of the substrate (cellulose crystallinity, degree of polymerization of cellulose, surface area, and lignin content).

Cellulase: Concentration of cellulase (up to a certain amount, the hydrolysis rate can be improved, but this significantly increases the cost of the process); retroactive inhibition due to accumulation of glucose and cellobiose in the reaction medium and thermal denaturation of the enzymes.

Compared to the chemical routes, the enzyme presents as a more adequate alternative to the ethanol production from the bagasse, from the technical point of view, due to the greater possibility of obtaining high glycosidic yields of the order of 90%, with enzymatic charges of about 7.5 FPU/g cellulose, at the same time as hydrolysates with reduced toxicity to the fermentation microorganisms are obtained. However, in spite of these advantages, there is the challenge of making the enzymatic process economically feasible.

In short, the enzymatic hydrolysis process has an excellent potential to produce fermentable sugars from lignocellulosic biomass and can be adapted to different cases. For sugarcane, this technology is already close to becoming commercial and will be a strategic point of importance for the next generations of cellulosic ethanol. Considering the current situation, commercial viability can be expected to be close to being achieved. The development of such technology will open the way for enzymes to be used and/or the raw material modified to obtain even higher yields (Buckeridge et al., 2010). Several studies have reported an improvement in cellulase production both by solid-state fermentation (Marín, 2018), and by genetic engineering of microorganisms, where cellulase production genes can be placed inside bacteria with high growth potential and expression of these enzymes (Darabzadeh et al., 2019; Rantasalo, 2019; Yang et al., 2019). Thus there may be a decrease in enzyme values, which is ultimately the determining factor for the high price of second-generation ethanol. Currently it is possible to find prices for the purchase of cellulases in the industrial market of enzymes with an average cost of it being $2.71/gallon of ethanol produced, reaching 48% of the minimum sales value of the gallon of ethanol. With theoretical literature yields, these values can reach up to $1.47/gallon of ethanol, reaching a huge potential of price competitiveness (Farinas, 2018). These results are extremely motivating for the future second-generation ethanol market, where new technologies in the biorefinery industries can bring more savings to the processes, also generating a more competitive product in the market.

Fermentation

Fermentation is the third step in the second-generation ethanol production process. In fermentation there is the metabolization of sugars, glucose, and xylose obtained in the previous stage by the microorganisms having then as a result of the metabolization, the ethanol.

The main sugars resulting from the hydrolysis of biomass are glucose and xylose. In the organism of the yeast *Sacharomyces cerevisiae*, the glucose is captured by transporters codified by genes of the family hexoses transporters (HXT), unlike xylose for which the yeast does not have mechanism of high expressiveness. After being transported, the glucose is phosphorylated and transformed into pyruvate in the glycolytic pathway. Pyruvate is then converted to ethanol by ethanolic fermentation. Xylose, in turn, follows a different pathway in which there is little affinity for metabolization. Yeast *S. cerevisiae* does not have a specific carrier for this sugar, so in the cell, xylose is transported by the HXT. However, the affinity of these transporters for glucose is greater, making it difficult for xylose uptake by the cell in the presence of glucose. Inside the cell, xylose is transformed into xylitol and then into xylulose. The xylulose formed then enters the nonoxidative phase of the pentose phosphate pathway. The product of this pathway follows the glycolytic pathway, being transformed into pyruvate, which is converted into ethanol. *S. cerevisiae* has the genes that encode enzymes that metabolize xylose; however, its expression is very low, which prevents the conversion of xylose into ethanol (Matsushika et al., 2009; Rubio-Texeira et al., 2009; Oreb et al., 2012).

Fermentative microorganisms

Researches have been developed aiming the isolation of strains of yeasts capable of growing and fermenting at high temperatures. Among the isolated organisms, the most important are *S. cerevisiae*, *Zymomonas mobilis*, and *Kluyveromyces marxianus*.

Yeast *S. cerevisiae* is the main organism used in ethanolic fermentation in Brazil and in the world. Such yeast is capable of fermenting sucrose to ethanol with high yields, high productivity, and tolerates high concentrations of ethanol compared to other yeast strains. It also has the ability to grow at low pH, tolerance to some inhibitors, and is genetically and physiologically well-known. *S. cerevisiae* are heterotrophic yeasts, unicellular, aclorophyllates, and examples of facultative organisms, since they have aerobic and anaerobic metabolism. Therefore in anaerobiosis they "ferment," that is, they use the sugar in the medium to metabolize, generating ethanol. Besides being widely used in the fuel industry, it is also widely used in food industry, mainly for baking and production of alcoholic beverages such as beer and wine (Santos et al., 2015). As it undergoes natural processes of genetic mutation, its maintenance is extremely necessary when it is desired to obtain optimal productions of ethanol. Therefore it is always recommended that a verification of the fermentation yield of the must is made, observing when there is a decrease in productivity, and maintenance of the organism in the process is necessary. Nowadays it is also possible to make the fermentation process using the microorganisms immobilized on inert substrates. As advantages of this technique, we have low cost of production, lower cost of separating the cells from the fermented wort, and high productivity. On the other hand, it can present problems with the support due to the mass transfer in large-scale processes, abrasion of the supports, and until their breakup by the evolution of CO_2 (Xia et al., 2019).

In recent years, the species *K. marxianus* has been adopted by the industry, mainly because it has characteristics that are desirable for biotechnological applications, which include thermotolerance. In this case, some strains are able to grow at temperatures above 45°C. The ability to assimilate a wide variety of substrates is also mentioned. Many of the sugars that this species is able to assimilate are present in the lignocellulosic biomass. These characteristics, coupled with the ability of some strains to ferment xylose in second-generation ethanol, with high yields, have aroused interest in their application in ethanol production process from lignocellulosic biomass (Lane et al., 2011).

Ways of conducting the fermentative process

Enzymatic hydrolysis and fermentation may be performed separately, in the so-called SHF process, or combined, in so-called simultaneous saccharification and fermentation (SSF). Regardless of the way of conduction, it is important to maintain high carbohydrate concentration in the hydrolysis step, in order to obtain a high concentration of ethanol in the fermentation process.

Separate saccharification and fermentation

SHF is the traditional design, in which the hydrolysis of the cellulose, after pretreatment of the raw material, is made prior to fermentation. In this form of conduction, hydrolysis of the biomass is initially performed. After hydrolysis, the medium is centrifuged, the unhydrolyzed solid is discarded, and in the liquid medium the fermentative microorganism is inoculated. The process continues until the sugars in the medium are consumed. An advantage of this process is the possibility of performing each step, hydrolysis and fermentation, under the optimal conditions of cellulase action and growth of the microorganism (Olofsson et al., 2008). In this case, the hydrolysis is performed near 50ºC, which is the optimal temperature of cellulase performance, and the fermentation at the optimum temperature of growth of the microorganism around 37ºC. The disadvantage that can occur is the partial inhibition of cellulases by the accumulation of cellobiose and glucose. This inhibition negatively influences the conversion of biomass to glucose (Wang et al., 2012). Despite this, in a comparison between SHF and SSF, using corn cob biomass, very similar ethanol yield values were achieved, where the SHF process had a slightly higher yield (76.47%) to the yield of SSF (73.45%), using the same biomass loading with enzymes produced by *Trichoderma reesei* (Li et al., 2018).

Simultaneous saccharification and fermentation

It has been shown that higher yields of ethanol could be obtained by SSF compared to separate hydrolysis and fermentation processes (SHF) (Tomás-Pejó et al., 2008; Santos et al., 2010). In this conduction form, the hydrolysis and the fermentation are carried out simultaneously, in the same container. Thus cellulases and microorganisms are inoculated in the same container at the same time.

In this process, the glucose released in the hydrolysis is concomitantly used by the yeast, preventing its accumulation in the reaction medium and, consequently, decreasing the inhibition of cellulases by hydrolysis products, cellobiose, and glucose (Olofsson et al., 2008). Several studies have shown that ethanol yield is higher when conducted by SSF than when using SHF, both for long fiber wood and agricultural residues, for sugarcane bagasse and straw (Ohgren et al., 2007).

Compared to the two-stage process, with separate hydrolysis and fermentation, SSF has several advantages: (1) it improves the rate of hydrolysis by the simultaneous conversion of sugars that inhibit cellulase activity; (2) decreases the requirement for enzymes; (3) increases yield; (4) decreases the requirement for sterile conditions, since glucose is immediately removed from the medium; and (5) decreases total process time. The main disadvantage that needs to be considered in the SSF process is the incompatibility of temperature of hydrolysis and fermentation, which makes it necessary to use strains of thermotolerant microorganisms (Sun and Cheng, 2002; Olofsson et al., 2008; Saha et al., 2011).

Presaccharification

The SSF process may be preceded by a presaccharification step. The function of this step is to provide an initial carbon source for the yeast, since the biomass is in the form of a polymer which is not a directly fermentable carbon source. It has been shown that presaccharification leads to rapid dissolution and good mixing of the substrate in the medium, resulting in higher yield of ethanol at the end of the process (Santos et al., 2010). De Souza et al. (2012) demonstrated that the initial glucose concentration is the main factor affecting the ethanol yield in the SSF process, hence the importance of this step in the process.

Simultaneous saccharification and cofermentation

For the economically viable production of ethanol, it is essential to use microorganisms that ferment not only the glucose released during the cellulose hydrolysis stage but also the xylose from the hemicellulose. However, this process lacks microorganisms that ferment xylose to ethanol in high yields. Research has focused on the isolation of nonconventional yeasts capable of naturally fermenting this sugar and/or on the genetic manipulation of microorganisms that ferment glucose to ethanol with high yields, such as *S. cerevisiae* yeast. Among the microorganisms that naturally ferment xylose, we highlight lineages of the genus *Spathaspora*. These organisms have been isolated from a variety of environments, such as decaying woods and tree fragments, and tested for the ability to ferment xylose in ethanol (Cadete et al., 2012, 2013).

The metabolic pathway of xylose to ethanol is present in bacteria, fungi, and yeasts that naturally use xylose. Genes that encode the enzymes of these pathways have been isolated and cloned in organisms that have high fermentative capacity, as is the case of *S. cerevisiae* (Cadete et al., 2012).

With the obtention of these microorganisms, it will be possible to apply them in the so-called saccharification and simultaneous cofermentation. In this process, the hydrolysis of the biomass is carried out simultaneously to the fermentation of the hexoses and pentoses generated in the cellulose and hemicellulose hydrolysis, respectively (Bacovsky et al., 2010). The xylose used in the process may be from the hydrolysis of the hemicelluloses by hemicellulases as well as from the xylose released in the biomass pretreatment step when using diluted acid pretreatment. This process allows full biomass utilization in order to increase ethanol yield and productivity (Bacovsky et al., 2010; Cardona et al., 2010).

Consolidated bioprocess

Nowadays the costs of pretreatment and enzymatic hydrolysis of cellulose are still the main economic obstacles to the commercialization of bioconversion technology of biomass in ethanol (Alvira et al., 2010; Hendriks and Zeeman, 2009; Li et al., 2018). Thus one way of reducing the costs of the hydrolysis process is the development of fermentative microorganisms capable of hydrolyzing the lignocellulosic biomass.

In this form of conduction, denominated consolidated bioprocess, enzymatic production, cellulose hydrolysis, and glucose fermentation take place in the same equipment (Lynd et al., 2002; Bacovsky et al., 2010). In this context, several works with genetically modified species expressing enzymes of the cellulolytic complex (Saha et al., 2011; Yuan et al., 2011) have appeared in the last years. This integrated management would mark a significant advance in terms of efficiency and simplicity of the process (Bacovsky et al., 2010).

Inclusion of second-generation ethanol production annex to standard distillery

With its already consolidated technology and highly competitive costs, Brazil is a major producer of sugarcane and ethanol (first-generation), and has a very favorable environment for bagasse and straw hydrolysis, as well as conversion of sugars into second-generation ethanol.

The inclusion model would be to integrate the hydrolysis process with a standard ethanol distillery, using bagasse and sugarcane as raw material. In this model, after pretreatment of bagasse with steam and sulfuric acid, enzymatic hydrolysis of the pretreated material in aqueous suspension and posttreatment of the hydrolysis liquor occurs (Fig. 11.9).

Processing would be continued from the standard distillery in which the liquor would be added to the treated cane juice and preconcentrated to form the wort. The processes of ethanol fermentation, distillation, rectification, and dehydration of ethanol and treatment of the vinasse would also be carried out in the distillery, now resized to meet the additional production of ethanol.

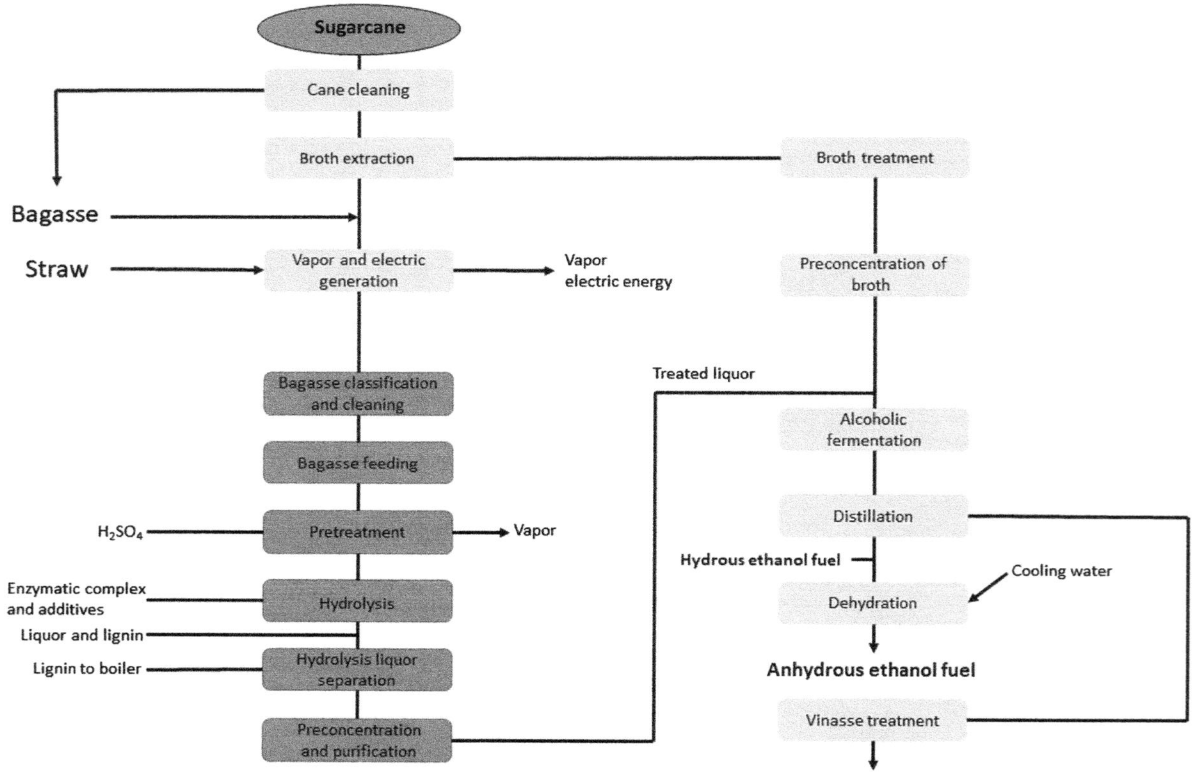

Figure 11.9 Hydrolysis process attached to the standard distillery. *Adapted from CGEE, 2009. Centro de Gestão e Estudos Estratégicos. Bioetanol combustível: uma oportunidade para o Brasil. CGEE, Brasília, 536 pp.*

The distillery would provide the necessary utilities for hydrolysis operation: process water, cooling, steam, and electric power, as well as storage facilities for raw materials and final ethanol (Rossel and Zainaghi, 2008). According to Bonomi (2010), the pretreatment processes with H_2SO_4 or hydrothermal pretreatment emerge as the most promising technological alternatives for the installation of hydrolysis units attached to the standard distillery.

Finally, it is well known that the Brazilian sugar and ethanol sector plays a fundamental role in the world market for biofuels, due to the massive investments that the sector has been receiving and also the unique geographical and climatic conditions. The second-generation technologies will bring to the entrepreneurs of the sector, greater variety in the portfolio of products. The current sugar and ethanol production plants will in fact be sugarcane biorefineries in the future. However, for this to happen, more studies and investments in research will be needed to achieve a greater and better understanding of the phenomena involved in biochemical conversion, which should happen in the coming years. Future results from the production of second-generation ethanol in conjunction with sugarcane plants are already in operation and will depend on a few key factors. The first of these is the reduction of the cost of enzymes and technologies of pretreatment of biomass, and until the use of modified cane varieties, at least the use of pretreatment is already considered (Fonseca et al., 2017). At the present time, the most profitable destination, as shown in the same study, is the use of about 15% of biomass of straw and bagasse for the production of second-generation ethanol, with the remainder being used in a cogeneration-integrated biomass with combined gasification cycle. It is known that there is a very attractive expectation in the processes of biorefinery for the production of energy and products with high added value; therefore second-generation ethanol must be a very targeted solution in the next years.

References

Alvira, P., Tomás-Pejó, E., Ballesteros, E.M., Negro, M.J., 2010. Pretreatment technologies for an efficient bioethanol production process based on enzymatic hydrolysis: a review. Bioresour. Technol. 101, 4851–4861.

Bacovsky, D., Dallos, M., Wörgetter, M., 2010 Status of 2nd generation biofuels demonstration facilities in June 2010. A report to bioenergy task 39.

Bahrani, S., Raeissi, S., Sarshar, M., 2015. Experimental investigation of ionic liquid pretreatment of sugarcane bagasse with 1-3-dimethylimadazolium dimethyl phosphate. Bioresour. Technol. 185, 411–415.

Bezerra, T.L., Ragauskas, A.J., 2016. A review of sugarcane bagasse for second-generation bioethanol and biopower production. Biofuel. Bioprod. Biorefin. 10 (5), 634–647.
Bian, J., Peng, F., Peng, X.-P., Xiao, X., Peng, P., Xu, F., 2014. Effect of [Emim]Ac pretreatment on the structure and enzymatic hydrolysis of sugarcane bagasse cellulose. Carbohydr. Polym. 100, 211–217.
Bonomi, A., 2010. Workshop hidrólise de materiais lignocelulósicos. In: Cortez, L.A.B. (Ed.). Bioetanol de cana-de-açúcar: P&D para produtividade sustentabilidade. Blucher, São Paulo, pp. 679–695.
Buckeridge, M.S., Silva, G.B., Cavalari, A., 2008. Parede celular. In: Kerbauy, G.B. (Ed.), Fisiologia vegetal, second ed. Guanabara Koogan, Rio de Janeiro, pp. 165–181.
Buckeridge, M.S., Santos, W.D., Souza, A.P., 2010. As rotas para o etanol celulósico no Brasil. In: Cortez, L.A.B. (Ed.), Bioetanol de cana-de-açúcar: P&D para Produtividade Sustentabilidade. Edgard Blücher Ltda, São Paulo, pp. 365–380.
Cadete, R.M., Melo, M.A., Dussán, K.J., Rodrigues, R.C., Silva, S.S., Zilli, J.E., et al., 2012. Diversity and physiological characterization of D-xylose-fermenting yeasts isolated from the Brazilian Amazonian Forest. PLoS One 7, e43135.
Cadete, R.M., Melo, M.A., Zilli, J.E., Vital, M.J., Mouro, A., Prompt, A.H., et al., 2013. *Spathaspora brasiliensis* sp. nov., *Spathaspora suhii* sp. nov., *Spathaspora roraimanensis* sp. nov. and *Spathaspora xylofermentans* sp. nov., four novel D-xylose-fermenting yeast species from Brazilian Amazonian forest. Antonie Van Leeuwenhoek. 103, 421–431.
Cardona, C.A., Quintero, J.A., Paz, I.C., 2010. Production of bioethanol from sugarcane bagasse: status and perspectives. Bioresour. Technol. 101, 4754–4766.
CGEE, 2009. Centro de Gestão e Estudos Estratégicos. Bioetanol combustível: uma oportunidade para o Brasil. CGEE, Brasília, 536 pp.
Chang, V.S., Holtzapple, M.T., 2000. Fundamental factors affecting biomass enzymatic reactivity. Appl. Biochem. Biotechnol. 84, 5–37.
Chen, W.-H., Tu, Y.-J., Sheen, H.-K., 2011. Disruption of sugarcane bagasse lignocellulosic structure by means of dilute sulfuric acid pretreatment with microwave-assisted heating. Appl. Energ. 88, 2726–2734.
Conway, G., 2003. Produção de alimentos no século XXI: Biotecnologia e meio ambiente. Estação Liberdade, São Paulo, 375 pp.
Cortez, L.A.B., Lora, E.E.S., Olivares Gómez, E., 2008. Biomassa para energia. Campinas, SP: Editora da Unicamp 733.
Darabzadeh, N., Hamidi-Esfahani, Z., Hejazi, P., 2019. Optimization of cellulase production under solid-state fermentation by a new mutant strain of *Trichoderma reesei*. Food Sci. Nutr 7, 572–578.
De Souza, C.J.A., Costa, D.A., Rodrigues, M.Q.R.B., Santos, A.F., Lopes, M.R., Abrantes, A.B.P., et al., 2012. The influence of presaccharification, fermentation temperature and yeast strain on ethanol production from sugarcane bagasse. Bioresour. Technol. 109, 63–69.
EMBRAPA – Empresa Brasileira de Pesquisa Agropecuária, 2019. Áreas cultivadas no Brasil e no Mundo. Embrapa . Available from: www.embrapa.br.
Farinas, C. S., Marconcini, J.M., & Mattoso, L.H.C., 2018. Enzymatic Conversion of Sugarcane Lignocellulosic Biomass as a Platform for the Production of Ethanol, Enzymes and Nanocellulose. Journal of Renewable Materials, 6(2), 203–216. Available from: https://doi.org/10.7569/jrm.2017.6341578.
Fonseca, G.C., Costa, C.B.B., Cruz, A.J.G., 2017. Superstructural economic optimization of sugarcane bagasse exploitation in an ethanol distillery connected to Rankine cycle, BIGCC system and second generation ethanol process. In: 27th European Symposium on Computer Aided Process Engineering, pp. 889–894.
Fuentes, L.L.G., 2009. Determinação de dados cinéticos da deslignificação do bagaço de cana-de-açúcar e da hidrólise enzimática no pré-tratamento com hidróxido de cálcio.

Dissertação (Mestrado) – Escola Engenharia Química, Universidade Estadual de Campinas, São Paulo.

Galbe, M., Zacchi, G., 2010. Produção de etanol a partir de materiais lignocelulósicos. In: Cortez, L.A.B. (Ed.). Bioetanol de cana-de-açúcar: P&D para produtividade sustentabilidade. Blucher, São Paulo, pp. 697–716.

Genomics. Department of Energy Genome Programs. Available at: <www.genomics.energy.gov> (accessed 10/01/12).

Gómez, E.O., Souza, R.T.G., Rocha, G.J.M., Almeida, E., Cortez, L.A.B., 2010. A palha de cana-de-açúcar como matéria-prima para processos de segunda geração. In: Cortez, L.A.B. (Ed.), Bioetanol de cana-de-açúcar: P&D para produtividade sustentabilidade. Edgard Blücher Ltda, São Paulo, pp. 637–659.

Hamelinck, C.N., Hooijdonk, G.V., Faaij, A.P.C., 2005. Ethanol from lignocellulosic biomass: techno-economic performance in short-, middle-and long-term. Biomass Bioenerg. 28, 384–410.

Hendriks, A.T.W.M., Zeeman, G., 2009. Pretreatments to enhance the digestibility of lignocellulosic biomass. Bioresour. Technol. 100, 10–18. n.

IPCC – International Panel on Climate Change. Climate change 2007: Impacts, Adaption and Vulnerability. Cambridge, 2007.

Lane, M.M., Burke, N., Karreman, R., Wolfe, K.H., O'byrne, C.P., Morrissey, J.P., 2011. Physiological and metabolic diversity in the yeast Kluyveromyces marxianus. Anton Leeuw 100, 507–519.

Li, Y.-H., Zhang, X.-Y., Zhang, F., Peng, L.-C., Zhang, D.-B., Kondo, A., et al., 2018. Optimization of cellulolytic enzyme components through engineering *Trichoderma reesei* and on-site fermentation using the soluble inducer for cellulosic ethanol production from corn stover. Biotechnol. Biofuels 11 (1).

Lynd, L.R., Weimer, P.J., Van zyl, W.H., Pretorius, I.S., 2002. Microbial cellulose utilization: fundamentals and biotechnology. Rev. Microbiol. Molecul. Biol 66, 506–577.

Magalhães, O.S.G., Braunbeck, O.A., 2010. Colheita de cana-de-açúcar e palha. In: Cortez, L.A.B. (Ed.), Bioetanol de cana-de-açúcar: P&D para produtividade sustentabilidade. Edgard Blücher Ltda, São Paulo, pp. 465–475.

Marín, M., Artola, A., & Sánchez, A., 2018. Optimization of Down-Stream for Cellulases Produced Under Solid-State Fermentation of Coffee Husk. Waste and Biomass Valorization. Available from: https://doi.org/10.1007/s12649-018-0327-5.

Matsuoka, S., Bressiani, J., Maccheroni, W., Fouto, I., 2012. Bioenergia da cana. In: Santos, F.A., Borém, A., Caldas, C. (Eds.). Cana-de-açúcar: bioenergia, açúcar e etanol – tecnologias e perspectivas, second ed. Revisada e ampliada. Viçosa, MG pp. 547–577.

Matsushika, A., Inoue, H., Kodaki, T., Sawaya, S., 2009. Ethanol production from xylose in engineered Saccharomyces cerevisiae strains: current state and perspectives. Appl. Microbiol. Biotechnol 84, 37–53.

Mesa L., et al., 2017. Desirability function for optimization of Dilute Acid pretreatment of sugarcane straw for ethanol production and preliminary economic analysis based in three fermentation configurations. Appl. Energy. Available from: http://dx.doi.org/10.1016/j.apenergy.2017.03.018.

Moro, M.K., et al., 2017. Continuous pretreatment of sugarcane biomass using a twin-screw extruder. Ind. Crop. Prod. 97, 509–517.

Moghaddam, L., Zhang, Z., Wellard, R.M., Bartley, J.P., O'hara, I.M., Doherty, W.O.S., 2014. Characterisation of lignins isolated from sugarcane bagasse pretreated with acidified ethylene glycol and ionic liquids. Biomass Bioenerg. 70, 498–512.

Ohgren, K., Bura, R., Lesnicki, G., Saddler, J., Zacchi, G., 2007. A comparison between simultaneous saccharification and fermentation and separate hydrolysis and fermentation using steam-pretreated corn stover. Process Biochem. 42, 834–839.

Olofsson, K., Bertilsson, M., Lidén, G., 2008. A short review on SSF – an interesting process option for ethanol production from lignocellulosic feedstocks. Biotechnol. Biofuels 1, 1–7.

Oreb, M., Dietz, H., Farwick, A., Boles, E., 2012. Novel strategies to improve co-fermentation of pentoses with d-glucose by recombinant yeast strains in lignocellulosic hydrolysates. Bioengineered 3, 347–351.

Palmqvist, E., et al., 2000. Fermentation of lignocellulosic hydrolysates. Inhibitors and mechanisms of inhibition. Bioresour. Tecnol. 74, 25–33.

Perdices, M.B., Arcaya, G.E.A., Coral, D.S.O., Mendoza, M.A.G., Nunes, D.C.L., 2012. Bioetanol a partir de materiais lignocelulósicos pela rota da hidrólise. In: Lora, E.E.S., Venturini, O.J. (Eds.), Biocombustíveis., 1. Interciência Ltda, Rio de Janeiro, pp. 335–588. , e2.

Rabelo, S.C. Costa, A.C., Rossel, C.E.V., 2012. Aproveitamento de resíduos industriais. In: Santos, F. A., Borém, A., Caldas, C. (Eds.). Cana-de-açúcar: bioenergia, açúcar e etanol – tecnologias e perspectivas, second ed. Revisada e ampliada. Viçosa, MG, pp. 515–536.

Rantasalo, A., Vitikainen, M., Paasikallio, T., Jäntti, J., Landowski, C.P., & Mojzita, D., 2019. Novel genetic tools that enable highly pure protein production in *Trichoderma reesei*. Scientific Reports, 9(1). Available from: https://doi.org/10.1038/s41598-019-41573-8.

Rossel, C.E.V., Zainaghi, G. Produção de etanol combustível pela hidrolise enzimática. Brasília: IEL/NC, 2008. p. 123–138. (Instituto Euvaldo Lodi. Núcleo Central. Álcool combustível/IEL. Núcleo Central).

Rubio-Texeira, M., Zeebroeck, G.V., Voordeckers, K., Thevelein, J.M., 2009. Saccharomyces cerevisiae plasma membrane nutrient sensors and their role in PKA signaling. FEMS Yeast Res. 10, 134–149.

Saha, B.C., Nichols, N.N., Qureshi, N., Cotta, M.A., 2011. Comparison of separate hydrolysis and fermentation and simultaneous saccharification and fermentation processes for ethanol production from wheat straw by recombinant *Escherichia coli* strain FBR5. Appl. Microbiol. Biotechnol. 92, 865–874.

Santos, J.R.A., Souto-Maior, A.M., Gouveia, E.R., 2010. Comparação entre processos em shf e em ssf de bagaço de cana-de-açúcar para a produção de etanol por *Saccharomyces cerevisiae*. Química Nova 3, 904–908.

Santos, F. et al., 2015. Sugarcane Agricultural Production, Bioenergy and Ethanol. Academic Press. Elsevier. San Diego.

Santos, F.A., Queiroz, J.H., Colodette, J.L., Fernandes, S.A., Guimarães, V.M., Rezende, S.T., 2012. Potencial da palha de cana-de-açúcar para produção de etanol. Química Nova 35 (5), 1004–1010.

Singh, J., Suhag, M., Dhaka, A., 2015. Augmented digestion of lignocellulose by steam explosion, acid and alkaline pretreatment methods: a review. Carbohydr. Polym. 117, 624–631.

Siripong, P., 2018. Potential of weed biomass for ethanol production. In: Basso, T.P., Basso L. C. Fuel and ethanol production from sugarcane. IntechOpen.

Sun, Y., Cheng, J., 2002. Hydrolysis of lignocellulosic materials for ethanol production: a review. Bioresour. Technol. 83, 1–11.

Tizazu, B.Z., & Moholkar, V.S., 2018. Kinetic and thermodynamic analysis of dilute acid hydrolysis of sugarcane bagasse. Bioresource Technology, 250, 197–203. Available from: https://doi.org/10.1016/j.biortech.2017.11.032.

Tomás-Pejó, E., García-Aparicio, M., Negro, M.J., Oliva, J.M., Ballesteros, M., 2008. Effect of different cellulase dosages on cell viability and ethanol productioby *Kluyveromyces marxianus* in SSF processes. Bioresour. Technol. 100, 890–895.

Wang, M., Li, Z., Fang, X., Wang, L., Qu, Y., 2012. Cellulolytic enzyme production and enzymatic hydrolysis for second-generation bioethanol production. Adv. Biochem. Engin./Biotechnol.

Yang, J., Deng, L., Zhao, C., Fang, H., 2019. Heterologous expression of *Neurospora crassa* cbh1 gene in *Pichia pastoris* resulted in production of a neutral cellobiohydrolase I. Biotechnol. Progress.

Yu, M., Li, J., Chang, S., Zhang, L., Mao, Y., Cui, T., et al., 2016. Bioethanol production using the sodium hydroxide pretreated sweet sorghum bagasse without washing. Fuel 175, 20–25.

Yuan, W.J., Chang, B.L., Ren, J.G., Liu, J.P., Bai, F.W., Li, Y.Y., 2011. Consolidated bioprocessing strategy for ethanol production from Jerusalem artichoke tubers by *Kluyveromyces marxianus* under high gravity conditions. J. Appl. Microbiol. 112, 38–44.

Xia, J., Yang, Y., Liu, C.-G., Yang, S., Bai, F.-W., 2019. Engineering *Zymomonas mobilis* for robust cellulosic ethanol production. Trends Biotechnol.

Zhang, Q., Hu, J., Lee, D.-J., 2017. Pretreatment of biomass using ionic liquids: research updates. Renew. Energ. 111, 77–84.

Further reading

Atalla, R.H., Hackney, J.M., Uhlin, I., 1993. Hemicelluloses as structure regulators in the aggregation of native cellulose. Int. J. Biol. Macromol. 15, 109–112.

BNDES, CGEE, 2008. Banco Nacional para o Desenvolvimento Social e Econômico; Centro de Gestão e Estudos Estratégicos. Bioetanol de cana-de-açúcar: Energia para o desenvolvimento sustentável. BNDES & CGEE, Rio de Janeiro, 316 pp.

D' Almeida, M.L.O., 1998. Composição química dos materiais lignocelulósicos, second ed., vol. 1. Instituto de Pesquisas do Estado de São Paulo, São Paulo (Celulose e Papel Tecnologia de Fabricação da Pasta Celulósica).

De Oliveira, Santos, Tabosa, Victor, et al., 2018. Role of hemicellulose removal during dilute acid pretreatment on the cellulose accessibility and enzymatic hydrolysis of compositionally diverse sugarcane hybrids. Ind. Crop. Prod. 111, 722–730.

Fengel, D., Wegener, G., 1989. Wood Chemistry Ultrastructure Reactions. Walter de Gruyter, Berlín, 610 pp.

Galbe, M., Zacchi, G., 2007. Pretreatment of lignocellulosic materials for efficient bioethanol production. Adv. Biochem. Engin/Biotechnol 108, 41–65.

Harifara Rabemanolontsoa, S.S., 2016. Various pretreatments of lignocellulosics. Bioresour. Technol. 199, 83–91.

Lehninger, A.L., 2006. Princípios de bioquímica, fourth ed. Coordenação de tradução: Arnaldo Antonio Simões e Wilson Roberto Navega Lodi. SARVIER, São Paulo.

Mussato, S.I., Roberto, I.C., 2004. Alternatives for detoxification of diluted-acid lignocellulosic hydrolyzates for use in fermentative processes: a review. Bioresour. Technol 93, 1–10.

Nosratpour, M.J., Karimi, K., Sadeghi, M., 2018. Improvement of ethanol and biogas production from sugarcane bagasse using sodium alkaline pretreatments. J. Environ. Manage. 226, 329–339.

Rabelo, S., 2007. Avaliação de desempenho do pré-tratamento com peróxido de hidrogênio alcalino para a hidrólise enzimática de bagaço de cana-de-açúcar. Tese (Mestrado) – Escola Engenharia Química. Universidade Estadual de Campinas, São Paulo.

Ramos, L.P., 2003. The chemistry involved in the pretreatment of lignocellulosic materials. Química Nova 26, 863–871.

Seabra, J.E.A., 2008. Avaliação técnico-econômico de opções para o aproveitamento integral da biomassa de cana no Brasil. 298 f. Tese (Doutorado) – Universidade Estadual de Campinas, Campinas, SP.
Sjöstrom, E., 1993. Wood Chemistry: Fundamentals and applications., second ed. Academic Press, San Diego, CA.
Teleman, A., 2009. Hemicelluloses and pectins. In: Gellerstedt, M.E.K.G., Henriksson, G. (Eds.), Wood Chemistry and Wood Biotechnology., 1. De Gruyter, Stockholm, pp. 102–120.
Wyman, C.E., 2003. Potential synergies and challenges in refining cellulosic biomass to fuels, chemicals, and power. Biotechnol. Prog. 19, 254–262.

CHAPTER 12

Biopolymers of sugarcane

Thalita Mendonça de Resende and Marcelo Moreira da Costa

Contents

Introduction 229
Sugarcane 231
Biopolymer 232
Sugarcane bagasse 233
Cellulose 236
Methylcellulose 236
Carboxymethylcellulose 237
Cellulose acetate 238
Hemicelluloses 238
Furfural resins 239
Phenolic resins 240
Lignin 241
Sugarcane juice 243
Polylactic acid 243
Acrylic acid 244
Polyhydroxybutyrate 246
Green polyethylene 247
Xanthan gum 249
Perspectives and challenges for the production of biopolymer from sugarcane 249
References 251
Further reading 254

Introduction

Recently environmental concerns have significantly increased regarding nonrenewable petrochemical utilization as raw material for chemical production. Among the key concerns and issues, at least three major points should be highlighted: (1) disposal in landfills or incineration of materials derived from nonrenewable resources with low biodegradability (Belloli, 2010; Yang et al., 2007; Yu et al., 2006) and heavy contamination of water streams and oceans, mainly with plastic microparticles; (2) production instability besides rise in oil prices (Figs. 12.1 and 12.2) will remain

Sugarcane Biorefinery, Technology and Perspectives
DOI: https://doi.org/10.1016/B978-0-12-814236-3.00012-3

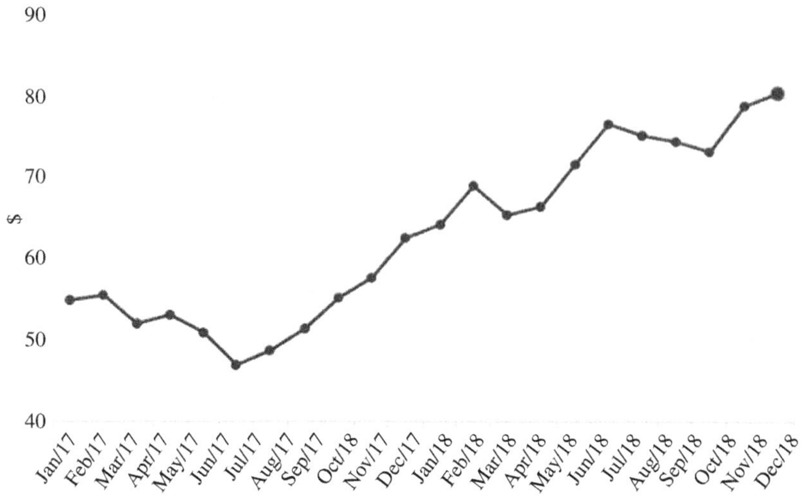

Figure 12.1 Evolution of the oil price: dollar per barrel of Brent crude. *Tendência Consultoria, 2018. Available from: <https://g1.globo.com/economia/noticia/2018/11/16/precos-do-petroleo-tem-alta-mas-devem-ter-6a-perda-semanal.ghtml> (accessed 15.03.19.).*

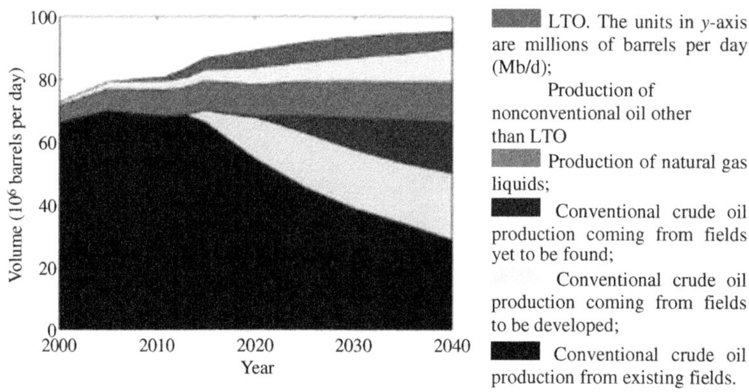

Figure 12.2 Evolution of the total oil liquids produced, according to the IEA in the "New Policies." *LTO*, Light tight oil. *WEO, 2014. Available from: <http://www.worldenergyoutlook.org/publications/weo-2014/> (accessed 15.03.19.).*

due to the limited availability of fossil raw materials (Tendências Consultoria, 2018; Solé et al., 2018); and (3) global warming concerns with pronounced public debate, looks like the major issue to be accomplished. There is a strong consensus that global surface temperatures have increased in recent decades (Fig. 12.3) and that the trend is caused by

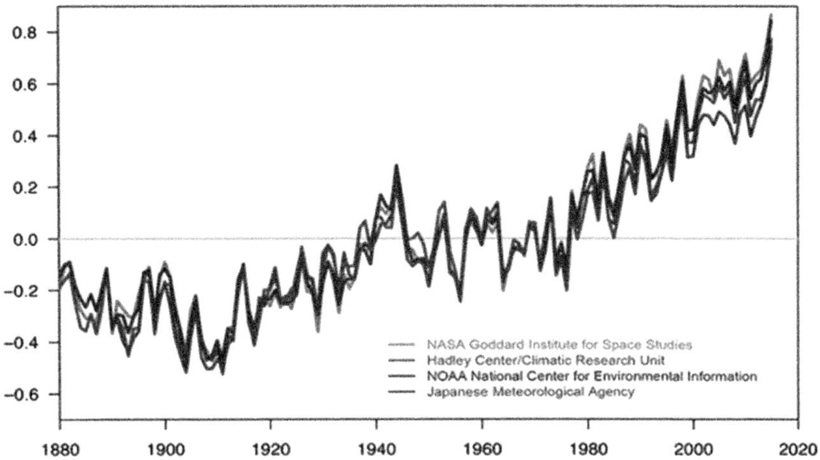

Figure 12.3 Temperature data (in degree Celsius) from four international science institutions. *NASA. NASA's Global Climate Change Website, 2019. NASA's Goddard Institute for Space Studies, NOAA National Climatic Data Center, Met Office Hadley Centre/Climatic Research Unit and the Japanese Meteorological Agency. (Graph produced by Earth Science Communications Team at NASA's Jet Propulsion Laboratory | California Institute of Technology). Available from <https://climate.nasa.gov> (accessed 15.03.19.).*

human-induced emissions of greenhouse gases (GHGs; e.g., CO_2) from nonrenewable resources utilization for fuel and chemical production (NASA's Global Climate Change Website, 2019). Fig. 12.3 revealed rapid warming in the past few decades and that the last decade has been the warmest on record.

Therefore, alternatives are required for mitigating these problems, assessing the total or partial replacement of nonrenewable resources utilization for renewable biomass with high biodegradability behavior from vegetable sources.

Sugarcane

Sugarcane is an important commercial crop worldwide, and one of the principal raw materials for the production of crystalized sucrose (sugar) and ethanol. The by-products of sugarcane harvesting are also used as feed for livestock in many countries. Brazil tops the list of sugarcane producers with an estimated annual production of 647.6 million metric tons for the 2017/18 season. South-Central Brazil accounts for greater than 85% of this national production. The sugar production forecast (for the 2017/18

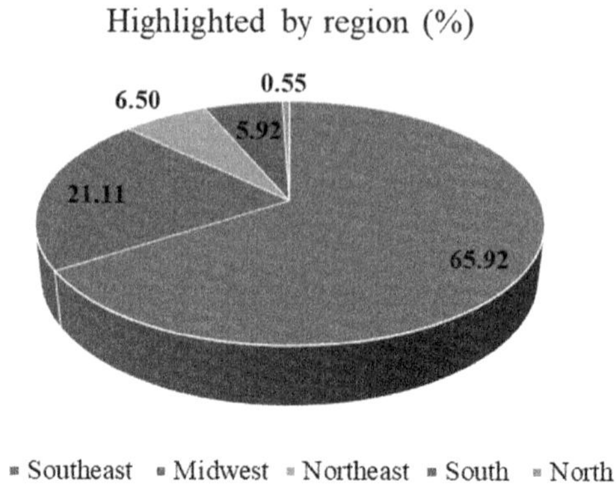

Figure 12.4 Sugarcane production statistics by region (%). *CONAB - Companhia Nacional de Abastecimento, 2018. Acompanhamento da safra brasileira de cana-de-açúcar. V. 4 - SAFRA 2017/18 N.3 - Terceiro levantamento (December 2017), pp. 2318–7921.*

season) is about 38.7 million tons and it is the main product derived from the sugarcane cultivated in Brazil (CONAB, 2018).

Sugarcane is one of the most efficient photosynthetic mechanisms, among the crops: it can convert around 2%–3% of radiant solar energy into green biomass. This high photosynthetic capacity also allows it to produce high coefficient of CO_2 fixation, comparable to the woods of moderate climates, contributing to the reduction of GHGs (McKendry, 2002).

Due to the land availability and proper conditions, Brazil finds itself in a privileged position as the major player worldwide in sugarcane production. As a result of large amount of sugarcane globally, it drives researchers to create alternatives to make feasible the biotechnological processes which employ the sugarcane as raw material (Guimarães et al., 2018). Fig. 12.4, shows the production per region in percentage (CONAB, 2018).

In this context, the objective of this chapter is to summarize the huge range of applications of polymeric products, that is, biopolymers for various applications, that can be obtained from basic constituents of sugarcane bagasse and juice and their products derivatives.

Biopolymer

Currently there are a large number of polymers derived from natural sources, which are available for various applications. Among them, cellulose, starch, lignin, and rubber are some of the most currently used.

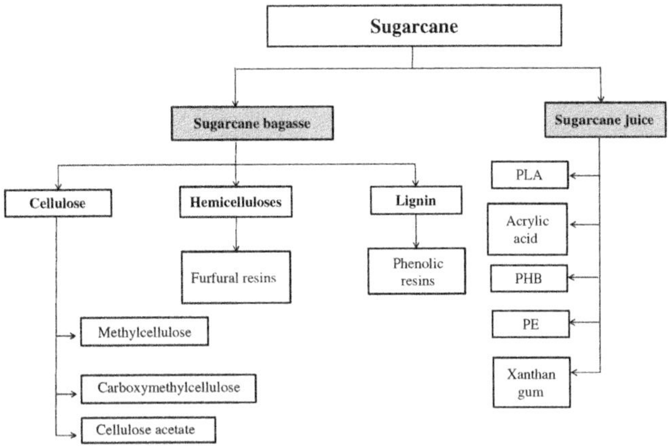

Figure 12.5 Examples of natural biopolymer that can be obtained from sugarcane. *PE*, Polyethylene; *PHB*, polyhydroxybutyrate; *PLA*, polylactic acid. *Adapted from Schuchardt, U., Ribeiro, M.L., Gonçalves, A.R., 2001. A indústria petroquímica no próximo século: como substituir o petróleo como matéria-prima? Quím. Nova 24 (2), 247–251.*

Fig. 12.5 shows some natural polymers and Fig. 12.6 shows materials that can be obtained from lignocellulosic biomass, highlighting the polymer.

The requirement of a variety of polymers that provide good performance and which are of low cost grows with the advancement of science and technology. In this sense, the difficulty of obtaining homopolymers leveraged new studies with composite materials and alloys (Sahoo, 2008, cited by Camargo et al., 2012; Reddy et al., 2003).

Sugarcane bagasse

The sugarcane plant, after being mechanically harvested, is transported to a mill and is crushed between the serrated rollers. The crushed sugarcane is then pressed to extract raw sugarcane juice, while the leftover fibrous material (10%–16% dry basis mass) sugarcane bagasse, is used mainly as a fuel to generate energy in biomass boilers (Lemos and Castro, 2010; Faria, 2005). On the other hand, many new applications and noble products with high added value have been proposed for sugarcane bagasse through the biorefinery concept.

These alternatives require raw materials at low cost and good availability. Consequently, agricultural residues have been considered as the main target in Europe, North America, and other places of temperate climate. In countries with a tropical climate, there is a great opportunity to use

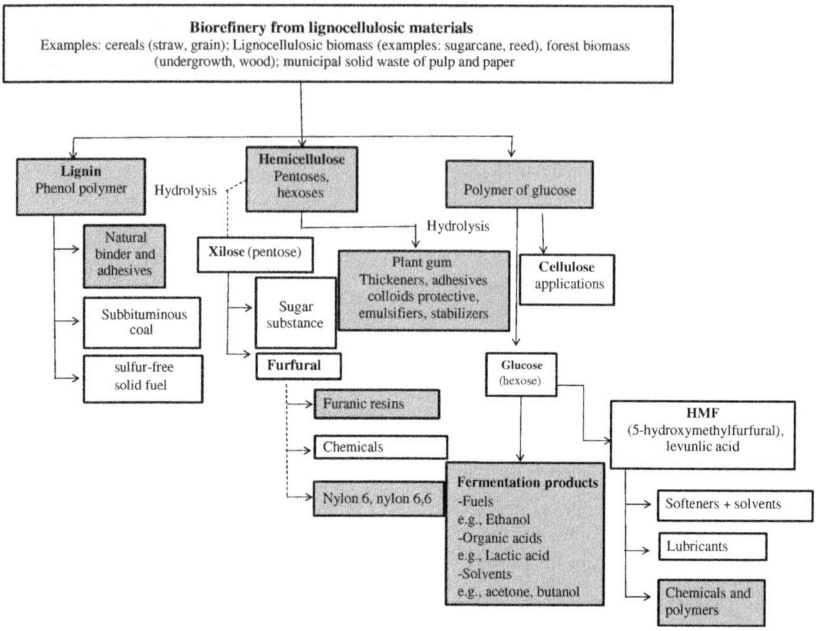

Figure 12.6 Potential products obtained in the biorefinery of lignocellulosic material. *Kamm, B., Gruber, P.R., Kamm, M., 2006. Biorefineries – Industrial Processes and Products. Wiley-VCH, Weinheim, p. 407.*

Table 12.1 Sugarcane bagasse chemical composition.

Sugarcane bagasse composition	% dry basis
Cellulose	41.7–46.6
Hemicellulose	25.2–34.0
Lignin	12.6–26.2
Organosoluble	2–3
Aquasoluble	2–3
Ashes	2–6
Humidity	47–52

sugarcane for high productivity, that is, 647.6 million metric tons for the 2017/18 to practice the concept of biorefinery (CONAB, 2018). The main chemical composition of sugarcane bagasse can be observed in Table 12.1 (Vieira et al., 2007; Sun et al., 2004, mentioned by Kumar et al., 2012; Triana et al., 1990)

Approximately, using the numbers mentioned above, annual sugarcane production and dry weight of the bagasse (Fig. 12.7A) on the total

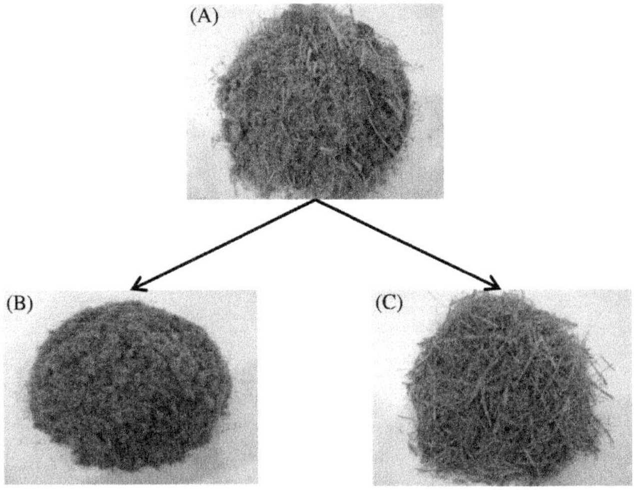

Figure 12.7 (A) sugarcane bagasse, (B) pith cells, and (C) fibrous material.

sugarcane weight (CONAB, 2018; Lemos and Castro, 2010; Faria, 2005). It will be possible to estimate the bagasse amount that exceeds annually 80 million dry tons. It is important to emphasize that 30%–35% of the bagasse crushed and milled at the industry sites are pith cells (Fig. 12.7B) with low content of cellulose. For the satisfactory production of pulp and other biopolymers from sugarcane bagasse, pith cells should be removed by using depithing process.

Depithing process should be applied to the bagasse, in order to prevent the waste of chemical products during the pulping process. The fibrous material of bagasse (Fig. 12.7C) is a rich source not only of cellulose, but also hemicelluloses (mainly xylans).

Together, cellulose and hemicelluloses represent around 70% of the sugarcane bagasse. Covalently linked to lignin, hemicelluloses are interlaced around these sugars. Lignin is an amorphous and highly branched polyphenolic polymer with phenylpropane units, which may be present in varying amounts in biomass, comprising around 8–30% of biomass weight (Table 12.2) (Honorato et al., 2017). Today the largest fraction of sugarcane bagasse (60%–90%) is burned to produce steam/energy to sugar and ethanol mills and to sell to the electrical grid. Table 12.2 shown the comparison among different biomass compositions vs. calorific values.

Despite the calorific power of sugarcane bagasse, the largest fraction of it consists of carbohydrates. Whereas the calorific power of carbohydrates is approximately half of the lignin (McKendry, 2002). It would be

Table 12.2 Biomass composition, calorific values and densities of compressed materials (Honorato et al., 2017)

Biomass	Cellulose	Hemicellulose	Lignin	Calorific value	Density of compressed
	(%)	(%)	(%)	(GJ/ton)	(kg/m³)
Wood *(Pinus ssp.)*	43.3	25.9	29.6	18.3	906.8
Agave bagasse *(Agave atrovirens)*	45.9	26.6	8.7	15.5	868
Sugarcane bagasse *(Saccharum officinarum)*	32.3	35.2	22.4	19.1	721.7
Corn stubble (*Zea mays)*	39.1	30.9	18.6	15.6	928.2
Wheat Straw *(Triticum* ssp.)	42.9	28.9	21.6	17.9	899

valuable to develop at least an effective path for the better use of sugarcane bagasse carbohydrates. The bagasse from sugarcane has several applications (Tita et al., 2002; Ripoli et al., 2000, mentioned by Kumar et al., 2012). From the basic constituents of bagasse, it is possible to produce a wide variety of biopolymers for various applications.

Cellulose

Cellulose presents semicrystalline behavior, which makes it insoluble in water and organic solvents. Despite its high crystallinity, it decomposes before merging and, thus to increase their applicability, it becomes necessary to convert it into its soluble derivatives, as the ethers and esters, etherification being one of the most important routes of derivatization of cellulose (Kumar et al., 2012). Among the products obtained by derivatization of cellulose are the carboxymethylcellulose (CMC) (Morais and Filho, 1999), the cellulose acetate, and the methylcellulose (Filho et al., 2007; Vieira et al., 2007). These products are the key application of dissolving pulp which could be a market for cellulose (pulp) from sugarcane bagasse.

Methylcellulose

Compared to cellulose, methylcellulose shows an increased thermal stability and solubility in water associated with the increase of the degree of substitution, characteristics that extend the commercial applicability of this biopolymer (Filho et al., 2007). Many products of commercial importance may be developed on the basis of methylcellulose, for example, a thickener for the food industry and additive for concrete in civil construction.

Kumar et al. (2012) studied the synthesis and characterization of a porous composite methylcellulose and polyvinyl alcohol (PVA), with the intention of adding to methylcellulose latter properties, such as biodegradability, biocompatibility, and nontoxicity, for biomedical applications. The methylation of purified cellulose was obtained by reaction between α-cellulose and the dimethyl sulfate (DMS) in acetone as a reactional state. The methylation of purified cellulose was obtained by reaction between α-cellulose and DMS in acetone as a reactional state. The porous composite was produced by freeze-drying of the mixture of methylcellulose and PVA. By three-dimensional analysis, the authors demonstrated the applicability of this biopolymer in the biomedical field.

Carboxymethylcellulose

CMC is an ether prepared from the replacement of a hydroxyl group of glucose by a group of monochloroacetic acid, with the previous swelling of the cellulose with alkali. The reactions [Eq. (12.1)—alkaline swelling/mercerization and Eq. (12.2)—etherification] for this preparation are described below (Heydarzadeh et al., 2009)

$$\left[C_6H_7O_2(OH)_3\right]_n + n\mathrm{NaOH} \rightarrow \left[C_6H_7O_2(OH)_2ONa\right]_n + nH_2O \quad (12.1)$$

$$\begin{gathered}\left[C_6H_7O_2(OH)_2ONa\right]_n + n\mathrm{ClCH_2COONa} \rightarrow \\ [C_6H_7O_2(OH_2)OCH_2COONa]_n + n\mathrm{NaCl}\end{gathered} \quad (12.2)$$

The sodium CMC is an anionic polyelectrolyte, soluble in water, whose properties allow applications such as thickening agent, emulsifier, adhesive binder, wetting, dispersant, etc. However, these properties, as well as its applications, depend greatly on the degree of substitution (DS—average number of carboxymethyl per unit of anhydrous glucose) of carboxyl groups along the chain. The applications of CMC are varied and among them stand the uses in the segments: pharmaceutical, food, cosmetic, and petroleum (Martinez, 1996).

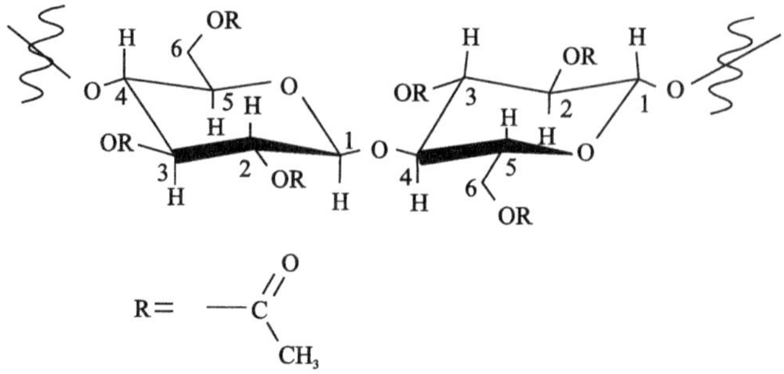

Figure 12.8 Structure of cellulose acetate. *Meireles, C.S., 2007. Síntese e caracterização de membranas de acetato de celulose, obtido do bagaço de cana-de-açúcar, e blendas de acetato de celulose com poliestireno de copos plásticos descartados (Dissertation (Master)). UFU, 80f.*

Cellulose acetate

Derived from cellulose of great commercial importance, cellulose acetate is an interesting alternative from an environmental point of view, the possibility of aggregating value to the bagasse, providing the removal of residue in the environment. Cellulose acetate is produced by substituting the hydroxyl groups of the cellulose acetyl units (Fig. 12.8). The obtaining of cellulose acetate is possible through the reaction of cellulose with a mixture of acetic acid and acetic anhydride, in the presence of sulfuric acid, perchloric acid, or as a catalyst (Sassi and Chanzy, 1995).

For cellulose acetate, there are several applications, such as in fibers, plastics, photographic films, and filters for cigarettes, among others (Steinmeir, 2004; Edgar et al., 2001).

Hemicelluloses

Among the hemicelluloses, xylans are more abundant in sugarcane bagasse. Xylans are a family of structurally diverse plant polysaccharides with a backbone composed of 1,4-linked β-D-xylosyl residues. Arabinoxylan is the predominant hemicelluloses of grasses. L-Arabinofuranoside is attached randomly by 1α→2 and/or 1α→3 linkages to the xylose units throughout the chain. Side chains containing arabinosyl, galactosyl, glucosyluronic acid, and 4-O-methyl glucosyluronic acid residues have been identified. Fig. 12.9 shows the structure of xylan from sugarcane bagasse. From xylan polymers it would be possible to get various chemicals, among them the furfuryl alcohol (FFA) and depressants for iron ore flotation process in order to replace food grade starch (Costa et al., 2017). In addition, xylans

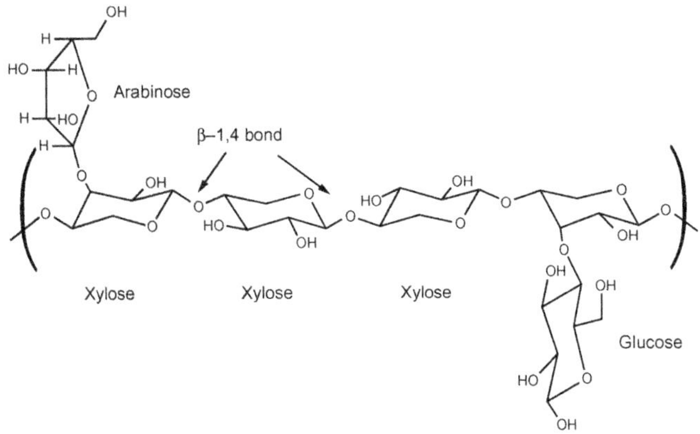

Figure 12.9 Simplified structure of the main bagasse hemicelluloses: arabinoxylan. *Zeinaly, F., Saraeian, A., Gabov, K. Fardim, P., 2017. Determination of carbohydrates in sugarcane bagasse pulp in different TCF bleaching sequences. Cell. Chem. Technol. 51 (1–2), 45–53.*

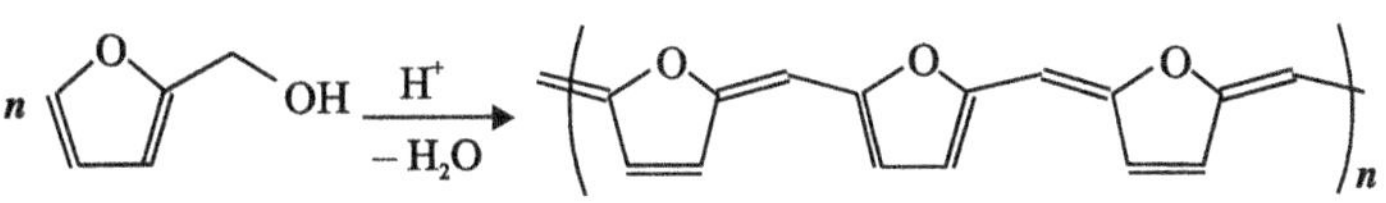

Figure 12.10 Polymerization of furfuryl alcohol. *Ibnyaich, A., 2010. Modification of the Properties of Biobased Thermoset Resin Using Cellulose Nano-Whiskers (CNW) as an Additive. 60f (Dissertation (Master)). Lulea University of Technology.*

are naturally soluble in alkali and can be used as biopolymers without the need for derivatization (Zeinaly et al., 2017).

Furfural resins

FFA is produced in two steps: first, the pentosans of hemicelluloses undergo an acid hydrolysis, producing the furfuraldehyde; second, the aldehyde group is reduced to alcohol by a catalyzed hydrogenation. In acidic conditions, the FFA goes through a very exothermic reaction of condensation, leading to the formation of a linear or branched polymer and water as products of reaction (Fig. 12.10). The resin, after the cure by cross-linking between the polymeric chains (Fig. 12.11), presents a framework which affords the material excellent chemical and fire resistance, typically justifying their application in foundry and refractory industries (Ibnyaich, 2010).

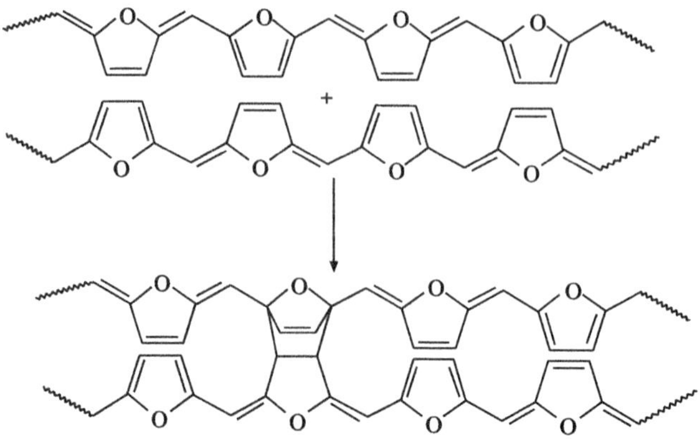

Figure 12.11 Chemical structure of furfuraldehyde cured resin. *Ibnyaich, A., 2010. Modification of the Properties of Biobased Thermoset Resin Using Cellulose Nano-Whiskers (CNW) as an Additive. 60f (Dissertation (Master)). Lulea University of Technology.*

At this point, it is worth mentioning that the biopolymers that are formed by chain carbohydrates, such as celluloses and hemicelluloses, have a chemical structure very similar to that of starch. The biopolymer starch, a constituent of the human food chain and/or animal, is widely used in industrial segments, such as pulp and paper, mining, textiles, and petroleum, for various purposes. The biopolymer celluloses (CMC, xylan, etc.), then become products with great potential to replace starch in industrial application, freeing this food source to decrease starvation and global hunger levels.

Phenolic resins

Phenolic resins can be prepared from the lignin extracted from sugarcane bagasse by organosolv process and replace phenol compounds from petrochemicals source (Santos and Curvelo, 1999). Phenolic resins exhibit dimensional stability, thermal stability, and chemical resistance at elevated temperatures. These resins present several industrial applications, particularly in the segments of abrasive, refractory, friction, molding compound or postmolding (Santos and Curvelo, 1999), plastics, automotive, and aerospace (Oliveira et al., 2008).

However, phenolic resins present high density of cross-linking, after curing, which gives them a low resistance to fracture, making the phenolic resins more fragile when compared to many other plastics. This fact prevents the phenolic resins implementation in some cases. Hence, in order to improve the impact resistance, it is necessary to strengthen these matrices by using composites (Tita et al., 2002).

Several research groups have been trying to use phenolic resins, formaldehyde (Doherty et al., 2007; Hoareau et al., 2006; Tita et al., 2002) or furfurylic (Oliveira et al., 2008), in combination with fibers of sugarcane bagasse and/or other fibrous sources, to produce lignocellulosic composites. Although these groups have reached several advances with respect to gains in physical–mechanical properties, is still needed achieve same properties, and other studies will be needed to release these composites to large-scale application.

Lignin

Usually lignin from sugarcane bagasse presents high content of *p*-hydroxyphenyl units lignin (H), therefore lower quantity of methoxyl groups are hardwood and softwood lignins. Final lignin structure depends not only on the raw material, but also the way it is extracted. Some advantages of lignin removed via organosolv process, for example, forward to the Kraft lignins, include high purity, relatively low molecular weight, and greater ability to be derivatized, due to the greater fraction of aromatic rings not replaced in position *ortho* (Doherty et al., 2007; Lora and Glasser, 2002).

A large amount of bagasse is currently burnt as a low-grade fuel for energy recovery; only a limited quantity has been used to make pulps and composite materials. The utilization of lignin from bagasse could be enhanced, if the industrial and scientific researchers' community can find better applications and products with high added value (Paiva and Frollini, 2002; Zarate et al., 2000).

The lignin from sugarcane bagasse has several applications:

1. Lignin is being used in the boilers of power plants due to their high energy content (Kleppe, 1970). Typically, the higher heating values (HHVs) of biomass fuels were correlated with their lignin content. There was a highly significant linear correlation ($r = 0.9504$) between the HHV of the biomass fuel and the lignin content explained by the following equation: HHV ($kJ\ g^{-1}$) $= 0.0889(L\ wt\%) + 16.8218$ (Demirbas, 2000).
2. Many studies suggest the use of lignin in mixtures containing polymers, aiming to improve various properties of these polymers, making them biodegradable, improving the mechanical properties, and reducing cost (Lora and Glasser, 2002). Lignin can be used as a sustainable

feedstock for production of a biobased version of Polyethylene terephthalate (PET), a thermoplastic polymer. This polymer is used in the manufacturing of a wide range of products like synthetic fibers and containers for food and plastic bobbles for beverage (Valmet Forward, 2017). PET is the most common thermoplastic polymer resin of the polyester family (Wikipedia, 2019). The world production of PET is according to some reports currently above 20 million tonnes per year, and rising. (Valmet Forward, 2017).

3. Production of vanillin, phenol, acetic acid, lignin oxidized, obtaining oils with characteristics similar to petroleum, the production of phenol-formaldehyde resins, dispersants, emulsifiers, builders of metal sequestering agent, adhesives, binders, obtaining of syngas used in the production of methanol (Lora and Glasser, 2002; Schuchardt et al., 2001).
4. Utilization of sugarcane bagasse lignin-based product for pesticide formulations. Lignin was first extracted from the black liquor of alkaline pulping of sugarcane bagasse, and then it was modified by oxidation, sulfonation, and copolymerization. Therefore after these steps a water-soluble product was obtained. The dispersing efficiency of the product being used as a dispersant for pesticide formulations was estimated via suspending rate measurement and was compared with some commercial dispersants (Zhili and Yuanyuan, 2011).
5. CMC lignin produced using sugarcane biomass was used as retarder additive for oil well cement slurry. It is established that CMC lignin production from sugarcane biomass can be an interesting process to focus on biorefinery associated to sugar industry. That process can reduce the disposal issues, avoiding its burning and generating more valuable products (Moreira et al., 2018).
6. Gregorová et al. (2005) showed that increased amount of lignin in polypropylene matrices enhanced some properties such as thermal stability and structural strength.
7. There are many studies on carbon fiber production using lignin extracted from sugarcane bagasse, and many properties of carbon fibers produced are still under study.

The range of work done with lignin as a raw material is quite wide and there is a tendency to increase, due to the availability of biomass generated in the mills and the necessity to use this material (Quinelato, 2016).

Sugarcane juice

The crushed sugarcane is then pressed to extract raw sugarcane juice from sugarcane bagasse. The juice represents approximately 80% dry basis mass of the sugarcane, being composed primarily of carbohydrates. Many studies are seeking to add value to the product, using it as a substrate in fermentation processes (Lemos and Castro, 2010; Faria, 2005). In fact, the composition of raw sugarcane juice is a mixture of carbohydrates, polyphenols, organic acids, and inorganic minerals as mentioned in Table 12.3 (Walford, 1996).

Polylactic acid

It is formed by the polymerization, via esterification, of units of lactic acid, which in turn can be obtained by fermentation of sugars. The spectrum of application of the polylactic acid (PLA) is expanding, and there is an opportunity to move several polymers produced from fossil fuel (Groot and Borém, 2010), in various industrial segments such as the pharmaceutical industry, biomedicine, food packaging, and plastics, in general (Shen et al., 2009; Serna et al., 2003). Fig. 12.12 illustrates the possible fields of application of lactic acid, highlighting the production of PLA and acrylic acid which will be discussed later in this chapter.

Fig. 12.13 shows the production of biopolymer from sugarcane.

Table 12.3 Sugarcane raw juice chemical composition.

Sugarcane juice	Chemical composition	Brix matter (%)
Sugars/carbohydrates	Sucrose	81–87
	Reducing sugars	03–06
	Oligosaccharides	0.06–0.6
	Polysaccharides (gums and dextrans)	0.2 – 0.8
Salts	Inorganic acids	1.5–3.7
Organic nonsugars	Organic acids	0.7–1.3
	Amino acids	0.5–2.5
	Dextrans	0.1–0.6
	Starch	0.11–0.5
	Gums	0.02–0.05
	Waxes, fats, phospholipids	0.05–0.15
	Colorants	0.1
Insolubles	Sand, bagasse, etc.	0.15–1

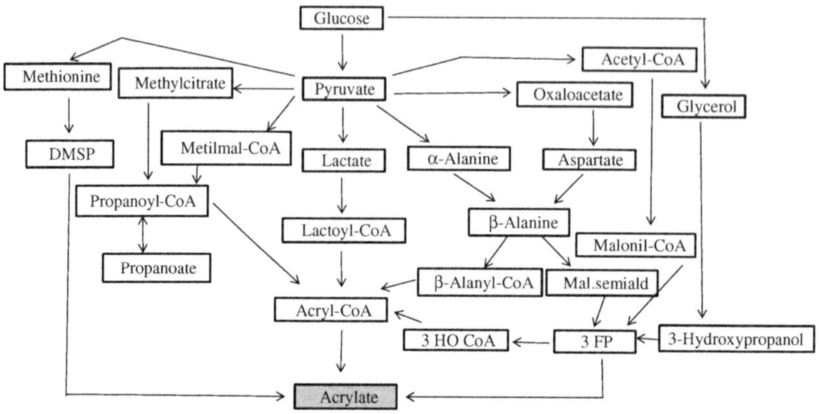

Figure 12.12 Commercial use and applications of lactic acid and its salts. *Wee, Y.J., Kim, J.N., Ryu, H.W., 2006. Biotechnological production of lactic acid and its recent applications. Food Technol. Biotechnol. 44 (2), 163–172.*

Acrylic acid

Acrylic acid and its esters are mainly used as raw materials in the production of polymer. The polymers produced from acrylic acid and its esters are characterized by transparency, easy adhesion, elasticity, and stability to heat and light.

They are used in the coating of surfaces and the production of textile sectors, adhesives, treatment of paper, fibers, detergents, superabsorbent materials, etc. Currently, all acrylic acids are produced from the petrochemical industry (Falbe, 1995, mentioned by Lunelli, 2010).

This acid has, in its structure, carboxylic group terminals that confer an anionic behavior to its molecule. This characteristic renders abilities of ionic exchange and complexation with ions of positive charge (Alpatova et al., 2004). This ability is maximized when the polymerized acrylic acid is used, for example, in the water and effluents treatment for the removal of heavy metals or as antiscaling agent to avoid low solubility salts deposition at several industrial processes (Costa et al., 2018).

The dehydration of lactic acid, produced by the fermentation of sugarcane juice, leading to acrylic acid, is considered an alternative process, independent of the petrochemical process (AIDA, 2009, mentioned by Lunelli, 2010). The dehydration of lactic acid for the production of acrylic acid is a viable process. Nevertheless, improvements are still needed to make it more competitive, as well as the development of new

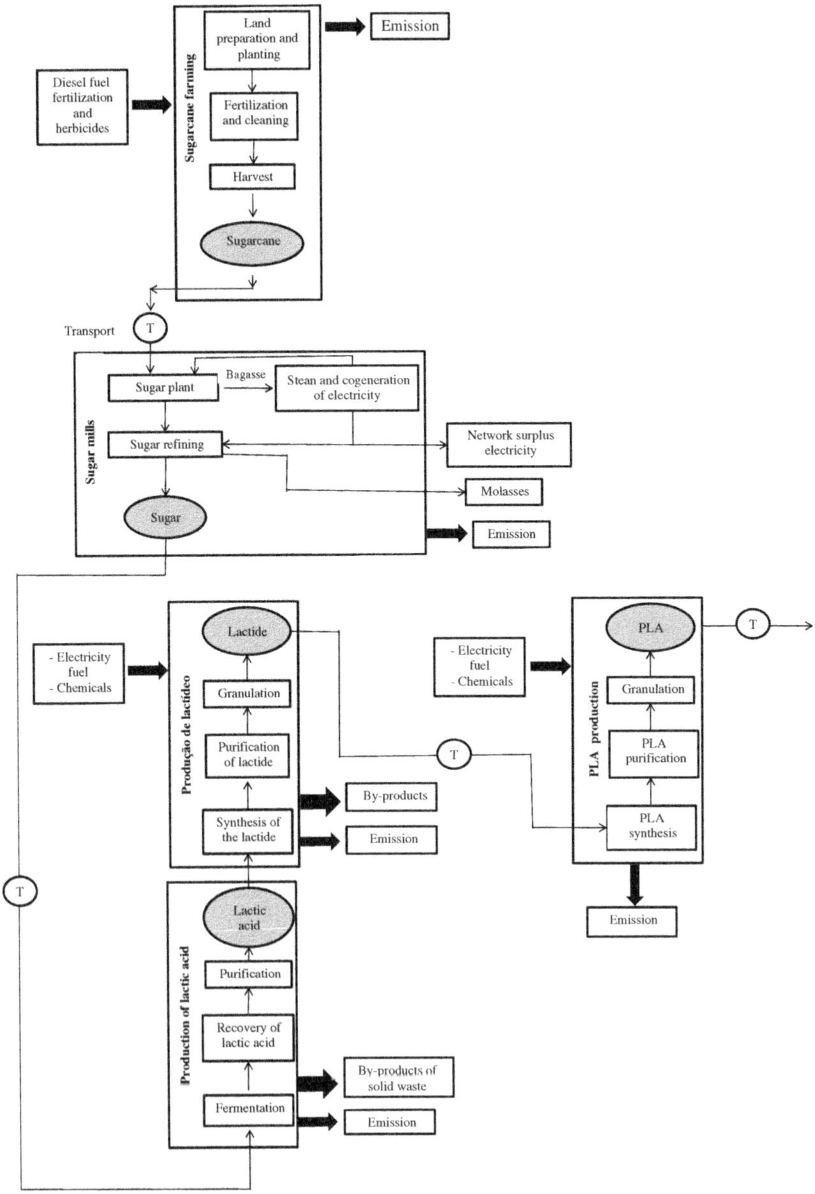

Figure 12.13 Productive chain of polylactic acid (PLA) via sugarcane biomass. *Groot, W., Borém, T., 2010. Life cycle assessment of the manufacture of lactides and PLA biopolymers form sugarcane in Thailand. Int. J. Life Cycle Assess. 15, 970–984.*

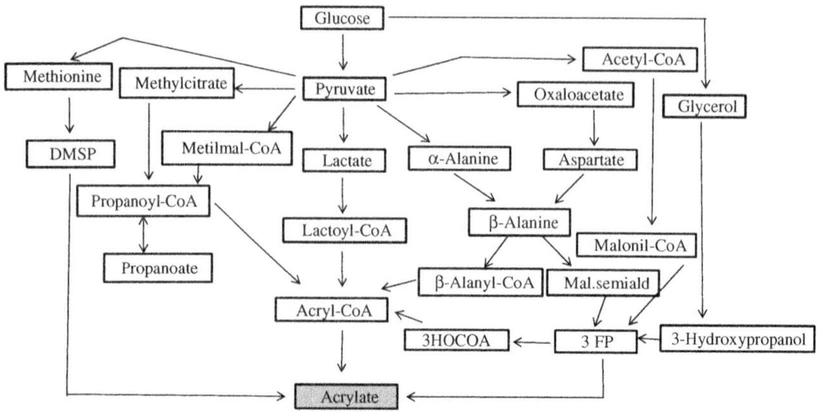

Figure 12.14 Routes for biosynthesis of acrylic acid. *Straathof, A.J.J., Sie, S., Franco, T. T., Van Der Wielen, L.A.M., 2005. Feasibility of acrylic acid production by fermentation. Appl. Microbiol. Biotechnol. 67, 727–734.*

fermentation processes. The direct dehydration of lactic acid for the acrylic acid is expressed by Eq. (12.3), as mentioned by Straathof et al. (2005).

$$C_3H_6O_3 \rightarrow C_3H_4O_2 + H_2O \tag{12.3}$$

Among the various metabolic routes presented in Fig. 12.14 for the production of acrylic acid from renewable sources, the most important one is via lactate. Obtaining acrylic acid polymers has gained focus from researches. The main reason is regarding the easy way to generate lactic acid and its precursor via fermentation of carbohydrates present in sugarcane juice.

Polyhydroxybutyrate

Fig. 12.15 shows the molecule of biopolymer polyhydroxybutyrate (PHB), which is a polyester family of polyhydroxyalkanoates (Fig. 12.16).

PHB presents white coloration and grainy appearance and can be obtained from the fermentation of sugarcane juice. The sucrose is initially inverted by enzymatic process, transforming it into a syrup, that is, fructose and glucose monomers. There is, then, the formation of the polymer inside of bacteria that feed the syrup (Fig. 12.17). PHB can be used in the manufacture of pots, spoons, and plastic bags, among other objects, and possess properties similar to polypropylene (Telles et al., 2011).

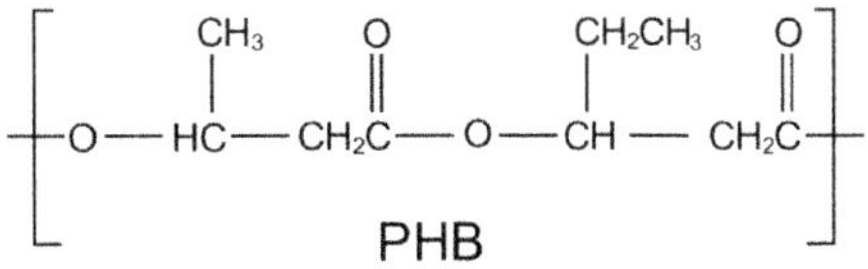

Figure 12.15 The biopolymer molecule polyhydroxybutyrate.

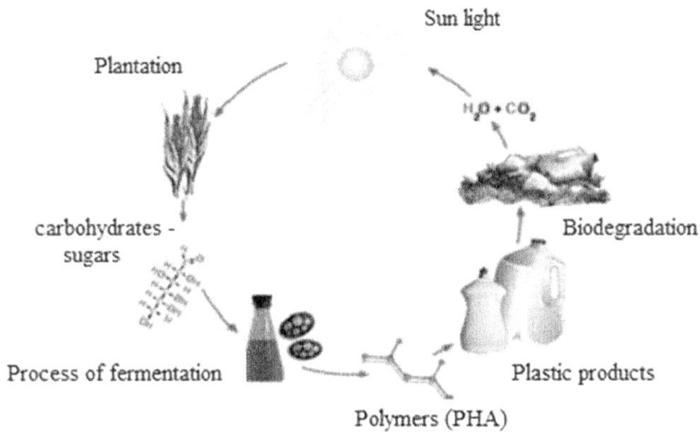

Figure 12.16 Polihidroxialcanoatos biocicle. *Adapted from Marangoni, C., 2000. Estudo de estratégias de produção de poli(3 hidroxibutirato-3-hidroxivalerato) por Ralstonia Eutropha utilizando substratos de baixo custo e ácido propiônico (Dissertation (Master)). UFSC, Florianópolis.191f.*

Rosa (2002), cited by Coutinho et al. (2004), proposed a cycle of PHB production in a sugar mill, where the effluent from that production would be used for fertigation and to produce 1 kg of PHB from 3 kg of sugar. Besides, this biopolymer has an average time of degradation of 6–12 months, compared with 40–50 years or up to 200 years in the case of synthetic polymers. In addition, to being biocompatible and easily absorbed by the human body, it can be used in medical and pharmaceutical industries, adequate to manufacture suture wires, bone implants, and capsules that gradually release medicines into the bloodstream (Coutinho et al., 2004).

Green polyethylene

Green polyethylene (GPE) is a polymer widely used in everyday applications in plastics in general, such as bags, films and packaging for food industries, and personal and domestic hygiene, but then again is a nonrenewable source. Thus the GPE which can be obtained from sugarcane juice carbohydrates via fermentation to ethanol, dehydration, and

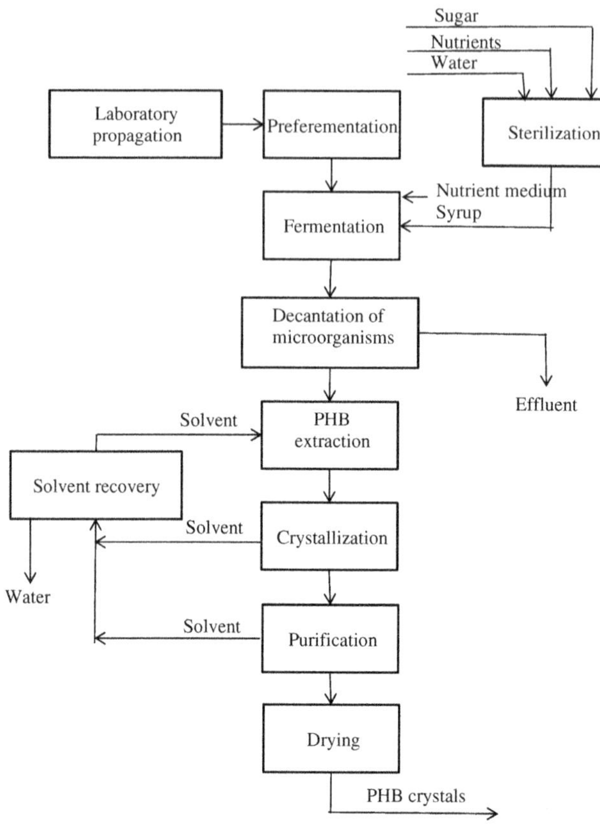

Figure 12.17 Production process of polyhydroxybutyrate (PHB). *Adapted from Bucci, D.Z., 2003. Avaliação de embalagens de PHB (poli (Ácido 3-Hidroxibutírico)) para alimentos (Dissertation (Master)). Universidade Federal de Santa Catariana, Florianópolis, 146f.*

polymerization [Eq. (12.4)] are an excellent cost-competitive alternative to renewable sources. If the biomass remained at gas phase, for example, methanol source, obtained from synthesis, gas can be converted into ethylene, propylene, and higher alcohols, by using proper acid zeolites.

$$CH_3CH_2OH \rightarrow CH_2CH_2 + H_2O \rightarrow (CH_2 - CH_2)n \qquad (12.4)$$

The GPE has the same technical characteristics and appearance as a conventional polyethylene from petrochemical sources. This means that there is no need of investment in equipment and/or processes retrofitting for using GPE. Consequently resulting in lower capital expenditures and higher speed of deployment of such more sustainable practices by

Figure 12.18 Green polyethylene (PE) production. *Adapted from Belloli, R., 2010. Polietileno Verde do Etanol da Cana-de-açúcar Brasileira: biopolímero de classe mundial. (Completion of Course Work). UFRS, Porto Alegre, 39 pp.*

companies. In Fig. 12.18, some important relationships in the production of GPE using sugarcane biomass are represented (Belloli, 2010).

Xanthan gum

It is a biopolymer classified as heteroexopolis-saccharide branched group that has the ability to form gels and viscous solutions in aqueous medium (Medeiros, 2000). Xanthans are high added value products and are not produced in Brazil. In this way, their local production would reduce the costs of importation as well.

Xanthans present themselves as an alternative to traditional gums due to their rheological properties, being widely used as thickener agents, gelling agents, suspension agents, colloid protectors, and stabilizers in the food industries, pharmaceutical, chemical, and petrochemical industries (Faria, 2005; Lima, 2001).

The sugarcane juice can be easily converted into xanthan, when in the presence of bacteria *Xanthomonas campestris* type. Sucrose has been used for conventional production of xanthans. However, using sugarcane juice, the yield average for production of gum is 10 times higher and the apparent viscosity five times smaller. Other tests should be performed for fermentation process optimization (Brandão et al., 2008).

Perspectives and challenges for the production of biopolymer from sugarcane

Certainly environmental problems faced today have driven effort of development of new technologies to answer them. Thus the biopolymers are an alternative in this scenario, but there are many challenges to present their positive environmental characteristics and subsequently to be able to insert them in a relevant industrial scale in the global market (Belloli, 2010).

The production of biopolymers from biomass, especially in a tropical country like Brazil, could easily replace many petrochemical products. Following, Schuchardt (2001) mentioned that about 1% of the biomass produced in Brazil would be necessary for total replacement of petrochemical source for chemicals products. Consequently, Brazil has a potential place as a player in this scenario.

Numerous researches with biopolymers started several decades ago, and these products are already on the European market for more than two decades. Besides, the current market is characterized by a high growth behavior with strong diversification as a bioplastic, for example, one of these biopolymers mentioned in this chapter. Fig. 12.19 shows the overall bioplastic world production capacity. It could be noted that bioplastic world production capacity has a growth of about 6.4% annually, from 2016 (4.16 million ton) to forecast for 2021 (6.11 million ton) mentioned by European Bioplastics (2012).

Currently bio-based building blocks are the core of the new bioeconomy. Fig. 12.20 illustrates the development of capacities for the main bio-based building blocks, used for the production of polymers. Between 2018 and 2023, the CAGR of 4.5% will be only slightly higher than that of bio-based polymers (4%) as a whole. The building blocks can be used in structural polymers as well as in functional polymers (for definition, see

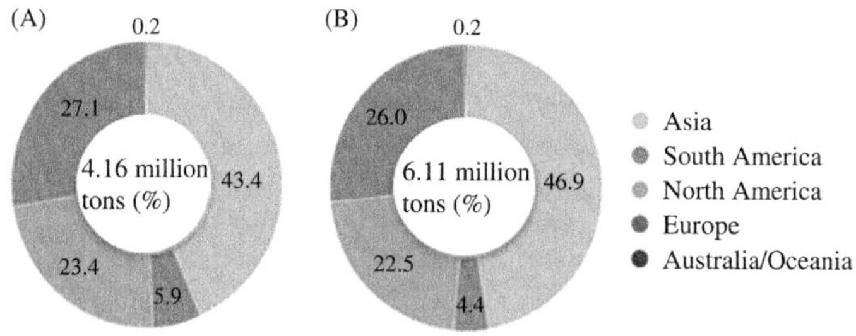

Figure 12.19 Overall bioplastic world production capacity by region for 2016 and forecast for 2021. (A) Global bioplastic production capacity in 2016 (by region) and (B) global bioplastic production capacity in 2021 (by region). *Adapted from European Bioplastics/University of Applied Sciences and Arts Hanover. 2012. Available from: <http://en.european-bioplastics.org/market/market-development/> (accessed 15.03.19.).*

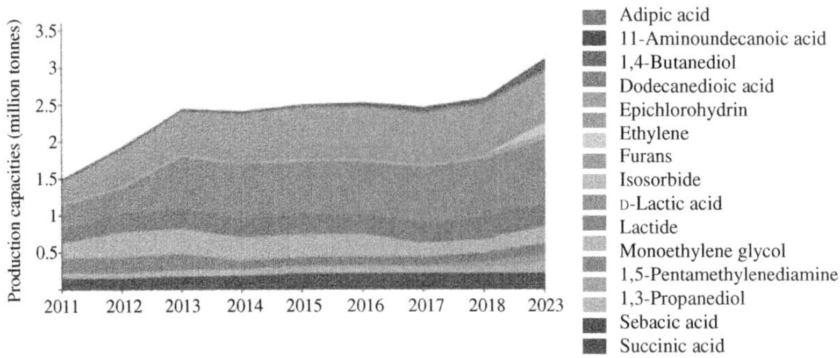

Figure 12.20 Bio-based building blocks: evolution of worldwide production capacities from 2011 to 2023. *Chinthapalli, R., Skoczinski, P., Carus, M., Baltus W., Guzman D., Kāb, H., et al., 2018. Bio-Based Building Blocks and Polymers – Global Capacities Production and Trends 2018–2023. Nova Institute for Ecology and Innovation, p. 8.*

below) and also in various other applications such as food, feed, cosmetics, or pharmaceuticals (Chinthapalli et al., 2018).

The overall production capacity of bio-based building blocks increased by about 5% (120,000 t/a) in 2018, although some pioneers went bankrupt. The overall forecast for bio-based building blocks evolution worldwide indicates a total growth by 4.5% until 2023.

References

Alpatova, A., Verbych, S., Bryk, M., Nigmatullin, R., Hilal, N., 2004. Ultrafiltration of water containing natural organic material: heavy metal removing in the hybrid complexation-ultrafiltration process. Sep. Purif. Technol. 40, 155–162.

Belloli, R., 2010. Polietileno Verde do Etanol da Cana-de-açúcar Brasileira: biopolímero de classe mundial. (Completion of course work). UFRS, Porto Alegre, 39 pp.

Brandão, L.V., Nery, T.B.R., Machado, B.A.S., Esperidião, M.C.A., Druzian, J.I., 2008. Produção de goma xantana obtida a partir do caldo de cana. Ciênc. Tecnol. Alimentos 28 (Suppl.), 217–222.

Bucci, D.Z., 2003. Avaliação de embalagens de PHB (Poli (Ácido 3-Hidroxibutírico)) para alimentos (Dissertation (Master)). Universidade Federal de Santa Catariana, Florianópolis, 146f.

Camargo, F.A., Innocentini-Mei, L.H., Lemes, A.P., Moraes, S.G., Durán, N., 2012. Processing and characterization of composites of poly(3-hydroxybutyrate-*co*-hydroxyvalerate) and lignin from sugarcane bagasse. J. Compos. Mater. 46 (4), 417–425.

Chinthapalli, R., Skoczinski, P., Carus, M., Baltus, W., Guzman, D., Kāb, H., et al., 2018. Bio-Based Building Blocks and Polymers – Global Capacities Production and Trends 2018–2023. Nova Institute for Ecology and Innovation, p. 8.

CONAB - Companhia Nacional de Abastecimento, 2018. Acompanhamento da safra brasileira de cana-de-açúcar. V. 4 - SAFRA 2017/18 N.3 - Terceiro levantamento | December 2017. ISSN: 2318-7921.

Costa, M.M., Hill, C., Freeland, C., Rodrigues, F., Hakas, A., 2018. KemTalo Scale 207 as antiscalant for sugarcane juice evaporation system. American Society of SugarCane Technologists. In: 48th Annual Joint Meeting Conference, Hyatt Regency Coconut Point Bonita Springs, FL. June 25–27, 2018.

Costa, M.M., Resende, T.M., Cabo, S., (2017). Novel Synthetic Depressant for Iron Ore Flotation. *19th International Symposium on Wood, Fibre and Puling Chemistry, Aug. 30 -Sept. 01, 2017. Porto Seguro, BA, Brazil.*

Coutinho, B.C., Miranda, G.B., Sampaio, G.R., De souza, L.B.S., Santana, W.J., Coutinho, H.D.M., 2004. A importância e as vantagens do polihidroxibutirato (plástico biodegradável). Holos 76–81.

Demirbas, A. (2000). Relationships between lignin contents and heating values of biomass. Department of Science Education, Karadeniz Technical University, 61365, Akçaabat, Trabzon, Turkey. Received 7 September 1999; accepted 14 March 2000.

Doherty, W., Halley, P., Edye, L., Rogers, D., Cardona, F., Park, Y., et al., 2007. Studies on polymers and composites from lignin and fibers derived from sugar cane. Polym. Adv. Technol. 18, 673–678.

Edgar, K.J., Buchanan, C.M., Debenham, J.S., Rundquist, P.A., Siller, B.D., Shelton, M. C., et al., 2001. Advance in cellulose ester performance and application progress. Polym. Sci. 26, 1605–1688.

European Bioplastics/University of Applied Sciences and Arts Hanover, 2012. Available from: <http://en.european-bioplastics.org/market/market-development/>. (accessed 15.03.19.).

Faria, S., 2005. Produção de goma xantana empregando caldo de cana por *Xanthomonas campestris* pv *campestris* (Dissertation (Master)). Universidade Federal de Uberlândia, Uberlândia, NRRL B-1459, 119f.

Filho, G.R., Toledo, L.C., Cerqueira, D.A., Assunção, R.M.N., Meireles, C.S., Otaguro, H., et al., 2007. Water flux, DSC, and cytotoxicity characterization of membranes of cellulose acetate produced from sugarcane bagasse, using PEG 600. Polym. Bull. 59, 73–81.

Gregorová, A., Cibulková, Z., Kosíková, B., Simon, P., 2005. Stabilization effect of lignin in polypropylene and recycled polypropylene. Polym. Degrad. Stabil. 89, 553–558.

Groot, W., Borém, T., 2010. Life cycle assessment of the manufacture of lactides and PLA biopolymers form sugarcane in Thailand. Int. J. Life Cycle Assess. 15, 970–984.

Guimarães, A.F., Silva, E.A., Colavite, A.P., 2018. Distribuição geográfica e escoamento da produção de biocombustíveis da região sul do brasil. Rev. Eletrôn. Cien. Inov. Tecnol. 9 (18).

Heydarzadeh, H.D., Najafpour, G.D., Nazari-Moghaddam, A.A., 2009. Catalyst-free conversion of alkali cellulose to fine carboxymethyl cellulose at mild conditions world applied. Sci. J. 6 (4), 564–569.

Honorato, A., Trejo, S., Aburto, J., 2017. Base de datos de materiales lignocelulósicos. CONACyT, Mexico, Project 151370.

Hoareau, W., Oliveira, F.B., Grelier, S., Siegmund, B., Frollini, E., Castellan, A., 2006. Fiberboards based on sugarcane bagasse lignin and fibers. Macromol. Mater. Eng. 29, 829–839.

Ibnyaich, A., 2010. Modification of the Properties Biobased Thermoset Resin Using Cellulose Nano-Whiskers (CNW) as an Additive (Dissertation (Master)). Lulea University of Technology, 60f.

Kamm, B., Gruber, P.R., Kamm, M., 2006. 407 pp Biorefineries – Industrial Processes and Products. Wiley-VCH, Weinheim.

Kleppe, P.J., 1970. Kraft pulping. TAPPI J. 53, 35–47.

Kumar, A., Negi, Y.S., Bhardwaj, N.K., Choudhary, V., 2012. Synthesis and characterization of methylcellulose/PVA based porous composite. Carbohydr. Polym. 88, 1364–1372.

Lemos, J.C.M., Castro, T.M.O., 2010. Uma análise comparativa entre o difusor e a moenda na extração da sacarose – (Completion of Course Work). Universidade de Franca, Franca, 67 pp.
Lima, U.A., 2001. Biotecnologia industrial: processos fermentativos e enzimáticos, vol. 3. Edgard Blucher Ltd, São Paulo, pp. 125–154.
Lora, J.H., Glasser, W.G., 2002. Recent industrial applications of lignina: a sustainable alternative to nonrenewable materials. J. Polym. 10, 39–48.
Lunelli, B.H., 2010. Produção e controle da síntese do éster de ácido acrílico através da fermentação do ácido láctico (Thesis (Doctorate)). UNICAMP, Campinas, 228f,.
Marangoni, C., 2000. Estudo de estratégias de produção de poli(3 hidroxibutirato-3-hidroxivalerato) por *Ralstonia Eutropha* utilizando substratos de baixo custo e ácido propiônico (Dissertation (Master)). UFSC, Florianópolis, 191f.
Martinez, C.M.F., 1996. Obtenção e caracterização de materiais absorventes através da carboximetilação de polpa etanol/água de medula de bagaço de cana-de-açúcar. (Completion of Course Work). USP, São Carlos, 146f.
McKendry, P., 2002. Energy production from biomass (part 1): overview of biomass. Bioresour. Technol. 83, 37–46.
Medeiros, I.S., 2000. Influência do método de recuperação e purificação de xantana produzida por *Xanthomonas campestris* pv. *pruni* cepa 06. In: Congresso brasileiro de ciência e tecnologia de alimentos, 17, Fortaleza. Anais… Fortaleza, vol. 3, pp. 9–29.
Meireles, C.S., 2007. Síntese e caracterização de membranas de acetato de celulose, obtido do bagaço de cana-de-açúcar, e blendas de acetato de celulose com poliestireno de copos plásticos descartados (Dissertation (Master)). UFU, 80f.
Morais, L.C., Filho, S.P.C., 1999. Carboximetilação de Polpas de Bagaço de Cana-de-Açúcar e Caracterização dos Materiais Absorventes Obtidos. Polímeros. Ciênc. Tecnol. 9, 46–51.
Moreira, P.H.S.S., Freiras, J.C.O., Braga, R.M., Araújo, R.M., Souza, M.A.F., 2018. Production of carboxymethyl lignin from sugarcane bagasse: a cement retarder additive for oilwell application. Ind. Crop. Prod. 116, 144–149.
NASA. NASA's Global Climate Change Website, 2019. NASA's Goddard Institute for Space Studies, NOAA National Climatic Data Center, Met Office Hadley Centre/Climatic Research Unit and the Japanese Meteorological Agency. (Graph produced by Earth Science Communications Team at NASA's Jet Propulsion Laboratory | California Institute of Technology). Available from <https://climate.nasa.gov> (accessed 15.3.19.).
Oliveira, F.B., Gardrat, C., Enjalbal, C., Frollini, E., Castellan, A., 2008. Phenol-furfural resins to elaborate composites reinforced with sisal fibers – molecular analysis of resin and properties of composites. J. Appl. Polym. Sci. 109, 2291–2303.
Paiva, J.M.F., Frollini, E., 2002. Sugarcane bagasse reinforced phenolic and lignophenolic composites. J. Appl. Polym. Sci. 83 (4).
Quinelato, C., 2016. Métodos de extração da lignina do bagaço da cana-de-açúcar da região noroeste do Estado de São Paulo. 2016. 95f. Dissertação (Mestrado em Química) – Instituto de Biociências, Letras e Ciências Exatas da Universidade Estadual Paulista "Júlio de Mesquita Filho", Campus de São José do Rio Preto.
Reddy, C.S.K., Ghai, R., Rashmi, Kalia, V.C., 2003. Polyhydroxyalkanoates: an overview. Bioresour. Technol. 87 (2), 137–146.
Santos, F., Curvelo, A.A.S., 1999. Utilização de ligninas em resinas fenólicas. 1. Preparação de Novolacas a partir de lignossulfonatos. Polímeros: Ciênc. Tecnol. 49–58.
Sassi, J.F., Chanzy, H., 1995. Ultrastructural aspects of the acetylation of cellulose. Cellulose 2, 111–127.
Schuchardt, U., Ribeiro, M.L., Gonçalves, A.R., 2001. A indústria petroquímica no próximo século: como substituir o petróleo como matéria-prima? Quím. Nova 24 (2), 247–251.

Serna, C.L., Rodríguez, S.A., Albán, A.F., 2003. Ácido Poliláctico (PLA): Propiedades y Aplicaciones. Ing. Compet. 5 (1).

Shen, L., Haufe, J., Patel, M.K., 2009. Overview and market projection of emerging bio-based plastics. Group Science, Technology and Society, Universiteit Utrecht, Pro-Bip Final Report.

Solé, J., Olivares, A.G., Turiel, A., Ballabrera-Poy, J., 2018. Renewable transitions and the net energy from oil liquids: a scenarios study. Renew. Energ. 116, 258–271.

Steinmeir, H., 2004. Acetate manufacturing, process and technology. Macromol. Symposia, 208, 49–60.

Straathof, A.J.J., Sie, S., Franco, T.T., Van Der Wielen, L.A.M., 2005. Feasibility of acrylic acid production by fermentation. Appl. Microbiol. Biotechnol. 67, 727–734.

Telles, M.R., Saran, L.M., Unêda-Trevisolli, S.H., 2011. Produção, propriedades e aplicações de bioplástico obtido a partir da cana-de-açúcar. Ciência & tecnologia: Fatec-jb, Jaboticabal 2 (1), 52–63.

Tendência Consultoria, 2018. Available from: <https://g1.globo.com/economia/noticia/2018/11/16/precos-do-petroleo-tem-alta-mas-devem-ter-6a-perda-semanal.ghtml> (accessed 15.03.19.).

Tita, S.P.S., Paiva, J.M.F., Frollini, E., 2002. Resistência ao Impacto e Outras Propriedades de Compósitos Lignocelulósicos: Matrizes Termofixas Fenólicas Reforçadas com Fibras de Bagaço de Cana-de-açúcar. Polímeros: Ciênc. Tecnol. 12 (4), 228–239.

Triana, O., et al., 1990. Atlas del Bagazo de la Cãna de Azucar, Cuba-9/GEPLACEA/PNUD/ICIDCA, México.

Vieira, R.G.P., Filho, G.R., Assunção, R.M.N., Meireles, C.S., Vieira, J.G., Oliveira, G. S., 2007. Synthesis and characterization of methycellulose from sugar bagasse celulose. Carbohydr. Polym. 67, 182–189.

Wee, Y.J., Kim, J.N., Ryu, H.W., 2006. Biotechnological production of lactic acid and its recent applications. Food Technol. Biotechnol. 44 (2), 163–172.

Yang, K., Wang, X., Wang, Y., 2007. Progress in nanocomposite of biodegradable polymer. J. Indust. Eng. Chem., Seoul, Republic of Korea 13, 485–500.

Yu, L., Dean, K., Li, L., 2006. Polymer blends and composites from renewable resources. Prog. Polym. Sci. 31 (6), 576–602.

Walford, S.N., 1996. Composition of Sugarcane Juice. Sugar Milling Research Institute, University of Natal, Durban. 266 ProcS. Afr. Sug. Technol. Ass., 70.

WEO, 2014. Available from: <http://www.worldenergyoutlook.org/publications/weo-2014/> (accessed 15.03.19.).

Zarate, C.N., Aranguren, M.I., Reboredo, M.M., 2000. Resol–vegetable fibers composites. J. Appl. Polym. Sci. 77 (8).

Zeinaly, F., Saraeian, A., Gabov, K., Fardim, P., 2017. Determination of carbohydrates in sugarcane bagasse pulp in different TCF bleaching sequences. Cell. Chem. Technol. 51 (1–2), 45–53.

Zhili, L.I., Yuanyuan, G.E., 2011. Extraction of lignin from sugar cane bagasse and its modification into a high performance dispersant for pesticide formulations. J. Braz. Chem. Soc., São Paulo 22 (10), 1866–1871.

Further reading

Braun, J.L., Holtman, K.M., Kadla, J.F., 2005. Lignin based carbon fibers: oxidative termo stabilization of kraft lignin. Carbon 43, 385–394.

CHAPTER 13

Bioelectricity of sugarcane: a case study from Brazil and perspectives

Zilmar José de Souza

Contents

Introduction 256
What is bioelectricity? 256
The electric matrix and power installed by bioelectricity 257
  The generation of electricity by the biomass source in general 257
São Paulo: responsible for 44% of bioelectricity in 2018 261
Productivity in the generation of sugarcane bioelectricity 263
  The commercialization of bioelectricity 264
  The system guarantees product quality and offer 264
  The free energy market and bioelectricity 265
  The regulated energy market and bioelectricity 266
  The year 2019 in the regulated market 268
One point of attention: the decontracting of bioelectricity in the coming years 269
  Evolution of investment in bioenergy 270
The potential of bioelectricity yet to enjoy 272
Expansion of biomass is far from indicated in the 10-year plan in Brazil 272
The potential of bioelectricity and the RenovaBio 273
An agenda for bioelectricity 275
  Efforts for regular and growing contracting for bioelectricity and biogas in regulated auctions and strengthening of the hiring free environment 275
  Equalization of the current judicialization in the financial settlements in the short-term market 275
  Improvement of the methodology of revision of the physical energy guarantee for biomass/biogas plants 277
References 278
Further reading 279

Sugarcane Biorefinery, Technology and Perspectives
DOI: https://doi.org/10.1016/B978-0-12-814236-3.00013-5

Introduction

This chapter aims to present an overview of the main topics related to the bioelectricity offered to the electricity grid in Brazil, focusing on the sugarcane industry. In order to do so, the general concepts and benefits of bioelectricity for the system, the installed capacity in the country, the evolution of the generation for the electric grid, the commercial environment and its challenges, the potential of bioelectricity and a guideline for its development will be presented.

What is bioelectricity?

Bioelectricity is a renewable and sustainable electricity produced from the biomass of sugarcane (bagasse and straw), wood waste, charcoal, rice hulls, elephant grass, and other biomasses.

In the specific case of the sugar-energy sector, each ton of sugarcane ground in sugar and ethanol production generates, on an average, 50% moisture, 250–270 kg of bagasse, and more than 200 kg of straw.

With high fiber content, sugarcane bagasse, since the industrial revolution, has been used in the production of steam and electric energy for the production of sugar and ethanol, guaranteeing the energy self-sufficiency of the plants during the harvest period.

But in addition to meeting the power needs of the mills, bagasse has allowed the generation of surplus electric power that is supplied to the Brazilian electric system.

In the 2018/2019 harvest, Brazil produced 620.8 million tons of sugarcane, resulting in 29 million tons of sugar and 33.1 billion liters of ethanol. A third product should also be remembered in the final results of the harvest, for its strategic role in guaranteeing the supply of electricity in 2018: bioelectricity generated through bagasse and sugarcane straw.

Since 1987, in addition to meeting the energy needs for sugar and ethanol production, the sugar-energy sector has been using part of this biomass to supply electricity to the National Interconnected System (NIS), reaching, in 2018, a total of 21.5 TWh, to account for more than 4% of Brazilian national consumption. In 2018, 82% of the bioelectricity that was supplied to the NIS came from the sugarcane sector.

The electric matrix and power installed by bioelectricity

In 2019 the installed capacity currently granted in the country by the National Electric Energy Agency (ANEEL) is 169,663 MW. The biomass source in general, with 14,889 MW, represents 9% of the power granted in the electric matrix of Brazil, occupying the fourth position in the matrix, behind the hydro source, natural gas, thermoelectric, and wind power plants (Fig. 13.1).

With regard to sugarcane bioelectricity, the sugarcane industry holds 11,424 MW, higher than the installed capacity at Belo Monte hydroelectric power plant, representing around 7% of the power granted in Brazil and 77% of the biomass source (Fig. 13.2).

In terms of the annual evolution of installed capacity, the biomass source had its record of expansion in 2010, with 1750 MW (equivalent to 12.5% of an Itaipu Plant), resulting from investment decisions prior to 2008, when the scenario was stimulating regarding the expansion of the sugar-energy sector.

Biomass, which already accounts for 32% of the country's annual installed capacity growth (in 2010), should only install 101 new MW in 2019, representing 3% of the installed capacity expansion in Brazil, as can be seen below (Figs. 13.3 and 13.4).

The generation of electricity by the biomass source in general

In 2018, 82% of the bioelectricity that was supplied to NIS came from the sugarcane sector (Fig. 13.5).

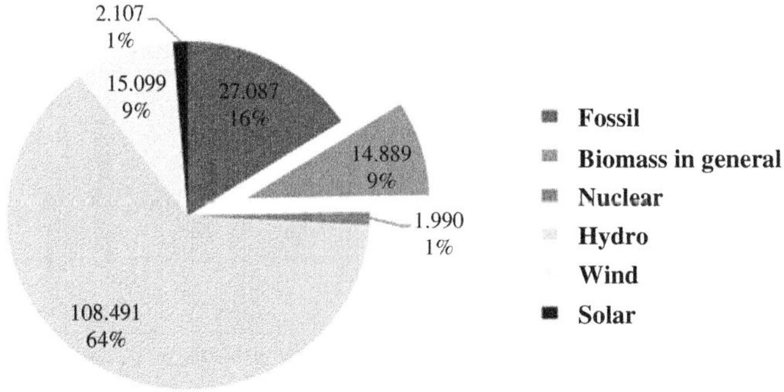

Figure 13.1 Power granted by source: in commercial operation (GW and %). *UNICA, 2019. Cartilha: a bioeletricidade da cana e o mercado livre de energia elétrica no brasil "Cartilha: The Bioelectricity of Sugarcane and the Free Market of Electric Energy in Brazil". União da Indústria de Cana-de-Açúcar (UNICA). July 2019. Available from: <http://www.unica.com.br/documentos/publicacoes/bioeletricidade/>, Data from ANEEL (2019).*

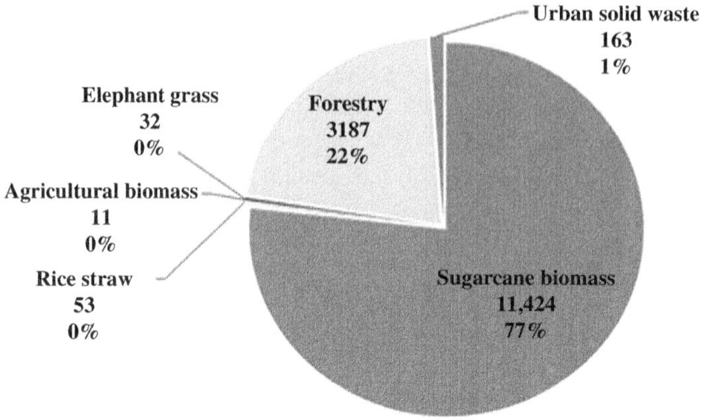

Figure 13.2 Power granted biomass source: in commercial operation (MW and %). *ÚNICA, 2019. Cartilha: a bioeletricidade da cana e o mercado livre de energia elétrica no brasil "Cartilha: The Bioelectricity of Sugarcane and the Free Market of Electric Energy in Brazil". União da Indústria de Cana-de-Açúcar (UNICA). July 2019. Available from: <http://www.unica.com.br/documentos/publicacoes/bioeletricidade/>, through ANEEL (2019).*

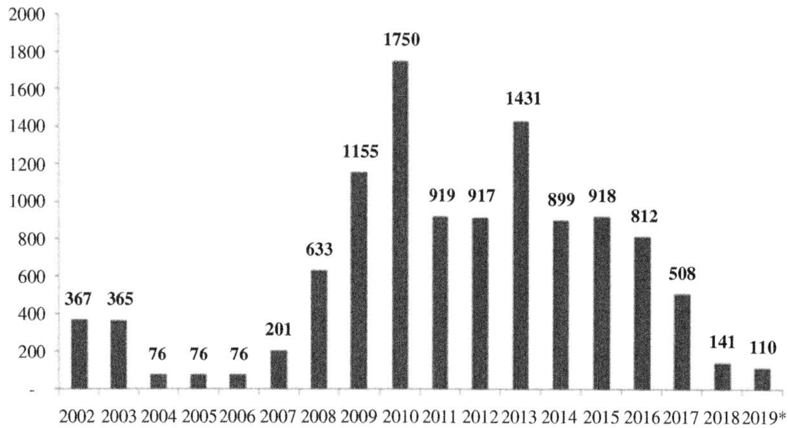

Figure 13.3 Annual increase in installed capacity by biomass, 2002–19, Brazil (MW). *Prediction, including projects with restriction in start of operations. Survey conducted on August 7, 2019. *ÚNICA, 2019. Cartilha: a bioeletricidade da cana e o mercado livre de energia elétrica no brasil "Cartilha: The Bioelectricity of Sugarcane and the Free Market of Electric Energy in Brazil". União da Indústria de Cana-de-Açúcar (UNICA). July 2019. Available from: <http://www.unica.com.br/documentos/publicacoes/bioeletricidade/>, database from ANEEL (2019).*

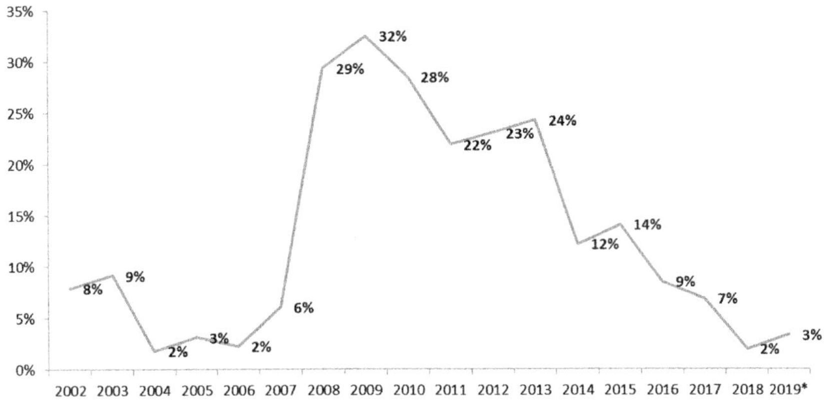

Figure 13.4 Representation of the annual increase of installed capacity by biomass in relation to the total increase in the electric energy matrix, 2002–19, Brazil (%). *Prediction, including projects with restriction in start of operations. Survey conducted on August 7, 2019. *ÚNICA, 2019. Cartilha: a bioeletricidade da cana e o mercado livre de energia elétrica no brasil "Cartilha: The Bioelectricity of Sugarcane and the Free Market of Electric Energy in Brazil". União da Indústria de Cana-de-Açúcar (UNICA). July 2019. Available from: <http://www.unica.com.br/documentos/publicacoes/bioeletricidade/>, database from ANEEL (2019).*

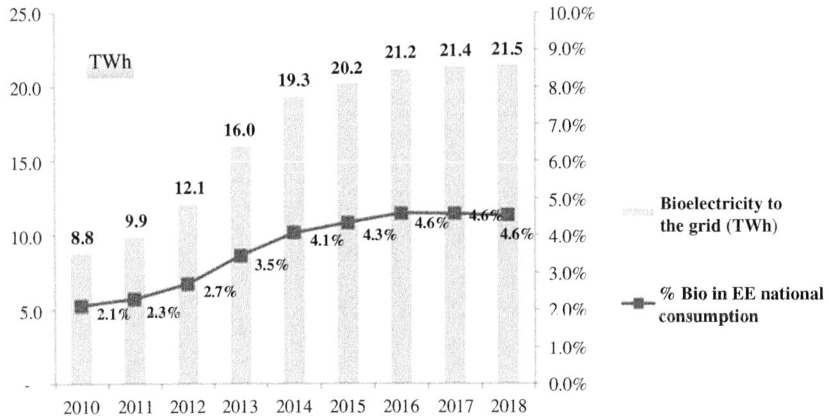

Figure 13.5 Bioenergy bioelectricity (GWh) offered for the network, 2010–18. *ÚNICA, 2019. Cartilha: a bioeletricidade da cana e o mercado livre de energia elétrica no brasil "Cartilha: The Bioelectricity of Sugarcane and the Free Market of Electric Energy in Brazil". União da Indústria de Cana-de-Açúcar (UNICA). July 2019. Available from: <http://www.unica.com.br/documentos/publicacoes/bioeletricidade/>, through CCEE, 2019. InfoMercado: Dados Individuais "InfoMercado: Individual Data". Câmara de Comercialização de Energia Elétrica, São Paulo, Brazil. Available from: <www.ccee.org.br>.*

In 2018, the bioelectricity offered to the grid by the sugar-energy sector was 21.5 thousand GWh. This is a generation equivalent to

- supply 11.4 million homes throughout the year;
- have avoided the emission of 6.4 million tons of CO_2, a mark that can only be achieved by cultivating 45 million native trees over 20 years;
- have saved 15% of the total stored energy in the reservoirs of the SouthEast (SE)/WestCenter (WC) submarket, due to the greater predictability and availability of bioelectricity in the dry period.

In addition, 83% of the bioelectricity offered to the grid in 2018 occurred in the dry period, when the tariff flag in the consumers' energy account was in the yellow or red mode, as shown below (Fig. 13.6).

When bioelectricity generation is added to self-consumption, bioelectricity closed 2018 as the third most important source in the country's internal energy supply offer (IESO), almost equalizing with natural gas.

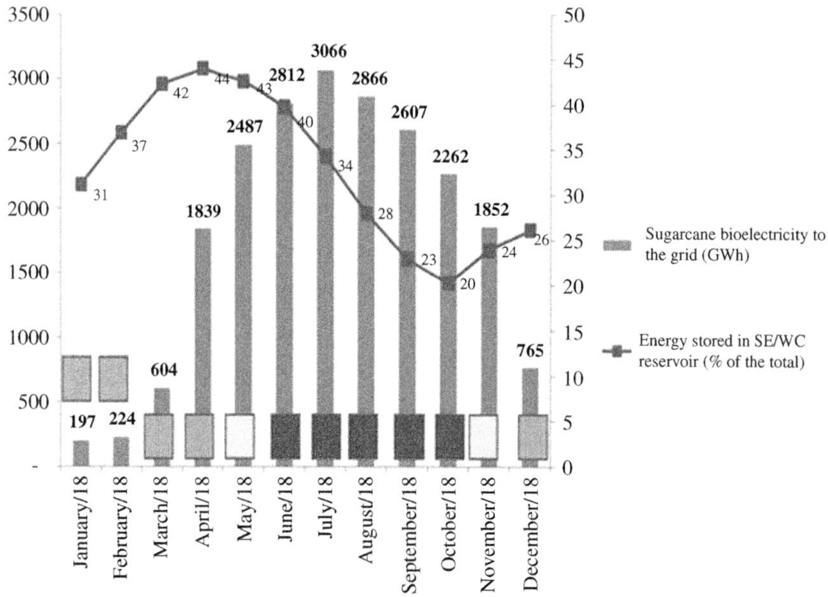

Figure 13.6 Sugarcane bioelectricity (GWh) offered to the grid, from January to December 2018 and tariff flag in the month. *ÚNICA, 2019. Cartilha: a bioeletricidade da cana e o mercado livre de energia elétrica no brasil "Cartilha: The Bioelectricity of Sugarcane and the Free Market of Electric Energy in Brazil". União da Indústria de Cana-de-Açúcar (UNICA). July 2019. Available from: <http://www.unica.com.br/documentos/publicacoes/bioeletricidade/>, through ONS, ANEEL e CCEE, 2019. InfoMercado: Dados Individuais "InfoMercado: Individual Data". Câmara de Comercialização de Energia Elétrica, São Paulo, Brazil. Available from: <www.ccee.org.br>.*

In 2018, the IESO was estimated at 632.1 thousand GWh, an increase of 1.3% over 2017. Hydropower generation remains the leader with 67% of the IESO total, followed by natural gas with 8.5%. The biomass source generated 52.5 thousand GWh, including the portion destined for self-consumption, representing 8.3% of the total domestic supply. Although the wind power continues to increase its share, it still ranked fourth, representing 7.7% of the IESO in 2018.

São Paulo: responsible for 44% of bioelectricity in 2018

In 2018 the State of São Paulo (SP) generated 11,608 GWh of bioelectricity for the NIS, a volume 3% lower than that of 2017, but still allowed the state to lead with, by far, the ranking of bioelectricity generation for the NIS, representing 44.1% of the total last year. The information is in a recent survey of the Sugarcane Industry Union (UNICA), based on data from the Electric Energy Trading Chamber or Câmara de Comercialização de Energia Elétrica (CCEE). The values include not only the generation of bioelectricity by the sugar-energy sector but also in other sectors of the economy and by different types of biomass. The second place remained with Mato Grosso do Sul (MS), with 4369 GWh of bioelectricity offered to the NIS, a significant increase of 19.8% in relation to 2017, leading the state to represent 16.6% of the total bioelectricity exported to NIS in 2018. The State of Minas Gerais (MG) ranked third, generating 3235 GWh for the grid in 2018, 8.3% higher than in 2017, representing 12.3% of the total bioelectricity generation offered to the NIS in 2018. The states of Goiás (GO) and Paraná (PR) closed the ranking of the most representative states in the generation of bioelectricity for the network, producing 2704 and 1632 GWh, respectively, each participating with 10.3% and 6.2% of the bioelectricity offered to the NIS in 2018.

The total bioelectricity production of these five states was 23,548 GWh, representing almost 90% of all the bioelectricity offered to the NIS in 2018. This shows the importance of the sugar-energy sector, as these are also the main sugarcane-producing states in the country. In the 2017/2018 harvest, the combined production of these states represented 90% of the country's sugarcane milling (Table 13.1).

This information also shows how significant the generation of bioelectricity is for the country and specifically for the State of São Paulo, which in 2018 imported 57% of the electricity consumed by its market. In 2018 the total consumption of electricity in the State of São Paulo was 131,978 GWh and the generation of energy in the state produced for the transmission system was 56,657 GWh. Considering that the bioelectricity produced in São Paulo for NIS was 11,608 GWh, this was equivalent to 20.5% of the total generation by the state for the system.

According to the Secretariat of Infrastructure and the Environment, with 18,917,329 consumer units, in December 2018 alone, the electricity distributors operating in the State of São Paulo met electricity consumption of 11,206 GWh, that is, the bioelectricity generated for NIS in São Paulo last year was equivalent to attending an entire month of electricity consumption in the state of São Paulo. Based on data from the Energy

Table 13.1 Bioelectricity in general offered for the network in 2017 and 2018 (in GWh).

State	Year		Variation		Total participation		
	2017	2018	GWh	%	2017 (%)	2018 (%)	Accumulated (%)
SP	11,969	11,608	− 361	− 3.0	47.0	44.1	
MS	3649	4370	721	19.8	14.3	16.6	60.6
MG	2986	3235	249	8.3	11.7	12.3	72.9
GO	2617	2704	87	3.3	10.3	10.3	83.2
PR	1497	1632	135	9.0	5.9	6.2	89.4
BA	973	953	− 20	− 2.1	3.8	3.6	93.0
MA	504	438	− 66	− 13.0	2.0	1.7	94.7
MT	239	231	− 8	− 3.2	0.9	0.9	95.5
RS	159	200	41	26.0	0.6	0.8	96.3
AL	151	190	39	25.6	0.6	0.7	97.0
PE	136	178	42	31.1	0.5	0.7	97.7
SC	137	146	10	7.1	0.5	0.6	98.3
ES	125	105	− 20	− 15.7	0.5	0.4	98.7
TO	117	98	− 19	− 16.6	0.5	0.4	99.0
RN	26	84	57	216.1	0.1	0.3	99.3
PA	98	57	− 41	− 42.2	0.4	0.2	99.6
SE	53	54	1	1.6	0.2	0.2	99.8
PB	36	46	11	29.9	0.1	0.2	99.9
PI	10	11	1	8.4	0.0	0.0	100.0
RJ	0	5	5	—	0.0	0.0	100.0
Total	25,482	26,345	863	3.4	100.0	100.0	

Source: UNICA, 2019. Cartilha: a bioeletricidade da cana e o mercado livre de energia elétrica no brasil "Cartilha: The Bioelectricity of Sugarcane and the Free Market of Electric Energy in Brazil". União da Indústria de Cana-de-Açúcar (UNICA). July 2019. Available from: <http://www.unica.com.br/documentos/publicacoes/bioeletricidade/>, through CCEE, 2019. InfoMercado: Dados Individuais "InfoMercado: Individual Data". Câmara de Comercialização de Energia Elétrica, São Paulo, Brazil. Available from: <www.ccee.org.br>.

Research Company or Empresa de Pesquisa Energética (EPE), it is estimated that we will only use 15% of the bioelectricity potential for the NIS. Therefore the improvement in the business environment for bioelectricity in the coming years, stimulating investment, can contribute to accelerate the development of this strategic generation source, reducing the gap between the actual production of bioelectricity and its technical potential of generation for the NIS.

Productivity in the generation of sugarcane bioelectricity

Since 2013 the sugar-energy sector has been generating more electricity for the NIS than for the own consumption of the plants, with 60% of energy for the grid and 40% for own consumption. Between 2010 and 2017, in terms of kilowatt hour indicator exported to the electricity grid per tons of processed cane, bioelectricity increased by greater than 110% (Table 13.2).

Table 13.2 Bioenergy bioelectricity generation (TWh) and indicators, 2013–18, Brazil.

	2013	2014	2015	2016	2017	2018
Sugar–energy bioelectricity						
Self-consumption (in TWh)[a]	13.9	13.3	14.0	14.0	14.2	14.2
Offered to grid (in TWh)	16.0	19.3	20.2	21.2	21.4	21.5
Total (in TWh)	29.9	32.6	34.2	35.2	35.6	35.7
Sugarcane (thousand tons)	651,294	633,927	666,824	651,841	641,066	620,832
Indicator kWh per tons of cane						
Self-consumption (kWh/tons)	21.34	20.98	21.00	21.42	22.15	22.87
Offered to grid (kWh/tons)	24.57	30.45	30.29	32.58	33.38	34.63
Total (kWh/tons)	45.91	51.43	51.29	54.00	55.69	57.50

[a]Self-consumption (in TWh) for the year 2018 is an estimate.
Note: Data on the volume of sugarcane based on crop year and the generation of energy in calendar year.
Source: UNICA, 2019. Cartilha: a bioeletricidade da cana e o mercado livre de energia elétrica no brasil "Cartilha: The Bioelectricity of Sugarcane and the Free Market of Electric Energy in Brazil". União da Indústria de Cana-de-Açúcar (UNICA). July 2019. Available from: <http://www.unica.com.br/documentos/publicacoes/bioeletricidade/>, through CCEE, 2019. InfoMercado: Dados Individuais "InfoMercado: Individual Data". Câmara de Comercialização de Energia Elétrica, São Paulo, Brazil. Available from: <www.ccee.org.br>.

The commercialization of bioelectricity

As of 2004 in relation to the commercialization of bioelectricity in the Brazilian electric sector, two possible environments were established to conclude purchase and sale contracts:

- regulated contracting environment (RCE), in which energy generation and distribution agents participate (in response to so-called captive consumers); and
- free contracting environment (FCE), which includes generation agents, traders, importers and exporters of energy, and free and special consumers of electricity.

Energy consumption in the free market was 18,406 MW (average) in February 2019, representing 29% of all electricity consumed in the country, a 1.1% increase in consumption in the last 12 months. In February 2019, 80% of the country's industrial electricity consumption was served through the free market. Producers deliver and receive energy to the system, in their center of gravity, assumed part of the losses between the point of generation and this center of gravity. Consumers, in an analogous way, deliver and receive energy to the system, in its center of gravity, assuming part of the losses between this center of gravity and the point of consumption.

The system guarantees product quality and offer

Differences between the contracted and the produced or consumed are settled by the settlement price of differences (SPD), defined in four submarkets, by the computational model. This settlement is made monthly by the CCEE in the so-called short-term market (STM). Contracts protect agents from the short-term price and are obligatory for 100% of the load, with no time constraints in the case of free market. Contracts can be registered after measuring actual consumption. Failure to prove, in addition to exposure to the payment of the SPD, implies the payment of penalties for lack of ballast.

The current criteria for migration to the free market were established in 1998 by Law no. 9.648/1998, which created two groups of consumers able to choose their electricity supplier. The first group consists of consumer units with a load greater than or equal to 2500 kW and serviced at any voltage. These consumers can purchase energy from any energy generation or commercialization agent. On January 1, 2020 the limit will be reduced to a load of 2000 kW. The second group, composed of consumer units with a demand greater than or equal to 500 kW served in any voltage, can also choose their supplier, but their range of choice is restricted to the energy coming from the so-called incentivized sources, namely small hydroelectric plants (SHPs), biomass plants, wind power plants, and qualified cogeneration systems. In February 2019 there were 5819 consumers in the free market, of which 4932 were special consumers (85%) and 887 were free consumers (15%), representing a 12% increase in the number of consumers in the last 12 months, that is, 627 new consumers migrated to the free market.

The free energy market and bioelectricity

In 2018, 69% of the generation by the biomass source was destined to the FCE and 31% to the RCE, as shown below (Fig. 13.7).

In February of this year (2019) 76% of the biomass generation for the grid was to the free market. In that month, bioelectricity led the ranking of how much of the energy generated by renewables is sold in the free market, followed by SHPs (65%) and then by wind power (45%).

According to EPE (2019), in relation to the sugar-energy sector, of the 369 sugarcane biomass plants in operation in 2018, 200 commercialized electricity (54%), three more than the previous year. A total of 169 plants produced only for self-consumption (46%). Of those that export energy to the NIS, part operate exclusively in the FCE (60% or 120 plants) or in the RCE (20% or 40 plants) and the rest (20% or 40 plants) sell in both contracting environments.

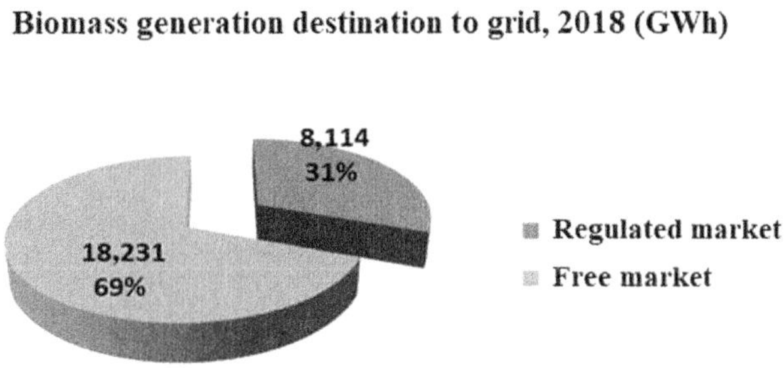

Figure 13.7 Destination of bioelectricity generation for the electricity grid in Brazil by commercialization environment (GWh and %). *ÚNICA, 2019. Cartilha: a bioeletricidade da cana e o mercado livre de energia elétrica no brasil "Cartilha: The Bioelectricity of Sugarcane and the Free Market of Electric Energy in Brazil". União da Indústria de Cana-de-Açúcar (UNICA). July 2019. Available from: <http://www.unica.com.br/documentos/publicacoes/bioeletricidade/>, through CCEE, 2019. InfoMercado: Dados Individuais "InfoMercado: Individual Data". Câmara de Comercialização de Energia Elétrica, São Paulo, Brazil. Available from: <www.ccee.org.br>.*

The regulated energy market and bioelectricity

The year 2018 was the third worst year of contracting new projects in the regulated auctions promoted by the federal government, since the implementation of this form of hiring in 2005, as shown below (Fig. 13.8).

In the so-called A-4 auction last year, biomass registered 28 projects, totaling 1422 MW, and ended up selling bioelectricity from only two projects. At the 2018 A-6 auction, biomass registered 25 projects totaling

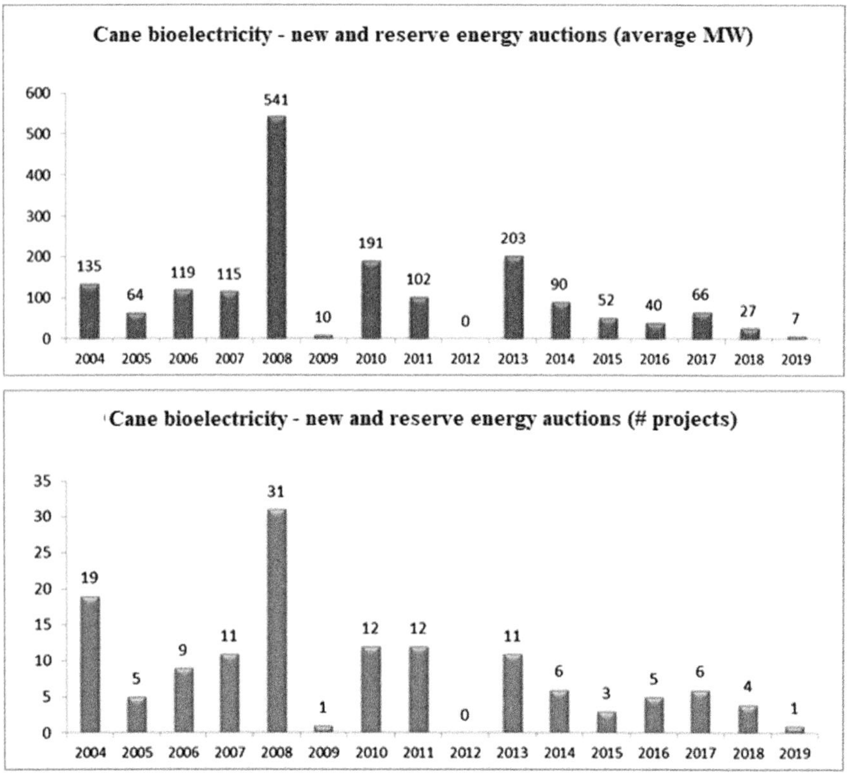

Figure 13.8 Projects and physical guarantee marketed by sugar–energy bioelectricity in the regulated contracting environment, from 2004 to 2019.
Note: The year 2019 considers only the A-4/2019 Auction, and the results of Auction A-5/2019 should be expected for the consolidation of this year. *ÚNICA, 2019. Cartilha: a bioeletricidade da cana e o mercado livre de energia elétrica no brasil "Cartilha: The Bioelectricity of Sugarcane and the Free Market of Electric Energy in Brazil". União da Indústria de Cana-de-Açúcar (UNICA). July 2019. Available from: <http://www.unica.com.br/documentos/publicacoes/bioeletricidade/>, through CCEE, 2019. InfoMercado: Dados Individuais "InfoMercado: Individual Data". Câmara de Comercialização de Energia Elétrica, São Paulo, Brazil. Available from: <www.ccee.org.br>.*

1040 MW and, at the end of the dispute, sold only two projects. The first regulated new energy auction promoted by the federal government took place in 2005, and since then the biomass source has sold 146 projects in new and reserve energy auctions, helping to make viable 7658 MW, 51% of the currently installed power by the source in the country (which is the fourth most important in the country). Since 2005 the new and reserve energy auctions have contracted a total of 418,707,111 MWh of the biomass source, in contracts of 15–25 years. This generation is equivalent to more than four times the annual production of Itaipu Power Plant, still the largest hydroelectric plant in production in the world and responsible for 21% of the national electricity consumption in Brazil in 2018.

In the last 14 years of regulated environment, the biomass source had 1203 average MW contracted in new and reserve energy auctions, with an annual average of 86 MW. However, last year, the two new energy auctions that took place contracted only four biomass projects totaling 26.9 average MW, well below the annual average contracting. In 2019 there is already the scheduling of two new A-4 and A-6 auctions. The hope is always that biomass can perform at least better than the average annual hiring of the last 14 years. But why is it important to encourage the contracting of this source to the NIS? The model of contracting via auctions is 14 years old and considers almost exclusively the production of energy and its price, to the detriment of the other attributes of the different sources of generation. It is time to improve it and evaluate the real costs and benefits of each generation source, considering the contribution of their respective attributes to the fulfillment of the various objectives of the energy supply in the NIS.

In this line, in October 2018, Instituto Escolhas presented a study with the real value and benefits of the various sources of energy generation, observing its attributes. According to the study, a bioelectricity plant, located in the SE, is the most cost-effective renewable energy for final consumers. The value of the bioelectricity energy in the SE would be R $168/MWh, of the hydroelectric plant would be R$286/MWh, of the wind power in the Northeast (NE) R$195/MWh and R$244/MWh in the South. The SHP in the SE would be R$285/MWh, the solar in the NE R$293/MWh, and R$328/MWh in the SE.

The year 2018 was the third worst year of contracting new projects in the regulated auctions promoted by the Federal Government. In addition to dismantling this important productive chain of the national economy, the results of the current contracting format in the regulated auctions have been bad for the biomass and also for the final consumer, considering the attributes that this source provides to the NIS. Brazil only uses 15% of the bioelectricity potential. If biomass was to be fully utilized in the sugarcane fields, bioelectricity would have the technical potential to reach 146 thousand GWh, almost seven times the volume offered in 2018, which would account for more than 30% of the energy consumption in the NIS. That is, there are several "Itaipu mills" still sleeping in the cane fields!

The commendable initiative of modernization of the electric sector, by the Ministry of Mines and Energy (MME), is very welcome in the aspect of seeking to assess the attributes of each generation source. We are sure that, at the end of this process, bioelectricity will be more competitive. Not forgetting the task of valuation of attributes, as part of the selection criteria for projects to be contracted, it is expected that the next regulated auctions will be able to evolve to at least the creation of a specific product for bioelectricity in A-6 Auctions. Thus separating it from noncomparable sources such as coal and natural gas, or even regional and specific auctions for this source, ideally within a long-term sector policy for bioelectricity, with clear guidelines and continuity, seeking to ensure full efficient use of this important renewable resource in the country's energy matrix.

The year 2019 in the regulated market

In 2019 the A-4/2019 Auction was held on June 28. The Auction contracted only 81.1 average MW, from 15 solar, wind, SHP, and biomass projects. The contracts are of 20 and 30 years, for beginning of supply from 2023. The solar source reached a new price record in Brazil, selling at R $67.48/MWh, against the ceiling of R$276.00/MWh, with a discount of 75.5%. The previous record was in the auction of 2018, which updating by inflation is at R$123.98/MWh. Wind power sold at R$79.99/MWh, against the ceiling of R$208.00/MWh, with a discount of 61.54%. The hydroelectric sold at R$198.12/MWh, against the ceiling of R$288.00/MWh, discount of 31.2%. The biomass source had registered 1039 MW, totaling 19 projects, but commercialized only one project at R$179.87/MWh, against the ceiling of R$311/MWh, with a discount of 42%.

The A-4/2019 Auction contracted the equivalent of 401.5 MW of installed capacity and 164.99 average MW of physical guarantee.

Table 13.3 Generation projects registered in the Auction A-6/2019 (in MW, per source).

Source	**A6 (2025) October 17, 2019**		
	Ceiling price for biomass: to be defined		
	Projects	**MW**	**% MW**
Biomass	25	1360	1.3
Hydroelectric	5	213	0.2
SHP	59	939	0.9
CHG	14	39	0.0
Coal	4	1667	1.7
Natural gas	52	41,718	41.4
Wind	845	25,158	24.9
Photovoltaic	825	29,780	29.5
Total	1829	100,874	100

CHG, Central hydroelectric generator; *SHPs*, small hydroelectric plants.
UNICA, 2019. Cartilha: a bioeletricidade da cana e o mercado livre de energia elétrica no brasil "Cartilha: The Bioelectricity of Sugarcane and the Free Market of Electric Energy in Brazil". União da Indústria de Cana-de-Açúcar (UNICA). July 2019. Available from: <http://www.unica.com.br/documentos/publicacoes/bioeletricidade/>, Data from ANEEL (2019).

The purchasing agents were the distributors "CPFL Santa Cruz" and "Light." The average sale price of the auction was R$151.15/MWh. The A-6/2019 Auction will take place on October 17, 2019. This auction will include hydroelectric projects, wind power and solar photovoltaic projects, and thermoelectric projects with biomass, national mineral coal, and natural gas.

In the A-6 Auction last year, wind energy accounted for the majority of the contracted volume (50.3%), followed by natural gas (40.3%), SHPs with 9.4% and, finally, biomass with only 1%. In A-6/2018, biomass competed in the so-called product availability, where a single thermal to natural gas took greater than 97% of the demand allocated to that product, displacing the bioelectricity projects. For the A-6/2019 Auction, the biomass source registered a total of 25 projects, according to Table 13.3.

One point of attention: the decontracting of bioelectricity in the coming years

By 2024 between new energy auctions, from alternative sources and from reserve, the biomass source will have almost 50% of its contracts maturing. There are 823 average MW. A generation equivalent to

- supply nearly 4 million homes throughout the year;
- avoiding the emission of 2.1 million tons of CO_2, a volume that can only be obtained by cultivating 15 million native trees over 20 years; and
- save 5% of the total stored energy in the reservoirs of the hydroelectric plants of the SE/WC submarket due to the greater predictability and availability of bioelectricity in the dry period.

If the plants' contracts for biomass are not renewed, there will be a need to put other plants in place. Biomass is renewable and is not an intermittent source, so it has attributes that are increasingly essential for the NIS. Most of the mills that have expiring contracts must invest significantly in the maintenance/refurbishment of the thermoelectric plants (TEPs) to continue serving the NIS. This requires clarity about what the Institutional Environment of the Electricity Sector will look like right now, not from 2024. Hiring capacity/reliability, separate from energy, is something for a distant future. The decision on the subject should come before 2024.

There is hope that the federal government can signal, from now on, how the treatment will be for this energy: (1) Will part of this energy be recontracted in the form of backup energy? (2) As it requires investment in maintenance/refurbishment of the TEPs, could we not rediscuss the concepts of old and new energy and let them participate in the upcoming new energy auctions? (3) Or make structuring auctions to rehire this energy?

Evolution of investment in bioenergy

There is a challenge in resuming the expansion of bioelectricity in the national electricity matrix and the level of investment in generation in the sugar-energy sector. From 2008 to 2018, the National Bank for Economic and Social Development (BNDES) had a total of R$7.5 billion of disbursement with the generation of electric energy in the sugar-energy sector. However, the trend has been declining: in 2008 the disbursement was R$1.9 billion (representing 29% of the total disbursement for the sugar-energy sector) and in 2017 only R$21 million (only 1% of the total disbursed for the sugar-energy sector). In 2018 the disbursement of BNDES with cogeneration in the sugar-energy sector increased to R $94 million (Figs. 13.9 and 13.10).

The decrease in BNDES' disbursements for bioenergy bioelectricity can be explained by the reduction of investment in the sugar and ethanol sector but also by the loss of competitiveness in the regulated auctions promoted by the Federal Government from 2009.

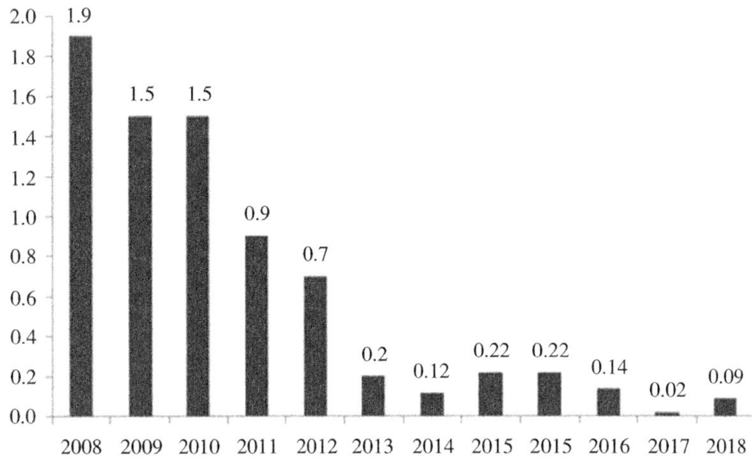

Figure 13.9 BNDES disbursements for bioenergy bioelectricity, 2008–18 (in R$ billion). *ÚNICA, 2019. Cartilha: a bioeletricidade da cana e o mercado livre de energia elétrica no brasil "Cartilha: The Bioelectricity of Sugarcane and the Free Market of Electric Energy in Brazil". União da Indústria de Cana-de-Açúcar (UNICA). July 2019. Available from: <http://www.unica.com.br/documentos/publicacoes/bioeletricidade/>, through BNDES e EPE, 2019. Empresa de Pesquisa Energética. Análise de Conjuntura dos Biocombustíveis "Analysis of the Biofuels Scenario". Available from: <http://www.epe.gov.br>.*

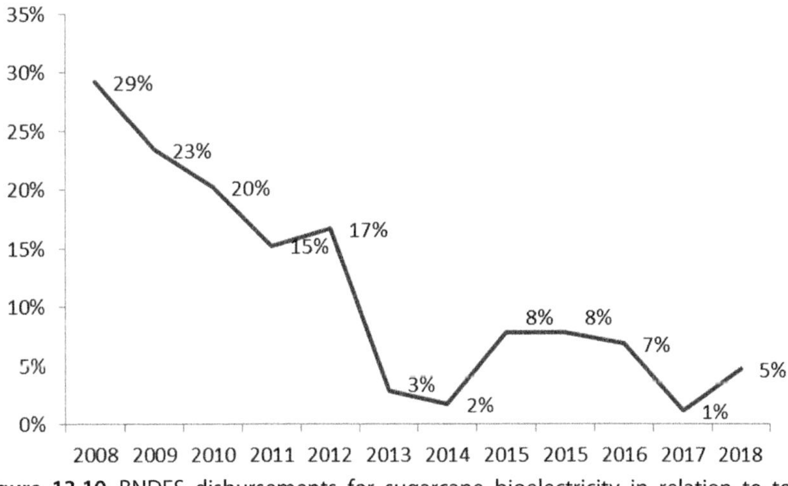

Figure 13.10 BNDES disbursements for sugarcane bioelectricity in relation to total disbursed to the sugar–energy sector, 2008–18 (%). *ÚNICA, 2019. Cartilha: a bioeletricidade da cana e o mercado livre de energia elétrica no brasil "Cartilha: The Bioelectricity of Sugarcane and the Free Market of Electric Energy in Brazil". União da Indústria de Cana-de-Açúcar (UNICA). July 2019. Available from: <http://www.unica.com.br/documentos/publicacoes/bioeletricidade/>, through BNDES e EPE, 2019. Empresa de Pesquisa Energética. Análise de Conjuntura dos Biocombustíveis "Analysis of the Biofuels Scenario". Available from: <http://www.epe.gov.br>.*

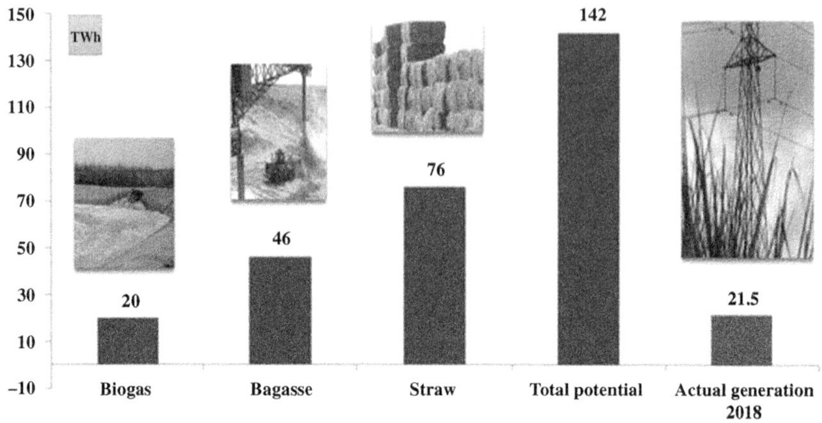

Figure 13.11 Technical potential of sugar-energy bioelectricity for the grid, 2018 (in TWh). *ÚNICA, 2019. Cartilha: a bioeletricidade da cana e o mercado livre de energia elétrica no brasil "Cartilha: The Bioelectricity of Sugarcane and the Free Market of Electric Energy in Brazil". União da Indústria de Cana-de-Açúcar (UNICA). July 2019. Available from: <http://www.unica.com.br/documentos/publicacoes/bioeletricidade/>, through EPE, 2019. Empresa de Pesquisa Energética. Análise de Conjuntura dos Biocombustíveis "Analysis of the Biofuels Scenario". Available from: <http://www.epe.gov.br>.*

The potential of bioelectricity yet to enjoy

The Brazilian electricity sector must undergo a modernization process, with market opening being one of the main guidelines of this sectoral reform. The expected growth for the free market should also stimulate the commercialization of new bioelectricity projects, due to the potential "dormancy" of this source in Brazilian sugarcane plantations. Only 15% of the potential of sugar-energy bioelectricity is used. If bioelectricity were to be fully utilized in sugarcane plantations, bioelectricity would have the technical potential to reach 142 thousand GWh, almost 7 times the volume offered in 2018, which would account for more than 30% of energy consumption in the NIS (Fig. 13.11).

Expansion of biomass is far from indicated in the 10-year plan in Brazil

In August 2018 ANEEL organized the Auction for the purchase of electric energy for new generation projects, called Auction A-6/2018, to assist the distributors' market from 2024 onwards. The sugar and ethanol sector have contracted only two projects, selling 9.8 average MW, which represents an increase of 28.5 MW by 2024. According to a survey by the UNICA, the last regulated auctions have already contracted 133 average MW, representing a power of 386 MW to be installed from 2021 to 2024. According to the UNICA survey, in the last 10-year Energy Expansion Plan (PDE 2026), is that the expansion to be contracted in the sugar-energy biomass source, from 2021 to 2024, is of 1868 MW.

To date, the regulated environment, which is still the main entry point into the electricity sector, contributed only 20% of the expansion indicated in the PDE for sugarcane biomass. Even without a free market expansion, low contracting at regulated auctions shows that we need to redesign the model and the form of participation of biomass in the regulated events to be adherent to the planning in the electric sector. At the last auction A-6/2018, the biomass source had registered 25 projects totaling 1040 MW, but sold only two projects, meaning an increase of 28.5 MW. The major winner was a natural gas thermoelectric, located in Maranhão, which took more than 97% of the demand allocated for the product availability, in which biomass, thermal, coal, and natural gas sources compete.

Direct competition between a renewable source such as biomass from nonrenewable sources such as coal and gas is not adequate. In addition to a very different scale of production and cost structure, putting to compete a sustainable source against one that can throw in the atmosphere almost a ton of CO_2 per MWh produced impairs the source of bioelectricity and distorts the correct valuation of the attributes that each source delivers to the electrical system. It is important that the upcoming regulated auctions can evolve to at least the creation of a specific product for bioelectricity, separating it from noncomparable sources such as coal and natural gas, ideally within a long-term sectoral policy for the sugar-energy sector, with clear guidelines and continuity, seeking to ensure the full efficient use of this important renewable energy resource in the country's electric energy matrix.

The potential of bioelectricity and the RenovaBio

According to the MME (2019), RenovaBio is a state policy that aims to draw up a joint strategy to recognize the strategic role of all types of biofuels in the Brazilian energy matrix, both for energy security and mitigation of emissions reduction of greenhouse gases (GHGs). This is the new National Biofuel Policy, instituted by Law 13, 576/2017.

The main objectives of the RenovaBio are

- to provide an important contribution to the fulfillment of Brazil's nationally determined commitments under the Paris Agreement;
- to promote the adequate expansion of biofuels in the energy matrix, with emphasis on the regularity of the fuel supply; and
- to ensure predictability for the fuel market, inducing gains in energy efficiency and reduction of GHG emissions in the production, commercialization and use of biofuels.

Unlike traditional measures, RenovaBio does not propose the creation of a carbon tax, subsidies, presumed credit, or volumetric mandates for the addition of biofuels to fuels. RenovaBio is designed to introduce market mechanisms to recognize the ability of each biofuel to reduce emissions individually by the producing unit. Basically, there are two main instruments:

1. Establishment of national emission reduction targets for the fuel matrix, defined for a period of 10 years. Targets are important to bring predictability in terms of volumetric fuel needs (fossil and renewable) over this time horizon, and thus enable private agents to make their investment planning and analysis in an environment with less uncertainty. The national targets will be broken down into individual annual targets, compulsory for all fuel distributors, according to their participation in the fossil fuel market.
2. Certification of the production of biofuels by private inspector firms, with different grades being assigned to each production unit (the higher the score for the producer that produces the highest amount of net energy with the lowest CO_{2eq} emissions in the life cycle). The note will accurately reflect the individual contribution of each producer agent to the mitigation of a specific amount of GHGs in relation to their fossil substitute (in terms of tons of CO_{2eq}). The process of certification of the production of biofuels under the RenovaBio will be under the responsibility of the National Agency of Oil, Natural Gas and Biofules (ANP).

The connection of these two instruments will occur with the creation of the biofuel decarbonization credit (CBIO). It will be a financial asset, traded on a stock exchange, issued by the biofuel producer, from the sale of its production (after the issuance of the invoice). Fuel distributors will meet the target by demonstrating the required amount of CBIOs on their property. Other agents (individuals and corporations) may buy and sell CBIOs on the stock exchange as a way to bring greater liquidity to this market. According to EPE (2017), with the operation of RenovaBio, the state policy to stimulate biofuels, the number of mills operating will increase from 360 to 390, while the supply of ethanol will increase from the current 31 billion liters to 49 billion, making available, as a consequence, a larger volume of biomass for the generation of energy.

Encouraged by RenovaBio and a favorable business environment in the electricity sector, sugar–energy bioelectricity for the grid has the potential to grow by almost 60% by 2030, rising from 21,500 GWh in 2018 to 34,000 GWh by 2030. Furthermore, the sugarcane sector will use less than 20% of the technical potential of this generation source in 2030, showing the possibility of positive response that bioelectricity can give to the also expected free market expansion.

An agenda for bioelectricity

In order to reduce the gap between the effective generation of bioelectricity and its potential, a long-term and stimulating sector policy with clear guidelines and continuity is important, among them:

Efforts for regular and growing contracting for bioelectricity and biogas in regulated auctions and strengthening of the hiring free environment

We need to promote auctions in the regulated environment for biomass (and continuity in contracting), with pricing models that incorporate bioelectricity externalities and the characteristics of each project (retrofit, greenfield, use of straw and bagasse, biogas generation, etc.). In the A-6 auctions there is a need to create a product for bioelectricity, just as there is for wind and photovoltaic sources. Direct competition with nonrenewable sources such as coal and natural gas is not adequate because there is no effective differentiation of the positive and negative attributes that each source delivers to the NIS. At the same time, the FCE must be strengthened so that it can also enable a number of ever-increasing bioelectricity projects.

Equalization of the current judicialization in the financial settlements in the short-term market

The strengthening of institutions such as the free market or technically called the FCE is necessarily a solution to institutional problems in the current environment of the electricity sector, without which an institutional scenario conducive to investment and capacity expansion of electricity generation will not be built. Among these institutional issues there is the long persistence of the serious problems related to financial liquidations in the STM of electric energy. In that market, basically, that generator that produces more energy than the energy volume established for its unit is entitled to receive the amount equivalent to the SPD, determined weekly by the CCEE. For those who generate less, it pays the difference because it generated less than the SPD also needed. The amounts are compensated and the transactions are managed by the CCEE, whose settlement occurs monthly. Thus initially no resources come from outside. The counterparties (deficit agent and surplus agent) do not have contact with each other due to the intermediation of the CCEE. The amount paid by the deficit agents is collected by the CCEE, which transfers it to the surplus agents in the financial settlements in the STM.

However, since 2015 electricity sector players who have credits in the STM are facing serious problems in receiving their revenues. The problem arose with the nonconformity of the water generators, participants in the so-called energy reallocation mechanism (ERM), who were in the position of debtors and generated less energy than expected to their commitments and their physical guarantee. The ERM is a kind of hydroelectric condominium and a way of sharing the hydrological risk among those agents. The indicator that measures the amount of energy produced in relation to the physical guarantee (a "certificate" of generation) of the hydropower plants belonging to the ERM is the ERM Adjustment Factor, or generation scaling factor (GSF). The GSF is calculated monthly by the CCEE and, since 2013, in the annual average, has been less than 1, that is, the hydraulic generation of plants belonging to the ERM is lower than the physical guarantee that is used as reference for commercialization of hydroelectric generation.

The adjustment of this little energy production should be assumed by the hydroelectric power plants participating in the ERM condominium, but this is not what happened. The lack of a proper administrative solution to the issue of low GSF and at the right time led such water generators to seek judicial protection, obtaining dozens of injunctions that paid them out of paying their debts in the STM. As a consequence of this situation, the energy generators, who were creditors in the STM, began to see their revenues decrease, being affected by judicial decisions in lawsuits in which they were not even part, regarding the injunctions of hydrological risk of hydro generators in the FCE. As a result, creditors also began to file lawsuits, requiring that the lawsuits they did not participate in would not affect the receipt of their claims.

This scenario of judicialization in the STM started to become more worrying around August 2015 and has been worsening. The financial settlement of July 2019, for example, referring to the April 2019 movement, when creditors without judicial protections in the electricity sector received a close payment of only 5%, accumulating a sector debt of R$7 billion referring to the judicialization of hydrological risk, in the share of energy traded in the free market.

Improvement of the methodology of revision of the physical energy guarantee for biomass/biogas plants

The physical energy guarantee (PEG) determines the amount of electricity that can be contracted in the market. On October 20, 2014, Ordinance no. 564/MME, dated October 17, 2014, which established the methodology for biomass-powered TEPs with variable unitary cost to be able to review their amounts of PEG based on the increase in the availability of fuel and/or energy efficiency.

In Article 7 of Ordinance no. 564/MME, it was established that the review of PEG, dealt with in the above item, will only be carried out for projects that cumulatively meet the following conditions: (1) projects with measurement data totaling a minimum of 12 months of records of electricity generated at CCEE, disregarded the first 12 months after the release of the commercial startup of the first generating unit; (2) ventures for which the average generation was greater than $0.95 \times$ PEG at the present moment; and (3) existing ventures participating in existing energy auctions, alternative energy auctions, or reserve power auctions.

Almost 5 years after the edition of Ordinance no. 564/MME, it is noted that the restriction of revision of PEG amounts based on increased fuel availability and/or energy efficiency, without an increase in installed capacity, only for those who are willing to market in the RCE is not in line with the current environment of the Brazilian Electricity Sector, where 30% of consumption occurs in the free market. It is therefore an unnecessary and unjustified barrier to achieving the abovementioned investments, which result in renewable and sustainable energy for the country.

In summary, the prohibition of the applicability of the methodology established in article Paragraph 1, item II, of Ordinance no. 564/MME for enterprises that market in the free environment does not favor the financing of projects involving fuel increase and/or energy efficiency, damaging the expansion of bioelectricity in the Brazilian electricity matrix and the isonomy between commercialization.

It is important that item I of Article 7 of Ordinance no. 564/MME, which obliges it to commercialize PEG from investments in energy efficiency and/or increase of fuel only in the regulated marketing environment, making it impossible to make the investment via sale directly to consumers, traders, or other agents, when the investor thinks it is more appropriate. It is estimated that the sugar and ethanol sector can increase generation by more than 10% in the short/medium term, without necessarily investing in installed capacity. The definition of PEG for biomass/biogas plants has not encouraged the generator to produce energy to the limit of its capacity, inhibiting a potential supply precisely at the critical moments of electric power shortage and high costs for the final energy consumer power.

The improvement in the business environment for bioelectricity is essential to contribute to the development and success of bioelectricity and the RenovaBio, which will translate into the expansion of ethanol production in Brazil. Ethanol and bioelectricity are self-respecting products and need a business environment that is attractive for its development and harnessing its incredible potential available to civil society.

References

ANEEL, 2019. BIG - Banco de Informações da Geração. Brasília – DF. 29 de maio de 2019. Available from: < http://www.aneel.gov.br > .

CCEE, 2019. InfoMercado: Dados Individuais "InfoMercado: Individual Data". Câmara de Comercialização de Energia Elétrica, São Paulo, Brazil. Available from: <www.ccee.org.br>.

EPE, 2017. Nota Técnica: Papel dos biocombustíveis na matriz "Technical Note: Role of Biofuels in the Matrix". Rio de Janeiro – RJ. 24 de fevereiro de 2017. Available from: <http://www.epe.gov.br>.

EPE, 2019. Empresa de Pesquisa Energética. Análise de Conjuntura dos Biocombustíveis "Analysis of the Biofuels Scenario". Available from: <http://www.epe.gov.br>.

MME, 2019. Ministério de Minas e Energia (Energy and Mines Ministry). Boletim Mensal de Energia de dezembro de 2018 "Monthly Energy Bulletin December 2018". Departamento de Informações e Estudos Energéticos – DIE. March 2019. Available from: <http://www.mme.gov.br>.

ÚNICA, 2019. Cartilha: a bioeletricidade da cana e o mercado livre de energia elétrica no brasil "Cartilha: The Bioelectricity of Sugarcane and the Free Market of Electric Energy in Brazil". União da Indústria de Cana-de-Açúcar (UNICA). July 2019. Available from: <http://www.unica.com.br/documentos/publicacoes/bioeletricidade/>.

Further reading

ABRACEEL, 2019. Associação Brasileira dos Comercializadores de Energia. Boletim. Brazilian Association of Energy Commercialization. February 2019. Available from: <http://www.abraceel.com.br/zpublisher/secoes/home.asp>.

EPE – Empresa de Pesquisa Energética, 2019. RenovaBio: Biocombustíveis 2030 "RenovaBio: Biofuels 2030". Available from: <http://www.epe.gov.br>.

MME, EPE, 2018. Ministério de Minas e Energia (Energy and Mines Ministry). Plano Decenal de Expansão de Energia 2027 "Ten Year Energy Expansion Plan 2027" Empresa de Pesquisa Energética. MME/EPE, Brasília, Brazil.

UNICA – União da Indústria de Cana-de-Açúcar. UNICADATA, São Paulo, Brazil. Available from: <http://unicadata.com.br/>.

UNICA, 2011. União da Indústria de Cana-de-Açúcar. Cartilha da Bioeletricidade "Bioelectricity Handbook". UNICA, São Paulo, Brazil. Available from: <http://www.unica.com.br/documentos/publicacoes/bioeletricidade/>.

Index

Note: Page numbers followed by "*f*" and "*t*" refer to figures and tables, respectively.

A

Acid hydrolysis, 41, 131–132, 213–214
Acid pretreatment, 207
Acinetobacter, 174
Acrylic acid, 244–246
 routes for biosynthesis of, 246*f*
Actinobacillus succinogenes, 154
Agroecological Zoning of sugarcane, 80–83
Agroenergy, 72–75
 bioenergy, 73–75
 ethanol and anhydrous alcohol, 72–73
Agroindustry, sugarcane-based, 168
Alcohol, 33–34, 71, 179, 239
 ethanol and anhydrous alcohol, 72–73
 sugar fermentation for production of, 8
Alkaline solutions, 188, 207–208
α-cellulose, 237
Aluminum oxide, 119–120
Ammonia fiber explosion (AFEX), 208–209
Anaerobic biodigestion, 184–188
 biogas composition, 185–188
 biogas desulfurization, 188
Anaerobic digestion process, 41, 185
Anaerobiospirillum succiniciproducens, 175
Arabinoxylan, 198, 238–239
Arthrobacter, 174
Ash content, of straw, 53–58, 57*f*
Aspergillus niger, 36
Aureobasidium pullulans, 154

B

Bacillus, 174
Bagasse, 12–13, 22, 26–29, 67–68, 73–74, 118–119, 122–127, 144, 149, 180, 202–203, 212, 233–236
 bioelectricity from, 144
 chemical composition, 234*t*
Basidiomycete, 39
β-glycosides, 214–215
Bi-fuel car, 72–73
Binders, 148
Bio-based building blocks, 250–251, 251*f*
Bio-based products, 153
Biochemical oxygen demand (BOD), 181
Biochemical routes, 174–175
Biodegradable polymer, 35–36
Biodigested vinasse, 184, 187
 concentration of, 190–191
Bioeconomy, 166
Bioelectricity of sugarcane, 144, 255–256
 agenda, 275–278
  efforts for regular and growing contracting for bioelectricity and biogas, 275
  equalization of current judicialization in the financial settlements, 275–277
  revision of physical energy guarantee for biomass/biogas plants, 277–278
 decontracting of bioelectricity in the coming years, 269–271
  evolution of investment in bioenergy, 270–271
 electric matrix and power installed by, 257–261
  generation of electricity by the biomass source, 257–261
 expansion of biomass, 272–273
 potential of
  commercialization, 272
  and RenovaBio, 273–275
 productivity in the generation of, 263–269
  commercialization, 264
  free energy market and bioelectricity, 265
  regulated energy market and bioelectricity, 266–268

Bioelectricity of sugarcane (*Continued*)
system guarantees product quality and offer, 264–265
the year 2019 in the regulated market, 268–269
São Paulo, 261–263
from sugarcane bagasse and straw, 144
Bioenergy, 73–75, 100, 117, 135–136
cultivation of sugarcane for, 121
evolution of investment in, 270–271
main conversion routes for, 130–134
electricity, production of, 130–131
hydrolysis, 131–132
Bioethanol, 135–136
production, in Brazil, 179–180
Biofuels, 39, 41–42, 51–53, 146–147, 151–152, 273–274
production, in Ecuador, 96–97
Biogas, 180, 184–185
from by-products, 145
composition, 185–188
desulfurization, 188
Biohydrogen, 39
Biomass, 16, 51–53, 55–56, 58–59, 62, 103–104, 109, 136, 144, 196, 196*f*, 197*t*, 257
electricity generation by biomass source, 257–261
expansion of, 272–273
lignocellulosic. *See* Lignocellulosic biomass
Biomass-derived carbon materials (BDCMs), 148
Bio-oil, 133
Bioplastic, 35–36, 250
Biopolymers of sugarcane, 229, 232–233
cellulose, 236–249
acrylic acid, 244–246
carboxymethylcellulose (CMC), 237
cellulose acetate, 238
furfural resins, 239–240
green polyethylene (GPE), 247–249
hemicelluloses, 238–239
lignin, 241–242
methylcellulose, 236–237
phenolic resins, 240–241
polyhydroxybutyrate (PHB), 246–247
polylactic acid (PLA), 243
sugarcane juice, 243
xanthan gum, 249
perspectives and challenges, 249–251
sugarcane bagasse, 233–236
Bioproducts from sugarcane, 170–175, 172*t*
Brandy, sugarcane, 71
Brazil, 3–4, 21, 51–52
agroenergy system, 72–75
average productivity of sugarcane in, 50
electricity sector, 272
ethanol production in, 14–16
internal energy supply in, 15*f*
liquor production in, 71
methanol production in, 21
products, production, export, and international trade, 69*t*
sugar, ethanol, and energy sectors in, 14–16
sugarcane as main Brazilian energy crop, 118–122
sugarcane as the main agricultural product of, 68–69
sugarcane crop and bioethanol production in, 179–180
sugarcane production in, 4–6, 8–9, 21
estimated production for 2018/19 crop, 14–15
sugar production in, 5–12, 21
Breeding, sugarcane, 6, 104–105
Brevbacterium, 174
Brix index, 189
Broth, 22–24
Brown sugar, 7–8
1,4-Butanediol, 152
2,3-Butanodiol, 39
Butanol, 39, 154
By-products of the sugarcane industry, 21, 34–36
bagasse and straw, 26–29
chemical potential of sugarcane residues, 36–42
filter cake, 31–32, 31*t*

future perspectives for sugarcane biorefinery, 42–45
molasses, 29
sugar and ethanol production process, 22–26, 23*f*
vinasse, 30
yeasts, 33–34, 33*t*

C

C4 photosynthetic cycle, 1
Calcium oxide, 119–120
Câmara de Comercialização de Energia Elétrica (CCEE), 261, 275–277
Canary Islands, 3
Carbohydrates, chemicals from, 149–152
Carbon dioxide, 34
Carbon fiber composites, 158–159
Carbon footprint, 97
Carboxymethylcellulose (CMC), 236–237
Catalytic cracking, 41–42
Cellobiohydrolases, 214–215
Cellobiose, 214–215
Cellulose, 148–149, 197–198, 235–249
acrylic acid, 244–246
-based carbon fibers, 158–159
carboxymethylcellulose (CMC), 237
cellulose acetate, 238
furfural resins, 239–240
green polyethylene (GPE), 247–249
hemicelluloses, 238–239
lignin, 241–242
methylcellulose, 236–237
phenolic resins, 240–241
polyhydroxybutyrate (PHB), 246–247
polylactic acid (PLA), 243
schematic representation of cellulose molecule, 198*f*
sugarcane juice, 243
xanthan gum, 249
Cellulose acetate, 238
structure of, 238*f*
Cellulose–hemicelluloses domain, 202
Cellulosic ethanol, 26, 29, 91, 110, 192, 216
Chemical oxygen demand (COD), 186–187
Chemical potential of sugarcane, 148–159
advanced bio-based polymers and derived materials, 155–159
carbon fiber composites, 158–159
nanomaterials, 155–158
and its residues, 36–42
microbial metabolic pathway and sugarcane-derived products, 153–155
technology platforms, 149–153
carbohydrates, chemicals from, 149–152
lignin, chemicals from, 152–153
Chemical routes, 173–174
China, sugarcane production in, 3, 7, 11–12
Circular economy of sugarcane, 141, 145, 155
Citric acid, 36
Cogeneration process, 144, 175
Coimbatore Research Institute in India, 6
Combustion, 147
Composition of sugarcane plant, 1
Countries, sugarcane-producing, 70*t*
Crabtree effect, 33
Creation of the biofuel decarbonization credit (CBIO), 274–275
Crude oil, 13
Cuba, sugarcane production in, 6
Cultivars, 104–108
Cyclohexane, 25–26
Cytogenetics of sugarcane, 104–105

D

Depithing process, 235
Depolymerization methods, 152–153
Dextran, 39
Diffusion, 123
Dilute acid process, 213
Dimethyl sulfate (DMS), 237
Diseases, sugarcane, 107
Distillation, 25–26
Dry sugarcane straw cleaning process, 56, 56*f*

E

Economical importance, 49

Ecuador
agricultural production, 96
biofuels production in, 96–97
Electric Energy Trading Chamber, 261
Electricity, production of, 130–131
Electricity generation by the biomass source, 257–261
Electricity sector, 272–273, 275–276
Electric matrix and power installed by bioelectricity, 257–261
electricity generation by the biomass source, 257–261
Endoglucanases, 214–215
Energy auction, 266–267
Energy cane, 43–44, 108–115
advantages and challenges of, 114–115
an opportunity to improve agricultural productivity, 135
hybridization, 111–113
procedure for obtaining energy cane cultivars, 112*f*
selection, 113–114
strategies to select, 110–111
Energy cogeneration, 49–50
Energy consumption, 264, 267–268
Energy from sugarcane, 117
energy cane, 135
future markets, 135–136
main Brazilian energy crop, 118–122
main conversion routes for sugarcane bioenergy, 130–134
electricity, production of, 130–131
hydrolysis, 131–132
pyrolysis, fast, 132–134
residues energy from sugar/ethanol mills, 122–130
bagasse, 122–127
straw, 127–128
vinasse, 128–130
Energy matrix in Brazil, 15–16
Energy potential of sugarcane, 142–148
combustion, 147
energy storage applications of sugarcane-derived materials, 147–148
gasification, 146–147
pyrolysis, 146
routes, 143–147
1G and 2G ethanol production, 143–144
bioelectricity from sugarcane bagasse and straw, 144
biogas from by-products, 145
thermochemical route, 145–147
Energy purposes, sustainability of sugarcane for, 89
flex-fuel vehicle technologies and well-to-wheels, 97–99
Energy reallocation mechanism (ERM), 276
Energy-valued waste, 145
Environment, sugarcane and, 79–85
Enzymatic hydrolysis, 132, 214–216
Erianthus, 113
Ethanol, 8, 39, 181–182
and anhydrous alcohol, 72–73
dehydration, 25–26
-fueled vehicles, 67–68, 73
production, 14, 50
in Africa, 12–14
in Brazil, 14–16
production process, 22–26
second-generation (2G), 91
1-Ethyl-3-methyl imidazolium acetate, 148
Eucalyptus, 42–43
Eucalyptus lignin, schematic representation of, 199*f*
Evaporation, 189–190
Exoglucanases, 214–215
Exopolysaccharide, 39

F

Fermentation, 30, 217–221
consolidated bioprocess, 221
fermentative microorganisms, 217–218
presaccharification, 220
separate saccharification and, 218–219
simultaneous saccharification and, 219–220
simultaneous saccharification and cofermentation, 220–221
Fertirrigation in natura, 182–184
Filter cake, 31–32, 31*t*
Fischer–Tropsch (FT) synthesis, 41, 146–147, 175

Flex-fuel vehicle technologies, 97–99
Food versus energy, 16
Fossil feedstocks, 142
Fractionation, sugarcane, 148–149
Free contracting environment (FCE), 264, 275–276
Fructose, 1
Funneling, 152–153
Furanics, 151–152
Furfural, 26, 41, 149
Furfuraldehyde cured resin, chemical structure of, 240*f*
Furfural resins, 239–240
Furfuryl alcohol (FFA), 238–239
  polymerization of, 239*f*
Future perspectives for sugarcane biorefinery, 42–45

G

Gasification, 146–147
Gasoline, 8
Generation scaling factor (GSF), 276
Germplasm bank and sugarcane hybridization, 105–106
Global warming potentials (GWP), differences in, 90*t*
Gluconic acid, 154
Gluconobacter oxydans, 154
Glucose, 1, 39
Glycerol, 151–152
Goias State, 50
Green chemistry, 166–167
Greenhouse gases (GHGs), 196–197, 229–231, 273–274
  emissions, 135, 141
Green polyethylene (GPE), 247–249, 249*f*
GREET model, 97–99

H

Harvest
  approximate values for harvest in 2016/17, 11–12
  approximate values for harvest in 2017/18, 12
Hemicelluloses, 149, 198, 202, 235, 238–239
High-density polyethylene, 151–152
Higher heating values (HHVs), 145–146
Hispaniola Island, 3
Historical itinerary of sugarcane, 3–6
History of the main products of sugarcane, 7–8
"Holy Land" of caravans, 7
Hybridization, sugarcane, 105–106
Hydrolysis, 131–132, 212–216
  acid hydrolysis, 213–214
  enzymatic hydrolysis, 214–216
  hydrolysis process attached to the standard distillery, 222*f*
Hydropowered mill, 5*f*
Hydropower generation, 261
Hydro-processing, 41–42
3-Hydroxypropionic acid, 171
5-Hydroxymethyl-furfural, 149

I

India, sugarcane production in, 3, 11–12
Indonesia, cane plantations in, 6
Integral utilization of sugarcane residues, 44–45
Inter- and intramolecular hydrogen bonds, 202
Intergovernmental Panel on Climate Change (IPCC) assessment report, 89–90
Internal energy supply offer (IESO), 260–261
Internal return rate (IRR), 94–95
Ionic liquids, 208
Iron oxide, 119–120
Itaconic acid, 154

J

Juice, 179–180, 243
  chemical composition, 243*t*
  produced after cane crushing, 120

K

Kluyveromyces marxianus, 217–218
Kraft lignins, 241

L

Lactic acid, 152, 154
  commercial use and applications of, 244*f*
Lactobacillus brevis, 39
Lactobacillus delbrueckii, 175
Lactobacillus plantarum, 154
Land use change (LUC), 94
l-Arabinofuranoside, 238–239
Leuconostoc mesenteroides, 39
Levulinic acid, 149, 152, 171
Life cycle assessment (LCA) framework, 89–90, 94–95
Lignin, 149, 157–158, 198–199, 202, 235, 241–242
  chemicals from, 152–153
Lignin–carbohydrate bonds, 207–208
Lignocellulose fractionation, 149
Lignocellulosic biomass
  effect of pretreatment on, 206*f*
  recalcitrance of, 200–202
  structural organization of, 197–200
Lignocellulosic material of sugarcane, 120
Lignocellulosic recalcitrant structure, sugar release from, 38
Liquor, sugarcane (Cachaça), 71–72
Lithium ion batteries (LIBs), 147
Low-density polyethylene, 151–152

M

Madeira Islands, 3
Medicinal purposes, cane sugar for, 7
Mediterranean region, sugarcane production in, 3
Melle–Boinot ethanolic fermentation process, 24, 25*f*
Methanol, 146–147
Methylcellulose, 236–237
Microbial metabolic pathway and sugarcane-derived products, 153–155
Middle East, use of cane sugar in, 7
Milling, 123
Minas Gerais State, 50
Ministry of Mines and Energy (MME), 268
Miscanthus, 113
Molasses, 24, 29, 179–180
Monascus ruber, 155
Monoethyleneglycol (MEG), 26
Monosaccharides, 38

N

Nanomaterials, 155–158
National Alcohol Program, 50, 75
National Association of Automobile Manufacturers, 50
National Association of Components for Automotive Vehicles, 50
National Bank for Economic and Social Development (BNDES), 270
National Electric Energy Agency (ANEEL), 257, 272–273
National Interconnected System (NIS), 256, 267
National policy for sugarcane, 80–83
Native cellulose, 148
Natural biopolymer, 233*f*
Noble cane, 104

O

1G ethanol production, 143
1G2G3G ethanol biorefinery integration systems, 96
Organic acids, 152

P

Papua New Guinea, 3
Penicillium chrysogenum, 155
Pentose fermentation, 149
Permanent preservation areas (PPA), 80–83
Phenolic resins, 240–241
Physical energy guarantee (PEG), 277–278
Plug-in hybrid flex-fuel technology, 99
Pneumatic classifier equipment, 124
Polyethylene terephthalate (PET), 35–36, 151–152, 241–242
Polyhydroxyalkanoates (PHAs), 39
Polyhydroxybutyrate (PHB), 246–247, 247*f*
  production process of, 248*f*
Polylactic acid (PLA), 39, 243, 245*f*
Polymalic acid, 154
Polyols, 151–152

Polysaccharides, 38
Polyvinyl alcohol (PVA), 237
Polyvinyl chloride, 151–152
Poor syrup, 179–180
Portugal, sugarcane production in, 4–5
Potassium oxide, 119–120
Potential yield of sugarcane, 108
Presaccharification, 220
Pró-Álcool program, 90
Production of sugarcane, 3–4, 16, 22, 50
in Brazil, 4–6, 8–9, 12, 14–15
demand for, 50
Productivity of sugarcane per hectare, 203–204
Prospective economic performance (PEP), 95
Pyrolysis, 146
fast, 132–134

R

Rankine steam power cycles, 147
Regulated contracting environment (RCE), 264
Renewable chemistry, sugarcane for, 165
biochemical routes, 174–175
challenges involved, 175–176
chemical routes, 173–174
perspectives for sugarcane bioproducts, 170–175
raw material, sugarcane as, 168–170
thermochemical routes, 175
Renewable raw materials, 167
RenovaBio, 273–275
Residual molasses, 29
Residues energy from sugar/ethanol mills, 122–130
bagasse, 122–127
straw, 127–128
vinasse, 128–130
Rind fraction, 123–124
Rural Development Office (RDO), 76–77

S

SAAKE, 130
Saccharification
presaccharification, 220
separate saccharification and fermentation, 218–219
simultaneous saccharification and cofermentation, 220–221
simultaneous saccharification and fermentation, 219–220
Saccharomyces cerevisiae, 24, 33, 174, 191
Saccharum, 3
Saccharum barberi, 3, 104, 110–111
Saccharum edule, 104, 110–111
Saccharum officinarum, 4, 104, 110–111
Saccharum robostun, 3, 104, 110–111
Saccharum sinense, 104, 110–111
Saccharum sinensis, 3
Saccharum spontaneum, 3, 104, 110–113
Sacharomyces cerevisiae, 217–218, 220
Sakkar, 3
São Paulo, 50, 75–79, 78*t*, 79*t*, 261–263
Second-generation ethanol (SGE), 49–51, 91, 192
Second-generation ethanol production from sugarcane, 168–170, 195, 202–204, 205*f*
fermentation, 217–221
consolidated bioprocess, 221
fermentative microorganisms, 217–218
presaccharification, 220
separate saccharification and fermentation, 218–219
simultaneous saccharification and cofermentation, 220–221
simultaneous saccharification and fermentation, 219–220
hydrolysis, 212–216
acid hydrolysis, 213–214
enzymatic hydrolysis, 214–216
inclusion to standard distillery, 221–223
pretreatments, 204–205, 210*t*
pretreatments, types of, 206–212
biological pretreatments, 208
chemical pretreatments, 206–208
combined pretreatments, 208–212
physical pretreatments, 206
recalcitrance of lignocellulosic biomass, 200–202

Second-generation ethanol production from sugarcane (*Continued*)
  structural organization of lignocellulosic biomass, 197–200
  theoretical yield, 203*f*
Separate hydrolysis and fermentation processes (SHF), 218–219
Separate saccharification and fermentation process, 218–219
Settlement price of differences (SPD), 264
Short-term market (STM), 264
Simultaneous saccharification and fermentation (SSF), 218–219
  advantages and disadvantages, 219–220
Sludge, 22–24
Small hydroelectric plants (SHPs), 264–265
Smut disease, 115
Socioeconomic and environmental benefits of using ethanol from sugarcane, 16–17
Sorbitol, 171
Southeast Asia, sugarcane extraction in, 7
Soybean production, 83
Spain, cane plantations in, 6
Spatial Dependency Index (SDI), 62
Spatial distribution of sugarcane, 77*f*
Sporisorium scitamineum, 115
Stirling engines, 147
Straw, 26–29, 56, 118–119, 127–128
  bioelectricity from, 144
Straw bales, sugarcane
  ash content, 53–58
  water content in, 58–64
Streptococcus mutants, 39
Structure and main components of sugarcane, 150*f*
Sub-Sahelian Africa, sugarcane production in, 11–12
Succinic acid, 152, 154
Sucrochemistry, 168–170
Sucrose, 1, 168–170
Sugar, 1, 70–71
  and ethanol production process, 22–26, 23*f*
  and ethanol sectors in Africa, 12–14
  largest sugar consumers in the world, 70–71
Sugar, ethanol, and energy sectors in Brazil, 14–16
Sugar beet, 7–8
Sugarcane culture, 4, 4*f*, 6, 42–43
Sugarcane Industry Union (UNICA), 261
Sugar degradation products, 149
Sugar-energy agro-industry products, 35*t*
Sugar-energy sector, 51–53, 256, 261, 265, 270
Sugar/ethanol mills, residues energy from, 122–130
  bagasse, 122–127
  straw, 127–128
  vinasse, 128–130
Sugar fermentation for alcohol production, 8
Sugar production, 9–10, 50
  in Brazil, 7–8
Sugar yield gain, 107–108
Sukkar, 3
Superior individuals, selection of, 106–107
Sustainability of sugarcane industry, 52
Sustainable chemistry, 165
Sweet salt, 7
Syngas, 146–147

T

Taxonomy and genetics of sugarcane, 104–105
Technoeconomic assessment (TEA), 89–90, 94–95, 95*t*
Thailand, sugarcane production in, 11–12
Thermochemical routes, 175
Thermoelectric plants (TEPs), 270
Traditional sugarcane, 109
Trametes hirsuta, 39
Trapiche, 5–6
Trash, 118
2G ethanol production, 94–95, 99, 143–144, 168–170
Type I energy cane, 109
Type II energy cane, 109

U

Upflow Anaerobic Sludge Blanket (UASB) reactor, 184–185

V

Vehicles, ethanol-fueled, 67–68, 73
Vinasse, 30, 74, 128–130, 168–170, 180–184
  alternatives for the use of, 182–191
    anaerobic biodigestion, 184–188
    fertirrigation in natura, 182–184
    vinasse concentration, 189–191
  anaerobic biodigestion of, 187–188
  biorefinery, 191*f*
  characterization and chemical composition of, 180–182
  concentration, 189–191
    biodigested vinasse, concentration of, 190–191
    evaporation, 189–190
  physicochemical characteristics of, 182*t*
Vinasse sugarcane use, perspectives of in Brazil, 191–192
Visible ash content in sugarcane straw bales, 56, 57*f*

W

Water content, 57
  in sugarcane straw bales, 58–64
Weissela confusa, 39
Well-to-tank (WTT), 97, 99
Well-to-wheels framework, 95, 97–99
Wine, distillation of, 8
Wood crops, 42–43
World scenario of sugarcane and its main products, 8–16
  approximate values for the 2016/17 harvest, 11–12
  approximate values for the 2017/18 harvest, 12
  sugar, ethanol, and energy sectors in Brazil, 14–16
  sugar and ethanol sectors in Africa, 12–14

X

Xanthan gum, 39, 249
Xanthomonas campestris, 249
Xylans, 238–239
Xylitol, 41
Xylose, 41, 217, 220–221

Y

Yeasts, 33–34, 33*t*

Z

Zymomonas mobilis, 217

CPI Antony Rowe
Eastbourne, UK
November 26, 2019

Contents

Preface xi

Acknowledgements xiii

Author Biographies xv

List of Figures xix

List of Tables xxv

Symbols xxvii

1 Introduction 1
- 1.1 Introduction 1
- 1.2 Fault Diagnosis 1
  - 1.2.1 Filter-Based Fault Diagnosis 1
  - 1.2.2 Data-Driven Fault Diagnosis 6
- 1.3 Remaining Useful Life Prediction 7
  - 1.3.1 Data-Driven Remaining Useful Life Prediction 9
  - 1.3.2 Filter-Based Remaining Useful Life Prediction 9
- 1.4 Outline of This Book 10

2 Filter/Estimator Design of Networked Multi-rate Sampled Systems with Network-Induced Phenomena 13
- 2.1 Estimator Design with Measurement Quantization and Sensor Failures 14
  - 2.1.1 Problem Formulation 14
  - 2.1.2 Variance-Constrained Estimator Design 19
  - 2.1.3 Illustrative Examples 27
- 2.2 Finite-Time Filter Design with Event-Based Relay and Fading Channels 29
  - 2.2.1 Problem Formulation 30
  - 2.2.2 Finite-Time Filter Design 35
  - 2.2.3 Illustrative Examples 44
- 2.3 Conclusion 50

3 **Fault Detection of Networked Multi-rate Systems with Filter-Based Methods** **51**
3.1 Fault Detection with Fading Measurements and Randomly Occurring Faults . . . 52
3.1.1 Problem Formulation . . . 52
3.1.2 Detection of Randomly Occurring Faults . . . 56
3.1.3 Illustrative Examples . . . 60
3.2 Fault Detection with Dynamic Quantization and Intermittent Faults 64
3.2.1 Problem Formulation . . . 64
3.2.2 Detection of Intermittent Faults . . . 69
3.2.3 Illustrative Example . . . 77
3.3 Conclusion . . . 82

4 **Fault Diagnosis of Multi-rate Time-Varying Systems with Filter-Based Methods** **85**
4.1 Event-Based Fault Diagnosis with Constrained Fault . . . 86
4.1.1 Problem Formulation . . . 86
4.1.2 Fault Detection and Fault Isolation . . . 87
4.1.3 Illustrative Examples . . . 98
4.2 Event-Based Fault Diagnosis with Bounded Unknown Fault . . . 102
4.2.1 Problem Formulation . . . 104
4.2.2 Fault Diagnosis and Fault Estimation . . . 105
4.2.3 Illustrative Examples . . . 116
4.3 Conclusion . . . 122

5 **Fault Diagnosis of Modular Multilevel Converters with Machine Learning Methods** **125**
5.1 Fault Diagnosis with Mixed Kernel Support Tensor Machine . . . 126
5.1.1 Operating Principles of Modular Multilevel Converters . . . 126
5.1.2 Mixed Kernel Support Tensor Machine . . . 127
5.1.3 Fault Diagnosis . . . 131
5.1.4 Illustrative Examples . . . 132
5.2 Fault Diagnosis with Synchrosqueezing Transform and Optimized Deep CNN . . . 139
5.2.1 Synchrosqueezing Transform . . . 139
5.2.2 Optimized Deep Convolutional Neural Network . . . 140
5.2.3 Fault Diagnosis . . . 142
5.2.4 Illustrative Examples . . . 143
5.3 Conclusion . . . 151

6 **Remaining Useful Life Prediction of Industrial Components with Filter-Based Methods** **153**
6.1 Remaining Useful Life Prediction with Adaptive UKF and SVR . . 154
6.1.1 Genetic Algorithm Optimized Support Vector Regression . . 157
6.1.2 Remaining Useful Life Prediction of Lithium-Ion Batteries . 158

6.1.3 Illustrative Examples 159
6.2 Remaining Useful Life Prediction with ALF-Optimized PF and LSTM 165
6.2.1 Adaptive Levy Flight Optimized Particle Filter 166
6.2.2 Remaining Useful Life Prediction of Lithium-Ion Batteries . 169
6.2.3 Illustrative Examples 170
6.3 Remaining Useful Life Prediction with Degradation Point Detection and EKF 175
6.3.1 Degradation Point Detection 179
6.3.2 Health Indicator Construction 182
6.3.3 Remaining Useful Life Prediction of Bearings 183
6.3.4 Illustrative Examples 185
6.4 Conclusion 194

7 Remaining Useful Life Prediction of Industrial Components with Machine Learning Methods 195
7.1 Remaining Useful Life Prediction with WPT and Optimized SVR . 196
7.1.1 Degenerate Point Detection 196
7.1.2 Remaining Useful Life Prediction of Turbine Engines . . . 198
7.1.3 Illustrative Examples 201
7.2 Remaining Useful Life Prediction with Complete Ensemble EMD and GRU 208
7.2.1 Health Indicator Construction 208
7.2.2 Remaining Useful Life Prediction of Bearings 212
7.2.3 Illustrative Examples 215
7.3 Remaining Useful Life Prediction with PSR and Error Compensation 219
7.3.1 Health Indicator Construction 220
7.3.2 Remaining Useful Life Prediction of Lithium-Ion Batteries . 221
7.3.3 Illustrative Examples 226
7.4 Conclusion 231

8 Conclusions and Future Topics 233

Bibliography 235

Index 261

Preface

Modern industrial systems consist of industrial production equipment, monitoring devices, execution components, and communication networks. As the utilization time of these units increases, the safety of units will inevitably decline and eventually deteriorate. Once the anomalies or faults happen in industrial units, it may cause economic losses, major accidents, environmental pollution, and even human casualties. Therefore, safety analysis has become the prerequisite for industrial systems, that has received unremitting attention from both academics and practitioners. Generally speaking, the whole life state of the system can be divided into the healthy one, degraded one and faulty one. The corresponding research topics of safety analysis include state estimation/monitoring, anomaly detection, prediction of the remaining useful life, fault location, and fault estimation. The traditional safety analysis of industrial systems mainly investigates the fault detection and isolation problems in a single-rate sampling frame, with little attention on the influence of network-induced phenomena and the multi-rate sampling mechanism on diagnostic performance. It is worth mentioning that classical methods are mainly based on the signal and model, and the information of equipment operation is not fully utilized, which might lead to low accuracy of the fault diagnosis and remaining useful life prediction.

The objective of this book is to present the up-to-date research methodologies on the fault diagnosis and remaining useful life prediction of industrial systems and equipment. The content of this book is composed of two parts, where the first part (Chapters 2–4) presents the model-based state estimation and fault diagnosis methodologies of industrial systems, and the second part (Chapters 5–7) investigates the data-driven fault diagnosis and remaining useful life prediction of the industrial equipment. The compendious frame and description of the book are given as follows. Chapter 1 introduces the recent progress on filter-based fault diagnosis and remaining useful life prediction for systems and the equipment, and gives the outline of the book. Chapter 2 is concerned with the filter design of multi-rate sampled systems with unreliable data transmission. Chapter 3 deals with the problem of filter-based fault detection for networked multi-rate systems with network-induced phenomena. In Chapter 4, the filter-based fault diagnosis problems are investigated for multi-rate time-varying systems with the event-triggered mechanism. Chapter 5 discusses the fault diagnosis problem of modular multilevel converters by using machine learning methods. The remaining useful life prediction problem is investigated in Chapter 6 for lithium-ion batteries in terms of the filer-based methods. In Chapter 7, the machine-learning-assisted remaining useful life prediction is investigated of the industrial equipment. Chapter 8 gives the conclusion and some possible future research topics.

This book is a research monograph whose intended audience is graduate and postgraduate students as well as researchers. The background required of the reader is knowledge of basic control system theory, basic Lyapunov stability theory, Kalman/particle filter theory and machine learning theory.

Yong Zhang
Wuhan, China

Zidong Wang
London, U.K.

Ye Yuan
Wuhan, China

Acknowledgements

The authors would like to express their deep appreciation to those who have been directly involved in various aspects of the research leading to this book. Special thanks go to Professor Zhenxing Liu from Wuhan University of Science and Technology, Professor Huajing Fang and Professor Ying Zheng from Huazhong University of Science and Technology, for their valuable suggestions, constructive comments and support. The authors also extend our thanks to many colleagues who have offered support and encouragement throughout this research effort. In particular, we would like to acknowledge the contributions from Lifeng Ma, Lei Zou, Sai Li, Xiujuan Zheng, Cheng Cheng, Junyang Jin, and Guijun Ma. Last but not the least, the authors are especially grateful to their families for their encouragement and never-ending support when it was most required. The writing of this book was supported by the National Natural Science Foundation of China under Grants 61873148, 61873197, 61933007, and 92167201.

Author Biographies

Prof. Yong Zhang Yong Zhang is currently full Professor of Automatic Control, at School of Information Science and Engineering, Wuhan University of Science and Technology since 2019. Prior to this, he was a Post-doctoral researcher in the Control Science and Engineering, Wuhan University of Science and Technology. He received the BSc degree in Mathematics Education in 2001 from Jiangsu Normal University, Xuzhou, China, the MSc degree in Applied Mathematics in 2007 from Three Gorges University, Yichang, China, and the PhD degree in Control Science and Engineering in 2010 from Huazhong University of Science and Technology, Wuhan, China. From 2014 to 2015, he was a visitor with the Department of Information Systems and Computing, Brunel University London, Uxbridge, U.K. He has published over 30 papers in refereed international journals. His current research interests include fault diagnosis and prognosis, robust control and filtering. He is a very active reviewer for several international journals.

Prof. Zidong Wang Zidong Wang is Professor of Dynamical Systems and Computing at Brunel University London, West London, United Kingdom. He was born in 1966 in Yangzhou, Jiangsu, China. He received the BSc degree in Mathematics in 1986 from Suzhou University, Suzhou, the MSc degree in Applied Mathematics in 1990 and the PhD degree in Electrical and Computer Engineering in 1994, both from Nanjing University of Science and Technology, Nanjing.

He was appointed as Lecturer in 1990 and Associate Professor in 1994 at Nanjing University of Science and Technology. From January 1997 to December 1998, he was an Alexander von Humboldt research fellow with the Control Engineering Laboratory, Ruhr-University Bochum, Germany. From January 1999 to February 2001, he was a Lecturer with the Department of Mathematics,

University of Kaiserslautern, Germany. From March 2001 to July 2002, he was a University Senior Research Fellow with the School of Mathematical and Information Sciences, Coventry University, U.K. In August 2002, he joined the Department of Computer Science, Brunel University London, U.K., as a Lecturer, and was then promoted to a Reader in September 2003 and to a Chair Professor in July 2007.

Professor Wang's research interests include dynamical systems, signal processing, bioinformatics, control theory and applications. He has published more than 600 papers in refereed international journals. He was awarded the Humboldt research fellowship in 1996 from Alexander von Humboldt Foundation, the JSPS Research Fellowship in 1998 from Japan Society for the Promotion of Science, and the William Mong Visiting Research Fellowship in 2002 from the University of Hong Kong. He was a recipient of the State Natural Science Award from the State Council of China in 2014 and the Outstanding Science and Technology Development Awards (once in 2005 and twice in 1997) from the National Education Committee of China.

Professor Wang is currently serving or has served as the Editor-in-Chief for International Journal of Systems Science, the Editor-in-Chief for Neurocomputing, Executive Editor for Systems Science and Control Engineering, Subject Editor for Journal of The Franklin Institute, an Associate Editor for IEEE Transactions on Automatic Control, IEEE Transactions on Control Systems Technology, IEEE Transactions on Systems, Man, and Cybernetics - Systems, Asian Journal of Control, Science China Information Sciences, IEEE/CAA Journal of Automatica Sinica, Control Theory and Technology, an Action Editor for Neural Networks, an Editorial Board Member for Information Fusion, IET Control Theory & Applications, Complexity, International Journal of Systems Science, Neurocomputing, International Journal of General Systems, Studies in Autonomic, Data-driven and Industrial Computing, and a member of the Conference Editorial Board for the IEEE Control Systems Society. He served as an Associate Editor for IEEE Transactions on Neural Networks, IEEE Transactions on Systems, Man, and Cybernetics - Part C, IEEE Transactions on Signal Processing, Circuits, Systems & Signal Processing, and an Editorial Board Member for International Journal of Computer Mathematics.

Professor Wang is a Member of the Academia Europaea (section of Physics and Engineering Sciences), a Fellow of the IEEE (for contributions to networked control and complex networks), a Fellow of the Chinese Association of Automation, a Member of the IEEE Press Editorial Board, a Member of the EPSRC Peer Review College of the UK, a Fellow of the Royal Statistical Society, a member of program committee for many international conferences, and a very active reviewer for many international journals. He was nominated an appreciated reviewer for IEEE Transactions on Signal Processing in 2006-2008 and 2011, an appreciated reviewer for IEEE Transactions on Intelligent Transportation Systems in 2008, an outstanding reviewer for IEEE Transactions on Automatic Control in 2004 and for the journal Automatica in 2000.

Prof. Ye Yuan Ye Yuan is currently full Professor of Automatic Control, at School of Artificial Intelligence & Automation, Huazhong University of Science and Technology since 2016. Prior to this, he was a Post-doctoral researcher in the Hybrid Systems Lab and BAIR Lab, UC Berkeley with Prof. Claire J. Tomlin, a Junior Research Fellow (JRF) at Darwin College, University of Cambridge. He received his BSc degree under the supervision of Prof. Yugeng Xi from Department of Automation, Shanghai Jiao Tong University in 2008 and M. Phil., PhD from Control Group, Department of Engineering, University of Cambridge in 2009, 2012 under the supervision of Prof. Jorge M. Goncalves. He has been holding visiting researcher positions at CDS, Caltech (2011, 2014, 2015), LIDS, MIT (2013), and Imperial College London (2011-2015).

He has served as an Associate Editor of the IEEE Transactions on Control of Network Systems (2019-2020) and IEEE Control Systems Society Conference Editorial Board (2017-2020). He is the recipient of China National Recruitment Program of 1000 Talented Young Scholars, Dorothy Hodgkin Postgraduate Award, Microsoft Research PhD Scholarship, Cambridge Overseas Student Award, Chinese Government Award for Outstanding Students Abroad, Henry Lester PhD Scholarship and a number of best paper awards in IEEE conferences. His current research interests include System Identification, Control, Optimization and Machine Learning with applications to the understanding and (re-)design of natural (biology) and man-made systems (manufacturing, robotics and power systems). He has published more than 60 papers in refereed international journals.

List of Figures

2.1 An example of multi-rate sampled-data systems with $b = 3$. 15
2.2 Estimation error $e_z(t_k)$ for different b. 29
2.3 The actual steady-state estimation error variance for $e_1(t_k)$ for different b. 30
2.4 The actual steady-state estimation error variance for $e_2(t_k)$ for different b. 30
2.5 The triggering instants for networked multi-rate systems with $\delta = 0.2$. 48
2.6 The trajectory of $\mathbb{E}\{\eta^T(t_k)R\eta(t_k)\}$ for networked multi-rate systems with $\delta = 0.2$. 49
2.7 The output $\bar{z}(t_k)$ and its estimation $\hat{z}(t_k)$ for networked multi-rate systems with $\delta = 0.1$. 49
2.8 Filtering error $\tilde{z}(t_k)$ for networked multi-rate systems with $\delta = 0.1$. 50

3.1 Residual signal $r(t_k)$ for networked multi-rate systems. 62
3.2 Evolution of residual evaluation function $J(t_k)$ for networked multi-rate systems. 62
3.3 Residual signal $r(t_k)$ for networked single-rate systems. 63
3.4 Evolution of residual evaluation function $J(t_k)$ for networked single-rate systems. 64
3.5 DTS200 setup. 78
3.6 Zooming variable and its region of networked multi-rate systems with nonuniformly sampled measurements. 81
3.7 Residual-based fault detection of networked multi-rate systems with nonuniformly sampled measurements. 81
3.8 Residual-evaluation-function-based fault detection of networked multi-rate systems with nonuniformly sampled measurements. . . . 82

4.1 The minimum H_∞ performance level with given variance constraint matrix $\Theta_{t_k} = \text{diag}\{0.4, 0.3\}$ and upper bound of triggering torus $\delta_2 = 10^{-5}$. 98
4.2 The minimum upper bound of variance $\text{E}\{e_1(t_k)e_1^T(t_k)\}$, $\text{E}\{e_2(t_k)e_2^T(t_k)\}$ with given H_∞ performance level $\gamma = 1.8$ and upper bound of triggering torus $\delta_2 = 10^{-5}$. 99

4.3 The minimum upper bound δ_2 of triggering torus with given H_∞ performance level $\gamma = 1.8$ and variance constraint matrix $\Theta_{t_k} = \text{diag}\{0.4, 0.3\}$. 100
4.4 Fault detection with lower bound of triggering-torus $\delta_1 = 10^{-5}$. . . 101
4.5 Fault detection with lower bound of triggering-torus $\delta_1 = 10^{-4}$. . . 102
4.6 Fault isolation with lower bound of triggering-torus $\delta_1 = 10^{-5}$. . . . 103
4.7 Fault isolation with lower bound of triggering-torus $\delta_1 = 10^{-4}$. . . . 103
4.8 The minimum upper bound of estimation error for annulus I and annulus II: (a) The minimum upper bound of $e_1^T(t_k)e_1(t_k)$; (b) The minimum upper bound of $e_2^T(t_k)e_2(t_k)$; and (c) The minimum upper bound of $e_3^T(t_k)e_3(t_k)$. 118
4.9 The minimum triggering annulus for the given estimation error constraint and the lower bound of triggering annulus. 119
4.10 Fault detection for the given annulus-event-triggered scheme: (a) Annulus-event-triggered scheme and (b) Evolution of $J(t_k)$ and J_{th}. 119
4.11 Fault isolation for the given annulus-event-triggered scheme: (a) Annulus-event-triggered scheme and (b) Evolution of $Y^{(1)}(t_k)$ and $Y_{th}^{(1)}$. 120
4.12 Fault isolation for the given annulus-event-triggered scheme: (a) Annulus-event-triggered scheme and (b) Evolution of $Y^{(2)}(t_k)$ and $Y_{th}^{(2)}$. 121
4.13 Fault isolation for the given annulus-event-triggered scheme: (a) Annulus-event-triggered scheme and (b) Evolution of $Y^{(3)}(t_k)$ and $Y_{th}^{(3)}$. 121
4.14 Fault signal and its estimation for the given annulus-event-triggered scheme: (a) Annulus-event-triggered scheme and (b) Fault signal and its estimate. 122

5.1 Three-phase topological structure of modular multilevel converter. . 126
5.2 Sub-module structure of modular multilevel converter. 127
5.3 Rank-one mapping of the third-order tensor. 129
5.4 Chordal distance visual computing of the tensor A and B. 130
5.5 Modular multilevel converter under normal operation. 133
5.6 Open-circuit fault occurs in sub-module of A-phase upper bridge arm. 134
5.7 201-level system under normal condition and open-circuit fault occurs in sub-module of A-phase upper bridge arm. 135
5.8 The classification accuracy of different kernel function. 136
5.9 Modular multilevel converter with upper bridge sub-module fault in A-phase. 137
5.10 The fault identification label of modular multilevel converter with sudden load under normal condition and upper bridge sub-module fault in A-phase. 137
5.11 Modular multilevel converter with upper bridge sub-module fault in B-phase. 138

5.12 Modular multilevel converter with upper bridge sub-module fault in C-phase. 138
5.13 Time-frequency representations of faulty sub-module in modular multilevel converter: (a) Raw signal; (b) Continuous wavelet transform result; and (c) Synchrosqueezing transform result. 141
5.14 The architecture of deep convolutional neural network model in this section. 142
5.15 The flowchart of fault diagnosis for modular multilevel converter with the proposed SST-GA-DCNN algorithm. 144
5.16 Normal operation of modular multilevel converter: (a) Three-phase AC currents and (b) Three-phase circuiting currents. 145
5.17 Three-phase AC currents and inner circuiting currents of modular multilevel converter under fault operation: (a) A-phase upper bridge arm sub-module fault and (b) A-phase lower bridge arm sub-module fault. 146
5.18 Three-phase AC currents and inner circuiting currents of modular multilevel converter under fault operation: (a) B-phase upper bridge arm sub-module fault and (b) B-phase lower bridge arm sub-module fault. 146
5.19 Three-phase AC currents and inner circuiting currents of modular multilevel converter under fault operation: (a) C-phase upper bridge arm sub-module fault and (b) C-phase lower bridge arm sub-module fault. 147
5.20 Time-frequency representations of synchrosqueezing transform for modular multilevel converter with normal signals. 149
5.21 Time-frequency representations of synchrosqueezing transform for modular multilevel converter with A-phase up/low arm fault. 149
5.22 Time-frequency representations of synchrosqueezing transform for modular multilevel converter with B-phase up/low arm fault. 149
5.23 Time-frequency representations of synchrosqueezing transform for modular multilevel converter with C-phase up/low arm fault. 150
5.24 Visualization of intermediate activation for DCNN. 150
5.25 Performance evolution of deep convolutional neural network: (a) Accuracy and (b) Loss. 151
5.26 Confusion matrix of fault diagnosis. 151

6.1 Flowchart of the integrated algorithm. 159
6.2 The capacity decay curves of four batteries. 160
6.3 Ten-step-prediction for battery B0005 with start point 80 cycle. . . . 161
6.4 Ten-step-prediction for battery B0005 with start point 100 cycles. . 161
6.5 Twenty-step-prediction for battery B0006 with start point 60 cycle. . 162
6.6 Twenty-step-prediction for battery B0006 with start point 80 cycle. . 162
6.7 Twenty-step-prediction for battery B0007 with start point 60 cycle. . 162
6.8 Twenty-step-prediction for battery B0007 with start point 80 cycle. . 163
6.9 Five-step-prediction for battery B0018 with start point 40 cycle. . . 163

6.10 Five-step-prediction for battery B0018 with start point 60 cycle. . . 163
6.11 Weight degeneracy and particle impoverishment phenomena of standard particle filter algorithm. 166
6.12 Illustration of the evolution process of adaptive Levy flight-based particle filter. 168
6.13 Network architecture of LSTM. 169
6.14 Flowchart of proposed method for remaining useful life prediction. . 171
6.15 The capacity decay curves of three batteries. 171
6.16 Remaining useful life prediction of batteries 05, 06, and 07 with four different methods. 173
6.17 Box plot of the predicted remaining useful life distribution for batteries 05, 06, and 07 with 50 particles. 176
6.18 Histogram plot of the predicted remaining useful life distribution for batteries 05, 06, and 07 with 50 particles, and red lines are the real remaining useful life. 177
6.19 Box plot of the predicted remaining useful life distribution for batteries 05, 06, and 07 with 150 particles. 178
6.20 Histogram plot of the predicted remaining useful life distribution for batteries 05, 06, and 07 with 150 particles, and red lines are the real remaining useful life. 179
6.21 The procedure of the proposed ensemble prognostic methodology. . 181
6.22 Overview of PRONOSTIA platform. 187
6.23 Bearing 1–3: (a) Vibration signal of the whole lifetime; (b) Vibration signal at TSP; (c) Degradation Point Detection; and (d) Envelope spectrum of vibration signal at TSP. 188
6.24 The variance and cumulative features of bearing 1–3. 189
6.25 RUL prediction results comparison of different models based on EKF with bearing B1. 191
6.26 RUL prediction results comparison of EKF and PF based on XJTU-SY bearing datasets. 192
6.27 RUL prediction results comparison of EKF and PF based on double exponential model with bearing B1. 192
6.28 Bearing data sampling platform of XJTU-SY. 193

7.1 Wavelet basis function selection process. 198
7.2 The flowchart of proposed two-stages approach. 199
7.3 Different characteristics of the original signal distribution. 203
7.4 Wavelet transform approximation coefficient. 203
7.5 Partial sensor recording signals. 204
7.6 Reconstructed signal after WPT. 205
7.7 (a) Second sensor reconstructed signal of all units and (b) Life span distribution of all units. 205
7.8 Actual versus predicted RULs of full cycle. 206
7.9 RUL prediction results of test_FD001. 207
7.10 RUL prediction results of test_FD003. 208

7.11 Integrated architecture for abnormal point detection and RUL prediction of REBs. . . . 209
7.12 The flowchart of GRU. . . . 213
7.13 Abnormal point time detection: (a) Bearing 1–1; (b) Bearing 1–3; (c) Bearing 1–5; and (d) Bearing 1–7. . . . 216
7.14 Cri of the extracted features in condition 1. . . . 217
7.15 Feature selection with mean of Cri in condition 1. . . . 217
7.16 HIs of the test bearing: (a) The HI with M1 and (b) The HI with BC-CEEMDAN-FC and GRU. . . . 218
7.17 RUL prediction: (a) Bearing 1–3 and (b) Bearing 1–7. . . . 219
7.18 Flowchart of proposed approach. . . . 220
7.19 Evolution trend of $H1$. . . . 227
7.20 Evolution trend of $H2$ for the 90th cycles. . . . 228
7.21 (a) Decomposed series of original H1 and (b) Decomposed series of original H2. . . . 229
7.22 (a) Reconstructed H1 and (b) Reconstructed H2. . . . 230
7.23 (a) Prediction for battery 05, (b) Prediction for battery 06, (c) Prediction for battery 07, and (d) Prediction for battery 18. . . . 231

List of Tables

1.1 Typical machine learning methods for fault diagnosis and remaining useful life prediction. 8

2.1 The permitted minimum γ and corresponding estimator gains H_ϱ $(\varrho = 1, 2, \cdots, b)$. 28
2.2 The minimum variance values σ_r $(r = 1, 2)$ and corresponding estimator gains H_ϱ $(\varrho = 1, 2, \cdots, b)$. 29
2.3 Model parameters and Main operating point. 45
2.4 The actual $\bar{c}_1$ and $\bar{\mu}$. 47
2.5 The permitted minimum c_2. 48
2.6 The permitted minimum γ. 48

4.1 The value of residual evaluation function and its threshold with different δ_1. 101
4.2 The value of residual matching function and its threshold with $\delta_1 = 10^{-5}$. 102
4.3 The value of residual matching function and its threshold with $\delta_1 = 10^{-4}$. 102
4.4 The minimum upper bound with given lower bound $\delta^- = 10^{-5}$. . . . 118
4.5 Fault isolation functions and their thresholds. 120

5.1 Parameters of the operating environment and the modular multilevel converter prototype. 132
5.2 Label value of fault classification. 135
5.3 The optimal parameters of grid search and the average of classification accuracy based on the support tensor machine model. 135
5.4 The optimal parameters of grid search and average of classification accuracy based on the SVM model. 136
5.5 Main parameters of modular multilevel converter prototype. 143
5.6 Parameters of deep convolutional neural network model. 148
5.7 Fault types and label values. 148

6.1 Prediction effect comparison of B0005 battery. 164
6.2 Prediction effect comparison of B0006 battery. 164
6.3 Comparison of four algorithms for battery 05. 174
6.4 Comparison of four algorithms for battery 06. 174
6.5 Comparison of four algorithms for battery 07. 175

6.6 Comparison of remaining useful life distribution with particles count of 50 and 150. 180
6.7 RMSE, MAPE, and elapsed time of the PF-LSTM using different order Markov processes based on battery 05. 180
6.8 Comparison of prediction performance with the existed results. . . . 181
6.9 Observed TTF values for bearings (in seconds). 187
6.10 Time to start prediction (TSPs). 188
6.11 Features trendability value comparison. 189
6.12 RUL prediction error comparison of PHM2012 dataset. 190
6.13 Comparison of RUL prediction errors with some machine learning methods. 191
6.14 XJTU-SY bearing datasets. 193
6.15 RUL prediction error comparison of XJTU-SY bearing datasets. . . 193

7.1 The main properties of commonly wavelets. 199
7.2 Comparison of different optimization algorithms. 207
7.3 The statistical features in time domain. 210
7.4 Comparison of prediction performance with three different algorithms. 230
7.5 Comparison of prediction performance with the existed results. . . . 231

Symbols

$\mathbb{R}^n$	The n-dimensional Euclidean space.
$\mathbb{R}^{n\times m}$	The set of all $n \times m$ real matrices.
$L_2[0,\infty)$	The space of square-integrable vector functions over $[0,\infty)$.
$X \geq Y$	The $X - Y$ is positive semi-definite, where X and Y are symmetric matrices.
$X > Y$	The $X - Y$ is positive definite, where X and Y are symmetric matrices.
$\mathrm{Prob}\{\cdot\}$	The occurrence probability of the event "·".
$\mathbb{E}\{\cdot\}$	The expectation of the stochastic variable "·" with respect to the given probability measure Prob.
0	The zero matrix of compatible dimension.
I	The identity matrix of compatible dimension.
$*$	The term induced by symmetry, in symmetric block matrices or complex matrix expressions.
$\mathrm{diag}\{\cdots\}$	The block-diagonal matrix.
$\mathrm{col}\{\cdots\}$	The column vector composed of elements.
$\Vert \bullet \Vert$	The Euclidean norm for vectors.
$\mathbb{Z}_{n\geq 0}$	The set of n component real vectors.
$\lfloor \star \rfloor$	The floor function which is the largest integer not greater than $\star$.
$\lambda_{\max}(Q)$	The maximum eigenvalue of matrix $Q \in \mathbb{R}^{n\times m}$.
$\lambda_{\min}(Q)$	The minimum eigenvalue of matrix $Q \in \mathbb{R}^{n\times m}$.
M^{T}	The transpose matrix of M.
$\Vert A \Vert$	The norm of matrix $A \in \mathbb{R}^{n\times m}$ defined by $\Vert A \Vert \triangleq \sqrt{\lambda_{\max}(A^T A)}$.
$\Vert x \Vert$	The norm of vector $x \in \mathbb{R}^n$, $\Vert x \Vert \triangleq \sqrt{x^T x}$.
$\otimes$	The Kronecker product.

$\mathrm{diag}_1^n\{A_i\}$ A block-diagonal matrix where the square matrices A_i are in the corresponding main diagonal blocks, and $\mathrm{diag}_n\{A\}$ denotes n diagonal blocks A.

$\mathrm{col}_1^n\{x_i\}$ Stacks the vectors as $[x_1^T\ x_2^T\ \cdots\ x_n^T]^T$.

1

Introduction

1.1 Introduction

Along with the rapid development of the economy and society, safety has become the primary concern in production. Small-scale safety accidents cause economic losses in factories, and large accidents cause casualties and environmental events. Therefore, it is necessary to monitor the status of the key production equipment and production process, detect anomalies and faults, predict their evolution trends, and take measures to avoid accidents before their occurrence. At the same time, the location and severity of faults need to be traced quickly and accurately. The timely maintenance should be carried out to restore the production quickly. In a word, these problems all boil down to the study of fault diagnosis [1,2] and remaining useful life (RUL) prediction [3,4] of the equipment or systems. In order to perform fault diagnosis and remaining useful life prediction, many advanced methods have been proposed, which can be roughly divided into model-based ones and data-driven ones [5–8]. Among them, the filter-based methods have attracted wide attention [9–11].

1.2 Fault Diagnosis

As the main parts of the safety analysis field, fault diagnosis consists of fault detection, fault isolation, and fault estimation. Specially speaking, the purpose of fault detection is to check the anomalies of equipment; fault isolation aims to locate and classify the fault; and fault estimation can provide the estimates of the fault signals.

1.2.1 Filter-Based Fault Diagnosis

The model-based fault diagnosis depends on the dynamics of systems where the most widely studied systems are linear time-invariant systems with the following form

$$\begin{cases} x(k+1) = Ax(k) + Bu(k) + E_\omega \omega(k) + E_f f(k) \\ y(k) = Cx(k) + Du(k) + F_\omega \omega(k) + F_f f(k) \end{cases} \tag{1.1}$$

DOI: 10.1201/9781003330998-1

where $x(k) \in \mathbb{R}^{n_x}$ represents the state vector, $u(k) \in \mathbb{R}^{n_u}$ is the input vector, $y(k) \in \mathbb{R}^{n_y}$ denotes the measurement output, $\omega(k) \in \mathbb{R}^{n_\omega}$ is the exogenous disturbance signal, and $f(k) \in \mathbb{R}^{n_f}$ is the fault signal. The matrices A, B, C, D, E_ω, E_f, F_ω, and F_f are known real matrices with appropriate dimensions.

From the point of view of system analysis, the types of disturbances/noises in model (1.1) directly affect the characteristics of the goal system, leading to several typical fault detection methods as follows: 1) the H_∞ filtering technique [12], parity space approach [12], unknown input observer method [13] as well as the optimization-based method [14] have been adopted to design the fault detection filters for systems with energy-bounded disturbances/noises; 2) the Kalman filter-based method [15] has proven to be particularly effective for Gaussian stochastic disturbances/noises; and 3) the set-membership estimation approach [16–18] has the capability of dealing with the ellipsoidal-set-constrained disturbances/noises. In order to integrally formulate the theory of fault diagnosis, we divide the process of fault diagnosis into fault detection, fault isolation, and fault estimation.

As the first step of fault diagnosis, successful fault detection plays a key role in the diagnosis of faults. Based on the basic model (1.1), the observer-based residual generator [19] can be constructed as follows

$$\begin{cases} \hat{x}(k+1) = A\hat{x}(k) + Bu(k) + L[y(k) - \hat{y}(k)] \\ \hat{y}(k) = C\hat{x}(k) + Du(k) \\ r(k) = V[y(k) - \hat{y}(k)] \end{cases} \tag{1.2}$$

where $\hat{x}(k)$ and $\hat{y}(k)$ are the estimation of $x(k)$ and $y(k)$, respectively. $r(k)$ is the residual signal. L and V are the observer gain matrices to be designed that have appropriate dimensions.

By denoting the state estimation error $e(k) \triangleq \hat{x}(k) - x(k)$ and combining (1.1)–(1.2), we can obtain the following fault detection error system

$$\begin{cases} e(k+1) = \bar{A}e(k) + \bar{E}_\omega \omega(k) + \bar{E}_f f(k) \\ r(k) = \bar{C}e(k) + \bar{F}_\omega \omega(k) + \bar{F}_f f(k) \end{cases} \tag{1.3}$$

where $\bar{A} \triangleq A - LC$, $\bar{E}_\omega \triangleq E_\omega - LF_\omega$, $\bar{E}_f \triangleq E_f - LF_f$, $\bar{C} \triangleq VC$, $\bar{F}_\omega \triangleq VF_\omega$, and $\bar{F}_f \triangleq VF_f$.

With the constructed filter structure (1.2), the gains of the fault detection filter are designed for the system (1.1) with bounded-energy disturbances/noises according to the following two performance indices: one is the H_∞ performance [12, 20], which reveals the influence of disturbances/noises on the residual; and another is the trade-off optimization by utilizing the H_∞/H_∞ and H_-/H_∞ performance indices [19, 21], where the H_∞ performance quantifies the robustness of residuals against the disturbance, and the H_- performance reflects the sensitivity of the residuals with respect to the fault.

After the fault detection filters are designed by employing the above two kinds of performance indices, the corresponding residual can be obtained. Consequently, the occurrence of a fault can be determined by comparing the residual-related function with their threshold. Obviously, constructing a reasonable and effective residual is the basis of fault detection. Here, a common method is introduced with the following residual evaluation function $J(k)$ and its threshold J_{th}:

$$J(k) \triangleq \Big[\sum_{h=k_0}^{k} r^T(h) r(h) \Big]^{\frac{1}{2}}, \quad J_{th} \triangleq \sup_{\substack{\omega(k)\in \ell_2 \\ f(k)=0}} J(k) \tag{1.4}$$

where k_0 represents the initial time of the residual evaluation.

The occurrence of fault can be detected by comparing $J(k)$ with J_{th} according to the following test rule:

$$\begin{cases} J(k) \geq J_{th} \Longrightarrow & \text{alarm for fault} \\ J(k) < J_{th} \Longrightarrow & \text{no fault} \end{cases} \tag{1.5}$$

and the fault detection time k_d is defined as $k_d \triangleq \min_k \{J(k) \geq J_{th}\}$.

After the fault is detected with (1.5), a natural problem is how to judge the performance of the selected fault detection method. In this aspect, some outstanding results have been acquired [7, 16, 22, 23], where the fault detection performance can be assessed with several indices including the false alarm rate, missing alarm rate, fault detection rate, averaged detection delay, and so on. Fundamentally speaking, an excellent fault detection method should possess a low false alarm rate and missing alarm rate, a high fault detection rate as well as less averaged detection delay. A widely used strategy is to give an acceptable false alarm rate for fault detection and then, the main goal is to maximize the fault detection rate. For instance, the parity relation-based offline residual generator and evaluator have been integrally designed to maximize the fault detection rate with the predefined false alarm rate in [16]. The authors in [23] have proposed the new concepts of false alarm rate and fault detection rate in the framework of the norm, where the detection can be realized by comparing concurrently the fault detection rate and false alarm rate with their thresholds. Furthermore, in order to reduce the false alarm rate and missing alarm rate, the adaptive threshold selection methods have been proposed by authors in [7, 24].

After the fault has been detected, the fault isolation module will be activated. Based on the empirical knowledge, it is assumed that there are q types of possible faults. Specifically speaking, it is assumed that $f(k)$ belongs to the following finite fault sets

$$\mathscr{F} \triangleq \{f_I^{(1)}(k), f_I^{(2)}(k), \cdots, f_I^{(q)}(k)\} \tag{1.6}$$

Similar to [25], the following fault isolation estimators are adopted

$$\begin{cases} \hat{x}_I^{(\ell)}(k+1) = A\hat{x}_I^{(\ell)}(k) + Bu(k) + E_f f_I^{(\ell)}(k) \\ \qquad\qquad\quad + L_I^{(\ell)}\big[y(k) - \hat{y}_I^{(\ell)}(k)\big] \\ r_I^{(\ell)}(k) = V_I^{(\ell)}\big[y(k) - \hat{y}_I^{(\ell)}(k)\big] \\ \hat{y}_I^{(\ell)}(k) = C\hat{x}_I^{(\ell)}(k) + Du(k) + F_f f_I^{(\ell)}(k) \end{cases} \tag{1.7}$$

where $\hat{x}_I^{(\ell)}(k)$ is the estimate of state $x(k)$ in the ℓth estimator, $L_I^{(\ell)}$ and $V_I^{(\ell)}$ are the gain matrices to be designed, $\ell = 1, 2, \cdots, q$.

By denoting $e_I^{(\ell)}(k) \triangleq x(k) - \hat{x}_I^{(\ell)}(k)$ and $\tilde{f}_I^{(\ell)}(k) \triangleq f(k) - f_I^{(\ell)}(k)$ as the corresponding estimation error. Together (1.1) with (1.7), one has the following fault isolation error system:

$$\begin{cases} e_I^{(\ell)}(k+1) = \bar{A}_I^{(\ell)} e_I^{(\ell)}(k) \\ \qquad\qquad + \bar{E}_{I,\omega}^{(\ell)} \omega(t_k) + \bar{E}_{I,f}^{(\ell)} \tilde{f}_I^{(\ell)}(k) \\ r_I^{(\ell)}(k) = \bar{C}_I^{(\ell)} e_I^{(\ell)}(k) \\ \qquad\qquad + \bar{F}_{I,\omega}^{(\ell)} \omega(t_k) + \bar{F}_{I,f}^{(\ell)} \tilde{f}_I^{(\ell)}(k) \end{cases} \tag{1.8}$$

where $\bar{A}_I^{(\ell)} \triangleq A - L_I^{(\ell)} C$, $\bar{E}_{I,\omega}^{(\ell)} \triangleq E_\omega - L_I^{(\ell)} F_\omega$, $\bar{E}_{I,f}^{(\ell)} \triangleq E_f - L_I^{(\ell)} F_f$, $\bar{C}_I^{(\ell)} \triangleq V_I^{(\ell)} C$, $\bar{F}_{I,\omega}^{(\ell)} \triangleq V_I^{(\ell)} F_\omega$, and $\bar{F}_{I,f}^{(\ell)} \triangleq V_I^{(\ell)} F_f$.

After the gain matrices of the fault isolator are obtained, the following residual estimation function $Y^{(\ell)}(k)$ and its threshold $Y_{th}^{(\ell)}$ are adopted:

$$Y^{(\ell)}(k) \triangleq \Big[\sum_{i=k_0}^{k} (r_I^{(\ell)}(i))^T r_I^{(\ell)}(i)\Big]^{\frac{1}{2}} \tag{1.9}$$

$$Y_{th}^{(\ell)} \triangleq \sup_{d(k)\in\mathfrak{W}_k \bar{f}(k)=0} Y^{(\ell)}(k) \tag{1.10}$$

where k_0 represents the initial time of the estimation.

Consequently, it can be concluded that fault $f(k)$ is the ℓth type fault $f_I^{(\ell)}(k)$ if the following decision scheme holds:

$$\begin{cases} Y^{(\ell)}(k) \le Y_{th}^{(\ell)} \\ Y^{(s)}(k) > Y_{th}^{(s)}, \quad s \in \{1, 2, \cdots, q\}/\ell \end{cases} \tag{1.11}$$

The absolute fault isolation time k_I is defined as $k_I \triangleq \max_k\{Y^{(s)}(k) > Y_{th}^{(s)}\}$.

To realize effective fault isolation, two representative methods are widely used [26, 27]: 1) dedicated observer scheme (DOS), under which the ℓth residual is only sensitive to the ℓth fault and is decoupled from all other faults; 2) the generalized observer scheme (GOS), which shows that the ℓth residual is sensitive to all faults except the ℓth one. By using the similar structure of the generalized observer scheme, the fault isolation problems have been investigated in [18, 28, 29], respectively, for nonlinear systems subject to sensor bias faults, incomplete measurements and the event-triggering-based multi-rate sampling mechanism. Particularly, to improve the fault isolation rate and reduce the isolation delay, the adaptive threshold has been introduced in [28].

If the detected fault cannot be isolated with the aforementioned fault isolation schemes, the fault estimation unit will be activated to identify both the amplitude and the evolutionary dynamics of the faults. Then, the fault dataset $\mathscr{F}$ can be updated and will become more complete. In terms of the model-based fault estimation, the authors in [30, 31] have assumed the fault dynamics satisfies $f(k+1) = A_k f(k)$ with the known time-varying matrix A_k. Such kinds of faults include incipient faults and constant faults as special cases. In [32, 33], another typical fault dynamics has been considered with the following model:

$$f(k+1) = f(k) + \phi(k) \tag{1.12}$$

where $\phi(k)$ is the fault increment.

To estimate the fault effectively, the following estimators are adopted:

$$\begin{cases} \hat{x}_E(k+1) = A\hat{x}_E(k) + Bu(k) + E_f\hat{f}_E(k) \\ \qquad\qquad\qquad + L_E\big[y(k) - \hat{y}_E(k)\big] \\ \hat{f}_E(k+1) = \hat{f}_E(t_k) + V_E\big[y(k) - \hat{y}_E(k)\big] \\ \qquad \hat{y}_E(k) = C\hat{x}_E(k) + Du(k) + F_f\hat{f}_E(k) \end{cases} \tag{1.13}$$

where $\hat{f}_E(k)$ and $\hat{x}_E(k)$ are the estimates of fault $f(k)$ and state $x(k)$, respectively. L_E and V_E are the estimator gain matrices to be designed.

By denoting $e_E(k) \triangleq x(k) - \hat{x}_E(k)$, $\tilde{f}_E(k) \triangleq f(k) - f_E(k)$, $\bar{\omega}(k) \triangleq \mathrm{col}\{\phi(k), \omega(k)\}$, and utilizing (1.1), (1.12), and (1.13), one can obtain the following estimation error systems:

$$\begin{cases} e_E(k+1) = \bar{A}_E e_E(k) + \bar{E}_{E,\omega}\bar{\omega}(k) + \bar{E}_{E,f}\tilde{f}_E(k) \\ \tilde{f}_E(k+1) = \bar{C}_E e_E(k) + \bar{F}_{E,\omega}\bar{\omega}(k) + \bar{F}_{E,f}\tilde{f}_E(k) \end{cases} \tag{1.14}$$

where $\bar{A}_E \triangleq A - L_E C$, $\bar{E}_{E,\omega} \triangleq [0 \quad E_\omega - L_E F_\omega]$, $\bar{E}_{E,f} \triangleq E_f - L_E F_f$, $\bar{C}_E \triangleq -V_E C$, $\bar{F}_{E,\omega} \triangleq [I \quad -V_E F_\omega]$, $\bar{F}^{(\ell)}_{E,f} \triangleq I - V_E F_f$.

In order to facilitate design, we define $\xi(k) \triangleq \mathrm{col}\{e_E(k), \tilde{f}_E(k)\}$. Then, the system (1.14) can be rewritten as follows

$$\xi(k+1) = \mathcal{A}\xi(k) + \mathcal{B}\bar{\omega}(k) \tag{1.15}$$

where $\mathcal{A} \triangleq \begin{bmatrix} \bar{A}_E & \bar{E}_{E,f} \\ \bar{C}_E & \bar{F}_{E,f} \end{bmatrix}$, $\mathcal{B} \triangleq \begin{bmatrix} \bar{E}_{E,\omega} \\ \bar{F}_{E,\omega} \end{bmatrix}$. Along similar lines of [18], the fault estimator can be designed based on the augmented system (1.15).

In addition to the aforementioned diagnosis performance indices (such as false alarm rate, missing alarm rate, fault detection rate, and averaged detection delay), fault detectability, isolability, and diagnosability have also attracted much research attention [1, 7, 34], since the later three indices are capable of revealing some essential problems in the fault diagnosis. In particular, this topic can be divided into the system-based method and performance-based technique. For example, from the fault detection and isolation point of view, the structural properties of a system have been discussed in system-based method [1]. The performance-based one has been discussed in [7, 34] where the conditions have been obtained to show how a fault can be detected, isolated and diagnosed.

In the above parts, we summarize the methods of fault detection, fault isolation and fault estimation for linear time-invariant systems from the perspective of methodology. Actually, the complexity of systems will directly affect the fault diagnosis process and thus, more and more research attention has been paid to fault diagnosis problems for other more complex systems, including time-varying systems, nonlinear systems, multi-rate sampling systems, and so on.

The time-varying characteristic exists in almost all real-time systems. For time-invariant systems [1], the fault diagnosis mainly focuses on the steady-state behaviours of systems that are discussed in the infinite time domain. In contrast, the fault diagnosis for time-varying systems usually emphasizes the transient performance of systems in the finite time domain [35–37]. For observer-based methods, there are three kinds of typical fault diagnosis design methods for time-varying systems, including Kalman-type recursive filter method [29], recursive Riccati difference equations method [37], and recursive linear matrix inequality method [18].

Nonlinearity is another common phenomenon in reality and a large number of industrial systems are essentially nonlinear. In the past decades, the fault diagnosis problems for nonlinear systems have attracted considerable attention with two typical fault diagnosis techniques being developed. One is the Takagi–Sugeno fuzzy model based method [38, 39], which has been recognized as a powerful tool to describe the global behaviours of nonlinear systems. Since the Takagi–Sugeno fuzzy model can be represented by several linear submodels connected via membership functions, some mature fault diagnosis methods for linear systems can be easily extended to the Takagi–Sugeno fuzzy models. Another typical approach is the extension of the existing linear system approach by adding some constraints on the considered nonlinearities, such as the sector-bounded condition [40] and Taylor series expansion [41].

Multi-rate sampled-data system is a kind of system, whose sensors have different sampling periods with the system state. The multi-rate sampled-data models are capable of describing industrial systems that have kinds of sensors employed for different monitoring tasks with different measurement updating requirements. Compared to the fault diagnosis for single-rate sampled-data systems, the fault diagnosis for multi-rate sampled-data systems is more challenging, as the traditional methods may fail to deal with the multi-rate phenomenon. In recent years, some interesting results have been reported concerning the fault diagnosis for multi-rate sampled-data systems with multi-rate uniform sampling [19] and nonuniform sampling [18].

1.2.2 Data-Driven Fault Diagnosis

With the rapid development in sensors, data storage, network transmissions and other new technologies, numerous data are generated to monitor the key manufacturing equipment. The research of fault diagnosis and remaining useful life prediction mainly focuses on mining the deterioration information from monitoring data and developing effective algorithms to detect the abnormality and predict their evolution trend.

Data-driven modelling and analysis have attracted considerable attention [42], on multivariable statistical process monitoring [43, 44], principal component analysis (PCA), and partial least squares (PLS). The subspace-aided approach and Fisher discriminant analysis (FDA) can be used to detect the fault of the linear process with Gaussian distribution. The independent component analysis (ICA) can investigate the fault detection of non-Gaussian linear processes. Many improved algorithms for nonlinear and time-varying process characteristics have been proposed such as the kernel and dynamic methods, which are abbreviated as K/DPCA, K/DPLS, K/DICA, and so on [45–47]. The process fault with Gaussian distribution can be detected by comparing squared prediction error (SPE) with its threshold (Hotelling's T^2 statistic) for PCA/PLS/FDA series. The kernel density estimation is chosen to calculate the control limit of the independent component analysis method for data with non-Gaussian distribution. After the fault is detected, contribution-based methods are usually used to isolate the fault [45,46]. For example, the relative reconstruction-based contribution and the minimum risk Bayesian decision theory are adopted in [48].

Machine learning-based fault diagnosis techniques have been utilized widely in many interesting fields including machinery equipment [49], electrical equipment [50], energy systems [51], and so on. According to the depth of model structures, machine learning methods for fault diagnosis can be classified into shallow learning method and deep learning method. Shallow machine learning methods mainly involve support vector machine (SVM) [52], extreme learning machine (ELM) [53], hidden Markov process (HMM) [54], and deep belief network (DBN) [55], which have no or few network units. In contrast, deep learning method adopts many neurons to form a network structure with a certain number of layers and width to realize the learning function. Typical methods involve convolutional neural networks (CNN) [56], generative adversarial networks (GAN) [57], long short-term memory (LSTM) [58], gated recurrent unit (GRU) [59], transfer learning (TL) [60], and graph neural networks (GNN) [61]. With the development of sensors and big data technology, deep learning methods have received constantly increasing attention and become the mainstream. Such popularity mainly benefits from the stronger analysis and learning ability and the excellent diagnostic performance of the deep learning methods. The advantages and disadvantages of the above typical shallow machine learning and deep learning methods, as well as typical application scenarios, are summarized in Table 1.1.

1.3 Remaining Useful Life Prediction

Remaining useful life prediction, which aims to determine the effective remaining use time or its probability distribution, is achieved by establishing the degradation mapping of equipment based on the degradation mechanism model or the monitoring data [95,96]. In recent years, the remaining useful life prediction-related research topic has attracted considerable research attention. More specifically, the popular implementation methods of remaining useful life prediction can be divided into two cat-

Table 1.1: Typical machine learning methods for fault diagnosis and remaining useful life prediction.

Model	Advantage	Disadvantage	Fault diagnosis	RUL prediction
SVM	⋆ Good robustness to noise. ⋆ Convenient to process high dimensional data. ⋆ Fast training speed.	∗ Difficult to train large-scale data. ∗ Unable for Multiple classifications. ∗ Sensitive to selection of parameters and kernel functions.	• [52] Bearings • [62] Steering actuator	• [63] Battery • [64] Bearing
ELM	⋆ Strong generalization. ⋆ Fast learning rate. ⋆ Few training parameters.	∗ Low accuracy. ∗ Easy to overfitting.	• [53] Turbofan engine • [65] Rolling bearing	• [66] Turbofan engines • [67] Rolling bearings
HMM	⋆ Easy to learn inherent structure by experimental data. ⋆ Good ability to deal with noise. ⋆ Consider the potential evolution of the state.	∗ Time-invariant state transition probability. ∗ Limited by assumptions and unable to predict directly. ∗ Complex calculation process.	• [54] Motor drive system • [68] PEM fuel cell	• [69] Tool wearing • [70] Tool wearing
DBN	⋆ Fast training speed and less convergence time. ⋆ Easy expansion and good performance. ⋆ Parallel computing is possible.	∗ Process one-dimensional data. ∗ Need a labelled sample set. ∗ Slow learning process.	• [55] Gearbox • [71] Reciprocating	• [72] Gas leakage • [73] Power grid
CNN	⋆ Strong ability of image processing. ⋆ Less training parameters. ⋆ Automatic feature extraction.	∗ Parameters need to be adjusted manually. ∗ Large amount of calculation. ∗ May lose useful information.	• [74] Bearing • [56] MMC	• [75] Turbofan engine • [76] Bearing
GAN	⋆ Generate more samples for training ⋆ Avoid the use of Markov Chain ⋆ Can train any generating network.	∗ Poor interpretability. ∗ Exist the problem of non-convergence. ∗ Not applicable to discrete data.	• [57] Chiller • [77] Bearing	• [78] Aero-engine • [79] Wind turbines
RNN	⋆ Fast training speed and less convergence time. ⋆ Easy expansion and good performance. ⋆ Parallel computing is possible.	∗ Process one-dimensional data. ∗ Need a labelled sample set. ∗ Slow learning process.	• [80] Bearing • [81] Bearing	• [82] Turbofan engine • [83] Ion etching
LSTM	⋆ Strong ability of time series data processing. ⋆ Solve the problem of long sequence dependence. ⋆ Small error and high accuracy.	∗ The problem of gradient explosion. ∗ Limited processing power for long sequences. ∗ Slow training speed.	• [84] Wind turbine • [58] Bearing	• [85] Battery • [86] Turbofan engine
GRU	⋆ Strong ability of time series data processing. ⋆ Solve the problem of long sequence dependence. ⋆ Small error and high accuracy.	∗ The problem of gradient explosion. ∗ Limited processing power for long sequences. ∗ Slow training speed.	• [87] Spur gear • [59] TE	• [88] Bearing • [89] Bearing
TL	⋆ Improve the generalization ability. ⋆ Save computing resources and time. ⋆ Reduce training data required for the model.	∗ Rely on the pretrained weight. ∗ Negative transfer due to network structure. ∗ Difficult to converge.	• [60] Bearing • [90] Gear	• [91] Bearings • [92] Battery
GCN	⋆ Ability to work with graph structure. ⋆ Suitable for large-scale graph data sets. ⋆ Better stability and invariance.	∗ Limitations on the relevance of spatial information. ∗ Poor generalization ability on other graph structures.	• ADR • Event detection • Detecting anomalies	• [93] Traffic • Chemical poisoning • [94] Crowd counting

egories, that is data-driven remaining useful life prediction method and filter-based remaining useful life prediction method.

1.3.1 Data-Driven Remaining Useful Life Prediction

Data-driven remaining useful life prediction techniques can be summarized into statistical methods [97] and machine learning-based remaining useful life prediction approaches [4,70]. Statistical methods utilize the theory of statistics to establish evaluation indices first, and then, assess the health status of equipment. However, the applications of these methods are limited by the data quality and the strict preconditions of statistical theory. Compared to statistical methods, machine learning-based methods, which are more flexible and practical, have long been popular techniques utilized for remaining useful life prediction in recent years. The machine learning-based remaining useful life prediction process consists of four steps (i.e. feature extraction, health index establishment, feature selection, and remaining useful life prediction). These steps are achieved differently with the shallow and the deep learning techniques.

Various shallow learning techniques have been adopted to obtain more accurate remaining useful life prediction results, which include, but are not limited to support vector regression [98], deep belief network [72], support vector machine [64], and so on. With the help of deep learning techniques, better results can be obtained by the data-based remaining useful life prediction method from the perspective of trend, scale similarity, and monotony [60, 61, 92]. The implementation of the remaining useful life prediction method with deep learning techniques relies on the employment of some representative deep learning algorithms, which include, but are not limited to convolutional neural networks [75, 76], graph neural networks [57, 99], recurrent neural network (RNN) [82], long short-term memory [85], gated recurrent unit [88], and so on.

1.3.2 Filter-Based Remaining Useful Life Prediction

Filter-based remaining useful life prediction, which takes stochastic filtering as a bridge to achieve prediction, is a typical data-model hybrid method [6]. The prediction accuracy can be significantly improved since this kind of method integrates the advantages of data-driven and model-based methods. Specifically, this approach can be divided into three parts: i) establish the health index (HI) based on the feature engineering and select the appropriate degradation model; ii) determine the model parameters by stochastic filtering methods; and iii) obtain the prediction results through model extrapolation. Typical filter-based methods include, but are not limited to Kalman filter (KF)-based method and particle filter (PF)-based method .

Kalman filter has long been used as a helpful tool to deal with the optimal filtering problems of a class of linear systems with Gaussian noise [100]. However, in engineering practice, the degradation processes of equipments are usually nonlinear, which brings significant challenges to the implementation of the filtering task. To overcome this challenge, the extended Kalman filter (EKF) algorithm [101] and the unscented Kalman filter (UKF) algorithm [63, 102], which are improved filtering methods of the Kalman filter, have been proposed. To be specific, the extended Kalman filter algorithm utilizes the partial derivative of the nonlinear degradation process to establish the Jacobian determinant, which realizes linearization. Unscented Kalman filter algorithm employs the unscented transformation method to approximate the nonlinear degradation process. The core idea of the Kalman filter-based method can be summarized into three steps: i) assume that the equipment degradation conforms to a specific degradation mechanism model, ii) integrate the mathematical model and key parameters into a discrete-time state-space model by establishing an expansion vector, and iii) utilize KF/EKF/UKF algorithm to implement the state update and prediction task. When it comes to the research of remaining useful life predictions of lithium-ion batteries and rolling bearings, the degradation mechanism model often is selected as the exponential function and its extended model, quadratic function model, and fusion curve model [103, 104].

Particle filter, which avoids the problem of massive data storage and recalculation, is a recursive Bayesian algorithm based on the Monte Carlo method [105]. Since it utilizes the concept of sequential importance sampling to approximate the approximate solution of the Bayesian optimal solution, the accuracy of Bayesian prediction largely depends on the sample size and diversity. However, as the number of iterations increases, the particle diversity decreases (i.e. the weight of some particles decreases or even can be ignored), which gives rise to particle depletion problems and finally leads to the loss of the sample diversity. Accordingly, optimizing the selections of sampled particles during resampling, which is one of the most effective approaches, has long been utilized to solve this problem [106].

1.4 Outline of This Book

This book is divided into eight chapters, where Chapters 2–4 mainly focus on model-based state estimation and fault diagnosis problems, and Chapters 5–7 are concerned with data-driven fault diagnosis and remaining useful life prediction problems. To be specific, the outline of this book is given as follows:

- Chapter 1 presents some concepts in safety analysis, reviews the research progress of filter-based fault diagnosis and remaining useful life prediction problems with model and data methods, and lists the outline of the book.

- Chapter 2 investigates the variance-constrained H_∞ and finite-time filter design problem for networked multi-rate systems. In the first half of the chapter, the estimation strategy is proposed for a class of networked multi-rate systems with measurement quantization and probabilistic sensor failures. The second half of this chapter designs a finite-time filter for a class of networked multi-rate systems with fading measurements in an event-triggered relay communication framework.

- Chapter 3 is concerned with the fault detection problem for uniformly and nonuniformly sampled multi-rate systems with randomly occurring faults, fading measurements, intermittent faults, and dynamic quantization. In the first half of the chapter, the fault detection issue is analyzed for a class of networked multi-rate systems with network-induced fading channels and randomly occurring faults. The remaining part of this chapter investigates the fault detection problem for a class of networked multi-rate sensor systems with nonuniform sampling and dynamic quantization.

- Chapter 4 discusses the fault detection and isolation problems for a class of time-varying multi-rate systems. In the first half of this chapter, the fault detection and isolation problem has been investigated for a class of multi-rate time-varying systems with ellipsoid-constrained fault and torus-event-triggering communication scheme. The second half of this chapter deals with the annulus-event-based fault detection, isolation, and estimation issue for multi-rate time-varying systems with sensor degradation as well as unknown-but-bounded-disturbances and faults.

- Chapter 5 deals with the fault diagnosis and location problems of the open-circuit fault for the modular multilevel converter. In the first section of this chapter, with the help of the mixed kernel support tensor machine, the fault diagnosis and location of the modular multilevel converter are achieved. The second section of this chapter investigates the fault diagnosis and location problem in terms of the deep convolutional neural networks algorithm. The last section deals with the fault diagnosis and location problem with the help of synchrosqueezing transform and genetic algorithm optimized convolutional neural networks.

- Chapter 6 is concerned with the remaining useful life prediction problems using hybrid algorithms. In the first subsection of this chapter, the adaptive unscented Kalman filtering algorithm is utilized to renovate dynamically the noise covariances, and the support vector regression algorithm and the genetic algorithm are employed to compensate for unknown observations and to realize multi-step prediction. In the second sub section of this chapter, a hybrid prognosis framework based on particle filter is put forward to optimize the performance of the particle filter and improve the prediction accuracy with the help of adaptive Levy flight and long short-term memory algorithm. In the last sub section of this chapter, with the help of the extended Kalman filter and the optimized long short-term memory, the remaining useful life problem is analyzed. In terms of the remaining useful life prediction of rolling bearings, the cumulative functions combined with noise-reduced time-domain features are used to establish trend health indicators. After the time to start prediction and health indicators have been determined, the fusion between double exponential degrade model and extended Kalman filter is desired to predict the remaining useful life.

- Chapter 7 is concerned with the remaining useful life prediction problem for the rolling element bearings and the lithium batteries with machine learning methods. For the purpose of achieving the remaining useful life of bearings, wavelet modulus maximum is used to confirm the degenerate point, and then genetic algorithm optimized support vector machine is proposed to achieve remaining useful life prediction in the first section. In the second section, the remaining useful life prediction of the rolling element bearings is addressed by constructing a framework based on a trend-reconstruct-based features selection and gated recurrent unit. In the third half of this chapter, the remaining useful life prediction of lithium-ion batteries is investigated by employing a hybrid data-driven method based on support vector regression and error compensation.

- Chapter 8 draws some conclusions on the book, and points out some potential research directions related to the work done in this book.

In addition, the illustrative numerical simulation examples, which are employed to validate the effectiveness of the results, are provided within the individual chapters.

2

Filter/Estimator Design of Networked Multi-rate Sampled Systems with Network-Induced Phenomena

The past few decades have witnessed an ever-increasing research interest in networked control systems because of their advantages of decreasing the need for hardwiring and reducing the costs of installation as well as implementation. So far, a great number of research results have been available in the literature (see e.g. [107, 108]), where various network-induced phenomena (e.g. communication delays, packet dropouts [109], signal quantization [110], and randomly occurring nonlinearities) have been thoroughly investigated.

In computer-based control systems, the interface between the plant and the estimator is often connected via analogue-to-digital and digital-to-analogue devices, which normally leads to the quantization process [111]. In particular, it is often unrealistic or sometimes impossible for large-scale networked control systems to sample all physical signals uniformly at a single rate. Consequently, the scheme of multi-rate sampled-data with signal quantization arises naturally and has become a research focus for many years.

State estimation or filtering has long been a research topic of fundamental importance in signal processing, communications and control applications [112]. Among a variety of existing approaches, the H_∞ method has gained particular research attention due to its capability of providing a bound for the worst-case estimation error without the need for knowledge of noise statistics. Recently, fading measurements and event-triggered mechanism have received initial attention in the context of filter/controller design. Different from conventional time-triggered communication schemes, the event-triggered strategy alleviates the unnecessary waste of computation and communication resources while maintaining the guaranteed filtering performance. As such, some pioneering works have been carried out on networked control systems with either fading channels [113–115] or event-triggered mechanism [116–119]. Nonetheless, up to now, there have been few results concerning the more practical problems with consideration of both the event-triggered communication schemes and the channel fading issues.

Inspired by the above discussion, we will investigate the estimator/filter design of multi-rate sampled-data systems with network-induced phenomena in the H_∞ framework. Section 2.1 is concerned with the state estimation problem with measurement quantization and sensor failures. In Section 2.2, a finite-time H_∞ filter is

DOI: 10.1201/9781003330998-2

derived with fading measurements in an event-triggered wireless relay communication framework. Section 2.3 gives our conclusions.

2.1 Estimator Design with Measurement Quantization and Sensor Failures

It is common in practical engineering that the estimation performance requirements are naturally expressed as the upper bounds on estimation error variances [120]. So, the variance-constrained theory has been widely applied in solving multi-objective control problems as well as filtering problems, see, for example [121, 122].

In this section, the variance-constrained H_∞ state estimation problem is investigated for networked multi-rate systems with measurement quantization and sensor failures.

2.1.1 Problem Formulation

Consider the following class of discrete-time systems:

$$x(T_{k+1}) = Ax(T_k) + B_1\omega(T_k) + B_2\nu(T_k) \tag{2.1}$$

$$z(T_k) = Lx(T_k), \quad k = 0, 1, 2, \cdots \tag{2.2}$$

where $x(T_k) \in \mathbb{R}^{n_x}$ represents the state vector, $z(T_k) \in \mathbb{R}^{n_z}$ is the signal to be estimated, $\omega(T_k) \in \mathbb{R}^{n_\omega}$ is a disturbance input with bounded energy which belongs to $\ell_2[0,\infty)$, and $\nu(T_k) \in \mathbb{R}^{n_\nu}$ is a zero mean Gaussian white noise sequence with covariance $R > 0$.

The measurement with probabilistic sensor failures is described by

$$y(t_k) = \Xi(t_k)Cx(t_k) + D\xi(t_k) = \sum_{s=1}^{m} \beta_s(t_k)C_s x(t_k) + D\xi(t_k) \tag{2.3}$$

where $y(t_k) \in \mathbb{R}^m$ is the measured output vector, $\Xi(t_k)$ is a diagonal matrix governing the probabilistic sensor failures described by

$$\Xi(t_k) \triangleq \text{diag}\{\beta_1(t_k), \beta_2(t_k), \cdots, \beta_m(t_k)\}$$

with $\beta_s(t_k)$ $(s = 1, ..., m)$ being m independent random variables which are also independent of $\nu(T_k)$, and the matrix C_s is defined by

$$C_s \triangleq \text{diag}\{\underbrace{0, \cdots, 0}_{s-1}, 1, \underbrace{0, \cdots, 0}_{m-s}\}C \ (s = 1, 2, \cdots, m).$$

It is assumed that $\beta_s(t_k)$ has the probabilistic density function $f(\beta_s)$ on the interval $[0, 1]$ with known mathematical expectations $\bar{\beta}_s$ and variances $\tilde{\bar{\beta}}_s^2$. In the sequel, we

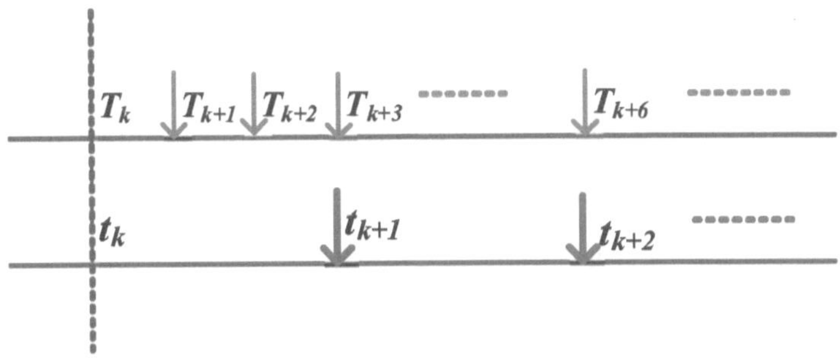

Figure 2.1: An example of multi-rate sampled-data systems with $b = 3$.

denote $\bar{\Xi} \triangleq \mathbb{E}[\Xi(t_k)] = \text{diag}\{\bar{\beta}_1, \bar{\beta}_2, \cdots, \bar{\beta}_m\}$. $\xi(t_k) \in \mathbb{R}^{n_\xi}$ is the measurement noise which belongings to $\ell_2[0, \infty)$. A, B_1, B_2, L, C, and D are known matrices with appropriate dimensions.

Remark 2.1. *Note that an increasing number of sensors have been installed to structures for monitoring and control, and sensor faults become more frequent compared to the structure's lifetime [123]. For small autonomous helicopters, the fault detection problem is investigated in [124], where several sensor failures were considered such as total sensor failure, stuck with constant bias sensor failure, drift or additive-type sensor failure, multiplicative-type sensor failure and outlier data sensor failure. In Eq. (2.3), the random variable $\beta_s(t_k)$ which takes value on the interval $[0\ ,1]$ is introduced to describe the $s-th$ ($s = 1, ..., m$) sensor failure, and the kind of measurement model can include the Bernoulli distribution model [125] as its special case. Actually, this sensor fault model is multiplicative-type sensor failure which has been discussed in [124].*

For a given frame period h, we make the following assumptions about the sampling period for (2.1)–(2.3):

A1. The system state $x(T_k)$ and its estimation $z(T_k)$ are updated at instants T_k, and $T_{k+1} - T_k \triangleq h,\ k = 0, 1, 2, \cdots$.

A2. The measurement $y(t_k)$ from the system is sampled at instants t_k, then $t_{k+1} - t_k \triangleq bh,\ k = 0, 1, 2, \cdots$, where b is a positive integer.

It can be seen that (2.1) and (2.2) evolve with a fast sampling period h, while the measurement dynamics (2.3) is generated with a slower period bh, that is the measurement sampling periods $t_{k+1} - t_k$ are integer multiples of the fast period h. Accordingly, (2.1), (2.2) and (2.3) are essentially a multi-rate sampled-data system model. An illustration of multi-rate sampled-data is shown in Fig. 2.1 with $b = 3$, where the system state and estimation signal are updated with period h, and the measurements are sampled with period $3h$.

In this section, the quantization effect on measurement $y(t_k)$ is considered with the map of the quantization process given by

$$\bar{y}(t_k) = q(y(t_k)) = \mathrm{col}\{q_1(y^{(1)}(t_k)), q_2(y^{(2)}(t_k)), \cdots, q_m(y^{(m)}(t_k))\}$$

where $\bar{y}(t_k)$ is the signal after quantization. The quantizer $q(\cdot)$ is assumed to be of the logarithmic type, that is, for each $q_s(y^{(s)}(t_k))(s = 1, 2, \cdots, m)$, the set of quantization levels is described by

$$\Im = \{\pm u_i^{(s)}, u_i^{(s)} = (\chi^{(s)})^i u_0^{(s)}, i = 0, \pm 1, \pm 2, \cdots\} \bigcup \{0\}$$

where $0 < \chi^{(s)} < 1, u_0^{(s)} > 0$, and $\chi^{(s)}(s = 1, 2, \cdots, m)$ is called the quantization density. Each of the quantization level corresponds to a segment such that the quantizer maps the whole segment to this quantization level. The logarithmic quantizer $q_s(\cdot)$ is defined as

$$q_s(y^{(s)}(t_k)) = \begin{cases} u_i^{(s)}, & \frac{1}{1+\delta_s} u_i^{(s)} < y^{(s)}(t_k) < \frac{1}{1-\delta_s} u_i^{(s)} \\ 0, & y^{(s)}(t_k) = 0 \\ -q_s(-y^{(s)}(t_k)), & y^{(s)}(t_k) < 0 \end{cases}$$

where $\delta_s = \frac{1-\chi^{(s)}}{1+\chi^{(s)}}$. It can be easily observed from the above definition that $q_s(y^{(s)}(t_k)) = (1 + \triangle^{(s)}(t_k))y^{(s)}(t_k)$ holds for certain $\triangle^{(s)}(t_k)$ satisfying $\triangle^{(s)}(t_k) \leq \delta_s$.

According to the above transformation, the quantization effects have been transformed into sector-bounded uncertainties [126]. By defining

$$\triangle(t_k) \triangleq \mathrm{diag}\{\triangle^{(1)}(t_k), \triangle^{(2)}(t_k), \cdots, \triangle^{(m)}(t_k)\},$$

the measurement after quantization can be expressed as

$$\bar{y}(t_k) = (I + \triangle(t_k))y(t_k) \tag{2.4}$$

Denoting $\Lambda \triangleq \mathrm{diag}\{\delta_1, \delta_2, \cdots, \delta_m\}$ and setting $F(t_k) \triangleq \triangle(t_k)\Lambda^{-1}$, we can know that $F(t_k)$ is a real-valued time-varying matrix satisfying $F(t_k)^T F(t_k) = F(t_k)F(t_k)^T \leq I$.

Note that it is mathematically difficult to handle the variance-constrained state estimation problem directly for such kind of multi-rate sampled-data system. In the next section, we are going to convert the resulting multi-rate sampled-data system into a single-rate system for technical convenience.

By applying the relation (2.1) recursively, one obtains the following equations with time scale t_k:

$$\begin{cases} x(t_{k+1}) = A^b x(t_k) + \bar{B}_{1,1}\bar{\omega}(t_k) + \bar{B}_{2,1}\bar{\nu}(t_k) \\ x(t_{k+1} - h) = x(t_k + (b-1)h) = A^{b-1}x(t_k) + \bar{B}_{1,2}\bar{\omega}(t_k) + \bar{B}_{2,2}\bar{\nu}(t_k) \\ \vdots \\ x(t_{k+1} - (b-1)h) = x(t_k + h) = Ax(t_k) + \bar{B}_{1,b}\bar{\omega}(t_k) + \bar{B}_{2,b}\bar{\nu}(t_k) \end{cases} \tag{2.5}$$

where

$$
\begin{aligned}
\bar{\omega}(t_k) &\triangleq \text{col}\{\omega(t_k), \omega(t_k+h), \cdots, \omega(t_k+(b-1)h)\}, \\
\bar{\nu}(t_k) &\triangleq \text{col}\{\nu(t_k), \nu(t_k+h), \cdots, \nu(t_k+(b-1)h)\}, \\
\bar{B}_{j,1} &\triangleq [A^{b-1}B_j \;\; A^{b-2}B_j \;\; \cdots \;\; AB_j \;\; B_j], \\
\bar{B}_{j,2} &\triangleq [A^{b-2}B_j \;\; A^{b-3}B_j \;\; \cdots \;\; B_j \;\; 0], \cdots, \\
\bar{B}_{j,b-1} &\triangleq [AB_j \;\; B_j \;\; \underbrace{0 \cdots 0}_{b-2}], \;\; \bar{B}_{j,b} \triangleq [B_j \;\; \underbrace{0 \cdots 0}_{b-1}], (j=1,2).
\end{aligned}
$$

Based on the quantized measurement signal $\bar{y}(t_k)$ and system (2.5), the following estimator is constructed:

$$
\begin{cases}
\hat{x}(t_{k+1}) = A^b \hat{x}(t_k) + H_1\Big(\bar{y}(t_k) - \bar{\Xi}C\hat{x}(t_k)\Big) \\
\hat{x}(t_{k+1}-h) = A^{b-1}\hat{x}(t_k) + H_2\Big(\bar{y}(t_k) - \bar{\Xi}C\hat{x}(t_k)\Big) \\
\vdots \\
\hat{x}(t_{k+1}-(b-1)h) = A\hat{x}(t_k) + H_b\Big(\bar{y}(t_k) - \bar{\Xi}C\hat{x}(t_k)\Big) \\
\hat{z}(t_k - ih) = L\hat{x}(t_k - ih), \; (i = 0, 1, 2, \cdots, b-1)
\end{cases} \tag{2.6}
$$

where $\hat{x}(t_k - ih) \in \mathbb{R}^{n_x}$ $(i = 0, 1, 2, \cdots, b-1)$ are the estimated state, $\hat{z}(t_k - ih) \in \mathbb{R}^{n_z}$ $(i = 0, 1, 2, \cdots, b-1)$ are the estimated output, and H_ϱ $(\varrho = 1, 2, \cdots, b)$ are the estimator gains to be designed.

Denoting

$$
\begin{aligned}
\eta(t_k) &\triangleq \text{col}\{x(t_k), x(t_k-h), \cdots, x(t_k-(b-1)h)\}, \\
e(t_k - ih) &\triangleq x(t_k - ih) - \hat{x}(t_k - ih), \\
e_\eta(t_k) &\triangleq \text{col}\{e(t_k), e(t_k-h), \cdots, e(t_k-(b-1)h)\}, \\
e_z(t_k - ih) &\triangleq z(t_k - ih) - \hat{z}(t_k - ih), \\
\tilde{z}_e(t_k) &\triangleq \text{col}\{e_z(t_k), e_z(t_k-h), \cdots, e_z(t_k-(b-1)h)\}, \\
\tilde{\beta}_s(t_k) &\triangleq \beta_s(t_k) - \bar{\beta}_s, \; d(t_k) \triangleq \text{col}\{\xi(t_k), \bar{\omega}(t_k)\}, \\
\bar{A}_1 &\triangleq \text{col}\{A^b, A^{b-1}, \cdots, A\}, \;\; \hat{A} \triangleq [\bar{A}_1 \;\; \underbrace{0 \cdots 0}_{b-1}], \\
\hat{H} &\triangleq \text{col}\{H_1, H_2, \cdots, H_b\}, \; \mathcal{I} \triangleq [I \;\; \underbrace{0 \cdots 0}_{b-1}], \\
\mathcal{A} &\triangleq \hat{A} - \hat{H}\bar{\Xi}C\mathcal{I}, \; \mathcal{C} \triangleq -\hat{H}F(t_k)\Lambda\bar{\Xi}C\mathcal{I}, \\
\bar{\mathcal{A}}_s &\triangleq -\hat{H}C_s\mathcal{I}, \; \bar{\mathcal{C}}_s \triangleq -\hat{H}F(t_k)\Lambda C_s\mathcal{I}, \\
\hat{B}_j &\triangleq \text{col}\{\bar{B}_{j,1}, \bar{B}_{j,2}, \cdots, \bar{B}_{j,b-1}, \bar{B}_{j,b}\}, \\
\mathcal{B}_1 &\triangleq \begin{bmatrix} -\hat{H}D & \hat{B}_1 \end{bmatrix}, \\
\bar{\mathcal{B}}_1 &\triangleq \begin{bmatrix} -\hat{H}F(t_k)\Lambda D & 0 \end{bmatrix}, \; \mathcal{B}_2 \triangleq \hat{B}_2,
\end{aligned}
$$

$$\hat{L} \triangleq \text{diag}\{\underbrace{L, L, \cdots, L}_{b}\},$$
$$(i = 0, 1, 2, \cdots, b-1;\ j = 1, 2;\ s = 1, 2, \cdots, m),$$

and using the lifting technique, the dynamics of estimation error can be obtained from (2.5) and (2.6) as follows:

$$\begin{cases} e_\eta(t_{k+1}) = \mathcal{A}e_\eta(t_k) + \Big\{\mathcal{C} + \sum_{s=1}^{m} \tilde{\beta}_s(t_k)(\bar{\mathcal{A}}_s + \bar{\mathcal{C}}_s)\Big\}\eta(t_k) \\ \qquad\qquad + (\mathcal{B}_1 + \bar{\mathcal{B}}_1)d(t_k) + \mathcal{B}_2\bar{\nu}(t_k) \\ \tilde{z}_e(t_k) = \hat{L}e_\eta(t_k) \end{cases} \tag{2.7}$$

and $\bar{\nu}(t_k)$ satisfies the following relationship:

$$\begin{aligned} &\mathbb{E}\{\bar{\nu}(t_k)\} = 0,\ \ \mathbb{E}\{\bar{\nu}(t_k)\bar{\nu}^T(t_i)\} = 0\ (k \neq i), \\ &\mathbb{E}\{\bar{\nu}(t_k)\bar{\nu}^T(t_k)\} = \text{diag}\Big\{\underbrace{R, \cdots, R}_{b}\Big\} \triangleq \mathcal{R}. \end{aligned} \tag{2.8}$$

By denoting $\tilde{e}(t_k) \triangleq \text{col}\{e_\eta, \eta(t_k)\}$, $\bar{H} \triangleq \text{col}\{-\hat{H}, 0\}$, $\bar{C} \triangleq \begin{bmatrix} 0 & \Lambda\bar{\Xi}C\mathcal{I} \end{bmatrix}$, $\bar{C}_s \triangleq \begin{bmatrix} 0 & C_s\mathcal{I} \end{bmatrix}$ and $\bar{D} \triangleq \begin{bmatrix} \Lambda D & 0 \end{bmatrix}$ $(s = 1, 2, \cdots, m)$, we have the following augmented system:

$$\begin{cases} \tilde{e}(t_{k+1}) = \Big\{(\mathscr{A} + \mathscr{C}) + \sum_{s=1}^{m} \tilde{\beta}_s(t_k)(\bar{\mathscr{A}}_s + \bar{\mathscr{C}}_s)\Big\}\tilde{e}(t_k) \\ \qquad\qquad + (\mathscr{B}_1 + \bar{\mathscr{B}}_1)d(t_k) + \mathscr{B}_2\bar{\nu}(t_k) \\ \tilde{z}_e(t_k) = \mathscr{L}\tilde{e}(t_k) \end{cases} \tag{2.9}$$

where

$$\begin{aligned} &\mathscr{A} \triangleq \text{diag}\{\mathcal{A}, \hat{A}\},\ \mathscr{C} \triangleq \bar{H}F(t_k)\bar{C}, \\ &\bar{\mathscr{A}}_s \triangleq \bar{H}\bar{C}_s,\ \bar{\mathscr{C}}_s \triangleq \bar{H}F(t_k)\Lambda\bar{C}_s\ (s = 1, 2, \cdots, m), \\ &\mathscr{B}_1 \triangleq \begin{bmatrix} -\hat{H}D & \hat{B}_1 \\ 0 & \hat{B}_1 \end{bmatrix},\ \mathscr{B}_2 \triangleq \begin{bmatrix} \hat{B}_2 \\ \hat{B}_2 \end{bmatrix}, \\ &\bar{\mathscr{B}}_1 \triangleq \bar{H}F(t_k)\bar{D},\ \mathscr{L} \triangleq [\hat{L}\ \ \underbrace{0 \cdots 0}_{b}]. \end{aligned}$$

Remark 2.2. *So far, by using the lifting technique, the model (2.9) for networked multi-rate systems has been obtained. Comparing with the models of networked multi-rate systems in [127], model (2.9) exhibits two distinguished features: i) both the quantization and probabilistic sensor failures are considered and therefore the model (2.9) is quite comprehensive to better reflect the networked environment; ii) the coefficients in model (2.3) are governed by individual random variables taking value on* $[0, 1]$ *and such representations include the widely studied Bernoulli distribution (see e.g. [127]) as a special case.*

Before proceeding further, we introduce the following definition.

Definition 2.1. *The augmented system (2.9) is said to be exponentially mean-square stable if, with $d(t_k) = 0$ and $\bar{v}(t_k) = 0$, there exist constants $\alpha \geq 1$ and $\hbar \in (0, 1)$ such that*

$$\mathbb{E}\{\|\tilde{e}(t_k)\|^2\} \leq \alpha \hbar^{t_k} \mathbb{E}\{\|\tilde{e}(t_0)\|^2\}$$

The main purpose of this section is to design the estimator in the form of (2.6) such that the following requirements are satisfied simultaneously:

(a) the augmented system (2.9) is exponentially mean-square stable;

(b) under zero-initial condition, the estimation error $\tilde{z}_e(t_k)$ with respect to the energy bounded disturbance $d(t_k)$ satisfies

$$\sum_{k=0}^{\infty} \mathbb{E}\left\{\|\tilde{z}_e(t_k)\|^2\right\} < \gamma^2 \sum_{k=0}^{\infty} \mathbb{E}\left\{\|d(t_k)\|^2\right\} \tag{2.10}$$

where γ is a given disturbance attenuation level; and

(c) the individual steady-state estimation error variance $\mathcal{E}_e^{(r)}$ satisfies

$$\mathcal{E}_e^{(r)} \triangleq \lim_{k \to \infty} \mathbb{E}\left\{e_r(t_k) e_r^T(t_k)\right\} \leq \sigma_r^2 \ (r = 1, 2, \cdots, n_x) \tag{2.11}$$

where $e_r(t_k)$ is the rth entry of the vector $e(t_k)$, $\mathcal{E}_e^{(r)}$ stands for the steady-state variance of the rth state estimation error, and $\sigma_r^2 > 0$ denotes the prespecified variance constraint on steady-state estimation error $e_r(t_k)$ $(r = 1, 2, \cdots, n_x)$.

2.1.2 Variance-Constrained Estimator Design

A. H_∞ Performance

The following theorem gives a sufficient condition for the exponential mean-square stability as well as H_∞ performance constraint of the augmented system (2.9).

Theorem 2.1. *For the given disturbance attenuation level $\gamma > 0$ and estimator gain H, the augmented system (2.9) is exponentially mean-square stable and simultaneously satisfies the H_∞ performance constraint (2.10) if there exists a positive definite matrix $\mathcal{P}$ such that the following matrix inequality*

$$\Phi \triangleq \begin{bmatrix} \Gamma + \mathscr{L}^T \mathscr{L} & (\mathscr{A} + \mathscr{C})^T \mathcal{P} (\mathscr{B}_1 + \bar{\mathscr{B}}_1) \\ * & (\mathscr{B}_1 + \bar{\mathscr{B}}_1)^T \mathcal{P} (\mathscr{B}_1 + \bar{\mathscr{B}}_1) - \gamma^2 I \end{bmatrix} < 0 \tag{2.12}$$

holds, where $\Gamma \triangleq (\mathscr{A} + \mathscr{C})^T \mathcal{P} (\mathscr{A} + \mathscr{C}) + \sum_{s=1}^{m} \tilde{\bar{\beta}}_s^2 (\bar{\mathscr{A}}_s + \bar{\mathscr{C}}_s)^T \mathcal{P} (\bar{\mathscr{A}}_s + \bar{\mathscr{C}}_s) - \mathcal{P}$.

Proof. Choose the following Lyapunov function:

$$V(\tilde{e}(t_k)) = \tilde{e}^T(t_k) \mathcal{P} \tilde{e}(t_k). \tag{2.13}$$

By calculating the difference of $V(\tilde{e}(t_k))$ along the trajectory of the augmented system (2.9) with $d(t_k) = 0$ and $\bar{v}(t_k) = 0$, and taking the mathematical expectation, one has

$$\begin{aligned}
\mathbb{E}\Big\{\triangle V(\tilde{e}(t_k))\Big\} &= \mathbb{E}\Big\{\tilde{e}^T(t_{k+1})\mathcal{P}\tilde{e}(t_{k+1}) - \tilde{e}^T(t_k)\mathcal{P}\tilde{e}(t_k)\Big\} \\
&= \mathbb{E}\Big\{\tilde{e}^T(t_k)\Big[\Big((\mathscr{A}+\mathscr{C}) + \sum_{s=1}^{m}\tilde{\beta}_s(t_k)(\bar{\mathscr{A}}_s + \bar{\mathscr{C}}_s)^T\mathcal{P} \\
&\quad \times\Big((\mathscr{A}+\mathscr{C}) + \sum_{s=1}^{m}\tilde{\beta}_s(t_k)(\bar{\mathscr{A}}_s + \bar{\mathscr{C}}_s)\Big) - \mathcal{P}\Big]\tilde{e}(t_k)\Big\} \\
&= \tilde{e}^T(t_k)\Big\{(\mathscr{A}+\mathscr{C})^T\mathcal{P}(\mathscr{A}+\mathscr{C}) \\
&\quad + \sum_{s=1}^{m}\tilde{\bar{\beta}}_s^2(\bar{\mathscr{A}}_s + \bar{\mathscr{C}}_s)^T\mathcal{P}(\bar{\mathscr{A}}_s + \bar{\mathscr{C}}_s) - \mathcal{P}\Big\}\tilde{e}(t_k) \\
&= \tilde{e}^T(t_k)\Gamma\tilde{e}(t_k). \qquad (2.14)
\end{aligned}$$

We can obtain from (2.12) that $\Gamma < 0$ and, subsequently,

$$\mathbb{E}\Big\{\triangle V(\tilde{e}(t_k))\Big\} \leq -\lambda_{\min}(-\Gamma)\|\tilde{e}(t_k)\|^2.$$

Hence, by following the similar analysis in [12], the augmented system (2.9) is exponentially mean-square stable.

Next, based on the zero initial condition, let us establish the H_∞ performance constraint of augmented system (2.9) with $\bar{v}(t_k) = 0$ by the following derivation:

$$\begin{aligned}
\mathbb{E}\{\triangle V(\tilde{e}(t_k))\} &+ \mathbb{E}\{\tilde{z}_e^T(t_k)\tilde{z}_e(t_k)\} - \gamma^2\mathbb{E}\{d^T(t_k)d(t_k)\Big\} \\
&= \tilde{e}^T(t_k)\{\Gamma + \mathscr{L}^T\mathscr{L}\}\tilde{e}(t_k) \\
&\quad +2\tilde{e}^T(t_k)\{(\mathscr{A}+\mathscr{C})^T\mathcal{P}(\mathscr{B}_1 + \bar{\mathscr{B}}_1)\}d(t_k) \\
&\quad +d^T(t_k)\{(\mathscr{B}_1 + \bar{\mathscr{B}}_1)^T\mathcal{P}(\mathscr{B}_1 + \bar{\mathscr{B}}_1) - \gamma^2 I\}d(t_k) \\
&= \vartheta^T(t_k)\Phi\vartheta(t_k) \qquad (2.15)
\end{aligned}$$

where $\vartheta(t_k) \triangleq \mathrm{col}\{\tilde{e}(t_k), d(t_k)\}$. Furthermore, by using the Schur Complement Lemma to (2.12), we have $\Phi < 0$ implying

$$\mathbb{E}\{\triangle V(\tilde{e}(t_k))\} + \mathbb{E}\{\tilde{z}_e^T(t_k)\tilde{z}_e(t_k)\} - \gamma^2\mathbb{E}\{d^T(t_k)d(t_k)\} < 0$$

for all nonzero $d(t_k)$.

By considering the zero initial condition, the above inequality indicates that

$$\sum_{k=0}^{\infty}\mathbb{E}\{\tilde{z}_e^T(t_k)\tilde{z}_e(t_k)\} < \gamma^2\sum_{k=0}^{\infty}\mathbb{E}\{d^T(t_k)d(t_k)\}$$

which is equivalent to (2.10). The proof of this theorem is now complete. □

B. Variance Analysis

The following theorem presents sufficient conditions that guarantee the exponential mean-square stability of the augmented system (2.9) and, at the same time, enforce the individual steady-state estimation error variance constraints.

Theorem 2.2. *For the given steady-state variance upper bounds σ_r^2 ($r = 1, 2, \cdots, n_x$) and estimator gain H, the augmented system (2.9) is exponentially mean-square stable and simultaneously satisfies the steady-state variance constraint (2.11) if there exists a positive definite matrix $\mathcal{Q}$ such that the following matrix inequalities*

$$(\mathscr{A}+\mathscr{C})\mathcal{Q}(\mathscr{A}+\mathscr{C})^T + \sum_{s=1}^{m} \tilde{\bar{\beta}}_s^2 (\bar{\mathscr{A}}_s + \bar{\mathscr{C}}_s)\mathcal{Q}(\bar{\mathscr{A}}_s + \bar{\mathscr{C}}_s)^T - \mathcal{Q} + \mathscr{B}_2 \mathcal{R} \mathscr{B}_2^T < 0 \tag{2.16}$$

$$\underline{\mathcal{I}}_r \bar{\mathcal{I}}_\ell \mathcal{Q} \bar{\mathcal{I}}_\ell^T \underline{\mathcal{I}}_r^T \le \sigma_r^2 \quad (r = 1, 2, \cdots, n_x; \ell = 1, 2, \cdots, b) \tag{2.17}$$

hold, where $\bar{\mathcal{I}}_\ell = [\underbrace{0 \ \cdots \ 0}_{\ell-1} \ I \ \underbrace{0 \ \cdots \ 0}_{2b-\ell}]$ *and* $\underline{\mathcal{I}}_r = [\underbrace{0 \ \cdots \ 0}_{r-1} \ 1 \ \underbrace{0 \ \cdots \ 0}_{n_x-r}]$.

Proof. First of all, it follows from (2.16) that

$$(\mathscr{A}+\mathscr{C})\mathcal{Q}(\mathscr{A}+\mathscr{C})^T + \sum_{s=1}^{m} \tilde{\bar{\beta}}_s^2 (\bar{\mathscr{A}}_s + \bar{\mathscr{C}}_s)\mathcal{Q}(\bar{\mathscr{A}}_s + \bar{\mathscr{C}}_s)^T - \mathcal{Q} < -\mathscr{B}_2 \mathcal{R} \mathscr{B}_2^T < 0. \tag{2.18}$$

Based on (2.18), it can be inferred from [128] that the augmented system (2.9) is exponentially mean-square stable and, subsequently, the steady-state covariance $\hat{\mathcal{Q}}$ defined by

$$\hat{\mathcal{Q}} \triangleq \lim_{k\to\infty} \mathbb{E}\left\{\tilde{e}(t_k)\tilde{e}^T(t_k)\right\}. \tag{2.19}$$

exists and satisfies the following discrete-time modified Lyapunov equation:

$$(\mathscr{A}+\mathscr{C})\hat{\mathcal{Q}}(\mathscr{A}+\mathscr{C})^T + \sum_{s=1}^{m} \tilde{\bar{\beta}}_s^2 (\bar{\mathscr{A}}_s + \bar{\mathscr{C}}_s)\hat{\mathcal{Q}}(\bar{\mathscr{A}}_s + \bar{\mathscr{C}}_s)^T - \hat{\mathcal{Q}} + \mathscr{B}_2 \mathcal{R} \mathscr{B}_2^T = 0. \tag{2.20}$$

Subtracting (2.20) from (2.18) gives

$$(\mathscr{A}+\mathscr{C})(\mathcal{Q}-\hat{\mathcal{Q}})(\mathscr{A}+\mathscr{C})^T + \sum_{s=1}^{m} \tilde{\bar{\beta}}_s^2 (\bar{\mathscr{A}}_s + \bar{\mathscr{C}}_s)(\mathcal{Q}-\hat{\mathcal{Q}})(\bar{\mathscr{A}}_s + \bar{\mathscr{C}}_s)^T - (\mathcal{Q}-\hat{\mathcal{Q}}) < 0 \tag{2.21}$$

which indicates from [128] that $\mathcal{Q} - \hat{\mathcal{Q}} \ge 0$.

Finally, considering the definitions of (2.11) and (2.19), we can obtain that

$$\begin{aligned}\mathcal{E}_e^{(r)} &\triangleq \lim_{k\to\infty}\mathbb{E}\{e_r(t_k)e_r^T(t_k)\} \triangleq \lim_{k\to\infty}\mathbb{E}\{\underline{\mathcal{I}}_r e(t_k)e^T(t_k)\underline{\mathcal{I}}_r^T\} \\ &\triangleq \underline{\mathcal{I}}_r\Big\{\lim_{k\to\infty}\mathbb{E}\{\bar{\mathcal{I}}_\ell\tilde{e}(t_k)\tilde{e}^T(t_k)\bar{\mathcal{I}}_\ell^T\}\Big\}\underline{\mathcal{I}}_r^T \\ &\triangleq \underline{\mathcal{I}}_r\bar{\mathcal{I}}_\ell\hat{\mathcal{Q}}\bar{\mathcal{I}}_\ell^T\underline{\mathcal{I}}_r^T \le \underline{\mathcal{I}}_r\bar{\mathcal{I}}_\ell\mathcal{Q}\bar{\mathcal{I}}_\ell^T\underline{\mathcal{I}}_r^T \end{aligned} \tag{2.22}$$

Therefore, matrix inequality (2.17) indicates that the requirement (c) is also met and the proof is now complete. □

To conclude the above analysis, we present a theorem which intends to take both the H_∞ performance and the variance constraint into consideration in a unified framework. Before giving our main result, we introduce the following well-known lemma.

Lemma 2.1. *[129] Let $\Omega = \Omega^T$, S and U be real matrices with appropriate dimensions, and matrix $F(\cdot)$ satisfies $F(\cdot)F^T(\cdot) \le I$, then*

$$\Omega + UF(\cdot)M + M^TF^T(\cdot)U^T < 0 \tag{2.23}$$

if and only if there exists a positive scalar ε such that

$$\Omega + \frac{1}{\varepsilon}UU^T + \varepsilon M^TM < 0 \tag{2.24}$$

or equivalently

$$\begin{bmatrix}\Omega & U & \varepsilon M^T \\ * & -\varepsilon I & 0 \\ * & * & -\varepsilon I\end{bmatrix} < 0 \tag{2.25}$$

For the convenience of later development, we denote

$$\begin{aligned}\breve{\mathcal{P}} &\triangleq \operatorname{diag}\Big\{P_1, P_2, \cdots, P_{2b}\Big\},\ \breve{\mathscr{P}} \triangleq \operatorname{diag}\{\underbrace{\breve{\mathcal{P}}, \cdots, \breve{\mathcal{P}}}_{m}\}, \\ \mathcal{W} &\triangleq \operatorname{diag}\{\underbrace{W, W, \cdots, W}_{b}\},\ \mathscr{W} \triangleq \operatorname{diag}\{\mathcal{W}, \mathcal{W}\}, \\ \hat{\mathcal{H}} &\triangleq \operatorname{col}\{\mathcal{H}_1, \cdots, \mathcal{H}_b\},\ \underline{\mathcal{H}}_s \triangleq \operatorname{col}\{\underbrace{0, \cdots, 0}_{s-1}, -\hat{\mathcal{H}}, \underbrace{0, \cdots, 0}_{m-s}\}, \\ \underline{\mathcal{H}} &\triangleq \operatorname{col}\{0, 0, 0, -\hat{\mathcal{H}}\}, \\ \mathscr{W}\mathscr{A} &= \operatorname{diag}\Big\{\mathcal{W}\hat{A} - \hat{\mathcal{H}}\Xi C\mathcal{I}, \mathcal{W}\hat{A}\Big\},\ \mathscr{W}\bar{\mathscr{A}}_s = \begin{bmatrix} 0 & -\hat{\mathcal{H}}C_s\mathcal{I} \\ 0 & 0\end{bmatrix}, \\ \mathscr{W}\mathscr{B}_1 &= \begin{bmatrix} -\hat{\mathcal{H}}D & \mathcal{W}\hat{B}_1 \\ 0 & \mathcal{W}\hat{B}_1\end{bmatrix}, \\ \hat{\mathcal{X}}^T &\triangleq [\tilde{\bar{\beta}}_1(\mathscr{W}\bar{\mathscr{A}}_1)^T\ \tilde{\bar{\beta}}_2(\mathscr{W}\bar{\mathscr{A}}_2)^T \cdots \tilde{\bar{\beta}}_m(\mathscr{W}\bar{\mathscr{A}}_m)^T],\end{aligned}$$

$$\begin{aligned}
\hat{\bar{\bar{\Sigma}}}_1 &\triangleq \begin{bmatrix} \hat{\bar{\Sigma}}_1 & \hat{\bar{\Sigma}}_2 \\ * & \breve{\mathcal{P}} - 2\mathscr{W} \end{bmatrix}, \ \hat{\bar{\Sigma}}_1 \triangleq \begin{bmatrix} -\breve{\mathcal{P}} & 0 & \mathscr{L}^T \\ * & -\gamma^2 I & 0 \\ * & * & -I \end{bmatrix}, \\
\hat{\bar{\Sigma}}_2 &\triangleq \mathrm{col}\{(\mathscr{W}\mathscr{A})^T, (\mathscr{W}\mathscr{B}_1)^T, 0\}, \\
\hat{\bar{\bar{\Sigma}}}_3 &\triangleq \mathrm{col}\{\hat{\bar{\mathcal{X}}}^T, 0, 0, 0\}, \ \hat{\bar{\bar{\Sigma}}}_4 \triangleq [\underline{\mathcal{H}} \ \varepsilon_1^{(1)} \underline{C}^T \ \underline{\mathcal{H}} \ \varepsilon_2^{(1)} \underline{D}^T], \\
\bar{\bar{\Sigma}}_5 &\triangleq [\bar{\bar{\Sigma}}_{5,1} \ \bar{\bar{\Sigma}}_{5,2} \ \cdots \ \bar{\bar{\Sigma}}_{5,m}], \ \bar{\bar{\Sigma}}_{5,s} \triangleq [0 \ \epsilon_s^{(1)} \underline{C}_s^T], \\
\hat{\bar{\bar{\Sigma}}}_6 &\triangleq [\hat{\bar{\bar{\Sigma}}}_{6,1} \ \hat{\bar{\bar{\Sigma}}}_{6,2} \ \cdots \ \hat{\bar{\bar{\Sigma}}}_{6,m}], \ \hat{\bar{\bar{\Sigma}}}_{6,s} \triangleq [\bar{\bar{\beta}}_s \underline{\mathcal{H}}_s \ 0], \ \breve{\mathscr{W}} \triangleq \mathrm{diag}\{\underbrace{\mathscr{W}, \cdots, \mathscr{W}}_{m}\},
\end{aligned}$$

$$\begin{aligned}
\hat{\bar{\Pi}}_2 &\triangleq \begin{bmatrix} \breve{\mathcal{P}} - 2\mathscr{W} & \mathscr{W}\mathscr{A} & \hat{\bar{\mathcal{Y}}} & \mathscr{W}\mathscr{B}_2 \\ * & -\breve{\mathcal{P}} & 0 & 0 \\ * & * & -\breve{\mathscr{P}} & 0 \\ * & * & * & -\mathcal{R}^{-1} \end{bmatrix}, \\
\hat{\bar{\mathcal{Y}}} &\triangleq [\bar{\bar{\beta}}_1(\mathscr{W}\bar{\mathscr{A}}_1) \ \bar{\bar{\beta}}_2(\mathscr{W}\bar{\mathscr{A}}_2) \ \cdots \ \bar{\bar{\beta}}_m(\mathscr{W}\bar{\mathscr{A}}_m)], \\
\bar{M} &\triangleq \mathrm{col}\{0, \bar{C}, \underbrace{0, \cdots, 0}_{m+1}\}, \ \bar{M}_s \triangleq \mathrm{col}\{\underbrace{0, \cdots, 0}_{s+1}, \Lambda\bar{C}_s, \underbrace{0, \cdots, 0}_{m+1-s}\}, \\
\underline{\hat{U}} &\triangleq \mathrm{col}\{-\hat{\mathcal{H}}, \underbrace{0, \cdots, 0}_{m+2}\}, \\
\hat{\bar{\Pi}}_3 &\triangleq [\underline{\hat{U}} \ \varepsilon_1^{(2)} \bar{M}^T], \ \hat{\bar{\Pi}}_4 \triangleq [\hat{\bar{\Pi}}_{4,1} \ \hat{\bar{\Pi}}_{4,2} \ \cdots \ \hat{\bar{\Pi}}_{4,m}], \ \hat{\bar{\Pi}}_{4,s} \triangleq [\bar{\bar{\beta}}_s \underline{\hat{U}} \ \epsilon_s^{(2)} \bar{M}_s^T], \\
\breve{\varepsilon}^{(1)} &\triangleq \mathrm{diag}\{\varepsilon_1^{(1)}, \varepsilon_1^{(1)}, \varepsilon_2^{(1)}, \varepsilon_2^{(1)}\}, \ \breve{\varepsilon}^{(2)} \triangleq \mathrm{diag}\{\varepsilon_1^{(2)}, \varepsilon_1^{(2)}\}, \\
\breve{\epsilon}^{(j)} &\triangleq \mathrm{diag}\{\bar{\epsilon}_1^{(j)}, \bar{\epsilon}_2^{(j)}, \cdots, \bar{\epsilon}_m^{(j)}\}, \\
\mathcal{R}^{-1} &\triangleq \mathrm{diag}\{\underbrace{R^{-1}, \ \cdots, \ R^{-1}}_{b}\}, \ (s = 1, 2, \cdots, m; \ j = 1, 2).
\end{aligned}$$

Theorem 2.3. *For the given disturbance attenuation level $\gamma > 0$ and steady-state variance upper bounds σ_r^2 $(r = 1, 2, \cdots, n_x)$, the augmented system (2.9) is exponentially mean-square stable while achieving the H_∞ performance constraint (2.10) for any nonzero $d(t_k)$ and the steady-state variance constraint (2.11) for $\bar{v}(t_k)$, if there exist matrices $\mathcal{H}_\varrho$ $(\varrho = 1, 2, \cdots, b)$, $W > 0$ and $P_h > 0$ $(h = 1, 2, \cdots, 2b)$ such that the following linear matrix inequalities hold:*

$$\begin{bmatrix} \hat{\bar{\bar{\Sigma}}}_1 & \hat{\bar{\bar{\Sigma}}}_3 & \hat{\bar{\bar{\Sigma}}}_4 & \bar{\bar{\Sigma}}_5 \\ * & \breve{\mathscr{P}} - 2\breve{\mathscr{W}} & 0 & \hat{\bar{\bar{\Sigma}}}_6 \\ * & * & -\breve{\varepsilon}^{(1)} & 0 \\ * & * & * & -\breve{\epsilon}^{(1)} \end{bmatrix} < 0 \tag{2.26}$$

$$\begin{bmatrix} \hat{\bar{\Pi}}_2 & \hat{\bar{\Pi}}_3 & \hat{\bar{\Pi}}_4 \\ * & -\breve{\varepsilon}^{(2)} & 0 \\ * & * & -\breve{\epsilon}^{(2)} \end{bmatrix} < 0 \tag{2.27}$$

$$\begin{bmatrix} -\sigma_r^2 & \mathcal{I}_r \\ * & -P_\ell \end{bmatrix} < 0 \; (r = 1, \cdots, n_x;\; \ell = 1, 0 \cdots, b) \quad (2.28)$$

Furthermore, if above inequalities are feasible, the desired estimator gains can be determined by

$$H_\varrho = W^{-1}\mathcal{H}_\varrho \; (\varrho = 1, 2, \cdots, b). \quad (2.29)$$

Proof. By using the Schur Complement Lemma, (2.12) is equivalent to the following inequality:

$$\Sigma \triangleq \begin{bmatrix} \Sigma_1 & \Sigma_2 & \Sigma_3 \\ * & -\mathcal{P}^{-1} & 0 \\ * & * & -\mathscr{P}^{-1} \end{bmatrix} < 0 \quad (2.30)$$

where

$$\Sigma_1 \triangleq \begin{bmatrix} -\mathcal{P} & 0 & \mathscr{L}^T \\ * & -\gamma^2 I & 0 \\ * & * & -I \end{bmatrix}, \; \Sigma_2 \triangleq \mathrm{col}\{\mathscr{A}^T + \mathscr{C}^T, \mathscr{B}_1^T + \bar{\mathscr{B}}_1^T, 0\},$$

$$\Sigma_3 \triangleq \mathrm{col}\{\mathcal{X}^T, 0, 0\},$$

$$\mathcal{X}^T \triangleq [\tilde{\bar{\beta}}_1(\bar{\mathscr{A}}_1^T + \bar{\mathscr{C}}_1^T) \; \tilde{\bar{\beta}}_2(\bar{\mathscr{A}}_2^T + \bar{\mathscr{C}}_2^T) \; \cdots \; \tilde{\bar{\beta}}_m(\bar{\mathscr{A}}_m^T + \bar{\mathscr{C}}_m^T)],$$

$$\mathscr{P}^{-1} \triangleq \mathrm{diag}\{\underbrace{\mathcal{P}^{-1}, \cdots, \mathcal{P}^{-1}}_{m}\}.$$

In order to cope with the uncertainty factor $F(\bullet)$, we rewrite (2.30) in the form of (2.23) as follows:

$$\begin{aligned} \bar{\Sigma} \; &+ \; \hat{\underline{H}}F(t_k)\hat{C} + \hat{C}^T F^T(t_k)\hat{\underline{H}}^T + \hat{\underline{H}}F(t_k)\hat{D} + \hat{D}^T F^T(t_k)\hat{\underline{H}}^T \\ &+ \; \sum_{s=1}^{m} \tilde{\bar{\beta}}_s \hat{\underline{H}}_s F(t_k)\hat{C}_s + \sum_{s=1}^{m} \tilde{\bar{\beta}}_s \hat{C}_s^T F^T(t_k)\hat{\underline{H}}_s^T < 0 \end{aligned} \quad (2.31)$$

where

$$\begin{aligned} \bar{\Sigma} \; &\triangleq \; \begin{bmatrix} \Sigma_1 & \bar{\Sigma}_2 & \bar{\Sigma}_3 \\ * & -\mathcal{P}^{-1} & 0 \\ * & * & -\mathscr{P}^{-1} \end{bmatrix}, \; \bar{\Sigma}_2 \triangleq \mathrm{col}\{\mathscr{A}^T, \mathscr{B}_1^T, 0\}, \\ \bar{\Sigma}_3 \; &\triangleq \; \mathrm{col}\{\bar{\mathcal{X}}^T, 0, 0\}, \; \bar{\mathcal{X}}^T \triangleq [\tilde{\bar{\beta}}_1 \bar{\mathscr{A}}_1^T \; \tilde{\bar{\beta}}_2 \bar{\mathscr{A}}_2^T \; \cdots \; \tilde{\bar{\beta}}_m \bar{\mathscr{A}}_m^T], \\ \underline{H} \; &\triangleq \; \mathrm{col}\{0, 0, 0, \bar{H}\}, \; \hat{\underline{H}} \triangleq \mathrm{col}\{\underline{H}, \underbrace{0, \cdots, 0}_{m}\}, \\ \underline{C} \; &\triangleq \; [\bar{C} \; 0 \; 0 \; 0], \; \hat{C} \triangleq [\underline{C} \; \underbrace{0 \cdots 0}_{m}], \\ \underline{D} \; &\triangleq \; [0 \; \bar{D} \; 0 \; 0], \; \hat{D} \triangleq [\underline{D} \; \underbrace{0 \cdots 0}_{m}], \end{aligned}$$

$$\underline{H}_s \triangleq \text{col}\{\underbrace{0,\cdots,0}_{s-1},\bar{H},\underbrace{0,\cdots,0}_{m-s}\},\ \hat{\bar{H}}_s \triangleq \text{col}\{0,0,0,0,\underline{H}_s\},$$

$$\underline{C}_s \triangleq [\Lambda\bar{C}_s\ 0\ 0\ 0],\ \hat{C}_s \triangleq [\underline{C}_s\ \underbrace{0\cdots 0}_{m}],\ (s=1,2,\cdots,m).$$

Applying Lemma 2.1 to (2.31), it follows that (2.31) holds if and only if there exist positive scalars $\varepsilon_1^{(1)}, \varepsilon_2^{(1)}, \epsilon_s^{(1)}$ $(s = 1, 2, \cdots, m)$ such that the following matrix inequality holds

$$\bar{\bar{\Sigma}} \triangleq \begin{bmatrix} \bar{\bar{\Sigma}}_1 & \bar{\bar{\Sigma}}_3 & \bar{\bar{\Sigma}}_4 & \bar{\bar{\Sigma}}_5 \\ * & -\mathscr{P}^{-1} & 0 & \bar{\bar{\Sigma}}_6 \\ * & * & -\breve{\bar{\varepsilon}}^{(1)} & 0 \\ * & * & * & -\breve{\epsilon}^{(1)} \end{bmatrix} < 0 \tag{2.32}$$

where

$$\bar{\bar{\Sigma}}_1 \triangleq \begin{bmatrix} \bar{\Sigma}_1 & \bar{\Sigma}_2 \\ * & -\mathcal{P}^{-1} \end{bmatrix},\ \bar{\bar{\Sigma}}_3 \triangleq \text{col}\{\bar{\Sigma}_3, 0\},\ \bar{\bar{\Sigma}}_4 \triangleq [\underline{H}\ \varepsilon_1^{(1)}\underline{C}^T\ \underline{H}\ \varepsilon_2^{(1)}\underline{D}^T],$$

$$\bar{\bar{\Sigma}}_6 \triangleq [\bar{\bar{\Sigma}}_{6,1}\ \bar{\bar{\Sigma}}_{6,2}\ \cdots\ \bar{\bar{\Sigma}}_{6,m}],\ \bar{\bar{\Sigma}}_{6,s} \triangleq [\tilde{\bar{\beta}}_s\underline{H}_s\ 0],\ (s=1,2,\cdots,m).$$

After using Schur Complement Lemma to (2.16), we have

$$\Pi_1 \triangleq \begin{bmatrix} -\mathcal{Q} & \mathscr{A}+\mathscr{C} & \mathcal{Y} & \mathscr{B}_2 \\ * & -\mathcal{Q}^{-1} & 0 & 0 \\ * & * & -\mathscr{Q}^{-1} & 0 \\ * & * & * & -\mathcal{R}^{-1} \end{bmatrix} < 0 \tag{2.33}$$

where $\mathscr{Q}^{-1} \triangleq \text{diag}\{\underbrace{\mathcal{Q}^{-1},\cdots,\mathcal{Q}^{-1}}_{\ell+1}\}$, $\mathcal{Y} \triangleq [\tilde{\bar{\beta}}_1(\bar{\mathscr{A}}_1 + \bar{\mathscr{C}}_1)\ \ \tilde{\bar{\beta}}_2(\bar{\mathscr{A}}_2 + \bar{\mathscr{C}}_2)\ \cdots\ \tilde{\bar{\beta}}_m(\bar{\mathscr{A}}_m + \bar{\mathscr{C}}_m)]$.

Similarly, by denoting $\mathcal{Q} = \mathcal{P}^{-1}$, we rewrite (2.33) in the form of (2.23) as follows:

$$\Pi_2 + \underline{U}F(t_k)\bar{M} + \bar{M}^T F^T(t_k)\underline{U}^T + \sum_{s=1}^{m}\tilde{\bar{\beta}}_s\underline{U}F(t_k)\bar{M}_s + \sum_{s=1}^{m}\tilde{\bar{\beta}}_s\bar{M}_s^T F^T(t_k)\underline{U}^T < 0 \tag{2.34}$$

where

$$\Pi_2 \triangleq \begin{bmatrix} -\mathcal{P}^{-1} & \mathscr{A} & \bar{\mathcal{Y}} & \mathscr{B}_2 \\ * & -\mathcal{P} & 0 & 0 \\ * & * & -\mathscr{P} & 0 \\ * & * & * & -\mathcal{R}^{-1} \end{bmatrix},$$

$$\bar{\mathcal{Y}} \triangleq [\tilde{\bar{\beta}}_1\bar{\mathscr{A}}_1\ \tilde{\bar{\beta}}_2\bar{\mathscr{A}}_2\ \cdots\ \tilde{\bar{\beta}}_m\bar{\mathscr{A}}_m],\ \underline{U} \triangleq \text{col}\{\bar{H},\underbrace{0,\cdots,0}_{m+2}\}.$$

By applying Lemma 2.1 again to (2.34), we know that (2.34) holds if and only if there exist positive scalars $\varepsilon_1^{(2)}, \epsilon_s^{(2)}$ $(s = 1, 2, \cdots, m)$ such that the following LMI holds:

$$\bar{\Pi}_2 \triangleq \begin{bmatrix} \Pi_2 & \Pi_3 & \Pi_4 \\ * & -\breve{\varepsilon}^{(2)} & 0 \\ * & * & -\breve{\epsilon}^{(2)} \end{bmatrix} < 0 \tag{2.35}$$

where

$$\Pi_3 \triangleq [\underline{U} \ \ \varepsilon_1^{(2)} \bar{M}^T], \ \ \Pi_4 \triangleq [\Pi_{4,1} \ \Pi_{4,2} \ \cdots \ \Pi_{4,m}],$$
$$\Pi_{4,s} \triangleq [\tilde{\bar{\beta}}_s \underline{U} \ \ \epsilon_s^{(2)} \bar{M}_s^T], \ (s = 1, 2, \cdots, m).$$

To this end, in order to design the estimator by Matlab LMI Toolbox to effectively, we assume $\breve{\mathcal{P}} \triangleq \mathrm{diag}\{P_1, P_2, \cdots, P_{2b}\}$ and let $\mathcal{H}_\varrho \triangleq WH_\varrho$ $(\varrho = 1, 2, \cdots, b)$. By noting $W > 0$ and $P_h > 0$ $(h = 1, 2, \cdots, 2b)$, we have $(P_h - W)P_h^{-1}(P_h - W) \geq 0$, which is equivalent to

$$-WP_h^{-1}W \leq P_h - 2W \ \ (h = 1, 2, \cdots, 2b)$$

Applying the congruence transformation $\mathrm{diag}\Big\{I, I, I, \underbrace{\mathscr{W}, \cdots, \mathscr{W}}_{m+1}, \underbrace{I, \cdots, I}_{2m+4}\Big\}$ to (2.32), we get (2.26). Further applying the congruence transformations $\mathrm{diag}\Big\{\mathscr{W}, \underbrace{I, \cdots, I}_{3m+b+2}\Big\}$ to (2.35), we obtain (2.27). At the same time, the estimator gain can be expressed as (2.29).

On the other hand, from $\mathcal{Q} = \mathcal{P}^{-1}$, we assume $\breve{\mathcal{Q}} \triangleq \breve{\mathcal{P}}^{-1} \triangleq \mathrm{diag}\{P_1^{-1}, P_2^{-1}, \cdots, P_{2b}^{-1}\}$, and rewrite (2.17) as follows:

$$\begin{aligned} \underline{\mathcal{I}}_r \bar{\mathcal{I}}_\ell \breve{\mathcal{Q}} \bar{\mathcal{I}}_\ell^T \underline{\mathcal{I}}_r^T &= \underline{\mathcal{I}}_r \bar{\mathcal{I}}_\ell \breve{\mathcal{P}}^{-1} \bar{\mathcal{I}}_\ell^T \underline{\mathcal{I}}_r^T = \underline{\mathcal{I}}_r P_\ell^{-1} \underline{\mathcal{I}}_r^T \\ &\leq \sigma_r^2 \ (r = 1, 2, \cdots, n_x; \ \ell = 1, 2, \cdots, b) \end{aligned} \tag{2.36}$$

By using Schur Complement Lemma to (2.36), we have (2.28), which concludes the proof from Theorems 2.1 and 2.2. □

Remark 2.3. *In this section, the variance-constrained state estimation problem is investigated for a class of networked multi-rate systems with quantization and probabilistic sensor failures. The main features of our results are twofold: i) the quantified relationships have been established among the H_∞ performance level, the upper bounds on the steady-state variances of the estimation errors, the quantizer parameters, the sensor failure probabilities and multi-rate multiple b of the sampling period h; and ii) the proposed approach has offered much flexibility in making compromise between the steady-state variances and the H_∞ performance, while the essential multiple objectives can all be achieved simultaneously in the framework of networked multi-rate systems.*

In order to show the combined effect of the considered variance constraints, quantizer parameters, sensor failure probabilities and the multi-rate sampling, we now discuss the following two optimization problems for given quantization density $\chi^{(s)}$ $(s = 1, 2, \cdots, m)$, multi-rate multiple b, sensor failure parameters $\bar{\beta}_s$ and $\bar{\tilde{\beta}}_s^2 (s = 1, 2, \cdots, m)$.

P1: For given steady-state estimation error variance-constrained bounds $\sigma_1^2, \cdots, \sigma_{n_x}^2$, the optimal H_∞ estimator design problem:

$$\min_{W,\mathcal{H}_1,\cdots,\mathcal{H}_b,P_1,\cdots,P_{2b}} \gamma^2 \text{ subject to } (2.26) - (2.28). \tag{2.37}$$

P2: For given H_∞ performance level γ, the minimum weighted variance-constrained estimator design problem:

$$\min_{W,\mathcal{H}_1,\cdots,\mathcal{H}_b,P_1,\cdots,P_{2b}} \sum_{r=1}^{n_x} c_r \sigma_r^2 \text{ subject to } (2.26) - (2.28). \tag{2.38}$$

where c_r $(r = 1, 2, \cdots, n_x)$ are given weighting coefficients for variances and satisfy $\sum_{r=1}^{n_x} c_r = 1$.

2.1.3 Illustrative Examples

In this section, similar to [130], a manoeuvring target tracking system is presented to demonstrate the effectiveness of the proposed design scheme, and the involved system has the following state-space model:

$$x(T_{k+1}) = \begin{bmatrix} 0.3 & h \\ 0 & 0.4 \end{bmatrix} x(T_k) + \begin{bmatrix} \frac{h}{2} \\ 0.1 \end{bmatrix} \omega(T_k) + \begin{bmatrix} 0.1 \\ \frac{h^2}{2} \end{bmatrix} \nu(T_k) \tag{2.39}$$

$$z(T_k) = \begin{bmatrix} 0.5 & 0.4 \end{bmatrix} x(T_k) \tag{2.40}$$

where h is the sampling period. $x(T_k) = \text{col}\{x_p(T_k), x_v(T_k)\}$ is the system state, $x_p(T_k)$ and $x_v(T_k)$ are the position and velocity of the target at time T_k, respectively.

In this networked manoeuvring target tracking system where the sensor signal is transmitted through communication network, it is often the case that the measurement outputs are quantized before being transmitted to the estimator. In the same time, the measurements received by sensor could be neither completely missing nor completely successful, but only part of the information can go through. Suppose that only the position of the manoeuvring target is measurable and the sensor is utilized to measure the position, then we use the following equation to model the measurements with quantization effects and probabilistic sensor failures at time t_k:

$$\bar{y}(t_k) = q\Big(\beta(t_k) \begin{bmatrix} 1 & 0 \end{bmatrix} x(t_k) + 0.5\xi(t_k)\Big) \tag{2.41}$$

The parameters of the logarithmic quantizer $q(\cdot)$ are chosen as $u_0 = 2, \chi = 0.4$, and the probability density function of sensor failure coefficient is taken as

$$f(\beta(t_k)) = \begin{cases} 10\beta(t_k), & 0 \le \beta(t_k) \le 0.20; \\ -2.50(\beta(t_k) - 1), & 0.20 < \beta(t_k) \le 1. \end{cases}$$

then the mathematical expectation $\bar{\beta}$ and the variance $\bar{\bar{\beta}}^2$ can be calculated as 0.4000 and 0.0467, respectively.

Here, the sampling period h is set as $0.5s$ and the variance of Gaussian white noise $\nu(T_k)$ is taken as $R = 0.3$. The disturbance input $\omega(T_k)$ and the measurement noise $\xi(t_k)$ are chosen as following:

$$\omega(T_k) = 0.1\mathrm{e}^{-0.05T_k} \sin(T_k),\ \xi(t_k) = \frac{\mathrm{e}^{-0.05t_k}}{0.2t_k + 1}.$$

Actually, the considered state estimation problem in this example are two-rate sampled-data systems, that is the state estimation in both position and velocity at a fast rate with the period h, while the sensor samples the target position at a slow one with the period bh. We aim to design the estimator, by using the quantized measurement, to estimate the state (position) of the manoeuvring target which is subjected to bounded energy disturbance and Gaussian white noise. Now, let us examine the following two cases.

The variance constraints on the steady-state estimation error are set as $\sigma_1 = 0.6$ and $\sigma_2 = 0.4$. By using the MATLAB LMI toolbox and considering the optimization problem (P1), we obtain the minimum disturbance attenuation level γ and corresponding estimator gains H_ϱ $(\varrho = 1, 2, \cdots, b)$ in Table 2.1 with different multi-rate multiple b. Take the initial state of (2.1) and its estimation as $x(T_0) = \mathrm{col}\{-0.1, 0.1\}$ and $\hat{x}(t_0) = \mathrm{col}\{-0.2, 0.2\}$, respectively. The estimated error $e_z(t_k)$ for the position of the manoeuvring target is plotted in Fig. 2.2.

Table 2.1: The permitted minimum γ and corresponding estimator gains H_ϱ $(\varrho = 1, 2, \cdots, b)$.

	γ	H_ϱ
$b = 2$	0.7734	$H_1 = \begin{bmatrix} 0.0288 \\ -0.0770 \end{bmatrix}, H_2 = \begin{bmatrix} -0.0040 \\ 0.0822 \end{bmatrix}$
$b = 3$	0.8073	$H_2 = \begin{bmatrix} 0.0009 \\ -0.0026 \end{bmatrix}, H_3 = \begin{bmatrix} 0.0724 \\ 0.0715 \end{bmatrix}$
$b = 4$	0.8472	$H_2 = \begin{bmatrix} 0.0007 \\ -0.0019 \end{bmatrix}, H_3 = \begin{bmatrix} 0.0428 \\ -0.0055 \end{bmatrix}, H_4 = \begin{bmatrix} 0.0229 \\ 0.0346 \end{bmatrix}$
$b = 5$	0.8728	$H_1 = \begin{bmatrix} 0.0096 \\ -0.0271 \end{bmatrix}, H_2 = \begin{bmatrix} 0.0004 \\ -0.0011 \end{bmatrix}, H_3 = \begin{bmatrix} 0.0266 \\ -0.0159 \end{bmatrix}, H_4 = \begin{bmatrix} 0.0190 \\ 0.0070 \end{bmatrix}, H_5 = \begin{bmatrix} 0.0146 \\ 0.0121 \end{bmatrix}$

For the given H_∞ performance level $\gamma = 0.95$ and weighting coefficients $c_1 = 0.4, c_1 = 0.6$, we now deal with the problem (P2). Solving the optimization problem (2.38), we obtain the minimum individual variance values σ_r $(r = 1, 2)$ and corresponding estimator gains H_ϱ $(\varrho = 1, 2, \cdots, b)$ in Table 2.2. Choosing the same initial values as Case I, the simulation results are shown in Figs. 2.3-2.4,

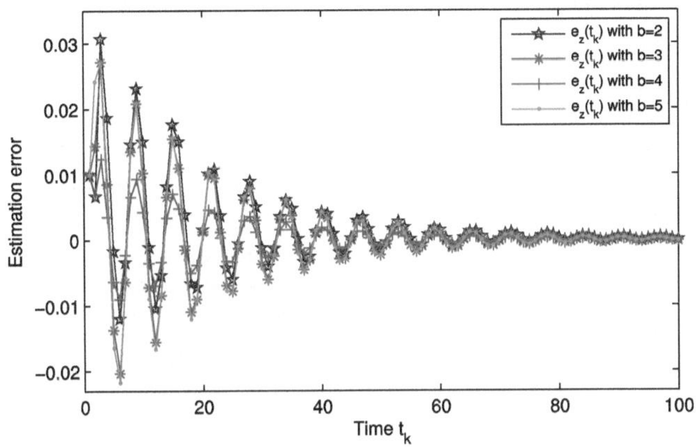

Figure 2.2: Estimation error $e_z(t_k)$ for different b.

which display the actual steady-state variance for $e_1(t_k) = x_1(t_k) - \hat{x}_1(t_k)$ and $e_2(t_k) = x_2(t_k) - \hat{x}_2(t_k)$, respectively.

Table 2.2: The minimum variance values σ_r ($r = 1, 2$) and corresponding estimator gains H_ϱ ($\varrho = 1, 2, \cdots, b$).

	σ_1	σ_2	H_ϱ
$b=2$	0.2368	0.0341	$H_1 = \begin{bmatrix} 0.0814 \\ -0.2221 \end{bmatrix}, H_2 = \begin{bmatrix} -0.0221 \\ 0.1561 \end{bmatrix}$
$b=3$	0.2759	0.0389	$H_1 = \begin{bmatrix} 0.0323 \\ -0.0907 \end{bmatrix}, H_2 = \begin{bmatrix} 0.0018 \\ -0.0051 \end{bmatrix}, H_3 = \begin{bmatrix} 0.0497 \\ 0.0913 \end{bmatrix}$
$b=4$	0.3011	0.0418	$H_1 = \begin{bmatrix} 0.0252 \\ -0.0712 \end{bmatrix}, H_2 = \begin{bmatrix} 0.0008 \\ -0.0022 \end{bmatrix}, H_3 = \begin{bmatrix} 0.0254 \\ 0.0245 \end{bmatrix}, H_4 = \begin{bmatrix} 0.0194 \\ 0.0336 \end{bmatrix}$
$b=5$	0.3166	0.0436	$H_1 = \begin{bmatrix} 0.0223 \\ -0.0631 \end{bmatrix}, H_2 = 10^{-3} \times \begin{bmatrix} 0.2863 \\ -0.8104 \end{bmatrix}, H_3 = \begin{bmatrix} 0.0154 \\ 0.0083 \end{bmatrix}, H_4 = \begin{bmatrix} 0.0135 \\ 0.0142 \end{bmatrix}, H_5 = \begin{bmatrix} 0.0108 \\ 0.0168 \end{bmatrix}$

Tables 2.1–2.2 demonstrate the relationship between H_∞ performance level γ and variance upper bounds σ_r ($r = 1, 2$) as well as the multi-rate multiple b of the sampling period h. It can be observed from Table 2.1 and Table 2.2 that, with increased b, the disturbance attenuation performance deteriorates and the variance upper bounds become bigger, and these observations can also been confirmed from Figs. 2.3 to 2.4, which are in agreement with the engineering practice.

2.2 Finite-Time Filter Design with Event-Based Relay and Fading Channels

For wireless networked systems, long-distance communication would increase the energy consumption and degrade the signal strength. Until now, almost results on filter/controller design problems for networked control systems have been based on

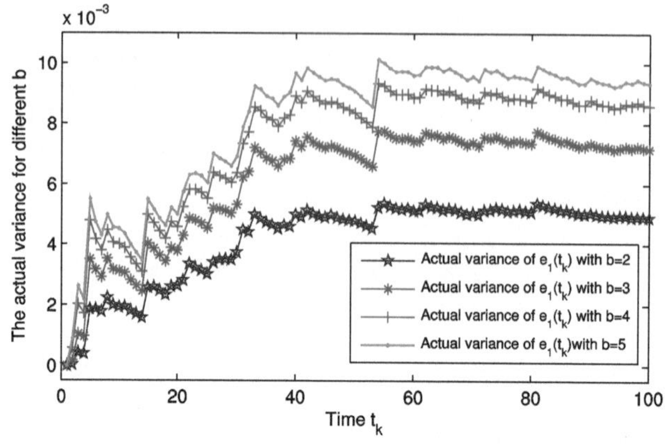

Figure 2.3: The actual steady-state estimation error variance for $e_1(t_k)$ for different b.

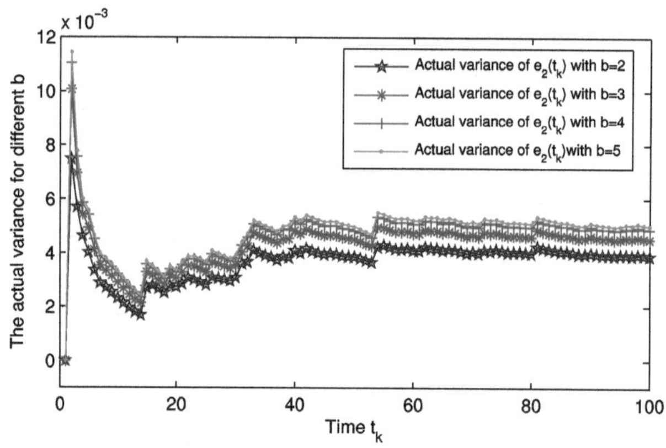

Figure 2.4: The actual steady-state estimation error variance for $e_2(t_k)$ for different b.

the traditional Lyapunov asymptotic stability [109, 118], and the finite-time H_∞ filtering problem for networked multi-rate systems has not been investigated yet.

In this section, a finite-time H_∞ filter for a class of networked multi-rate systems with fading measurements in an event-triggered wireless relay communication framework is investigated.

2.2.1 Problem Formulation

A. Multi-Rate Sampled-Data System

Consider a discrete-time system described by the following state-space model:

$$x(T_{k+1}) \quad = \quad Ax(T_k) + B_1\omega(T_k) \tag{2.42}$$

$$\bar{y}(t_k) = Cx(t_k) + B_2\nu(t_k) \tag{2.43}$$
$$z(T_k) = Dx(T_k), \quad k = 0, 1, 2, \cdots \tag{2.44}$$

where $x(T_k) \in \mathbb{R}^{n_x}$ represents the state vector, $\bar{y}(t_k) \in \mathbb{R}^{n_y}$ is the measurement output, and $z(t_k) \in \mathbb{R}^{n_z}$ is the signal to be estimated. $\omega(T_k) \in \mathbb{R}^{n_\omega}$ and $\nu(t_k) \in \mathbb{R}^{n_\nu}$ are the exogenous disturbance signals. The matrices A, B_1, C, B_2, and D are known real matrices with appropriate dimensions.

The sampling period for (2.42) and (2.44) is denoted by $h \triangleq T_{k+1} - T_k$ ($k = 0, 1, 2, \cdots$). For simplicity, it is assumed that the measurement period is integer multiples of the sampling period for (2.42) and (2.44), that is $t_{k+1} - t_k = bh$ ($k = 0, 1, 2, \cdots$), where b is a positive integer. It can be seen that (2.42) and (2.44) evolve with a constant period h, while the measurement dynamics (2.43) is generated with a slower period bh. Accordingly, the system (2.42)-(2.44) is essentially a multi-rate sampled-data system.

By applying the relation (2.42) recursively, one obtains the following equations with time scale t_k:

$$\begin{cases} x(t_{k+1}) = A^b x(t_k) + \sum_{j=0}^{b-1} A^{b-1-j} B_1 \omega(t_k + jh) \\ x(t_{k+1} - h) = x(t_k + (b-1)h) = A^{b-1} x(t_k) + \\ \qquad \sum_{j=0}^{b-2} A^{b-2-j} B_1 \omega(t_k + jh) \\ \qquad \vdots \\ x(t_{k+1} - (b-1)h) = Ax(t_k) + B_1\omega(t_k) \end{cases} \tag{2.45}$$

Denoting

$$\begin{aligned}
\bar{x}(t_k) &\triangleq \mathrm{col}\{x(t_k), x(t_k - h), \cdots, x(t_k - (b-1)h)\}, \\
\bar{\omega}(t_k) &\triangleq \mathrm{col}\{\omega(t_k), \omega(t_k + h), \cdots, \omega(t_k + (b-1)h)\}, \\
\bar{z}(t_k) &\triangleq \mathrm{col}\{z(t_k), z(t_k - h), \cdots, z(t_k - (b-1)h)\}, \\
\bar{A}_1 &\triangleq \mathrm{col}\{A^b, A^{b-1}, \cdots, A\}, \quad \hat{A} \triangleq [\bar{A}_1 \ \underbrace{0 \cdots 0}_{b-1}], \\
\bar{B}_{1,1} &\triangleq [A^{b-1}B_1 \ A^{b-2}B_1 \ \cdots \ AB_1 \ B_1], \\
\bar{B}_{1,2} &\triangleq [A^{b-2}B_1 \ A^{b-3}B_1 \ \cdots \ B_1 \ 0], \cdots, \\
\bar{B}_{1,b-1} &\triangleq [AB_1 \ B_1 \ \underbrace{0 \cdots 0}_{b-2}], \quad \bar{B}_{1,b} \triangleq [B_1 \ \underbrace{0 \cdots 0}_{b-1}], \\
\hat{\bar{B}}_1 &\triangleq \mathrm{col}\{\bar{B}_{1,1}, \bar{B}_{1,2}, \cdots, \bar{B}_{1,b}\}, \\
\bar{D} &\triangleq \mathrm{diag}\{\underbrace{D, \cdots, D}_{b}\},
\end{aligned}$$

an augmented system can be obtained from (2.44) and (2.45) as follows:

$$\begin{cases} \bar{x}(t_{k+1}) = \hat{A}\bar{x}(t_k) + \hat{\bar{B}}_1\bar{\omega}(t_k) \\ \bar{z}(t_k) = \bar{D}\bar{x}(t_k) \end{cases} \tag{2.46}$$

B. Event-Based Relay Communication Mechanism

The entire communication between the sensor nodes and the filter consists of two phases. In the first phase, the sensor nodes transmit the measurement signal to the relay node, where the signal might be fading during the transmission. In the second phase, the relay node would forward the received signal to the filter based on the event-triggered transmission scheme. First, let's introduce the fading channels between sensors and the relay node. The following Rice fading model is utilized to describe the actually received measurement signal at the relay node after going through the fading channels:

$$y(t_k) = \sum_{s=0}^{\ell(t_k)} \alpha_s(t_k)\bar{y}(t_k - sbh) + B_3\xi(t_k) \tag{2.47}$$

where $\ell(t_k) = \min\{\ell, \lfloor \frac{t_k}{bh} \rfloor\}$ with ℓ being a given positive scalar denoting the number of paths, $y(t_k)$ is the real measured output, and $\alpha_s(t_k)$ $(s = 0, 1, ..., \ell(t_k))$ are the mutually independent channel coefficients having probability density functions $\psi(s)$ on the interval $[0, 1]$ with mathematical expectations $\bar{\alpha}_s$ and variances $\tilde{\bar{\alpha}}_s^2$. $\xi(t_k)$ is an external disturbance and B_3 is a constant matrix.

Remark 2.4. *In wireless communications, fading is the deviation of the attenuation affecting a signal over certain propagation media which might be due to multipath propagation. Fading phenomena might vary with time, geographical position or radio frequency. Hence, fading effects are always modelled as certain random process (e.g. Rayleigh fading model, Rice fading model and Weibull fading model [114]). In this section, we adopt the ℓth-order Rice fading model [113] which is capable of accounting for channel fading, time delay and data dropout simultaneously. In such a model, $\alpha(t_k)$ denotes the attenuation of the signal strength, $\bar{y}(t_k - sbh)$ represents the delay effects caused by the path (or channel) s. Note that, the channel coefficients in (2.47) is assumed to obey any probability density function (not necessarily Gaussian) on [0,1], which is more suitable for practical engineering application.*

An objective of this section is to take the event-based relay communication mechanism into consideration to reduce the communication frequency. For this purpose, we consider the following event generator functions $f(.,.)$ as [117]

$$f(\sigma(t_k), \delta) = \sigma^T(t_k)\Omega\sigma(t_k) - \delta y^T(t_k)\Omega y(t_k) \tag{2.48}$$

where $y(t_{k_i})$ is the measurement at the latest event time k_i, $y(t_k)$ is the current measurement, and $\sigma(t_k) \triangleq y(t_{k_i}) - y(t_k)$. Ω is a symmetric positive-definite weighting matrix and $\delta \geq 0$ is the threshold.

At the second phase, the relay node acts as the event generator, and the execution is triggered as long as the condition

$$f(\sigma(t_k), \delta) > 0 \tag{2.49}$$

is satisfied. Therefore, the sequence of event-triggered instants $0 \leq t_{k_0} \leq t_{k_1} \leq \cdots \leq t_{k_i} \leq \cdots$ is determined iteratively by

$$t_{k_{i+1}} = \inf \{t_k \in \mathbb{Z}_{k \geq 0} \mid t_k > t_{k_i}, f(\sigma(t_k), \delta) > 0\} \tag{2.50}$$

Accordingly, only the received fading measurement signal satisfying the event condition (2.49) will be transmitted to the filter.

Remark 2.5. *Event-based transmission scheme has attracted considerable research interest in the past few years [116–118]. Such a transmission scheme is capable of improving the efficiency in resource utilization. So far, a growing number of results have been reported on the filtering problems with event-based transmission strategies. For example, in [117], the finite-horizon filtering problem has been investigated for time-varying systems with event-based transmission mechanism. In this section, we focus our attention on the event-based filtering problem for time-invariant systems with multi-rate sampling and fading channels. It is worth mentioning that the event-based transmission mechanism under consideration is equipped in the relay node, whose received signal is the measurement signal after going through the fading channels.*

C. Structure of the Filter

In this section, the zero-order holder is applied to hold the signal transmitted to the filter until a new measurement is transmitted. Let $\bar{y}(t_k)$ be the signal received by the filter. It is easy to see that

$$\bar{y}(t_k) = y(t_{k_i}), \quad t_{k_i} \leq t_k < t_{k_{i+1}}.$$

Based on the received signal $\bar{y}(t_k)$, we adopt the following event-based filter for system (2.46):

$$\begin{cases} \hat{\bar{x}}(t_{k+1}) = \hat{A}\hat{\bar{x}}(t_k) + \hat{H}\left[\bar{y}(t_k) - \hat{y}(t_k)\right] = \hat{A}\hat{\bar{x}}(t_k) + \hat{H}\left[y(t_{k_i}) - \hat{y}(t_k)\right] \\ \hat{y}(t_k) = \sum_{s=0}^{\ell} \bar{\alpha}_s C \hat{\bar{x}}(t_k - sbh) \\ \hat{\bar{z}}(t_k) = \bar{D}\hat{\bar{x}}(t_k) \end{cases} \tag{2.51}$$

where $\hat{\bar{x}}(t_k)$ represents the estimate of the state $\bar{x}(t_k)$ and $\hat{\bar{z}}(t_k)$ is the estimate of $\bar{z}(t_k)$. $\hat{H} \triangleq \text{col}\{H_1, H_2, \cdots, H_b\}$ and H_r $(r = 1, 2, \cdots, b)$ are the filter parameters to be determined.

Throughout the section, the following assumptions are made.

Assumption 2.1. *The external disturbance signals $\bar{\omega}(t_k)$, $\nu(t_k)$ and $\xi(t_k)$ are time-varying and satisfy the following constraint*

$$\sum_{t_k=0}^{N}\left[\bar{\omega}^T(t_k)\bar{\omega}(t_k)+\nu^T(t_k)\nu(t_k)+\xi^T(t_k)\xi(t_k)\right]\leq\mu$$

where μ is a given positive scalar and N is the given positive integer.

Assumption 2.2. *For $-\ell\leq i\leq -1$, we have $y(i)=0$ and $\mathrm{col}\{\omega(i),\nu(i),\xi(i)\}=0$. Without loss of generality, we also assume that $T_0=t_0=0$ and therefore $t_k-sbh=(k-s)bh=t_{k-s}$.*

By using Assumption 2.2 and letting

$$\begin{aligned}
d(k-s) &\triangleq \mathrm{col}\{\bar{\omega}(k-s),\nu(k-s),\xi(k-s)\}\\
&\triangleq col\{\bar{\omega}(t_k-sbh),\nu(t_k-sbh),\xi(t_k-sbh)\},\\
\hat{\bar{x}}(k) &\triangleq \mathrm{col}\{\hat{x}(t_k),\hat{x}(t_k-h),\cdots,\hat{x}(t_k-(b-1)h)\},\\
\bar{e}(t_k) &\triangleq \bar{x}(t_k)-\hat{\bar{x}}(t_k),\\
\eta(k-s) &\triangleq \mathrm{col}\{\bar{x}(t_k-sbh),\bar{e}(t_k-sbh)\},\\
\hat{\bar{z}}(k) &\triangleq \mathrm{col}\{\hat{z}(t_k),\hat{z}(t_k-h),\cdots,\hat{z}(t_k-(b-1)h)\},\\
\tilde{\bar{z}}(k) &\triangleq \bar{z}(t_k)-\hat{\bar{z}}(t_k),\sigma(k)\triangleq y(t_{k_i})-y(t_k),\\
\tilde{\alpha}_s(k) &\triangleq \alpha_s(t_k)-\bar{\alpha}_s\ (s=0,1,2,\cdots,\ell),\\
\bar{C} &\triangleq [C\ \underbrace{0\cdots 0}_{b-1}],
\end{aligned}$$

the combination of (2.46) and (2.51) yields the following augmented system:

$$\begin{cases}\eta(k+1)=\left(\mathcal{A}+\tilde{\alpha}_0(k)\tilde{\mathcal{A}}\right)\eta(k)+\mathcal{H}\sigma(k)+\sum_{s=1}^{\ell}\bar{\alpha}_s\bar{\mathcal{A}}\eta(k-s)\\
\qquad+\sum_{s=1}^{\ell}\tilde{\alpha}_s(k)\bar{\mathcal{A}}\eta(k-s)+\left(\mathcal{B}+\tilde{\alpha}_0(k)\bar{\mathcal{B}}\right)d(k)\\
\qquad+\sum_{s=1}^{\ell}\bar{\alpha}_s\bar{\mathcal{B}}d(k-s)+\sum_{s=1}^{\ell}\tilde{\alpha}_s(k)\bar{\mathcal{B}}d(k-s)\\
\tilde{\bar{z}}(k)=\mathcal{D}\eta(k)\end{cases}\tag{2.52}$$

where

$$\begin{aligned}
\mathcal{A} &\triangleq \begin{bmatrix}\hat{A} & 0\\ 0 & \hat{A}-\bar{\alpha}_0\hat{H}\bar{C}\end{bmatrix},\ \tilde{\mathcal{A}}\triangleq\begin{bmatrix}0 & 0\\ -\hat{H}\bar{C} & 0\end{bmatrix},\\
\bar{\mathcal{A}} &\triangleq \begin{bmatrix}0 & 0\\ 0 & -\hat{H}\bar{C}\end{bmatrix},\ \mathcal{H}\triangleq\begin{bmatrix}0\\ -\hat{H}\end{bmatrix},
\end{aligned}$$

$$\mathcal{B} \triangleq \begin{bmatrix} \hat{\bar{B}}_1 & 0 & 0 \\ 0 & -\bar{\alpha}_0 \hat{H} B_2 & -\hat{H} B_3 \end{bmatrix},$$

$$\bar{\mathcal{B}} \triangleq \begin{bmatrix} 0 & 0 & 0 \\ 0 & -\hat{H} B_2 & 0 \end{bmatrix}, \mathcal{D} \triangleq \begin{bmatrix} 0 & \bar{D} \end{bmatrix}.$$

Remark 2.6. *Comparing with the existing works on filtering problems of networked control systems [117, 130, 131], our work exhibits two distinguished features: i) this section represents one of the first attempts to consider the finite-time filtering problem of networked multi-rate systems; ii) both the fading measurements and event-based relay communication mechanism are considered and therefore it is quite comprehensive.*

Before proceeding further, we introduce the following definition as in [132–134].

Definition 2.2. *The augmented system (2.52) is said to be stochastically finite-time bounded (SFTB) with respect to (c_1, c_2, R, N, μ), if*

$$\begin{cases} \mathbb{E}\{\eta^T(k_1) R \eta(k_1)\} \leq c_1 \\ \sum_{i=0}^{N} d^T(i) d(i) \leq \mu \end{cases} \Rightarrow \mathbb{E}\{\eta^T(k_2) R \eta(k_2)\} < c_2,$$

$$\forall k_1 \in \{-\ell, -\ell+1, \cdots, -1\},\ k_2 \in \{0, 1, \cdots, N-1\} \tag{2.53}$$

where $0 < c_1 < c_2$, $\mu > 0$, $R > 0$ and $N \in \mathbb{Z}_{k \geq 0}$.

The main purpose of this section is to design the event-based filter of the form (2.51) such that the following requirements are satisfied simultaneously:

(a) The augmented system (2.52) is stochastically finite-time bounded with respect to (c_1, c_2, R, N, μ).

(b) Under zero-initial condition, the estimation error output $\tilde{\bar{z}}(k)$ with respect to all nonzero $d(k)$ satisfies

$$\mathbb{E}\Big\{\sum_{k=0}^{N} \tilde{\bar{z}}^T(k) \tilde{\bar{z}}(k)\Big\} < \gamma^2 \sum_{k=0}^{N} d^T(k) d(k) \tag{2.54}$$

where $\gamma > 0$ is a given finite-time H_∞ performance index.

2.2.2 Finite-Time Filter Design

In this subsection, we investigate the event-based finite-time H_∞ filtering problem for networked multi-rate systems (2.52). First, we propose the following finite-time filtering performance analysis results for networked multi-rate systems with fading measurements and event-based relay communication mechanism.

Theorem 2.4. *For given finite-time bounded parameters c_1, c_2, R, N, μ and filter gain $\hat{H}$, the augmented system (2.52) is stochastically finite-time bounded with respect to (c_1, c_2, R, N, μ) if there exist positive definite matrices $\mathcal{P}$, $\mathcal{Q}_s$ $(s = 1, 2, \cdots, \ell)$ and a scalar $\beta > 1$ such that the following inequalities*

$$\Gamma + \bar{\Gamma} - \tilde{\Gamma} < 0 \tag{2.55}$$

$$(\lambda_{\max}(\tilde{\mathcal{P}}) + \frac{\ell(\ell+1)}{2} \max_{1 \le s \le \ell} \{\lambda_{\max}(\tilde{\mathcal{Q}}_s)\})c_1 + \frac{\mu}{\beta - 1} < \beta^{-N} \lambda_{\min}(\tilde{\mathcal{P}}) c_2 \tag{2.56}$$

hold, where

$$\begin{aligned}
\tilde{\Gamma} &\triangleq \mathrm{diag}\{\beta\mathcal{P}, 0, 0, \beta I, \beta I\},\ I_\alpha \triangleq \mathrm{diag}\{\tilde{\bar{\alpha}}_1^2,\ \tilde{\bar{\alpha}}_2^2, \cdots, \tilde{\bar{\alpha}}_\ell^2\},\\
\hat{\bar{\alpha}} &\triangleq \begin{bmatrix} \bar{\alpha}_1 I & \bar{\alpha}_2 I & \cdots & \bar{\alpha}_\ell I \end{bmatrix},\ \mathcal{C} \triangleq \begin{bmatrix} \bar{C} & 0 \end{bmatrix},\\
\mathcal{B}_2 &\triangleq [0\ B_2\ 0],\ \mathcal{B}_3 \triangleq [0\ 0\ B_3],\ \tilde{\mathcal{P}} \triangleq R^{-\frac{1}{2}} \mathcal{P} R^{-\frac{1}{2}},\\
\tilde{\mathcal{Q}}_s &\triangleq R^{-\frac{1}{2}} \mathcal{Q}_s R^{-\frac{1}{2}},\ \hat{\mathcal{Q}} \triangleq \mathrm{diag}\{\mathcal{Q}_1, \cdots, \mathcal{Q}_\ell\},\\
\Gamma &\triangleq \begin{bmatrix} \Gamma_{11} & \Gamma_{12} & \mathcal{A}^T\mathcal{P}\mathcal{H} & \Gamma_{13} & \mathcal{A}^T\mathcal{P}\bar{\mathcal{B}}\hat{\bar{\alpha}} \\ * & \Gamma_{22} & \Gamma_{23} & \Gamma_{24} & \Gamma_{25} \\ * & * & \mathcal{H}^T\mathcal{P}\mathcal{H} & \mathcal{H}^T\mathcal{P}\mathcal{B} & \mathcal{H}^T\mathcal{P}\bar{\mathcal{B}}\hat{\bar{\alpha}} \\ * & * & * & \Gamma_{33} & \mathcal{B}^T\mathcal{P}\bar{\mathcal{B}}\hat{\bar{\alpha}} \\ * & * & * & * & \Gamma_{44} \end{bmatrix},\\
\Gamma_{11} &\triangleq \mathcal{A}^T\mathcal{P}\mathcal{A} + \tilde{\bar{\alpha}}_0^2 \tilde{\mathcal{A}}^T\mathcal{P}\tilde{\mathcal{A}} + \sum_{s=1}^{\ell} \mathcal{Q}_s,\ \Gamma_{12} \triangleq \mathcal{A}^T\mathcal{P}\bar{\mathcal{A}}\hat{\bar{\alpha}},\\
\Gamma_{13} &\triangleq \mathcal{A}^T\mathcal{P}\mathcal{B} + \tilde{\bar{\alpha}}_0^2 \tilde{\mathcal{A}}^T\mathcal{P}\bar{\mathcal{B}}, \Gamma_{23} \triangleq \hat{\bar{\alpha}}^T \bar{\mathcal{A}}^T\mathcal{P}\mathcal{H},\\
\Gamma_{22} &\triangleq \hat{\bar{\alpha}}^T \bar{\mathcal{A}}^T\mathcal{P}\bar{\mathcal{A}}\hat{\bar{\alpha}} + I_\alpha \otimes (\tilde{\mathcal{A}}^T\mathcal{P}\tilde{\mathcal{A}}) - \hat{\mathcal{Q}},\\
\Gamma_{24} &\triangleq \hat{\bar{\alpha}}^T \bar{\mathcal{A}}^T\mathcal{P}\mathcal{B},\ \Gamma_{25} \triangleq \hat{\bar{\alpha}}^T \bar{\mathcal{A}}^T\mathcal{P}\bar{\mathcal{B}}\hat{\bar{\alpha}} + I_\alpha \otimes (\tilde{\mathcal{A}}^T\mathcal{P}\bar{\mathcal{B}}),\\
\Gamma_{33} &\triangleq \mathcal{B}^T\mathcal{P}\mathcal{B} + \tilde{\bar{\alpha}}_0^2 \bar{\mathcal{B}}^T\mathcal{P}\bar{\mathcal{B}},\\
\Gamma_{44} &\triangleq \hat{\bar{\alpha}}^T \bar{\mathcal{B}}^T\mathcal{P}\bar{\mathcal{B}}\hat{\bar{\alpha}} + I_\alpha \otimes (\bar{\mathcal{B}}^T\mathcal{P}\bar{\mathcal{B}}),\\
\bar{\Gamma} &\triangleq \begin{bmatrix} \bar{\Gamma}_{11} & \bar{\Gamma}_{12} & 0 & \bar{\Gamma}_{13} & \delta\bar{\alpha}_0 \mathcal{C}^T\Omega\mathcal{B}_2\hat{\bar{\alpha}} \\ * & \bar{\Gamma}_{22} & 0 & \bar{\Gamma}_{23} & \bar{\Gamma}_{24} \\ * & * & -\Omega & 0 & 0 \\ * & * & * & \bar{\Gamma}_{33} & \bar{\Gamma}_{34} \\ * & * & * & * & \bar{\Gamma}_{44} \end{bmatrix},\\
\bar{\Gamma}_{11} &\triangleq \delta(\bar{\alpha}_0^2 + \tilde{\bar{\alpha}}_0^2)\mathcal{C}^T\Omega\mathcal{C},\ \bar{\Gamma}_{12} \triangleq \delta\bar{\alpha}_0\mathcal{C}^T\Omega\mathcal{C}\hat{\bar{\alpha}},\\
\bar{\Gamma}_{13} &\triangleq \delta\bar{\alpha}_0\mathcal{C}^T\Omega(\bar{\alpha}_0\mathcal{B}_2 + \mathcal{B}_3) + \tilde{\bar{\alpha}}_0^2\mathcal{C}^T\Omega\mathcal{B}_2,\\
\bar{\Gamma}_{22} &\triangleq \delta\hat{\bar{\alpha}}^T\mathcal{C}^T\Omega\mathcal{C}\hat{\bar{\alpha}} + \delta I_\alpha \otimes (\mathcal{C}^T\Omega\mathcal{C}),\\
\bar{\Gamma}_{23} &\triangleq \delta\hat{\bar{\alpha}}^T\mathcal{C}^T\Omega(\bar{\alpha}_0\mathcal{B}_2 + \mathcal{B}_3),\\
\bar{\Gamma}_{24} &\triangleq \delta\hat{\bar{\alpha}}^T\mathcal{C}^T\Omega\mathcal{B}_2\hat{\bar{\alpha}} + \delta I_\alpha \otimes (\mathcal{C}^T\Omega\mathcal{B}_2),
\end{aligned}$$

$$
\begin{aligned}
\bar{\Gamma}_{33} &\triangleq \delta(\bar{\alpha}_0\mathcal{B}_2+\mathcal{B}_3)^T\Omega(\bar{\alpha}_0\mathcal{B}_2+\mathcal{B}_3)+\tilde{\bar{\alpha}}_0^2\mathcal{B}_2^T\Omega\mathcal{B}_2,\\
\bar{\Gamma}_{34} &\triangleq \delta(\bar{\alpha}_0\mathcal{B}_2^T+\mathcal{B}_3^T)\Omega\mathcal{B}_2\hat{\bar{\alpha}},\\
\bar{\Gamma}_{44} &\triangleq \delta\hat{\bar{\alpha}}^T\mathcal{B}_2^T\Omega\mathcal{B}_2\hat{\bar{\alpha}}+\delta I_\alpha\otimes(\mathcal{B}_2^T\Omega\mathcal{B}_2).
\end{aligned}
$$

Proof. Choose the following Lyapunov-like functional:

$$
V(k)\triangleq\eta^T(k)\mathcal{P}\eta(k)+\sum_{s=1}^{\ell}\sum_{r=k-s}^{k-1}\eta^T(r)\mathcal{Q}_s\eta(r). \tag{2.57}
$$

By taking the mathematical expectation of $V(k)$ along the trajectory of the augmented system (2.52), one has

$$
\begin{aligned}
\mathbb{E}\{V(k+1)\} &= \eta^T(k)\Big[\mathcal{A}^T\mathcal{P}\mathcal{A}+\tilde{\bar{\alpha}}_0^2\tilde{\mathcal{A}}^T\mathcal{P}\tilde{\mathcal{A}}\Big]\eta(k)+2\tilde{\bar{\alpha}}_0^2\eta^T(k)\mathcal{A}^T\mathcal{P}\bar{\mathcal{B}}d(k)\\
&+2\eta^T(k)\mathcal{A}^T\mathcal{P}\Big[\sum_{s=1}^{\ell}\bar{\alpha}_s\bar{\mathcal{A}}\eta(k-s)+\mathcal{H}\sigma(k)+\mathcal{B}d(k)+\sum_{s=1}^{\ell}\bar{\alpha}_s\bar{\mathcal{B}}d(k-s)\Big]\\
&+\Big[\sum_{s=1}^{\ell}\bar{\alpha}_s\bar{\mathcal{A}}\eta(k-s)\Big]^T\mathcal{P}\Big[\sum_{s=1}^{\ell}\bar{\alpha}_s\bar{\mathcal{A}}\eta(k-s)\Big]\\
&+2\Big[\sum_{s=1}^{\ell}\bar{\alpha}_s\bar{\mathcal{A}}\eta(k-s)\Big]^T\mathcal{P}\Big[\mathcal{H}\sigma(k)+\mathcal{B}d(k)+\sum_{s=1}^{\ell}\bar{\alpha}_s\bar{\mathcal{B}}d(k-s)\Big]\\
&+\sigma^T(k)\mathcal{H}^T\mathcal{P}\mathcal{H}\sigma(k)+\sum_{s=1}^{\ell}\tilde{\bar{\alpha}}_s^2\eta^T(k-s)\tilde{\mathcal{A}}^T\mathcal{P}\Big[\tilde{\mathcal{A}}\eta(k-s)+\bar{\mathcal{B}}d(k-s)\Big]\\
&+2\sigma^T(k)\mathcal{H}^T\mathcal{P}\mathcal{B}d(k)+2\sum_{s=1}^{\ell}\bar{\alpha}_s\sigma^T(k)\mathcal{H}^T\mathcal{P}\bar{\mathcal{B}}d(k-s)\\
&+d^T(k)\Big[\mathcal{B}^T\mathcal{P}\mathcal{B}+\tilde{\bar{\alpha}}_0^2\bar{\mathcal{B}}^T\mathcal{P}\bar{\mathcal{B}}\Big]d(k)\\
&+2\sum_{s=1}^{\ell}\bar{\alpha}_s d^T(k)\mathcal{B}^T\mathcal{P}\bar{\mathcal{B}}d(k-s)+\sum_{s=1}^{\ell}\tilde{\bar{\alpha}}_s^2 d^T(k-s)\bar{\mathcal{B}}^T\mathcal{P}\bar{\mathcal{B}}d(k-s)\\
&+\sum_{s=1}^{\ell}\sum_{r=k-s}^{k-1}\eta^T(r)\mathcal{Q}_s\eta(r)+\sum_{s=1}^{\ell}\Big[\eta^T(k)\mathcal{Q}_s\eta(k)-\eta^T(k-s)\mathcal{Q}_s\eta(k-s)\Big]\\
&=\vartheta^T(k)\Gamma\vartheta(k)+\sum_{s=1}^{\ell}\sum_{r=k-s}^{k-1}\eta^T(r)\mathcal{Q}_s\eta(r)
\end{aligned} \tag{2.58}
$$

where

$$
\begin{aligned}
\eta_k^\ell &\triangleq \mathrm{col}\{\eta(k-1),\eta(k-2),\cdots,\eta(k-\ell)\},\\
d_k^\ell &\triangleq \mathrm{col}\{d(k-1),d(k-2),\cdots,d(k-\ell)\},\\
\vartheta(k) &\triangleq \mathrm{col}\{\eta(k),\eta_k^\ell,\sigma(k),d(k),d_k^\ell\}.
\end{aligned}
$$

Moreover, substituting (2.43) into (2.47), we have

$$\begin{aligned}
\mathbb{E}\{y^T(k)\Omega y(k)\} &= \eta^T(k)\Big[(\bar{\alpha}_0^2+\tilde{\bar{\alpha}}_0^2)\mathcal{C}^T\Omega\mathcal{C}\Big]\eta(k)+2\tilde{\bar{\alpha}}_0^2\eta^T(k)\mathcal{C}^T\Omega\mathcal{B}_2 d(k)\\
&+2\eta^T(k)\bar{\alpha}_0\mathcal{C}^T\Omega\Big[\sum_{s=1}^{\ell}\bar{\alpha}_s\mathcal{C}\eta(k-s)+(\bar{\alpha}_0\mathcal{B}_2+\mathcal{B}_3)d(k)+\sum_{s=1}^{\ell}\bar{\alpha}_s\mathcal{B}_2 d(k-s)\Big]\\
&+\Big[\sum_{s=1}^{\ell}\bar{\alpha}_s\mathcal{C}\eta(k-s)\Big]^T\Omega\Big[\sum_{s=1}^{\ell}\bar{\alpha}_s\mathcal{C}\eta(k-s)\Big]+2\Big[\sum_{s=1}^{\ell}\bar{\alpha}_s\mathcal{C}\eta(k-s)\Big]^T\Omega\\
&\Big[(\bar{\alpha}_0\mathcal{B}_2+\mathcal{B}_3)d(k)+\sum_{s=1}^{\ell}\bar{\alpha}_s\mathcal{B}_2 d(k-s)\Big]+\sum_{s=1}^{\ell}\tilde{\bar{\alpha}}_s^2\eta^T(k-s)\mathcal{C}^T\Omega\mathcal{C}\eta(k-s)\\
&+2\sum_{s=1}^{\ell}\tilde{\bar{\alpha}}_s^2\eta^T(k-s)\mathcal{C}^T\Omega\mathcal{B}_2 d(k-s)\\
&+d^T(k)\Big[(\bar{\alpha}_0\mathcal{B}_2+\mathcal{B}_3)^T\Omega(\bar{\alpha}_0\mathcal{B}_2+\mathcal{B}_3)+\tilde{\bar{\alpha}}_0^2\mathcal{B}_2^T\Omega\mathcal{B}_2\Big]d(k)\\
&+2d^T(k)(\bar{\alpha}_0\mathcal{B}_2+\mathcal{B}_3)^T\Omega\Big[\sum_{s=1}^{\ell}\bar{\alpha}_s\mathcal{B}_2 d(k-s)\Big]\\
&+\Big[\sum_{s=1}^{\ell}\bar{\alpha}_s\mathcal{B}_2 d(k-s)\Big]\Omega\Big[\sum_{s=1}^{\ell}\bar{\alpha}_s\mathcal{B}_2 d(k-s)\Big]\\
&+\sum_{s=1}^{\ell}\tilde{\bar{\alpha}}_s^2 d^T(k-s)\mathcal{B}_2^T\Omega\mathcal{B}_2 d(k-s)
\end{aligned}\tag{2.59}$$

Considering the event-triggered condition (2.49) and combining (2.58) and (2.59), one obtains

$$\begin{aligned}
\mathbb{E}\{V(k+1)\} &\le \vartheta^T(k)\Gamma\vartheta(k)+\sum_{s=1}^{\ell}\sum_{r=k-s}^{k-1}\eta^T(r)\mathcal{Q}_s\eta(r)-\\
&\quad\ \sigma(k)\Omega\sigma(k)+\delta\mathbb{E}\{y^T(k)\Omega y(k)\}\\
&= \vartheta^T(k)(\Gamma+\bar{\Gamma})\vartheta(k)+\sum_{s=1}^{\ell}\sum_{r=k-s}^{k-1}\eta^T(r)\mathcal{Q}_s\eta(r)
\end{aligned}\tag{2.60}$$

It follows from (2.55) that $\Gamma+\bar{\Gamma}<\tilde{\Gamma}$, and therefore

$$\begin{aligned}
\mathbb{E}\{V(k+1)\} &\le \beta\eta^T(k)\mathcal{P}\eta(k)+\beta\sum_{s=1}^{\ell}\sum_{r=k-s}^{k-1}\eta^T(r)\mathcal{Q}_s\eta(r)+\\
&\quad\ \beta\sum_{s=0}^{\ell}d^T(k-s)d(k-s)\\
&= \beta\mathbb{E}\{V(k)\}+\beta\sum_{s=0}^{\ell}d^T(k-s)d(k-s)
\end{aligned}\tag{2.61}$$

According to $\sum_{i=1}^{N} d^T(i)d(i) \leq \mu$ and applying above relation (2.61) recursively, we have the following inequality:

$$\begin{aligned}
\mathbb{E}\{V(k)\} &\leq \beta^k \mathbb{E}\{V(0)\} + \sum_{i=0}^{k-1}\sum_{s=0}^{\ell} \beta^{i+1} d^T(k-i-s)d(k-i-s) \\
&\leq \beta^k \mathbb{E}\{\eta^T(0)\mathcal{P}\eta(0)\} + \beta^k \sum_{s=1}^{\ell}\sum_{r=-s}^{-1} \mathbb{E}\{\eta^T(r)\mathcal{Q}_s\eta(r)\} \\
&\quad + \sum_{i=0}^{k-1}\sum_{s=0}^{\ell} \beta^{i+1} d^T(k-i-s)d(k-i-s) \\
&\leq \beta^k \lambda_{\max}(\tilde{\mathcal{P}})\mathbb{E}\{\eta^T(0)R\eta(0)\} + \sum_{i=0}^{k-1}\beta^{i+1}\sum_{s=0}^{\ell} d^T(s)d(s) \\
&\quad + \beta^k \max_{1\leq s\leq \ell}\{\lambda_{\max}(\tilde{\mathcal{Q}}_s)\} \sum_{s=1}^{\ell}\sum_{r=-s}^{-1} \mathbb{E}\{\eta^T(r)R\eta(r)\} \\
&\leq \beta^N \left[(\lambda_{\max}(\tilde{\mathcal{P}}) + \frac{\ell(\ell+1)}{2} \max_{1\leq s\leq \ell}\{\lambda_{\max}(\tilde{\mathcal{Q}}_s)\})c_1 + \frac{\mu}{\beta-1} \right]
\end{aligned} \tag{2.62}$$

On the other hand, for $V(k)$, we have

$$\mathbb{E}\{V(k)\} \;\geq\; \mathbb{E}\{\eta^T(k)\mathcal{P}\eta(k)\} \geq \lambda_{\min}(\tilde{\mathcal{P}})\mathbb{E}\{\eta^T(k)R\eta(k)\} \tag{2.63}$$

Then, the combination of (2.62) and (2.63) results in

$$\begin{aligned}
&\mathbb{E}\{\eta^T(k)R\eta(k)\} \\
&< \frac{\beta^N}{\lambda_{\min}(\tilde{\mathcal{P}})} \left[\frac{\mu}{\beta-1} + (\lambda_{\max}(\tilde{\mathcal{P}}) + \frac{\ell(\ell+1)}{2} \max_{1\leq s\leq \ell}\{\lambda_{\max}(\tilde{\mathcal{Q}}_s)\})c_1 \right]
\end{aligned} \tag{2.64}$$

Noticing (2.64), it follows from (2.55) that $\mathbb{E}\{\eta^T(k)R\eta(k)\} < c_2$ for all $k = 1, 2, \cdots, N$. Therefore, by using Definition 2.2, the stochastically finite-time bounded of the augmented system (2.52) with respect to (c_1, c_2, R, N, μ) is ensured, and this completes the proof. □

The following theorem presents sufficient conditions that guarantee the finite-time bounded of the augmented system (2.52) and, at the same time, satisfy the finite-time H_∞ performance constraint (2.54).

Theorem 2.5. *For given finite-time bounded parameters c_1, c_2, R, N, μ and filter gain $\hat{H}$, the augmented system (2.52) is stochastically finite-time bounded with respect to (c_1, c_2, R, N, μ) and simultaneously satisfies the finite-time H_∞ performance constraint (2.54) if there exist positive definite matrices $\mathcal{P}$, $\mathcal{Q}_s$ $(s =$*

$1, 2, \cdots, \ell$) *and a scalar* $\beta > 1$ *such that the following inequalities*

$$\beta(\Gamma + \bar{\Gamma}) + \bar{\bar{\Gamma}} < 0 \tag{2.65}$$

$$c_1\lambda_{\max}(\tilde{\mathcal{P}}) + \frac{\ell(\ell+1)}{2}c_1 \max_{1\leq s\leq \ell}\{\lambda_{\max}(\tilde{\mathcal{Q}}_s)\} + \frac{\gamma^2\mu}{(\ell+1)(\beta-1)} < \beta^{-N}c_2\lambda_{\min}(\tilde{\mathcal{P}}) \tag{2.66}$$

hold, where $\bar{\bar{\Gamma}} \triangleq \mathrm{diag}\{-\beta\mathcal{P} + \mathcal{D}^T\mathcal{D}, 0, 0, -\beta\frac{\gamma^2}{\ell+1}I, -\beta\Upsilon\}$,
$\Upsilon \triangleq \mathrm{diag}\{\underbrace{\frac{\gamma^2}{\ell+1}I, \cdots, \frac{\gamma^2}{\ell+1}I}_{\ell}\}$, *and other corresponding matrices are defined in Theorem 2.4.*

Proof. Inequality (2.65) can be written as

$$\beta(\Gamma + \bar{\Gamma}) + \tilde{\bar{\Gamma}} + \bar{\mathcal{D}}^T\bar{\mathcal{D}} < 0 \tag{2.67}$$

where

$$\tilde{\bar{\Gamma}} \triangleq \mathrm{diag}\{-\beta\mathcal{P}, 0, 0, -\beta\frac{\gamma^2}{\ell+1}I, -\beta\Upsilon\},\ \bar{\mathcal{D}} \triangleq \mathrm{col}\{\mathcal{D}, 0, 0, 0, 0\}.$$

Noting the fact $\bar{\mathcal{D}}^T\bar{\mathcal{D}} \geq 0$, (2.67) means that

$$\beta(\Gamma + \bar{\Gamma}) + \tilde{\bar{\Gamma}} < 0 \tag{2.68}$$

Combining (2.60) and (2.68), we have

$$\beta\mathbb{E}\{V(k+1)\} \leq \beta\mathbb{E}\{V(k)\} + \beta\frac{\gamma^2}{\ell+1}\sum_{s=0}^{\ell} d^T(k-s)d(k-s) \tag{2.69}$$

On the other hand, for $\beta > 1$, it follows from (2.69) that

$$\mathbb{E}\{V(k+1)\} \leq \beta\mathbb{E}\{V(k)\} + \beta\frac{\gamma^2}{\ell+1}\sum_{s=0}^{\ell} d^T(k-s)d(k-s) \tag{2.70}$$

Obviously, by following the similar analysis of obtaining Theorem 2.4, we have

$$\mathbb{E}\{\eta^T(k)R\eta(k)\} < \frac{\beta^N}{\lambda_{\min}(\tilde{\mathcal{P}})}\left[c_1\left(\lambda_{\max}(\tilde{\mathcal{P}}) + \frac{\ell(\ell+1)}{2}\max_{1\leq s\leq \ell}\{\lambda_{\max}(\tilde{\mathcal{Q}}_s)\}\right) + \frac{\gamma^2\mu}{(\ell+1)(\beta-1)}\right] \tag{2.71}$$

Hence, it can be verified from (2.66) that the augmented system (2.52) is stochastically finite-time bounded.

Next, let us consider the finite-time H_∞ performance. For this purpose, we introduce the following index:

$$\mathcal{J}_N \triangleq \mathbb{E}\Big\{\sum_{k=0}^{N} \tilde{\bar{z}}^T(k)\tilde{\bar{z}}(k) - \gamma^2 \sum_{k=0}^{N} d^T(k)d(k)\Big\}. \tag{2.72}$$

Under the zero-initial condition, due to $\{d(k)\}_{-\ell \le k \le -1} = 0$, it follows from (2.52) and (2.60) that

$$\begin{aligned}
\mathcal{J}_N &\triangleq \mathbb{E}\Big\{\sum_{k=0}^{N} \tilde{\bar{z}}^T(k)\tilde{\bar{z}}(k) - \frac{\gamma^2}{\ell+1}\sum_{k=0}^{N}\sum_{s=0}^{\ell} d^T(k)d(k)\Big\} \\
&\le \mathbb{E}\Big\{\sum_{k=0}^{N} \tilde{\bar{z}}^T(k)\tilde{\bar{z}}(k) - \frac{\gamma^2}{\ell+1}\sum_{k=0}^{N}\sum_{s=0}^{\ell} d^T(k-s)d(k-s)\Big\} \\
&\le \sum_{k=0}^{N} \mathbb{E}\Big\{\tilde{\bar{z}}^T(k)\tilde{\bar{z}}(k) - \frac{\gamma^2}{\ell+1}\sum_{s=0}^{\ell} d^T(k-s)d(k-s) + [V(k+1) - V(k)]\Big\} \\
&\quad -\mathbb{E}\{V(N+1)\} \\
&\le \sum_{k=0}^{N} \mathbb{E}\Big\{\tilde{\bar{z}}^T(k)\tilde{\bar{z}}(k) - \frac{\gamma^2}{\ell+1}\sum_{s=0}^{\ell} d^T(k-s)d(k-s) + \beta[V(k+1) - V(k)]\Big\} \\
&= \sum_{k=0}^{N} \vartheta^T(k)\Big[\beta(\Gamma + \bar{\Gamma}) + \bar{\bar{\Gamma}}\Big]\vartheta(k)
\end{aligned} \tag{2.73}$$

Inequality (2.65) guarantees that $\beta(\Gamma + \bar{\Gamma}) + \bar{\bar{\Gamma}} < 0$, that is, $\mathcal{J}_N < 0$, and then we have

$$\mathbb{E}\Big\{\sum_{k=0}^{N} \tilde{\bar{z}}^T(k)\tilde{\bar{z}}(k)\Big\} < \gamma^2 \sum_{k=0}^{N} d^T(k)d(k)$$

which means that the finite-time H_∞ performance constraint (2.54) is satisfied. The proof of this theorem is now complete. □

Having established the analysis results, we are now ready to deal with the filter design problem. In the following theorem, sufficient conditions are provided for the existence of the desired event-based filter. For the convenience of later development, we denote

$$\begin{aligned}
\bar{I}_\alpha &\triangleq \operatorname{diag}\{\tilde{\bar{\alpha}}_1, \cdots, \tilde{\bar{\alpha}}_\ell\},\ \breve{\mathcal{P}} \triangleq \operatorname{diag}\Big\{P_1, \cdots, P_{2b}\Big\}, \\
\breve{\mathscr{P}} &\triangleq \operatorname{diag}\{\underbrace{\breve{\mathcal{P}}, \cdots, \breve{\mathcal{P}}}_{\ell}\},\ \mathcal{W}_1 \triangleq \operatorname{diag}\{\underbrace{W, \cdots, W}_{b}\}, \\
\mathscr{W} &\triangleq \operatorname{diag}\{\underbrace{\mathcal{W}, \cdots, \mathcal{W}}_{\ell}\},\ \mathcal{W} \triangleq \operatorname{diag}\{\underbrace{W, \cdots, W}_{2b}\}, \\
\breve{\mathcal{Q}}_s &\triangleq \operatorname{diag}\Big\{Q_{s,1}, Q_{s,2}, \cdots, Q_{s,2b}\Big\}\ (s = 1, 2, \cdots, \ell),
\end{aligned}$$

$$
\begin{aligned}
\hat{\breve{\mathcal{Q}}} &\triangleq \operatorname{diag}\{\breve{\mathcal{Q}}_1, \cdots, \breve{\mathcal{Q}}_\ell\},\ \breve{\mathcal{H}} \triangleq \operatorname{col}\{\bar{H}_1, \bar{H}_2, \cdots, \bar{H}_b\},\\
\breve{\breve{\mathcal{H}}} &\triangleq \begin{bmatrix} 0 \\ -\breve{\mathcal{H}} \end{bmatrix},\ \breve{\mathcal{A}} \triangleq \begin{bmatrix} \mathcal{W}_1\hat{A} & 0 \\ 0 & \mathcal{W}_1\hat{A} - \bar{\alpha}_0\breve{\mathcal{H}}\bar{C} \end{bmatrix},\\
\breve{\bar{\mathcal{A}}} &\triangleq \begin{bmatrix} 0 & 0 \\ 0 & -\breve{\mathcal{H}}\bar{C} \end{bmatrix},\ \breve{\tilde{\mathcal{A}}} \triangleq \begin{bmatrix} 0 & 0 \\ -\breve{\mathcal{H}}\bar{C} & 0 \end{bmatrix},\\
\breve{\mathcal{B}} &\triangleq \begin{bmatrix} \mathcal{W}_1\hat{\hat{B}}_1 & 0 & 0 \\ 0 & -\bar{\alpha}_0\breve{\mathcal{H}}B_2 & -\breve{\mathcal{H}}B_3 \end{bmatrix},\\
\breve{\bar{\mathcal{B}}} &\triangleq \begin{bmatrix} 0 & 0 & 0 \\ 0 & -\breve{\mathcal{H}}B_2 & 0 \end{bmatrix},\\
\tilde{\bar{\Xi}}_1 &\triangleq \operatorname{diag}\Big\{\sum_{s=1}^{\ell}\beta\breve{\mathcal{Q}}_s + \mathcal{D}^T\mathcal{D} - \beta\breve{\mathcal{P}}, -\beta\hat{\breve{\mathcal{Q}}}, -\beta\Omega, -\beta\frac{\gamma^2}{\ell+1}I, -\beta\Upsilon\Big\},\\
\bar{\Xi}_2 &\triangleq \begin{bmatrix}\sqrt{\beta}\bar{\Xi}_{21} & \sqrt{\beta}\bar{\Xi}_{22} & \sqrt{\beta}\bar{\Xi}_{23}\end{bmatrix},\\
\bar{\Xi}_{21} &\triangleq \operatorname{col}\{\tilde{\bar{\alpha}}_0\breve{\tilde{\mathcal{A}}}^T, 0, 0, \tilde{\bar{\alpha}}_0\breve{\bar{\mathcal{B}}}^T, 0\},\\
\bar{\Xi}_{22} &\triangleq \operatorname{col}\{0, (\bar{I}_\alpha \otimes \breve{\bar{\mathcal{A}}})^T, 0, 0, (\bar{I}_\alpha \otimes \breve{\bar{\mathcal{B}}})^T\},\\
\bar{\Xi}_{23} &\triangleq \operatorname{col}\{\breve{\mathcal{A}}^T, \hat{\bar{\alpha}}^T\breve{\bar{\mathcal{A}}}^T, \breve{\breve{\mathcal{H}}}^T, \breve{\mathcal{B}}^T, \hat{\bar{\alpha}}^T\breve{\bar{\mathcal{B}}}^T\},\\
\Xi_3 &\triangleq \begin{bmatrix}\sqrt{\beta}\Xi_{31} & \sqrt{\beta}\Xi_{32} & \sqrt{\beta}\Xi_{33}\end{bmatrix},\\
\Xi_{31} &\triangleq \operatorname{col}\{\sqrt{\delta}\tilde{\bar{\alpha}}_0\mathcal{C}^T\Omega, 0, 0, \sqrt{\delta}\tilde{\bar{\alpha}}_0\mathcal{B}_2^T\Omega, 0\},\\
\Xi_{32} &\triangleq \operatorname{col}\{0, \sqrt{\delta}\Big(\bar{I}_\alpha \otimes (\mathcal{C}\Omega)\Big)^T, 0, 0, \sqrt{\delta}\Big(\bar{I}_\alpha \otimes (\mathcal{B}_2\Omega)\Big)^T\},\\
\Xi_{33} &\triangleq \sqrt{\delta}\operatorname{col}\{\bar{\alpha}_0\mathcal{C}^T\Omega, \hat{\bar{\alpha}}^T\mathcal{C}^T\Omega, 0, (\bar{\alpha}_0\mathcal{B}_2^T + \mathcal{B}_3^T)\Omega, \hat{\bar{\alpha}}^T\mathcal{B}_2^T\Omega\},\\
\bar{\Xi}_4 &\triangleq \operatorname{diag}\Big\{\breve{\mathscr{P}} - 2\mathscr{W}, \breve{\mathscr{P}} - 2\mathscr{W}, \breve{\mathcal{P}} - 2\mathcal{W}\Big\},\\
\Xi_5 &\triangleq \operatorname{diag}\Big\{-\Omega, \underbrace{-\Omega, \cdots, -\Omega}_{\ell}, -\Omega\}.
\end{aligned}
$$

Theorem 2.6. *Let the disturbance attenuation level $\gamma > 0$ and the finite-time bounded parameters c_1, c_2, R, N, μ be given. The augmented system (2.52) is stochastically finite-time bounded with respect to (c_1, c_2, R, N, μ) and simultaneously satisfies the finite-time H_∞ performance constraint (2.54) if there exist matrices $\bar{H}_r$, $W > 0$, $P_j > 0$, $Q_{s,j} > 0$ ($r = 1, 2, \cdots, b$; $s = 1, 2, \cdots, \ell$; $j = 1, 2, \cdots, 2b$) and scalars $\beta > 1$, λ_i ($i = 1, 2$) such that the following linear matrix inequalities hold:*

$$
\begin{bmatrix} \tilde{\bar{\Xi}}_1 & \bar{\Xi}_2 & \Xi_3 \\ * & \bar{\Xi}_4 & 0 \\ * & * & \Xi_5 \end{bmatrix} < 0 \tag{2.74}
$$

$$R \leq \mathcal{P} \leq \lambda_1 R,\ 0 < \mathcal{Q}_s \leq \lambda_2 R \tag{2.75}$$

$$c_1\lambda_1 + \frac{\ell(\ell+1)}{2}c_1\lambda_2 + \frac{\gamma^2\mu}{(\ell+1)(\beta-1)} < \beta^{-N}c_2 \tag{2.76}$$

Furthermore, if the above inequalities are feasible, the desired event-based filters can be determined by

$$H_r = W^{-1}\bar{H}_r\ (r = 1, 2, \cdots, b) \tag{2.77}$$

Proof. By using the Schur Complement Lemma, (2.65) is equivalent to the following inequality:

$$\Xi = \begin{bmatrix} \Xi_1 & \Xi_2 & \Xi_3 \\ * & \Xi_4 & 0 \\ * & * & \Xi_5 \end{bmatrix} < 0 \tag{2.78}$$

where

$$\begin{aligned}
\Xi_1 &\triangleq \mathrm{diag}\Big\{\sum_{s=1}^{\ell}\beta\mathcal{Q}_s + \mathcal{D}^T\mathcal{D} - \beta\mathcal{P}, -\beta\hat{\mathcal{Q}}, -\beta\Omega, -\beta\frac{\gamma^2}{\ell+1}I, -\beta\Upsilon\Big\},\\
\Xi_2 &\triangleq \Big[\sqrt{\beta}\Xi_{21} \quad \sqrt{\beta}\Xi_{22} \quad \sqrt{\beta}\Xi_{23}\Big],\\
\Xi_{21} &\triangleq \mathrm{col}\{\tilde{\bar{\alpha}}_0\tilde{\mathcal{A}}^T, 0, 0, \tilde{\bar{\alpha}}_0\bar{\mathcal{B}}^T, 0\},\\
\Xi_{22} &\triangleq \mathrm{col}\{0, (\bar{I}_\alpha \otimes \tilde{\mathcal{A}})^T, 0, 0, (\bar{I}_\alpha \otimes \bar{\mathcal{B}})^T\},\\
\Xi_{23} &\triangleq \mathrm{col}\{\mathcal{A}^T, \hat{\bar{\alpha}}^T\bar{\mathcal{A}}^T, \mathcal{H}, \mathcal{B}, \hat{\bar{\alpha}}^T\bar{\mathcal{B}}^T\},\\
\Xi_4 &\triangleq \mathrm{diag}\Big\{-\mathcal{P}^{-1}, \underbrace{-\mathcal{P}^{-1}, \cdots, -\mathcal{P}^{-1}}_{\ell}, -\mathcal{P}^{-1}\Big\}.
\end{aligned}$$

Let $\mathcal{P} \triangleq \mathrm{diag}\{P_1, P_2, \cdots, P_{2b}\}$ and $\mathcal{P}^{-1} \triangleq \mathrm{diag}\{P_1^{-1}, P_2^{-1}, \cdots, P_{2b}^{-1}\}$. Considering $W > 0$ and $P_j > 0$, we have $(P_j - W)P_j^{-1}(P_j - W) \geq 0$, which is equivalent to $-WP_j^{-1}W \leq P_j - 2W$ $(j = 1, 2, \cdots, 2b)$.

Applying congruence transformation $\mathrm{diag}\Big\{\underbrace{\mathscr{W}, \cdots, \mathscr{W}}_{\ell+2}\Big\}$ to (2.78) and denoting $\bar{H}_r \triangleq WH_r$ $(r = 1, 2, \cdots, b)$, (2.74) can be obtained and the filters can be expressed as (2.77).

Finally, for $\tilde{\mathcal{P}} \triangleq R^{-\frac{1}{2}}\mathcal{P}R^{-\frac{1}{2}}$ and $\tilde{\mathcal{Q}}_s \triangleq R^{-\frac{1}{2}}\mathcal{Q}_sR^{-\frac{1}{2}}$ in Theorem 2.4, there exist scalars λ_i $(i = 1, 2)$ satisfying

$$R \leq \lambda_{\min}(\tilde{\mathcal{P}})R \leq \mathcal{P} \leq \lambda_{\max}(\tilde{\mathcal{P}})R \leq \lambda_1 R,\ 0 < \mathcal{Q}_s \leq \max_{1\leq s\leq \ell}\{\lambda_{\max}(\tilde{\mathcal{Q}}_s)\}R \leq \lambda_2 R \tag{2.79}$$

By substituting (2.79) into (2.66), we obtain (2.76). Therefore, the proof of Theorem 2.6 is complete. □

Remark 2.7. *In this section, we examine how the event-triggered strategy and multi-rate sampled-data pattern influence the H_∞ performance over finite-time interval. Compared with existing literatures, our results have the following three distinguishing features: 1) a more comprehensive networked control systems model is introduced that takes the multi-rate sampling, the event-triggered mechanism and the fading measurements into account; 2) a more practical analysis framework for finite-time H_∞ is established to investigate the filter design problem; and 3) quantitative relationships are investigated between the H_∞ performance index, the event-triggering threshold, the coefficients of the fading measurements, the parameters of the finite-time boundedness and the multi-rate multiple b of the sampling period h.*

In order to show the multiobjective flavour of the considered event-triggered mechanism, fading measurements and multi-rate sampling on the finite-time filter design, we now formulate the following two optimization problems for given channel fading parameters $\bar{\alpha}_s, \bar{\tilde{\alpha}}_s (s = 1, 2, \cdots, \ell)$, event-triggered threshold δ, finite-time bounded parameters (c_1, R, N, μ), and multi-rate multiple b.

P1: For the given H_∞ performance level γ, the minimum boundedness-constrained filter design problem for the networked multi-rate systems with event-triggered mechanism and fading measurements:

$$\min_{W,\bar{H}_r,P_j,Q_{s,j}\ (r=1,2,\cdots,b;\ s=1,2,\cdots,\ell;\ j=1,2,\cdots,2b)} c_2 \quad \text{subject to (2.74)} - \text{(2.76)} \tag{2.80}$$

P2: For the given boundedness-constrained parameter c_2, the optimal H_∞ filter design problem for the networked multi-rate systems with event-triggered mechanism and fading measurements:

$$\min_{W,\bar{H}_r,P_j,Q_{s,j}\ (r=1,2,\cdots,b;\ s=1,2,\cdots,\ell;\ j=1,2,\cdots,2b)} \gamma^2 \quad \text{subject to (2.74)} - \text{(2.76)} \tag{2.81}$$

Note that the above two optimization problems can be readily solved via the existing semi-definite programming. The flexibility of our developed strategy is shown through the addressed two optimization problems, and such kind of flexibility allows us to make compromise between filtering performance and triggering frequency to achieve a balance between accuracy and cost.

2.2.3 Illustrative Examples

In this section, a numerical example is presented to demonstrate the effectiveness of the proposed filter design scheme with fading measurements and event-based relay communication mechanism for networked multi-rate systems.

Consider a continuous stirred tank reactor (CSTR) as [127] with the following balance equations:

$$\frac{dx_A}{dt} = \frac{F}{V}(x_{A0} - x_A) - k_1 x_A - k_3 x_A^2 \tag{2.82}$$

$$\frac{dx_B}{dt} = -\frac{F}{V}x_B + k_1 x_A - k_2 x_B \tag{2.83}$$

$$\frac{dx_\vartheta}{dt} = \frac{F}{V}(x_{\vartheta 0} - x_\vartheta) + \frac{k_W A_R}{\varrho C_P V}(x_\psi - x_\vartheta) - \frac{k_1 x_A H_{AB} + k_2 x_B H_{BC} + k_3 x_A^2 H_{AD}}{\varrho C_P} \tag{2.84}$$

$$\frac{dx_\psi}{dt} = \frac{1}{m_K C_{PK}}[Q_K + k_W A_R(x_\vartheta - x_\psi)] \tag{2.85}$$

where x_A and x_B represent the concentrations of product A and the desired product B within the reactor respectively, x_ϑ denotes the reactor temperature and x_ψ the coolant temperature. The rate coefficients k_1, k_2, and k_3 depend exponentially on the reactor temperature x_ϑ via Arrhenius' law

$$k_i = k_{0_i} e^{\frac{-E_{Ai}}{x_\vartheta + 273.15}}, \ (i = 1, 2, 3).$$

Furthermore, F, V, ϱ, C_P, and $H_{AB,AD,BC}$ are the normalized process stream inflow, volume flow, the density, the heat capacity and the reaction enthalpy, respectively. The model parameters and main operating point of the CSTR (2.82)–(2.84) are given in Table 2.3.

Table 2.3: Model parameters and Main operating point.

Model parameters	
$k_{0_{1,2}} = 1.287 \times 10^{12}\ \mathrm{h}^{-1}$	$k_{0_3} = 9.043 \times 10^{9}\ 1/\mathrm{mol}$
$E_{A_{1,2}}/R = 9758.3\ \mathrm{K}$	$E_{A_3}/R = 8560.0\ \mathrm{K}$
$H_{AB} = 4.2\ \mathrm{kJ/mol}$	$H_{AD} = -41.85\ \mathrm{kJ/mol}$
$H_{BC} = -11.0\ \mathrm{kJ/mol}$	$\frac{F}{V} = 18.83\ \mathrm{h}^{-1}$
$\varrho = 0.9342 \times 10^{-4}\ \mathrm{kg/l}$	$C_p = 3.01\ \mathrm{kJ/kgK}$
$m_K = 0.9342 \times 10^{-4}\ \mathrm{kg/l}$	$C_{pK} = 2.0\ \mathrm{kJ/kgK}$
$A_R = 0.215\ \mathrm{m}^2$	$k_W = 4032\ \mathrm{kJ/kgK}$
Main operating point	
$x_{A_s} = 1.235\ mol/l$	$x_{B_s} = 0.9\ mol/l$
$x_{\vartheta_s} = 134.14\ ^0C$	$x_{\psi_s} = 128.95\ ^0C$

Letting $x(t) = \text{col}\{x_A(t), x_B(t), x_\vartheta(t), x_\psi(t)\}$, the linearized state-space model of the CSTR (2.82)–(2.84) near the operating point $x_s = \text{col}\{x_{As}, x_{Bs}, x_{\vartheta s}, x_{\psi s}\}$ is given by

$$\dot{x}(t) = \mathscr{A}x(t) + \mathscr{B}\omega(t) \tag{2.86}$$

where $\omega(t)$ is the exogenous disturbance signal belonging to $L_2[0,\infty)$, $\mathscr{B} = [0\ 0\ 1\ 0]^T$, and the linearized system matrix $\mathscr{A}$ is obtained by

$$\mathscr{A} = \begin{bmatrix} a_{11} & 0 & a_{13} & 0 \\ k_1 & -\frac{F}{V} - k_2 & a_{23} & 0 \\ a_{31} & -\frac{k_2 H_{BC}}{\varrho C_P} & a_{33} & \frac{k_W A_R}{\varrho C_P V} \\ 0 & 0 & \frac{k_W A_R}{m_K C_{PK}} & -\frac{k_W A_R}{m_K C_{PK}} \end{bmatrix}_{|x_s}$$

$$= \begin{bmatrix} -86.0948 & 0 & -4.2075 & 0 \\ 50.6138 & -69.4438 & 0.9974 & 0 \\ 172.2205 & 197.9984 & -36.7414 & 30.7978 \\ 0 & 0 & 86.6880 & -86.6880 \end{bmatrix}$$

where

$$a_{11} = -\frac{F}{V} - k_1 - 2k_3 x_A,\ a_{13} = -\frac{E_{A1}k_1 x_A + E_{A3}k_3 x_A^2}{R(x_\vartheta + 273.15)^2},$$

$$a_{23} = \frac{E_{A1}k_1 x_A - E_{A2}k_2 x_B}{R(x_\vartheta + 273.15)^2},\ a_{31} = -\frac{k_1 H_{AB} + 2k_3 x_A H_{AD}}{\varrho C_P},$$

$$a_{33} = -\frac{E_{A1}k_1 x_A H_{AB} + E_{A2}k_2 x_B H_{BC}}{R\varrho C_P(x_\vartheta + 273.15)^2} - \frac{E_{A3}k_3 x_A^2 H_{AD}}{R\varrho C_P(x_\vartheta + 273.15)^2}\cdots$$
$$-\frac{F}{V} - \frac{k_W A_R}{\varrho C_P V}.$$

For the CSTR, it is usually expensive to know the concentration of the desired product x_B by using traditional chemical approaches. An alternative yet inexpensive approach is to use signal processing approaches to estimate the concentration. In this example, only the measurements of the reactor temperature x_ϑ are used to estimate the concentration of the desired product. Here, the estimation is performed by a remote estimator collecting the measurement information through a wireless channel. Wireless sensors are deployed to monitor the reactor temperature with $C = [0\ 0\ 1\ 0]$, $D = [0\ 1\ 0\ 0]$, $B_2 = 1$, and $B_3 = 1$. Discretizing system (2.86) with period $h = 0.05$min, one obtains

$$A = \begin{bmatrix} -0.0752 & -0.0700 & -0.0226 & -0.0088 \\ 0.0228 & 0.0050 & -0.0111 & -0.0035 \\ 1.7660 & 1.3396 & 0.3882 & 0.1640 \\ 1.8159 & 1.4322 & 0.4618 & 0.1938 \end{bmatrix},$$

$$B_1 = \begin{bmatrix} -0.0025 & 0.0135 & 0.0671 & 0.0506 \end{bmatrix}.$$

The exogenous disturbance inputs are selected as

$$\omega(t_k) = \quad 0.2\cos(5t_k),\ \xi(t_k) = 0.2\mathrm{e}^{-0.1t_k}\sin(2t_k),\ \nu(t_k) = \tfrac{\sin(1.2t_k)}{0.2t_k+0.8}. \quad (2.87)$$

Due to the multi-path transmission and shadowing problem, channel fading usually takes place in the form of (2.47). Assume that the order of the fading model is $\ell = 1$ and the probability density functions of channel coefficients are

$$\begin{cases} \psi(\vartheta_0) = 0.0005(\mathrm{e}^{9.89\vartheta_0} - 1), & 0 \le \vartheta_0 \le 1; \\ \psi(\vartheta_1) = \begin{cases} 10\vartheta_1, \\ -2.50(\vartheta_1 - 1), \end{cases} & \begin{matrix} 0 \le \vartheta_1 \le 0.20; \\ 0.20 < \vartheta_1 \le 1. \end{matrix} \end{cases}$$

The mathematical expectations $\bar{\alpha}_s$ $(s = 0, 1)$ can be calculated as 0.8991 and 0.4000, and the variances $\tilde{\bar{\alpha}}_s^2$ $(s = 0, 1)$ are 0.0133 and 0.0467, respectively.

On the other hand, in order to reduce the network resource occupancy while maintaining the guaranteed filtering performance, the event-based wireless relay communication mechanism (2.48) is also adopted. The objective here is to design the event-based filter (2.51) for networked multi-rate systems (2.42)-(2.44) with fading measurements (2.47) so that the augmented system (2.52) is stochastically finite-time bounded with respect to (c_1, c_2, R, N, μ) and simultaneously satisfies the finite-time H_∞ performance constraint (2.54). In this numerical example, choose the event-weighted matrix as $\Omega = I$ and the finite-time bounded-parameters as $R = \mathrm{diag}\{0.5, 0.2, 1, 0.4\}$, $N = 21$, $\beta = 1.02$. The initial state and its estimation are taken as $x(T_0) = \mathrm{col}\{0.02, -0.03, -0.02, 0.01\}$, $\hat{x}(T_0) = \mathrm{col}\{-0.05, 0.02, -0.01, 0.02\}$, respectively. According to Definition 2.2 and above parameters, we have $\bar{c}_1 = \mathbb{E}\{\eta^T(k_0)R\eta(k_0)\}$ and $\bar{\mu} = \sum_{i=0}^{20} d^T(i)d(i)$, then actual $\bar{c}_1$ and $\bar{\mu}$ is listed in Table 2.4.

Table 2.4: The actual $\bar{c}_1$ and $\bar{\mu}$.

	$b = 2$	$b = 3$	$b = 4$
$\bar{\mu}$	6.9006	7.7001	8.4997
$\bar{c}_1$	0.004	0.0056	0.0072

Case 1. In this case, we wish to design filters to solve the problem (P1), that is, we are interested in minimizing the finite-time bounded parameter c_2 under the given $\gamma = 1.3$, $\delta = 0.2$, $c_1 = 0.01 > \bar{c}_1$ and $\mu = 9 > \bar{\mu}$. Solving the optimization problem (2.80) using LMI Toolbox yields the minimum values c_2 in Table 2.5 with different multi-rate multiple b. At the same time, the desired optimal filters can be calculated by using Theorem 2.6. Based on the obtained optimal filter parameters, the triggering instants and the evolution of $\mathbb{E}\{\eta^T(t_k)R\eta(t_k)\}$ for different multi-rate multiple b are shown in Figs. 2.5-2.6, respectively.

Case 2. Without loss of generality, for the given $\mu = 8$ and $c_2 = 7$, we now deal with the problem (P2) for $b = 2$. By solving the optimization problem (2.81),

Table 2.5: The permitted minimum c_2.

	$b=2$	$b=3$	$b=4$
$\delta=0.2$	0.7265	0.7280	0.7283

the minimum finite-time H_∞ performance level can be obtained in Table 2.6 with different initial finite-time bounded parameter c_1 and event-triggered parameter δ, and $\star$ means that there is no feasible solution.

Table 2.6: The permitted minimum γ.

	$\delta=0.1$	$\delta=0.3$	$\delta=0.5$	$\delta=0.7$
$c_1=0.01$	0.6352	1.1053	1.4350	1.7091
$c_1=0.05$	0.6431	1.1515	1.5449	1.9106
$c_1=0.1$	0.6612	1.2674	1.9118	$\star$

If we assume $c_1=0.1$, $\delta=0.1$ and $\gamma=1.5$, then the filters can be designed by Matlab as follows:

$$\begin{aligned} H_1 &= \text{col}\{0.0001,-0.0007,-0.0019,0.0021\},\\ H_2 &= \text{col}\{-0.0010,-0.0093,-0.0233,0.0262\}. \end{aligned}$$

The simulation results are shown in Figures 2.7 and 2.8. Figure 2.7 shows the output $z(t_k)$ and its estimation $\hat{z}(t_k)$, and the filtering error $\tilde{z}(t_k)$ is drawn in Figure 2.8. The simulation results have confirmed that the filter performs very well.

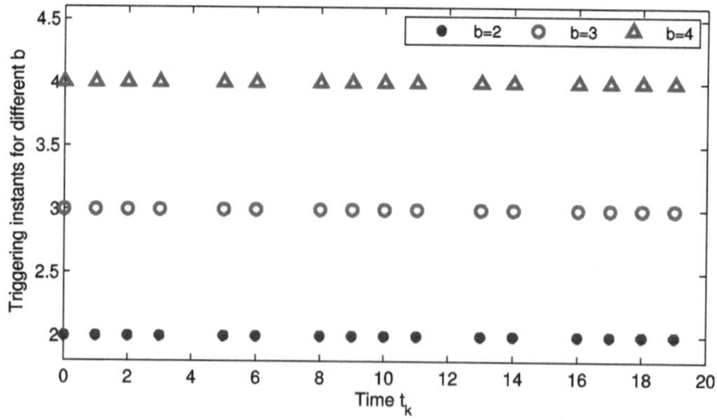

Figure 2.5: The triggering instants for networked multi-rate systems with $\delta=0.2$.

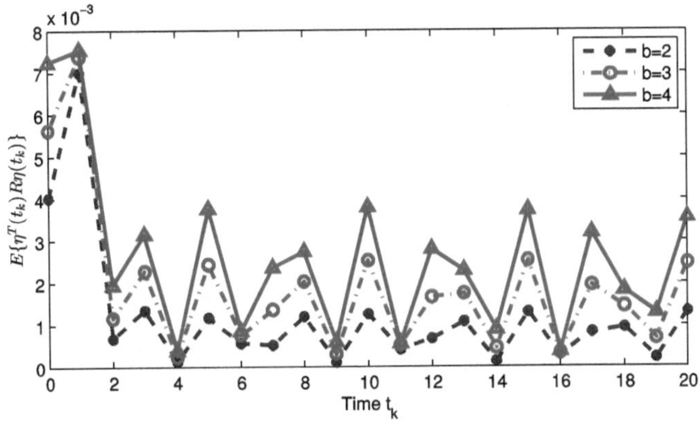

Figure 2.6: The trajectory of $\mathbb{E}\{\eta^T(t_k)R\eta(t_k)\}$ for networked multi-rate systems with $\delta = 0.2$.

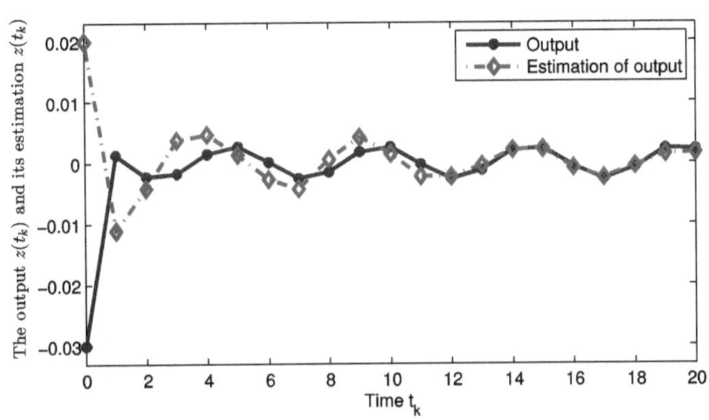

Figure 2.7: The output $\bar{z}(t_k)$ and its estimation $\hat{z}(t_k)$ for networked multi-rate systems with $\delta = 0.1$.

Remark 2.8. *For the given fading channel number ℓ, Tables 2.5 and 2.6 demonstrate the quantitative relationship between H_∞ performance level γ, finite-time bounded-parameters c_1, c_2 and event-triggered parameter δ as well as the multi-rate multiple b of the sampling period h. It can be observed from Tables 2.5 and 2.6 that, with the increased δ or b, the disturbance attenuation performance deteriorates or the finite-time boundedness become bigger, which are in agreement with the engineering practice.*

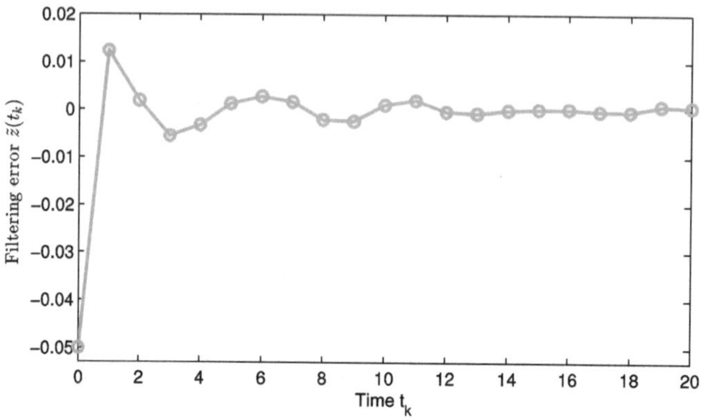

Figure 2.8: Filtering error $\tilde{z}(t_k)$ for networked multi-rate systems with $\delta = 0.1$.

2.3 Conclusion

In this chapter, both variance-constrained H_∞ estimator and finite-time filter have been designed for networked multi-rate systems. The system under consideration involves network-induced probabilistic sensor failures, signal quantization, fading measurements, and event-triggered communication mechanisms. With the help of lifting technique, the augmented multi-rate systems models have been established, and sufficient conditions have been achieved in the form of matrix inequalities by utilizing stochastic analysis techniques. The obtained results have revealed the relationship among the H_∞ performance index, finite-time bounded parameters, event-triggered threshold, fading channel parameters, multi-rate multiples, and other system or network parameters. Finally, several simulation examples have been given to verify the effectiveness of the proposed analysis and design schemes. It is worth noting that the design approaches in this chapter provide support for improving the state estimation technology of cyber physical systems.

3

Fault Detection of Networked Multi-rate Systems with Filter-Based Methods

Networked control systems (NCSs), which have been successfully applied to a wide range of field, have attracted persistently increasing research interests [135]. Up to now, a rich body of literatures focusing on dealing with the estimation, filtering and control problems for various networked control systems against network-related phenomena have been published [122, 136]. Recently, the fault detection problems of networked control systems [12, 137–139] have attracted a great deal of research attention. The T-S fuzzy-model-based fault detection problem for networked control systems with Markov delays has been studied in [140]. In [141], an fault detection framework has been proposed for a class of nonlinear networked control systems via a shared communication medium.

The single-rate sampled-data setting is an underlying assumption for most available literature concerning networked control systems. However, in real applications, especially when it comes to large-scale networked systems, system components may be structured distributively, which means that the sensors, actuators, and controllers are connected by communication networks. For such kind of networked control systems, faster analogue-to-digital (A/D) and digital-to-analogue (D/A) conversions would contribute to better performance. However, higher implementation cost of faster A/D and D/A conversions is a serious consideration owing to the limited economic budget. In this case, the scheme of multi-rate sampled-data, which allows different speeds for A/D and D/A conversions and achieves a satisfactory balance between the performance and implementation cost, arises naturally and have become a research focus for many years, see [142–144]. The fault detection problem of multi-rate sampled-data systems has been investigated in [145, 146]. To date, the fault detection problem has not been adequately examined for networked multi-rate systems, not to mention the cases when fading measurements, dynamic quantization, missing measurements and randomly occurring faults are simultaneously presented in nonuniform sampling scenario.

Inspired by the above discussion, we propose the reasonable fault detection methods for networked multi-rate systems. The rest of this chapter is outlined as follows. Section 3.1 is concerned with the fault detection problem with fading measurements and randomly occurring faults. In Section 3.2, under the influence of network-induced phenomena and dynamic quantization, fault detection filters with nonuniform sampling are constructed. Section 3.3 gives our conclusions.

DOI: 10.1201/9781003330998-3

3.1 Fault Detection with Fading Measurements and Randomly Occurring Faults

In this section, we aim to investigate the fault detection problem for a class of networked multi-rate systems with fading measurements and randomly occurring faults.

3.1.1 Problem Formulation

Consider the following class of discrete-time systems with randomly occurring faults:

$$x(T_{k+1}) = Ax(T_k) + B_1\omega(T_k) + B_2 f(T_k) \tag{3.1}$$

$$y(t_k) = Cx(t_k), \quad k = 0, 1, 2, \cdots \tag{3.2}$$

where $x(T_k) \in \mathbb{R}^{n_x}$ represents the state vector, $y(t_k) \in \mathbb{R}^{n_y}$ is the ideal measurement, $\omega(T_k) \in \mathbb{R}^{n_\omega}$ is the disturbance input which belongs to $\ell_2[0, \infty)$, and $f(T_k) \in \mathbb{R}^{n_f}$ is the fault signal to be detected. A, B_1, B_2, and C are constant matrices with appropriate dimensions.

The sampling period of system (3.1) is denoted by $h \triangleq T_{k+1} - T_k$. For simplicity, it is assumed that the measurement period is integer multiples of the system (3.2), that is $t_{k+1} - t_k \triangleq bh$, where b is a positive integer.

In comparison with the wired networked control systems, the wireless networked control systems are susceptible to fading effect because of multipath propagation or shadowing from obstacles affecting the wave propagation [147]. In this section, the actually received measurement signal with probabilistic fading channels is described by

$$\bar{y}(t_k) = \sum_{s=0}^{\ell(t_k)} \beta_s(t_k) y(t_k - sbh) \tag{3.3}$$

where $\ell(t_k) = \min\{\ell, \lfloor \frac{t_k}{bh} \rfloor\}$ with ℓ being a given positive scalar denoting the number of paths. $\bar{y}(t_k) \in \mathbb{R}^{n_y}$ is the measurement output through fading channels. $\beta_s(t_k)$ $(s = 0, 1, ..., \ell(t_k))$ are assumed to be mutually independent channel coefficients having probability density functions on the interval $[0, 1]$ with known mathematical expectations $\bar{\beta}_s$ and variances $\tilde{\bar{\beta}}_s^2$.

Remark 3.1. *In a networked environment, the faults could occur in a random way due to a variety of reasons such as limited bandwidth of the communication channels, random fluctuation of the network load, unreliability of the wireless links with large distances, as well as the fading measurement signals. The network-induced fault can be modelled in (3.1) whose probability distribution information can be specified a prior through statistical tests. Note that both the time delays and packet dropouts can be described by this kind of fading model.*

It can be seen that (3.1) evolves with a constant period h, while the fading measurement dynamics (3.3) is generated with a slower period bh. Accordingly, (3.1) and

(3.3) are essentially multi-rate sampled-data system models. Note that it is mathematically difficult to handle the fault detection problem directly for such kind of multi-rate sampled-data system. In the next section, we are going to convert the resulting multi-rate sampled-data system into a single-rate system for technical convenience.

By applying the relation (3.1) recursively, one obtains a new system with time scale t_k as follows:

$$\begin{aligned} x(t_{k+1}) &= A^b x(t_k) + \sum_{i=0}^{b-1} A^{b-1-i} B_1 \omega(t_k + ih) + \\ &\quad \sum_{i=0}^{b-1} A^{b-1-i} B_2 f(t_k + ih) \end{aligned} \tag{3.4}$$

where

$$\begin{aligned} \bar{\omega}(t_k) &\triangleq \text{col}\{\omega(t_k), \omega(t_k + h), \cdots, \omega(t_k + (b-1)h)\}, \\ \bar{f}(t_k) &\triangleq \text{col}\{f(t_k), f(t_k + h), \cdots, f(t_k + (b-1)h)\}, \\ \tilde{\alpha}(t_k + ih) &\triangleq \alpha(t_k + ih) - \bar{\alpha} \ (i = 0, 1, 2, \cdots, b-1), \\ \bar{A}_{11} &\triangleq [A^{b-1}B_1 \ A^{b-2}B_1 \cdots AB_1 \ B_1], \\ \bar{A}_{12} &\triangleq [\bar{\alpha}A^{b-1}B_2 \ \bar{\alpha}A^{b-2}B_2 \cdots \bar{\alpha}AB_2 \ \bar{\alpha}B_2]. \end{aligned}$$

Consider the following observer-based fault detection filter

$$\begin{cases} \hat{x}(t_{k+1}) = A^b \hat{x}(t_k) + L(y(t_k) - \hat{y}(t_k)) \\ r(t_k) = V(y(t_k) - \hat{y}(t_k)) \\ \hat{y}(t_k) = \bar{\Xi} C \hat{x}(t_k) \end{cases} \tag{3.5}$$

where $\hat{x}(t_k) \in \mathbb{R}^{n_{\hat{x}}}$ is the estimated state, $r(t_k) \in \mathbb{R}^{n_r}$ is the residual that is compatible with the fault vector, and the L and V are the appropriately dimensioned fault detection filter gain matrices to be designed. In our present work, it is intended to make the error between the residual signal $r(t_k)$ and the fault signal $f(t_k)$ as small as possible in H_∞ framework.

Letting $e(t_k) \triangleq x(t_k) - \hat{x}(t_k)$, $\bar{x}(t_k) \triangleq \text{col}\{x(t_k), x(t_k - h), \cdots, x(t_k - (b-1)h)\}$ and $\tilde{\beta}_s(t_k) \triangleq \beta_s(t_k) - \bar{\beta}_s$, the error dynamics for the fault detection filter can be obtained from (3.4)-(3.5) and Assumption 3.2 as follows:

$$\begin{cases} e(t_{k+1}) = (A^b - LC)e(t_k) + \bar{A}_{11}\bar{\omega}(t_k) + \bar{A}_{12}\bar{f}(t_k) + LCx(t_k) \\ \qquad - \sum_{s=0}^{\ell} \tilde{\beta}_s(t_k) LCx(t_k - sbh) - \sum_{s=0}^{\ell} \bar{\beta}_s LCx(t_k - sbh) \\ \qquad + \sum_{i=0}^{b-1} \tilde{\alpha}(t_k + ih) A^{b-1-i} B_2 f(t_k + ih) \\ r(t_k) = VCe(t_k) - VCx(t_k) + \sum_{s=0}^{\ell} \bar{\beta}_s VCx(t_k - sbh) \\ \qquad + \sum_{s=0}^{\ell} \tilde{\beta}_s(t_k) VCx(t_k - sbh) \end{cases} \tag{3.6}$$

On the other hand, with similar procedure for obtaining (3.4), we have

$$\begin{cases} x(t_{k+1}-h) = A^{b-1}x(t_k) + \bar{A}_{21}\bar{\omega}(t_k) + \bar{A}_{22}\bar{f}(t_k) \\ \qquad + \sum_{i=0}^{b-2} \tilde{\alpha}(t_k+ih)A^{b-2-i}B_2 f(t_k+ih) \\ \qquad \cdots \qquad \cdots \\ x(t_{k+1}-(b-1)h) = Ax(t_k) + \bar{A}_{b1}\bar{\omega}(t_k) + \bar{A}_{b2}\bar{f}(t_k) + \tilde{\alpha}(t_k)B_2 f(t_k) \end{cases} \tag{3.7}$$

where

$$\begin{aligned} \bar{A}_{21} &\triangleq [A^{b-2}B_1 \ A^{b-3}B_1 \cdots B_1 \ 0], \\ \bar{A}_{(b-1)1} &\triangleq [AB_1 \ B_1 \cdots 0 \ 0], \\ \bar{A}_{b1} &\triangleq [B_1 \ 0 \cdots 0 \ 0], \\ \bar{A}_{22} &\triangleq [\bar{\alpha}A^{b-2}B_2 \ \bar{\alpha}A^{b-3}B_2 \cdots \bar{\alpha}B_2 \ 0], \\ \bar{A}_{(b-1)2} &\triangleq [\bar{\alpha}AB_2 \ \bar{\alpha}B_2 \cdots 0 \ 0], \\ \bar{A}_{b2} &\triangleq [\bar{\alpha}B_2 \ 0 \cdots 0 \ 0]. \end{aligned}$$

For convenience of later analysis, we denote

$$\begin{aligned} \eta(t_k) &\triangleq \text{col}\{e(t_k), \bar{x}(t_k), \bar{x}(t_k-bh), \cdots, \bar{x}(t_k-\ell bh)\}, \\ r_e(t_k) &\triangleq r(t_k) - f(t_k), \\ \mathcal{I} &\triangleq \text{col}\Big\{I, \underbrace{0, \cdots, 0}_{(\ell+1)b}\Big\}, \\ \bar{A} &\triangleq \text{col}\{(1-\bar{\beta}_0)LC, A^b, A^{b-1}, \cdots, A, \underbrace{0, \cdots, 0}_{\ell b}\}, \\ \bar{B}_1 &\triangleq \text{col}\Big\{A^{b-1}B_2, \underbrace{A^{b-1}B_2, A^{b-2}B_2, \cdots, AB_2, B_2}_{b}, \underbrace{0, \cdots, 0}_{\ell b}\Big\}, \\ \bar{B}_2 &\triangleq \text{col}\Big\{A^{b-2}B_2, \underbrace{A^{b-2}B_2, A^{b-3}B_2, \cdots, B_2, 0}_{b}, \underbrace{0, \cdots, 0}_{\ell b}\Big\}, \cdots, \\ \bar{B}_b &\triangleq \text{col}\{B_2, \underbrace{B_2, 0, \cdots, 0, 0}_{b}, \underbrace{0, \cdots, 0}_{\ell b}\}. \end{aligned}$$

Then, by using the lifting technique, the augmented system resulting from (3.4), (3.6), and (3.7) can be written as

$$\begin{cases} \eta(t_{k+1}) = \Big(\mathcal{A} + \sum_{s=0}^{\ell} \tilde{\beta}_s(t_k)\tilde{\mathcal{A}}_s\Big)\eta(t_k) + \mathcal{B}_1\bar{\omega}(t_k) + \ldots \\ \qquad \Big(\mathcal{D} + \sum_{i=0}^{b-1} \tilde{\alpha}(t_k+ih)\tilde{\mathcal{D}}_i\Big)\bar{f}(t_k) \\ r_e(t_k) = \Big(\mathcal{C} + \sum_{s=0}^{\ell} \tilde{\beta}_s(t_k)\tilde{\mathcal{C}}_s\Big)\eta(t_k) + \mathcal{B}_2\bar{f}(t_k) \end{cases} \tag{3.8}$$

where

$$\mathcal{A} \triangleq [(A^b - LC)\mathcal{I}\ \bar{A}\ -\bar{\beta}_1 LC\mathcal{I}\ -\bar{\beta}_2 LC\mathcal{I}\ \cdots\ -\bar{\beta}_\ell LC\mathcal{I}\ \underbrace{0\ \cdots\ 0}_{\ell b}],$$

$$\tilde{\mathcal{A}}_s \triangleq [\underbrace{0\ \cdots\ 0}_{s+1}\ LC\mathcal{I}\ \underbrace{0\ \cdots\ 0}_{(\ell+1)b-s-1}],$$

$$\mathcal{B}_1 \triangleq \mathrm{col}\{\bar{A}_{11}, \bar{A}_{11}, \bar{A}_{21}, \cdots, \bar{A}_{(b-1)1}, \bar{A}_{b1}, \underbrace{0, \cdots, 0}_{\ell b}\},$$

$$\mathcal{D} \triangleq \mathrm{col}\{\bar{A}_{12}, \bar{A}_{12}, \bar{A}_{22}, \cdots, \bar{A}_{(b-1)2}, \bar{A}_{b2}, \underbrace{0, \cdots, 0}_{\ell b}\},$$

$$\tilde{\mathcal{D}}_i \triangleq [\underbrace{0\ \cdots\ 0}_{i}\ \bar{B}_{i+1}\ \underbrace{0\ \cdots\ 0}_{b-i-1}],$$

$$\mathcal{C} \triangleq [VC\ -(1-\bar{\beta}_0)VC\ \bar{\beta}_1 VC\ \bar{\beta}_2 VC\ \cdots\ \bar{\beta}_\ell VC\ \underbrace{0\ \cdots\ 0}_{\ell b}],$$

$$\tilde{\mathcal{C}}_s \triangleq [\underbrace{0\ \cdots\ 0}_{s+1}\ VC\ \underbrace{0\ \cdots\ 0}_{(\ell+1)b-s-1}],\ \mathcal{B}_2 \triangleq [-I\ \underbrace{0\ \cdots\ 0}_{b-1}],$$

$$(s = 0, 1, \cdots, \ell;\ i = 0, 1, \cdots b-1).$$

Remark 3.2. *By using the lifting technique, the model (3.8) for networked multi-rate systems is obtained. Comparing with the fault detection models of the multi-rate sampled-data system in [145, 146], the model (3.8) exhibits two distinguished features: i) both the fading measurements and randomly occurring faults are considered and therefore the model (3.8) is quite comprehensive to better reflect the networked environment; ii) the introduction of the stochastic coefficients in model (3.3) results in significant delays in the overall dynamics governed by (3.8). Note that the communication delay issues have not been considered in [145, 146].*

Before proceeding further, we introduce the following definition.

Definition 3.1. *The augmented system (3.8) is said to be exponentially mean-square stable if, with $\bar{\omega}(t_k) = 0$ and $\bar{f}(t_k) = 0$, there exist scalars $\delta > 0$ and $\varrho \in (0, 1)$ such that*

$$\mathbb{E}\{\|\eta(t_k)\|^2\} \leq \delta\varrho^{t_k}\mathbb{E}\{\|\eta(t_0)\|^2\}, \quad \forall \eta(t_0) \in \mathbb{R}^{(b+1)n_x}$$

The observer-based fault detection filters is designed to simultaneously meet the following requirements:

(a) the augmented system (3.8) is exponentially mean-square stable;

(b) under the zero-initial condition, the error $r_e(t_k)$ between the residual and the fault estimate satisfies

$$\sum_{k=0}^{\infty}\mathbb{E}\{\|r_e(t_k)\|^2\} < \gamma^2\sum_{k=0}^{\infty}(\|\bar{\omega}(t_k)\|^2 + \|\bar{f}(t_k)\|^2) \tag{3.9}$$

for any nonzero $\bar{\omega}(t_k)$ or $\bar{f}(t_k)$, where scalar $\gamma > 0$ is a given disturbance attenuation level.

For the fault detection purpose, we adopt the threshold J_{th} and the residual evaluation function $J(t_k)$ as follows:

$$J(t_k) = \left\{ \sum_{h=t_{k0}}^{t_k} r^T(h)r(h) \right\}^{\frac{1}{2}}, \quad J_{th} = \sup_{\substack{\bar{\omega}(t_k)\in\ell_2 \\ \bar{f}(t_k)=0}} \mathbb{E}\{J(t_k)\}$$

where t_{k0} denotes the initial evaluation time instant and $t_k - t_{k0}$ denotes the evaluation time steps.

The occurrence of faults can be detected by comparing $J(t_k)$ with J_{th} according to the following test rule:

$$\begin{cases} J(t_k) \geq J_{th} \Longrightarrow \text{alarm for fault} \\ J(t_k) < J_{th} \Longrightarrow \text{no fault} \end{cases} \tag{3.10}$$

Remark 3.3. *As discussed in [148], depending on the type of the system under consideration, there exist two residual evaluation strategies, that is the statistic testing and norm-based residual evaluation. For the norm-based residual evaluation, the well-established robust control theory can be used to compute the threshold, therefore, it is widely adopted. On the other hand, from the engineering viewpoint, the determination of a threshold is to find out the tolerant limit for disturbances and model uncertainties under fault-free operation conditions. There are some factors such as the dynamics of the residual generator as well as the bounds of the unknown inputs and model uncertainties, they all significantly influence this procedure. As a result, false alarm and missed detection are two common phenomenon in fault diagnosis.*

3.1.2 Detection of Randomly Occurring Faults

By resorting to the stochastic analysis techniques, we shall provide the H_∞ performance analysis result for the augmented system (3.8) and then proceed with the subsequent fault detection filter design stage.

Theorem 3.1. *Let the disturbance attenuation level $\gamma > 0$ and the fault detection filter parameters L and V be given. The augmented system (3.8) is exponentially mean-square stable while achieving the H_∞ performance constraint (3.9) if there exists matrix P such that the following matrix inequality holds:*

$$\hat{\Phi} \triangleq \begin{bmatrix} \bar{\Phi}_{11} & \bar{\Phi}_{12} & \bar{\Phi}_{13} \\ * & -I & 0 \\ * & * & -I \end{bmatrix} < 0 \tag{3.11}$$

where

$$\bar{\Phi}_{11} \triangleq \begin{bmatrix} \Gamma & \mathcal{A}^T P \mathcal{B}_1 & \mathcal{A}^T P \mathcal{D} + \mathcal{C}^T \mathcal{B}_2 \\ * & \mathcal{B}_1^T P \mathcal{B}_1 - \gamma^2 I & \mathcal{B}_1^T P \mathcal{D} \\ * & * & \Phi_{33} \end{bmatrix},$$

$$\begin{aligned}
\Gamma &\triangleq \sum_{s=0}^{\ell} \bar{\tilde{\beta}}_s^2 \tilde{\mathcal{A}}_s^T P \tilde{\mathcal{A}}_s + \mathcal{A}^T P \mathcal{A} - P, \ \bar{\Phi}_{12} \triangleq \mathrm{col}\{\mathcal{C}^T, 0, 0\}, \\
\bar{\Phi}_{13} &\triangleq \mathrm{col}\{\hat{\tilde{\mathcal{C}}}^T, 0, 0\}, \ \hat{\tilde{\mathcal{C}}}^T \triangleq [\bar{\tilde{\beta}}_0 \tilde{\mathcal{C}}_0^T \ \bar{\tilde{\beta}}_1 \tilde{\mathcal{C}}_1^T \ \cdots \ \bar{\tilde{\beta}}_\ell \tilde{\mathcal{C}}_\ell^T], \\
\Phi_{33} &\triangleq \sum_{i=0}^{b-1} \breve{\alpha}^2 \tilde{\mathcal{D}}_i^T P \tilde{\mathcal{D}}_i + \mathcal{D}^T P \mathcal{D} + \mathcal{B}_2^T \mathcal{B}_2 - \gamma^2 I.
\end{aligned}$$

Proof. Choose the following Lyapunov function:

$$V(\eta(t_k)) = \eta^T(t_k) P \eta(t_k) \tag{3.12}$$

By calculating the difference of $V(\eta(t_k))$ along the trajectory of the augmented system (3.8) with $\bar{\omega}(t_k) = 0$ and $\bar{f}(t_k) = 0$, and taking the mathematical expectation, one has

$$\begin{aligned}
&\mathbb{E}(\Delta V(\eta(t_k))) \\
= \ &\mathbb{E}\{\eta^T(t_{k+1}) P \eta(t_{k+1}) - \eta^T(t_k) P \eta(t_k)\} \\
= \ &\mathbb{E}\{\eta^T(t_k)((\mathcal{A} + \sum_{s=0}^{\ell} \tilde{\beta}_s(t_k) \tilde{\mathcal{A}}_s)^T P (\mathcal{A} + \sum_{s=0}^{\ell} \tilde{\beta}_s(t_k) \tilde{\mathcal{A}}_s) - P) \eta(t_k)\} \\
= \ &\eta^T(t_k) \Big(\mathcal{A}^T P \mathcal{A} - P + \sum_{s=0}^{\ell} \tilde{\beta}_s^2 \tilde{\mathcal{A}}_s^T P \tilde{\mathcal{A}}_s \Big) \eta(t_k) \\
= \ &\eta^T(t_k) \Gamma \eta(t_k)
\end{aligned} \tag{3.13}$$

It follows from (3.11) that $\Gamma < 0$ and, subsequently,

$$\mathbb{E}\big(\Delta V(\eta(t_k))\big) \le -\lambda_{\min}(-\Gamma) \|\eta(t_k)\|^2$$

By following the similar analysis in [125], the augmented system (3.8) is exponentially mean-square stable.

Finally, let us consider the H_∞ performance of the overall estimation dynamics. For this purpose, we introduce the following index:

$$J_n \triangleq \mathbb{E}\Big\{ \sum_{k=0}^{n} \|r_e(t_k)\|^2 - \sum_{k=0}^{n} \gamma^2 (\|\bar{\omega}(t_k)\|^2 + \|\bar{f}(t_k)\|^2) \Big\} \tag{3.14}$$

Under the zero-initial condition, it follows from (3.14) that

$$\begin{aligned}
J_n &\triangleq \mathbb{E}\Big\{ \sum_{k=0}^{n} \|r_e(t_k)\|^2 - \sum_{k=0}^{n} \gamma^2 (\|\bar{\omega}(t_k)\|^2 + \|\bar{f}(t_k)\|^2) \Big\} \\
&\le \sum_{k=0}^{n} \mathbb{E}\Big\{ \|r_e(t_k)\|^2 - \gamma^2 (\|\bar{\omega}(t_k)\|^2 + \|\bar{f}(t_k)\|^2) + \Delta V(\eta(t_k)) \Big\} - ... \\
&\quad \mathbb{E}\{V(\eta(t_{n+1}))\}
\end{aligned}$$

$$
\begin{aligned}
&\leq \sum_{k=0}^{n}\mathbb{E}\Big\{\|r_e(t_k)\|^2-\gamma^2(\|\bar{\omega}(t_k)\|^2+\|\bar{f}(t_k)\|^2)+\Delta V(\eta(t_k))\Big\}\\
&= \sum_{k=0}^{n}\Big\{\eta^T(t_k)[\sum_{s=0}^{\ell}\tilde{\bar{\beta}}_s^2\tilde{\mathcal{A}}_s^T P\tilde{\mathcal{A}}_s+\mathcal{A}^T P\mathcal{A}+\sum_{s=0}^{\ell}\tilde{\bar{\beta}}_s^2\tilde{\mathcal{C}}_s^T\tilde{\mathcal{C}}_s\\
&\quad +\mathcal{C}^T\mathcal{C}-P]\eta(t_k)+2\eta^T(t_k)[\mathcal{A}^T P\mathcal{D}+\mathcal{C}^T\mathcal{B}_2]\bar{f}(t_k)\\
&\quad +2\eta^T(t_k)\mathcal{A}^T P\mathcal{B}_1\bar{\omega}(t_k)+2\bar{\omega}^T(t_k)\mathcal{B}_1^T P\mathcal{D}\bar{f}(t_k)\\
&\quad +\bar{f}^T(t_k)[\sum_{i=0}^{b-1}\breve{\alpha}^2\tilde{\mathcal{D}}_i^T P\tilde{\mathcal{D}}_i+\mathcal{D}^T P\mathcal{D}+\mathcal{B}_2^T\mathcal{B}_2-\gamma^2 I]\bar{f}(t_k)\\
&\quad +\bar{\omega}^T(t_k)[\mathcal{B}_1^T P\mathcal{B}_1-\gamma^2 I]\bar{\omega}(t_k)\Big\}\\
&= \sum_{k=0}^{n}\Big\{\vartheta^T(t_k)\Phi\vartheta(t_k)\Big\}\\
&= \sum_{k=0}^{n}\Big\{\vartheta^T(t_k)(\bar{\Phi}_{11}+\tilde{\Phi})\vartheta(t_k)\Big\}
\end{aligned}
\tag{3.15}
$$

where

$$
\vartheta(t_k)\triangleq \mathrm{col}\{\eta(t_k),\bar{\omega}(t_k),\bar{f}(t_k)\},\ \mathbb{E}\{\tilde{\alpha}^2(t_k+ih)\}=(\sqrt{\bar{\alpha}(1-\bar{\alpha})})^2\triangleq\breve{\alpha}^2,
$$

$$
\Phi\triangleq\bar{\Phi}_{11}+\tilde{\Phi},\ \tilde{\Phi}\triangleq\mathrm{diag}\{\sum_{s=0}^{\ell}\tilde{\bar{\beta}}_s^2\tilde{\mathcal{C}}_s^T\tilde{\mathcal{C}}_s+\mathcal{C}^T\mathcal{C},0,0\}=\bar{\Phi}_{12}\bar{\Phi}_{12}^T+\bar{\Phi}_{13}\bar{\Phi}_{13}^T.
$$

By using the Schur Complement Lemma to (3.11), we have

$$
\hat{\Phi}=\bar{\Phi}_{11}+\bar{\Phi}_{12}\bar{\Phi}_{12}^T+\bar{\Phi}_{13}\bar{\Phi}_{13}^T<0 \tag{3.16}
$$

that is $\bar{\Phi}_{11}+\tilde{\Phi}<0$, therefore, we obtain the following relation from (3.15)

$$
\mathbb{E}(\Delta V(\eta(t_k)))+\mathbb{E}(\|r_e(t_k)\|^2)-\gamma^2(\|\bar{\omega}(t_k)\|^2+\|\bar{f}(t_k)\|^2)<0 \tag{3.17}
$$

for all nonzero $\bar{\omega}(t_k)$ and $\bar{f}(t_k)$. Considering zero initial condition, the inequality (3.17) implies that

$$
\sum_{k=0}^{n}\mathbb{E}\{\|r_e(t_k)\|^2\}<\gamma^2\sum_{k=0}^{n}(\|\bar{\omega}(t_k)\|^2+\|\bar{f}(t_k)\|^2)
$$

Letting $n\to\infty$, it follows from the aforementioned inequality that

$$
\sum_{k=0}^{\infty}\mathbb{E}\{\|r_e(t_k)\|^2\}<\gamma^2\sum_{k=0}^{\infty}(\|\bar{\omega}(t_k)\|^2+\|\bar{f}(t_k)\|^2)
$$

which is (3.9). The proof is now complete. □

Having established the analysis results, we are now ready to deal with the filter design problem. In the following theorem, a sufficient condition is provided for the existence of the desired H_∞ multi-rate fault detection filter. For the technical convenience, we denote

$$
\begin{aligned}
\bar{\mathcal{A}}_{10}^T &\triangleq \text{col}\Big\{-\bar{\beta}_1 C^T\bar{L}^T, -\bar{\beta}_2 C^T\bar{L}^T, \cdots, -\bar{\beta}_\ell C^T\bar{L}^T\Big\},\\
\bar{\mathcal{A}}_1^T &\triangleq \text{col}\Big\{(A^b)^T P_1 - C^T\bar{L}^T, (1-\bar{\beta}_0)C^T\bar{L}^T, \bar{\mathcal{A}}_{10}^T, \underbrace{0,\cdots,0}_{\ell b}\Big\},\\
\bar{\mathcal{A}}_i^T &\triangleq \text{col}\Big\{(A^{b+2-i})^T P_i, \underbrace{0,\cdots,0}_{(\ell+1)b}\Big\}, \hat{\mathcal{A}}^T \triangleq \Big[\bar{\mathcal{A}}_1^T\ \bar{\mathcal{A}}_2^T \cdots \bar{\mathcal{A}}_{b+1}^T\ \underbrace{0\ \cdots\ 0}_{\ell b}\Big],\\
\mathcal{X}_j^T &\triangleq \text{col}\Big\{\underbrace{0,\cdots,0}_{j+1}, -C^T\bar{L}^T, \underbrace{0,\cdots,0}_{b-1-j}\Big\}, \bar{\tilde{\mathcal{A}}}_j^T \triangleq \Big[\mathcal{X}_j^T\ \underbrace{0\ \cdots\ 0}_{b}\Big],\\
\hat{P} &\triangleq \text{diag}\{P_1, P_2, \cdots, P_{(\ell+1)b+1}\}, \hat{\bar{\mathcal{A}}}^T \triangleq \Big[\bar{\bar{\beta}}_0\bar{\tilde{\mathcal{A}}}_0^T\ \bar{\bar{\beta}}_1\bar{\tilde{\mathcal{A}}}_1^T\ \cdots \bar{\bar{\beta}}_\ell\bar{\tilde{\mathcal{A}}}_\ell^T\Big],\\
\hat{\bar{\mathcal{D}}}^T &\triangleq \Big[\breve{\alpha}\tilde{\mathcal{D}}_0^T\hat{P}\ \breve{\alpha}\tilde{\mathcal{D}}_1^T\hat{P}\ \cdots \breve{\alpha}\tilde{\mathcal{D}}_{b-1}^T\hat{P}\Big], \hat{\bar{\mathcal{C}}}^T \triangleq [\bar{\bar{\beta}}_0\tilde{\mathcal{C}}_0^T\ \bar{\bar{\beta}}_1\tilde{\mathcal{C}}_1^T\ \cdots\ \bar{\bar{\beta}}_\ell\tilde{\mathcal{C}}_\ell^T],\\
&\quad (i = 2, 3, \cdots, b+1;\ j = 0, 1, 2, \cdots, \ell).
\end{aligned}
$$

Theorem 3.2. *For the given disturbance attenuation level $\gamma > 0$, the augmented system (3.8) is exponentially mean-square stable while achieving the performance constraint (3.9) for any nonzero $\bar{\omega}(t_k)$ and $\bar{f}(t_k)$ if there exist matrices $\bar{L}$, $\bar{V}$ and $P_i > 0$ $(i = 1, 2, \cdots, (\ell+1)b+1)$ such that the following linear matrix inequality (LMI) holds:*

$$
\bar{\Xi} \triangleq \begin{bmatrix} \bar{\Xi}_{11} & \Xi_{12} & \bar{\Xi}_{13} \\ * & \Xi_{22} & 0 \\ * & * & \bar{\Xi}_{33} \end{bmatrix} < 0 \tag{3.18}
$$

where

$$
\begin{aligned}
\bar{\Xi}_{11} &\triangleq \text{diag}\Big\{-\hat{P}, -\gamma^2 I, -\gamma^2 I\Big\},\\
\Xi_{12} &\triangleq \begin{bmatrix} \hat{\bar{\mathcal{C}}}^T & \mathcal{C}^T \\ 0 & 0 \\ 0 & \mathcal{B}_2^T \end{bmatrix}, \bar{\Xi}_{13} \triangleq \begin{bmatrix} \hat{\bar{\mathcal{A}}}^T & 0 & \hat{\mathcal{A}}^T \\ 0 & 0 & \mathcal{B}_1^T\hat{P} \\ 0 & \hat{\bar{\mathcal{D}}}^T & \mathcal{D}^T\hat{P} \end{bmatrix},\\
\Xi_{22} &\triangleq \text{diag}\{-I, \cdots, -I\}, \bar{\Xi}_{33} \triangleq \text{diag}\{-\hat{P}, \cdots, -\hat{P}\},
\end{aligned}
$$

and other corresponding matrices are defined in Theorem 3.1. Furthermore, if the inequality (3.18) is feasible, the desired fault detection filter gain can be determined by

$$
L = P_1^{-1}\bar{L},\ V = \bar{V}. \tag{3.19}
$$

Proof. By using the Schur Complement Lemma, (3.11) is equivalent to the following inequality:

$$\Xi = \begin{bmatrix} \Xi_{11} & \Xi_{12} & \Xi_{13} \\ * & \Xi_{22} & 0 \\ * & * & \Xi_{33} \end{bmatrix} < 0 \tag{3.20}$$

where

$$\Xi_{11} \triangleq \mathrm{diag}\{-P, -\gamma^2 I, -\gamma^2 I\},\ \Xi_{13} \triangleq \begin{bmatrix} \breve{\tilde{\mathcal{A}}}^T & 0 & \mathcal{A}^T P \\ 0 & 0 & \mathcal{B}_1^T P \\ 0 & \breve{\tilde{\mathcal{D}}}^T & \mathcal{D}^T P \end{bmatrix},$$

$$\Xi_{33} \triangleq \mathrm{diag}\{\underbrace{-P, \cdots, -P}_{b+\ell+2}\},\ \breve{\tilde{\mathcal{A}}}^T \triangleq \begin{bmatrix} \tilde{\bar{\beta}}_0 \tilde{\mathcal{A}}_0^T P & \tilde{\bar{\beta}}_1 \tilde{\mathcal{A}}_1^T P & \cdots & \tilde{\bar{\beta}}_\ell \tilde{\mathcal{A}}_\ell^T P \end{bmatrix},$$

$$\breve{\tilde{\mathcal{D}}}^T \triangleq \begin{bmatrix} \breve{\alpha} \tilde{\mathcal{D}}_0^T P & \breve{\alpha} \tilde{\mathcal{D}}_1^T P & \cdots & \breve{\alpha} \tilde{\mathcal{D}}_{b-1}^T P \end{bmatrix}.$$

In order to utilize the Matlab LMI Toolbox to design the fault detection filter effectively, we assume P as $\hat{P} = \mathrm{diag}\{P_1, P_2, \cdots, P_{(\ell+1)b+1}\}$, let $\bar{L} = P_1 L$ and $\bar{V} = V$, then (3.18) can be obtained and the fault detection filter can be expressed as (3.19). The proof of this theorem is now complete. □

Remark 3.4. *We first establish a comprehensive model that covers multi-rate sampled-data dynamics, network-induced fading measurements and randomly occurring faults, thereby better reflecting the reality of networked control systems. In this case, sufficient conditions are given in Theorem 3.1-3.2 which make sure that the augmented system (3.8) is exponentially mean-square stable and H_∞ criterion in (3.9) is satisfied. Note that, at this stage, the designed fault detection filter which shows the combined effects of fading parameters, fault occurrence probability as well as multi-rate multiple. Next, in the special case of networked multi-rate systems, that is $b = 1$, the general networked single-rate systems with fading measurements and randomly occurring faults are taken into account, and corresponding fault detection filter is also designed in Corollary.*

3.1.3 Illustrative Examples

In this section, two numerical examples are presented to demonstrate the effectiveness of the proposed fault detection filter design scheme with fading measurements and randomly occurring faults for networked multi-rate systems and networked single-rate systems, respectively.

Example 1 In this numerical example, the system parameters of (3.1) and (3.2) are chosen as follows:

$$A = \begin{bmatrix} 0.8 & h \\ 0 & 0.6 \end{bmatrix}, B_1 = \begin{bmatrix} \frac{h^2}{2} \\ h \end{bmatrix}, B_2 = \begin{bmatrix} \frac{3h}{2} \\ 0.6 \end{bmatrix}, C = \begin{bmatrix} 0 & 0.3 \end{bmatrix}.$$

Here, the sampling period h of system (3.1) is $0.5s$, the measurement updating period is $1.5s$ (i.e. $b = 3$), the number of paths is $\ell = 1$, the probability of the randomly occurring faults is $\bar{\alpha} = 0.6$, and the probability density functions of channel coefficients are

$$\begin{cases} q(\beta_0) = 0.0005(\mathrm{e}^{9.89\beta_0} - 1), & 0 \le \beta_0 \le 1; \\ q(\beta_1) = \begin{cases} 10\beta_1, & 0 \le \beta_1 \le 0.20; \\ -2.50(\beta_1 - 1), & 0.20 < \beta_1 \le 1; \end{cases} \end{cases} \tag{3.21}$$

The mathematical expectations $\bar{\beta}_s$ can be calculated as 0.8991 and 0.4000, and the variances $\tilde{\bar{\beta}}_s^2$ $(s = 0, 1)$ are 0.0133 and 0.0467, respectively. By using the MATLAB LMI toolbox, for the augmented system (3.8), we obtain the minimum disturbance attenuation level as $\gamma_* = 1.0094$. The sub-optimal FD filter can then be obtained as following:

$$L = \begin{bmatrix} 2.1427 \\ -1.0263 \end{bmatrix}; V = -0.0389.$$

Letting the initial state of (3.1) be $x(T_0) = \mathrm{col}\{0.1, -0.1\}$ and its estimation be $\hat{x}(t_0) = \mathrm{col}\{0.1, 0\}$. To further illustrate the effectiveness of the designed fault detection filter, for $t_k = 0, 1, 2, \cdots, 100$, let the fault signal and the disturbance input be given as

$$f(t_k) = \begin{cases} 0.1, & 30 \le t_k \le 50 \\ 0, & \text{else} \end{cases}, \quad \omega(t_k) = e^{-0.01t_k}\sin(2t_k).$$

The residual response $r(t_k)$ and evolution of residual evaluation function $J(t_k) = \left\{\sum_{h=t_{k_0}}^{t_k} r^T(h)r(h)\right\}^{\frac{1}{2}}$ for networked multi-rate systems are shown in Figs. 3.1∼3.2, respectively. After 200 runs of the simulations, we get an average value of $J_{th} = 0.0369$. From Fig. 3.2, it can be shown that $0.0275 = J(29) < J_{th} < J(30) = 0.1090$, which means that the fault can be detected as soon as its occurrence.

Example 2 As the special case of networked multi-rate systems, in this example, an internet-based three-tank system is introduced to illustrate the effectiveness of our proposed networked single-rate systems.

With the variables defined in [149], the system model (3.1) and (3.2) with following parameters are adopted:

$$A = \begin{bmatrix} 0.9974 & 0 & 0.0026 \\ 0 & 0.9951 & 0.0024 \\ 0.0026 & 0.0024 & 0.9950 \end{bmatrix}, B_1 = \begin{bmatrix} 16.2190 & 0 \\ 0 & 16.2007 \\ 0.0212 & 0.0193 \end{bmatrix},$$

$$B_2 = \begin{bmatrix} 0.0212 \\ 0.0193 \\ 16.1997 \end{bmatrix}, C = \begin{bmatrix} 1 & 0 & 0 \\ 0 & 1 & 0 \end{bmatrix}.$$

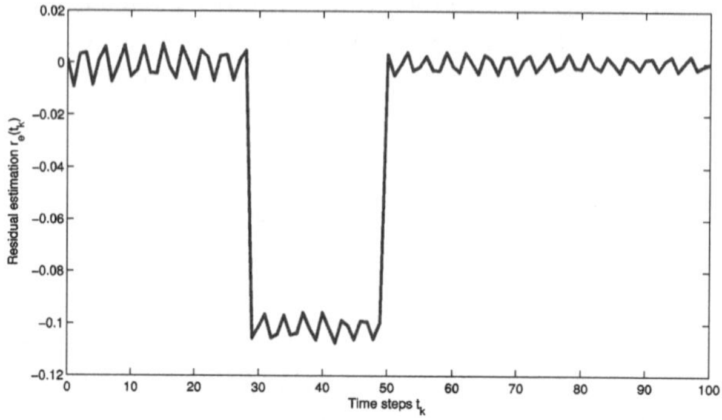

Figure 3.1: Residual signal $r(t_k)$ for networked multi-rate systems.

where $x(T_k) \in R^3$ is the system state representing the liquid levels of the three tanks; similar to [149], $\omega(T_k) \in R^2$ is the disturbance used to model the unknown disturbance and input, and $f(T_k) \in R$ is the fault signal reflecting the leakages in tank 3, $y(T_k) \in R^2$ is the measurement output describing the height measurements of tank 1 and tank 2. Here, we mainly investigate the internet-based fault detection problem, the measurement signal will obtain through remote network, thus, due to the multi-path transmission and shadowing problem, network-induced channel fading and randomly occurring faults usually take place, then the actually received measurement signal through network is $\bar{y}(T_k) \in R^2$, which satisfies (3.3).

$$\begin{cases} q(\beta_0) = 0.0005(e^{9.89\beta_0} - 1), & 0 \leq \beta_0 \leq 1; \\ q(\beta_1) = \begin{cases} 10\beta_1, & 0 \leq \beta_1 \leq 0.20; \\ -2.50(\beta_1 - 1), & 0.20 < \beta_1 \leq 1; \end{cases} \end{cases} \tag{3.22}$$

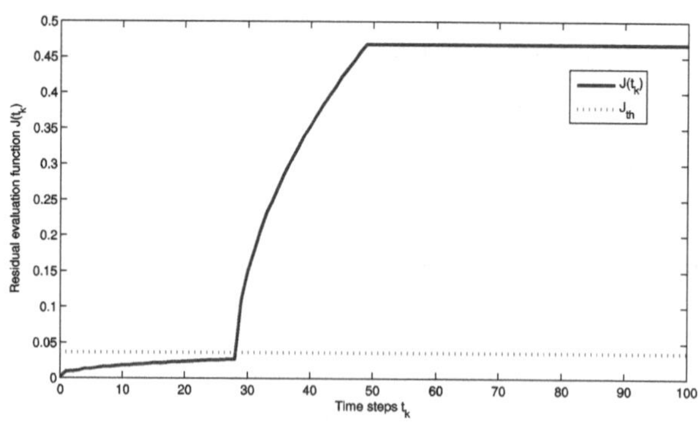

Figure 3.2: Evolution of residual evaluation function $J(t_k)$ for networked multi-rate systems.

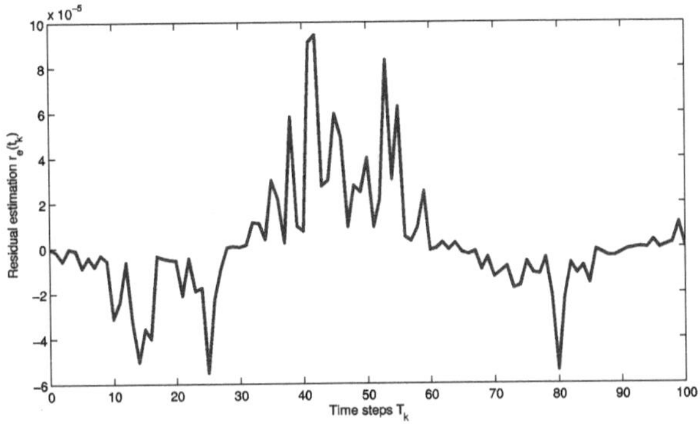

Figure 3.3: Residual signal $r(t_k)$ for networked single-rate systems.

Our aim here is to detect the faults by using the established mathematic model of the system (3.1) as well as the measurement signals (3.2) through network in the presence of a leakage in tank 3. In order to discuss simply the fault detection problem with fading measurement, we choose the fading parameters as (3.22). Choosing the faults occurrence probability as $\bar{\alpha} = 0.6$, similar to Example 1, by using Corollary (3.1), the sub-optimal fault detection filter and the minimum H_∞ attenuation level can be obtained as follows: $\gamma_* = 1.0023$,

$$L = \begin{bmatrix} -0.0042 & 0.0100 \\ -0.0030 & 0.0069 \\ -0.0095 & 0.0212 \end{bmatrix}, \quad V = 10^{-5} \times \begin{bmatrix} 0.1185 & -0.2856 \end{bmatrix}.$$

The initial value of (3.8) is chosen as $\bar{e}(T_0) = \text{col}\{0.1, -0.1, 0, 0.2, 0, -0.6, 0, 0.3, 0\}$, for $T_k = 0, 1, 2, \cdots, 100$, the fault signal and exogenous disturbance input signal are set as

$$f(T_k) = \begin{cases} 0.5, & 30 \le T_k \le 50 \\ 0, & \text{else} \end{cases}, \quad \omega(T_k) = \begin{bmatrix} e^{-0.02T_k} \sin(0.2T_k) \\ e^{-0.01T_k} \cos(0.1T_k) \end{bmatrix}.$$

The residual response $r(T_k)$ and evolution of residual evaluation function $J(T_k) = \left\{\sum_{h=T_0}^{T_k} r^T(h)r(h)\right\}^{\frac{1}{2}}$ for networked single-rate systems are shown in Figs. 3.3-3.4, respectively. After 200 runs of the simulations, we get an average value of $J_{th} = 1.4582 \times 10^{-4}$. From Fig. 3.4, it can be shown that $1.3526 \times 10^{-4} = J(41) < J_{th} < J(42) = 1.6304 \times 10^{-4}$, which means that the fault can be detected within 11 time steps after the fault occurred at $T_k = 30$.

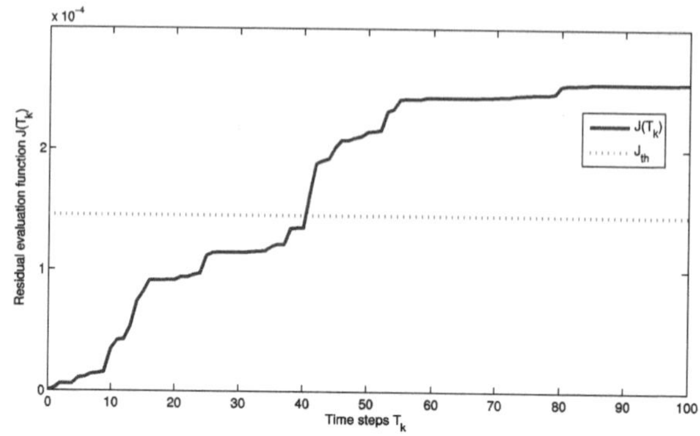

Figure 3.4: Evolution of residual evaluation function $J(t_k)$ for networked single-rate systems.

3.2 Fault Detection with Dynamic Quantization and Intermittent Faults

For the networked control systems with digital-communication scheme, the data is usually quantized before they are transmitted among sensors, controller/actuators and filter/estimator. As long as the limited communication capacity (e.g. shared network bandwidth) is a concern under resource constraints, the quantization error would never vanish and the resulting signal quantization has now widely recognized as a source that might degrade the system performance if not adequately dealt with.

In the following part, the fault detection problem is addressed for a class of networked multi-rate systems where the sampling period of measurement output is allowed to be nonuniform, where a Markovian jumping system model with incomplete transition probabilities is established.

3.2.1 Problem Formulation

Consider the following class of discrete-time systems:

$$x(T_{k+1}) = Ax(T_k) + B_1\omega(T_k) + \alpha(T_k)B_2f(T_k) \tag{3.23}$$

$$y(t_k) = Cx(t_k) + D_1\omega(t_k) + D_2f(t_k), \quad k = 0, 1, 2, \cdots \tag{3.24}$$

where $x(T_k) \in \mathbb{R}^{n_x}$ is the state vector, $y(t_k) \in \mathbb{R}^{n_y}$ is the process output, $\omega(T_k) \in \mathbb{R}^{n_\omega}$ is the disturbance input which belongs to $\ell_2[0, \infty)$, and $f(T_k) \in \mathbb{R}^{n_f}$ is the fault signal to be detected. A, B_1, B_2, C, D_1, and D_2 are known matrices with appropriate dimensions. The Bernoulli distributed stochastic variable $\alpha(T_k)$ in (3.23) is introduced to describe the phenomenon of intermittent faults satisfying

$$\text{Prob}\{\alpha(T_k) = 1\} = \bar{\alpha}, \ \text{Prob}\{\alpha(T_k) = 0\} = 1 - \bar{\alpha} \tag{3.25}$$

where $\bar{\alpha} \in [0, 1]$ is a known constant.

For a given frame period $h > 0$, we make the following assumptions on the sampling feature for systems (3.23)–(3.24):

Assumption 3.1. *The system state $x(T_k)$ is updated periodically at instants T_k with $h \triangleq T_{k+1} - T_k,\ k = 0, 1, 2, \cdots$.*

Assumption 3.2. *The measurement $y(t_k)$ from (3.24) is sampled at time t_k where the sampling interval is time-varying and random, that is $\mu_k h \triangleq t_{k+1} - t_k$, $k = 0, 1, 2, \cdots$, where the random variable $\mu_k \in \mathcal{Z} \triangleq \{1, 2, ..., b\}$ with b being a positive integer.*

It is clear that, for the considered systems (3.23)–(3.24), the state is updated with a uniform updating period h while the measurement output is sampled with a nonuniform sampling interval $\mu_k h$. As such, (3.23)–(3.24) are essentially nonuniformly sampled multi-rate systems. By applying the relation (3.23) recursively with some k satisfying $T_k = t_k$, one obtains the following equations with time scale t_k:

$$\begin{aligned} x(t_{k+1}) &= A^{\mu_k} x(t_k) + \bar{B}_{1,\mu_k} \bar{\omega}(t_k) + \\ &\quad \big[\bar{B}_{2,\mu_k} + \sum_{s=0}^{\mu_k - 1} \tilde{\alpha}(t_k + sh) \tilde{B}_{s,\mu_k,0}\big] \bar{f}(t_k) \end{aligned} \tag{3.26}$$

where

$$\begin{aligned} \tilde{\alpha}(t_k + sh) &\triangleq \alpha(t_k + sh) - \bar{\alpha}, \\ \bar{\omega}(t_k) &\triangleq \mathrm{col}_0^{b-1}\{\omega(t_k + sh)\},\ \bar{f}(t_k) \triangleq \mathrm{col}_0^{b-1}\{f(t_k + sh)\}, \\ \bar{B}_{1,\mu_k} &\triangleq \big[A^{\mu_k - 1} B_1 \ \cdots \ B_1 \ \underbrace{0 \cdots 0}_{b-\mu_k}\big], \\ \bar{B}_{2,\mu_k} &\triangleq \big[\bar{\alpha} A^{\mu_k - 1} B_2 \ \cdots \ \bar{\alpha} B_2 \ \underbrace{0 \cdots 0}_{b-\mu_k}\big], \\ \tilde{B}_{s,\mu_k,0} &\triangleq \big[\underbrace{0 \cdots 0}_{s} \ A^{\mu_k - 1 - s} B_2 \ \underbrace{0 \cdots 0}_{b-s-1}\big]. \end{aligned}$$

Let μ_k be a homogeneous Markov chain that takes value in $\mathcal{Z}$ with transition probability matrix (TPM) $\mathfrak{R} \triangleq [\rho_{ij}]_{b \times b}$ given by

$$\mathrm{Prob}\{\mu_{k+1} = j | \mu_k = i\} = \rho_{ij} \quad \forall i, j \in \mathcal{Z}$$

where $\rho_{ij} > 0$ is the transition probability from mode i to j and $\sum_{j=0}^{b} \rho_{ij} = 1$ for all $i, j \in \mathcal{Z}$.

It is assumed that $\mathfrak{R}$ is uncertain and belongs to a given convex-bounded polyhedral domain with vertices $\mathfrak{R}^{(\ell)}$, that is $\mathfrak{R} \in \Pi_{\mathfrak{R}}$, where $\Pi_{\mathfrak{R}}$ is described by the following set:

$$\Pi_{\mathfrak{R}} \triangleq \{\mathfrak{R} | \mathfrak{R} = \sum_{\ell=1}^{N} c_\ell \mathfrak{R}^{(\ell)};\ c_\ell \geq 0,\ \sum_{\ell=1}^{N} c_\ell = 1\}. \tag{3.27}$$

Here, $\mathfrak{R}^{(\ell)} \triangleq [\rho_{ij}^{(\ell)}]$ $\big(i, j \in \mathcal{Z}; \ell \in \mathcal{N} \triangleq \{1, 2, \cdots, N\}\big)$ are given transition probability matrices.

By synthesizing the features of both uncertain transition probabilities [150] and partly unknown transition probabilities [151], we focus on more general Markovian jump systems with partly unknown and uncertain transition probabilities. For notation clarity, for any $i \in \mathcal{Z}$, we denote $\mathcal{Z} \triangleq \mathcal{Z}_{\mathcal{K}}^i \bigcup \mathcal{Z}_{\mathcal{UC}}^i \bigcup \mathcal{Z}_{\mathcal{UK}}^i$, where $\mathcal{Z}_{\mathcal{K}}^i \triangleq \{j : \rho_{ij} \text{ is known}\}$, $\mathcal{Z}_{\mathcal{UC}}^i \triangleq \{j : \rho_{ij} \text{ is uncertain}\}$ and $\mathcal{Z}_{\mathcal{UK}}^i \triangleq \{j : \rho_{ij} \text{ is unknown}\}$. Also, we define $\rho_{\mathcal{K}}^i \triangleq \sum_{j \in \mathcal{Z}_{\mathcal{K}}^i} \rho_{ij}$, $\rho_{\mathcal{UC}}^i \triangleq \sum_{j \in \mathcal{Z}_{\mathcal{UC}}^i} \sum_{\ell \in \mathcal{N}} c_\ell \rho_{ij}^{(\ell)}$.

Remark 3.5. *In the past decades, Markov chain has been extensively employed to model the network-induced phenomena including communication delay [152] and data missing (also called packet dropout) [153]. On the other hand, both deterministic model [154] and stochastic model [142, 155] have been utilized to describe the non-uniformly sampled multi-rate systems. For example, by introducing a white binary-valued Bernoulli sequence, the nonuniform estimation rates have been modelled as random switch model [155]. According to the probability that a new measurement is not sampled, the process of nonuniform measurement sampling has been modelled in [142] model as a special Markov chain, that is two elements are known in every row of the TPM and others are all zero. Here, we adopt the Markovian jump model to describe the nonuniform sampling, thereby reflecting the statistical correlation between adjacent sampling intervals.*

In an NCS, the signal is quantized before transmitted to the observer through communication network. Let $z = \text{col}\{z_1, z_2, \cdots, z_{n_z}\} \in \mathbb{R}^{n_z}$ be the vector being quantized. As in [156], the quantizer is piecewise constant function $q_\hbar : \mathbb{R}^{z_\hbar} \rightarrow \mathcal{D}_\hbar$ $(\hbar = 1, 2, \cdots, n_z)$, where $\mathcal{D}_\hbar$ is a finite subset of $\mathbb{R}^{z_\hbar}$. It is assumed that there exist positive real numbers $M_\hbar$ and $\Delta_\hbar$ such that the following two conditions hold:

$$
\begin{aligned}
a).\ \text{If } \| z_\hbar \| &\leq M_\hbar \text{ then } \| q^{(\hbar)}(z_\hbar) - z_\hbar \| \leq \Delta_\hbar; && (3.28)\\
b).\ \text{If } \| z_\hbar \| &> M_\hbar \text{ then } \| q^{(\hbar)}(z_\hbar) \| > M_\hbar - \Delta_\hbar\,. && (3.29)
\end{aligned}
$$

where $M_\hbar$ and $\Delta_\hbar$ are the quantization range and quantization error bound, respectively.

The following sampling-interval-dependent dynamic quantizer is adopted to quantize the measurements:

$$
q_{\eta_{\mu_k},k}^{(\hbar)}\big(y^{(\hbar)}(t_k)\big) \triangleq \eta_{\mu_k,k} q^{(\hbar)}\Big(\frac{y^{(\hbar)}(t_k)}{\eta_{\mu_k,k}}\Big) \tag{3.30}
$$

where zoom variable $\eta_{\mu_k,k}$ is a time-varying positive scalar jumping synchronously with the Markov chain μ_k. Based on (3.28) and (3.29), the dynamic quantizer $q_{\eta_{\mu_k},k}^{(\hbar)}(*)$ satisfies the following conditions:

$$
\begin{aligned}
a').\quad & \text{If } \| y^{(\hbar)}(t_k) \| \leq \eta_{\mu_k,k} M_\hbar \\
& \text{then } \| q_{\eta_{\mu_k},k}^{(\hbar)}(y^{(\hbar)}(t_k)) - y^{(\hbar)}(t_k) \| \leq \eta_{\mu_k,k}\, \Delta_\hbar && (3.31)\\
b').\quad & \text{If } \| y^{(\hbar)}(t_k) \| > \eta_{\mu_k,k} M_\hbar \\
& \text{then } \| q_{\eta_{\mu_k},k}^{(\hbar)}(y^{(\hbar)}(t_k)) \| > \eta_{\mu_k,k}(M_\hbar - \Delta_\hbar). && (3.32)
\end{aligned}
$$

Remark 3.6. *Since the zooming variable $\eta_{\mu_k,k}$ is time-varying with sampling instants t_k, the quantization range $\eta_{\mu_k,k}M_\hbar$ and quantization error bound $\eta_{\mu_k,k}\ \Delta_\hbar$ of the quantizer $q^{(\hbar)}_{\eta_{\mu_k,k}}(*)$ are adjusted dynamically. It is worth pointing out that, for given $M_\hbar$ and $\Delta_\hbar$, the range of zooming variable $\eta_{\mu_k,k}$ can be used to reflect the characteristics of the dynamic quantizer $q^{(\hbar)}_{\eta_{\mu_k,k}}(*)$. Therefore, sampling-interval-dependent zooming regions become one of our main focuses in the current study.*

When the quantized signal is transmitted through communication network from the sensor to the observer, the data missing (or packet dropout) phenomenon frequently occurs owing to the limited bandwidth of the communication channel. In this case, the received signal with data missing in the observer can be described by

$$\bar{y}(t_k) \triangleq \beta(t_k) q_{\eta_{\mu_k,k}}(y(t_k)) \tag{3.33}$$

where $q_{\eta_{\mu_k,k}}(y(t_k)) = \text{col}_0^{n_y}\{q^{(s)}_{\eta_{\mu_k,k}}(y^{(s)}(t_k))\}$, $\beta(t_k)$ is a Bernoulli distributed stochastic variable satisfying

$$\text{Prob}\{\beta(t_k) = 1\} = \bar{\beta},\ \text{Prob}\{\beta(t_k) = 0\} = 1 - \bar{\beta}$$

where $\bar{\beta} \in [0, 1]$ is a known constant.

Throughout this section, the following assumption is also needed.

Assumption 3.3. *The random variables $\alpha(t_k)$ and $\beta(t_k)$ are mutually independent, and are also uncorrelated to μ_k.*

For the purpose of residual generation, the following observer-based fault detection filter is adopted:

$$\begin{cases} \hat{x}(t_{k+1}) = A^{\mu_k}\hat{x}(t_k) + L_{\mu_k}\Big(\bar{y}(t_k) - \hat{y}(t_k)\Big) \\ r(t_k) = V_{\mu_k}\Big(\bar{y}(t_k) - \hat{y}(t_k)\Big) \\ \hat{y}(t_k) = C\hat{x}(t_k) \end{cases} \tag{3.34}$$

where $\hat{x}(t_k)$ and $\hat{y}(t_k)$ denote the estimates of $x(t_k)$ and $y(t_k)$, respectively. $r(t_k)$ is the residual that is compatible with the fault vector $f(t_k)$. For such a sampling-interval-dependent fault detection filter (SIDFDF), L_{μ_k} and V_{μ_k} are the filter gains with appropriate dimension to be designed.

For presentation convenience, for each $\mu_k = i \in \mathcal{Z}$, matrices $A^{\mu_k - 1 - s}$ $(s \in \mathcal{S})$, L_{μ_k} and V_{μ_k} are denoted by A^{i-1-s}, L_i and V_i, respectively. By introducing

$$\begin{aligned} \xi(t_k) &\triangleq \text{col}\{x(t_k), \tilde{x}(t_k)\}, \\ d(t_k) &\triangleq \text{col}\{\bar{\omega}(t_k), \bar{f}(t_k)\},\ \tilde{\beta}(t_k) \triangleq \beta(t_k) - \bar{\beta}, \\ e^{(\hbar)}_{\eta_{i,k}}(y^{(\hbar)}(t_k)) &\triangleq \eta_{i,k}\big[q^{(\hbar)}(\frac{y^{(\hbar)}(t_k)}{\eta_{i,k}}) - \frac{y^{(\hbar)}(t_k)}{\eta_{i,k}}\big], \\ e_{\eta_{i,k}}(t_k) &\triangleq \text{col}_0^{n_y}\{e^{(s)}_{\eta_{i,k}}(t_k)\},\ \tilde{r}(t_k) \triangleq r(t_k) - \bar{f}(t_k), \end{aligned}$$

we have the overall fault detection dynamics governed by the following augmented systems:

$$\begin{cases} \xi(t_{k+1}) = [\bar{A}_i + \tilde{\beta}(t_k)\tilde{A}_i]\xi(t_k) + [\bar{E}_i + \tilde{\beta}(t_k)\tilde{E}_i]e_{\eta_{i,k}}(t_k) \\ \qquad + [\bar{\mathcal{B}}_{1,i} + \tilde{\beta}(t_k)\tilde{\mathcal{B}}_{1,i}]\bar{\omega}(t_k) \\ \qquad + [\bar{\mathcal{B}}_{2,i} + \tilde{\beta}(t_k)\tilde{\mathcal{B}}_{2,i} + \sum_{s=0}^{i-1} \tilde{\alpha}(t_k + sh)\tilde{\mathfrak{B}}_{2,s,i}]\bar{f}(t_k) \\ \tilde{r}(t_k) = [\bar{C}_i + \tilde{\beta}(t_k)\tilde{C}_i]\xi(t_k) + [\bar{\beta}V_i + \tilde{\beta}(t_k)V_i]e_{\eta_{i,k}}(t_k) \\ \qquad + [\bar{\mathcal{D}}_{1,i} + \tilde{\beta}(t_k)\tilde{\mathcal{D}}_{1,i}]\bar{\omega}(t_k) + [\bar{\mathcal{D}}_{2,i} + \tilde{\beta}(t_k)\tilde{\mathcal{D}}_{2,i}]\bar{f}(t_k) \end{cases} \tag{3.35}$$

where

$$\begin{aligned}
\bar{A}_i &\triangleq \begin{bmatrix} A^i & 0 \\ (1-\bar{\beta})L_iC & A^i - L_iC \end{bmatrix}, \ \tilde{A}_i \triangleq \begin{bmatrix} 0 & 0 \\ -L_iC & 0 \end{bmatrix}, \\
\tilde{E}_i &\triangleq \begin{bmatrix} 0 \\ -L_i \end{bmatrix}, \ \bar{E}_i \triangleq \bar{\beta}\tilde{E}_i, \\
\bar{\mathcal{B}}_{\iota,i} &\triangleq \begin{bmatrix} \bar{B}_{\iota,i} \\ -\bar{\beta}L_i \otimes \bar{D}_\iota \end{bmatrix}, \ \tilde{\mathcal{B}}_{\iota,i} \triangleq \begin{bmatrix} 0 \\ -\bar{\beta}L_i \otimes \bar{D}_\iota \end{bmatrix}, \\
\tilde{\mathfrak{B}}_{2,s,i} &\triangleq \begin{bmatrix} \tilde{B}_{2,s,i} \\ \tilde{B}_{2,s,i} \end{bmatrix}, \ \bar{C}_i \triangleq \begin{bmatrix} (\bar{\beta}-1)V_iC & V_iC \end{bmatrix}, \ \tilde{C}_i \triangleq \begin{bmatrix} V_iC & 0 \end{bmatrix}, \\
\bar{\mathcal{D}}_{1,i} &\triangleq \bar{\beta}V_i \otimes \bar{D}_1, \ \bar{D}_2 \triangleq \Big[\bar{\beta}V_iD_2 - I \ \underbrace{-I \ \cdots \ -I}_{b-1}\Big], \\
\bar{D}_\iota &\triangleq [D_\iota \ \underbrace{0 \ \cdots \ 0}_{b-1}] \ (\iota = 1,2), \tilde{\mathcal{D}}_{\iota,i} \triangleq V_i \otimes \bar{D}_\iota,
\end{aligned}$$

Before proceeding further, we introduce the following definition as in [151].

Definition 3.2. *The augmented system (3.35) with $\omega(t_k) = 0$, $f(t_k) = 0$ and $e_{\eta_{i,k}}(t_k) = 0$ is said to be stochastically stable if*

$$\mathbb{E}\Big\{\sum_{k=0}^{\infty} \|\xi(t_k)\|^2 | (\xi(t_0), \mu_0)\Big\} < \infty$$

for any initial conditions $(\xi(t_0), \mu_0)$.

Similar to [1, 157–159], the approach of H_-/H_∞ is introduced to design the gain matrices L_i, V_i $(i \in \mathcal{Z})$ such that the following requirements are met simultaneously:

(a) The augmented system (3.35) is stochastically stable.

(b) Under the zero-initial condition, the residual estimation error $\tilde{r}(t_k)\Big|_{\bar{f}(t_k)=0}$ with respect to all nonzero $\bar{\omega}(t_k)$ satisfies

$$\sum_{t_k=0}^{\infty} \mathbb{E}\big[\tilde{r}^T(t_k)\tilde{r}(t_k)\big]\Big|_{\bar{f}(t_k)=0} \leq \gamma^2 \sum_{t_k=0}^{\infty} \bar{\omega}^T(t_k)\bar{\omega}(t_k) \tag{3.36}$$

where $\gamma > 0$ is a given H_∞ performance index.

(c) Under the zero-initial condition, the residual estimation error $\tilde{r}(t_k)\Big|_{\bar{\omega}(t_k)=0}$ with respect to all nonzero $\bar{f}(t_k)$ satisfies

$$\sum_{t_k=0}^{\infty} \mathbb{E}\big[\tilde{r}^T(t_k)\tilde{r}(t_k)\big]\Big|_{\bar{\omega}(t_k)=0} \geq \lambda^2 \sum_{t_k=0}^{\infty} \bar{f}^T(t_k)\bar{f}(t_k) \tag{3.37}$$

where $\lambda > 0$ is a given H_- performance index.

Remark 3.7. *Different from usual H_∞-based fault detection, we adopt the H_-/H_∞ approach to detect the fault in this section. It should be pointed out that the H_∞ performance index γ quantifies the robustness of residuals with respect to the disturbance, and the H_- performance index λ reflects the sensitivity of the residuals against the fault. Therefore, H_-/H_∞ approach can be used to achieve a satisfactory trade-off between the robustness against the disturbances and the sensitivity to the faults.*

In order to detect the intermittent fault, we introduce the residual evaluation function $J(t_k)$ and the threshold J_{th} with the following form:

$$J(t_k) \triangleq \Big[\sum_{h=t_{k0}}^{t_k} r^T(h)r(h)\Big]^{\frac{1}{2}}, \quad J_{th} \triangleq \sup_{\substack{\bar{\omega}(t_k)\in\ell_2 \\ \bar{f}(t_k)=0}} \mathbb{E}\big[J(t_k)\big]$$

where t_{k_0} represents the initial time of the residual evaluation.

The occurrence of intermittent fault can be detected by comparing $J(t_k)$ with J_{th} according to the following test rule:

$$\begin{cases} J(t_k) \geq J_{th} \Longrightarrow \text{alarm for fault} \\ J(t_k) < J_{th} \Longrightarrow \text{no fault} \end{cases} \tag{3.38}$$

and the fault detection time t_{k_d} is defined as $t_{k_d} \triangleq \min_{t_k}\{J(t_k) \geq J_{th}\}$.

3.2.2 Detection of Intermittent Faults

In this section, by resorting to the stochastic analysis techniques, we will provide the analysis result of both the H_-/H_∞ performance and the dynamic zooming region for the augmented system (3.35). For presentation convenience, we denote

$$
\begin{aligned}
\bar{\Sigma}_{1,i} &\triangleq \begin{bmatrix} \bar{\Sigma}_{1,1,i} & \bar{\Sigma}_{1,2,i} \\ * & \bar{\Sigma}_{1,3,i} \end{bmatrix}, \\
\bar{\Sigma}_{1,1,i} &\triangleq \begin{bmatrix} (1-\bar{\beta})C^T U_i C - P_{1,i} & (\bar{\beta}-1)C^T U_i C \\ * & C^T U_i C - P_{2,i} \end{bmatrix}, \\
\bar{\Sigma}_{1,2,i} &\triangleq \begin{bmatrix} 0 & 0 & 0 & 0 \\ \bar{\beta} C^T U_i D_1 & 0 & 0 & 0 \end{bmatrix}, \\
\bar{\Sigma}_{1,3,i} &\triangleq \mathrm{diag}\{\bar{\beta} D_1^T U_i D_1 - \gamma^2 I, -\gamma^2 I, -\gamma^2 I, -\gamma^2 I\}, \\
\bar{\Sigma}_{2,i} &\triangleq \begin{bmatrix} \bar{\Sigma}_{2,1,i} \\ \bar{\Sigma}_{2,2,i} \end{bmatrix}, \ \bar{\Sigma}_{2,1,i} \triangleq \begin{bmatrix} (A^i)^T R_1^T & (1-\bar{\beta})C^T \bar{L}_i^T \\ 0 & (A^i)^T R_2^T - C^T \bar{L}_i^T \end{bmatrix}, \\
U_i &\triangleq V_i^T V_i, \\
\bar{\Sigma}_{2,2,i} &\triangleq \begin{bmatrix} \bar{\Sigma}_{2,2,1,i} & \bar{\Sigma}_{2,2,2,i} \end{bmatrix}, \\
\bar{\Sigma}_{2,2,1,i} &\triangleq \mathrm{col}\Big\{ \underbrace{B_1^T (A^{i-1})^T R_1^T, \cdots, B_1^T R_1^T}_{i}, \underbrace{0, \cdots, 0}_{b-i} \Big\}, \\
\bar{\Sigma}_{2,2,2,i} &\triangleq \mathrm{col}\Big\{ -\bar{\beta} D_1^T \bar{L}_i^T, \underbrace{0, \cdots, 0}_{b-1} \Big\}, \ \bar{\Sigma}_{3,i} \triangleq \begin{bmatrix} \bar{\Sigma}_{3,1,i} \\ \bar{\Sigma}_{3,2,i} \end{bmatrix}, \\
\bar{\Sigma}_{3,1,i} &\triangleq \begin{bmatrix} 0 & -\tilde{\bar{\beta}} C^T \bar{L}_i^T \\ 0 & 0 \end{bmatrix}, \\
\bar{\Sigma}_{3,2,i} &\triangleq \begin{bmatrix} 0 & \mathrm{col}\{ -\tilde{\bar{\beta}} D_1^T \bar{L}_i^T, \underbrace{0, \cdots, 0}_{b-1} \} \end{bmatrix}, \\
\bar{\Sigma}_{4,i} &\triangleq \mathrm{diag}\{\Upsilon_{1,j} - R_1 - R_1^T, \Upsilon_{2,j} - R_2 - R_2^T\}, \\
\hat{\bar{\Sigma}}_{1,i} &\triangleq \begin{bmatrix} \hat{\bar{\Sigma}}_{1,1,i} & \bar{\Sigma}_{1,2,i} \\ * & \hat{\bar{\Sigma}}_{1,3,i} \end{bmatrix}, \\
\hat{\bar{\Sigma}}_{1,1,i} &\triangleq \begin{bmatrix} (1-\bar{\beta})C^T U_i C - \delta_{1,1}^2 I & (\bar{\beta}-1)C^T U_i C \\ * & C^T U_i C - \delta_{1,1}^2 I \end{bmatrix}, \\
\bar{\Sigma}_{1,3,i} &\triangleq \mathrm{diag}\{\bar{\beta} D_1^T U_i D_1 - \delta_{1,1}^2 I, -\delta_{1,1}^2 I, -\delta_{1,1}^2 I, -\delta_{1,1}^2 I\}, \\
\Sigma_{5,i} &\triangleq \begin{bmatrix} 0 & -\bar{L}_i^T \end{bmatrix}, \\
\bar{a}_{1,2,i} &\triangleq \frac{\delta_{1,1}\delta_{1,2} + \sqrt{(\delta_{1,1}\delta_{1,2})^2 + \lambda_{\min}(-\bar{\Phi}_i)\delta_{1,2}^2}}{\lambda_{\min}(-\bar{\Phi}_i)} \bar{\triangle}, \ \bar{\triangle} \triangleq (\sum_{\hbar=1}^{n_y} \triangle_{\hbar}^2)^{\frac{1}{2}}, \\
\tilde{\bar{\alpha}} &\triangleq \sqrt{\bar{\alpha}(1-\bar{\alpha})}, \ \tilde{\bar{\beta}} \triangleq \sqrt{\bar{\beta}(1-\bar{\beta})}, \ \bar{M} \triangleq (\sum_{\hbar=1}^{n_y} M_{\hbar}^2)^{\frac{1}{2}}.
\end{aligned}
$$

Theorem 3.3. *For the given performance* $\gamma > 0$, $\lambda > 0$ *and the positive scalars* $\delta_{1,1}, \delta_{1,2}, M_{\hbar}, \triangle_{\hbar}$ $(\hbar = 1, 2, ..., n_y), \bar{\alpha}$ *and* $\bar{\beta}$, *if there exist matrices* $R_1, R_2, \bar{L}_i, U_i, V_i$ *and* $P_{i,1}$, $P_{i,2}$ $(i \in \mathcal{Z})$ *such that the following linear matrix inequalities (LMIs) hold:*

$$\bar{\Phi}_i \triangleq \begin{bmatrix} \bar{\Sigma}_{1,i} & \bar{\Sigma}_{2,i} & \bar{\Sigma}_{3,i} \\ * & \bar{\Sigma}_{4,i} & 0 \\ * & * & \bar{\Sigma}_{4,i} \end{bmatrix} < 0 \tag{3.39}$$

$$\begin{bmatrix} \hat{\bar{\Sigma}}_{1,i} & \bar{\Sigma}_{2,i} & \bar{\Sigma}_{3,i} \\ * & \bar{\Sigma}_{4,i} & 0 \\ * & * & \bar{\Sigma}_{4,i} \end{bmatrix} < 0 \tag{3.40}$$

$$\begin{bmatrix} \bar{\beta} U_i - \delta_{1,2}^2 I & \bar{\beta}\Sigma_{5,i} & \bar{\bar{\beta}}\Sigma_{5,i} \\ * & \bar{\Sigma}_{4,i} & 0 \\ * & * & \bar{\Sigma}_{4,i} \end{bmatrix} < 0 \tag{3.41}$$

$$\bar{a}_{1,2,i}\|C\| < \bar{M} \tag{3.42}$$

where

$$\Upsilon_j \triangleq \begin{cases} \dfrac{1}{\rho_{\mathcal{K}}^i} \sum\limits_{j \in \mathcal{Z}_{\mathcal{K}}^i} \rho_{ij} P_j, & \forall j \in \mathcal{Z}_{\mathcal{K}}^i \\ \dfrac{1}{\rho_{\mathcal{UC}}^i} \sum\limits_{j \in \mathcal{Z}_{\mathcal{UC}}^i} \sum\limits_{\ell \in \mathcal{N}} c_\ell \rho_{ij}^{(\ell)} P_j, & \forall j \in \mathcal{Z}_{\mathcal{UC}}^i \\ P_j, & \forall j \in \mathcal{Z}_{\mathcal{UK}}^i \end{cases} \tag{3.43}$$

then the augmented system (3.35) is stochastically stable with the H_∞ *performance constraint (3.36), and the zoom variable* $\eta_{i,k}$ *meets the following sampling-interval-dependent regions:*

$$\left[\frac{\|y(t_k)\|}{\bar{M}}, \frac{\|y(t_k)\|}{\bar{a}_{1,2,i}\|C\|}\right]. \tag{3.44}$$

Especially, if the above inequalities are feasible, the filter parameters of the desired SIDFDF can be determined by

$$L_i = R_2^{-1}\bar{L}_i. \tag{3.45}$$

Proof. Choose the following Lyapunov function:

$$V(\xi(t_k), \mu_k) = \xi^T(t_k) P_i \xi(t_k). \tag{3.46}$$

Then, calculating the difference of $V(\xi(t_k), \mu_k)$ along the trajectory of (3.35) with $\bar{\omega}(t_k) = 0$, $\bar{f}(t_k) = 0$ and $e_\eta(t_k) = 0$, and taking the mathematical expectation, one has

$$\begin{aligned}
\mathbb{E}\{\triangle V(\xi(t_k),\mu_k)\} &= \xi^T(t_{k+1})\sum_{j\in\mathcal{Z}}\rho_{ij}P_j\xi(t_{k+1})-\xi^T(t_k)P_i\xi(t_k) \\
&= \xi^T(t_k)\big(\bar{A}_i^T\bar{P}_i\bar{A}_i+\tilde{\bar{\beta}}^2\tilde{A}_i^T\bar{P}_i\tilde{A}_i-P_i\big)\xi(t_k) \\
&\triangleq \xi^T(t_k)\Gamma_i\xi(t_k) < -\varrho\xi^T(t_k)\xi(t_k) \qquad (3.47)
\end{aligned}$$

where $\varrho \triangleq \inf_{i\in\mathcal{Z}}\{\lambda_{\min}(-\Gamma_i)\}$. Inequality (3.47) implies that

$$\mathbb{E}\Big\{\sum_{k=0}^{\infty}\xi^T(t_k)\xi(t_k)\Big\} < \frac{1}{\varrho}\mathbb{E}\Big\{V(\xi(t_0),\mu_0)\Big\} < \infty$$

and, according to Definition 3.2, the augmented system (3.35) is stochastically stable.

To establish the H_∞ performance constraint of augmented system (3.35) with $\bar{f}(t_k)=0$, we assume zero initial conditions and introduce the following index:

$$\begin{aligned}
J &\triangleq \mathbb{E}\Big\{\triangle V(\xi(t_k),\mu_k)\Big\}+\mathbb{E}\Big\{\tilde{r}_{\bar{\omega}}^T(t_k)\tilde{r}_{\bar{\omega}}(t_k)-\gamma^2\bar{\omega}^T(t_k)\bar{\omega}(t_k)\Big\} \\
&= \xi^T(t_k)\Big\{\Gamma_i+\bar{C}_i^T\bar{C}_i+\tilde{\bar{\beta}}^2\tilde{C}_i^T\tilde{C}_i\Big\}\xi(t_k) \\
&\quad + 2\xi^T(t_k)\Big\{\bar{A}_i^T\bar{P}_i\bar{\mathcal{B}}_{1,i}+\bar{C}_i^T\bar{\mathcal{D}}_{1,i}+\tilde{\bar{\beta}}^2\tilde{A}_i^T\bar{P}_i\tilde{\mathcal{B}}_{1,i}+\tilde{\bar{\beta}}^2\tilde{C}_i^T\tilde{\mathcal{D}}_{1,i}\Big\}\bar{\omega}(t_k) \\
&\quad + 2\xi^T(t_k)\Big\{\bar{A}_i^T\bar{P}_i\bar{E}_i+\tilde{\bar{\beta}}^2\tilde{A}_i^T\bar{P}_i\tilde{E}_i+\bar{\beta}\bar{C}_i^TV_i+\tilde{\bar{\beta}}^2\tilde{C}_i^TV_i\Big\}e_{\eta_{i,k}}(t_k) \\
&\quad + e_{\eta_{i,k}}^T(t_k)\Big\{\bar{E}_i^T\bar{P}_i\bar{E}_i+\tilde{\bar{\beta}}^2\tilde{E}_i^T\bar{P}_i\tilde{E}_i+\bar{\beta}^2V_i^TV_i+\tilde{\bar{\beta}}^2V_i^TV_i\Big\}e_{\eta_{i,k}}(t_k) \\
&\quad + 2e_{\eta_{i,k}}^T(t_k)\Big\{\bar{E}_i^T\bar{P}_i\bar{\mathcal{B}}_{1,i}+\bar{\beta}V_i^T\bar{\mathcal{D}}_{1,i}+\tilde{\bar{\beta}}^2\tilde{E}_i^T\bar{P}_i\tilde{\mathcal{B}}_{1,i}+\tilde{\bar{\beta}}^2V_i^T\tilde{\mathcal{D}}_{1,i}\Big\}\bar{\omega}(t_k) \\
&\quad + \bar{\omega}^T(t_k)\Big\{\bar{\mathcal{B}}_{1,i}^T\bar{P}_i\bar{\mathcal{B}}_{1,i}+\bar{\mathcal{D}}_{1,i}^T\bar{\mathcal{D}}_{1,i}+\tilde{\bar{\beta}}^2\tilde{\mathcal{B}}_{1,i}^T\bar{P}_i\tilde{\mathcal{B}}_{1,i}+\tilde{\bar{\beta}}^2\tilde{\mathcal{D}}_{1,i}^T\tilde{\mathcal{D}}_{1,i}-\gamma^2I\Big\}\bar{\omega}(t_k) \\
&\triangleq \zeta^T(t_k)\Phi_i\zeta(t_k)+2\zeta^T(t_k)\Psi_ie_{\eta_{i,k}}(t_k)+e_{\eta_{i,k}}^T(t_k)\Omega_ie_{\eta_{i,k}}(t_k) \qquad (3.48)
\end{aligned}$$

where

$$\begin{aligned}
\zeta(t_k) &\triangleq \mathrm{col}\{\xi(t_k),\bar{\omega}(t_k)\},\ \tilde{r}_{\bar{\omega}}(t_k)\triangleq\tilde{r}(t_k)\Big|_{\bar{f}(t_k)=0},\ \bar{P}_i\triangleq\sum_{j\in\mathcal{Z}}\rho_{ij}P_j, \\
\Phi_i &\triangleq \begin{bmatrix}\Phi_{1,i} & \Phi_{2,i}\\ * & \Phi_{3,i}\end{bmatrix},\ \Gamma_i\triangleq\bar{A}_i^T\bar{P}_i\bar{A}_i+\tilde{\bar{\beta}}^2\tilde{A}_i^T\bar{P}_i\tilde{A}_i-P_i, \\
\Phi_{1,i} &\triangleq \Gamma_i+\bar{C}_i^T\bar{C}_i+\tilde{\bar{\beta}}^2\tilde{C}_i^T\tilde{C}_i, \\
\Phi_{2,i} &\triangleq \bar{A}_i^T\bar{P}_i\bar{\mathcal{B}}_{1,i}+\bar{C}_i^T\bar{\mathcal{D}}_{1,i}+\tilde{\bar{\beta}}^2\tilde{A}_i^T\bar{P}_i\tilde{\mathcal{B}}_{1,i}+\tilde{\bar{\beta}}^2\tilde{C}_i^T\tilde{\mathcal{D}}_{1,i}, \\
\Phi_{3,i} &\triangleq \bar{\mathcal{B}}_{1,i}^T\bar{P}_i\bar{\mathcal{B}}_{1,i}+\bar{\mathcal{D}}_{1,i}^T\bar{\mathcal{D}}_{1,i}+\tilde{\bar{\beta}}^2\tilde{\mathcal{B}}_{1,i}^T\bar{P}_i\tilde{\mathcal{B}}_{1,i}+\tilde{\bar{\beta}}^2\tilde{\mathcal{D}}_{1,i}^T\tilde{\mathcal{D}}_{1,i}-\gamma^2I, \\
\Psi_i &\triangleq \begin{bmatrix}\bar{A}_i^T\bar{P}_i\bar{E}_i+\tilde{\bar{\beta}}^2\tilde{A}_i^T\bar{P}_i\tilde{E}_i+\bar{\beta}\bar{C}_i^TV_i+\tilde{\bar{\beta}}^2\tilde{C}_i^TV_i\\ \bar{\mathcal{B}}_{1,i}^T\bar{P}_i\bar{E}_i+\bar{\mathcal{D}}_{1,i}^T\bar{\beta}V_i+\tilde{\bar{\beta}}^2\tilde{\mathcal{B}}_{1,i}^T\bar{P}_i\tilde{E}_i+\tilde{\bar{\beta}}^2\tilde{\mathcal{D}}_{1,i}^TV_i\end{bmatrix}, \\
\Omega_i &\triangleq \bar{E}_i^T\bar{P}_i\bar{E}_i+\tilde{\bar{\beta}}^2\tilde{E}_i^T\bar{P}_i\tilde{E}_i+\bar{\beta}^2V_i^TV_i+\tilde{\bar{\beta}}^2V_i^TV_i.
\end{aligned}$$

By utilizing (3.31) and the factorization method, we have the following inequality

$$\begin{aligned} J &\le -\lambda_{\min}(-\Phi_i)\|\zeta(t_k)\|^2 + 2\eta_{i,k}\bar{\triangle}\|\Psi_i\|\|\zeta(t_k)\| + (\eta_{i,k}\bar{\triangle})^2\|\Omega_i\| \\ &= -\lambda_{\min}(-\Phi_i)\big(\|\zeta(t_k)\| - a_{1,1,i}\eta_{i,k}\big)\big(\|\zeta(t_k)\| - a_{1,2,i}\eta_{i,k}\big) \end{aligned} \tag{3.49}$$

where
$a_{1,1,i} \triangleq \frac{\|\Psi_i\| - \sqrt{\|\Psi_i\|^2 + \lambda_{\min}(-\Phi_i)\|\Omega_i\|}}{\lambda_{\min}(-\Phi_i)}\bar{\triangle}$, $a_{1,2,i} \triangleq \frac{\|\Psi_i\| + \sqrt{\|\Psi_i\|^2 + \lambda_{\min}(-\Phi_i)\|\Omega_i\|}}{\lambda_{\min}(-\Phi_i)}\bar{\triangle}$.

If Φ_i satisfies the following equality

$$\Phi_i \triangleq \begin{bmatrix} \Phi_{1,i} & \Phi_{2,i} \\ * & \Phi_{3,i} \end{bmatrix} < 0 \tag{3.50}$$

then we have $\lambda_{\min}(-\Phi_i) > 0$, $a_{1,1,i} < 0$ and $\|\zeta(t_k)\| - a_{1,1,i}\eta_{i,k} > 0$.

Furthermore, it can be found that $\|x(t_k)\| \le \|\xi(t_k)\| \le \|\zeta(t_k)\|$ and $\|y(t_k)\| \le \|C\|\|x(t_k)\|$, which results in

$$J \le -\lambda_{\min}(-\Phi_i)\big(\|\zeta(t_k)\| - a_{1,1,i}\eta_{i,k}\big)\Big(\frac{\|y(t_k)\|}{\|C\|} - a_{1,2,i}\eta_{i,k}\Big) \tag{3.51}$$

and further leads to $J < 0$ by considering (3.42). Therefore, we have

$$\mathbb{E}\Big\{\triangle V(\xi(t_k), \mu_k)\Big\} + \mathbb{E}\Big\{\tilde{r}^T(t_k)\tilde{r}(t_k) - \gamma^2\bar{f}^T(t_k)\bar{f}(t_k)\Big\} < 0$$

for all nonzero $d(t_k)$. By noticing the zero initial condition, (3.36) is achieved.

Next, we proceed with the design stage of SIDFDF, by using the Schur Complement Lemma, (3.50) holds if and only if the following inequalities are true:

$$\begin{bmatrix} \Sigma_{1,i} & \Sigma_{2,i}^{(\bar{P}_i)} & \Sigma_{3,i}^{(\bar{P}_i)} \\ * & -\bar{P}_i & 0 \\ * & * & -\bar{P}_i \end{bmatrix} < 0 \ \ (i \in \mathcal{Z}) \tag{3.52}$$

where

$$\begin{aligned} \Sigma_{1,i} &\triangleq \begin{bmatrix} \bar{C}_i^T\bar{C}_i + \bar{\tilde{\beta}}^2\tilde{C}_i^T\tilde{C}_i - P_i & \bar{C}_i^T\bar{\mathcal{D}}_{1,i} + \bar{\tilde{\beta}}^2\tilde{C}_i^T\tilde{\mathcal{D}}_{1,i} \\ * & \bar{\mathcal{D}}_{1,i}^T\bar{\mathcal{D}}_{1,i} + \bar{\tilde{\beta}}^2\tilde{\mathcal{D}}_{1,i}^T\tilde{\mathcal{D}}_{1,i} - \gamma^2 I \end{bmatrix}, \\ \Sigma_{2,i}^{(\bar{P}_i)} &\triangleq \text{col}\{\bar{A}_i^T\bar{P}_i,\ \bar{\mathcal{B}}_{1,i}^T\bar{P}_i\},\ \Sigma_{3,i}^{(\bar{P}_i)} \triangleq \text{col}\{\bar{\tilde{\beta}}\tilde{A}_i^T\bar{P}_i,\ \bar{\tilde{\beta}}\tilde{\mathcal{B}}_{1,i}^T\bar{P}_i\}. \end{aligned}$$

According to the classification of transition probabilities, we give the following decomposition:

$$\bar{P}_i \triangleq \sum_{j\in\mathcal{Z}_{\mathcal{K}}^i} \rho_{ij}P_j + \sum_{j\in\mathcal{Z}_{\mathcal{UC}}^i}\sum_{\ell=1}^{N} c_\ell\rho_{ij}^{(\ell)}P_j + \sum_{j\in\mathcal{Z}_{\mathcal{UK}}^i} \rho_{ij}P_j \tag{3.53}$$

and rewrite (3.52) from (3.43) and (3.53) as follows:

$$\begin{bmatrix} \Sigma_{1,i} & \Sigma_{2,i}^{(\Upsilon_j)} & \Sigma_{3,i}^{(\Upsilon_j)} \\ * & -\Upsilon_j & 0 \\ * & * & -\Upsilon_j \end{bmatrix} < 0 \ \ (\forall j \in \mathcal{Z}) \tag{3.54}$$

where the matrices $\Sigma_{2,i}^{(\Upsilon_j)}$ and $\Sigma_{3,i}^{(\Upsilon_j)}$ can be obtained by replacing $\bar{P}_i$ with Υ_j in (3.52).

In order to design effectively the filter parameters for SIDFDF by using Matlab LMI Toolbox, we assume $P_i \triangleq \mathrm{diag}\{P_{i,1}, P_{i,2}\}$, $R \triangleq \mathrm{diag}\{R_1, R_2\}$, and denote $\bar{L}_i \triangleq R_2 L_i$ $(i = 1, 2, \cdots, b)$. Due to $R > 0$ and $\Upsilon_j > 0$ $(\forall j \in \mathcal{Z})$, we have $(\Upsilon_j - R)\Upsilon_j^{-1}(\Upsilon_j - R) \geq 0$, which is equivalent to $-R\Upsilon_j^{-1}R \leq \Upsilon_j - 2R$ $(\forall j \in \mathcal{Z})$. Pre- and post-multiplying inequality (3.54) by $\mathrm{diag}\{I, \Upsilon_j^{-1}R, \Upsilon_j^{-1}R\}$ result in (3.57).

On the other hand, by considering the following facts:

$$\begin{aligned}
\bar{\Psi}_i &\triangleq \begin{bmatrix} \bar{A}_i^T\Upsilon_j\bar{E}_i + \tilde{\bar{\beta}}^2\tilde{A}_i^T\Upsilon_j\tilde{E}_i + \bar{\beta}\bar{C}_i^T V_i + \tilde{\bar{\beta}}^2\tilde{C}_i^T V_i \\ \bar{\mathcal{B}}_{1,i}^T\Upsilon_j\bar{E}_i + \bar{\mathcal{D}}_{1,i}^T\bar{\beta}V_i + \tilde{\bar{\beta}}^2\tilde{\mathcal{B}}_{1,i}^T\Upsilon_j\tilde{E}_i + \tilde{\bar{\beta}}^2\tilde{\mathcal{D}}_{1,i}^T V_i \end{bmatrix} = \Pi_i^T\Xi_i, \\
\bar{\Omega}_i &\triangleq \bar{E}_i^T\Upsilon_j\bar{E}_i + \tilde{\bar{\beta}}^2\tilde{E}_i^T\Upsilon_j\tilde{E}_i + \bar{\beta}^2V_i^TV_i + \tilde{\bar{\beta}}^2V_i^TV_i = \Xi_i^T\Xi_i \\
\|\bar{\Psi}_i\| &\triangleq \|\Pi_i^T\Xi_i\| \leq \|\Pi_i\|\|\Xi_i\|
\end{aligned}$$

where $\Pi_i^T \triangleq \begin{bmatrix} \bar{A}_i^T\Upsilon_j^{\frac{1}{2}} & \tilde{\bar{\beta}}\tilde{A}_i^T\Upsilon_j^{\frac{1}{2}} & \bar{C}_i^T & \tilde{\bar{\beta}}\tilde{C}_i^T \\ \bar{\mathcal{B}}_{1,i}^T\Upsilon_j^{\frac{1}{2}} & \tilde{\bar{\beta}}\tilde{\mathcal{B}}_{1,i}^T\Upsilon_j^{\frac{1}{2}} & \bar{\mathcal{D}}_{1,i}^T & \tilde{\bar{\beta}}\tilde{\mathcal{D}}_{1,i}^T \end{bmatrix}$,
$\Xi_i^T \triangleq \begin{bmatrix} \bar{E}_i^T\Upsilon_j^{\frac{1}{2}} & \tilde{\bar{\beta}}\tilde{E}_i^T\Upsilon_j^{\frac{1}{2}} & \bar{\beta}V_i^T & \tilde{\bar{\beta}}V_i^T \end{bmatrix}$, we know that $\|\Pi_i\| < \delta_{1,1}$ and $\|\Xi_i\| < \delta_{1,2}$ are equivalent to the following inequalities:

$$\Pi_i^T\Pi_i < \delta_{1,1}^2 I \tag{3.55}$$

$$\Xi_i^T\Xi_i < \delta_{1,2}^2 I \tag{3.56}$$

By employing the Schur Complement Lemma to (3.55) and (3.56), respectively, (3.40) and (3.41) can be achieved. Furthermore, if the condition (3.42) for quantizer range $\bar{M}$ is satisfied, then the stability and H_∞ performance constraint (3.36) are guaranteed, and the SIDFDF parameters can be expressed as (3.62). The proof of this theorem is now complete. □

Now, we will supply the H_- performance constraint (3.37) for the augmented system (3.35).

Theorem 3.4. *For the given performance $\gamma > 0$, $\lambda > 0$ and the positive scalars $\delta_{2,1}, \delta_{2,2}, M_\hbar, \Delta_\hbar$ $(\hbar = 1, 2, ..., n_y), \bar{\alpha}$ and $\bar{\beta}$, if there exist matrices $R_1, R_2, \bar{L}_i, U_i, V_i$ and $P_{i,1}$, $P_{i,2}$ $(i \in \mathcal{Z})$ such that the following LMIs hold:*

$$\bar{\Gamma}_i \triangleq \begin{bmatrix} \Gamma_{1,i} & \Gamma_{2,i} & \Gamma_{3,i} & \Gamma_{4,i} \\ * & \bar{\Sigma}_{4,i} & 0 & 0 \\ * & * & \bar{\Sigma}_{4,i} & 0 \\ * & * & * & \Gamma_{5,i} \end{bmatrix} < 0 \tag{3.57}$$

$$\begin{bmatrix} \bar{\Gamma}_{1,i} & \Gamma_{2,i} & \Gamma_{3,i} \\ * & \bar{\Sigma}_{4,i} & 0 \\ * & * & \bar{\Sigma}_{4,i} \end{bmatrix} < 0 \tag{3.58}$$

$$\begin{bmatrix} \bar{\beta}U_i - \delta_{2,2}^2 I & \bar{\beta}\Sigma_{5,i} & \tilde{\bar{\beta}}\Sigma_{5,i} \\ * & \bar{\Sigma}_{4,i} & 0 \\ * & * & \bar{\Sigma}_{4,i} \end{bmatrix} < 0 \tag{3.59}$$

$$\bar{a}_{2,2,i}\|C\| < \bar{M} \tag{3.60}$$

then the augmented system (3.35) is stochastically stable with the H_- performance constraint (3.37), and the zoom variable $\eta_{i,k}$ meets the following sampling-interval-dependent regions:

$$\left[\frac{\|y(t_k)\|}{\bar{M}}, \frac{\|y(t_k)\|}{\bar{a}_{2,2,i}\|C\|}\right]. \tag{3.61}$$

Especially, if above inequalities are feasible, the filter parameters of the desired SIDFDF can be determined by

$$L_i = R_2^{-1}\bar{L}_i. \tag{3.62}$$

where

$$\begin{aligned}
\Gamma_{1,i} &\triangleq \begin{bmatrix} \bar{\Sigma}_{1,1,i} & \Gamma_{1,2,i} \\ * & \Gamma_{1,3,i} \end{bmatrix}, \\
\Gamma_{1,2,i} &\triangleq \begin{bmatrix} \Gamma_{1,2,1,i} & \Gamma_{1,2,1,i} & \Gamma_{1,2,1,i} & \Gamma_{1,2,1,i} \\ \bar{\beta}C^T U_i D_2 - C^T V_i & -C^T V_i & -C^T V_i & -C^T V_i \end{bmatrix}, \\
\Gamma_{1,2,1,i} &\triangleq (1-\bar{\beta})C^T V_i,\ \Gamma_{1,3,1,i} \triangleq \bar{\beta}D_2^T U_i D_2 + (\lambda^2-1)I + 2\Gamma_{1,3,2,i}, \\
\Gamma_{1,3,2,i} &\triangleq \bar{\beta}D_2^T V_i, \\
\Gamma_{1,3,i} &\triangleq \begin{bmatrix} \Gamma_{1,3,1,i} & \Gamma_{1,3,2,i}-I & \cdots & \Gamma_{1,3,2,i}-I & \Gamma_{1,3,2,i}-I \\ * & (\lambda^2-1)I & \cdots & -I & -I \\ \vdots & \vdots & \ddots & \vdots & \vdots \\ * & * & \cdots & (\lambda^2-1)I & -I \\ * & * & \cdots & * & (\lambda^2-1)I \end{bmatrix}, \\
\Gamma_{2,i} &\triangleq \begin{bmatrix} \bar{\Sigma}_{2,1,i} \\ \Gamma_{2,2,i} \end{bmatrix}, \\
\Gamma_{2,2,i} &\triangleq \begin{bmatrix} \Gamma_{2,2,1,i} & \bar{\beta}\Gamma_{2,2,2,i} \end{bmatrix}, \\
\Gamma_{2,2,1,i} &\triangleq \mathrm{col}\Big\{\underbrace{\bar{\alpha}B_2^T(A^{i-1})^T R_1^T, \cdots, \bar{\alpha}B_2^T R_1^T}_{i}, \underbrace{0, \cdots, 0}_{b-i}\Big\}, \\
\Gamma_{2,2,2,i} &\triangleq \mathrm{col}\Big\{-D_2^T\bar{L}_i^T, \underbrace{0, \cdots, 0}_{b-1}\Big\}, \\
\Gamma_{3,i} &\triangleq \begin{bmatrix} \bar{\Sigma}_{3,1,i} \\ \Gamma_{3,2,i} \end{bmatrix},\ \Gamma_{3,2,i} \triangleq \begin{bmatrix} 0 & \tilde{\bar{\beta}}\Gamma_{2,2,2,i} \end{bmatrix},
\end{aligned}$$

$$
\begin{aligned}
\Gamma_{4,i} &\triangleq \Big[\underbrace{\tilde{\bar{\alpha}}\hat{\tilde{B}}_{1,i}^T \ \cdots \ \tilde{\bar{\alpha}}\hat{\tilde{B}}_{i-1,i}^T}_{i}\Big], \ \hat{\tilde{B}}_{s,i} \triangleq \begin{bmatrix} R_1\tilde{B}_{s,i,1} \\ R_2\tilde{B}_{s,i,1} \end{bmatrix}, \\
\Gamma_{5,i} &\triangleq \Big[\underbrace{\bar{\Sigma}_{4,i} \ \cdots \ \bar{\Sigma}_{4,i}}_{i}\Big], \\
\bar{\Gamma}_{1,i} &\triangleq \begin{bmatrix} \hat{\bar{\Sigma}}_{1,1,i} & \Gamma_{1,2,i} \\ * & \bar{\Gamma}_{1,3,i} \end{bmatrix}, \\
\hat{\bar{\Sigma}}_{1,1,i} &\triangleq \begin{bmatrix} (1-\bar{\beta})C^TU_iC - \delta_{2,1}^2 I & (\bar{\beta}-1)C^TU_iC \\ * & C^TU_iC - \delta_{2,1}^2 I \end{bmatrix}, \\
\bar{\Gamma}_{1,3,i} &\triangleq \begin{bmatrix} \bar{\Gamma}_{1,3,1,i} & I-\Gamma_{1,3,2,i} & \cdots & I-\Gamma_{1,3,2,i} & I-\Gamma_{1,3,2,i} \\ * & (1-\delta_{2,1}^2)I & \cdots & I & I \\ \vdots & \vdots & \ddots & \vdots & \vdots \\ * & * & \cdots & (1-\delta_{2,1}^2)I & I \\ * & * & \cdots & * & (1-\delta_{2,1}^2)I \end{bmatrix}, \\
\bar{\Gamma}_{1,3,1,i} &\triangleq \bar{\beta}D_2^TU_iD_2 + (1-\delta_{2,1}^2)I - 2\Gamma_{1,3,2,i}, \\
\bar{a}_{2,2,i} &\triangleq \frac{\delta_{2,1}\delta_{2,2} + \sqrt{(\delta_{2,1}\delta_{2,2})^2 + \lambda_{\min}(-\bar{\Gamma}_i)\delta_{2,2}^2}}{\lambda_{\min}(-\bar{\Gamma}_i)} \bar{\Delta}.
\end{aligned}
$$

and the other symbols are the same as defined in Theorem 3.3. The proof of Theorem 3.4 follows a similar line in obtaining Theorem 3.3, and therefore it is omitted.

Based on the above analysis of either the H_∞ performance or H_- performance, we are now ready to provide the robust fault detection with H_-/H_∞ performance constraint for augmented system (3.35) with the incomplete knowledge of transition probabilities.

Theorem 3.5. *For the given performance $\gamma > 0$, $\lambda > 0$ and the positive scalars $\delta_{\iota,1}, \delta_{\iota,2}, M_\hbar, \Delta_\hbar$ ($\iota = 1, 2$; $\hbar = 1, 2, ..., n_y$), $\bar{\alpha}$ and $\bar{\beta}$, if there exist matrices R_1, R_2, $\bar{L}_i, U_i, V_i$ and $P_{i,1}$, $P_{i,2}$ ($i \in \mathcal{Z}$) such that the LMIs (3.39)-(3.44) and (3.57)-(3.61) hold, then the augmented system (3.35) is stochastically stable with both the H_∞ performance constraint (3.36) and H_- performance constraint (3.37). Meanwhile, the zoom variable satisfies $\eta_{i,k} \in \left[\frac{\|y(t_k)\|}{M}, \frac{\|y(t_k)\|}{\max\{\bar{a}_{1,2,i},\bar{a}_{2,2,i}\}\|C\|}\right]$ Especially, if above LMIs are feasible, the desired SIDFDF can be determined by $L_i = R_2^{-1}\bar{L}_i$.*

Theorem 3.5 can be utilized to obtain the SIDFDF with under partially unknown transition probabilities. Obviously, the designed SIDFDF reflect the effect both the robustness from exogenous disturbance on the residual and the sensitivity of faults on the residual. In order to demonstrate the flexibility of our purposed design method, the algorithm is provided in **Algorithm 3.1**, which handles the problems for not only SIDFDF design but also fault detection.

Algorithm 3.1 Fault detection algorithm with optimal SIDFDF.

Step 1. Given parameters $M_{\hbar}, \triangle_{\hbar}$ $(\hbar = 1, 2, ..., n_y)$,$\bar{\alpha}$ and $\bar{\beta}$. Set the initial value for $\gamma, \lambda, \delta_{\iota,1}, \delta_{\iota,2}$ $(\iota = 1, 2)$ which satisfy conditions (3.39)-(3.44) and (3.57)-(3.61).

Step 2. Solve Theorem 3.5 to obtain matrices $R_1, R_2, P_{1,i}, P_{2,i}, \bar{L}_i, V_i$ $(i \in \mathcal{Z})$, respectively.

Step 3. If the obtained matrices $R_1, R_2, P_{1,i}, P_{2,i}, \bar{L}_i$,$V_i$ $(i \in \mathcal{Z})$ satisfy $\lambda_{\min}(-\bar{\Phi}_i) > 0$ and $\lambda_{\min}(-\bar{\Gamma}_i) > 0$, then employ (3.62) to design the SIDFDF L_i, V_i $(i \in \mathcal{Z})$, else go to Step 1 and adjust $\gamma, \lambda, \delta_{\iota,1}, \delta_{\iota,2}$ $(\iota = 1, 2)$.

Step 4. Utilize the designed SIDFDF in Step 3 to produce residual evaluation function $J(t_k)$ and appropriate threshold J_{th}. Meanwhile, $\bar{a}_i = \max\{\bar{a}_{1,2,i}, \bar{a}_{2,2,i}\}$ can be computed and the zooming regions $[\frac{\|y(t_k)\|}{M}, \frac{\|y(t_k)\|}{\bar{a}_i\|C\|}]$ is achieved ultimately.

Step 5. Compare $J(t_k)$ with J_{th} to determine whether there is a fault by using the test rule (3.38), and determine the fault detection time t_{k_d} according to $J(t_k) > J_{th}$.

Remark 3.8. *So far, we have discussed the designing problem of SIDFDF with partially unknown transition probabilities. The solvability of the addressed problem is cast into the feasibility of a set of LMIs. By means of the algorithm in* ***Algorithm 3.1****, we can obtain the tradeoff between the H_∞ performance, H_- performance and zooming region thereby demonstrating the flexibility of our purposed design method.*

Remark 3.9. *Compared to existing literatures such as fault detection [12, 29, 140, 160] or fault estimation [161, 162] in single-rate systems, our results have the following two distinguishing features: 1) a more practical system framework for multi-rate sampled data is introduced which can contain the single-rate systems as its special case; 2) the H_-/H_∞ approach is adopted to detect the intermittent fault, the performance of both the robustness against disturbances and the sensitivity to faults can be obtained.*

3.2.3 Illustrative Example

Three-tank system DTS200 has typical characteristics of tanks, pipelines and pumps which are widely used in chemical industry, thus it often serves as a benchmark process in laboratories. Especially, the precise mathematical model of three-tank system can be established and the leakage/sensor/actuator faults can be easily described. Therefore, the leakage detection of three-tank system DTS200 is adopted to verify the effectiveness of the proposed fault detection scheme in this section, and the profile of DTS 200 is shown in Fig. 3.5.

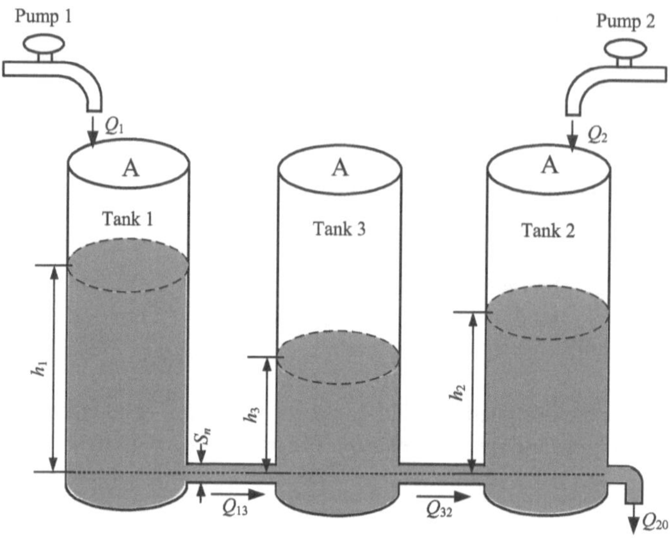

Figure 3.5: DTS200 setup.

Applying the incoming and outgoing mass flows with Torricellies law, the dynamics of DTS200 as [1] is modelled by

$$\begin{cases} A\dot{h}_1 = Q_1 - Q_{13}, \\ A\dot{h}_2 = Q_3 + Q_{32} - Q_{20}, \\ A\dot{h}_3 = Q_{13} - Q_{32} \\ Q_{20} = a_2 s_0 \sqrt{2gh_2} \\ Q_{13} = a_1 s_{13} sgn(h_1 - h_3)\sqrt{2g|h_1 - h_3|} \\ Q_{32} = a_3 s_{23} sgn(h_3 - h_2)\sqrt{2g|h_3 - h_2|} \end{cases} \tag{3.63}$$

where Q_1, Q_2 represent are incoming mass flow (cm^3/s), Q_{ij} denotes the mass flow (cm^3/s) from the $i-th$ tank to the $j-th$ tank, $h_i(t)$ $(i = 1, 2, 3)$ are the water level (cm) of each tank. Furthermore, the parameters are chosen as [1].

Letting $x(t) = y(t) \triangleq \text{col}\{h_1(t), h_2(t), h_3(t)\}$, $u(t) \triangleq \text{col}\{Q_1, Q_2\}$, the linearized state-space model of the three-tank system (3.63) near the operating point $x_0 = \text{col}\{45cm, 30cm, 15cm\}$ is given by

$$\begin{cases} \dot{x}(t) = \begin{bmatrix} -0.0085 & 0 & 0.0085 \\ 0 & -0.0195 & 0.0084 \\ 0.0085 & 0.0084 & -0.0169 \end{bmatrix} x(t) + \begin{bmatrix} 0.0065 & 0 \\ 0 & 0.0065 \\ 0 & 0 \end{bmatrix} u(t), \\ y(t) = \begin{bmatrix} 1 & 0 & 0 \\ 0 & 1 & 0 \\ 0 & 0 & 1 \end{bmatrix} x(t) \end{cases} \tag{3.64}$$

In this example, two types of faults are considered, that is component faults and sensor faults. One kind of component fault is the leaks in the three tanks, which can be modelled as additional mass flows out of tanks. Another is plugging between two tanks and in the letout pipe by tank 2, which cause changes in Q_{13}, Q_{32}, and Q_{20}. After choosing the appropriate controller, the linearized closed-loop model with faults can be expressed as

$$\begin{cases} \dot{x}(t) = \begin{bmatrix} -0.0025 & 0 & 0.0085 \\ 0 & -0.0013 & 0.0084 \\ 0.0085 & 0.0084 & -0.0169 \end{bmatrix} x(t) + \begin{bmatrix} 0.0019 & 0 \\ 0 & -0.0054 \\ 0 & 0 \end{bmatrix} \omega(t) \\ \quad + \begin{bmatrix} -0.0168 & 0 & -0.0085 \\ 0 & 0.1297 & 0.0084 \\ 0 & 0 & 0 \end{bmatrix} f(t), \\ y(t) = Cx(t) + D_2 f(t) \end{cases} \tag{3.65}$$

where $f(t) = \text{col}\{f_1(t), f_2(t), f_3(t)\}$ and $\omega(t) = \text{col}\{\omega_1(t), \omega_2(t)\}$ are the fault signal and disturbance signals, respectively. Discretizing system (3.65) with period $h = 0.5min$, one obtains

$$A = \begin{bmatrix} -0.7510 & 0.0058 & -0.0852 \\ 0.0058 & -0.4049 & 0.0242 \\ -0.0852 & 0.0242 & -0.8294 \end{bmatrix}, \ B_1 = \begin{bmatrix} 0.0513 & 0.0032 \\ -0.0004 & -0.2534 \\ 0.0063 & 0.0109 \end{bmatrix},$$

$$B_2 = \begin{bmatrix} -0.0123 & -0.0008 & -0.0063 \\ 0.0001 & 0.0605 & 0.0040 \\ -0.0015 & -0.0026 & -0.0009 \end{bmatrix}, C = D_2 = \text{diag}\{1,1,1\}, \ D_1 = 0.$$

From Assumption 3.2, the measurements in (3.24) are updated nonuniformly and randomly, that is $\mu_k h \triangleq t_{k+1} - t_k$ with Markov chain $\mu_k \in \mathcal{Z} \triangleq \{1,2,3,4\}$. According to the statistical information of μ_k, we have following case:

$$\Re_1 = \begin{bmatrix} 0.1 & ? & ? & ? \\ ? & \clubsuit & ? & \clubsuit \\ ? & 0.1 & 0.2 & ? \\ 0.6 & ? & ? & 0.1 \end{bmatrix},$$

where ? is the unknown element and ♣ represents the uncertain one. Actually, $\Re_1$ marks the Markov chains with partly unknown and partly uncertain transition probabilities. Assume that the TPM $\Re_1$ comprises three vertices $\Re_1^{(\ell)}$ $(\ell = 1,2,3)$, and the second lines of $\Re_1$ belong to a three-vertices polytope $\Pi_{\Re_{1,2}}$, that is

$$\Re_{1,2}^{(0)} \in \Pi_{\Re_{1,2}} \triangleq \{\Re_{1,2} \big| \Re_{1,2} = \sum_{\ell=1}^{3} c_\ell \Re_{1,2}^{(\ell)}, c_1 = 0.3, c_2 = 0.5, c_3 = 0.2\}$$

where

$$\begin{aligned}\Re_{1,2}^{(0)} &\triangleq \begin{bmatrix} ? & \clubsuit & ? & \clubsuit \end{bmatrix}, \Re_{1,2}^{(1)} \triangleq \begin{bmatrix} ? & 0.3 & ? & 0.2 \end{bmatrix}, \\ \Re_{1,2}^{(2)} &\triangleq \begin{bmatrix} ? & 0.2 & ? & 0.3 \end{bmatrix}, \Re_{1,2}^{(3)} \triangleq \begin{bmatrix} ? & 0.4 & ? & 0.1 \end{bmatrix}.\end{aligned}$$

For the given robustness performance level $\gamma = 2$ and sensitivity performance index $\lambda = 0.9$, by using the Theorem 3.5, we achieve following SIDFDF:

$$L_1 = \begin{bmatrix} -0.3839 & -0.0000 & 0.0000 \\ -0.0000 & -0.3313 & -0.0000 \\ -0.0445 & -0.0331 & -0.4081 \end{bmatrix}, V_1 = \begin{bmatrix} 0.1224 & 0.0000 & 0.0000 \\ 0.0000 & 0.1224 & 0.0000 \\ 0.0000 & 0.0000 & 0.1224 \end{bmatrix};$$

$$L_2 = \begin{bmatrix} 0.2948 & -0.0000 & -0.0000 \\ 0.0000 & 0.2195 & -0.0000 \\ 0.0705 & 0.0489 & 0.3330 \end{bmatrix}, V_2 = \begin{bmatrix} 0.1204 & -0.0000 & 0.0000 \\ -0.0000 & 0.1204 & 0.0000 \\ 0.0000 & 0.0000 & 0.1204 \end{bmatrix};$$

$$L_3 = \begin{bmatrix} -0.2264 & 0.0000 & 0.0000 \\ -0.0000 & -0.1455 & 0.0000 \\ -0.0838 & -0.0544 & -0.2718 \end{bmatrix}, V_3 = \begin{bmatrix} 0.1204 & -0.0000 & 0.0000 \\ -0.0000 & 0.1204 & 0.0000 \\ 0.0000 & 0.0000 & 0.1204 \end{bmatrix};$$

$$L_4 = \begin{bmatrix} 0.1739 & -0.0000 & -0.0000 \\ -0.0000 & 0.0964 & 0.0000 \\ 0.0885 & 0.0540 & 0.2218 \end{bmatrix}, V_4 = \begin{bmatrix} 0.1204 & -0.0000 & 0.0000 \\ -0.0000 & 0.1204 & 0.0000 \\ 0.0000 & 0.0000 & 0.1204 \end{bmatrix}.$$

Similar to [163], the following dynamic quantizer is chosen as

$$q_{\eta_{i,k}}^{(\hbar)}(y^{(\hbar)}(t_k)) = \begin{cases} M_\hbar\, \eta_{i,k}\, \mathrm{sgn}(y^{(\hbar)}(t_k)), \text{if } \|y^{(\hbar)}(t_k)\| > M_\hbar\, \eta_{i,k} \\ \eta_{i,k} \left\lfloor \dfrac{y^{(\hbar)}(t_k)}{\eta_{i,k}} + \triangle_\hbar \right\rfloor, \text{if } \|y^{(\hbar)}(t_k)\| \le M_\hbar\, \eta_{i,k} \end{cases} \tag{3.66}$$

where $(i = 1, 2, 3, 4; \hbar = 1, 2, 3)$, the quantization ranges are given by $M_1 = M_2 = M_3 = 1$ and the quantization error bounds are selected as $\triangle_1 = \triangle_2 = \triangle_3 = 0.01$. Then, the zooming variables are taken to be:

$$\eta_{i,k} = \frac{1}{2}\left(\frac{\|y(t_k)\|}{\bar{M}} + \frac{\|y(t_k)\|}{\bar{a}_{2,i}\|C\|}\right) \tag{3.67}$$

Based on above parameters, $\bar{a}_i = \max\{\bar{a}_{1,2,i}, \bar{a}_{2,2,i}\}$ can be computed as $\bar{a}_1 = 3.9167 \times 10^3$, $\bar{a}_2 = 2.9898 \times 10^3$, $\bar{a}_3 = 6.6175 \times 10^2$, $\bar{a}_4 = 2.2547 \times 10^2$. For $t_k = 0, 1, 2, \cdots, 100$, the disturbance is uniformly distributed over $[-0.05, 0.05]$ and fault signals are taken as

$$f_1(t_k) = f_2(t_k) = f_3(t_k) = \begin{cases} 0.01\ ,\ t_k \in [20, 40] \\ 0\ ,\ else \end{cases} \tag{3.68}$$

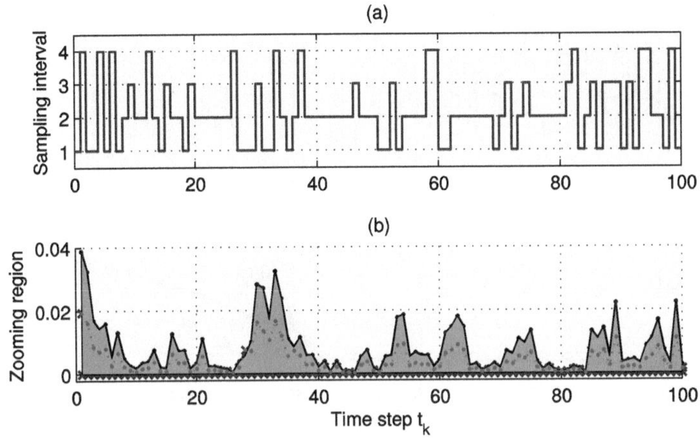

Figure 3.6: Zooming variable and its region of networked multi-rate systems with nonuniformly sampled measurements.

Combining the obtained $L_i, V_i, \bar{a}_i$ with quantizer (3.66) and zooming variable (3.67), and utilizing Algorithm 1, Figs. 3.6–3.8 can be acquired. With the stochastic jumping feature of nonuniform sampling, Fig. 3.6 shows the range of zooming variable for dynamic quantizer (3.30), and Fig. 3.7 offers the residual signals $r(t_k) = \mathrm{col}\{r_1(t_k), r_2(t_k), r_3(t_k)\}$. The evolution of the residual evaluation function $J(t_k)$ is presented in Fig. 3.8. Determine the threshold by using 200 Monte Carlo simulations as $J_{th} = 0.0110$, we have $0.0084 = J(57) < J_{th} < J(58) = 0.0128$,

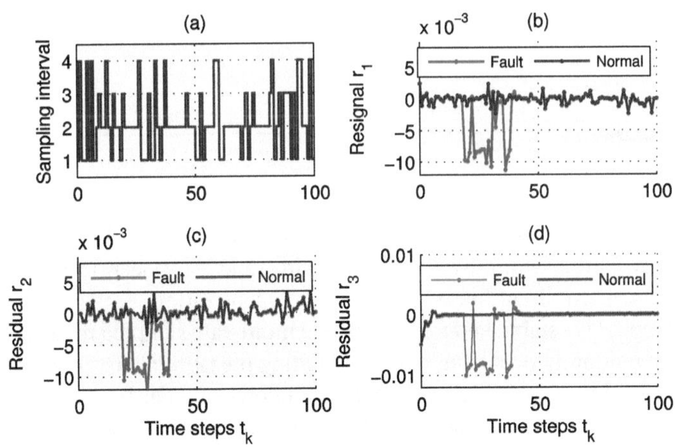

Figure 3.7: Residual-based fault detection of networked multi-rate systems with nonuniformly sampled measurements.

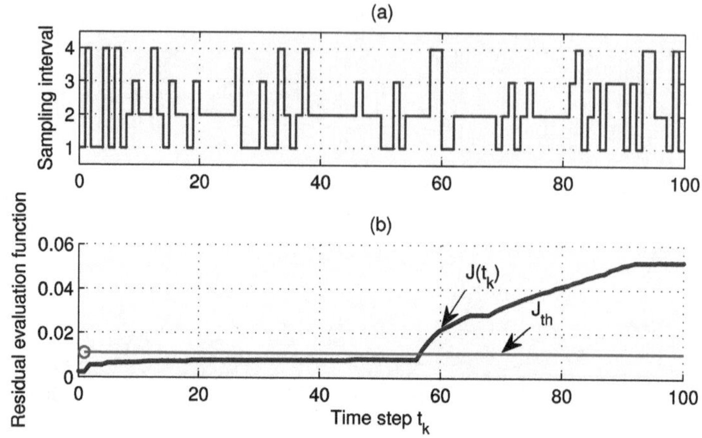

Figure 3.8: Residual-evaluation-function-based fault detection of networked multi-rate systems with nonuniformly sampled measurements.

then the alarm signal is generated for the first time and the fault detection time is $t_{k_d} = 58$.

Remark 3.10. *Figs. 3.7 and 3.8 represent two common fault detection methods including residual signal and residual evaluation function. Compared to the residual estimation function in this simulation verification, residual signal method can be used to detect the intermittent fault, which has distinguishing advantage over both effectiveness and quickness. The superiority of residual estimation function lies in that it supply quantitative results on the fault detection time, but its detection delays are often unavoidable.*

3.3 Conclusion

In this chapter, we have dealt with the fault detection problem for uniformly/non-uniformly sampled multi-rate systems with randomly occurring faults, fading measurements, intermittent faults, and dynamic quantization. In the first section, the augmented model is established to describe the multi-rate sampled networked control systems with randomly occurring faults and fading measurements. In the second section, Markovian jumping models with partly unknown and partly uncertain transition probabilities have been established, which represent the networked control systems with nonuniform sampling, intermittent fault, dynamic quantization, and data missing. By choosing linear matrix inequality technique and convex optimization tool, we have employed Matlab LMI Toolbox to design the fault detection filter effectively.

Several examples have been used to highlight the effectiveness of the proposed fault detection technology.

It's worth emphasizing that, intermittent fault described in this chapter is a new description, which is more consistent with the actual situation of the industrial fault. At the same time, nonuniform multi-rate sampling is a more effective sampling method in line with industrial practice. Therefore, the results of this chapter not only promote the theory of fault diagnosis but also provide meaningful solutions for industrial data collection methods.

4

Fault Diagnosis of Multi-rate Time-Varying Systems with Filter-Based Methods

The past decade has witnessed an ever-growing interest in the study of stochastic systems, such as CDMA systems [164], wind-thermal power systems [165], rail vehicle passive suspension systems [166], multi-agent systems [122], and networked systems [119, 167], where the variance-constrained theory [168] has gained particular research attention. On the other hand, note that in engineering practice, it is extremely difficult to adopt the single-rate sampling strategy for different kinds of devices [169]. In large-scale practical systems, the majority of systems have certain time-varying parameters and the transient performance of such time-varying systems is often more appealing/realistic than its steady-state counterpart (which is often suitable for time-invariant systems) [170]. As such, the analysis and design problems for time-varying and/or multi-rate sampled-data systems [143, 171, 172] have recently received considerable research attention.

On another research frontier, as compared with the traditional time-triggered communication mechanism, the event-triggered communication scheme has advantages of improving the usage efficiency of network bandwidth and saving energy under resource constraints [122]. In the past few years, a growing number of results have been reported on the applications of event-based strategies to various complex systems. Unfortunately, all the event-triggering conditions are assumed to be unilateral, which might be difficult to meet the engineering practice in the higher security systems. When it comes to the fault diagnosis problems, the corresponding results have been relatively few, although the event-triggered fault diagnosis problem is of particular importance for time-varying multi-rate sampled-data systems with limited communication capacity.

Inspired by the above discussion, we aim to handle the fault diagnosis problem for multi-rate time-varying systems with event-triggering communication scheme. Section 4.1 investigates the design of torus-event-based fault detection filter and fault isolation estimator, in the framework of variance-constrained H_∞ estimation with ellipsoid-constrained faults. In Section 4.2, the annulus-event-based fault detection estimator, fault isolation estimator as well as fault estimator by developing a new set-membership estimation method are investigated. Section 4.3 gives our conclusions.

DOI: 10.1201/9781003330998-4

4.1 Event-Based Fault Diagnosis with Constrained Fault

The utilization efficiency of communication network is of utmost importance in networked systems [143, 173–175], in which the communication protocol and event-triggering communication scheme are the research focuses. In this subsection, the multi-rate sampled-data time-varying systems are transformed into the single-rate model via the lifting technique, where the ellipsoid-constrained fault and a new torus-event-triggering communication scheme are taken into account. Then, a torus-event-based fault detection filter and fault isolation estimators are designed in the framework of variance-constrained H_∞ estimation.

4.1.1 Problem Formulation

Consider a discrete time-varying stochastic system described by the following state-space model:

$$x(T_{k+1}) = A_{T_k}x(T_k) + B_{1,T_k}\omega(T_k) + B_{2,T_k}f(T_k) \tag{4.1}$$

$$y(t_k) = C_{t_k}x(t_k) + D_{t_k}\nu(t_k) \tag{4.2}$$

where $x(T_k) \in \mathbb{R}^{n_x}$ is the system state, $y(t_k) \in \mathbb{R}^{n_y}$ is the measured output. External disturbances $\omega(T_k) \in \mathbb{R}^{n_\omega}$ and $\nu(t_k) \in \mathbb{R}^{n_\nu}$ are uncorrelated zero-mean white sequences with $\mathrm{E}[\omega(T_k)\omega^T(T_k)] = W_{T_k}$ and $\mathrm{E}[\nu(t_k)\nu^T(t_k)] = V_{t_k}$, respectively. A_{T_k}, B_{1,T_k}, B_{2,T_k}, C_{t_k}, and D_{t_k} are real-valued time-varying matrices with appropriate dimensions. $f(T_k) \in \mathbb{R}^{n_f}$ is the fault signal satisfying the following assumption.

Assumption 4.1. *The fault sequences $f(T_k)$ are confined to the following ellipsoidal set:*

$$\mathfrak{F}_{T_k} \triangleq \{f(T_k) :\ f^T(T_k)F_{T_k}^{-1}f(T_k) \leq 1\} \tag{4.3}$$

where F_{T_k} are known positive definite matrix with compatible dimension characterizing the scope of the ellipsoid.

Remark 4.1. *The fault is almost assumed as energy-bounded for either time-invariant systems [176–178] or time-varying systems [29, 179]. Actually, it's a broad and impractical assumption because most of faults have a limited amplitude in engineering. Therefore, the ellipsoid-constrained fault (4.3) in this section is meaningful and comprehensive.*

The sampling period of (4.1) is set as $h \triangleq T_{k+1} - T_k$, and the measurement (4.2) is assumed as $t_{k+1} - t_k = bh$, where b is a positive integer. As a consequence, (4.1) and (4.2) constitute a class of *multi-rate sampled-data systems*.

Let us now take the torus-event-based communication mechanism into consideration to reduce the communication frequency. Suppose that the sequence of triggering instants $0 \leq t_{k_0} \leq t_{k_1} \leq \cdots \leq t_{k_i} \leq \cdots$ is determined iteratively by

$$t_{k_{i+1}} = \inf\{t_k \in \mathbb{Z}_{k\geq 0} \mid t_k > t_{k_i}, \delta_1 I \leq \sigma^T(t_k)\Omega_{t_k}^{-1}\sigma(t_k) \leq \delta_2 I\} \tag{4.4}$$

where $0 < \delta_1 < \delta_2$, $\sigma(t_k) \triangleq y(t_{k_i}) - y(t_k)$, $y(t_{k_i})$ is the measurement at the latest event time k_i and $y(t_k)$ is the current measurement. Ω_{t_k} is referred to as the triggering threshold matrix, and δ_1, δ_2 are the boundness of torus.

Remark 4.2. *In comparison with conventional unilateral-event-triggering strategy, the torus-event-triggering condition defined in (4.4) has the following two distinguished features: 1) the triggering pattern (4.4) accords with practical demand because the measurement error $\sigma(t_k)$ usually resides in some torus for a moderately stable system; 2) triggering condition (4.4) is general that covers the unilateral triggering [122, 143, 180] as special cases with $\Omega_{t_k} = I$, $\delta_2 = +\infty$.*

Accordingly, the triggered measurement signal will be transmitted to the filter/estimator, and the zero-order holder is applied to hold the signal until a new measurement is achieved. Let $\bar{y}(t_k)$ be the signal received by the filter/estimator, we have

$$\bar{y}(t_k) = y(t_{k_i}), \quad t_{k_i} \leq t_k < t_{k_{i+1}} \tag{4.5}$$

4.1.2 Fault Detection and Fault Isolation

By applying the relation (4.1) recursively with some $t_k = T_k$, one obtains the following equations with time scale t_k:

$$x(t_{k+1}) = \bar{A}_{t_k}x(t_k) + \bar{B}_{1,t_k}d(t_k) + \bar{B}_{2,t_k}\bar{f}(t_k) \tag{4.6}$$

with

$$\mathcal{W}_{t_k} \triangleq \mathrm{E}\{d(t_k)d^T(t_k)\} = \mathrm{diag}\{V_{t_k}, W_{t_k}, W_{t_k+h}, \cdots, W_{t_k+(b-1)h}\}, \tag{4.7}$$

where

$$\begin{aligned}
\bar{\omega}(t_k) &\triangleq \mathrm{col}_0^{b-1}\{\omega(t_k+ih)\},\ d(t_k) \triangleq \mathrm{col}\{\nu(t_k), \bar{\omega}(t_k)\},\\
\bar{f}(t_k) &\triangleq \mathrm{col}_0^{b-1}\{f(t_k+ih)\},\ \bar{A}(t_k) \triangleq \prod_{i=0}^{b-1} A_{t_k+ih},\\
\bar{B}_{1,t_k} &\triangleq [0\ \prod_{i=1}^{b-1} A_{t_k+ih}B_{1,t_k}\ \prod_{i=2}^{b-1} A_{t_k+ih}B_{1,t_k+h}\ \cdots\\
&\quad \prod_{i=b-1}^{b-1} A_{t_k+ih}B_{1,t_k+(b-2)h}\ B_{1,t_k+(b-1)h}],\\
\bar{B}_{2,t_k} &\triangleq [\prod_{i=1}^{b-1} A_{t_k+ih}B_{2,t_k}\ \prod_{i=2}^{b-1} A_{t_k+ih}B_{2,t_k+h}\ \cdots\\
&\quad \prod_{i=b-1}^{b-1} A_{t_k+ih}B_{2,t_k+(b-2)h}\ B_{2,t_k+(b-1)h}].
\end{aligned}$$

Based on the received signal $\bar{y}(t_k)$, we adopt the following torus-event-based fault detection filter structure:

$$\begin{cases} \hat{x}(t_{k+1}) = \bar{A}_{t_k}\hat{x}(t_k) + L_{D,t_k}\left[\bar{y}(t_k) - \hat{y}(t_k)\right] \\ r(t_{k+1}) = H_{D,t_k}\left[\bar{y}(t_k) - \hat{y}(t_k)\right] \\ \quad \hat{y}(t_k) = C_{t_k}\hat{x}(t_k) \end{cases} \tag{4.8}$$

where $\hat{x}(t_k)$ is the estimate of system state $x(t_k)$, $r(t_k)$ is the residual signal. L_{D,t_k}, and H_{D,t_k} are the torus-event-based fault detection filter parameters to be determined.

By denoting $e(t_k) \triangleq x(t_k) - \hat{x}(t_k)$, combining (4.3), (4.6) and (4.8), one has the following fault detection systems:

$$\begin{cases} e(t_{k+1}) = \mathcal{A}_{t_k}e(t_k) + \mathcal{L}_{t_k}\sigma(t_k) + \mathcal{B}_{1,t_k}d(t_k) + \bar{B}_{2,t_k}\bar{f}(t_k) \\ r(t_{k+1}) = \mathcal{C}_{t_k}e(t_k) + H_{D,t_k}\sigma(t_k) + \mathcal{D}_{t_k}d(t_k) \end{cases} \tag{4.9}$$

with the following constraint

$$\bar{f}^T(t_k)\mathcal{F}_{t_k}^{-1}\bar{f}(t_k) \ \leq \ b \tag{4.10}$$

where

$$\begin{aligned} \mathcal{A}_{t_k} &\triangleq \bar{A}_{t_k} - L_{D,t_k}C_{t_k}, \ \mathcal{L}_{t_k} \triangleq -L_{D,t_k}, \\ \mathcal{B}_{1,t_k} &\triangleq \bar{B}_{1,t_k} - L_{D,t_k}\bar{D}_{t_k}, \ \mathcal{C}_{t_k} \triangleq H_{D,t_k}C_{t_k}, \\ \mathcal{D}_{t_k} &\triangleq H_{D,t_k}\bar{D}_{t_k}, \ \bar{D}_{t_k} \triangleq [D_{t_k} \ \underbrace{0 \cdots 0}_{b}], \\ \mathcal{F}_{t_k} &\triangleq \operatorname{diag}_0^{b-1}\{F_{t_k+ih}\}. \end{aligned}$$

Before proceeding further, we need the following assumption.

Assumption 4.2. *The initial state $x(0)$ and its estimate $\hat{x}(0)$ satisfy*

$$\mathrm{E}\{e(0)e^T(0)\} \leq \Theta_0 \tag{4.11}$$

where $e(0) \triangleq x(0) - \hat{x}(0)$, Θ_0 is the given positive definite matrix.

The purpose of this section is to design the torus-event-based fault detection filter parameters L_{D,t_k}, H_{D,t_k} for the multi-rate time-varying systems (4.1) and (4.2) such that the following two requirements are satisfied simultaneously.

1) For the given disturbance attenuation level $\gamma > 0$, the positive definite matrices U_d, $U_{\bar{f}}$, and U_e, and the initial state $x(t_0)$, the residual signal $r(t_k)$ satisfy the following H_∞-type performance constraint:

$$\begin{aligned} J \ \triangleq \ & \mathrm{E}\Big\{\sum_{t_k=0}^{N-1}\left[\|r(t_k)\|^2 - \gamma^2\|d(t_k)\|_{U_d}^2 - \gamma^2\|\bar{f}(t_k)\|_{U_{\bar{f}}}^2\right]\Big\} - \\ & \gamma^2 e^T(t_0)U_e e(t_0) < 0 \end{aligned} \tag{4.12}$$

where $\|\phi(t_k)\|_\Sigma^2 \triangleq \phi^T(t_k)\Sigma\phi(t_k)$ $(\phi(t_k) = d(t_k), \bar{f}(t_k);\ \Sigma = U_d, U_{\bar{f}})$.

2) The estimation error covariances satisfy the following constraints:

$$\mathrm{E}\{e(t_k)e^T(t_k)\} \leq \Theta_{t_k} \tag{4.13}$$

where Θ_{t_k} are the sequence of given matrices specifying the acceptable estimation accuracy according to the practical requirements.

Once a residual $r(t_k)$ is generated by (4.8), the following residual evaluation function $J(t_k)$ and corresponding threshold J_{th} are given by:

$$J(t_k) \triangleq \mathrm{E}\Big[\sum_{s=t_{k0}}^{t_k} r^T(i)r(i)\Big]^{\frac{1}{2}}, \quad J_{th} \triangleq \sup_{\substack{d(t_k)\in\bar{\mathfrak{W}}_{t_k}\\ \bar{f}(t_k)=0}} J(t_k)$$

where t_{k0} represents the initial time of the residual evaluation.

The occurrence of fault can be detected by comparing $J(t_k)$ with J_{th} according to the following test rule:

$$\begin{cases} J(t_k) \geq J_{th} \Longrightarrow \text{ alarm for fault} \\ J(t_k) < J_{th} \Longrightarrow \text{ no fault} \end{cases} \tag{4.14}$$

and the fault detection time t_D can be defined as $t_D \triangleq \min_{t_k}\{J(t_k) \geq J_{th}\}$.

Let us begin with the analysis of the H_∞ performance over the finite-horizon, that is, establish the sufficient conditions that guarantee the requirement (4.12).

Theorem 4.1. *Consider the multi-rate time-varying systems (4.1) and (4.2) with constrained fault (4.3). Let the disturbance attenuation level γ, the positive definite weighted matrices $U_d, U_{\bar{f}}, U_e$, and the triggering parameters $\delta_1, \delta_2, \Omega_{t_k}$ be given. The performance criterion (4.12) for the fault detection systems (4.9) is guaranteed for all nonzero $\omega(t_k), \nu(t_k), f(t_k)$, if there exist sequences of real-valued matrices $P_{t_k}, L_{D,t_k}, H_{D,t_k}$ and sequences of non-negative scalars ε_{i,t_k} $(i = 1,2,3)$ satisfying the following recursive matrix inequalities*

$$\Phi_{t_k} + \bar{\Phi}_{t_k} \leq 0 \tag{4.15}$$

for all $0 \leq t_k \leq N-1$ with initial condition

$$P_0 \leq \gamma^2 U_e \tag{4.16}$$

where

$$\Phi_{t_k} \triangleq \begin{bmatrix} \Phi_{1,t_k} & \Phi_{2,t_k}^T & \Phi_{3,t_k}^T \\ 0 & -I & 0 \\ 0 & 0 & -P_{t_{k+1}}^{-1} \end{bmatrix},$$

$$\Phi_{1,t_k} \triangleq \mathrm{diag}\Big\{\Phi_{1,1,t_k}, -P_{t_k}, (\varepsilon_{2,t_k} - \varepsilon_{1,t_k})\Omega_{t_k}^{-1}, -\gamma^2 U_d, \ldots$$
$$-\gamma^2 U_{\bar{f}} - \frac{\varepsilon_{3,t_k}}{b}\mathcal{F}_{t_k}^{-1}\Big\},$$

$$\begin{aligned}
\Phi_{1,1,t_k} &\triangleq \varepsilon_{1,t_k}\delta_1 - \varepsilon_{2,t_k}\delta_2 + \varepsilon_{3,t_k},\\
\Phi_{2,t_k}^T &\triangleq \mathrm{col}\Big\{0, \mathcal{C}_{t_k}^T, H_{D,t_k}^T, \mathcal{D}_{t_k}^T, 0\Big\},\\
\Phi_{3,t_k}^T &\triangleq \mathrm{col}\Big\{0, \mathcal{A}_{t_k}^T, \mathcal{L}_{t_k}^T, \mathcal{B}_{1,t_k}^T, \bar{B}_{2,t_k}^T\Big\},\\
\bar{\Phi}_{t_k} &\triangleq \mathrm{diag}\Big\{\frac{1}{N}e^T(t_0)(P_0 - \gamma^2 U_e)e(t_0), 0, 0, 0, 0\Big\}.
\end{aligned}$$

Proof. Defining

$$J_{t_k} \triangleq e^T(t_{k+1})P_{t_{k+1}}e(t_{k+1}) - e^T(t_k)P_{t_k}e(t_k) \tag{4.17}$$

and taking (4.9) into consideration, we have

$$\begin{aligned}
\mathrm{E}\{J_{t_k}\} = \mathrm{E}\Big\{&e^T(t_k)\big[\mathcal{A}_{t_k}^T P_{t_{k+1}}\mathcal{A}_{t_k} - P_{t_k}\big]e(t_k) + 2e^T(t_k)\big[\mathcal{A}_{t_k}^T P_{t_{k+1}}\mathcal{L}_{t_k}\sigma(t_k)\\
&+ \mathcal{A}_{t_k}^T P_{t_{k+1}}\mathcal{B}_{1,t_k}d(t_k)\big]\\
&+ 2e^T(t_k)\big[\mathcal{A}_{t_k}^T P_{t_{k+1}}\bar{B}_{2,t_k}\big]\bar{f}(t_k) + \sigma^T(t_k)\big[\mathcal{L}_{t_k}^T P_{t_{k+1}}\mathcal{L}_{t_k}\big]\sigma(t_k)\\
&+ 2\sigma^T(t_k)\big[\mathcal{L}_{t_k}^T P_{t_{k+1}}\mathcal{B}_{1,t_k}d(t_k) + \mathcal{L}_{t_k}^T P_{t_{k+1}}\bar{B}_{2,t_k}\bar{f}(t_k)\big]\\
&+ d^T(t_k)\big[\mathcal{B}_{1,t_k}^T P_{t_{k+1}}\mathcal{B}_{1,t_k}d(t_k) + 2\mathcal{B}_{1,t_k}^T P_{t_{k+1}}\bar{B}_{2,t_k}\bar{f}(t_k)\big]\\
&+ \bar{f}^T(t_k)\big[\bar{B}_{2,t_k}^T P_{t_{k+1}}\bar{B}_{2,t_k}\big]\bar{f}(t_k)\Big\}
\end{aligned} \tag{4.18}$$

Recurring to the following fact:

$$\begin{aligned}
&r^T(t_k)r(t_k) - \gamma^2 d^T(t_k)U_d d(t_k) - \gamma^2\bar{f}^T(t_k)U_{\bar{f}}\bar{f}(t_k) - r^T(t_k)r(t_k)\\
&+ \gamma^2 d^T(t_k)U_d d(t_k) + \gamma^2\bar{f}^T(t_k)U_{\bar{f}}\bar{f}(t_k) = 0
\end{aligned} \tag{4.19}$$

then, the following relation from (4.19) and (4.18) can be obtained

$$\begin{aligned}
\mathrm{E}\{J_{t_k}\} \;=\; &\mathrm{E}\Big\{\zeta^T(t_k)\Phi_{t_k}\zeta(t_k) - r^T(t_k)r(t_k) + \gamma^2 d^T(t_k)U_d d(t_k) +\\
&\gamma^2\bar{f}^T(t_k)U_{\bar{f}}\bar{f}(t_k)\Big\}
\end{aligned} \tag{4.20}$$

where $\zeta(t_k) \triangleq \mathrm{col}\{1, e(t_k), \sigma(t_k), d(t_k), \bar{f}(t_k)\}$.

Summing up (4.20) on both sides from 0 to $N-1$ with respect to t_k results in

$$\begin{aligned}
\sum_{t_k=0}^{N-1}\mathrm{E}\{J_{t_k}\} &= \mathrm{E}\{e^T(N)P_N e(N)\} - e^T(0)P_0 e(0)\\
&= \sum_{t_k=0}^{N-1}\mathrm{E}\{\zeta^T(t_k)\Phi_{t_k}\zeta(t_k)\} - \sum_{t_k=0}^{N-1}\mathrm{E}\Big\{r^T(t_k)r(t_k)\\
&\quad - \gamma^2 d^T(t_k)U_d d(t_k) - \gamma^2\bar{f}^T(t_k)U_{\bar{f}}\bar{f}(t_k)\Big\}
\end{aligned} \tag{4.21}$$

Hence, the H_∞ performance index J defined in (4.12) can be rewritten as follows:

$$J \triangleq \sum_{t_k=0}^{N-1} \mathrm{E}\{\zeta^T(t_k)[\Phi_{t_k} + \bar{\Phi}_{t_k}]\zeta(t_k)\} - \mathrm{E}\{e^T(N)P_N e(N)\} \tag{4.22}$$

Considering the triggering inequality (4.4) and fault constraint (4.10), one has

$$\begin{cases} \sigma^T(t_k)\Omega_{t_k}^{-1}\sigma(t_k) \leq \delta_1 \\ -\sigma^T(t_k)\Omega_{t_k}^{-1}\sigma(t_k) \leq -\delta_2 \\ \bar{f}^T(t_k)\mathcal{F}_{t_k}^{-1}\bar{f}(t_k) \leq b \end{cases} \tag{4.23}$$

which can be rearranged by means of $\zeta(t_k)$ as follows:

$$\begin{cases} \zeta^T(t_k)\mathrm{diag}\{-\delta_1, 0, \Omega_{t_k}^{-1}, 0, 0\}\zeta(t_k) \leq 0 \\ \zeta^T(t_k)\mathrm{diag}\{\delta_2, 0, -\Omega_{t_k}^{-1}, 0, 0\}\zeta(t_k) \leq 0 \\ \zeta^T(t_k)\mathrm{diag}\{-1, 0, 0, 0, \dfrac{1}{b}\mathcal{F}_{t_k}^{-1}\}\zeta(t_k) \leq 0 \end{cases} \tag{4.24}$$

where $\zeta(t_k) \triangleq \mathrm{col}\{1, e(t_k), \sigma(t_k), d(t_k), \bar{f}(t_k)\}$.

On the other hand, if there exist non-negative scalars $\varepsilon_{j,t_k} \geq 0$ $(j = 1, 2, 3)$, by employing the well-known Schur Complement to (4.15), the following inequality is true:

$$\begin{aligned} \Phi_{t_k} + \bar{\Phi}_{t_k} - \varepsilon_{1,t_k}\mathrm{diag}\{-\delta_1, 0, \Omega_{t_k}^{-1}, 0, 0\} - \varepsilon_{2,t_k}\mathrm{diag}\{\delta_2, 0, -\Omega_{t_k}^{-1}, 0, 0\} \\ -\varepsilon_{3,t_k}\mathrm{diag}\{-1, 0, 0, 0, \tfrac{1}{b}\mathcal{F}_{t_k}^{-1}\} \leq 0 \end{aligned} \tag{4.25}$$

Consequently, $\Phi_{t_k} + \bar{\Phi}_{t_k} < 0$ can be obtained by utilizing S-procedure to (4.25). Taking $P_{N+1} > 0$ and the initial condition (4.16) into consideration, we acquire $J < 0$. □

Now, we are in the position to analyze the variance-constrained estimation problem for the addressed stochastic multi-rate systems.

Theorem 4.2. *Consider the multi-rate time-varying systems (4.1) and (4.2) with constrained fault (4.3). Let the constraint matrices Θ_{t_k} and the triggering parameters $\delta_1, \delta_2, \Omega_{t_k}$ be given. The variance-constrained performance (4.13) for the fault detection systems (4.9) is guaranteed, if there exist sequences of real-valued matrices $P_{t_k}, L_{D,t_k}, H_{D,t_k}$ and sequences of non-negative scalars λ_{t_k}, ϵ_{j,t_k} $(j = 1, 2, 3, 4)$ satisfying the following recursive matrix inequalities*

$$\begin{bmatrix} \Xi_{1,t_k} & \Xi_{t_k}^T \\ 0 & -\Theta_{t_{k+1}} \end{bmatrix} \leq 0 \tag{4.26}$$

$$\begin{bmatrix} -\lambda_{t_k} & \Psi_{t_k}^T \\ 0 & -\breve{\Theta}_{t_{k+1}} \end{bmatrix} \leq 0 \tag{4.27}$$

where

$$\begin{aligned}
\breve{\Theta}_{t_{k+1}} &\triangleq \text{diag}_\kappa\{\Theta_{t_{k+1}}\}, \\
\Xi_{1,t_k} &\triangleq \text{diag}\Big\{\Xi_{1,1,t_k}, -\epsilon_{3,t_k}, (\epsilon_{2,t_k} - \epsilon_{1,t_k})\Omega_{t_k}^{-1}, -\frac{\epsilon_{3,t_k}}{b}\mathcal{F}_{t_k}^{-1}\Big\}, \\
\Xi_{1,1,t_k} &\triangleq \epsilon_{1,t_k}\delta_1 - \epsilon_{2,t_k}\delta_2 + \epsilon_{3,t_k} + \epsilon_{4,t_k} + \lambda_{t_k} - 1, \\
\bar{\Xi}_{t_k} &\triangleq \begin{bmatrix} 0 & \mathcal{A}_{t_k}U_{t_k} & \mathcal{L}_{t_k} & \bar{B}_{2,t_k} \end{bmatrix}, \\
\Psi_{t_k}^T &\triangleq \begin{bmatrix} \vartheta_{l,t_k}\mathcal{B}_{1,t_k}^T & \vartheta_{2,t_k}\mathcal{B}_{1,t_k}^T & \cdots & \vartheta_{\kappa,t_k}\mathcal{B}_{1,t_k}^T \end{bmatrix}.
\end{aligned}$$

Moreover, the matrix $\mathcal{W}_{t_k}$ can be decomposed by $\mathcal{W}_{t_k} \triangleq \sum_{l=1}^{\kappa}\vartheta_{l,t_k}\vartheta_{l,t_k}^T$ ($\vartheta_{l,t_k} \in \mathbb{R}^\kappa$, $\kappa = n_\nu + bn_\omega$), and Θ_{t_k} can be factorized with $\Theta_{t_k} = U_{t_k}U_{t_k}^T$.

Proof. We will prove this theorem by induction. Firstly, it can be known directly from Assumption 4.2 that (4.13) is right with $t_k = 0$.

Secondly, we assume that $\mathrm{E}\{e(t_k)e^T(t_k)\} \leq \Theta_{t_k}$ is true at time instant t_k, then there exits $z(t_k)$ satisfies

$$\mathrm{E}\{z(t_k)z^T(t_k)\} \leq I \tag{4.28}$$

with $\Theta_{t_k} = U_{t_k}U_{t_k}^T$ and $e(t_k) = U_{t_k}z(t_k)$. By employing Schur Complement to (4.28), we have the following inequality:

$$\mathrm{E}\{z^T(t_k)z(t_k)\} \leq 1 \tag{4.29}$$

Denoting $\xi(t_k) \triangleq \text{col}\{1, z(t_k), \sigma(t_k), \bar{f}(t_k)\}$, then (4.29) can be rewritten with $\xi(t_k)$ as follows

$$\mathrm{E}\{\xi^T(t_k)\text{diag}\{-1, I, 0, 0\}\xi(t_k)\} \leq 0 \tag{4.30}$$

Next, we proceed to prove that $\mathrm{E}\{e(t_{k+1})e^T(t_{k+1})\} \leq \Theta_{t_{k+1}}$ holds for t_{k+1}. The estimation error $e(t_{k+1})$ in (4.9) can be further expressed as

$$e(t_{k+1}) = (\bar{\Xi}_{t_k} + \vec{\Xi}_{t_k})\xi(t_k) = \Xi_{t_k}\xi(t_k) \tag{4.31}$$

where $\Xi_{t_k} \triangleq \bar{\Xi}_{t_k} + \vec{\Xi}_{t_k}$, $\vec{\Xi}_{t_k} \triangleq \begin{bmatrix} \mathcal{B}_{1,t_k}d(t_k) & 0 & 0 & 0 \end{bmatrix}$.

Meanwhile, taking triggering inequality (4.4) and fault constraint (4.10) into account, (4.23) can be represented with the following equalities:

$$\begin{cases} \xi^T(t_k)\text{diag}\{-\delta_1, 0, \Omega_{t_k}^{-1}, 0\}\xi(t_k) \leq 0 \\ \xi^T(t_k)\text{diag}\{\delta_2, 0, -\Omega_{t_k}^{-1}, 0\}\xi(t_k) \leq 0 \\ \xi^T(t_k)\text{diag}\{-1, 0, 0, \frac{1}{b}\mathcal{F}_{t_k}^{-1}\}\xi(t_k) \leq 0 \end{cases} \tag{4.32}$$

Considering (4.30), (4.32), and utilizing Schur Complement to (4.26), there exist non-negative scalars $\epsilon_{j,t_k} \geq 0$ $(j = 1, 2, 3, 4)$ satisfying the following inequality:

$$\begin{aligned}&\bar{\Xi}_{t_k}^T \Theta_{t_{k+1}}^{-1} \bar{\Xi}_{t_k} + \Sigma_{t_k} - \epsilon_{1,t_k} \mathrm{diag}\{-\delta_1, 0, \Omega_{t_k}^{-1}, 0\} \\ &-\epsilon_{2,t_k} \mathrm{diag}\{\delta_2, 0, -\Omega_{t_k}^{-1}, 0\} - \epsilon_{3,t_k} \mathrm{diag}\{-1, I, 0, 0\} \\ &-\epsilon_{4,t_k} \mathrm{diag}\{-1, 0, 0, \frac{1}{b}\mathcal{F}_{t_k}^{-1}\} \leq 0 \end{aligned} \tag{4.33}$$

where $\Sigma_{t_k} \triangleq \mathrm{diag}\{\lambda_{t_k} - 1, 0, 0, 0\}$. With the aid of S-procedure, one has:

$$\xi_{t_k}^T \big[\bar{\Xi}_{t_k}^T \Theta_{t_{k+1}}^{-1} \bar{\Xi}_{t_k}\big] \xi_{t_k} \leq 1 - \lambda_{t_k} \tag{4.34}$$

On the other hand, in virtue of Schur Complement, inequality (4.27) holds if and only if

$$\sum_{l=1}^{\kappa} \vartheta_{l,t_k}^T \mathcal{B}_{1,t_k}^T \Theta_{t_{k+1}}^{-1} \mathcal{B}_{1,t_k} \vartheta_{l,t_k} \leq \lambda_{t_k} \tag{4.35}$$

In the meantime, by setting $\mathcal{W}_{t_k} \triangleq \sum_{l=1}^{\kappa} \vartheta_{l,t_k} \vartheta_{l,t_k}^T$ and using the properties of matrix trace, (4.35) is equivalent to

$$\mathrm{tr}\big[\mathcal{B}_{1,t_k}^T \Theta_{t_{k+1}}^{-1} \mathcal{B}_{1,t_k} \mathcal{W}_{t_k}\big] \leq \lambda_{t_k} \tag{4.36}$$

Combining (4.7) and (4.36) with the properties of matrix trace, we have

$$\begin{aligned} \mathrm{E} \quad & \{\xi^T(t_k) \vec{\Xi}_{t_k}^T \Theta_{t_{k+1}}^{-1} \vec{\Xi}_{t_k} \xi(t_k)\} \\ = \quad & \xi^T(t_k)\mathrm{diag}\Big\{\mathrm{E}\{d^T(t_k)\mathcal{B}_{1,t_k}^T \Theta_{t_{k+1}}^{-1} \mathcal{B}_{1,t_k} d(t_k)\}, 0, 0, 0\Big\}\xi(t_k) \\ = \quad & \xi^T(t_k)\mathrm{diag}\Big\{\mathrm{tr}\Big[\mathrm{E}(d^T(t_k)\mathcal{B}_{1,t_k}^T \Theta_{t_{k+1}}^{-1} \mathcal{B}_{1,t_k} d(t_k))\Big], 0, 0, 0\Big\}\xi(t_k) \\ = \quad & \xi^T(t_k)\mathrm{diag}\Big\{\mathrm{tr}\{\mathcal{B}_{1,t_k}^T \Theta_{t_{k+1}}^{-1} \mathcal{B}_{1,t_k} \mathcal{W}_{t_k}\}, 0, 0, 0\Big\}\xi(t_k) \leq \lambda_{t_k} \end{aligned} \tag{4.37}$$

Therefore, the inequality can be verified from (4.31), (4.35), (4.36) and (4.37) as follows:

$$\begin{aligned} &\mathrm{E}\{e^T(t_{k+1})\Theta_{t_{k+1}}^{-1} e(t_{k+1})\} = \mathrm{E}\{\xi^T(t_k)\Xi_{t_k}^T \Theta_{t_{k+1}}^{-1} \Xi_{t_k} \xi(t_k)\} \\ &= \mathrm{E}\{\xi^T(t_k)\bar{\Xi}_{t_k}^T \Theta_{t_{k+1}}^{-1} \bar{\Xi}_{t_k} \xi(t_k)\} + \mathrm{E}\{\xi^T(t_k)\vec{\Xi}_{t_k}^T \Theta_{t_{k+1}}^{-1} \vec{\Xi}_{t_k} \xi(t_k)\} \leq 1 \end{aligned} \tag{4.38}$$

Finally, by applying Schur Complement to inequalities (4.38), we acquire $\mathrm{E}\{e(t_{k+1})e^T(t_{k+1})\} \leq \Theta_{t_{k+1}}$. □

To conclude the above analysis, we present a theorem which intends to take both the H_∞ performance index and the variance-constraint into account in a multiobjective analysis framework.

Theorem 4.3. *Consider the multi-rate systems (4.1) and (4.2) with constrained fault (4.3). Let the disturbance attenuation level $\gamma > 0$, the positive definite weighted* *matrices $U_d, U_{\bar{f}}, U_e$, the constraint matrices Θ_{t_k}, and the triggering parameters $\delta_1, \delta_2, \Omega_{t_k}$ be given. Then, for the fault detection systems (4.9), the H_∞ performance criterion (4.12) and the variance-constrained performance (4.13) are guaranteed simultaneously, if there exist sequences of real-valued matrices $P_{t_k}, L_{D,t_k}, H_{D,t_k}$ and sequences of non-negative scalars $\lambda_{t_k}, \varepsilon_{i,t_k}, \epsilon_{j,t_k}$ $(i = 1,2,3;\ j = 1,2,3,4)$ satisfying matrix inequalities (4.15), (4.16), (4.26) and (4.27).*

Theorems 4.1-4.3 outline the principles of designing the torus-event-based fault detection filter by solving the corresponding recursive linear matrix inequalities. It should be pointed out that, however, the proposed methodologies provide merely a feasible solution. In the following stage, three optimization problems based on Theorem 4.3 will be solved to demonstrate the optimal selections of the filter parameters.

Corollary 4.3.1. *Consider the multi-rate systems (4.1) and (4.2) with constrained fault (4.3). Let the positive definite weighted matrices $U_d, U_{\bar{f}}, U_e$, the constraint matrices Θ_{t_k}, and the triggering parameters $\delta_1, \delta_2, \Omega_{t_k}$ be given. The minimal H_∞ performance level γ can be guaranteed if there exist sequences of real-valued matrices $P_{t_k}, L_{D,t_k}, H_{D,t_k}$ and sequences of non-negative scalars $\lambda_{t_k}, \varepsilon_{i,t_k}, \epsilon_{j,t_k}$ $(i = 1,2,3;\ j = 1,2,3,4)$ by solving the following optimization problem:*

$$\begin{aligned} &\min_{P_{t_k}, L_{D,t_k}, H_{D,t_k}, \lambda_{t_k},\ \varepsilon_{i,t_k}, \epsilon_{j,t_k}} \quad \gamma^2 \\ &\text{subject to } (4.15), (4.16), (4.26), (4.27) \end{aligned} \tag{4.39}$$

Corollary 4.3.2. *Consider the multi-rate systems (4.1) and (4.2) with constrained fault (4.3). Let the disturbance attenuation level $\gamma > 0$, the positive definite weighted matrices $U_d, U_{\bar{f}}, U_e$, and the triggering parameters $\delta_1, \delta_2, \Omega_{t_k}$ be given. The minimal upper bound of estimation error variance Θ_{t_k} can be guaranteed (in the sense of matrix trace) if there exist sequences of real-valued matrices $P_{t_k}, L_{D,t_k}, H_{D,t_k}$ and sequences of non-negative scalars $\lambda_{t_k}, \varepsilon_{i,t_k}, \epsilon_{j,t_k}$ $(i = 1,2,3;\ j = 1,2,3,4)$ by solving the following optimization problem:*

$$\begin{aligned} &\min_{P_{t_k}, L_{D,t_k}, H_{D,t_k}, \lambda_{t_k},\ \varepsilon_{i,t_k}, \epsilon_{j,t_k}} \quad \mathrm{tr}\left[\Theta_{t_k}\right] \\ &\text{subject to } (4.15), (4.16), (4.26), (4.27) \end{aligned} \tag{4.40}$$

Corollary 4.3.3. *Consider the multi-rate systems (4.1) and (4.2) with constrained fault (4.3). Let the disturbance attenuation level $\gamma > 0$, the positive definite weighted matrices $U_d, U_{\bar{f}}, U_e$, the constraint matrices Θ_{t_k} and the triggering parameters δ_1, Ω_{t_k} be given. The minimal upper bound of triggering torus δ_2 can be guaranteed if there exist sequences of real-valued matrices $P_{t_k}, L_{D,t_k}, H_{D,t_k}$ and sequences of non-negative scalars $\lambda_{t_k}, \varepsilon_{i,t_k}, \epsilon_{j,t_k}$ $(i = 1,2,3;\ j = 1,2,3,4)$ by solving the following optimization problem:*

$$\begin{aligned} &\min_{P_{t_k}, L_{D,t_k}, H_{D,t_k}, \lambda_{t_k},\ \varepsilon_{i,t_k}, \epsilon_{j,t_k}} \quad \delta_2 \\ &\text{subject to } (4.15), (4.16), (4.26), (4.27) \end{aligned} \tag{4.41}$$

Remark 4.3. *Corollaries 4.3.1, 4.3.2 and 4.3.3 can be utilized to design the optimal torus-event-based fault detection estimator with the minimal H_∞ performance level, the minimal upper bound of estimation error variance or the minimal triggering-torus, respectively. Especially, it should be pointed out that there exists certain trade-off among the H_∞ performance level, the upper bound of estimation error variance and the triggering-torus, and such a tradeoff could provide much flexibility in making compromise among these performance parameters. Furthermore, in the design process of torus-event-based fault isolation estimators, similar optimization problems can be found, we omit the corresponding narrations due to the space.*

After the fault has been detected, the fault isolation module will be activated and the torus-event-based fault isolation estimators can be designed with recursive linear matrix inequalities method. Based on the empirical knowledge, it is assumed that there are q types of possible fault, specifically, $f(T_k)$ belongs to the finite set of functions given by

$$\mathscr{F} \triangleq \{\hat{f}^{(1)}(T_k), \hat{f}^{(2)}(T_k), \cdots, \hat{f}^{(q)}(T_k)\} \tag{4.42}$$

By using the received signal $\bar{y}(t_k)$, we employ the following q torus-event-based fault isolation estimators structure:

$$\begin{cases} \hat{x}^{(\ell)}(t_{k+1}) = \bar{A}_{t_k}\hat{x}^{(\ell)}(t_k) + \bar{B}_{2,t_k}\hat{\bar{f}}^{(\ell)}(t_k) + L_{I,t_k}^{(\ell)}\left[\bar{y}(t_k) - \hat{y}^{(\ell)}(t_k)\right] \\ r^{(\ell)}(t_k) = H_{I,t_k}^{(\ell)}\left[\bar{y}(t_k) - \hat{y}^{(\ell)}(t_k)\right] \\ \hat{y}^{(\ell)}(t_k) = C_{t_k}\hat{x}^{(\ell)}(t_k), \quad (\ell = 1, 2, \cdots, q) \end{cases} \tag{4.43}$$

where $\hat{\bar{f}}^{(\ell)}(t_k) \triangleq \text{col}_0^{b-1}\{\hat{f}^{(\ell)}(t_k + ih)\}$, $\hat{x}^{(\ell)}(t_k)$ is the estimate of state $x(t_k)$, $\hat{f}^{(\ell)}(t_k+ih)$ is the estimate of fault $f^{(\ell)}(t_k+ih)$, $L_{I,t_k}^{(\ell)}$ and $H_{I,t_k}^{(\ell)}$ are the ℓ estimator parameters to be designed.

Before proceeding further, the following assumption is needed.

Assumption 4.3. *The sequences $\tilde{f}^{(\ell)}(t_k)$ satisfy*

$$[\tilde{f}^{(\ell)}(t_k + ih)]^T[G_{t_k+ih}^{(\ell)}]^{-1}\tilde{f}^{(\ell)}(t_k + ih) \le 1 \tag{4.44}$$

where $\tilde{f}^{(\ell)}(t_k + ih) \triangleq f(t_k + ih) - \hat{f}^{(\ell)}(t_k + ih)$, matrix sequences $G_{t_k+ih}^{(\ell)} > 0$ characterize the scope of the fault estimation error.

Setting $e^{(\ell)}(t_k) \triangleq x(t_k) - \hat{x}^{(\ell)}(t_k)$ and $\tilde{\bar{f}}^{(\ell)}(t_k) \triangleq \text{col}_0^{b-1}\{\tilde{f}^{(\ell)}(t_k + ih)\}$, synthesizing (4.6) and (4.43), one has the following fault isolation systems:

$$\begin{cases} e^{(\ell)}(t_{k+1}) = \mathcal{A}_{t_k}^{(\ell)}e^{(\ell)}(t_k) + \mathcal{L}_{t_k}^{(\ell)}\sigma(t_k) + \mathcal{B}_{1,t_k}^{(\ell)}d(t_k) + \bar{B}_{2,t_k}\tilde{\bar{f}}^{(\ell)}(t_k) \\ r^{(\ell)}(t_k) = \mathcal{C}_{t_k}^{(\ell)}e^{(\ell)}(t_k) + H_{I,t_k}^{(\ell)}\sigma(t_k) + \mathcal{D}_{t_k}^{(\ell)}d(t_k) \end{cases} \tag{4.45}$$

with following constraints

$$[\tilde{\bar{f}}^{(\ell)}(t_k)]^T[\mathcal{G}_{t_k}^{(\ell)}]^{-1}\tilde{\bar{f}}^{(\ell)}(t_k) \le b \tag{4.46}$$

where

$$\begin{aligned}
\mathcal{A}_{t_k}^{(\ell)} &\triangleq \bar{A}_{t_k} - L_{I,t_k}^{(\ell)} C_{t_k},\ \mathcal{L}_{t_k}^{(\ell)} \triangleq -L_{I,t_k}^{(\ell)},\\
\mathcal{B}_{1,t_k}^{(\ell)} &\triangleq \bar{B}_{1,t_k} - L_{I,t_k}^{(\ell)} \bar{D}_{t_k},\ \mathcal{C}_{t_k}^{(\ell)} \triangleq H_{I,t_k}^{(\ell)} C_{t_k},\\
\mathcal{D}_{t_k}^{(\ell)} &\triangleq H_{I,t_k}^{(\ell)} \bar{D}_{t_k},\ \mathcal{G}_{t_k}^{(\ell)} \triangleq \mathrm{diag}_0^{b-1}\{G_{t_k+ih}^{(\ell)}\}.
\end{aligned}$$

For convenience of later analysis, we denote

$$\begin{aligned}
\Xi_{1,t_k}^{(\ell)} &\triangleq \mathrm{diag}\Big\{\Xi_{1,1,t_k}^{(\ell)}, -\epsilon_{3,t_k}^{(\ell)}, (\epsilon_{2,t_k}^{(\ell)} - \epsilon_{1,t_k}^{(\ell)})\Omega_{t_k}^{-1}, -\frac{\epsilon_{3,t_k}^{(\ell)}}{b}(\mathcal{G}_{t_k}^{(\ell)})^{-1}\Big\},\\
\Xi_{1,1,t_k}^{(\ell)} &\triangleq \epsilon_{1,t_k}^{(\ell)}\delta_1 - \epsilon_{2,t_k}^{(\ell)}\delta_2 + \epsilon_{3,t_k}^{(\ell)} + \epsilon_{4,t_k}^{(\ell)} + \lambda_{t_k}^{(\ell)} - 1,\\
\bar{\Xi}_{t_k}^{(\ell)} &\triangleq \begin{bmatrix} 0 & \mathcal{A}_{t_k}^{(\ell)} R_{t_k}^{(\ell)} & \mathcal{L}_{t_k}^{(\ell)} & \bar{B}_{2,t_k} \end{bmatrix},\\
\breve{\Upsilon}_{t_{k+1}}^{(\ell)} &\triangleq \mathrm{diag}\{\underbrace{\Upsilon_{t_{k+1}}^{(\ell)}, \cdots, \Upsilon_{t_{k+1}}^{(\ell)}}_{\kappa}\}\\
(\Psi_{t_k}^{(\ell)})^T &\triangleq \Big[\vartheta_{1,t_k}(\mathcal{B}_{1,t_k}^{(\ell)})^T\ \vartheta_{2,t_k}(\mathcal{B}_{1,t_k}^{(\ell)})^T\ \cdots \vartheta_{\kappa,t_k}(\mathcal{B}_{1,t_k}^{(\ell)})^T\Big].\\
\Phi_{1,t_k}^{(\ell)} &\triangleq \mathrm{diag}\{\Phi_{1,1,t_k}^{(\ell)}, -Q_{t_k}^{(\ell)}, (\varepsilon_{2,t_k}^{(\ell)} - \varepsilon_{1,t_k}^{(\ell)})\Omega_{t_k}^{-1}, -\gamma^2 U_d, \ldots\\
&\qquad -\gamma^2 U_{\bar{f}} - \frac{\varepsilon_{3,t_k}^{(\ell)}}{b}(\mathcal{G}_{t_k}^{(\ell)})^{-1}\Big\},\\
\Phi_{1,1,t_k}^{(\ell)} &\triangleq \varepsilon_{1,t_k}^{(\ell)}\delta_1 - \varepsilon_{2,t_k}^{(\ell)}\delta_2 + \varepsilon_{3,t_k}^{(\ell)},\\
(\Phi_{2,t_k}^{(\ell)})^T &\triangleq \mathrm{col}\Big\{0, (\mathcal{C}_{t_k}^{(\ell)})^T, (H_{I,t_k}^{(\ell)})^T, (\mathcal{D}_{t_k}^{(\ell)})^T, 0\Big\},\\
(\Phi_{3,t_k}^{(\ell)})^T &\triangleq \mathrm{col}\Big\{0, (\mathcal{A}_{t_k}^{(\ell)})^T, (\mathcal{L}_{t_k}^{(\ell)})^T, (\mathcal{B}_{1,t_k}^{(\ell)})^T, \bar{B}_{2,t_k}^T\Big\}.
\end{aligned}$$

and $R_{t_k}^{(\ell)}$ is the factorization of $\Upsilon_{t_k}^{(\ell)}$ (i.e. $\Upsilon_{t_k}^{(\ell)} \triangleq R_{t_k}^{(\ell)}(R_{t_k}^{(\ell)})^T$).

Theorem 4.4. *Consider the multi-rate systems (4.1) and (4.2) with constrained fault (4.3) and (4.44). Let the disturbance attenuation level $\gamma > 0$, the positive definite weighted matrices $U_d, U_{\bar{f}}, U_e$, the constraint matrices $\Upsilon_{t_k}^{(\ell)}$ and the triggering parameters $\delta_1, \delta_2, \Omega_{t_k}$ be given. Then, for the fault isolation systems (4.45), the H_∞ performance criterion (4.12) and the variance-constraint (4.13) are guaranteed simultaneously, if there exist sequences of real-valued matrices $Q_{t_k}^{(\ell)}, L_{I,t_k}^{(\ell)}, H_{I,t_k}^{(\ell)}$ and sequences of non-negative scalars $\lambda_{t_k}^{(\ell)}$, $\varepsilon_{i,t_k}^{(\ell)}, \epsilon_{j,t_k}^{(\ell)}$ ($i = 1,2,3;\ j = 1,2,3,4$), for all $0 \le t_k \le N-1$ with initial condition (4.16), satisfying the following recursive matrix inequalities*

$$\bar{\Phi}_{t_k} + \begin{bmatrix} \Phi_{1,t_k}^{(\ell)} & (\Phi_{2,t_k}^{(\ell)})^T & (\Phi_{3,t_k}^{(\ell)})^T \\ 0 & -I & 0 \\ 0 & 0 & -(Q_{t_{k+1}}^{(\ell)})^{-1} \end{bmatrix} \le 0 \tag{4.47}$$

$$\begin{bmatrix} \Xi_{1,t_k}^{(\ell)} & (\bar{\Xi}_{t_k}^{(\ell)})^T \\ 0 & -\Upsilon_{t_{k+1}}^{(\ell)} \end{bmatrix} \le 0 \tag{4.48}$$

$$\begin{bmatrix} -\lambda_{t_k}^{(\ell)} & (\Psi_{t_k}^{(\ell)})^T \\ 0 & -\breve{\Upsilon}_{t_{k+1}}^{(\ell)} \end{bmatrix} \leq 0 \tag{4.49}$$

Proof. Following the similar line of getting Theorem 4.3, we can complete the proof of this theorem and omit it here. □

After achieving the torus-event-based fault isolation estimators by utilizing Theorem 4.4, the residual matching function $J^{(\ell)}(t_k)$ and its threshold $J_{th}^{(\ell)}$ corresponding to $\ell-$ type fault are chosen as follows:

$$J^{(\ell)}(t_k) \triangleq \mathrm{E}\Big\{\sum_{i=t_{k_0}}^{t_k}[r^{(\ell)}(i)]^T r^{(\ell)}(i)\Big\}^{\frac{1}{2}}, J_{th}^{(\ell)} \triangleq \sup_{\substack{d(t_k)\in\mathfrak{W}_{t_k} \\ \bar{f}(t_k)=0}} J^{(\ell)}(t_k)$$

where t_{k_0} represents the initial time.

The fault-isolation decision scheme of the $\ell-$type fault $\hat{f}^{(\ell)}(t_k)$ is based on the following principle:

$$\begin{cases} J^{(\ell)}(t_k) \leq J_{th}^{(\ell)} \\ J^{(s)}(t_k) > J_{th}^{(s)}, \quad s \in \{1,2,\cdots,q\}/\ell \end{cases} \tag{4.50}$$

then the occurrence of the fault $\hat{f}^{(\ell)}(t_k)$ can be deduced. The absolute fault isolation time t_I is defined as $t_I^{(\ell)} \triangleq \max_{t_k}\{J^{(s)}(t_k) > J_{th}^{(s)}, \ s \in \{1,2,\cdots,q\}/\ell\}$.

From what has been discussed above, the fault detection and isolation scheme for the multi-rate time-varying systems can be handled by employing the following **Algorithm 4.1**.

Algorithm 4.1 The algorithm of fault detection and isolation.

Step 1. Design the fault detection filter by using Theorem 4.3.
Step 2. Employ the fault detection scheme (4.14) with the obtained torus-event-based fault detection estimator in step 1 to construct the residual evaluation function and achieve its threshold.
Step 3. After detecting the fault in step 2, activate the fault isolation estimators to devise the torus-event-based fault isolation estimators by using Theorem 4.4.
Step 4. Recur to the obtained torus-event-based fault isolation estimators in step 3, isolate the fault by using the fault isolation scheme (4.50).

Remark 4.4. *Compared to the traditional fault isolation with generalized observer scheme [7], our results have the following three significant characteristics: 1) a more comprehensive system is investigated that takes the multi-rate sampling, the event-triggered mechanism and the time-varying fashion into account; 2) a more practical analysis framework with variance-constrain and H_∞ performance is adopted to design the torus-event-based fault isolation estimators with ellipsoid-constrained fault; and 3) the residual matching function is chosen to isolate the fault by fully considering the fault information of experts. Therefore, the fault isolation strategy in this section is more in line with the engineering practice.*

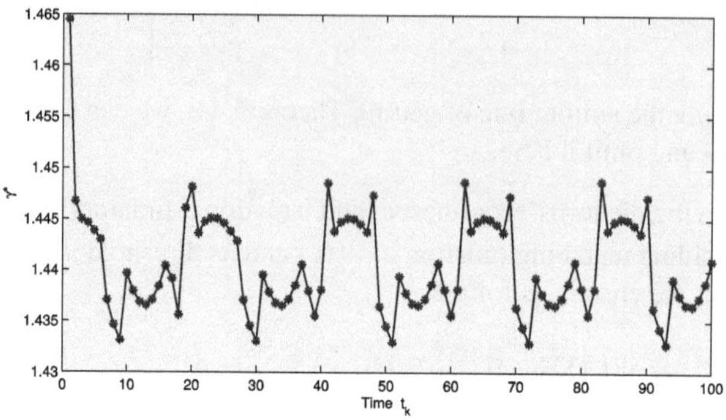

Figure 4.1: The minimum H_∞ performance level with given variance constraint matrix $\Theta_{t_k} = \text{diag}\{0.4, 0.3\}$ and upper bound of triggering torus $\delta_2 = 10^{-5}$.

4.1.3 Illustrative Examples

In this section, two numerical examples are presented to demonstrate the effectiveness of the proposed fault detection and isolation scheme for multi-rate time-varying systems.

Example 1 Consider the stochastic multi-rate sampled-data systems with the following parameters:

$$
\begin{aligned}
A_{T_k} &= \begin{bmatrix} 0.5 & 0.3 \\ 0.3 & 0.4 \end{bmatrix} + \Delta_{T_k},\ B_{1,T_k} = \begin{bmatrix} 0.7 & 0.1 \\ 0.1 & 0.9 \end{bmatrix} + \Delta_{T_k}, \\
B_{2,T_k} &= \begin{bmatrix} 0.6 & 0.2 \\ 0.2 & 0.4 \end{bmatrix} + \Delta_{T_k},\ C_{t_k} = D_{t_k} = I, \\
\Delta_{T_k} &= 10^{-2} \times \text{diag}\{\sin(0.1T_k), \cos(0.1T_k)\}, \\
F_{T_k} &= 5.0 \times 10^{-3},\ \Omega_{t_k} = \text{diag}\{1, 0.8\}, \\
U_\nu &= \text{diag}\{0.9, 1\} = U_\omega = U_f,\ U_e = \text{diag}\{2, 1\} = P_0, \\
U_d &= \text{diag}\{U_v, U_\omega\},\ h = T_{k+1} - T_k = 1, \\
W_{T_k} &= 10^{-2}I,\ V_{t_k} = 10^{-4}I,\ \delta_1 = 2.0 \times 10^{-5}, \\
\Theta_{t_k} &\triangleq \begin{bmatrix} \Theta_{1,t_k} & \Theta_{3,t_k} \\ \Theta_{3,t_k} & \Theta_{2,t_k} \end{bmatrix},\ e(t_k) \triangleq \text{col}\{e_1(t_k), e_2(t_k)\},
\end{aligned}
$$

the initial state and its estimation are set as $x(0) = \text{col}\{0.2, 0.4\}$, $\hat{x}(0) = \text{col}\{0.25, 0.35\}$, respectively. By solving the optimization problems (4.39), (4.40) and (4.41), we achieve separately the law of dynamic evolution for the minimal H_∞ performance level, the minimal upper bound of estimation error variance and the minimal upper bound of triggering-torus as Figs. 4.1– 4.3.

Example 2 In this example, the experiment of the three-tank system in the framework of multi-rate sampled data is considered, that is, the plant (4.1) is sampled with period $T_p = h$, and the sampling period of measurement (4.2) is assumed as $T_m = 3h$, other system parameters from [29] are listed as follows:

$$\begin{aligned} A &= \begin{bmatrix} 0.9908 & 0 & 0.0091 \\ 0 & 0.9856 & 0.0072 \\ 0.0091 & 0.0072 & 0.9836 \end{bmatrix}, \ C = \begin{bmatrix} 1 & 0 & 0 \\ 0 & 1 & 0 \end{bmatrix}, \\ B_1 &= B_2 = \begin{bmatrix} 64.6627 & 0.0007 & 0.2978 \\ 0.0007 & 64.4908 & 0.2358 \\ 0.2978 & 0.2358 & 64.4271 \end{bmatrix}, \\ D &= \text{diag}\{1, 1\}, \ \Omega_{t_k} = \text{diag}\{1, 0.8\} = U_\nu, \\ U_\omega &= \text{diag}\{0.9, 1, 0.8\} = U_f, \ U_d = \text{diag}\{U_\nu, U_\omega\}, \\ U_e &= \text{diag}\{2, 1, 1.5\} = P_0, \ \delta_2 = 0.1. \end{aligned}$$

Considering the existence of unmodeled dynamic and the perturbation of working point, we choose the following time-varying parameters:

$$\begin{aligned} \triangle A_{T_k} &\triangleq \sin(0.1 \times T_k) \times \begin{bmatrix} 0 & 0 & 0 \\ 0 & 0 & -0.001 \\ 0 & 0 & 0.001 \end{bmatrix}, \\ \triangle B_{1,T_k} &= \triangle B_{2,T_k} \triangleq \sin(0.1 \times T_k) \times \begin{bmatrix} 0 & 0 & 0 \\ 0 & 0 & 0.001 \\ 0 & 0 & -0.003 \end{bmatrix}, \\ A_{T_k} &\triangleq \triangle A_{T_k} + A, \ B_{1,T_k} = B_{2,T_k} \triangleq \triangle B_{1,T_k} + B_1, \\ C_{t_k} &\triangleq C, \ D_{t_k} \triangleq D. \end{aligned}$$

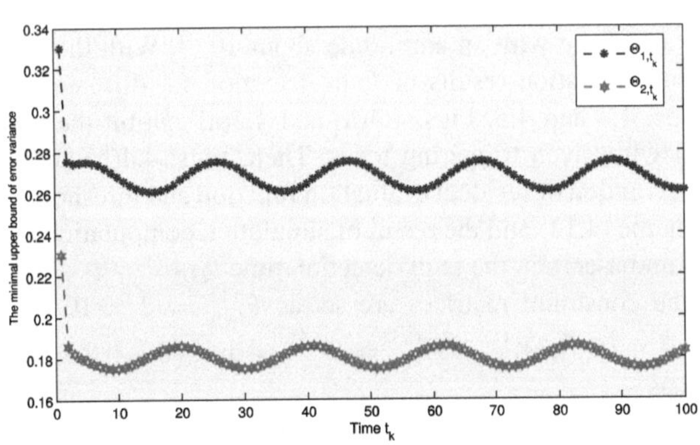

Figure 4.2: The minimum upper bound of variance $\mathrm{E}\{e_1(t_k)e_1^T(t_k)\}$, $\mathrm{E}\{e_2(t_k)e_2^T(t_k)\}$ with given H_∞ performance level $\gamma = 1.8$ and upper bound of triggering torus $\delta_2 = 10^{-5}$.

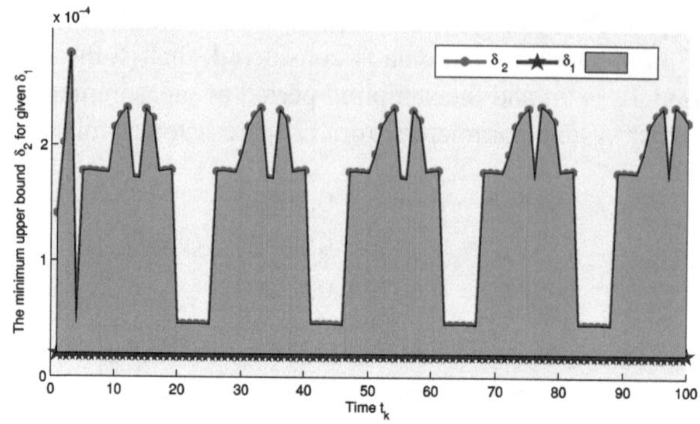

Figure 4.3: The minimum upper bound δ_2 of triggering torus with given H_∞ performance level $\gamma = 1.8$ and variance constraint matrix $\Theta_{t_k} = \text{diag}\{0.4, 0.3\}$.

Here, $x(T_k) \in \mathbb{R}^3$ represents the liquid levels of the three tanks; $y(t_k) \in \mathbb{R}^2$ describes the height measurements of tank $\mathcal{T}_1$ and tank $\mathcal{T}_2$; $w(T_k), v(T_k)$ are the unknown disturbance; $f(T_k) \in \mathbb{R}^3$ is the fault signal reflecting the leakages in the three tanks. Our aim is to detect and isolate the faults for multi-rate systems (4.1) and (4.2) in the presence of a leakage in tank $\mathcal{T}_1$, $\mathcal{T}_2$, or $\mathcal{T}_3$. It is assumed that the initial liquid level is $x(T_0) = \text{col}\{0.6813,\ 0.3321,\ 0.5534\}$ and its estimation is $\hat{x}(T_0) = \text{col}\{0.68, 0.34, 0.55\}$, and the uncorrelated zero-mean white external disturbances are set as $W_{T_k} = 10^{-8}I$ and $V_{t_k} = 10^{-6}I$, respectively.

Case I. The constraint matrix of fault is set as $F_{t_k} = 10^{-6}$ and the variance-constraint matrix is chosen as $\Theta_{t_k} = \text{diag}\{0.1, 0.2, 0.4\}$. The fault $f(T_k)$ is selected as

$$f_0(t_k) \triangleq \begin{cases} -10^{-4}, & 30 \le t_k \le 50; \\ 0, & otherwise. \end{cases} \tag{4.51}$$

that is there is a leakage with an amplitude about 10^{-4}. With the help of step 2 in Algorithm I, the simulation results of fault detection for different triggering-torus is shown in Figs. 4.4 and 4.5. Figs. 4.4(a) and 4.5(a) exhibit the release time and release interval with given triggering torus. Then, Figs. 4.4(b) and 4.5(b) provide corresponding evolution of residual evaluation function and threshold. By employing the decision scheme (4.14) and the result of simulation computation in Table 4.1, we detect the fault and ascertain the fault detection time t_D.

Case II. The constraint matrices are set as $F_{t_k}^{(1)} = 1 \times 10^{-6}$, $F_{t_k}^{(2)} = 5 \times 10^{-6}$, $F_{t_k}^{(3)} = 9 \times 10^{-6}$, $\Theta_{t_k}^{(1)} = \Theta_{t_k}^{(2)} = \Theta_{t_k}^{(3)} = \text{diag}\{0.1, 0.2, 0.4\}$, and the fault sets are chosen as

$$\mathscr{F} \triangleq \quad \{\hat{f}^{(1)}(t_k), \hat{f}^{(2)}(t_k), \hat{f}^{(3)}(t_k)\}$$

Table 4.1: The value of residual evaluation function and its threshold with different δ_1.

	J_{th}	$J(t_k)$	$J(t_k+1)$	t_D
$\delta_1 = 10^{-5}$	2.8815×10^{-4}	2.7918×10^{-4}	2.9561×10^{-4}	37
$\delta_1 = 10^{-4}$	9.1618×10^{-4}	9.1162×10^{-4}	9.1653×10^{-4}	39

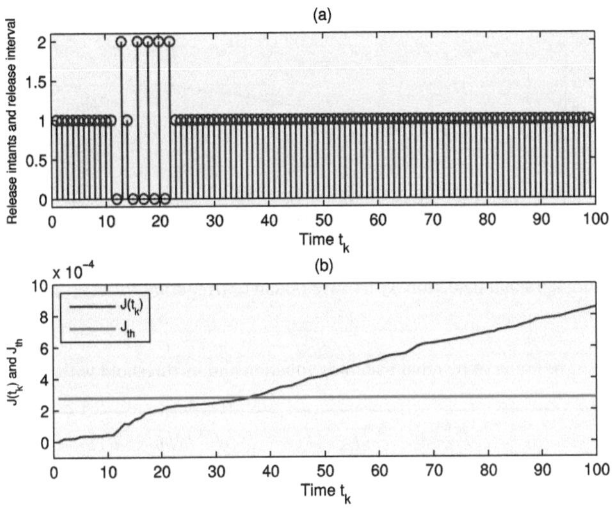

Figure 4.4: Fault detection with lower bound of triggering-torus $\delta_1 = 10^{-5}$.

$$\triangleq \left\{ \begin{bmatrix} f_O(t_k) \\ 0 \\ 0 \end{bmatrix}, \begin{bmatrix} 0 \\ f_O(t_k) \\ 0 \end{bmatrix}, \begin{bmatrix} 0 \\ 0 \\ f_O(t_k) \end{bmatrix} \right\} \tag{4.52}$$

Recur to Theorem 4.4, the simulation results of fault isolation for different triggering-torus can be achieved in Figs. 4.6 and 4.7. Figs. 4.6(a) and 4.7(a) show the release time and release interval, Figs. 4.6(b)–(d) and 4.7(b)–(d) provide the evolution of residual matching function and its threshold for estimators $\ell = 1, 2, 3$, respectively. It can be seen from that the residual matching function of estimator 3 always remains below its threshold, whereas corresponding estimators 1, 2 exceed their thresholds. Therefore, the emerging fault can be confirmed as 3-type fault by employing the fault isolation scheme (4.50). Furthermore, the values of residual matching function and its threshold for different lower bound δ_1 are listed in Tables 4.2 and 4.3, respectively. As a result, the fault isolation time is judged to be $t_I = 45$ for $\delta_1 = 10^{-5}$ and $t_I = 53$ for $\delta_1 = 10^{-4}$, respectively.

Remark 4.5. *It's concluded from Theorems 4.3 and 4.4 that the bound of triggering-torus δ_1, δ_2 is directly related to the design of torus-event-based fault detection estimator and torus-event-based fault isolation estimators. Actually, as one of the major*

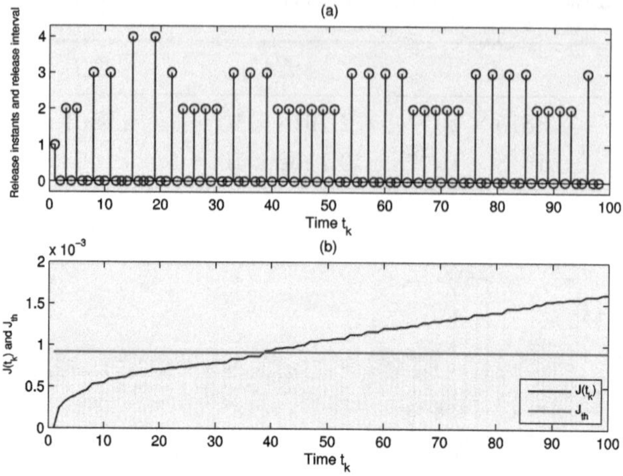

Figure 4.5: Fault detection with lower bound of triggering-torus $\delta_1 = 10^{-4}$.

Table 4.2: The value of residual matching function and its threshold with $\delta_1 = 10^{-5}$.

	$J_{th}^{(\ell)}$	$J^{(\ell)}(t_k)$	$J^{(\ell)}(t_k+1)$	t_k
$\ell = 1$	0.0017	0.0016	0.0018	44
$\ell = 1$	0.0017	0.0016	0.0018	44
$\ell = 2$	0.0036	0.0035	0.0039	39

Table 4.3: The value of residual matching function and its threshold with $\delta_1 = 10^{-4}$.

	$J_{th}^{(\ell)}$	$J^{(\ell)}(t_k)$	$J^{(\ell)}(t_k+1)$	t_k
$\ell = 1$	0.0030	0.0029	0.0031	52
$\ell = 2$	0.0040	0.0039	0.0040	39

performance of fault detection and isolation, the fault detection time t_D and fault isolation time t_I are affected by the bound of triggering-torus, and it can be seen by comparing the simulation Figs. 4.4–4.7 and Tables 4.1–4.3.

4.2 Event-Based Fault Diagnosis with Bounded Unknown Fault

In engineering practice, the disturbance or fault signals are often persistent but reside within certain elliptic sets. Such kind of ellipsoidal state estimation problem cannot

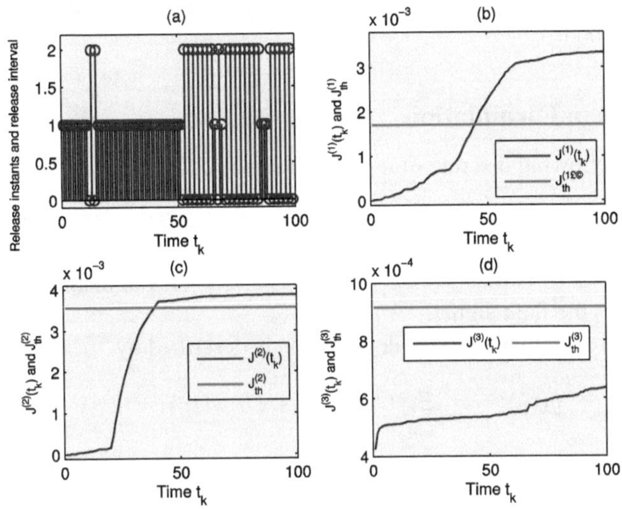

Figure 4.6: Fault isolation with lower bound of triggering-torus $\delta_1 = 10^{-5}$.

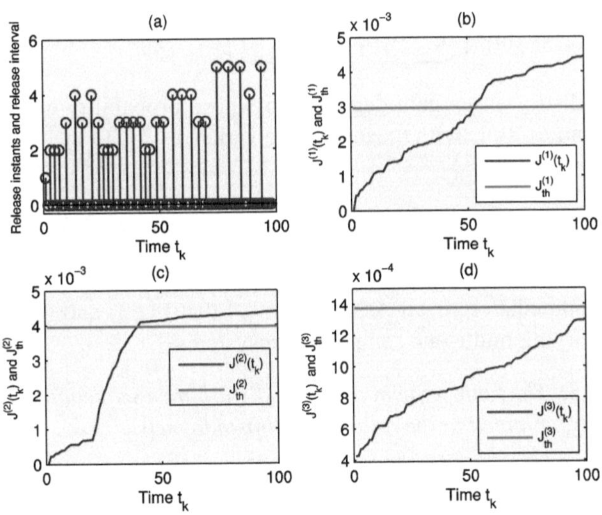

Figure 4.7: Fault isolation with lower bound of triggering-torus $\delta_1 = 10^{-4}$.

be properly handled by the conventional constrained vector method [143,162]. In this subsection, an annulus-event-based fault detection, isolation and estimation scheme for time-varying multi-rate systems with sensor degradation as well as unknown but bounded disturbances/faults are addressed by developing a new set-membership estimation method.

4.2.1 Problem Formulation

Consider the following discrete time-varying system:

$$x(T_{k+1}) = A_{T_k}x(T_k) + B_{1,T_k}\omega(T_k) + B_{2,T_k}f(T_k) \tag{4.53}$$

where $x(T_k) \in \mathbb{R}^{n_x}$ is the state vector, $\omega(T_k) \in \mathbb{R}^{n_\omega}$ is the disturbance input, and $f(T_k) \in \mathbb{R}^{n_f}$ is the fault signal.

The measurement with sensor degradation is described by

$$\begin{aligned} y(t_k) &= \Xi_{t_k}C_{t_k}x(t_k) + D_{t_k}\nu(t_k) \\ &= \sum_{s=1}^{m}\alpha_s(t_k)C_{s,t_k}x(t_k) + D_{t_k}\nu(t_k) \end{aligned} \tag{4.54}$$

where $y(t_k) \in \mathbb{R}^m$ is the measured output, $\nu(t_k) \in \mathbb{R}^{n_\nu}$ is the measurement noise. A_{T_k}, B_{1,T_k}, B_{2,T_k}, C_{t_k}, and D_{t_k} are real-valued time-varying matrices with appropriate dimensions. $\Xi_{t_k} \triangleq \text{diag}_1^m\{\alpha_s(t_k)\}$ where $\alpha_s(t_k)$ $(s = 1, ..., m)$ are m unrelated random variables taking values on the interval $[0, 1]$, and the matrix C_{s,t_k} is defined by $C_{s,t_k} \triangleq \text{diag}\{\underbrace{0, \cdots, 0}_{s-1}, 1, \underbrace{0, \cdots, 0}_{m-s}\}C_{t_k}$. The variable $\alpha_s(t_k)$ accounts for the probabilistic sensor gain degradation whose probability density function is $p_s(\cdot)$ on the annulus $[0, 1]$ with means $\bar{\alpha}_s$ and variances $\tilde{\bar{\alpha}}_s^2$. In the sequel, we denote $\bar{\Xi} \triangleq \mathbb{E}(\Xi_{t_k}) = \text{diag}_1^m\{\bar{\alpha}_s\}$.

Assumption 4.4. *The sampling period of (4.53) is set as $h \triangleq T_{k+1} - T_k$, and the measurement (4.54) is assumed to be $t_{k+1} - t_k = bh$, where b is a positive integer.*

It follows immediately from Assumption 4.4 that (4.53) and (4.54) constitute a class of time-varying multi-rate sampled-data systems.

Assumption 4.5. *The fault sequences $f(t_k)$, disturbance sequences $\omega(t_k)$ and $\nu(t_k)$ are, respectively, confined to the following ellipsoidal sets:*

$$\begin{cases} \mathfrak{W}_{t_k} \triangleq \{\omega(t_k) : \omega^T(t_k)W_{t_k}^{-1}\omega(t_k) \le 1\} \\ \mathfrak{F}_{t_k} \triangleq \{f(t_k) : \ f^T(t_k)F_{t_k}^{-1}f(t_k) \le 1\} \\ \mathfrak{M}_{t_k} \triangleq \{\nu(t_k) : \ \nu^T(t_k)M_{t_k}^{-1}\nu(t_k) \le 1\} \end{cases} \tag{4.55}$$

where W_{t_k}, F_{t_k}, and M_{t_k} are known positive definite matrices with compatible dimensions characterizing the scope of the ellipsoid.

Remark 4.6. *In most existing literature concerning model-based fault diagnosis, the external disturbances (or noises) have been assumed to be deterministic or stochastic. For deterministic disturbances/noises, the H_∞ filtering technique ([38]), the parity-equation approach [140] as well as the optimization-based method*

([181, 182]) have been utilized to deal with the disturbances/noises with bounded-energy (i.e. $w_k \in \ell_2[0, \infty)$). As for the Gaussian-type stochastic disturbances/noises, the Kalman filtering approach has proven to be particularly effective [183]. In practical engineering, due to the man-made electromagnetic interference as well as other natural sources, the disturbances are sometimes neither stochastic nor energy-bounded. Rather, the disturbances/noises are unknown but bounded within certain ellipsoidal set, for which the aforementioned estimation methods are no longer applicable. Fortunately, the set-membership estimation method is known to be suitable for the ellipsoidal-set-constrained disturbances/noises, and has therefore been investigated in this section.

the annulus-event-based communication mechanism is taken into consideration to reduce the communication frequency. Suppose that the sequence of the triggering instants is t_{k_i} satisfying $0 \leq t_{k_0} \leq t_{k_1} \leq \cdots \leq t_{k_i} \leq \cdots$.

Define

$$\sigma(t_k) \triangleq y(t_{k_i}) - y(t_k) \tag{4.56}$$

where $y(t_{k_i})$ is the measurement at the latest event time t_{k_i}, $y(t_k)$ is the current measurement. Then, the sequence of annulus-event-triggering instants is determined iteratively by

$$\begin{aligned} t_{k_{i+1}} \quad &\triangleq \quad \inf\{t_k \in \mathbb{Z}_{k \geq 0} \mid t_k > t_{k_i}, \\ &\qquad \delta^- < \sigma^T(t_k)\sigma(t_k) < \delta^+\} \end{aligned} \tag{4.57}$$

where the triggering threshold satisfies $0 < \delta^- < \delta^+$.

Accordingly, only the received measurement signal satisfying the event condition (4.57) will be transmitted to the filer/estimator, and the zero-order holder is applied to hold the signal transmitted to the filter/estimator until a new measurement is transmitted. Let $\bar{y}(t_k)$ be the signal received by the estimator, we have

$$\bar{y}(t_k) = y(t_{k_i}), \quad t_{k_i} \leq t_k < t_{k_{i+1}}. \tag{4.58}$$

Remark 4.7. *In comparison with the conventional time-triggered communication, the event-triggering strategy alleviates the network burden by avoiding unnecessary waste of computation/communication resources while maintaining desired system performance. The annulus triggering condition defined in (4.57) exhibits two distinguished features as follows. i) From an engineering viewpoint, it is often the case that the measurement error $\sigma(t_k)$ resides within certain limit annulus. In an extreme case, a largely abnormal measurement may trigger the action (e.g. monitoring and maintenance). ii) The triggering condition (4.57) is quite general that covers a well-studied triggering condition as the special case with $\delta^+ = +\infty$.*

4.2.2 Fault Diagnosis and Fault Estimation

A. Design of Fault Detection Filters

By applying the relation (4.53) recursively with some $t_k = T_k$, one obtains the following equation with time scale t_k:

$$x(t_{k+1}) = \bar{A}_{t_k} x(t_k) + \bar{B}_{1,t_k} d(t_k) + \bar{B}_{2,t_k} \bar{f}(t_k) \tag{4.59}$$

where

$$
\begin{aligned}
\bar{\omega}(t_k) &\triangleq \text{col}_0^{b-1}\{\omega(t_k+ih)\},\ d(t_k) \triangleq \text{col}\{\nu(t_k),\bar{\omega}(t_k)\},\\
\bar{f}(t_k) &\triangleq \text{col}_0^{b-1}\{f(t_k+ih)\},\\
\bar{A}_{t_k} &\triangleq \underbrace{A_{t_k+(b-1)h}A_{t_k+(b-2)h}\cdots A_{t_k+h}A_{t_k}}_{b},\\
\bar{B}_{1,t_k} &\triangleq \Big[0\ \ \underbrace{A_{t_k+(b-1)h}A_{t_k+(b-2)h}\cdots A_{t_k+h}}_{b-1}B_{1,t_k}\\
&\quad \underbrace{A_{t_k+(b-1)h}A_{t_k+(b-2)h}\cdots A_{t_k+2h}}_{b-2}B_{1,t_k+h}\\
&\quad \cdots\ \ A_{t_k+(b-1)h}B_{1,t_k+(b-2)h}\ \ B_{1,t_k+(b-1)h}\Big],\\
\bar{B}_{2,t_k} &\triangleq \Big[\underbrace{A_{t_k+(b-1)h}A_{t_k+(b-2)h}\cdots A_{t_k+h}}_{b-1}B_{2,t_k}\\
&\quad \underbrace{A_{t_k+(b-1)h}A_{t_k+(b-2)h}\cdots A_{t_k+2h}}_{b-2}B_{2,t_k+h}\\
&\quad \cdots\ \ A_{t_k+(b-1)h}B_{2,t_k+(b-2)h}\ \ B_{2,t_k+(b-1)h}\Big].
\end{aligned}
$$

Based on the received signal $\bar{y}(t_k)$, we adopt the following annulus-event-based fault detection estimator:

$$
\begin{cases}
\hat{x}(t_{k+1}) = \bar{A}_{t_k}\hat{x}(t_k) + L_{t_k}\left[\bar{y}(t_k) - \hat{y}(t_k)\right]\\
\hat{y}(t_k) = \bar{\Xi}C_{t_k}\hat{x}(t_k)
\end{cases}
\tag{4.60}
$$

where $\hat{x}(t_k)$ is the estimate of state $x(t_k)$ and L_{t_k} is the estimator parameter to be determined.

Setting $e(t_k) \triangleq x(t_k) - \hat{x}(t_k)$ and $\xi(t_k) \triangleq \text{col}\{e(t_k), x(t_k)\}$, the augmented systems is derived from (4.55), (4.59) and (4.60) as follows:

$$
\begin{aligned}
\xi(t_{k+1}) &= (\mathcal{A}_{t_k} + \sum_{s=1}^{m}\tilde{\alpha}_s(t_k)\tilde{\mathcal{A}}_{s,t_k})\xi(t_k) + \mathcal{L}_{t_k}\sigma(t_k) +\\
&\quad \mathcal{B}_{1,t_k}d(t_k) + \mathcal{B}_{2,t_k}\bar{f}(t_k)
\end{aligned}
\tag{4.61}
$$

with following constraints

$$
d^T(t_k)\mathcal{W}_{t_k}^{-1}d(t_k) \leq b+1 \tag{4.62}
$$

$$
\bar{f}^T(t_k)\mathcal{F}_{t_k}^{-1}\bar{f}(t_k) \leq b \tag{4.63}
$$

where

$$
\mathcal{A}_{t_k} \triangleq \begin{bmatrix} \bar{A}_{t_k} - L_{t_k}\bar{\Xi}C_{t_k} & 0\\ 0 & \bar{A}_{t_k}\end{bmatrix},\ \mathcal{L}_{t_k} \triangleq \begin{bmatrix} -L_{t_k}\\ 0\end{bmatrix},
$$

$$\begin{aligned}
\tilde{\mathcal{A}}_{s,t_k} &\triangleq \begin{bmatrix} 0 & -L_{t_k}C_{s,t_k} \\ 0 & 0 \end{bmatrix}, \ \mathcal{B}_{2,t_k} \triangleq \begin{bmatrix} \bar{B}_{2,t_k} \\ \bar{B}_{2,t_k} \end{bmatrix}, \\
\mathcal{B}_{1,t_k} &\triangleq \begin{bmatrix} \bar{B}_{1,t_k} - L_{t_k}\bar{D}_{t_k} \\ \bar{B}_{1,t_k} \end{bmatrix}, \ \bar{D}_{t_k} \triangleq [D_{t_k} \ \underbrace{0 \cdots 0}_{b}], \\
\mathcal{W}_{t_k} &\triangleq \mathrm{diag}\{M_{t_k}, \bar{W}_{t_k}\}, \ \bar{W}_{t_k} \triangleq \mathrm{diag}_0^{b-1}\{W_{t_k+ih}\}, \\
\mathcal{F}_{t_k} &\triangleq \mathrm{diag}_0^{b-1}\{F_{t_k+ih}\}.
\end{aligned}$$

Before proceeding further, we give the following assumption and lemma.

Assumption 4.6. *The initial state $x(t_0)$ and its estimate $\hat{x}(t_0)$ satisfy*

$$\xi^T(t_0)P_{t_0}^{-1}\xi(t_0) \le 1 \tag{4.64}$$

where $\xi(t_0) \triangleq \mathrm{col}\{x(t_0) - \hat{x}(t_0), x(t_0)\}$ and $P_{t_0} > 0$ is a given positive definite matrix.

Lemma 4.1. *(S-Procedure) Let $\psi_0(*), \psi_1(*), \cdots, \psi_p(*)$ be quadratic functions of the variable $\varsigma \in \mathbb{R}^n$: $\psi_j(\varsigma) \triangleq \varsigma^T X_j \varsigma$ $(j = 0, 1, 2, \cdots, p)$, where $X_j^T = X_j$. If there exist $\varepsilon_j \ge 0$ $(j = 0, 1, 2, \cdots, p)$ such that*

$$X_0 - \sum_{j=0}^{p} \varepsilon_j X_j \le 0 \tag{4.65}$$

then the following is true

$$\psi_1(\varsigma) \le 0, \cdots, \psi_p(\varsigma) \le 0 \Longrightarrow \psi_0(\varsigma) \le 0 \tag{4.66}$$

For convenience of later analysis, we denote

$$\begin{aligned}
\Phi_{1,t_k} &\triangleq \mathrm{diag}\{\Phi_{1,1,t_k}, -\varepsilon_{1,t_k}I, -\varepsilon_{2,t_k}I + \varepsilon_{3,t_k}I, \\
&\quad -\frac{\varepsilon_{4,t_k}}{b+1}\mathcal{W}_{t_k}^{-1}, -\frac{\varepsilon_{5,t_k}}{b}\mathcal{F}_{t_k}^{-1}\}, \\
\Phi_{1,1,t_k} &\triangleq -1 + \sum_{i=1,4,5} \varepsilon_{i,t_k} - \delta^+\varepsilon_{2,t_k} + \delta^-\varepsilon_{3,t_k}, \\
\Phi_{2,t_k}^T &\triangleq \mathrm{col}\Big\{0, (\mathcal{A}_{t_k}V_{t_k})^T, \mathcal{L}_{t_k}^T, \mathcal{B}_{1,t_k}^T, \mathcal{B}_{2,t_k}^T\Big\}, \\
\Phi_{3,t_k}^T &\triangleq \begin{bmatrix} -\tilde{\bar{\alpha}}_1(\tilde{\mathcal{A}}_{1,t_k}V_{t_k})^T & \cdots & -\tilde{\bar{\alpha}}_m(\tilde{\mathcal{A}}_{m,t_k}V_{t_k})^T \end{bmatrix}, \\
V_{t_k} &\triangleq \mathrm{diag}\{V_{1,t_k}, V_{2,t_k}\}, \ P_{t_k} \triangleq \mathrm{diag}\{P_{1,t_k}, P_{2,t_k}\},
\end{aligned}$$

and V_{t_k} is the factorization of P_{t_k} (i.e. $P_{t_k} = V_{t_k}V_{t_k}^T$).

Theorem 4.5. *Consider the time-varying multi-rate system (4.53)–(4.54) with constraints (4.62)–(4.63). If there exist sequences of real-valued matrices $P_{t_{k+1}}$, L_{t_k} and* *sequences of non-negative scalars* ε_{j,t_k} $(j = 1, 2, \cdots, 5)$ *satisfying the following recursive matrix inequalities*

$$\begin{bmatrix} \Phi_{1,t_k} & \Phi_{2,t_k}^T & \Phi_{3,t_k}^T \\ 0 & -P_{t_{k+1}} & 0 \\ 0 & 0 & -\mathrm{diag}_m\{P_{t_{k+1}}\} \end{bmatrix} \le 0, \tag{4.67}$$

then the state $x(t_k)$ *resides in its ellipsoidal set* $\mathcal{E} \triangleq \Big\{(x(t_k), \hat{x}(t_k)) : \xi(t_k)^T P_{t_k}^{-1} \xi(t_k) \le 1\Big\}$.

Proof. The proof is performed by induction. First, it is known immediately from Assumption 4.6 that $\xi^T(t_0)P_{t_0}^{-1}\xi(t_0) \le 1$ holds. Second, if $\xi^T(t_k)P_{t_k}^{-1}\xi(t_k) \le 1$ is true at instant t_k, then there exits $\eta(t_k)$ such that

$$\|\eta(t_k)\| \le 1 \tag{4.68}$$

with $P_{t_k} \triangleq V_{t_k}V_{t_k}^T$ and $\xi(t_k) \triangleq V_{t_k}\eta(t_k)$.

Next, we shall proceed to prove the following inequality:

$$\xi^T(t_{k+1})P_{t_{k+1}}^{-1}\xi(t_{k+1}) \le 1 \tag{4.69}$$

Denoting $\zeta(t_k) \triangleq \mathrm{col}\{1, \eta(t_k), \sigma(t_k), d(t_k), \bar{f}(t_k)\}$, (4.61) can be further rewritten as

$$\xi(t_{k+1}) = \Sigma_{t_k}\zeta(t_k) \tag{4.70}$$

where $\Sigma_{t_k} \triangleq \begin{bmatrix} 0 & \Sigma_{1,t_k} & \mathcal{L}_{t_k} & \mathcal{B}_{1,t_k} & \mathcal{B}_{2,t_k} \end{bmatrix}$, $\Sigma_{1,t_k} \triangleq \mathcal{A}_{t_k}V_{t_k} + \sum_{s=1}^{m}\tilde{\alpha}_s(t_k)\tilde{\mathcal{A}}_{s,t_k}V_{t_k}$.

It follows from (4.57), (4.62), (4.63), and (4.68) that the vectors $\eta(t_k)$, $\sigma(t_k)$, $d(t_k)$ and $\bar{f}(t_k)$ satisfy

$$\begin{cases} \|\eta(t_k)\| \le 1 \\ \sigma(t_k)^T\sigma(t_k) \le \delta^- \\ -\sigma(t_k)^T\sigma(t_k) \le -\delta^+ \\ d^T(t_k)\mathcal{W}_{t_k}^{-1}d(t_k) \le b+1 \\ \bar{f}^T(t_k)\mathcal{F}_{t_k}^{-1}\bar{f}(t_k) \le b \end{cases} \tag{4.71}$$

which can be rearranged by means of $\zeta(t_k)$ as follows:

$$\begin{cases} \zeta^T(t_k)\mathrm{diag}\{-1, I, 0, 0, 0\}\zeta(t_k) \le 0 \\ \zeta^T(t_k)\mathrm{diag}\{-\delta^-, 0, I, 0, 0\}\zeta(t_k) \le 0 \\ \zeta^T(t_k)\mathrm{diag}\{\delta^+, 0, -I, 0, 0\}\zeta(t_k) \le 0 \\ \zeta^T(t_k)\mathrm{diag}\{-1, 0, 0, \frac{1}{b+1}\mathcal{W}_{t_k}^{-1}, 0\}\zeta(t_k) \le 0 \\ \zeta^T(t_k)\mathrm{diag}\{-1, 0, 0, 0, \frac{1}{b}\mathcal{F}_{t_k}^{-1}\}\zeta(t_k) \le 0 \end{cases} \tag{4.72}$$

On the other hand, it can be seen from Lemma 4.1 that if there exist non-negative scalars $\varepsilon_{j,t_k} \geq 0$ $(j = 1, 2, 3, 4, 5)$ such that

$$\begin{aligned}
& \mathbb{E}\Big(\Sigma_{t_k}^T P_{t_{k+1}}^{-1} \Sigma_{t_k}\Big) - \mathrm{diag}\{1, 0, 0, 0, 0\} \\
- \ & \varepsilon_{1,t_k} \mathrm{diag}\{-1, I, 0, 0, 0, 0\} \\
- \ & \varepsilon_{2,t_k} \mathrm{diag}\{\delta^+, 0, I, 0, 0\} - \varepsilon_{3,t_k} \mathrm{diag}\{-\delta^-, 0, -I, 0, 0\} \\
- \ & \varepsilon_{4,t_k} \mathrm{diag}\{-1, 0, 0, \frac{1}{b+1} \mathcal{W}_{t_k}^{-1}, 0\} \\
- \ & \varepsilon_{5,t_k} \mathrm{diag}\{-1, 0, 0, 0, \frac{1}{b} \mathcal{F}_{t_k}^{-1}\} \leq 0,
\end{aligned} \tag{4.73}$$

then (4.69) is guaranteed and the induction is accomplished. By employing Schur Complement to (4.73), (4.67) is obtained and the proof of Theorem 4.5 is then complete. □

B. Fault Detection Decision Scheme

Once the sequences of annulus-event-based fault detection estimator are designed, we now employ them to detect the fault. In this section, the method of residual evaluation function is used to do fault detection, and the residual signal is chosen as:

$$r(t_k) \triangleq \bar{y}(t_k) - \hat{y}(t_k) = y(t_{k_i}) - \bar{\Xi} C_{t_k} \hat{x}(t_k) \tag{4.74}$$

Based on the generated residual signal $r(t_k)$ in (4.74), the residual evaluation function $J(t_k)$ and its threshold J_{th} are defined as:

$$J(t_k) \triangleq \Big[\sum_{s=t_{k_0}}^{t_k} r^T(i) r(i)\Big]^{\frac{1}{2}}, \quad J_{th} \triangleq \sup_{\substack{d(t_k) \in \bar{\mathfrak{W}}_{t_k} \\ \bar{f}(t_k) = 0}} J(t_k)$$

where t_{k_0} represents the initial time of the residual evaluation.

The occurrence of fault can be detected by comparing $J(t_k)$ with J_{th} according to the following test rule:

$$\begin{cases} J(t_k) \geq J_{th} \Longrightarrow \text{alarm for fault} \\ J(t_k) < J_{th} \Longrightarrow \text{no fault} \end{cases} \tag{4.75}$$

and the fault detection time t_{k_D} is defined as $t_{k_D} \triangleq \min_{t_k} \{J(t_k) \geq J_{th}\}$.

Remark 4.8. *In Theorem 4.5, we examine how the event-triggered strategy and the multi-rate sampled-data pattern influence the fault detection performance over a finite horizon. Compared to the traditional design of fault detection filter in [14, 19, 140, 181], our results have the following three distinguishing features: 1) a more comprehensive system model is introduced that takes the multi-rate sampling, the event-triggered mechanism, the time-varying fashion and the sensor degradation* *into account; 2) a more practical analysis framework for set-membership estimation is established to investigate the estimator design problem with unknown but bounded disturbances/faults; and 3) quantitative relationships are investigated between the estimation performance index P_{1,t_k}, the event-triggering threshold δ^-, δ^+, the statistical information of sensor degradation $\bar{\alpha}_s, \tilde{\bar{\alpha}}_s^2$, the parameters of the ellipsoidal constraint $W_{t_k}, F_{t_k}, M_{t_k}$, and the multi-rate multiple b of the sampling period h. Similarly, these advantages can be found in the design of annulus-event-based fault isolation estimator and annulus-event-based fault estimator to be carried out in Theorems 4.6 and 4.7, respectively.*

Theorem 4.5 outlines the principles of designing the annulus-event-based fault detection estimator by solving the corresponding recursive matrix inequalities. It should be pointed out that, however, the proposed methodologies provide merely a feasible solution without consideration of the performance enhancement. In the following stage, two optimization problems based on Theorem 4.5 will be used to demonstrate the optimal selection of the filter parameters.

Corollary 4.5.1. *Consider the time-varying multi-rate system (4.53)–(4.54) with constraints (4.62)–(4.63). Let the triggering threshold δ^-, δ^+ be given. The ellipsoid constraint $P_{1,t_{k+1}}$ on estimation error can be minimized (in the sense of matrix trace) if there exist sequences of real-valued matrices $P_{2,t_{k+1}}, L_{t_k}$ and sequences of nonnegative scalars ε_{j,t_k} $(j = 1, 2, \cdots, 5)$ solving the following optimization problem:*

$$\min_{P_{1,t_{k+1}}, P_{2,t_{k+1}}, L_{t_k}, \varepsilon_{j,t_k}} \operatorname{tr}\left[P_{1,t_{k+1}}\right] \text{ subject to } (4.67) \tag{4.76}$$

Corollary 4.5.2. *Consider the time-varying multi-rate system (4.53)–(4.54) with constraints (4.62)–(4.63). Let the lower bound of triggering threshold δ^- be given. The minimum upper bound δ^+ of triggering annulus can be guaranteed if there exist sequences of real-valued matrices $P_{1,t_{k+1}}, P_{2,t_{k+1}}, L_{t_k}$, sequences of nonnegative scalars ε_{j,t_k} $(j = 1, 2, \cdots, 5)$ solving the following optimization problem:*

$$\min_{P_{1,t_{k+1}}, P_{2,t_{k+1}}, L_{t_k}, \varepsilon_{j,t_k}} \delta^+ \text{ subject to } (4.67) \tag{4.77}$$

Remark 4.9. *Corollaries 4.5.1 and 4.5.2 can be utilized to design the optimal annulus-event-based fault detection estimator with minimum ellipsoid constraint P_{1,t_k} of estimation error or minimum triggering annulus $[\delta^-, \delta^+]$, respectively. Especially, it should be pointed out that there exists certain tradeoff between the constraint of estimation error and the triggering annulus, and such a tradeoff could provide much flexibility in making compromise between estimation performance and triggering frequency. Then, we can summarize the design of annulus-event-based fault detection estimator in* **Algorithm 4.2**. *Furthermore, in the design process of annulus-event-based fault isolation estimator and annulus-event-based fault estimator, similar optimization problems and design algorithms can be found in Sections 4.1.1, we omit the corresponding narration due to the space limit.*

Algorithm 4.2 The algorithm of annulus-event-based fault detection estimator.

Step 1. Given the time horizon N, the triggering annulus $[\delta^-, \delta^+]$. Set $t_0 = 0$, the initial state $x(t_0)$ and the initial constrain matrix P_{1,t_0}, P_{2,t_0} satisfying Assumption 4.6.
Step 2. Calculate the matrix factorization V_{t_0} according to $P_{t_0} = V_{t_0}V_{t_0}^T$, solve recursive matrix inequalities (4.67) to obtain P_{1,t_1}, P_{2,t_1} and fault detection estimator L_{t_0}.
Step 3. Set $t_k = t_{k+1}$, update the matrices V_{1,t_k}, V_{2,t_k}, solve (4.67) to achieve $P_{1,t_{k+1}}, P_{2,t_{k+1}}$ and fault detection estimator L_{t_k}.
Step 4. If $t_k < N$ then go to Step 3, else Stop.

After a fault has been detected, the fault isolation module is activated. Now, we supply the design of annulus-event-based fault isolation estimators which are based on set-membership estimation method.

A. Design of Fault Isolation Estimators

For isolation purposes, we assume that there are q types of possible fault function. Specifically, $f(T_k)$ belongs to a finite set of functions given by

$$\mathscr{F} \triangleq \{\hat{f}^{(1)}(T_k), \hat{f}^{(2)}(T_k), \cdots, \hat{f}^{(q)}(T_k)\} \tag{4.78}$$

By using the received signal $\bar{y}(t_k)$, the following q estimators are used as annulus-event-based fault isolation estimators for system (4.59):

$$\begin{cases} \hat{x}^{(\ell)}(t_{k+1}) = \bar{A}_{t_k}\hat{x}^{(\ell)}(t_k) + \bar{B}_{2,t_k}\hat{\bar{f}}^{(\ell)}(t_k) + K_{t_k}^{(\ell)}\left[\bar{y}(t_k) - \hat{y}^{(\ell)}(t_k)\right] \\ \quad \hat{y}^{(\ell)}(t_k) = \bar{\Xi}C_{t_k}\hat{x}^{(\ell)}(t_k), \quad (\ell = 1, 2, \cdots, q) \end{cases} \tag{4.79}$$

where $\hat{\bar{f}}^{(\ell)}(t_k) \triangleq \mathrm{col}_0^{b-1}\{\hat{f}^{(\ell)}(t_k + ih)\}$, $\hat{x}^{(\ell)}(t_k)$ is the estimate of state $x(t_k)$, $\hat{f}^{(\ell)}(t_k+ih)$ is the estimate of fault $f^{(\ell)}(t_k+ih)$, and $K_{t_k}^{(\ell)}$ are the estimator parameters to be determined.

Before proceeding further, we give the following assumption.

Assumption 4.7. *The sequences $\tilde{\bar{f}}^{(\ell)}(t_k)$ satisfy*

$$[\tilde{\bar{f}}^{(\ell)}(t_k)]^T[\bar{\mathcal{F}}^{(\ell)}]_{t_k}^{-1}\tilde{\bar{f}}^{(\ell)}(t_k) \le b \tag{4.80}$$

where $\tilde{f}^{(\ell)}(t_k+ih) \triangleq f(t_k+ih) - \hat{f}^{(\ell)}(t_k+ih)$, $\tilde{\bar{f}}^{(\ell)}(t_k) \triangleq \mathrm{col}_0^{b-1}\{\tilde{f}^{(\ell)}(t_k+ih)\}$, $\bar{\mathcal{F}}_{t_k}^{(\ell)} \triangleq \mathrm{diag}_0^{b-1}\{\bar{F}_{t_k+ih}^{(\ell)}\}$, $\bar{F}_{t_k+ih}^{(\ell)}$ are the constraint matrices characterizing the scope of the ellipsoid for $\tilde{f}^{(\ell)}(t_k+ih)$.

Denoting $\bar{e}^{(\ell)}(t_k) \triangleq x(t_k) - \hat{x}^{(\ell)}(t_k)$ and $\breve{f}^{(\ell)}(t_k) := \mathrm{col}\{\tilde{\bar{f}}^{(\ell)}(t_k), \bar{f}(t_k)\}$, the error system is obtained from (4.59), (4.79), and Assumption 4.7 as follows:

$$\begin{aligned}\bar{e}^{(\ell)}(t_{k+1}) &= \left(\bar{A}_{t_k} - K_{t_k}^{(\ell)}\bar{\Xi}C_{t_k}\right)\bar{e}^{(\ell)}(t_k) - K_{t_k}^{(\ell)}\sigma(t_k) \\ &\quad - \textstyle\sum_{s=1}^{m}\tilde{\alpha}_s(t_k)K_{t_k}^{(\ell)}C_{s,t_k}x(t_k) \\ &\quad + \left(\bar{B}_{1,t_k} - K_{t_k}^{(\ell)}\bar{D}_{t_k}\right)d(t_k) + \bar{B}_{2,t_k}\breve{f}^{(\ell)}(t_k)\end{aligned} \tag{4.81}$$

with following constraints

$$[\breve{f}^{(\ell)}(t_k)]^T[\mathcal{Z}_{t_k}^{(\ell)}]^{-1}\breve{f}^{(\ell)}(t_k) \le 2b \tag{4.82}$$

Then, setting $\bar{\xi}^{(\ell)}(t_k) \triangleq \text{col}\{\bar{e}^{(\ell)}(t_k), x(t_k)\}$, the following augmented system is derived from (4.59) and (4.81):

$$\begin{aligned}\bar{\xi}^{(\ell)}(t_{k+1}) &= \left(\bar{\mathcal{A}}_{t_k}^{(\ell)} + \textstyle\sum_{s=1}^{m}\tilde{\alpha}_s(t_k)\tilde{\bar{\mathcal{A}}}_{s,t_k}^{(\ell)}\right)\bar{\xi}^{(\ell)}(t_k) + \\ &\quad \mathcal{K}_{t_k}^{(\ell)}\sigma(t_k) + \bar{\mathcal{B}}_{1,t_k}^{(\ell)}d(t_k) + \bar{\mathcal{B}}_{2,t_k}\breve{f}^{(\ell)}(t_k)\end{aligned} \tag{4.83}$$

where

$$\begin{aligned}\bar{\mathcal{A}}_{t_k}^{(\ell)} &\triangleq \begin{bmatrix}\bar{A}_{t_k} - K_{t_k}^{(\ell)}\bar{\Xi}C_{t_k} & 0 \\ 0 & \bar{A}_{t_k}\end{bmatrix}, \ \mathcal{K}_{t_k}^{(\ell)} \triangleq \begin{bmatrix}-K_{t_k}^{(\ell)} \\ 0\end{bmatrix}, \\ \tilde{\bar{\mathcal{A}}}_{s,t_k}^{(\ell)} &\triangleq \begin{bmatrix}0 & -K_{t_k}^{(\ell)}C_{s,t_k} \\ 0 & 0\end{bmatrix}, \ \bar{\mathcal{B}}_{2,t_k} \triangleq \begin{bmatrix}\bar{B}_{2,t_k} & 0 \\ 0 & \bar{B}_{2,t_k}\end{bmatrix}, \\ \bar{\mathcal{B}}_{1,t_k}^{(\ell)} &\triangleq \begin{bmatrix}\bar{B}_{1,t_k} - K_{t_k}^{(\ell)}\bar{D}_{t_k} \\ \bar{B}_{1,t_k}\end{bmatrix}, \ \mathcal{Z}_{t_k}^{(\ell)} \triangleq \text{diag}\{\mathcal{F}_{t_k}^{(\ell)}, \bar{\mathcal{F}}_{t_k}^{(\ell)}\}.\end{aligned}$$

For presentation clarity, we denote

$$\begin{aligned}\Psi_{1,t_k}^{(\ell)} &\triangleq \text{diag}\{\Psi_{1,1,t_k}^{(\ell)}, -\epsilon_{1,t_k}^{(\ell)}I, -\epsilon_{2,t_k}^{(\ell)}I + \epsilon_{3,t_k}^{(\ell)}I, \\ &\quad -\frac{\epsilon_{4,t_k}^{(\ell)}}{b+1}\mathcal{W}_{t_k}^{-1}, -\frac{\epsilon_{5,t_k}^{(\ell)}}{2b}(\mathcal{Z}_{t_k}^{(\ell)})^{-1}\}, \\ \Psi_{1,1,t_k}^{(\ell)} &\triangleq -1 + \sum_{i=1,4,5}\epsilon_{i,t_k}^{(\ell)} - \delta^{+}\epsilon_{2,t_k}^{(\ell)} + \delta^{-}\epsilon_{3,t_k}^{(\ell)}, \\ (\Psi_{2,t_k}^{(\ell)})^T &\triangleq \text{col}\Big\{0, (\bar{\mathcal{A}}_{t_k}^{(\ell)}U_{t_k}^{(\ell)})^T, (\mathcal{K}_{t_k}^{(\ell)})^T, (\bar{\mathcal{B}}_{1,t_k}^{(\ell)})^T, \bar{\mathcal{B}}_{2,t_k}^T\Big\}, \\ (\Psi_{3,t_k}^{(\ell)})^T &\triangleq \begin{bmatrix}\tilde{\bar{\alpha}}_1(\tilde{\bar{\mathcal{A}}}_{1,t_k}^{(\ell)}U_{t_k}^{(\ell)})^T & \cdots & \tilde{\bar{\alpha}}_m(\tilde{\bar{\mathcal{A}}}_{m,t_k}^{(\ell)}U_{t_k}^{(\ell)})^T\end{bmatrix}, \\ Q_{t_k}^{(\ell)} &\triangleq \text{diag}\{Q_{1,t_k}^{(\ell)}, Q_{2,t_k}^{(\ell)}\}, \ U_{t_k}^{(\ell)} \triangleq \text{diag}\{U_{1,t_k}^{(\ell)}, U_{2,t_k}^{(\ell)}\}\end{aligned}$$

where $U_{t_k}^{(\ell)}$ is the factorization of $Q_{t_k}^{(\ell)}$ (i.e. $Q_{t_k}^{(\ell)} \triangleq U_{t_k}^{(\ell)}(U_{t_k}^{(\ell)})^T$).

Theorem 4.6. *Consider the time-varying multi-rate system (4.53)–(4.54) with constraints (4.62), (4.63), and (4.82). If there exist sequences of real-valued matrices*

$Q^{(\ell)}_{t_{k+1}}, K^{(\ell)}_{t_k}$ *and sequences of non-negative scalars* $\epsilon^{(\ell)}_{j,t_k}$ $(j = 1, 2, \cdots, 5,\ \ell = 1, 2, \cdots, q)$ *satisfying the following recursive matrix inequalities*

$$\begin{bmatrix} \Psi^{(\ell)}_{1,t_k} & (\Psi^{(\ell)}_{2,t_k})^T & (\Psi^{(\ell)}_{3,t_k})^T \\ 0 & -Q^{(\ell)}_{t_{k+1}} & 0 \\ 0 & 0 & -\mathrm{diag}_m\{Q^{(\ell)}_{t_{k+1}}\} \end{bmatrix} \leq 0\,, \tag{4.84}$$

then the state $x(t_k)$ *resides in its ellipsoidal sets* $\bar{\mathcal{E}}^{(\ell)} \triangleq \Big\{ \big(x(t_k), \hat{x}^{(\ell)}(t_k)\big) : [\bar{\xi}^{(\ell)}(t_k)]^T (Q^{(\ell)}_{t_k})^{-1} \bar{\xi}(t_k)^{(\ell)} \leq 1 \Big\}$.

B. Fault Isolation Decision Scheme

After acquiring the annulus-event-based fault isolation estimators by using Theorem (4.6), the estimation error function $Y^{(\ell)}(t_k)$ and its threshold $Y^{(\ell)}_{th}$ are set as follows:

$$\begin{aligned} Y^{(\ell)}(t_k) &\triangleq \Big[\sum_{i=t_{k_0}}^{t_k} (\bar{e}^{(\ell)}(i))^T \bar{e}^{(\ell)}(i) \Big]^{\frac{1}{2}}, \\ Y^{(\ell)}_{th} &\triangleq \sup_{d(t_k)\in \bar{\mathfrak{W}}_{t_k}, \bar{f}(t_k)=0} Y^{(\ell)}(t_k), \end{aligned}$$

where t_{k_0} represents the initial time of the estimation.

In order to isolate the detected fault $f(t_k)$, we activate the *isolation estimator bank* (4.79) and determine the fault type by employing the following decision scheme:

$$\begin{cases} Y^{(\ell)}(t_k) \leq Y^{(\ell)}_{th} \\ Y^{(s)}(t_k) > Y^{(s)}_{th}, \quad s \in \{1, 2, \cdots, q\}/\ell \end{cases} \tag{4.85}$$

Then, we can conclude that fault $f(t_k)$ is ℓ type fault $\hat{f}^{(\ell)}(t_k)$. The absolute fault isolation time t_I is defined as $t_I \triangleq \max_{t_k}\{Y^{(s)}(t_k) > Y^{(s)}_{th}\}$.

Remark 4.10. *For the generalized observer scheme [7, 27] of fault isolation, the estimation error* $\bar{e}(t_k)$ *is used to isolate directly the fault. In our section, the estimation error function* $Y^{(\ell)}(t_k)$ *are adopted to enhance the isolate effect. Especially, the set-membership estimation method is employed to investigate the annulus-event-based fault isolation of time-varying multi-rate systems with unknown but bounded disturbances/faults. Therefore, the fault isolation strategy with modified generalized observer scheme proposed in this section is more in line with the engineering practice.*

If the detected fault cannot be isolated with aforementioned fault isolation scheme, then the fault estimation unit will be started, and the fault database can be updated. The dynamic characteristics of the fault vector $f(T_k)$ in (4.53) can be described as:

$$f(T_{k+1}) = f(T_k) + \phi(T_k) \tag{4.86}$$

where $\phi(T_k)$ is the fault increment satisfying the following assumption.

Assumption 4.8. *The sequences $\phi(T_k)$ satisfy*

$$\phi^T(T_k)U_{T_k}^{-1}\phi(T_k) \leq 1 \tag{4.87}$$

where U_{T_k} are the constraint matrices characterizing the scope of the ellipsoid for $\phi(T_k)$.

By using (4.86) iteratively, we have the following equation

$$f(t_{k+1}) = f(t_k) + \mathbb{I}_\phi \bar{\phi}(t_k) \tag{4.88}$$

where $\bar{\phi}(t_k) \triangleq \mathrm{col}_0^{b-1}\{\phi(t_k + ih)\}$, $\mathbb{I}_\phi \triangleq [\underbrace{I \cdots I}_{b}]$. Then, the dynamics of augmented vector $\bar{f}(t_k)$ can be described as:

$$\bar{f}(t_{k+1}) = \bar{f}(t_k) + \hat{\mathcal{I}}_\phi d_\phi(t_k) \tag{4.89}$$

where

$$d_\phi(t_k) \triangleq \mathrm{col}\{\bar{\phi}(t_k), \bar{\phi}(t_{k+1})\}, \hat{\mathcal{I}}_\phi \triangleq [\ \mathcal{I}_\phi \quad \bar{\mathcal{I}}_\phi\ ],$$

$$\mathcal{I}_\phi \triangleq \begin{bmatrix} I & I & \cdots & I & I \\ 0 & I & \cdots & I & I \\ \vdots & \vdots & \ddots & \vdots & \vdots \\ 0 & 0 & \cdots & I & I \\ 0 & 0 & \cdots & 0 & I \end{bmatrix}_{(b\times b)}, \bar{\mathcal{I}}_\phi \triangleq \begin{bmatrix} 0 & 0 & \cdots & 0 & 0 \\ I & 0 & \cdots & 0 & 0 \\ \vdots & \vdots & \ddots & \vdots & \vdots \\ I & I & \cdots & 0 & 0 \\ I & I & \cdots & I & 0 \end{bmatrix}_{(b\times b)}.$$

Taking advantage of the received signal $\bar{y}(t_k)$, (4.59) and (4.89), the following annulus-event-based estimators are adopted:

$$\begin{cases} \hat{\bar{f}}(t_{k+1}) = \hat{\bar{f}}(t_k) + \bar{H}_{t_k}\left[\bar{y}(t_k) - \hat{y}(t_k)\right] \\ \hat{x}(t_{k+1}) = \bar{A}_{t_k}\hat{x}(t_k) + \bar{B}_{2,t_k}\hat{\bar{f}}(t_k) + R_{t_k}\left[\bar{y}(t_k) - \hat{y}(t_k)\right] \end{cases} \tag{4.90}$$

where $\hat{\bar{f}}(t_k)$ and $\hat{x}(t_k)$ are the estimate of the augmented fault $\bar{f}(t_k)$ and the state $x(t_k)$, respectively.
$\bar{H}_{t_k} \triangleq \mathrm{col}\{\underbrace{H_{1,t_k}, \cdots, H_{b,t_k+(b-1)h}}_{b}\}$, $H_{i,t_k+(i-1)h}$ $(i = 1, 2, \cdots, b)$ and R_{t_k} are the estimator parameters to be determined.

By denoting $e_{\bar{f}}(t_k) \triangleq \bar{f}(t_k) - \hat{\bar{f}}(t_k)$, $e_x(t_k) \triangleq x(t_k) - \hat{x}(t_k)$, $\varsigma(t_k) \triangleq \mathrm{col}\{x(t_k), \bar{f}(t_k), e_{\bar{f}}(t_k), e_x(t_k)\}$ and $\bar{d}(t_k) \triangleq \mathrm{col}\{d(t_k), d_\phi(t_k)\}$, synthesizing (4.59), (4.89) and (4.90), one has the following augmented systems:

$$\varsigma(t_{k+1}) = \Big(\mathscr{A}_{t_k} + \sum_{s=1}^{m} \tilde{\alpha}_s(t_k)\tilde{\mathscr{A}}_{s,t_k}\Big)\varsigma(t_k + \mathscr{H}_{t_k}\sigma(t_k) + \mathscr{B}_{t_k}\bar{d}(t_k) \quad (4.91)$$

with following constraint

$$\bar{d}^T(t_k)\mathscr{U}_{t_k}^{-1}\bar{d}(t_k) \leq 3b + 1 \quad (4.92)$$

where

$$\begin{aligned}
\mathscr{A}_{t_k} &\triangleq \begin{bmatrix} \vec{A}_{t_k} & 0 \\ 0 & \breve{A}_{t_k} \end{bmatrix}, \ \vec{A}_{t_k} \triangleq \begin{bmatrix} \bar{A}_{t_k} & \bar{B}_{2,t_k} \\ 0 & I \end{bmatrix}, \\
\breve{A}_{t_k} &\triangleq \begin{bmatrix} I & -\bar{H}_{t_k} \otimes (\bar{\Xi} C_{t_k}) \\ \bar{B}_{2,t_k} & \bar{A}_{t_k} - R_{t_k}\bar{\Xi} C_{t_k} \end{bmatrix}, \\
\tilde{\mathscr{A}}_{s,t_k} &\triangleq \begin{bmatrix} 0 & 0 \\ \vec{\mathcal{A}}_{s,t_k} & 0 \end{bmatrix}, \ \vec{\mathcal{A}}_{s,t_k} \triangleq \begin{bmatrix} -\bar{H}_{t_k} \otimes C_{s,t_k} & 0 \\ -R_{t_k} C_{s,t_k} & 0 \end{bmatrix}, \\
\mathscr{H}_{t_k} &\triangleq \mathrm{col}\{0, 0, -\bar{H}_{t_k}, -R_{t_k}\}, \\
\mathscr{B}_{t_k} &\triangleq \mathrm{col}\{\mathscr{B}_{1,t_k}, \mathscr{B}_{2,t_k}\}, \ \mathscr{B}_{1,t_k} \triangleq \mathrm{diag}\{\bar{B}_{1,t_k}, \hat{\mathcal{I}}_\phi\}, \\
\mathscr{B}_{2,t_k} &\triangleq \begin{bmatrix} -\bar{H}_{t_k} \otimes \bar{D}_{t_k} & \hat{\mathcal{I}}_\phi \\ \bar{B}_{1,t_k} - R_{t_k}\bar{D}_{t_k} & 0 \end{bmatrix}, \\
\mathscr{U}_{t_k} &\triangleq \mathrm{diag}\{\mathcal{W}_{t_k}, \mathcal{U}_{t_k}, \mathcal{U}_{t_{k+1}}\}, \ \mathcal{U}_{t_k} \triangleq \mathrm{diag}_0^{b-1}\{U_{t_k+ih}\}.
\end{aligned}$$

For convenience of later analysis, we denote

$$\begin{aligned}
\Upsilon_{1,t_k} &\triangleq \mathrm{diag}\{\Upsilon_{1,1,t_k}, -\vartheta_{1,t_k} I, -\vartheta_{2,t_k} I + \vartheta_{3,t_k} I, -\frac{\vartheta_{4,t_k}}{3b+1}\mathscr{U}_{t_k}^{-1}\}, \\
\Upsilon_{1,1,t_k} &\triangleq -1 + \sum_{i=1,4} \vartheta_{i,t_k} - \delta^+ \vartheta_{2,t_k} + \delta^- \vartheta_{3,t_k}, \\
\Upsilon_{2,t_k}^T &\triangleq \mathrm{col}\Big\{0, (\mathscr{A}_{t_k}\mathcal{G}_{t_k})^T, \mathscr{H}_{t_k}^T, \mathscr{B}_{t_k}^T\Big\}, \\
\Upsilon_{3,t_k}^T &\triangleq \begin{bmatrix} \tilde{\bar{\alpha}}_1(\tilde{\mathscr{A}}_{1,t_k}\mathcal{G}_{t_k})^T & \cdots & \tilde{\bar{\alpha}}_m(\tilde{\mathscr{A}}_{m,t_k}\mathcal{G}_{t_k})^T \end{bmatrix}, \\
\mathcal{G}_{t_k} &\triangleq \mathrm{diag}_1^8\{\mathcal{G}_{i,t_k}\}, \ \mathcal{O}_{t_k} \triangleq \mathrm{diag}_1^8\{\mathcal{O}_{i,t_k}\}
\end{aligned}$$

where $\mathcal{G}_{t_k}$ is the factorization of $\mathcal{O}_{t_k}$ (i.e. $\mathcal{O}_{t_k} \triangleq \mathcal{G}_{t_k}\mathcal{G}_{t_k}^T$).

Theorem 4.7. *Consider the time-varying multi-rate system (4.53)–(4.54) with constraint (4.62). If there exist sequences of real-valued matrices $\mathcal{O}_{t_{k+1}}, R_{t_k}, H_{i,t_k+(i-1)h} (i = 1, 2, \cdots, b)$ and sequences of non-negative scalars ϑ_{j,t_k} $(j = 1, 2, 3, 4)$ satisfying the following recursive matrix inequalities*

$$\begin{bmatrix} \Upsilon_{1,t_k} & \Upsilon_{2,t_k}^T & \Upsilon_{3,t_k}^T \\ 0 & -\mathcal{O}_{t_k} & 0 \\ 0 & 0 & -\mathrm{diag}_m\{\mathcal{O}_{t_k}\} \end{bmatrix} \leq 0, \quad (4.93)$$

then the augmented state $\varsigma(t_k)$ *resides in its ellipsoidal set*
$\mathscr{E} \triangleq \left\{(x(t_k), \bar{f}(t_k), e_{\bar{f}}(t_k), e_x(t_k)) : \varsigma^T(t_k)\mathcal{O}_{t_k}^{-1}\varsigma(t_k) \leq 1\right\}$.

Theorem 4.7 supplies a design method of annulus-event-based fault estimator which can be used to estimate simultaneously the state and the fault. Based on the obtained annulus-event-based fault estimator, we can solve the fault estimation problem. According to the previous discussions, the series of problems for fault detection, isolation and estimation for the time-varying multi-rate systems can be handled by employing **Algorithm 4.3**.

Algorithm 4.3 The algorithm of fault detection and isolation.

Step 1. Design a series of annulus-event-based fault detection estimator by using Theorem 4.5.
Step 2. Apply the obtained annulus-event-based fault detection estimator in step 1, and detect the fault by employing the fault detection scheme given in Section 4.2.2.
Step 3. After the fault has been detected in step 2, activate the fault isolation estimators to devise the annulus-event-based fault isolation estimators by using Theorem 4.6.
Step 4. Recur to the devised annulus-event-based fault isolation estimators in step 3, isolate the fault by using the fault isolation program.
Step 5. Supply the result when the fault can be isolated, otherwise, trigger the fault estimator to design the annulus-event-based fault estimator by using Theorem 4.7.
Step 6. With the aid of the acquired annulus-event-based fault estimator in step 5, estimate the fault signal and update fault database.

4.2.3 Illustrative Examples

To illustrate the effectiveness of the proposed methodology, a simulation experiment on the three-tank system is introduced. Three-tank system DTS200 has typical characteristics of tanks, pipelines and pumps which are widely used in chemical industry, thus it often serves as a benchmark process in laboratories. Especially, the precise mathematical model of three-tank system can be established and the leakage/sensor/actuator faults can be easily described.

Applying the incoming and outgoing mass flows with Torricellies law, the dynamics of DTS200 as [29] is modelled by

$$\begin{cases} A\dot{h}_1(t) = Q_1(t) - Q_{13}(t) + d_1(t) + v_1(t) + f_1(t) \\ A\dot{h}_2(t) = Q_2(t) + Q_{32}(t) - Q_{20}(t) + d_2(t) + v_2(t) + f_2(t) \\ A\dot{h}_3(t) = Q_{13}(t) - Q_{32}(t) + d_3(t) + f_3(t) \\ Q_{20}(t) = a_2 s_0 \sqrt{2gh_2(t)} \\ Q_{13}(t) = a_1 s_{13} sgn(h_1(t) - h_3(t))\sqrt{2g|h_1(t) - h_3(t)|} \\ Q_{32}(t) = a_3 s_{23} sgn(h_3(t) - h_2(t))\sqrt{2g|h_3(t) - h_2(t)|} \end{cases} \tag{4.94}$$

where $Q_1(t), Q_2(t)$ represent are incoming mass flow (cm^3/s), $Q_{ij}(t)$ denotes the mass flow (cm^3/s) from the ith tank to the jth tank, $h_i(t)$ $(i = 1, 2, 3)$ are the water level (cm) of each tank. Different from the single-rate sampling in [29], in our experiment, the plant (4.53) is sampled with period $T_p = h$ and the sampling period of measurement (4.54) is assumed as $T_m = 3h$. Under the assumption of $h_1(t) > h_3(t) > h_2(t)$, the sampling period of (4.95) is $T_p = 1\ s$, the equilibrium point of three-tank system is chosen as $x(T_0) = \text{col}\{0.6813,\ 0.3321,\ 0.5534\}$, and other parameters are adopted from [29]. By linearizing and discrediting, we have the discrete-time multi-rate sampled-data three-tank system:

$$x(T_{k+1}) = Ax(T_k) + B_1\omega(T_k) + B_2 f(T_k) \tag{4.95}$$

$$y(t_k) = \Xi_{t_k} Cx(t_k) + D\nu(t_k) \tag{4.96}$$

where $x(T_k) = \text{col}\{h_1(t),\ h_2(t),\ h_3(t)\}$ is the system state representing the liquid levels of the three tanks, $y(t_k) = \text{col}\{h_1(t),\ h_2(t)\}$ is the measurement output describing the height measurements of tank $\mathcal{T}_1$ and tank $\mathcal{T}_2$, $w(T_k) = \text{col}\{d_1(t),$ $d_2(t),\ d_3(t)\}$, $v(t_k) = \text{col}\{v_1(t),\ v_2(t)\}$ are used to model the unknown disturbance, and $f(T_k) = \text{col}\{f_1(t),\ f_2(t),\ f_3(t)\}$ is the fault signal reflecting the leakages in the three tanks. Other matrices can obtained as follows:

$$\begin{aligned}
A &= \begin{bmatrix} 0.9908 & 0 & 0.0091 \\ 0 & 0.9856 & 0.0072 \\ 0.0091 & 0.0072 & 0.9836 \end{bmatrix}, \\
B_1 &= B_2 = \begin{bmatrix} 64.6627 & 0.0007 & 0.2978 \\ 0.0007 & 64.4908 & 0.2358 \\ 0.2978 & 0.2358 & 64.4271 \end{bmatrix}, \\
C &= \begin{bmatrix} 1 & 0 & 0 \\ 0 & 1 & 0 \end{bmatrix}, \ D = \begin{bmatrix} 1 & 0 \\ 0 & 1 \end{bmatrix}.
\end{aligned}$$

Consider the existence of unmodeled dynamic and the perturbation of working point, the time-varying parameters are assumed as:

$$\begin{aligned}
A_{T_k} &\triangleq A + \sin(0.1 \times T_k) \times \begin{bmatrix} 0 & 0 & 0 \\ 0 & 0 & -0.001 \\ 0 & 0 & 0.001 \end{bmatrix}, \\
B_{1,T_k} &\triangleq B_{2,T_k} \triangleq B_1 + \sin(0.1 \times T_k) \times \begin{bmatrix} 0 & 0 & 0 \\ 0 & 0 & 0.001 \\ 0 & 0 & -0.003 \end{bmatrix}, \\
C_{t_k} &\triangleq C,\ D_{t_k} \triangleq D,\ \Xi_{t_k} \triangleq \text{diag}\{\alpha_{1,t_k}, \alpha_{2,t_k}\}, \\
\bar{\alpha}_1 &= 0.7,\ \bar{\alpha}_2 = 0.9,
\end{aligned}$$

and the disturbances are set as $w(t_k) = 3 \times 10^{-5} \times \cos(0.5 \times k)$, $v(t_k) = 5 \times 10^{-5} \times \frac{\sin(0.5 \times k)}{0.2 \times k + 0.8}$.

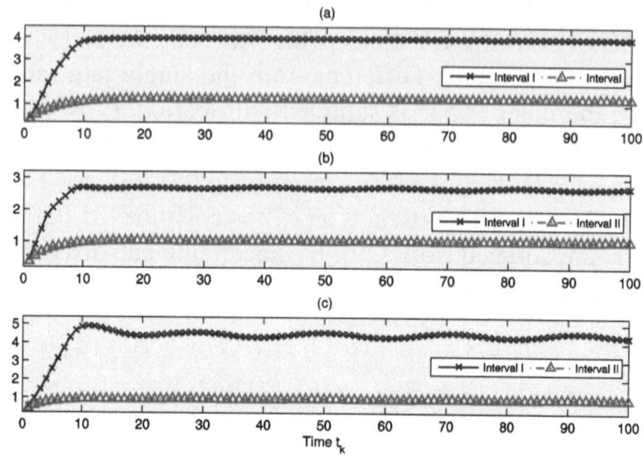

Figure 4.8: The minimum upper bound of estimation error for annulus I and annulus II: (a) The minimum upper bound of $e_1^T(t_k)e_1(t_k)$; (b) The minimum upper bound of $e_2^T(t_k)e_2(t_k)$; and (c) The minimum upper bound of $e_3^T(t_k)e_3(t_k)$.

Our aim here is to detect, isolate and estimate the faults by using the established mathematic model of the system as well as the measurement signals in the presence of a leakage in tank $\mathcal{T}_1$, $\mathcal{T}_2$ or $\mathcal{T}_3$.

Case I. Let the constraint matrices of disturbances and fault signal in (4.55) be $W_{t_k} = 10^{-7}$, $M_{t_k} = 3 \times 10^{-7}$, $F_{t_k} = 10^{-7}$. The initial constraint matrices are chosen as $P_{2,t_0} \triangleq \text{diag}\{1, 1.5, 2\}$. For given triggering **annulus I** ($[10^{-5}, 5 \times 10^{-2}]$) and **annulus II** ($[10^{-5}, 5 \times 10^{-1}]$), solving the optimization problem (4.76) by using Corollary 4.5.1 leads to the minimum constraint bound of estimation error in Fig. 4.8. On the other hand, when the bound of estimation error is supposed to be $P_{1,t_0} \triangleq \text{diag}\{0.1, 0.15, 0.2\}$, and the lower bound of triggering annulus is chosen as $\delta^- = 10^{-5}$, by solving the optimization problem (4.77), the minimum triggering region is shown in Fig. 4.9, and the specific values are shown in Table 4.4.

Table 4.4: The minimum upper bound with given lower bound $\delta^- = 10^{-5}$.

	$t_k = 1$	$t_k = 2$	$t_k = 3$	$\cdots$	$t_k = 98$	$t_k = 99$	$t_k = 100$
δ^+	1.0584	0.8020	0.6931	$\cdots$	0.5732	0.5722	0.5717

Suppose that the triggering annulus is $[\delta^-, \delta^+] \triangleq [10^{-5}, 0.9]$ and the fault is

$$f_0(t_k) \triangleq \begin{cases} -5 \times 10^{-3}, & 20 \le t_k \le 100; \\ 0, & \text{otherwise.} \end{cases} \tag{4.97}$$

Then, the fault signal is chosen as $f(t_k) \triangleq \text{col}\{0, 0, f_0(t_k)\}$, that is there is a leakage in Tank $\mathcal{T}_3$ with an amplitude about 5×10^{-3}. By using Algorithm I, the annulus-event-based fault detection estimator can be designed and the simulation results can be achieved in Fig. 4.10. Fig. 4.10(a) shows the release time and corresponding release annulus with given triggering annulus. Fig. 4.10(b) provides the evolution of residual evaluation function and threshold. By using simulation computation, we have $J_{th} = 2.7781$ and $2.5481 = J(30) < J_{th} < J(31) = 3.0156$. According to the fault detection decision scheme in Section 3.2, the fault detection time is $t_D = 31$, i.e, the fault can be detected in 11 time steps after its occurrence.

Case II. After the fault has been detected in Case I, we begin to isolate the fault. Firstly, the constraint matrices are chosen as

$$\begin{aligned} Q_{1,t_0}^{(1)} &= Q_{1,t_0}^{(2)} = Q_{1,t_0}^{(3)} = \text{diag}\{0.1, 0.15, 0.2\}, \\ Q_{2,t_0}^{(1)} &= Q_{2,t_0}^{(2)} = Q_{2,t_0}^{(3)} = \text{diag}\{1, 1.5, 2\}, \\ F_{t_k}^{(1)} &= 5 \times 10^{-7},\ F_{t_k}^{(2)} = 7 \times 10^{-7},\ F_{t_k}^{(3)} = 9 \times 10^{-7}; \\ \bar{F}_{t_k}^{(1)} &= 1 \times 10^{-7},\ \bar{F}_{t_k}^{(2)} = 3 \times 10^{-7},\ \bar{F}_{t_k}^{(3)} = 5 \times 10^{-7} \end{aligned}$$

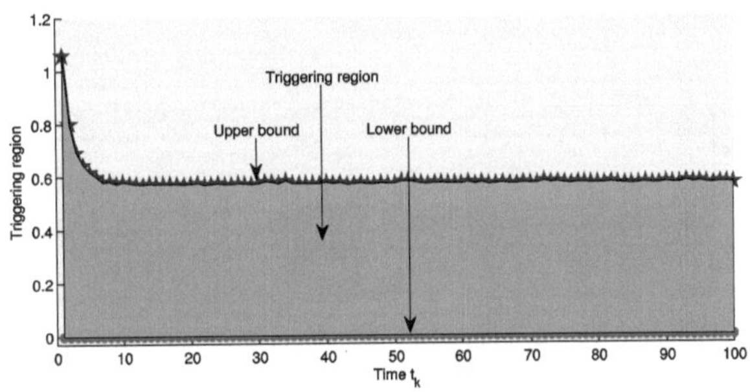

Figure 4.9: The minimum triggering annulus for the given estimation error constraint and the lower bound of triggering annulus.

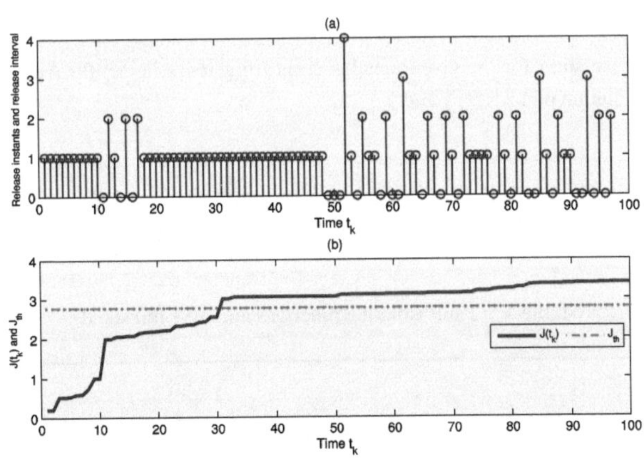

Figure 4.10: Fault detection for the given annulus-event-triggered scheme: (a) Annulus-event-triggered scheme and (b) Evolution of $J(t_k)$ and J_{th}.

and the triggering annulus is set as $[10^{-5}, 10^{-1}]$, the fault sets are chosen as

$$\begin{aligned}\mathscr{F} &\triangleq \{\hat{f}^{(1)}(t_k), \hat{f}^{(2)}(t_k), \hat{f}^{(3)}(t_k)\} \\ &\triangleq \left\{ \begin{bmatrix} f_O(t_k) \\ 0 \\ 0 \end{bmatrix}, \begin{bmatrix} 0 \\ f_O(t_k) \\ 0 \end{bmatrix}, \begin{bmatrix} 0 \\ 0 \\ f_O(t_k) \end{bmatrix} \right\}\end{aligned} \tag{4.98}$$

Similar to Algorithm I, by using Theorem 4.6, the annulus-event-based fault isolation estimators can be calculated and the result of fault isolation is shown in Figs. 4.11–4.13. Figs. 4.11(a), 4.12(a), and 4.13(a) offer the release time and corresponding release annulus with given triggering annulus. Figs. 4.11(b), 4.12(b), and 4.13(b) provide the estimation error functions and their thresholds corresponding to estimators $\ell = 1, 2, 3$. It can be seen from Table 4.5 that the estimator 3 always remains below its threshold, whereas the estimators $1, 2$ exceed their thresholds at $t^{(1)} = 52, t^{(2)} = 46$, respectively. According to the fault isolation program in Section 4.2, the fault is the 3-type fault and fault isolation time is $t_I = 52$.

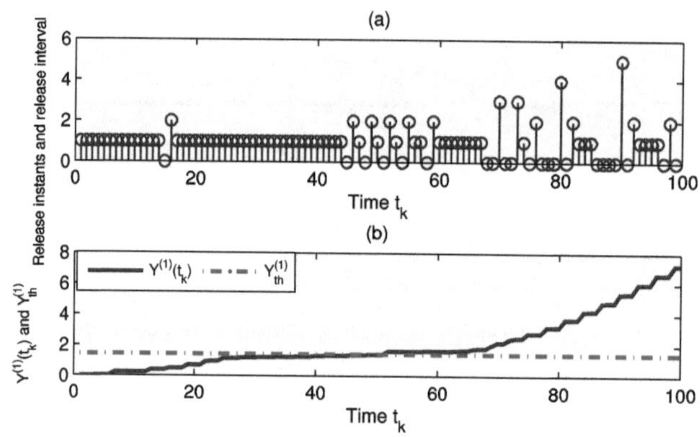

Figure 4.11: Fault isolation for the given annulus-event-triggered scheme: (a) Annulus-event-triggered scheme and (b) Evolution of $Y^{(1)}(t_k)$ and $Y_{th}^{(1)}$.

Case III. Suppose that the fault has not been isolated by using 4.2, the fault estimation scheme should be started. The fault evolution parameter is set as $A_{f,t_k} \triangleq$

Table 4.5: Fault isolation functions and their thresholds.

	$Y_{th}^{(\ell)}$	$Y^{(\ell)}(t_k)$
$1 - type$	1.3360	$1.3202 = Y^{(1)}(51) < Y_{th}^{(1)} < Y^{(1)}(52) = 1.3805$
$2 - type$	1.3209	$1.1079 = Y^{(2)}(45) < Y_{th}^{(2)} < Y^{(2)}(46) = 1.3270$
$3 - type$	2.0681	$Y^{(3)}(t_k) < Y_{th}^{(3)}$

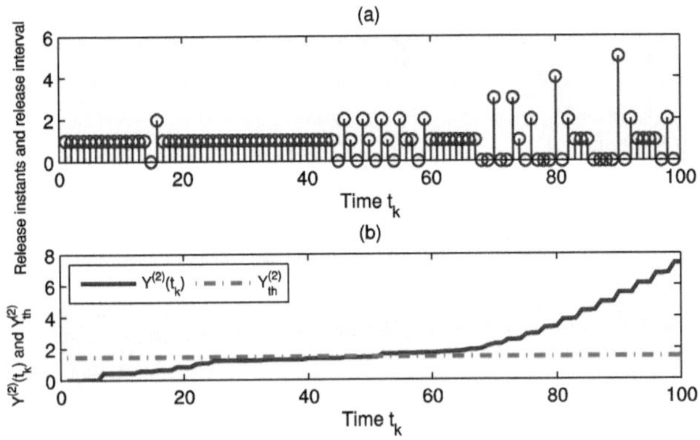

Figure 4.12: Fault isolation for the given annulus-event-triggered scheme: (a) Annulus-event-triggered scheme and (b) Evolution of $Y^{(2)}(t_k)$ and $Y_{th}^{(2)}$.

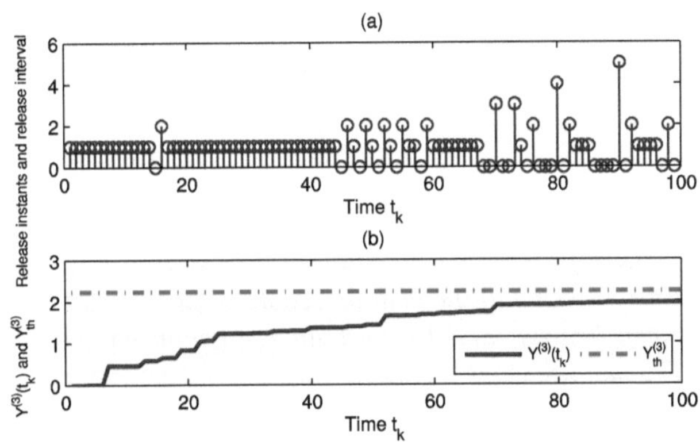

Figure 4.13: Fault isolation for the given annulus-event-triggered scheme: (a) Annulus-event-triggered scheme and (b) Evolution of $Y^{(3)}(t_k)$ and $Y_{th}^{(3)}$.

diag$\{1,1,1\}$, the constraint matrices are chosen as

$$\begin{aligned}
\mathcal{O}_{1,t_0} &= \mathcal{O}_{2,t_0} = \mathcal{O}_{3,t_0} = \mathcal{O}_{4,t_0} = \text{diag}\{1, 1.5, 2\},\\
\mathcal{O}_{5,t_0} &= \mathcal{O}_{6,t_0} = \mathcal{O}_{7,t_0} = \mathcal{O}_{8,t_0} = \text{diag}\{0.1, 0.15, 0.2\},\\
M_{t_k} &= 7\times 10^{-6},\ W_{t_k} = 2\times 10^{-5},\ U_{t_k} = 1\times 10^{-4},
\end{aligned}$$

and we assume that the triggering annulus is $[10^{-5}, 5\times 10^{-1}]$. By means of Theorem 4.7, the simulation results are supplied in Fig. 4.14. Fig. 4.14(a) plots the release time and corresponding release annulus with given triggering annulus, Fig. 4.14(b) shows the fault signal and its estimation.

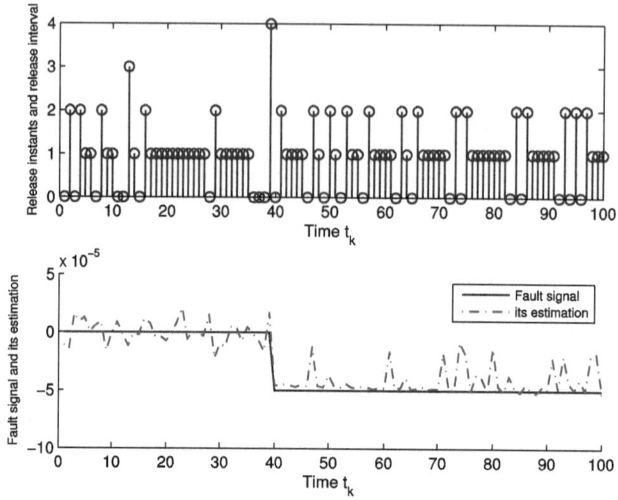

Figure 4.14: Fault signal and its estimation for the given annulus-event-triggered scheme: (a) Annulus-event-triggered scheme and (b) Fault signal and its estimate.

Remark 4.11. *The effect of fault detection, isolation, and estimation is shown in Figs. 4.10 to 4.14. In fact, as can be observed, there are certain issues with the fault diagnosis, for example, a) the fault detection/isolation is not instant (i.e. there appear to be some delays); and b) the fault estimation has certain errors (see Fig. 4.14(b)). The main reasons for these observations are outlined as follows: 1) the considered time-varying systems are quite complicated and are subject to disturbance input $\omega(T_k)$, measurement noise $\nu(t_k)$ and fault increment $\phi(T_k)$; 2) both the sensor degradation and the event-triggered communication mechanism would affect the integrity of the measurements which, in turn, impact on the fault detection/isolation/estimation; 3) although the set-membership constraint on disturbances/faults is suitable to reflect the practical engineering, such constraint could constitute another source of estimation errors.*

4.3 Conclusion

In this chapter, fault detection and isolation problems are investigated for a class of time-varying multi-rate systems. A novel annulus-event-triggering communication mechanism has been utilized to reduce the sensor data transmission rate and the energy consumption. Sufficient conditions have been established in the form of recursive linear matrix inequalities revealing the relationship between estimation performance index, the event-triggering threshold, the parameters of the ellipsoidal constraint and multi-rate multiples. Finally, simulation examples are utilized to show the effectiveness of the proposed schemes in this chapter.

5

Fault Diagnosis of Modular Multilevel Converters with Machine Learning Methods

In the past few years, increasing research interests have been attracted by the modular multilevel converter (MMC) for high-voltage direct-current transmission systems due primarily to their significant advantages such as easy construction, assembling, high efficiency, and high quality of the output voltages [184–186]. An inherent character of the modular multilevel converter is a large number of switching devices, which are all the potential failure points, and would lead to the occurrence of faults, and then, result in significant loss of life and property. Therefore, efficient fault diagnosis and location of the sub-module for modular multilevel converter can be used to quickly formulate maintenance plans and avoid more severe accidents [187, 188].

Fault diagnosis methods of modular multilevel converter [189, 190] can be divided into two categories, that is model-based approaches [18, 20, 191, 192] and data-driven methods [48, 193]. Model-based methods have been widely used in the fault diagnosis of modular multilevel converter [187, 188, 194]. However, these methods have one weakness, that is it is difficult to accurately represent the physical model in complex conditions [63, 193, 195]. Different from the model-based methods which utilize physical law to estimate the operation process of modular multilevel converter, data-driven methods firstly extract the features of the sampled data, and then realize fault classification and recognition based on the extracted features. Feature extraction, which can be achieved by time domain, frequency domain, and time-frequency representations methods, is one of the key steps in fault diagnosis [50, 88, 196]. After the features have been extracted, filtering is an important part in feature engineering. Compared with the traditional filtering methods [122, 197–199] such as continuous wavelet transform and short-time Fourier transform, synchrosqueezing transform (SST) is more effective in characterizing signals, and it's not limited by Heisenberg uncertainty principle. It should be noted that the recent research relating to the fault diagnosis of modular multilevel converter was mainly focused on model-based methods. However, the data-driven methods have not attracted enough attention yet [200]. Inspired by the above discussion, we aim to provide systematic approaches to diagnose the open-circuit fault of modular multilevel converter.

DOI: 10.1201/9781003330998-5

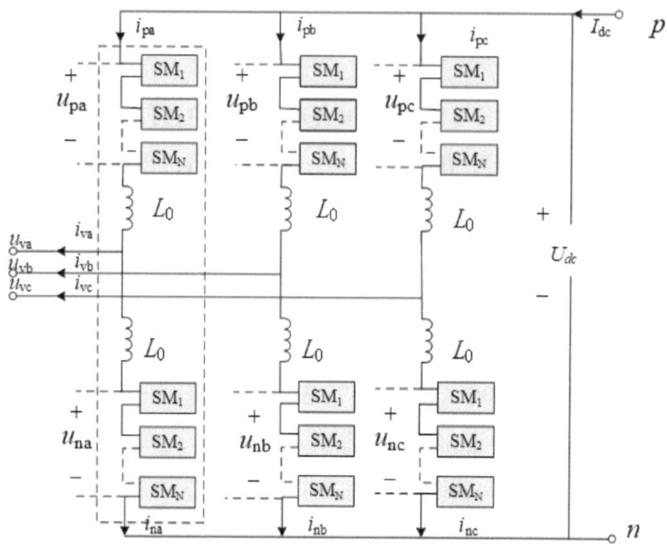

Figure 5.1: Three-phase topological structure of modular multilevel converter.

5.1 Fault Diagnosis with Mixed Kernel Support Tensor Machine

Support tensor machine (STM) has shown obvious advantages in dealing with small sample data, but its disadvantages in generalization function or learning ability with common kernel functions are obvious. In order to obtain higher classification accuracy, both the poly kernel function and the RBF kernel function are combined with mixed kernel support tensor machine (MKSTM), and in this way, more accurate fault diagnosis results can be obtained.

5.1.1 Operating Principles of Modular Multilevel Converters

As shown in Fig. 5.1, a typical modular multilevel converter topology is composed of three-phase. Each phase leg is divided into upper and lower arms. Each arm consists of N identical submodules and the arm inductor L_0. The circuit configuration of the sub-module is shown in Fig. 5.2, which contains a DC capacitor C, two insulated gate bipolar transistor (IGBT) devices (VT1 and VT2) and two anti-parallel diodes (VD1 and VD2). $U_{d,c}$ and $I_{d,c}$ are the DC side voltage and current, $u_{v,j}$ is the voltage and $i_{v,j}$ $(v = p, n;\ j = a, b, c)$ is the current of the AC side in upper and lower arm. C is the sub-module capacitance, $u_{s,m}$ is the sub-module voltage across the board.

According to [201], the arm current $i_{p,j}$ and $i_{n,j}$ can be expressed as

$$\begin{cases} i_{p,j} = i_{diff,j} + \dfrac{i_{v,j}}{2} \\ i_{n,j} = i_{diff,j} - \dfrac{i_{v,j}}{2} \end{cases} (j = a, b, c) \tag{5.1}$$

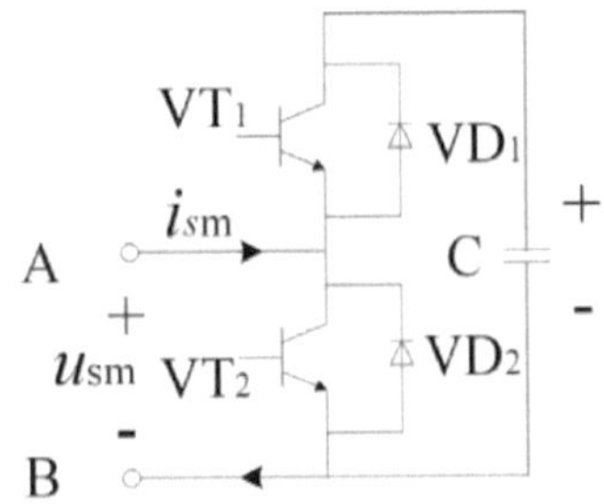

Figure 5.2: Sub-module structure of modular multilevel converter.

where $i_{diff,j}$ is the inner difference current of phase j. Taking (5.1) into account, $i_{diff,j}$ can be rewritten as

$$i_{diff,j} = \frac{i_{p,j} + i_{n,j}}{2} \tag{5.2}$$

In the normal case, the current of the DC side can be averagely divided into three phases. The inductance in each bridge arm is set equal, and the current of AC side is allocated equally to the upper and lower arms in each phase. Just as [202], the open circuit fault of sub-module in either VT1 or VT2 leads to the increase in output voltage. Therefore, the sub-module fault will cause adverse effects such as the asymmetric operation of the system in the upper and lower arm, current DC bias in AC side, asymmetric fundamental frequency, and double frequency circulation [187].

It is known that the circulation current can be restrained either by increasing the arm inductance or by using the suppression strategy, which cannot be completely eliminated. Actually, the variation of circulation current and the current of the AC side has a direct relationship with the fault in modular multilevel converter. Therefore, the fault can be located by analyzing the variation characteristics of circulation current and the current of the AC side. Since the circulation exists only inside the modular multilevel converter inverter, it is not affected by the external power supply and load, but will be significantly affected if the sub-module fails. Therefore, the change of the circulating current and the alternating current is taken as input data which improves the accuracy rating of the diagnostic system.

5.1.2 Mixed Kernel Support Tensor Machine

The collected observation data of open circuit fault in modular multilevel converter are expressed as the following fourth-order tensor data

$$X \in R^{I_1 \times I_2 \times I_3 \times I_4} \tag{5.3}$$

where I_i $(i = 1, 2, 3, 4)$ are, respectively, time series, the three-phase current value, the envelope mean value of the three-phase current value, and the three-phase circulating current value.

A. Support Tensor Machine

The fault diagnosis for modular multilevel converter with support tensor machine can be divided into two steps. The first step is classifying the high-dimensional samples. The second step is judging the fault based on the classification results. For the faulty modular multilevel converter, the type of the fault also can be acquired. Therefore, fault diagnosis of modular multilevel converter with support tensor machine is essentially a multi-classification problem. Now, let us supply the principle of multi-class problem and set that there are n following samples

$$\left\{x_i, y_i\right\}_1^n, \ x_i \in R^{I_1 \times I_2 \times \cdots \times I_M}, \ y_i \in \{1, 2, \cdots, n\} \tag{5.4}$$

where x_i is the ith input tensor sample, y_i is the corresponding label, and $I_1, I_2, \cdots, I_M$ are the samples of model space.

By using the outer product, the inner product, and the norm of the tensor [203], the optimization model with M vector spaces can be transformed into the following optimization model with tensor space:

$$\begin{cases} \min\limits_{W,b,\xi} \ \dfrac{1}{2}\|W\|_F^2 + C\sum\limits_{i=1}^{l}\xi_i, \\ s.t. \ \ y_i(< W, x_i > +b) + \xi_i \geq 1, \\ \qquad \xi_i \geq 0, \\ \qquad i = 1, 2, \cdots, I. \end{cases} \tag{5.5}$$

where the weight parameter W and the input sample x_i are the tensors, their scale is set as consistent. Based on the optimization problem (5.5), we construct the following *Lagrange* function:

$$\begin{aligned} L(W, b, \alpha_i, \beta_i, \xi_i) = &\frac{1}{2}\|W\|_F^2 + C\sum_{i=1}^{l}\xi_i - \sum_{i=1}^{i}\beta_i\xi_i \\ &- \sum_{i=1}^{l}\alpha_i[y_i(\langle W, \chi_i\rangle + b) + \xi_i - 1] \end{aligned} \tag{5.6}$$

where α_i, β_i $(\alpha_i \geq 0, \beta_i \geq 0)$ are the Lagrange multiplier, ξ_i is the slack variable and C is the punish coefficient. Then, the partial derivative of each variable can be expressed as

$$\begin{cases} \dfrac{\partial L}{\partial W} = 0 \Rightarrow W = \sum\limits_{i=1}^{l}\alpha_i y_i \chi_i \\ \dfrac{\partial L}{\partial b} = 0 \Rightarrow \sum\limits_{i=1}^{l}\alpha_i y_i = 0 \\ \dfrac{\partial L}{\partial \xi_i} = 0 \Rightarrow C - \alpha_i - \beta_i = 0 \end{cases} \tag{5.7}$$

where x_i is the sample of the support tensor, i satisfies the following set

$$I = \{i|\alpha_i \neq 0, i = 1, 2, \cdots, l\} \tag{5.8}$$

At the same time, W can rewritten as

$$W = \sum_{i=1}^{l} \alpha_i y_i \chi_i = \sum_{i \in I} \alpha_i y_i \chi_i \tag{5.9}$$

As far as the above support tensor model is concerned, the weight parameter of the tensor is limited to rank one. Essentially, a set of vector projections is employed to map the tensor data into the real space along the direction of each pattern, and the main ideas can be illustrated by Fig. 5.3. Actually, with the increase of the number of the sample points, corresponding structural risks are also bigger, which has negative impacts on the classification performance of the decision functions. Therefore, the following KSTM method is needed.

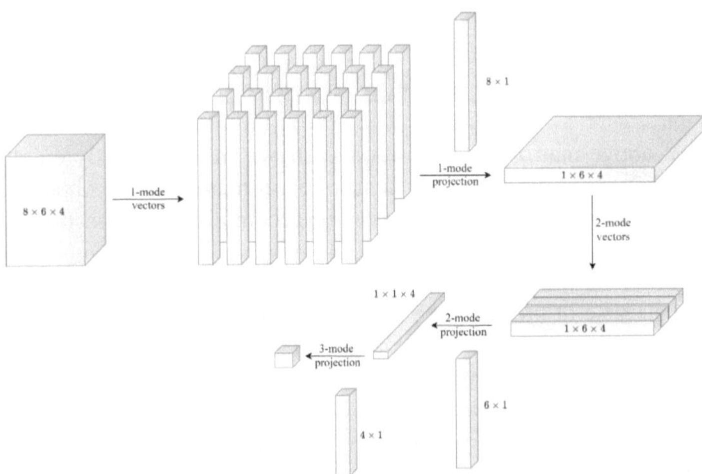

Figure 5.3: Rank-one mapping of the third-order tensor.

B. Kernel Support Tensor Machine

By utilizing nonlinear transformation, the original tensor can be mapped into a high-dimensional tensor feature space, in which the classification can be disposed in virtue of support tensor machine. One notable drawback of this method is that the kernel functions of SVM cannot be directly employed in support tensor machine. To deal with this problem, with the help of the idea of matrix similarity measure and Chrodal distance [204, 205], we adopt the following tensor kernel function which is modified based on the Gauss radial basis function (RBF):

$$K(A, B) = \prod_{i=1}^{M} \exp\left\{-\frac{\|D_{A,1}^{(i)} D_{A,2}^{T(i)} - D_{B,1}^{(i)} D_{B,2}^{T(i)}\|_F^2}{2\sigma^2}\right\} \tag{5.10}$$

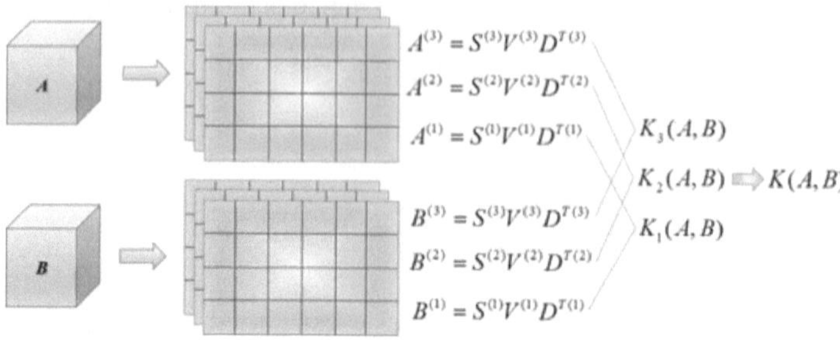

Figure 5.4: Chordal distance visual computing of the tensor A and B.

where $A, B \in R^{I_1 \times I_2 \times \cdots I_M}$ are the tensors and $D_{A,1}^{(i)}$, $D_{B,1}^{(i)}$ are the singular value decomposition with the ith partialization matrix for tensor A, B, respectively. More specifically, the ith partialization matrix can be described as follows:

$$A^{(i)} = S^{(i)}V^{(i)}D^{T(i)} = \begin{bmatrix} S_{A,1}^{(i)} & S_{A,2}^{(i)} \end{bmatrix} \begin{bmatrix} S_{A,1}^{(i)} & 0 \\ 0 & 0 \end{bmatrix} \begin{bmatrix} D_{A,1}^{(i)} \\ D_{A,1}^{(i)} \end{bmatrix} \tag{5.11}$$

The idea of above matrix decomposition is shown in Fig. 5.4. By employing the tensor kernel function $K(A, B)$, the tensor kernel matrix can be obtained, and the classification task based on tensor data can be accomplished by substituting the kernel matrix into support tensor machine.

C. Mixed Kernel Support Tensor Machine

For the purpose of imitating the multi-core support vector machine method, we propose the mixed kernel support tensor machine method to realize higher fault diagnosis. In order to achieve higher classification accuracy, we construct the following mixed tensor kernel function with convex combination:

$$K(A, B) = \sum_{i=1}^{N} \theta_i K_i(A, B), \ \sum_{i=1}^{N} \theta_i = 1. \tag{5.12}$$

where $K_i(A, B)$ is the i-type basic tensor kernel function, θ_i is the weight parameter corresponding to the i-type tensor kernel function, and N is the number of mixed kernel function. Since mixed kernel function satisfies the Mercer condition [206], it can be utilized by the support tensor machine to train and classify [207–209].

Taking the learning and the generalization ability into consideration, we adopt the convex combination of *RBF* kernel function K_{RBF} and *Polynomial* kernel function K_{Poly} to constitute the hybrid kernel function. The mathematical version of the convex combination is indicated as

$$\begin{cases} K(x_i, y_j) = \theta K_{Poly}(x_i, y_j) + (1-\theta) K_{RBF}(x_i, y_j) \\ K_{Poly}(x_i, y_j) = [x_i \times y_j + 1]^m \\ K_{RBF}(x_i, y_j) = \exp\{-\dfrac{\|x_i - y_j\|^2}{h^2}\} \end{cases} \tag{5.13}$$

where θ is the weight coefficient, m is the order of the $Poly$ kernel function $K_{Poly}(x_i, y_j)$, h is the kernel width of the RBF kernel function $K_{RBF}(x_i, y_j)$, Both the penalty coefficient C and the relaxation variable ξ_i can be found in (5.6). In order to ascertain the mixed kernel function (5.13), the above parameters need to be designed.

5.1.3 Fault Diagnosis

To overcome the subjectivity of the artificial selection on mixed kernel support tensor machine parameters (5.13), grid search and cross-validation are used to obtain optimal parameters of the kernel function, and the quantum genetic algorithm is selected to achieve the optimal combination coefficient. Specific steps can be found in **Algorithm 5.1**.

Algorithm 5.1 Optimal combination algorithm of mixed kernel support tensor machine.

Step 1. Load the tensor data as the training and testing set.
Step 2. Select the kernel function and initialize the corresponding parameters.
Step 3. Employ the combination algorithm of grid search and the cross validation to obtain the optimal parameters of kernel function in Step 2, and establish the support tensor machine model.
Step 4. Recur to the quantum genetic algorithm and achieve the optimal combination coefficient in (5.13), then construct the mixed kernel support tensor machine model.

Our objective is to diagnose the open-circuit fault by analyzing the characteristics of AC side current and three-phase internal circulation current, and then, locate the fault by using the mixed kernel support tensor machine. Corresponding method can be expressed as **Algorithm 5.2**.

Algorithm 5.2 Fault diagnosis algorithm of modular multilevel converter.

Step 1. Collect the AC current signal and the internal circulation signal in three phase for modular multilevel converter.
Step 2. Filter and normalize the current and circulating signals, and use the cubic spline interpolation method to obtain the envelope mean of the three-phase current.
Step 3. Select normal and faulty processed circulating data and three-phase current envelope mean data as mixed kernel support tensor machine training set, determine the decision output value of mixed kernel support tensor machine.
Step 4. Randomly set the open fault of the modular multilevel converter submodule, and diagnose and locate the fault through the decision output value of mixed kernel support tensor machine.

5.1.4 Illustrative Examples

In this section, a double-star modular multilevel converter topology is used to construct a 61-level modular multilevel converter system in the Matlab/Simulink platform with the nearest level approximation modulation strategy. The system parameters of the operating environment and the modular multilevel converter are shown in Table 5.1.

A. Fault Mechanism Analysis

The open-circuit fault of sub-module occurs in the upper or lower arm of j-phase, it will cause the positive and negative DC shunt current of the same phase, synchronously, corresponding DC components with different amplitude exists in other two phases. If the fault is diagnosed only by the current, it is easy to cause misjudgment due to the improper threshold and the load mutation. In particular, it is difficult to see the current changes directly for the sub-module fault in the high-level systems. Once the fault happens, the internal circulation current in three phases changes greatly, and the amplitude of double circulation current increases. Therefore, the characteristic of changes in both the three-phase current and the internal circulation current can be combined to determine the modular multilevel converter with fault.

B. Simulation Results

Both three-phase AC current and circulation current in normal operation are shown in Figs. 5.5(a) and (b), respectively, and their envelope means can be obtained by spline interpolation in Fig. 5.5(c). For the faulty case, the open-circuit fault is set at 2.5s, that is one of the three-phase is failure and the other two are normal. Taking the fault of upper bridge on A-phase, for example, the simulation of three-phase AC current, circulating current and current envelope mean values are shown in Fig. 5.6(a)–(c), respectively.

It can be seen from Fig. 5.5 with normal situation that the three-phase current in the AC side is symmetrical with coincident amplitude, the phase difference is 120°, and the three-phase envelope mean is 0. For the faulty case from Fig. 5.6, the open-circuit fault of sub-module leads to the DC component in AC side, for example, the open-circuit fault of sub-module occurs in the upper arm with A-phase, which gives rise to a positive DC bias. According to the Kirchhoff current law, the sum of

Table 5.1: Parameters of the operating environment and the modular multilevel converter prototype.

Parameters	Value
Number of regular SMs per arm n	60
Load resistance R (Ω)	10
Sub-module capacitor C (mF)	2
Three-phase input AC frequency f(Hz)	50
Phase difference E	120°
Inductance L (mH)	2

three-phase current is 0A in any time, the other two phases produce negative DC bias. In terms of circulation, the circulation amplitude of faulty phase decreases, the other two phase increases, and the double frequency circulation magnifies obviously.

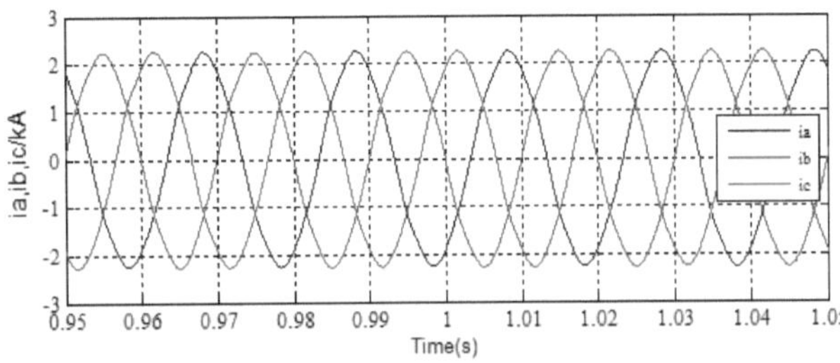

(a) Three-phase AC currents

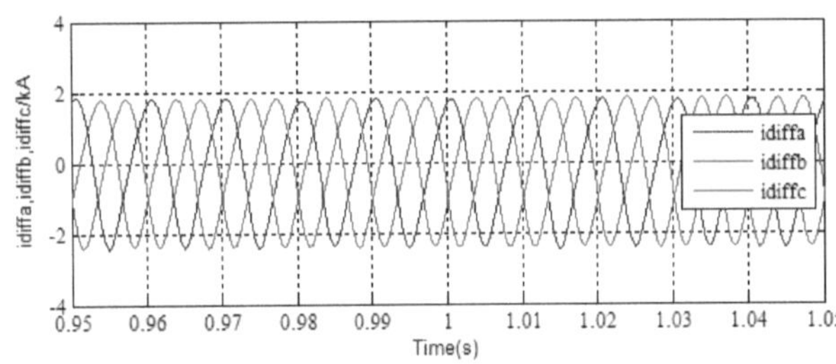

(b) Three-phase inner circulating current.

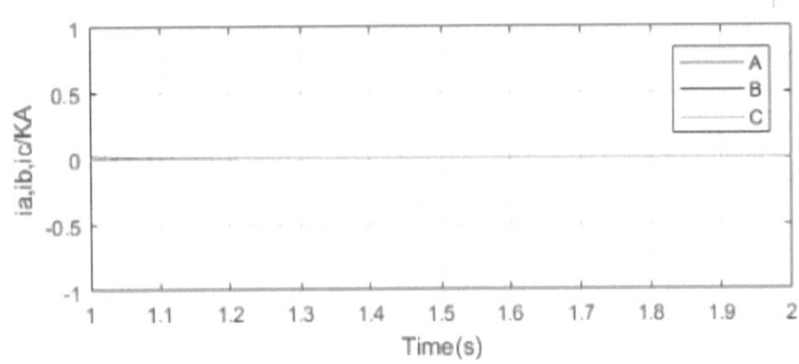

(c) Three-phase current envelope mean.

Figure 5.5: Modular multilevel converter under normal operation.

Due to the difference in a single sub-module fails between the fault and the normal three-phase current is not large for 201-level system Waveform, it is shown in Fig. 5.7 with three-phase current under normal level and sub-module open-circuit fault, therefore, the DC component of the current for sub-module fault is not suitable to the high-level modular multilevel converter system.

We also consider different faults of the three-phase bridge in modular multilevel converter and set the output value of mixed kernel support tensor machine as 7 categories. In each case, the number of the tensor sample groups is selected as 30, the

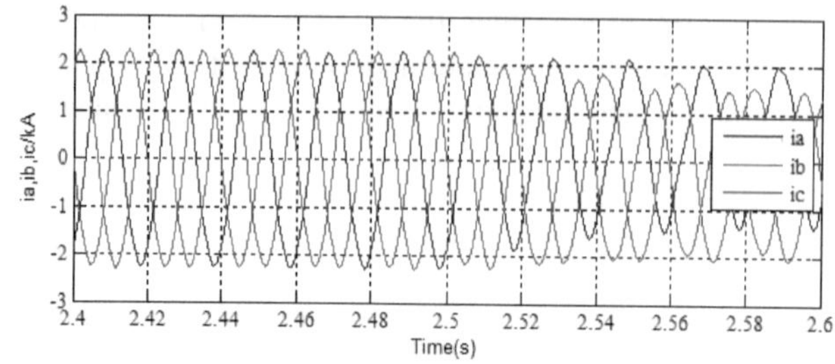

(a) Three-phase AC currents.

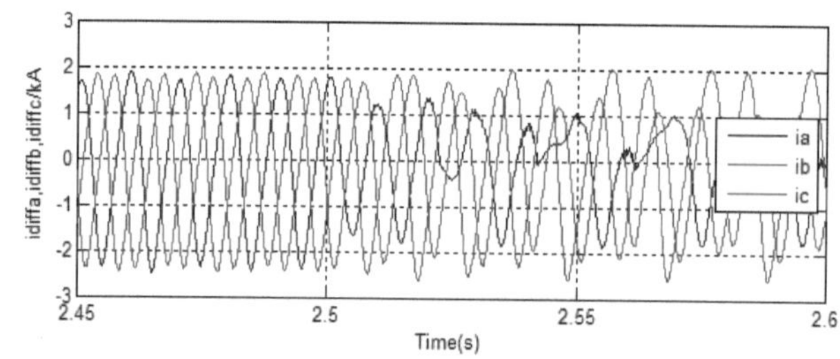

(b) Three-phase inner circulating current.

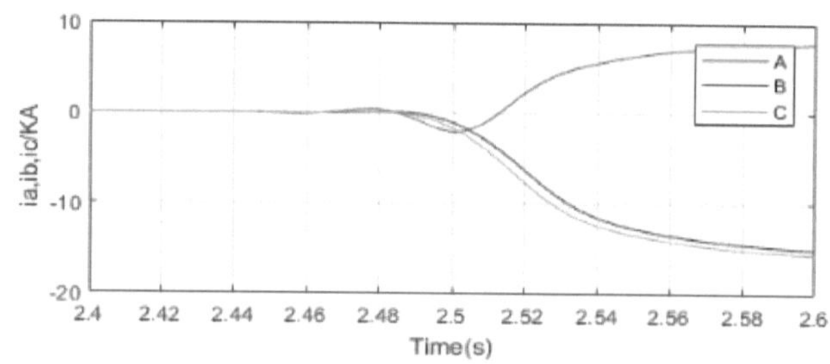

(c) Three-phase current envelope mean.

Figure 5.6: Open-circuit fault occurs in sub-module of A-phase upper bridge arm.

number of the training sample (NTRS) groups is set as 20, and the number of the testing samples (NTES) groups is set as 10. The classification results of label values can be shown in Table 5.2.

The parameters selected are shown in Table 5.3, in which θ is the weight coefficient in (5.13), m is the order of the kernel function, and h is the kernel width of the kernel function. Both the penalty coefficient C and the relaxation variable ξ_i can be found in (5.6), and K_{mix} is the mixed kernel function. The comparison with some kernel functions for the classification accuracy, is shown in Fig. 5.8. It can be seen from Table 5.3 and Fig. 5.8 that the mixed kernel function has the highest average classification accuracy, the performance of the RBF kernel function, and the polynomial kernel function are poorer, and the performance of the linear kernel function

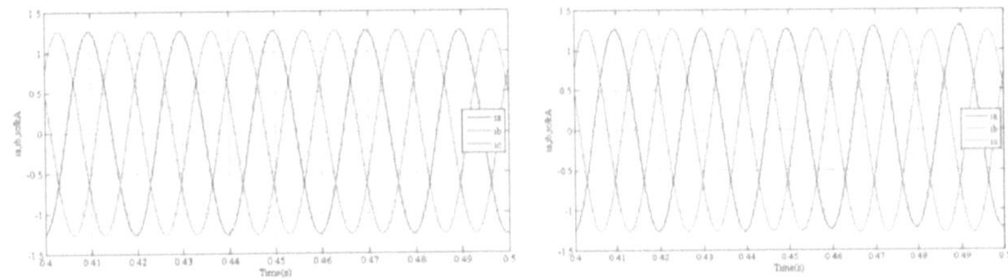

(a) Three-phase AC currents under normal condition.

(b) Three-phase AC currents under open-circuit fault condition.

Figure 5.7: 201-level system under normal condition and open-circuit fault occurs in sub-module of *A*-phase upper bridge arm.

is the lowest, which also proves that the classification accuracy of the mixed kernel support tensor machine algorithm is higher than the single kernel function support tensor machine algorithm.

In order to verify the effectiveness of the proposed methods, a comparison of different kernel functions is launched. The experimental results can be found in Table 5.4. It can be observed from Tables 5.3 and 5.4 that the classification precision

Table 5.2: Label value of fault classification.

$NTRS$	$NTES$	$Faulty\ bridge$	$Label\ value$
20	10	$Normal$	0
20	10	A-$phase\ upper$	1
20	10	B-$phase\ upper$	2
20	10	C-$phase\ upper$	3
20	10	A-$phase\ lower$	4
20	10	B-$phase\ lower$	5
20	10	C-$phase\ lower$	6

Table 5.3: The optimal parameters of grid search and the average of classification accuracy based on the support tensor machine model.

Supoort tensor machine	Parameters					Classification
model	C	$\xi \times 10^{-3}$	h	m	θ	accuracy (%)
Line	16					92.681
Poly	128	27.57		3		95.342
RBF	512	1.86	100			97.774
Sigmoid	64	6.47				94.471
K_{mix}	32	4.18	200	2	0.29	98.872

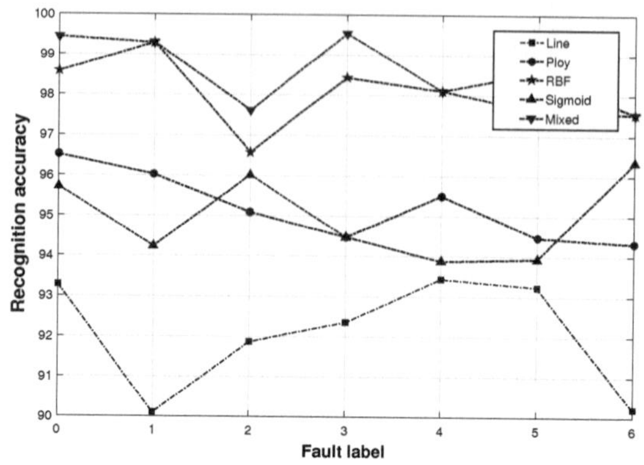

Figure 5.8: The classification accuracy of different kernel function.

depends on the choice of model and the utilization of mixed kernel function. The optimal combination of parameters of mixed kernel can be obtained by repeatedly training the experimental data, the classification accuracy based on the mixed kernel up to 98.872%, but the classification accuracy of other single kernel is 97.774%, and the classification accuracy of support vector machine based on mixed kernel is only 89.917%, which is lower than the classification accuracy based on support tensor machine single kernel. Obviously, the classification accuracy of support tensor machine based on mixed kernel is highest.

To realize the diagnosis of fault, the sub-module fault point of the upper and the lower bridge arm in each phase are set (i.e. one of the three phase fails, and the other two are normal). Specially, the open-circuit fault of sub-module is assumed to appear in the upper bridge arm of j-phase ($j = a, b, c$), after normalization processing, the AC side current, the circulation current, and the envelope mean can be found in Figs. 5.9, 5.11, and 5.12, respectively. Corresponding classification label values

Table 5.4: The optimal parameters of grid search and average of classification accuracy based on the SVM model.

SVM model	Parameters					Classification accuracy (%)
	C	$\xi \times 10^{-3}$	h	m	θ	
Line	16					82.249
Poly	128	9.42		8		86.612
RBF	5256	1.14	200			86.996
Sigmoid	32	4.67				85.519
K_{mix}	32	4.97	300	5	0.21	89.917

can be obtained by using mixed kernel support tensor machine, which is shown in Figs. 5.9(d), 5.11(d), and 5.12(d). Set the sub-module fault of upper bridge arm appears at 2.5 s. the classification tag value is 1 after 2.502 s, which is consistent with the results of the previous classification in Table 5.3. The accuracy of the classification is 98.8%, which indicates that the mixed kernel support tensor machine method achieves effective diagnosis of the fault.

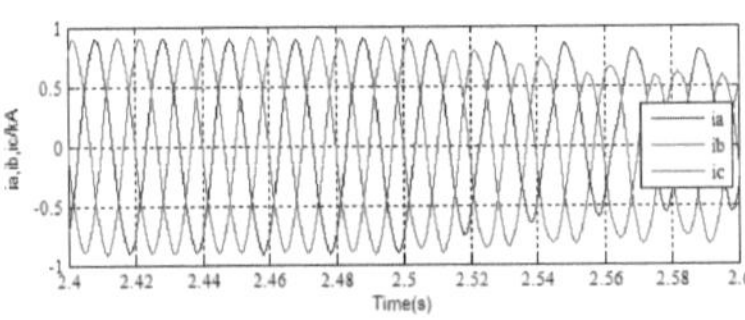

(a) Three-phase AC currents.

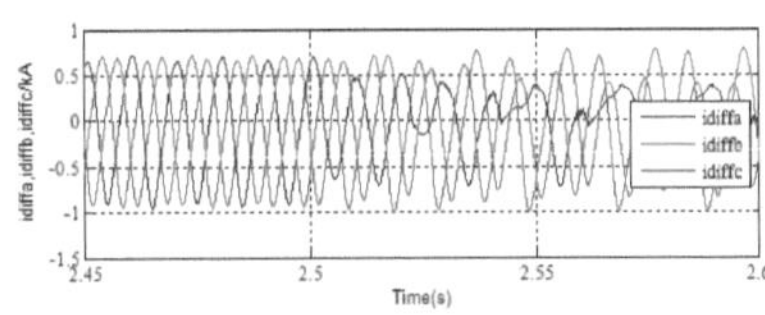

(b) Three-phase inner circulating current.

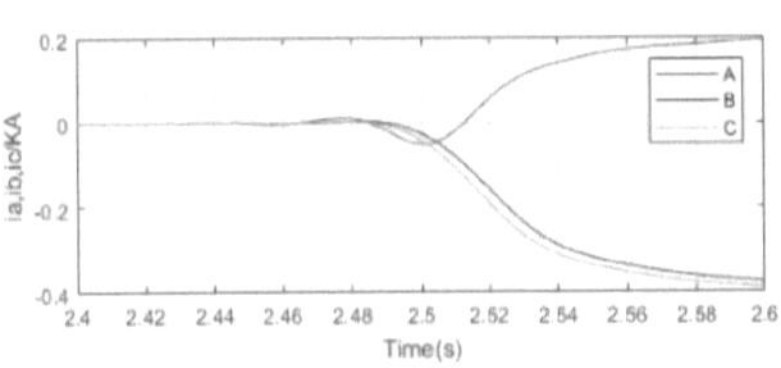

(c) Three-phase current envelope mean.

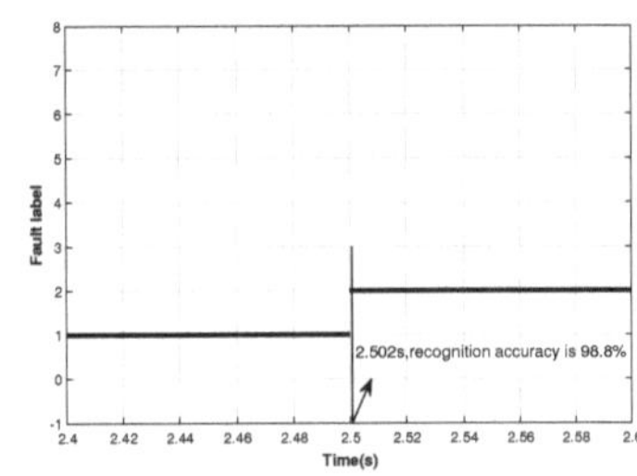

(d) The fault identification label with open-circuit fault.

Figure 5.9: Modular multilevel converter with upper bridge sub-module fault in A-phase.

(a) The fault identification label of modular multilevel converter with sudden load under normal condition.

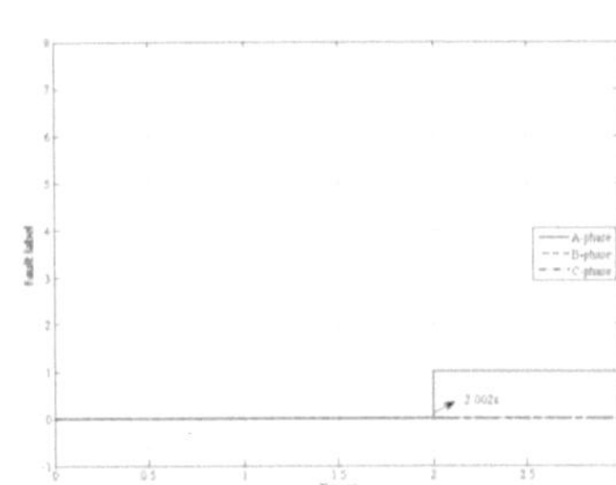

(b) The fault identification label of modular multilevel converter with sudden load under the upper bridge with open-circuit fault.

Figure 5.10: The fault identification label of modular multilevel converter with sudden load under normal condition and upper bridge sub-module fault in A-phase.

In addition, in order to verify the proposed method is still valid under sudden load conditions, at 2 s, we add 50% load to the normal and the A-phase fault system. As shown in Figs. 5.10(a) and (b), it is obvious that the operation efficiency is high

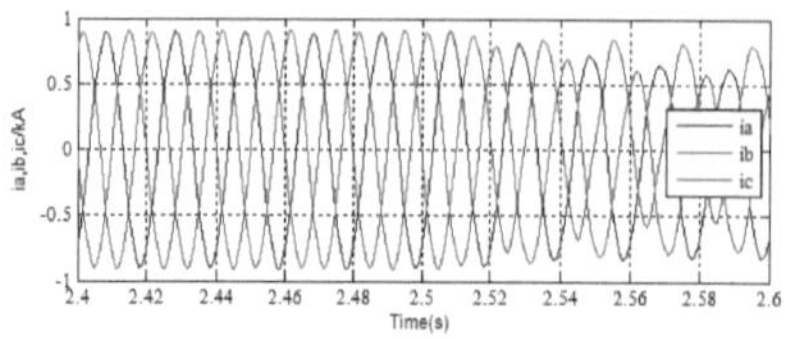

(a) Three-phase AC currents.

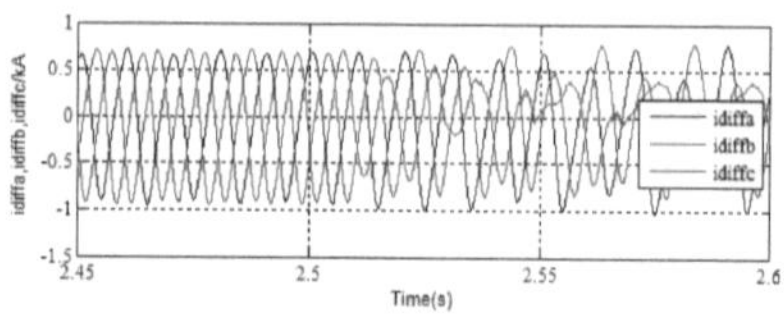

(b) Three-phase inner circulating current.

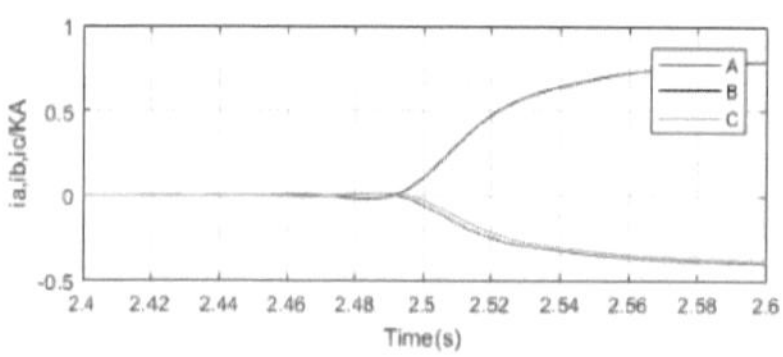

(c) Three-phase current envelope mean.

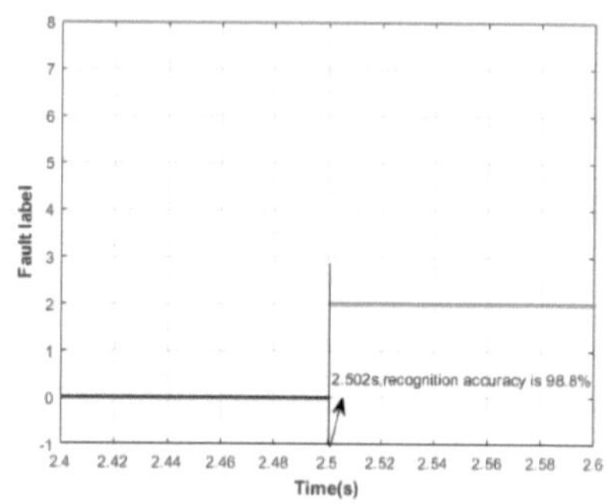

(d) The fault identification label with open-circuit fault.

Figure 5.11: Modular multilevel converter with upper bridge sub-module fault in B-phase.

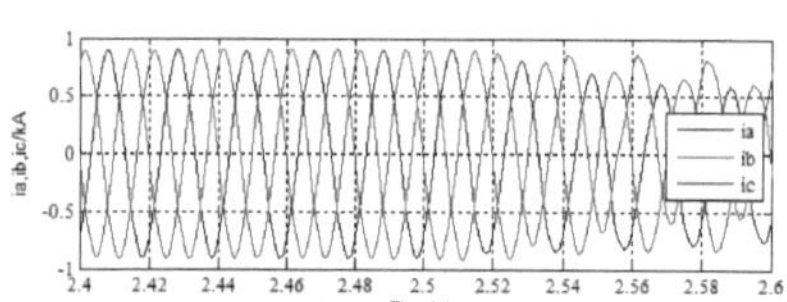

(a) Three-phase AC currents.

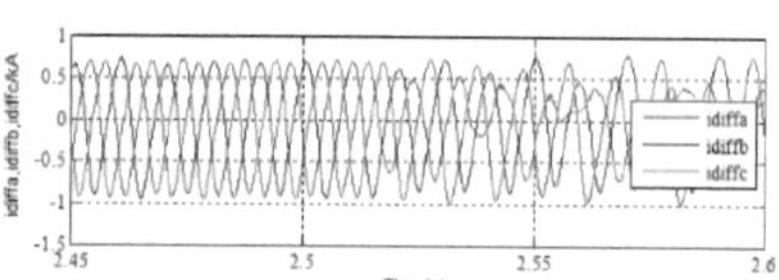

(b) Three-phase inner circulating current.

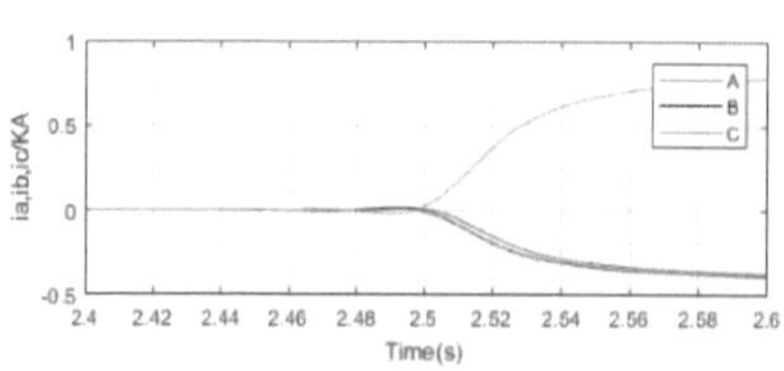

(c) Three-phase current envelope mean.

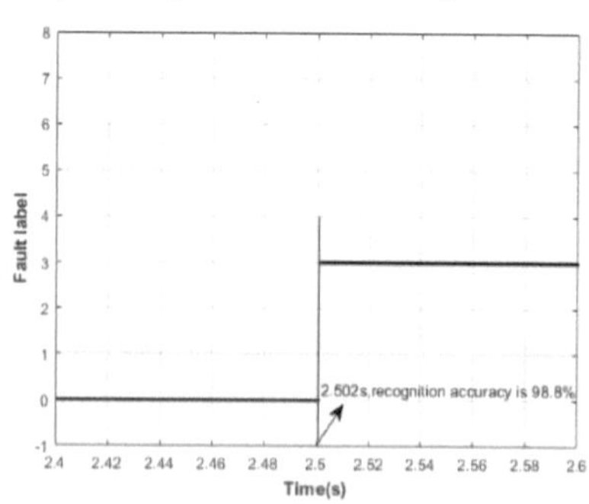

(d) The fault identification label with open-circuit fault.

Figure 5.12: Modular multilevel converter with upper bridge sub-module fault in C-phase.

and the result is not misjudged. Therefore, it can be proved that the proposed method can effectively detect the fault of modular multilevel converter and will not misjudge the fault under the influence of the suddenly imposed load.

5.2 Fault Diagnosis with Synchrosqueezing Transform and Optimized Deep CNN

In the following part of this chapter, a fault diagnosis framework with synchrosqueezing transform and optimized deep convolutional neural network (DCNN) is constructed, under which the frequency domain and time-frequency representations are calculated by synchrosqueezing transform. In addition, batch normalization (BN) and dropout technologies are introduced to prevent the deep convolutional neural network model from overfitting.

5.2.1 Synchrosqueezing Transform

Since wavelet transform is the base of synchrosqueezing transform, we first introduce the continuous wavelet transform. With the aid of the mother wavelet function ψ, the continuous wavelet transform of signal $x(t)$ can be defined as

$$W_x(a,b) = \int_{-\infty}^{+\infty} x(t)\frac{1}{\sqrt{a}}\overline{\psi}(\frac{t-b}{a})\mathrm{d}t \tag{5.14}$$

where $W_x(a,b)$ is the wavelet coefficients, $\overline{\psi}(\frac{t-b}{a})$ is the conjugate complex of wavelet function $\psi(\frac{t-b}{a})$, and a and b are scale and translation factors, respectively. By utilizing the mapping relationship between wavelet scale factor and signal frequency, the wavelet coefficients can be easily displayed on the time-frequency plane. For purely harmonic signal $x(t) = A\cos(\omega t)$, the continuous wavelet transform of $x(t)$ with plancherel theorem can be obtained

$$\begin{aligned} W_x(a,b) &= \frac{1}{2\pi}\int \hat{x}(\xi)\sqrt{a}\overline{\hat{\psi}(a\xi)}e^{ib\xi}\mathrm{d}\xi \\ &= \frac{A}{4\pi}\int [\delta(\xi-\omega)+\delta(\xi+\omega)]\sqrt{a}\overline{\hat{\psi}}(a\xi)e^{ib\xi}\mathrm{d}\xi \\ &= \frac{A}{4\pi}a^{\frac{1}{2}}\overline{\hat{\psi}}(a\omega)e^{ib\xi} \end{aligned} \tag{5.15}$$

where $\hat{\psi}(\xi) = \frac{1}{\sqrt{2\pi}}\int_{-\infty}^{+\infty}\psi(t)e^{-ib\xi}\mathrm{d}t$, $\hat{\psi}(\xi)$, and $\hat{x}(\xi)$ are the Mother wavelet function ψ and continuous wavelet transform, respectively.

Considering the fact that the wavelet coefficients $W_x(a,b)$ are distributed on all scales a, and taking into account the oscillation characteristics of the wavelet coefficients on the initial frequency ω, the initial estimation of the instantaneous frequency can be carried out as

$$[W_x(a,b)]^{-1}\partial b W_x(a,b) = i\omega \tag{5.16}$$

and the instantaneous frequency $\omega_x(a,b)$ of signal x can be expressed as

$$\omega_x(a,b) = \begin{cases} \frac{-i}{W_x(a,b)} \frac{\partial W_x(a,b)}{\partial b}, & |W_x(a,b)| > 0 \\ \infty, & |W_x(a,b)| = 0 \end{cases} \tag{5.17}$$

According to (5.17), the instantaneous frequency $\omega_x(a,b)$ can be estimated, and the wavelet coefficients can be transformed from time-scale domain (b,a) to time-frequency domain $(b, \omega_x(a,b))$. The wavelet transform coefficient $W_x[\omega_x(a,b), b]$ near any centre frequency ω_l of the interval $[\omega_l - \frac{\Delta\omega}{2}, \omega_l + \frac{\Delta\omega}{2}]$ can be used for squeezing calculation, that is, the whole time-frequency distribution is reassigned, and the synchrosqueezing results can be obtained

$$T_x(\omega_i, b) = \frac{1}{\Delta\omega} \cdot \sum_{a_j:|\omega(a_j,b)-\omega_i| \le 0} W_x(a,b) a_j^{-3/2} (\Delta a)_j \tag{5.18}$$

where a_j is the discrete point of all different scales in the time-frequency representations, ω_l denotes the discrete angular frequency, $a_j - a_{j-1} = (\Delta a)_j, \omega_l - \omega_{l-1} = \Delta\omega$.

Based on the above discussion, we found that the essence of synchrosqueezing transform is to enhance the local time-frequency characteristics by squeezing the wavelet time-frequency distribution along the frequency axis with continuous wavelet transform. From Fig. 5.13, it can be observed that synchrosqueezing transform has higher time-frequency resolution than commonly used time-frequency transform methods such as continuous wavelet transform. The time-frequency graph obtained by synchrosqueezing transform is more distinct, and the spectral lines are relatively concentrated. Therefore, the synchrosqueezing transform clearly reflects the fine changes of the spectral line colour corresponding to the current harmonic frequency components of different types of faults, and is more conducive to identifying the sub-module fault occurring in different bridge arms.

5.2.2 Optimized Deep Convolutional Neural Network

Deep convolutional neural network, which includes convolutional, pooling, and full-connected layers, is a typical deep learning algorithm. Convolution layers and pooling layers are alternately stacked for feature learning and feature selection. Full-connected layers following the convolutions and pooling layers are used to transform the 2D feature vector into a 1D feature vector, and then realize the classification. The specific architecture of deep convolutional neural network is shown in Fig. 5.14.

In order to improve the performance of deep convolutional neural network, two tricks, including dropout and batch normalization, are introduced. Specifically,

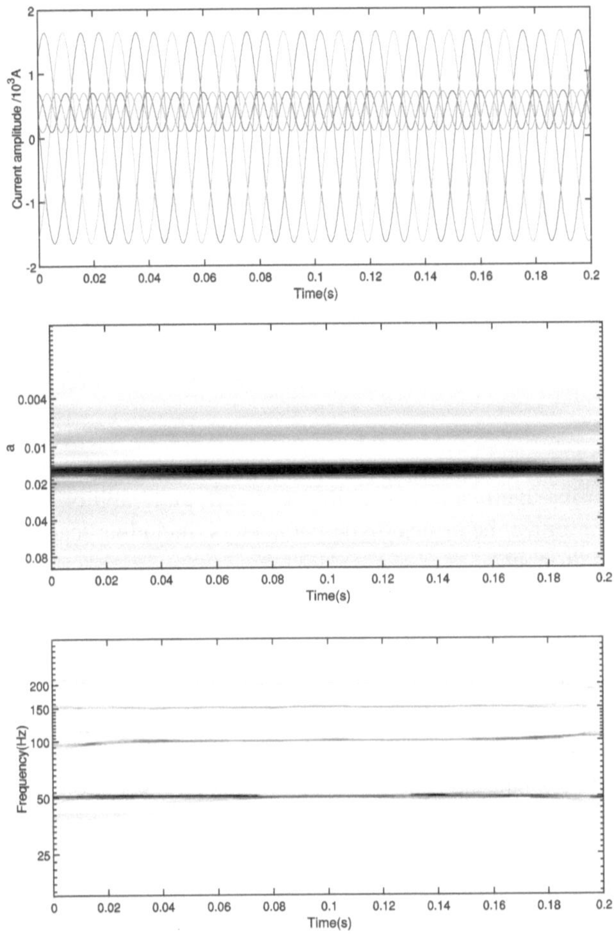

Figure 5.13: Time-frequency representations of faulty sub-module in modular multilevel converter: (a) Raw signal; (b) Continuous wavelet transform result; and (c) Synchrosqueezing transform result.

dropout is used to break the accidental correlation and prevent overfitting, and batch normalization is used to normalize the input of the hidden layer and make the network easier to train.

The architecture of deep convolutional neural network mainly contains weight parameters and hyperparameters. In the process of network training, the weight parameters are automatically updated by the error backpropagation algorithm. The network hyperparameters mainly include learning rate, sample batch size, dropout, and the number of neurons in each layer. Even small changes in hyperparameters will lead to significant impacts on network training. As a result, manual parameter tuning becomes an important step in deep learning training, which takes a lot of time and does not necessarily achieve good performance. Compared with traditional

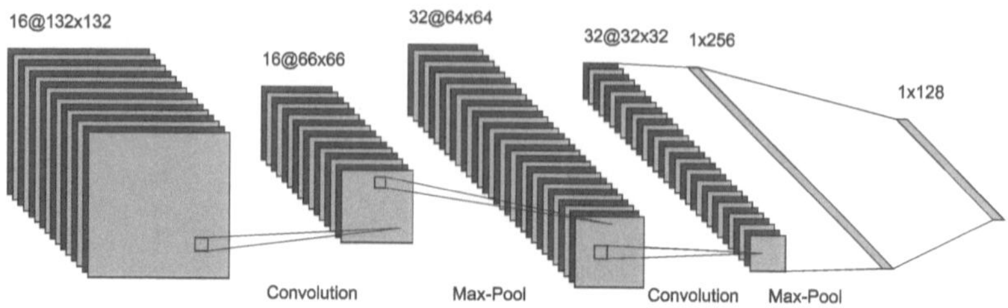

Figure 5.14: The architecture of deep convolutional neural network model in this section.

intelligent optimization algorithms, genetic algorithm (GA) has significant advantages owing to its significant adaptability and self-study ability. Accordingly, GA is adopted to optimize the critical hyperparameters, including learning rate, sample batch size, and dropout.

5.2.3 Fault Diagnosis

Since there are a large number of high-frequency harmonics in the output current of modular multilevel converter, the hard threshold method is employed to deal with it. However, the time-frequency representations of the denoised signal of modular multilevel converter contain the time-frequency ridges corresponding to different fault types. Fortunately, with the help of synchrosqueezing transform, the energy of signal is gathered near the real instantaneous frequency. Moreover, synchrosqueezing transform clearly reflects the fine colour changes of the spectral lines corresponding to the current harmonic frequency components for different types of faults, and it accurately expresses the energy distribution of signals and identifies efficaciously the faults of different bridge arm submodules. More specifically, the feature extraction with synchrosqueezing transform can be described in **Algorithm 5.3**.

Algorithm 5.3 Feature extraction algorithm with synchrosqueezing transform.

Step 1. Denote randomly a sub module with a open-circuit fault on a bridge arm of modular multilevel converter.
Step 2. Collect the raw signal of modular multilevel converter in both three-phase AC current and inner circulating current.
Step 3. Apply wavelet hard threshold method to denoise the collected raw signals and then normalize them.
Step 4. Utilize synchrosqueezing transform to process the three-phase AC current signals and circulating current signals, and then obtain the time-frequency representations of seven different fault types.

Then the next task is to diagnose the fault of modular multilevel converter. A hybrid fault diagnosis method, which combines the advantages of synchrosqueezing transform and GA-DCNN, can be used to implement the fault diagnosis of modular multilevel converter. Accordingly, this method is named as SST-GA-DCNN. The specific steps are summarized in **Algorithm 5.4**, and the flowchart of SST-GA-DCNN is shown in Fig. 5.15.

5.2.4 Illustrative Examples

A half-bridge modular multilevel converter topology is employed to construct a 31-level modular multilevel converter system in the Matlab/Simulink platform with the carrier phase-shifted modulation strategy. By using the operation data with the parameters shown in Table 5.5, the fault diagnosis method of modular multilevel converter is verified.

Table 5.5: Main parameters of modular multilevel converter prototype.

Parameter	Value
Rated DC voltage	8 kV
Number of submodules per arm	30
Line frequency	50 Hz
Carrier switching frequency	10,000 Hz
Sub-module capacitance	4.7 mF
Arm inductance	5 mH
Carrier ratio	30

Algorithm 5.4 Fault diagnosis algorithm with SST-GA-DCNN.

Step 1. Establish the deep convolutional neural network model: Select network optimization tricks and determine the hyperparameters of the network.
Step 2. Divide the data set: Divide the seven fault types time-frequency representations which are obtained from **Algorithm 5.3** into training set, validation set and test set.
Step 3. Data preprocessing: For the input image data of the deep convolutional neural network model, normalize all image data and then compress their size.
Step 4. Optimized deep convolutional neural network model: The training set is used to train and update the parameters of the network model. Employ the genetic algorithm to optimize network hyperparameters in **Step 1**, and then utilize 4-fold cross-validation method to evaluate the model after optimization of hyperparameters.
Step 5. Fault diagnosis: Employ the obtained GA-DCNN model in **Step 4** to test data, and obtain the final classification results of different fault types for modular multilevel converter.

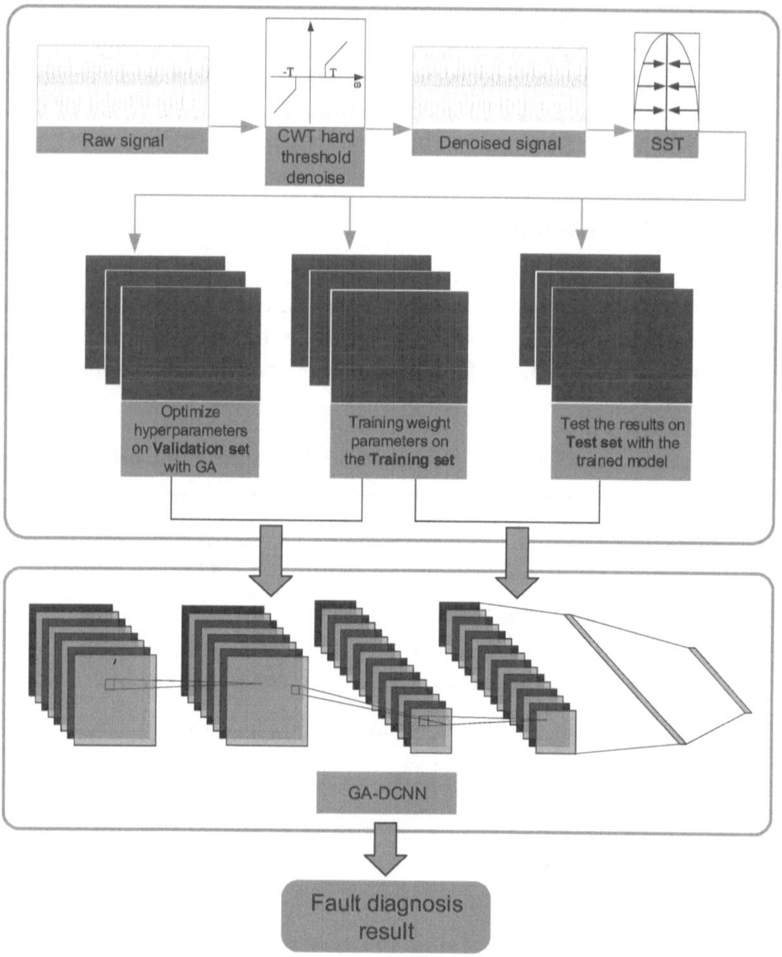

Figure 5.15: The flowchart of fault diagnosis for modular multilevel converter with the proposed SST-GA-DCNN algorithm.

IGBT faults are the main fault type of modular multilevel converter, and they can be divided into short-circuit faults and open-circuit faults. More specially, short-circuit faults have a relatively mature protection mechanism, while open circuit faults are not easy to detect and is more harmful to the system. Therefore, this section mainly studies the open-circuit fault of a single IGBT for modular multilevel converter.

Once the open-circuit fault of sub-module occurs [188], the DC side capacitance will be charged and discharged abnormally, and some values, such as the fluctuation of the capacitance voltage, and then the harmonic components of the three-phase output current and internal three-phase circulating current will change. Specifically, the faulty sub-module on a phase of modular multilevel converter generates corresponding positive and negative DC biases components in this phase, and other two-phase AC will also produce different DC bias. In addition, the three-phase internal circulation will also change. Therefore, this section selects the three-phase AC and three-phase internal circulating current as the raw signals of fault diagnosis.

For the healthy modular multilevel converter as Fig. 5.16, the three-phase AC current is symmetrical with the phase difference as 120, and the three-phase circulating currents are double power frequency circulation current with a relatively small amplitude and a larger positive DC bias.

In the case that the fault case is concerned, the open-circuit up/low arm fault of sub-module is set at 1.06s on the six arms of modular multilevel converter. The simulation results of three-phase AC current and internal circulating current are shown in Figs. 5.17–5.19, respectively. On the one hand, it can be seen from the waveform diagram that the three-phase AC current has a DC bias component, but it is not obvious since a higher modular multilevel converter level number accompany by a smaller DC bias. On the other hand, the inner circulating current changes are relatively obvious as the fault phase increased, and the circulating current of the non-fault phase is reduced.

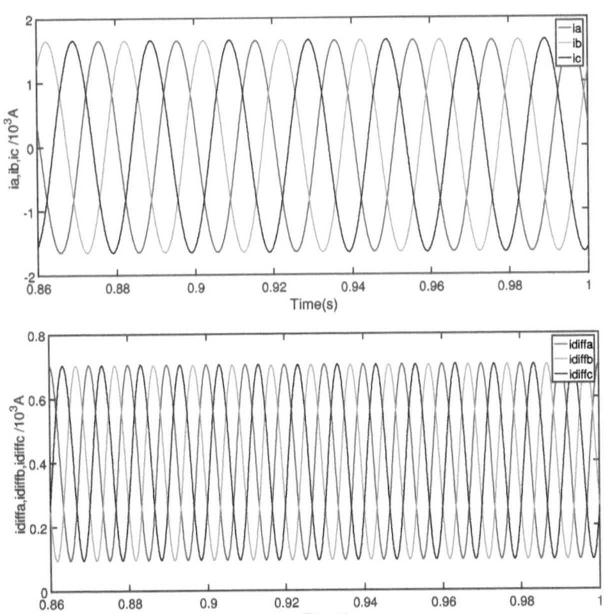

Figure 5.16: Normal operation of modular multilevel converter: (a) Three-phase AC currents and (b) Three-phase circuiting currents.

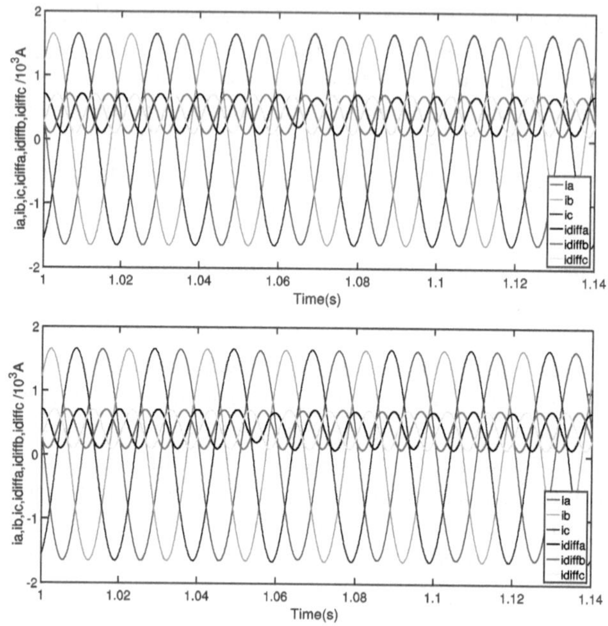

Figure 5.17: Three-phase AC currents and inner circuiting currents of modular multilevel converter under fault operation: (a) A-phase upper bridge arm sub-module fault and (b) A-phase lower bridge arm sub-module fault.

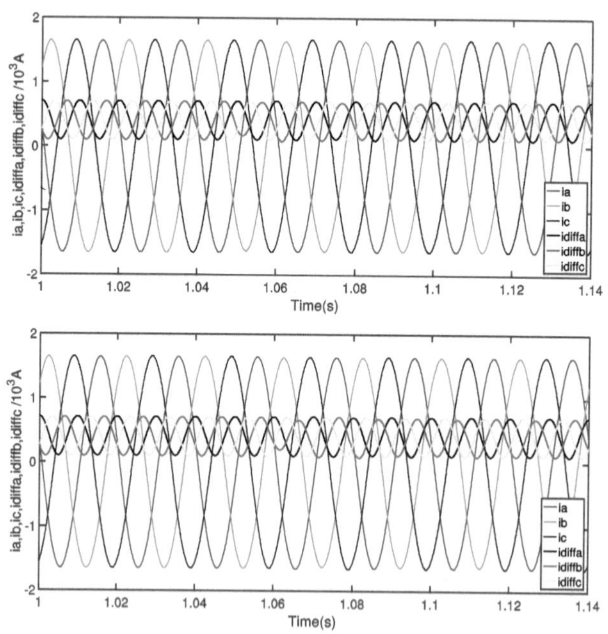

Figure 5.18: Three-phase AC currents and inner circuiting currents of modular multilevel converter under fault operation: (a) B-phase upper bridge arm sub-module fault and (b) B-phase lower bridge arm sub-module fault.

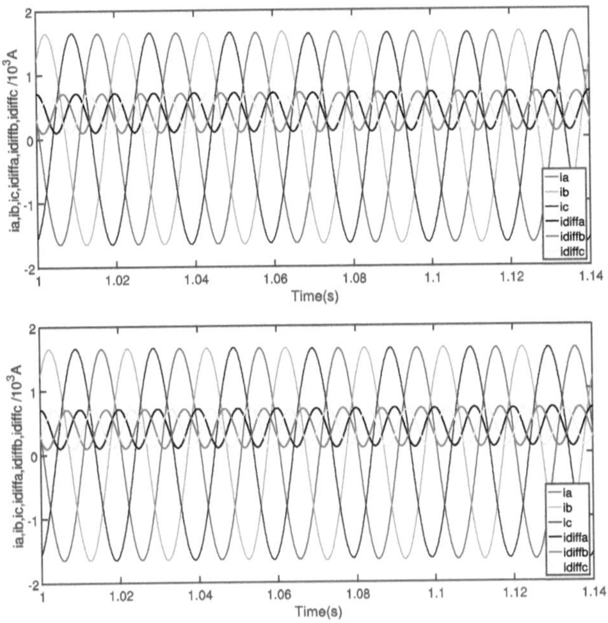

Figure 5.19: Three-phase AC currents and inner circuiting currents of modular multilevel converter under fault operation: (a) C-phase upper bridge arm sub-module fault and (b) C-phase lower bridge arm sub-module fault.

It is difficult to effectively diagnose the specific fault types with the time domain waveforms directly. However, the spectral components of the three-phase AC and inner circulating current with the faulty sub-module have undergone profound changes. Therefore, the three-phase AC current and internal circulating current are selected as the raw signals, and synchrosqueezing transform is adopted to extract the 3-D time-frequency representations.

Before diagnosing the fault of modular multilevel converter, the parameters of deep convolutional neural network model, the fault types, and label values are listed in Tables 5.6 and 5.7, respectively. The raw signals of seven fault types are collected with a sampling frequency of 10 kHz and sampling time of 22 s. In each fault case, the size of every time-frequency representations is compressed at 150 × 150. After 300 samples are randomly selected, the number of training samples (NTRS), validation samples (NVAS) and test samples (NTES) are set as 180, 60, and 60, respectively. After denoising, reconstructing, and synchrosqueezing transform, the corresponding time-frequency representations of seven type fault signals are shown in Figs. 5.20–5.23. Obviously, it is difficult to find the difference and diagnosis the fault type by manual visual inspection method.

With the aid of the time-frequency map, the complete visualization of three convolutional layers for convolutional neural network is shown in Fig. 5.24. Specially, the first convolutional layer has 32 channels, the second and third convolutional layers have 64 and 128 channels, respectively.

Table 5.6: Parameters of deep convolutional neural network model.

Layers	Parameters	
	Size of filter	Output shape
$Conv2d\_1$	3×3	(None,148,148,32)
$Maxpooling2d\_1$	2×2	(None,74,74,32)
$Conv2d\_2$	3×3	(None,72,72,128)
$Maxpooling2d\_2$	2×2	(None,36,36,64)
$Conv2d\_30$	3×3	(None,34,34,64)
$Conv2d\_31$	3×3	(None,32,32,128)
$Maxpooling2d\_3$	2×2	(None,16,16,128)
$Flatten$		(None,128)
$Fullyconnected\_1$		(None,128)
$Fullyconnected\_2$		(None,7)

Table 5.7: Fault types and label values.

NTRS	NVAS	NTES	Fault arm	Label
180	60	60	Normal	0
180	60	60	A-phase upper arm	1
180	60	60	A-phase lower arm	2
180	60	60	B-phase upper arm	3
180	60	60	B-phase lower arm	4
180	60	60	C-phase upper arm	5
180	60	60	C-phase lower arm	6

The optimization of deep convolutional neural network consists of two parts: one is the optimization of network hyperparameters, and the other is the deep convolutional neural network trick. Set initial values of the learning rate, dropout rate, and batch size as 0.01, 0.25, and 20, respectively, then these network hyperparameters are optimized with genetic algorithm and cross-validation algorithm. Considering parameter optimization for 1260 samples of the training set, 63 batches are required for the supplied 20 batch size, the training was carried out for 35 epochs in the training network. In view of deep convolutional neural network tricks, two deep convolutional neural network models are established, including with BN and dropout or without it.

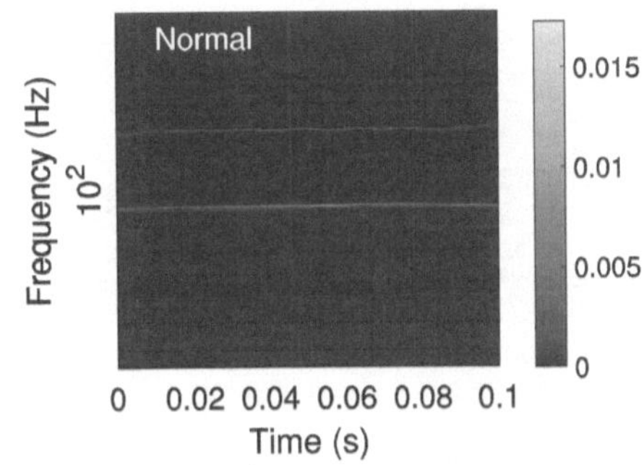

Figure 5.20: Time-frequency representations of synchrosqueezing transform for modular multilevel converter with normal signals.

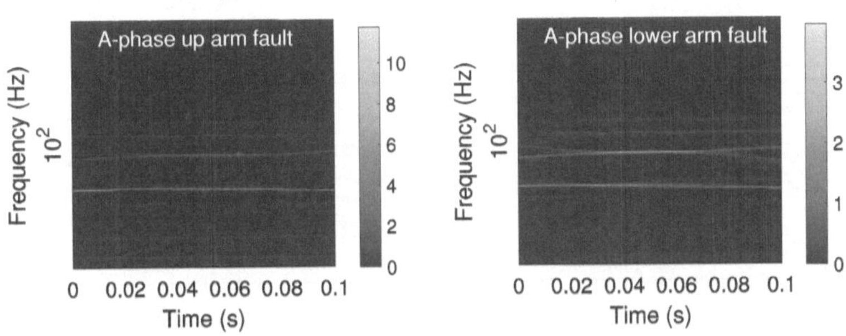

Figure 5.21: Time-frequency representations of synchrosqueezing transform for modular multilevel converter with A-phase up/low arm fault.

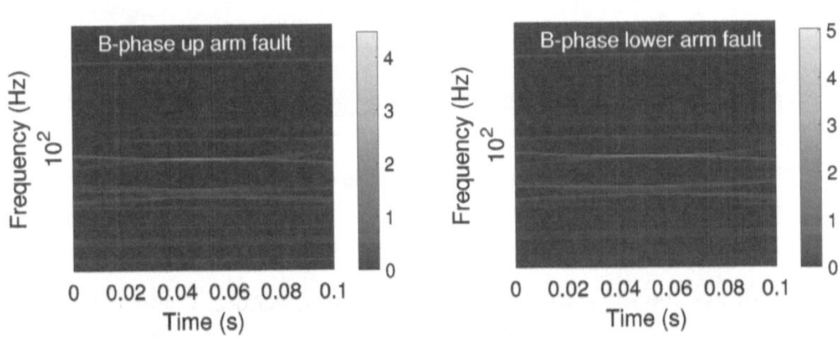

Figure 5.22: Time-frequency representations of synchrosqueezing transform for modular multilevel converter with B-phase up/low arm fault.

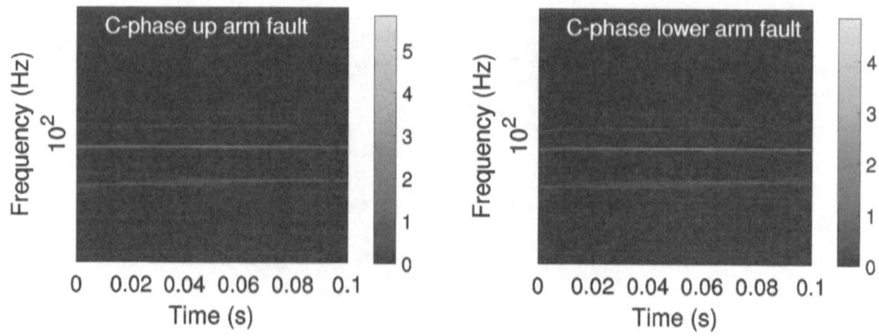

Figure 5.23: Time-frequency representations of synchrosqueezing transform for modular multilevel converter with C-phase up/low arm fault.

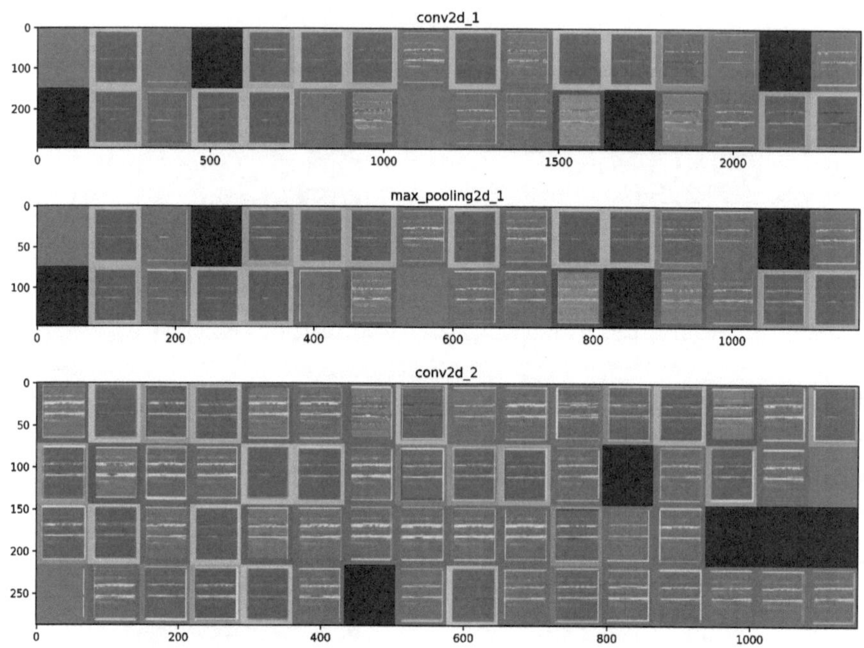

Figure 5.24: Visualization of intermediate activation for DCNN.

Fig. 5.25 shows the diagnosis effect of validation accuracy and validation loss, respectively. As can be seen that the accuracy with BN and dropout is obviously improved. Additionally, the BN model also converged more quickly than the other one with overfitting after 30 epochs. Moreover, the normalized confusion matrix is obtained in Fig. 5.26, and the experiment results demonstrate the effectiveness of deep convolutional neural network with BN and dropout.

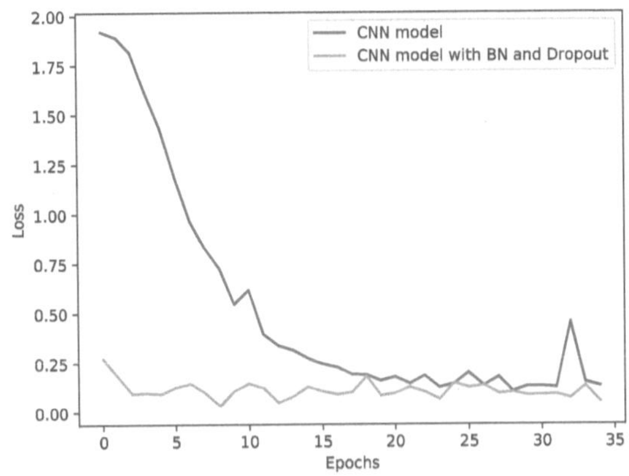

Figure 5.25: Performance evolution of deep convolutional neural network: (a) Accuracy and (b) Loss.

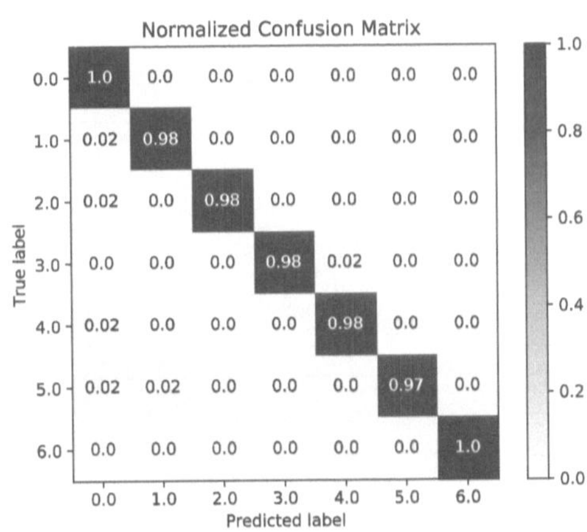

Figure 5.26: Confusion matrix of fault diagnosis.

5.3 Conclusion

In this chapter, the fault diagnosis and location problems of the open circuit fault for modular multilevel converter are investigated. Firstly, By extracting the characteristic data of AC current and internal circulation current in either normal operation or open-circuit fault, and training and classifying the obtained samples with mixed kernel support tensor machine, the fault diagnosis and location of modular multilevel converter can be achieved with the help of the supplied algorithm. Secondly, the time-frequency representations of the raw signals, which are synthesized by the AC and the inner circulating current of the modular multilevel converter, are calculated by synchrosqueezing transform. Then, deep convolutional neural network is introduced to learn the underlying features from the time-frequency representations, and their key hyperparameters are optimized with the genetic algorithm. Based on the obtained residual, both the residual evaluation function and its threshold are constructed to detect the fault.

6

Remaining Useful Life Prediction of Industrial Components with Filter-Based Methods

As a kind of lightweight and high energy density power source, lithium-ion batteries are widely used in spacecraft, aircraft, electric vehicles and portable electronic devices [210]. With regard to the practical application of lithium batteries, it inevitably degrades until the failure with the increase in consumption time. To avert the accidents caused by battery degradation, it is necessary to monitor the state of charge, evaluate the state of health and predict the remaining useful life of lithium-ion batteries [211].

Recently, a great deal of literature [211, 212] has reported the prediction of remaining useful life for lithium-ion batteries, where the mainly used approaches can be categorized into model-based ones, data-driven ones and data-model-fusion ones. For model-based methods, it is a key procedure to establish a mathematical model of battery life cycles with prior knowledge, which can describe the physical mechanism of a lithium battery. To do so, the Kalman filter (KF), extended Kalman filter (EKF), and unscented Kalman filter (UKF) are typically used techniques. Although the model-based method has achieved some successes in the remaining useful life prediction, an accurate mathematical model is usually difficult or even impossible to be obtained in practice, especially when the batteries operate in noisy and/or uncertain environments. Such a realistic situation would limit the applications of model-based methods. As an alternative, the data-driven methods need less system knowledge, that can extract typical features from the degradation data of lithium-ion batteries. Then, some techniques such as the machine learning method can be used to construct the mapping relationship between the degraded data and health state, which can be further used to estimate the battery degradation and forecast its remaining useful life [213]. The support vector regression (SVR), as a typical representative of artificial intelligence algorithm, has been widely utilized in the remaining useful life prediction of batteries ([214–217]). For data-driven prediction methods, the accuracy of prediction often depends on the established mapping relationships which are sensitive to the quality and quantity of the battery data.

To realize the complementary advantages of model-based methods and data-driven methods, the hybrid method was proposed by combining the mechanism characteristics of the degradation model and the machine learning strategy, which has received extensive research attention [218]. Different from the KF/UKF/UKF technique dealing with Gaussian noise, the particle filter (PF) method is derived from the sequential importance sampling and Bayesian theory [219, 220], which is

DOI: 10.1201/9781003330998-6

suitable for nonlinear degrade models with non-Gaussian noises. To be more specific, during the utilization of the particle filter for remaining useful life prediction, a number of weighted particles are adopted to approximate the posterior distribution of the state of degrade model. As the number of iterations increase, the big weight will be concentrated gradually and the small weights will fade away, which would result in the defects of weight degeneracy and particle impoverishment [221]. Such defects would bring serious side effects on the accuracy of the remaining useful life prediction. To address the above concerns of the standard particle filter, two typical techniques are adopted in the existing results to improve the performance of the particle filter [222]. One method is resampling technology [223], at which those particles with big weights are duplicated repeatedly and the small weights are abandoned. Then, the particle diversity decrease inevitably. The other is the particle distribution optimization technique [224], in which the intelligent algorithms (e.g. swarm optimization and genetic algorithm) are chosen to optimize the proposal distribution.

Inspired by the above discussion, we aim to provide hybrid prognosis frameworks to improve prediction accuracy. Section 6.1 is concerned with the remaining useful life prediction based on adaptive unscented Kalman filter and optimized support vector regression. In Section 6.2, the problem of remaining useful life prediction with Levy optimized particle filtering and long short-term memory network is investigated. Section 6.3 develops health assessment of model-data fused remaining useful life prediction of bearings with feature enhancement. Section 6.4 gives our conclusions.

6.1 Remaining Useful Life Prediction with Adaptive UKF and SVR

In order to update both the process noise covariance and the observation noise covariance, that can not be realized via the standard unscented Kalman filtering, the adaptive unscented Kalman filtering (AUKF) is introduced. On the other hand, the genetic algorithm (GA) is adopted to optimize the key parameters of kernel function in support vector regression, which is adopted to realize multi-step prediction of the remaining useful life.

As verified in [225], the exponential model usually has a better global regression performance than the polynomial model and other typical models. Therefore, we choose the following double exponential models as the capacity degradation pattern of lithium battery:

$$y_k = a * \exp(b \cdot k) + c * \exp(d \cdot k) \tag{6.1}$$

where y_k denotes battery capacity, k is the number of cycles, and a, b, c, d represent the model parameters which need to be initialized with actual data.

Due to the complexity and timeliness of battery degradation, the model parameters a_k, b_k, c_k, d_k in this section are assumed to be time-varying, which is more in line with the actual situation of battery degradation. By denoting $x_k \triangleq \mathrm{col}\{a_k, b_k, c_k, d_k\}$, and considering the influence of external noise and uncertainty, we establish the following state space model:

$$\begin{cases} x_{k+1} = x_k + \varpi_k \\ y_k = a_k * \exp(b_k \cdot k) + c_k * \exp(d_k \cdot k) + \upsilon_k \end{cases} \tag{6.2}$$

where ϖ_k and υ_k represent the process noise and observation noise, which are uncorrelated zero-mean white noise with covariance Q_k and R_k, respectively. By generalizing (6.2), one obtains the nonlinear system as follows:

$$\begin{cases} x_{k+1} = f(x_k) + \varpi_k \\ y_k = h(x_k) + \upsilon_k \end{cases} \tag{6.3}$$

where $f(*)$ and $g(*)$ are nonlinear functions endowing the mapping relation between capacity and the parameters of degrade model (6.1). Furthermore, we set P_0 as the initialization covariance of initial state x_0, Q_0 and R_0 as the initial covariance of noises, and $\bar{x}$ and P as the mean and variance of state $x(k)$ with four dimensions, respectively.

Different from the extended Kalman filter method which employs linearization to achieve the estimation of the nonlinear stochastic system, we adopt the unscented Kalman filter [226] because it utilizes unscented transform to deal with the nonlinear transfer of mean and variance, so that higher calculation accuracy and better stability can be obtained. The pseudocode of the adaptive unscented Kalman filter algorithm is shown in **Algorithm 6.1**.

The obtained data $\{e_k\}$ and its mapping $\{z_k\}$ make up the observation sample set $\mathbb{Z} = \{(e_1, z_1), (e_2, z_2), \cdots, (e_n, z_n)\}$ $(e_n \in \mathbf{R}^n, z_n \in \mathbf{R})$. The mapping relationship of observation set $\mathbb{Z}$ can be expressed as

$$z = g(e) = w^T e + q, w \in \mathbf{R}^n \tag{6.4}$$

where w and q are the weight vector and the offset constant, respectively.

The mapping (6.4) can be further transformed into the following optimization problem:

$$\min \left[\frac{\|w\|^2}{2} + C \cdot \sum_{i=1}^{n} (\xi_i + \xi_i^*)\right] \tag{6.5}$$

$$s.t. \begin{cases} z_i - w^T e_i - q \le \varepsilon + \xi_i \\ w^T e_i + q - z_i \le \varepsilon + \xi_i \\ \xi_i, \xi_i^* \ge 0, i = 1, ..., n \end{cases} \tag{6.6}$$

where C is a preset penalty coefficient, ε is the deviation between the training data and the actual observation data, ξ_i and ξ_i^* are slack variables. Usually, the penalty parameter C is used to adjust the balance between generalization ability and classification accuracy.

Algorithm 6.1 Adaptive Unscented Kalman Filter

Input:

Initialization parameters P_0, x_0, Q_0, R_0

Iteration:

while $k < N$ **do**

Step 1. Prediction process:

1.1). Using unscented transform to get sigma point set $\mathfrak{R}_x = \{x^{(i)}_{k-1|k-1}\}$, $i = 0, \cdots, 2n$

1.2). $x^{(i)}_{k|k-1} = f(x^{(i)}_{k-1|k-1})$

1.3). $\hat{x}_{k|k-1} = \sum_{i=0}^{2n} \omega_m^{(i)} x^{(i)}_{k|k-1}$

1.4). $e_x^{(i)} = x^{(i)}_{k|k-1} - \hat{x}_{k|k-1}$

1.5). $\boldsymbol{P}_{k|k-1} = \sum_{i=0}^{2n} \omega_c^{(i)} e_x^{(i)} (e_x^{(i)})^T + \boldsymbol{Q}_{k-1}$

1.6). Using new prediction point $\hat{x}_{k|k-1}$ and unscented transformation again. we can obtain the new sigma point sets $\mathfrak{R}_{\hat{x}} = \{x^{(i)}_{k|k-1}\}$, $i = 0, \cdots, 2n$

1.7). $y^{(i)}_{k|k-1} = h(x^{(i)}_{k|k-1})$

1.9). $\hat{y}_{k|k-1} = \sum_{i=0}^{2n} \omega_m^{(i)} y^{(i)}_{k|k-1}$

Step 2. Update process:

2.1). $e_y^{(i)} = y^{(i)}_{k|k-1} - \hat{y}_{k|k-1}$

2.2). $\boldsymbol{P}_{yy} = \sum_{i=0}^{2n} \omega_c^{(i)} e_y^{(i)} (e_y^{(i)})^T + r_{k-1}$

2.3). $\boldsymbol{P}_{xy} = \sum_{i=0}^{2n} \omega_c^{(i)} e_x^{(i)} (e_y^{(i)})^T$

2.4). $\boldsymbol{K}_k = \boldsymbol{P}_{xy} \boldsymbol{P}_{yy}^{-1}$

2.5). $e_k = y_k - \hat{y}_{k|k-1}$

2.6). $\hat{x}_{k|k} = \hat{x}_{k|k-1} + \boldsymbol{K}_k e_k$

2.7). $\boldsymbol{P}_{k|k} = \boldsymbol{P}_{k|k-1} - \boldsymbol{K}_k \boldsymbol{P}_{yy} \boldsymbol{K}_k^T$

2.8). $\hat{y}_{k|k} = h(\hat{x}_{k|k})$

Step 3. Covariance updating process:

3.1). $r_k^{(i)} = y^{(i)}_{k|k-1} - \hat{y}_{k|k}$

3.2). $\boldsymbol{G}_k = \frac{1}{L} \sum_{i=k-L+1}^{k} r_k^{(i)} \cdot (r_k^{(i)})^T$

3.3). $\boldsymbol{Q}_k = \boldsymbol{K}_k \boldsymbol{G}_k \boldsymbol{K}_k^T$

3.4). $\bar{r}_k = y_k - \hat{y}_{k|k}$

3.5). $\boldsymbol{R}_k = \sum_{i=0}^{2n} \omega_c^{(i)} \bar{r}_k \bar{r}_k^T + \boldsymbol{G}_k$

end while

Output:

$\hat{y}_{k|k}$– Capacity estimate at time k.

Similar to [227], by introducing the Lagrange multiplier α_i and α_i^*, equation (6.5) can be turned into a dual problem, and the optimal solution with parameters w, α_i, α_i^* can be written by

$$g(e) = \sum_{i=1}^{n} (\alpha_i - \alpha_i^*) K(e_i, e) + b \tag{6.7}$$

where $K(e_i, e)$ is a kernel function, which has a significant impact on the regression performance of support vector regression. The radial basis kernel function is often chosen as the following form

$$K(e_i, e) = \exp\{\frac{-|e_i - e|^2}{2\sigma^2}\} \tag{6.8}$$

where σ is the parameter of kernel function, which can affect the complexity of support vector regression algorithm.

In order to achieve more effective prognosis, the genetic algorithm [228] is introduced to optimize the parameters C and σ. genetic algorithm is an adaptive global optimization search method which imitates the evolutionary law of biology. Compared with traditional optimization algorithms, the genetic algorithm does not depend on specific mathematical equation and derivative expression and has advantages of strong global search ability, high efficiency and fast search speed. Therefore, genetic algorithm is often regarded as an appropriate choice to optimize the support vector regression for remaining useful life.

6.1.1 Genetic Algorithm Optimized Support Vector Regression

In this part, the multi-step prediction method with residual data is adopted to obtain higher prediction accuracy. Firstly, the starting point of prediction is supposed to be T and the residual data $e_{1:T}$ from 1 to T can be computed with the unscented Kalman filter algorithm. Next, the obtained residual data $e_{1:T}$ is utilized to build the support vector regression prediction model which can be expressed as follows:

$$\eta_t = g(\zeta_t) = \sum_{i=1}^{T-m} (\alpha_i - \alpha_i^*) K(\zeta_i, \zeta_t) + p \tag{6.9}$$

where $\zeta_t \triangleq [e_t, e_{t+1}, \cdots, e_{t+m-1}]$ and $\eta_t \triangleq e_{t+m}$ $(t = 1, 2, ..., T - m)$.

Relying on the past m sets of data $\zeta_{t+1-m} = [e_{t+1-m}, \cdots, e_t]$, the one-step prediction relation at time $t = T$ can be established as follows:

$$\tilde{e}_{t+1} \quad = \quad g(\zeta_{t+1-m}) = \sum_{i=1}^{T-m} (\alpha_i - \alpha_i^*) K(\zeta_i, \zeta_{t+1-m}) + p \tag{6.10}$$

where $\tilde{e}_{t+1}$ represents the residual. To improve the prognosis effect of the one-step prediction (6.10), the genetic algorithm is employed to optimize the kernel function $K(\zeta_i, \zeta_{t+1-m})$, where the optimization goal are the support vector regression parameters C and σ.

On the other hand, by performing the one-step prediction (6.10), the p-step prediction can be obtained:

$$\begin{cases} \hat{e}_{t+1} = g(\zeta_{t+1-m}) = g(e_{t+1-m}, \cdots, e_t) \\ \hat{e}_{t+2} = g(\zeta_{t+2-m}) = g(e_{t+2-m}, \cdots, e_t, \hat{e}_{t+1}) \\ \vdots \\ \hat{e}_{t+p} = g(\zeta_{t+n-m}) = g(e_{t+n-m}, \cdots, e_t, \hat{e}_{t+1}, \cdots, \hat{e}_{t+p-1}) \end{cases} \tag{6.11}$$

After the new prediction residual sequences $\{\hat{e}_{t+i}\ (i = 1, 2, ..., p)\}$ is obtained by the GA-SVR, the updated residual data is added to the update process via the adaptive unscented Kalman filtering algorithm. Consequently, the state vector value a_k, b_k, c_k, d_k can be updated each time and then, the adaptive character of the exponential model (6.1) can be acquired. Finally, the multi-step prediction can be realized.

6.1.2 Remaining Useful Life Prediction of Lithium-Ion Batteries

Based on the adaptive unscented Kalman filtering and GA-SVR algorithm discussed above, we propose a hybrid remaining useful life prediction method that combines adaptive unscented Kalman filtering and GA-SVR algorithm. The flowchart of AUKF-GA-SVR method is shown in Fig. 6.1, and the specific prediction process is summarized as follows:

Step 1. **State estimation**: Set the time point of starting prediction as T, the real capacity data $\{x_i, y_i\}_{i=1}^{T}$ of the lithium battery from 1 to T is used to determine the empirical model (6.2) parameters a_k, b_k, c_k, d_k with adaptive unscented Kalman filtering algorithm.

Step 2. **Residual data acquisition**: Utilize the empirical model obtained in **Step 1** and adaptive unscented Kalman filtering algorithm to acquire the residual data $\{e_i\}_{i=1}^{T}$.

Step 3. **Training the GA-SVR model**: Firstly, the residual data $\{e_i\}_{i=1}^{T}$ are filtered with adaptive unscented Kalman filtering algorithm, and the outliers are removed. Consequently, take the obtained data as a training set and recur to the genetic algorithm to optimize the support vector regression model (6.10), then acquire the multi-step prediction result $\{\hat{e}_{T+s}\}_{s=1}^{p}$ by the iterative calculation method (6.11).

Step 4. **State update**: Employ the obtained residual data $\{\hat{e}_{T+s}\}_{s=1}^{p}$ and the adaptive unscented Kalman filtering algorithm to produce the state estimation value $\{\hat{x}_{T+s}\}_{s=1}^{p}$. Utilize the measurement equation (6.2) to obtain the predicted capacity $\{\hat{y}_{T+s}\}_{s=1}^{p}$.

Step 5. **Remaining useful life prediction**: If the predicted battery capacity reaches the defined threshold, then output the remaining useful life by calculating the interval between the prediction starting cycle T and the battery failure threshold point. Else, go to **Step 3**.

Actually, the adaptive unscented Kalman filtering algorithm can not only achieve the adaptive update of process noise covariance and observation noise covariance, but also improve the filter accuracy. On the other hand, the support vector regression is adopted to implement multi-step prediction, and genetic algorithm is added to modify the parameters (C, σ) in the support vector regression so that the prediction accuracy can be improved.

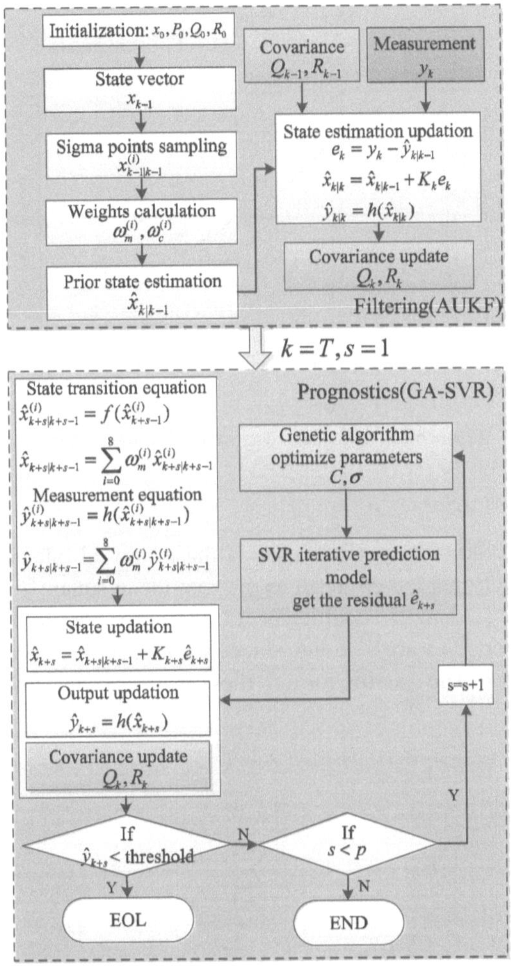

Figure 6.1: Flowchart of the integrated algorithm.

6.1.3 Illustrative Examples

A. Experimental Data Set Description

In this section, we utilize the National Aeronautics and Space Administration (NASA) battery dataset [229] to verify the effectiveness of the proposed method. By analyzing the characteristics of the battery capacity degradation data and experimental conditions, B0005, B0006, B0007, and B0018 are chosen and their capacity decay curves are shown in Fig. 6.2.

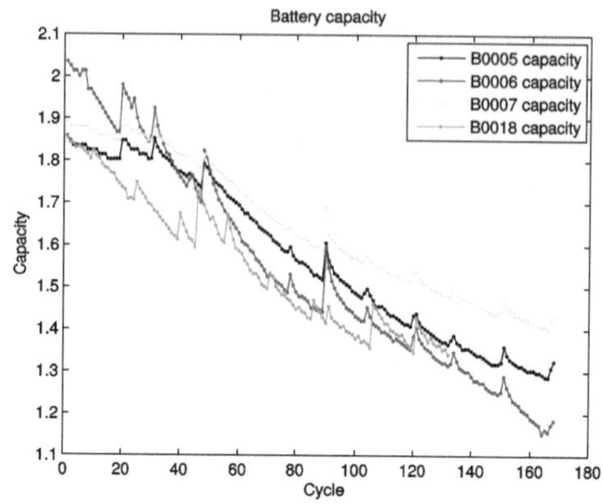

Figure 6.2: The capacity decay curves of four batteries.

B. Evaluation Matrices

To assess the prediction performance of the proposed integrated approach, we consider some traditional indices such as the root mean square error E_{RMSE}, mean absolute error E_{MAE} and R^2 coefficient. Furthermore, the prediction error E_{RUL} and its relative error E_{RA} are also introduced to indicate the accuracy of remaining useful life prediction. The specific form of these matrices can be described as follows

$$\begin{aligned} E_{RMSE} &= \sqrt{\frac{1}{n}\sum_{k=1}^{n}(y_k - \hat{y}_k)^2} \\ E_{MAE} &= \frac{1}{n}\sum_{k=1}^{n}\left|\frac{(\hat{y}_k - y_k)}{y_k}\right| \\ R^2 &= 1 - \frac{\sum_{k=1}^{n}(y_k - \hat{y}_k)^2}{\sum_{k=1}^{n}(y_k - \bar{y}_k)^2} \\ E_{RUL} &= |R_t - R_p| \\ E_{RA} &= 1 - \frac{|R_t - R_p|}{R_t} \end{aligned}$$

where y_k indicates the actual battery capacity, $\hat{y}_k$ denotes the predicted battery capacity, $\bar{y}_k$ represents the average of actual battery capacity, R_t is the true remaining useful life result, and R_p describes the remaining useful life prediction value.

Specifically, for indicators E_{RMSE}, E_{MAE}, and E_{RUL}, if they are closer to 0, then it means that the capacity prediction accuracy is higher. For parameters R^2 and E_{RA}, if they are closer to 1, then the more accurate remaining useful life prediction results can be achieved.

C. Simulation Result Analysis

In order to verify the superiority of the proposed technology, several widely-used methods are introduced as the comparison algorithms such as the unscented Kalman filter, extended Kalman filter, adaptive extended Kalman filter, adaptive unscented Kalman filtering, unscented Kalman filter and relevance vector regression (RVR-UKF) [103] and adaptive extended Kalman filter and genetic algorithm optimized support vector regression (AEKF-GASVR).

On the other hand, some key factors such as the number of data cycles, initial capacity and discharge conditions of each battery group are different, all of which have an important effect on the remaining useful life prediction. In this simulation part, we choose different parameters (including prediction start points, steps of multi-step prediction and threshold of battery failure) to verify the prediction effect of the developed approach. As a result, the remaining useful life prediction curves for four kinds of batteries with different starting prediction points are shown in Figs. 6.3-6.10, and their quantitative results of prediction accuracy are listed in Tables 6.1-6.2 with five kinds of performance matrices.

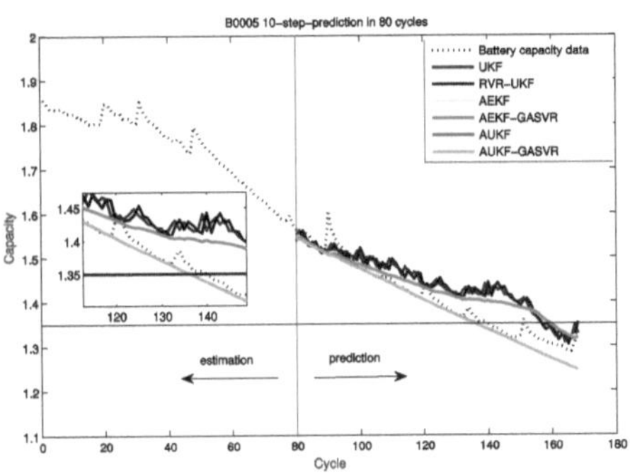

Figure 6.3: Ten-step-prediction for battery B0005 with start point 80 cycle.

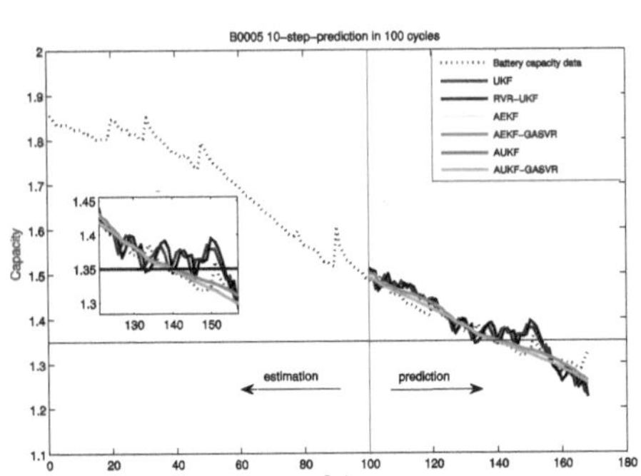

Figure 6.4: Ten-step-prediction for battery B0005 with start point 100 cycles.

Specifically, the degradation model of the battery capacity is established by using the real data before the prediction points, and the integrated algorithm proposed

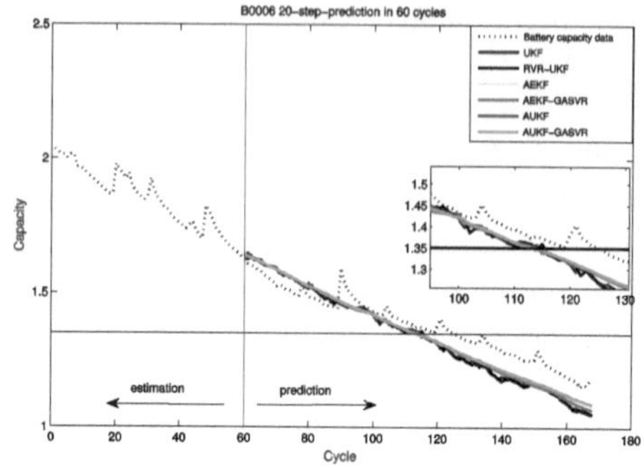

Figure 6.5: Twenty-step-prediction for battery B0006 with start point 60 cycle.

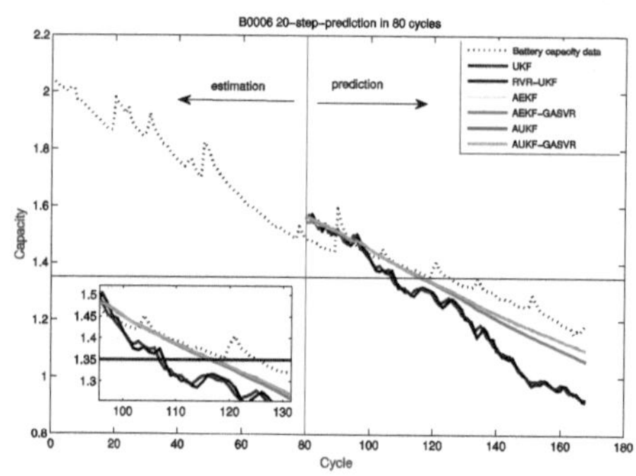

Figure 6.6: Twenty-step-prediction for battery B0006 with start point 80 cycle.

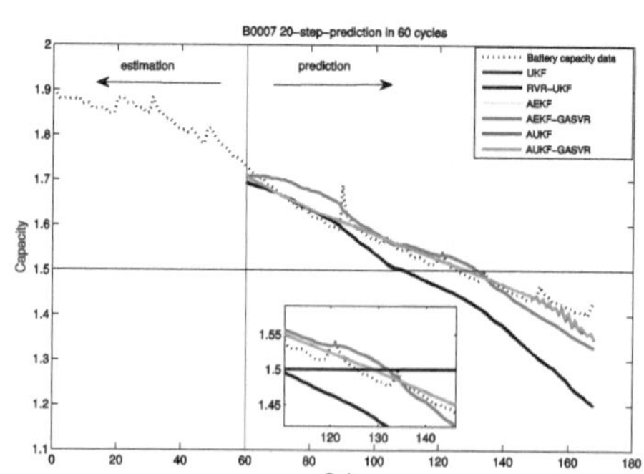

Figure 6.7: Twenty-step-prediction for battery B0007 with start point 60 cycle.

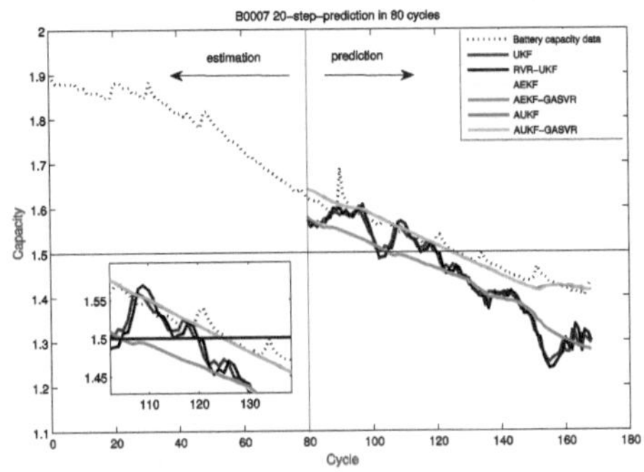

Figure 6.8: Twenty-step-prediction for battery B0007 with start point 80 cycle.

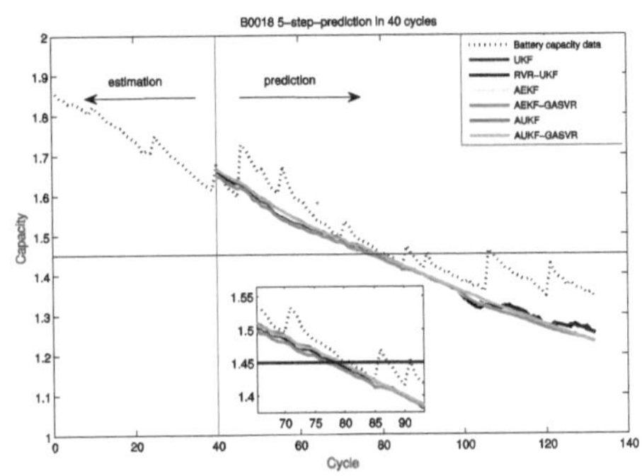

Figure 6.9: Five-step-prediction for battery B0018 with start point 40 cycle.

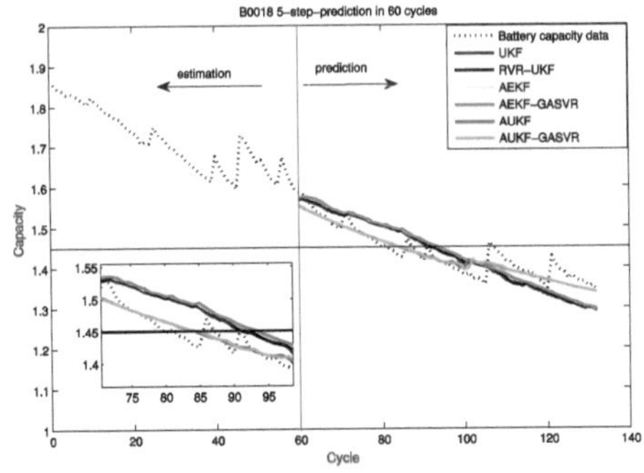

Figure 6.10: Five-step-prediction for battery B0018 with start point 60 cycle.

Table 6.1: Prediction effect comparison of B0005 battery.

Method	Cycle	R_t	R_p	E_{RUL}	E_{RMSE}	E_{MAE}	R_2	E_{RA}
UKF	80	60	76	16	0.0385	0.0279	0.7006	0.733
	100	40	44	4	0.0262	0.0192	0.7447	0.9
RVR-UKF	80	60	74	14	0.0381	0.0277	0.7116	0.767
	100	40	39	1	0.0238	0.0175	0.7552	0.975
AEKF	80	60	79	19	0.0306	0.0221	0.8127	0.683
	100	40	39	1	0.013	0.01	0.9176	0.975
AEKF-GASVR	80	60	78	18	0.0304	0.0218	0.8156	0.7
	100	40	40	0	0.0129	0.0095	0.9118	1
AUKF	80	60	56	4	0.0196	0.0136	0.9086	0.933
	100	40	39	1	0.0126	0.0094	0.9199	0.975
AUKF-GASVR	**80**	**60**	**57**	**3**	**0.0192**	**0.0125**	**0.9099**	**0.95**
	100	**40**	**40**	**0**	**0.0124**	**0.0092**	**0.9211**	**1**

Table 6.2: Prediction effect comparison of B0006 battery.

Method	Cycle	R_t	R_p	E_{RUL}	E_{RMSE}	E_{MAE}	R_2	E_{RA}
UKF	60	65	53	12	0.0617	0.0465	0.6613	0.8
	80	45	26	19	0.1275	0.0994	−1.1996	0.578
RVR-UKF	60	65	54	11	0.0603	0.0453	0.6651	0.817
	80	45	28	17	0.1265	0.0985	−1.1973	0.622
AEKF	60	65	53	12	0.0573	0.0429	0.7245	0.8
	80	45	36	9	0.0599	0.0468	0.5240	0.8
AEKF-GASVR	60	65	54	11	0.0565	0.0423	0.7259	0.817
	80	45	37	8	0.0593	0.0459	0.5265	0.822
AUKF	60	65	54	11	0.0527	0.0395	0.7731	0.817
	80	45	37	8	0.0489	0.0371	0.6885	0.822
AUKF-GASVR	**60**	**65**	**55**	**10**	**0.0510**	**0.0392**	**0.7743**	**0.833**
	80	**45**	**38**	**7**	**0.0483**	**0.0368**	**0.6888**	**0.844**

in this section is used to predict the remaining useful life after the starting prediction points. Compared with unscented Kalman filter, RVR-UKF, adaptive extended Kalman filter and AEKF-GASVR techniques, adaptive unscented Kalman filtering and AUKF-GASVR have better remaining useful life prediction performance from the above simulation results. Four methods, such as the unscented Kalman filter, can hardly complete the remaining useful life prediction task. It can be seen from Fig. 6.4 that adaptive extended Kalman filter, AEKF-GASVR, adaptive unscented Kalman filtering and AUKF-GASVR algorithms have much better filtering performance than unscented Kalman filter and RVR-UKF methods because the adaptive characteristic efficiently reduces the prediction and estimation error. On the other hand, the amount of degraded data also has an important effect on the prediction accuracy. For example, the 60th and 80th cycles are chosen as the prediction starting points for B0006 battery, where Figs. 6.5 and 6.6 and Table 6.2 verify that the later starting point results in higher prediction accuracy.

By analyzing the above simulation results, we can conclude that the higher prediction accuracy may benefit from the following reasons: 1) the unscented Kalman filter algorithm has not involved the calculation of Jacobian matrix since it ignores the high-order term; 2) the adaptive characteristics are used to dynamically adjust the estimation and prediction error; 3) the support vector regression algorithm is introduced to offset the shortcoming of the adaptive unscented Kalman filtering, which can only do one-step-predict; and 4) the genetic algorithm is employed to optimize parameters of support vector regression. Therefore, it is indeed effective to employ the AUKF-GA-SVR algorithm with the above advantages to predict the remaining useful life precisely.

6.2 Remaining Useful Life Prediction with ALF-Optimized PF and LSTM

In this subsection, a hybrid prognosis framework is established by fusing the adaptive Levy flight (ALF) optimized particle filter and long short term memory (LSTM) network. Under this fusion framework, the adaptive Levy flight algorithm is introduced to optimize the particle distribution, and the long short-term memory network network is adopted to learn about long-term dependencies.

Different from the extended Kalman filter or unscented Kalman filter with exponential model dealing with the Gaussian noises [63], the particle filter is adopted to investigate the nonlinear exponential degrade model with no-Gaussian noises. The remaining useful life prediction process of the particle filter with exponential model (PF-E) is shown in **Algorithm 6.2**, and some basic representations will be introduced in the next. Specially, we suppose that x_0^i $(i = 1, 2, ..., N)$ is generated from the prior probability density function $p(x_0)$. ω_k^i is the current weight. $y_{1:k+1}$ is defined as $y_{1:k+1} \triangleq [y_1, y_2, ..., y_{k+1}]$. N is the number of particles.

Although the resampling method can alleviate the weight degeneracy which is inevitable in the particle filter method [106, 221], the resample method would bring the new problem of particle impoverishment. To be specific, the large-weight particles are bound to be copied more times, more identical particles appear and thus, the diversity of particles gradually disappear. These phenomena of the particle filter algorithm are shown in Fig. 6.11. In Fig. 6.11, the circles filled with colour represent particles; the radiuses denote the size of the particle weight; and the particle position represents the particle states. It can be observed from Fig. 6.11 that the evolution process of the particles weight degeneracy, resampling, particle impoverishment in the particle filter algorithm. Meanwhile, some interesting results have revealed that the particle impoverishment causes inaccurate approximation of the posterior probability density function of the state, and resampling is not an effective solution to this problem [221, 222].

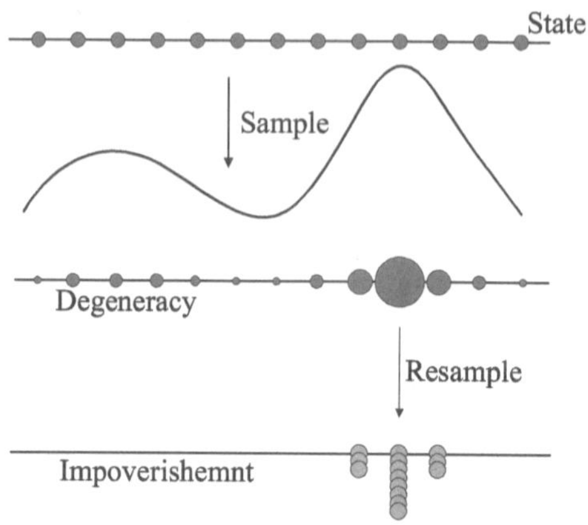

Figure 6.11: Weight degeneracy and particle impoverishment phenomena of standard particle filter algorithm.

6.2.1 Adaptive Levy Flight Optimized Particle Filter

Levy flight is a random search algorithm obeying Levy distribution, which is a non-Gaussian process with the characteristics of a short-distance local search or an occasionally long-distance jump to increase the search space and improve the global search ability, then accelerate the convergence speed. It has been verified from [230] that the Levy flight provides superior performance than several well-known meta-heuristic algorithms such as the simulated annealing (SA), differential evolution (DE), particle swarm optimization (PSO), elephant herding optimization (EHO), and genetic algorithm.

Considering its distinct advantages, the adaptive Levy flight algorithm is proposed to optimize the standard particle filter for the purpose of the performance improvement. Specifically speaking, the Levy flight algorithm is used to update the particle evolution with the following location update law:

$$Z_{\ell+1} = Z_\ell + \alpha \otimes \mathcal{L}(\beta) \tag{6.12}$$

where Z_ℓ represents the location of generation ℓ, α is the step size parameter, $\otimes$ is the dot product between two matrices, and $\mathcal{L}(\beta)$ denotes the path obeying Levy distribution. There are many ways to generate random steps that obey Levy distribution, where the common and efficient one is the well-known Mantegna algorithm (which can generate a symmetric and stable Levy distribution) [231], and the corresponding $\mathcal{L}(\beta)$ can be calculated as follows:

$$\mathcal{L}(\beta) \sim \frac{\beta\Gamma(\beta)\sin(\frac{\beta}{2}\pi)}{\pi}\frac{1}{s^{1+\beta}}\ (s \gg s_0 > 0) \tag{6.13}$$

where Γ is the standard Gamma function, s is the step length. By considering the maximum search efficiency in the unknown space at [231], we set $\beta = 1.5$.

Algorithm 6.2 Remaining useful life prediction algorithm with PF-E.

Step 1. Initialization: let $k = 0$, generate N initial particles and initial weights from $p(x_0)$, $x_0^i, w_0^i = 1/N$, $i = 1, 2, ..., N$, respectively.
Step 2. Particle Propagate: propagate each particle x_{k+1}^i by important density function $q(x_{k+1}^i|x_k^i, y_{1:k+1})$ usually expressed as state transition function $x_{k+1} = f(x_k) + \varsigma_k$ at k cycle, predict the state x_{k+1} by $x_{k+1} = \sum_{i=1}^{N} \omega_k^i x_{k+1}^i$.
Step 3. Particle Update: update the particles when the measurement y_{k+1} becomes available, calculate the weights by denoting $q(x_{k+1}^i|x_k^i, y_{1:k+1})$ as $p(x_{k+1}^i|x_k^i)$:

$$\begin{aligned}\omega_{k+1}^i &= \omega_k^i \frac{p(y_{k+1}|x_{k+1}^i)p(x_{k+1}^i|x_k^i)}{q(x_{k+1}^i|x_k^i, y_{1:k+1})} \\ &\propto \omega_k^i p(y_{k+1}|x_{k+1}^i)\end{aligned} \tag{6.14}$$

where $q(x_{k+1}^i|x_k^i, y_{1:k+1})$ is the important density function, $p(x_{k+1}^i|x_k^i)$ is the prior probability density, $\propto$ represents "in proportion to" and $p(y_{k+1}|x_{k+1}^i)$ is the likelihood function.

Normalize the updated particle weights as

$$\omega_{k+1}^i = \frac{\omega_{k+1}^i}{\sum_{j=1}^{N} \omega_{k+1}^j}$$

the posterior probability density function (PDF) $p(x_{k+1}|y_{1:k+1})$ can be approximated by those particles and associated weights:

$$p(x_{k+1}|y_{1:k+1}) \approx \sum_{i=1}^{N} \omega_{k+1}^i \delta(x_{k+1} - x_{k+1}^i) \tag{6.15}$$

where $\delta(\cdot)$ is the Dirac delta function.
Step 4. Particle Resampling: set a valid sample number N_t as the threshold, estimate the degree of degradation with $\hat{N}_e = \frac{1}{\sum_{i=1}^{N}(\omega_{k+1}^i)^2}$. If $\hat{N}_e < N_t$, a new set of particles $\{\hat{x}_{k+1}^i, 1/N\}$ are obtained with resample; otherwise, the resample is not performed.
Step 5. State Estimation: $\hat{x}_{k+1} = \sum_{i=1}^{N} \omega_{k+1}^i \hat{x}_{k+1}^i$.
Step 6. Remaining useful life Prediction: predict $\hat{y}_{k+l}$ recursively by extrapolating the degrade model $x_{k+1} = f(x_k) + \varsigma_k$ and $y_k = g(x_k) + \eta_k$ when the measurement y_{k+1} becomes unavailable, $l = 1, 2, 3, ..., T$. Once $\hat{y}_{k+l}$ reaches the threshold, the remaining useful life at time k can be achieved.

Similar to [231], the step length s is defined by:

$$s = \frac{u}{|v|^{\frac{1}{\beta}}}, \ u \sim N(0, \sigma^2), v \sim N(0, 1) \tag{6.16}$$

where u and v obey Gaussian distribution, β is a bounded index, σ satisfies

$$\sigma = \left\{ \frac{\Gamma(1+\beta) sin(\frac{\pi\beta}{2})}{\beta\Gamma(\frac{1+\beta}{2}) 2^{\frac{\beta-1}{2}}} \right\}^{\frac{1}{\beta}}.$$

To sum up, the ith particle update with Levy flight generation ℓ can be written as

$$x^i_{\ell+1,k} = x^i_{\ell,k} + \frac{\alpha}{\omega^i_{\ell,k}} s,\ \ell = 0, 1, ..., P \tag{6.17}$$

where $\omega^i_{\ell,k}$ denotes the weight of the ith particle at the ℓth generation Levy flight in k cycle. Especially, $\frac{a}{\omega^i_{\ell,k}}$ means that the smaller weight of current particle needs the larger search step size, that increases the search efficiency. Since the item $\frac{\alpha}{\omega^i_{\ell,k}}$ of (6.17) varies with i , ℓ, k, we call it adaptive Levy flight.

It has been verified from [232] that the Levy flight possesses some significant advantages including increased population diversity and expanded search area. Thus, it is adopted to improve the particle search efficiency and reduce resampling times. Especially, the Levy flight has the advantage of increasing the diversity of particles, that can be further used to solve the particle impoverishment problem. Therefore, the individual particle sampled from the important density function can be searched by adaptive Levy flight.

The optimization process of particle filter with adaptive Levy flight is illustrated in Fig. 6.12, in which N particles are sampled from the important density function, and K populations with P individuals are generated by P-times adaptive Levy flight. As a result, the individual with the largest weight in each population is picked out to form a new recommended distribution.

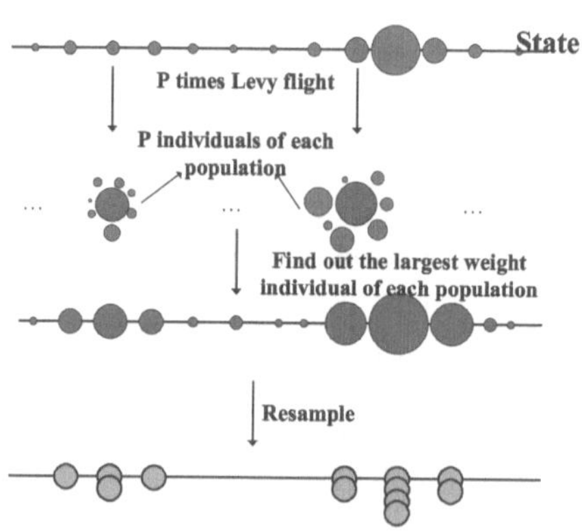

Figure 6.12: Illustration of the evolution process of adaptive Levy flight-based particle filter.

After using the ALF-PF algorithm, the weights of small-weight particles increase prominently than the large-weight particles due to the adaptive step. Several significant advantages are shown in Fig. 6.12 that the normalization of particle weights ω^i_k becomes more harmonious; more particle information is available for resampling; and the diversity of the resampled particles increases significantly. In other words, the weight degeneracy and particle impoverishment problem can be solved in this way.

Once the problem of weight degeneracy and particle impoverishment in the particle filter is solved, the subsequent task is to dispose the degradation model of LIBs. In fact, it is difficult to establish the accurate degradation model due to the internal

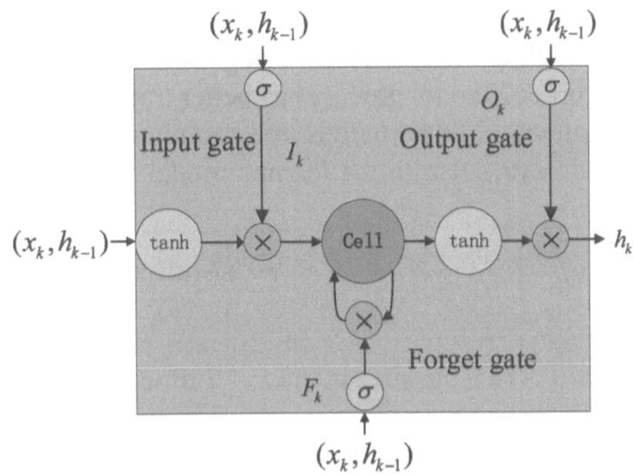

Figure 6.13: Network architecture of LSTM.

electrochemical reaction and complex external environment. To facilitate us to finish the modelling teaks, in this section, the well-known long short-term memory network is adopted to learn the battery degradation model.

A typical network structure of long short-term memory network is shown in Fig. 6.13, which contains a separate memory cell and three gates (i.e. the forgetting gate, input gate and output gate). The learning model can be described by

$$\begin{cases} F_k = \sigma(W_{Fx}x_k + W_{Fh}h_{k-1} + b_F) \\ I_k = \sigma(W_{Ix}x_k + W_{Ih}h_{k-1} + b_I) \\ O_k = \sigma(W_{Ox}x_k + W_{Oh}h_{k-1} + b_O) \\ \xi_k = F_k * \xi_{k-1} + I_k * \tanh(W_{\xi x}x_k + W_{\xi h}h_{k-1} + b_\xi) \\ h_k = O_k * \tanh(\xi_k) \end{cases} \tag{6.18}$$

where F_k is the forget gate designed to forget the redundant information, I_k is the input gate which is used to select the key information storing in the memory cell, and O_k is the output gate for determining the output information, which is calculated according to the current input x_k and last hidden state h_{k-1}. In each update moment k, the memory cell ξ_k updates by forget gate to forget some information and input gate to add some information, respectively. Concurrently, the current hidden state h_k is obtained by the current updated memory cell ξ_k and output gate O_k. Matrix weights $W_{Fx}/W_{Ix}/W_{Ox}/W_{\xi x}$ and $W_{Fh}/W_{Ih}/W_{Oh}/W_{\xi h}$, and biases $b_F/b_I/b_O/b_\xi$ need to be trained by (6.18). $\sigma(\cdot)$ (i.e. logistic sigmoid) and $\tanh$ (i.e. hyperbolic tangent) represents pointwise nonlinear activation functions.

6.2.2 Remaining Useful Life Prediction of Lithium-Ion Batteries

In the process of remaining useful life prediction with ALF-PF-LSTM, the state of each particle is assumed to propagate with a n-order Markov model

$$x_{k+1} = f(x_k, x_{k-1}, ..., x_{k-n+1}) + \varsigma_k \tag{6.19}$$

Generally speaking, higher-order Markov processes are thought to provide a better state transition model, at the cost of requiring more calculation. It is ascertained in Table 6.7 that a 4-order Markov process has better trade-off between the accuracy and calculation complexity. The particle is updated recursively by extrapolating the fixed transition function (6.19) with the learned model by long short-term memory network

$$\begin{aligned}\hat{x}^i_{k+l} &= f(\hat{x}^i_{k+l-1}, \hat{x}^i_{k+l-2}, \hat{x}^i_{k+l-3}, \hat{x}^i_{k+l-4}) \\ &\quad + \varsigma_k,\ l = 1, 2, 3, ..., T\end{aligned} \tag{6.20}$$

Based on ALF-PF-LSTM, the state at the $k+l$ time can be predicted as

$$\hat{x}_{k+l} = \sum_{i=1}^{N} \omega^i_k \hat{x}^i_{k+l} \tag{6.21}$$

Once the state of the ith particle at the end-of-life L^i_k extrapolated by (6.20) reaches the threshold, the remaining useful life of the ith particle can be achieved by $T^i_k = L^i_k - k$. At the same time, the posterior probability density function of remaining useful life can be approximated by

$$p(T_k|y_{1:k}) \approx \sum_{i=1}^{N} \omega^i_k \delta(T_k - T^i_k) \tag{6.22}$$

Especially, the approximated probability density function of remaining useful life indicates the probability of LIBs's failure at the future time, and the narrower probability density function width means the more reliable prediction result.

Finally, the remaining useful life can be estimated by:

$$T_k = \sum_{i=1}^{N} \omega^i_k T^i_k \tag{6.23}$$

Fig. 6.14 shows the flowchart of remaining useful life prediction with the proposed method. Each particle is propagated using the state transition relation $f(*)$ learned by long short-term memory network in the **Propagate** step. After the measurement y_k is available, the weights of particles are calculated with ALF-PF. Finally, if there is no new measurement, the remaining useful life prediction will be performed with the proposed method.

6.2.3 Illustrative Examples

A. Description of Experimental Dataset

In this section, the NASA battery dataset [229] is utilized to verify the effectiveness of the proposed method. Specially, batteries B05, B06 and B07 are taken as experimental data which has more obvious battery attenuation trend than other batteries. The evolution curves are shown in Fig. 6.15.

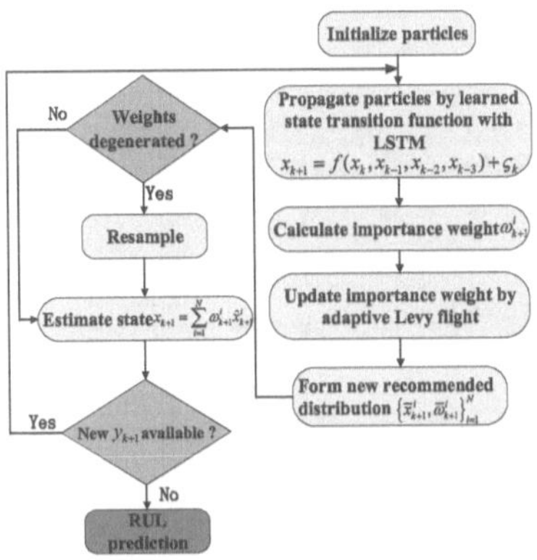

Figure 6.14: Flowchart of proposed method for remaining useful life prediction.

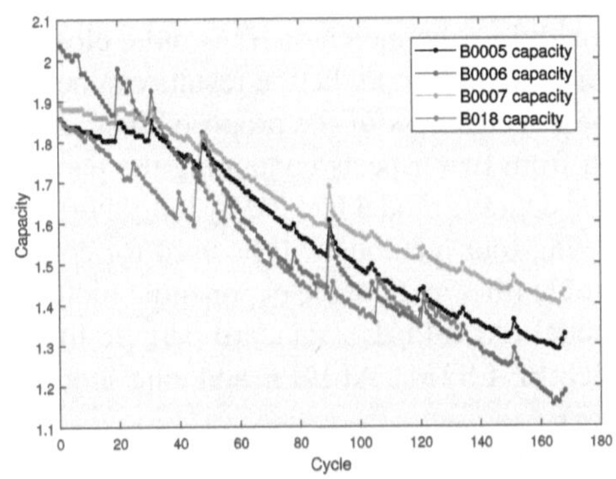

Figure 6.15: The capacity decay curves of three batteries.

The 18650 battery operates at 24°C in three conditions (charging, discharging, and impedance), and its parameters include current (charging and discharging), voltage (charging and discharging), impedance, ambient temperature, and so on. Similar to [63, 98], the battery capacity is chosen as the health indicator which plays an important role in the successful remaining useful life prediction. By considering the difference of experimental conditions, the failure thresholds of B05, B06, and B07 are set as 75%, 68%, and 80% of the rated capacity, respectively.

B. Evaluation Criteria

In order to prove the superiority of the proposed method, we adopt the following traditional indices such as root mean square error (E_{RMSE}), mean absolute percent error (E_{MAPE}), and R^2 coefficient as the evaluation criteria. Furthermore, the prediction error (E_{RUL}) and probability density function width are also applied to evaluate the accuracy of the remaining useful life prediction.

$$
\begin{aligned}
E_{RMSE} &= \sqrt{\frac{1}{n}\sum_{k=1}^{n}(y_k - \hat{y}_k)^2} \\
E_{MAPE} &= \frac{1}{N}\sum_{k}^{N}\left|\frac{y_k - \hat{y}_k}{y_k}\right| \\
R^2 &= 1 - \frac{\sum_{k=1}^{n}(y_k - \hat{y}_k)^2}{\sum_{k=1}^{n}(y_k - \bar{y}_k)^2} \\
E_{RUL} &= |R_t - R_p|
\end{aligned}
$$

where y_k represents the actual capacity, $\hat{y}_k$ indicates the predicted battery capacity, $\bar{y}_k$ denotes the average of actual data, R_t describes the true remaining useful life, and R_p is the prediction value of remaining useful life. It is worth noting that the indicators E_{RMSE}, E_{MAPE}, and E_{RUL} are close to 0, meaning that the prediction accuracy is higher. In addition, the indicator R^2 is more close to 1, showing that the more accurate remaining useful life prediction results can be achieved.

To demonstrate the effectiveness of the proposed method, we will carry out the simulation verification from two aspects containing the particle filter optimized by adaptive Levy flight and degrade model learned by long short-term memory network. Specifically, the following four methods will be used for comparative analysis, that cover the standard particle filter and double exponential model (PF-E), ALF-PF, and double exponential model (ALF-PF-E), standard particle filter and long short-term memory network model (PF-LSTM), ALF-PF, and long short-term memory network model (ALF-PF-LSTM).

It can be seen obviously from Fig. 6.16 that the prediction accuracy is greatly improved with the utilization of the Levy-fight-based particle filter or LSTM-based degradation model. Meanwhile, some key factors (such as the number of training data) have an important effect on the remaining useful life prediction. Thus, different prediction starting points are selected in the simulation to verify the prediction effect of the proposed approach. The quantitative results of prediction accuracy can be obtained in Tables 6.3–6.5 with five kinds of performance matrices. Compared with PF-E, ALF-PF-E and PF-LSTM, ALF-PF-LSTM has better prediction performance. In addition, the amount of degraded data also has an important effect on the prediction accuracy in the same approach, that is verified by Tables 6.3–6.5.

C. Simulation results analysis

As the important criteria to measure the prediction quality of the proposed method, the box plot and histogram of particles are provided in Figs. 6.17–6.20 which consist of the particle number of 50 and 150 at 100 cycle. Actually, the width of probability density function can be verified concisely with box plot, and the detailed distribution of particles can be visualized with histogram. In general, adaptive Levy flight leads to the narrower width of the predicted probability density function,

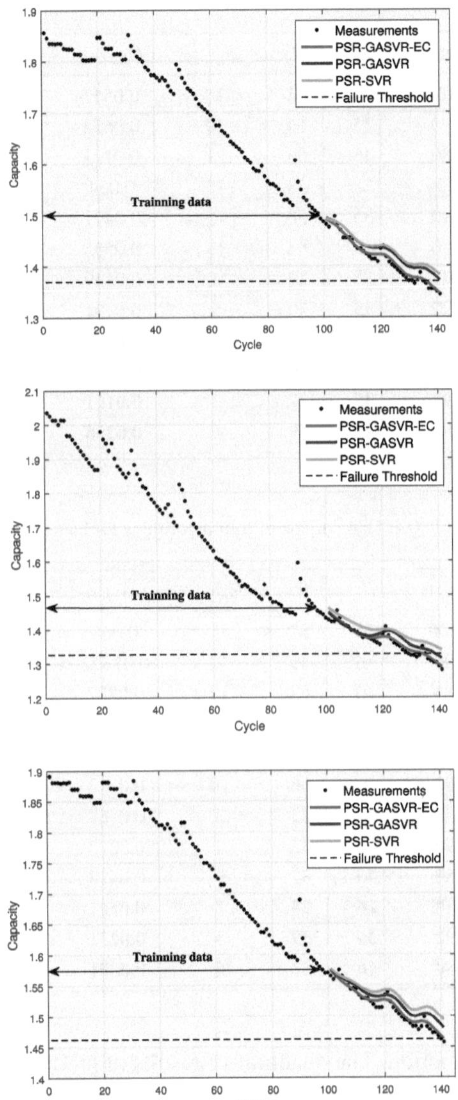

Figure 6.16: Remaining useful life prediction of batteries 05, 06, and 07 with four different methods.

and long short-term memory network leads to the more accurate prediction, which indicates that the real remaining useful life is in the high likelihood region of the predicted probability density function. Compared with the other three methods, the ALF-PF-LSTM not only has higher prediction accuracy but also narrower width of

Table 6.3: Comparison of four algorithms for battery 05.

Algorithm	Cycle	R_t	R_p	E_{RUL}	E_{RMSE}	E_{MAPE} (%)	R^2
	100	25	43	18	0.0567	3.7605	0.9245
PF-E	90	35	54	19	0.0652	3.923	0.9147
	80	45	66	21	0.0721	4.3625	0.9029
	100	25	38	13	0.034	2.2744	0.973
ALF-PF-E	90	35	50	15	0.0471	2.4214	0.9611
	80	45	64	19	0.0565	3.0224	0.9421
	100	25	29	4	0.0149	0.7988	0.9948
PF-LSTM	90	35	40	5	0.0191	0.8781	0.9881
	80	45	53	8	0.0289	0.9481	0.9641
	100	**25**	**26**	**1**	**0.0116**	**0.5635**	**0.9968**
ALF-PF-LSTM	**90**	**35**	**27**	**2**	**0.0161**	**0.6332**	**0.9891**
	80	**45**	**49**	**4**	**0.0256**	**0.7223**	**0.9721**

Table 6.4: Comparison of four algorithms for battery 06.

Algorithm	Cycle	R_t	R_p	E_{RUL}	E_{RMSE}	E_{MAPE} (%)	R^2
	100	26	37	11	0.068	4.2763	0.9278
PF-E	90	36	49	13	0.07322	4.3531	0.9163
	80	46	63	17	0.0882	4.6132	0.9001
	100	26	30	4	0.0503	3.3061	0.9605
ALF-PF-E	90	36	42	6	0.0591	3.5218	0.9519
	80	46	56	10	0.0723	4.4126	0.9211
	100	26	23	3	0.0164	0.8864	0.9958
PF-LSTM	90	36	32	4	0.0223	0.9183	0.9793
	80	46	37	9	0.0413	0.9015	0.9513
	100	**26**	**28**	**2**	**0.0161**	**0.9023**	**0.996**
ALF-PF-LSTM	**90**	**36**	**39**	**3**	**0.0231**	**0.9531**	**0.9823**
	80	**46**	**52**	**6**	**0.0381**	**1.0253**	**0.9619**

probability density function. The quantitative results of predicted probability density function at 100 cycle are shown in Table 6.6.

The particle number, as another kind of concerned parameter, has an impact in the particle filter technique. Generally speaking, the larger number of particles leads to the better prediction results, at the cost of requiring more calculation. The prediction effect of the proposed method is shown in Table 6.6 and Figs. 6.17–6.20 with 150 particles and 50 particles. The results mean that the proposed method is less affected by the particle number thereby having better robustness.

In the process of learning the degradation model with long short-term memory network, the order of Markov process is an important parameter that affects the remaining useful life performance. Normally, the higher-order Markov process leads to more elapsed time. The detailed comparison on RMSE, MAPE, and elapsed time

Table 6.5: Comparison of four algorithms for battery 07.

Algorithm	Cycle	R_t	R_p	E_{RUL}	E_{RMSE}	E_{MAPE} (%)	R^2
	100	25	38	13	0.0354	2.1302	0.9567
PF-E	90	35	51	16	0.0481	2.3136	0.9411
	80	45	67	22	0.0719	3.1921	0.9271
	100	25	34	9	0.0183	1.1128	0.9884
ALF-PF-E	90	35	45	10	0.0223	1.1934	0.9801
	80	45	60	15	0.0344	1.3236	0.9671
	100	25	26	1	0.0103	0.4948	0.9963
PF-LSTM	90	35	38	3	0.0191	0.5424	0.9843
	80	45	52	7	0.0313	0.7346	0.9763
	100	**25**	**25**	**0**	**0.011**	**0.5251**	**0.9958**
ALF-PF-LSTM	**90**	**35**	**36**	**1**	**0.0134**	**0.5532**	**0.9913**
	80	**45**	**48**	**3**	**0.0231**	**0.5665**	**0.9832**

for different orders on battery 05 is shown in Table 6.7. It can be concluded that the 4-order Markov process has the highest prediction accuracy. It is the reason that a 4-order Markov process was applied in the proposed ALF-PF-LSTM method.

Compared to other classical methods of machine learning such as Gaussian process regression with linear mean function (LM-GPR) [233], multiscale Gaussian process regression with squared exponential (SE-MGPR) [234], rest time-based prognostic framework (RTPF) [235], and support vector regression with particle swarm optimization (PSO-SVR) [236], the results in Table 6.8 show that the proposed hybrid method in this section enjoys higher precision.

By analyzing the above simulation results, we can conclude that the higher prediction accuracy may benefit from the following reasons: 1) the Levy flight with adaptive steps effectively increases the diversity of particles and solve the problem of particle impoverishment, which forces the particles to shift to areas with high likelihood; and 2) the degraded model learned by long short-term memory network has better prediction performance than the double exponential model.

6.3 Remaining Useful Life Prediction with Degradation Point Detection and EKF

As a key connecting unit of the rotation machine, bearing is widely employed to support the mechanical rotating body, reduce the friction coefficient during its movement and ensure its turning precision. With the passage of operating time, the bearing will continue to accumulate damage until it fails with the complexity and harshness of the operating conditions, and unexpected machine downtime and even catastrophic

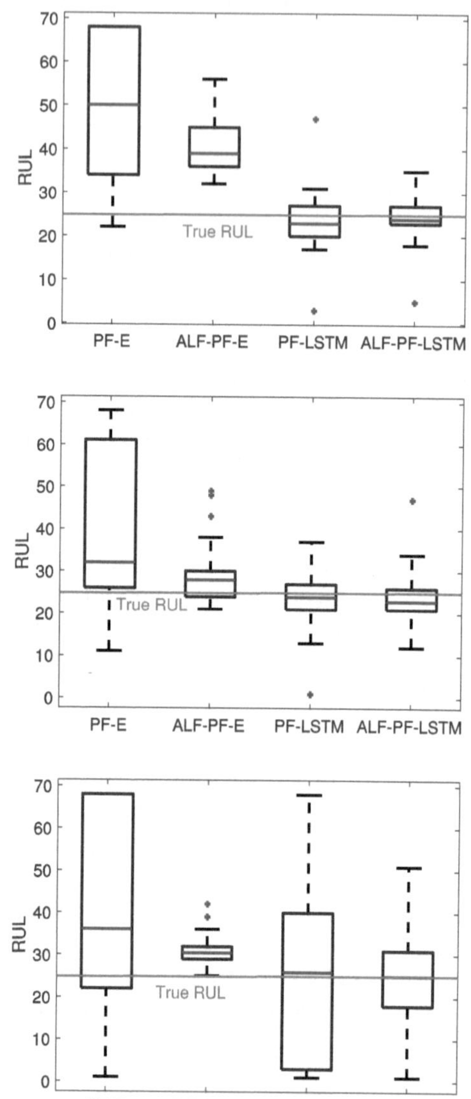

Figure 6.17: Box plot of the predicted remaining useful life distribution for batteries 05, 06, and 07 with 50 particles.

damage would happened. Actually, bearing failure is one of the most common causes of machine breakdown. Therefore, it is very meaningful to assess the health status and predict the remaining useful life of bearing during the working process so that a reasonable maintenance strategy can be timely adopted [4].

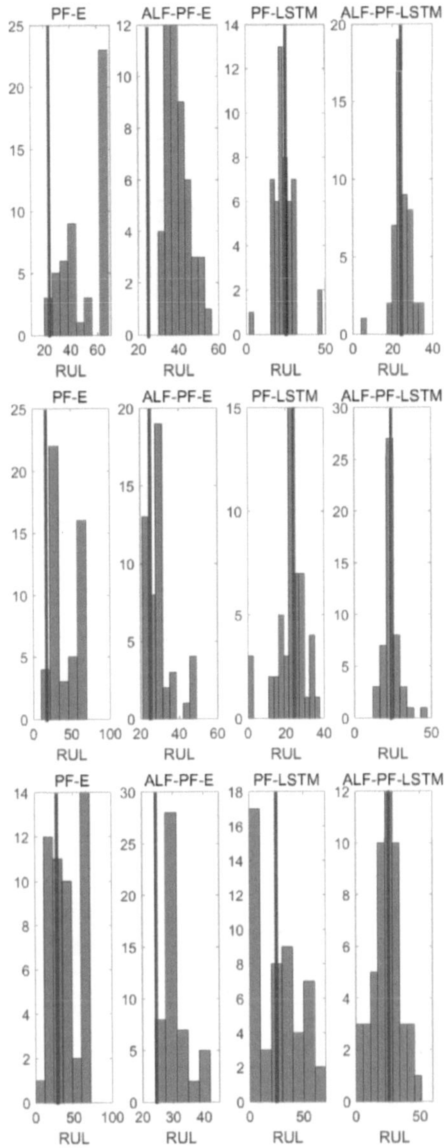

Figure 6.18: Histogram plot of the predicted remaining useful life distribution for batteries 05, 06, and 07 with 50 particles, and red lines are the real remaining useful life.

In this subsection, complete ensemble empirical mode decomposition with adaptive noise is introduced to reduce the noise and reconstruct the degrade trend of bearing. Next, the cumulative function of reconstructed trend is employed to realize the feature enhancement. After the time to start prediction and health indicators

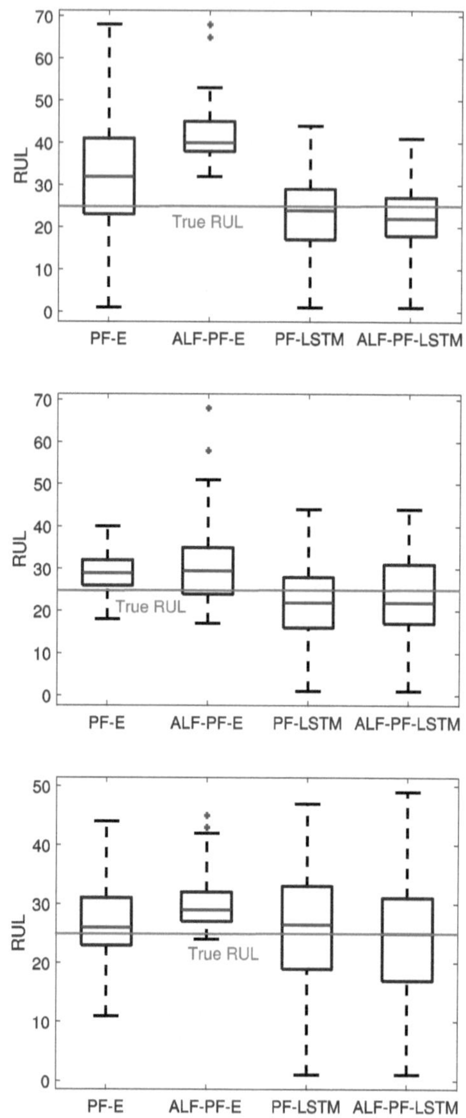

Figure 6.19: Box plot of the predicted remaining useful life distribution for batteries 05, 06, and 07 with 150 particles.

have been determined, the fusion between double exponential degrade model and extended Kalman filter is desired to predict the remaining useful life. The overall structure of the proposed method can be summarized in Fig. 6.21, which contains three aspects: 1) degradation point detection; 2) health index establishment; and 3) remaining useful life prediction.

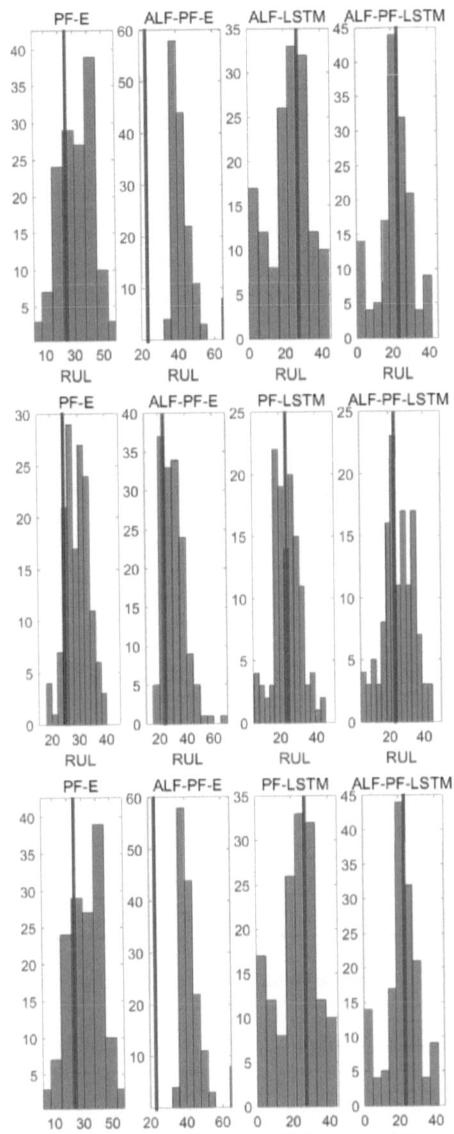

Figure 6.20: Histogram plot of the predicted remaining useful life distribution for batteries 05, 06, and 07 with 150 particles, and red lines are the real remaining useful life.

6.3.1 Degradation Point Detection

According to the accepted view [4], the whole life of the bearing can be usually divided into three stages: running-in period, working period and degradation period. Among them, the running-in period is very short due to the installation. Until the

Table 6.6: Comparison of remaining useful life distribution with particles count of 50 and 150.

Particles count	Algorithm	Battery number	E_{RUL}	PDF width
50	PF-E	5	18	48
		6	11	52
		7	13	67
	ALF-PF-E	5	13	23
		6	4	28
		7	9	15
	PF-LSTM	5	9	15
		6	4	14
		7	3	24
	ALF-PF-LSTM	5	1	17
		6	1	24
		7	2	45
150	PF-E	5	11	67
		6	4	233
		7	6	33
	ALF-PF-E	5	10	33
		6	3	36
		7	4	51
	ALF-PF-E	5	2	43
		6	2	43
		7	0	50
	ALF-PF-LSTM	5	1	40
		6	1	41
		7	0	48

Table 6.7: RMSE, MAPE, and elapsed time of the PF-LSTM using different order Markov processes based on battery 05.

Order	3	4	5	6	7	8
$RMSE$	0.0269	0.0108	0.1032	0.0216	0.0308	0.0452
$MAPE$ (%)	0.337	0.3137	0.8471	0.3606	0.3687	0.4265
Elapsed time (s)	61	157	223	351	467	671

degradation occurs, the bearing will be in a relatively stable working period, after which the bearing begins to degrade until the end of life. Especially, the detection of the degradation point will affect the prediction accuracy of remaining useful life. Therefore, it is important to determine timely the degradation point and ascertain it as the time to start prediction (TSP), which can significantly improve the accuracy of remaining useful life prediction [101].

Before determining the TSP, the variance feature of the original is selected because it can reflect efficiently the evolution of entire life of the bearing. Especially, a obviously facts of bearing can be found that the evolution trend is linear during the working period, nevertheless, the evolution trend gradually becomes nonlinear in the age stage due to the degradation occurs and intensifies. Therefore, the TSP point on the bearing can be transformed as discovering cut-off point on the evolution trend, where the turning point from linear to non-linear of evolution trend is an interesting choice. Fortunately, the Kalman Filter (KF) is a very effective linear filtering method which can be used to obtain the optimal estimation of linear systems. After the filter error of variance feature with KF is chosen as the abnormal factor, and the relative

Table 6.8: Comparison of prediction performance with the existed results.

Method	Battery 05		Battery 06		Battery 07	
	RMSE	MAPE (%)	RMSE	MAPE (%)	RMSE	MAPE (%)
LM-GPR [233]	1.36	1.6	6.86	10.2	1.73	1.7
SE-MGPR [234]	1.2	1.38	2.11	2.93	1.07	1.02
RTPF [235]	0.68	0.76	0.93	1.25	0.43	0.44
PSO-SVR [236]	0.75	0.82	0.93	1.25	0.97	1.02

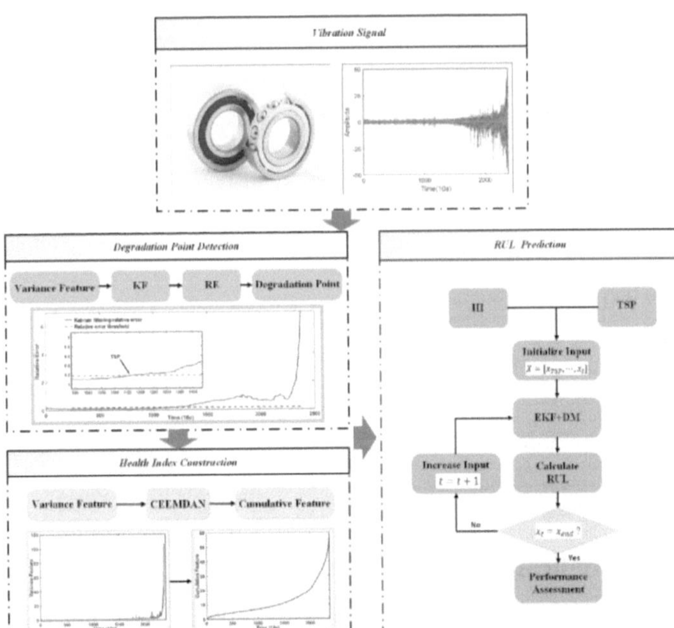

Figure 6.21: The procedure of the proposed ensemble prognostic methodology.

errors (RE) is calculated with sliding window method. Consequently, TSP can be ascertained when RE exceeds the preset threshold, and the algorithm of TSP detection is shown minutely in Algorithm 6.3.

Algorithm 6.3 Var-KF-RE

Input:

$\{S_i(k)\}_{i=1,\ldots,n,k=1,\ldots,K}$ represent the raw vibration signals of *i-th* sampling sequences at time *k*, n, K are the total number of sampling and sample length, respectively.

1 Calculate the variance sequences $V(k)$ of $S_i(k)$ with its mean value $\bar{S}(k)$ by
$$V(k) = \frac{1}{n}\sum_{i=1}^{n}(S_i(k) - \overline{S}(k))^2$$

2 Utilize the variance sequences $V(k)$ to obtain the filtered feature sequences $F(k)$ and filter error sequences $E(k)$ with KF.

3 Set the sliding window size m as $5\% * K$.

4 Compute the RE $R_E(k)$ by

$$R_E(k) = \begin{cases} \dfrac{|\sum_{i=1}^{k} E(i)|}{\sum_{i=1}^{k} F(i)} \times 100\%, & if\ k < m \\ \dfrac{|\sum_{i=k-m+1}^{k} E(i)|}{\sum_{i=k-m+1}^{k} F(i)} \times 100\%, & if\ k \geq m \end{cases}$$

5 Let the degradation threshold as

$$T_{th} = 2 * \overline{R_E}(j), \quad j = 10\% * K, \ldots, 20\% * K$$

6 Denote $t = 20\% * K$

7 **if** $R_E(t) > T_{th}$ **then**

8 **go to** step 10

9 **else** $t = t + 1$, **go to** step 7

10 **end if**

Output:

The degradation point *t*, and *TSP* = *t*.

6.3.2 Health Indicator Construction

After the starting point of degradation is ascertained as a TSP, the construction of health indicator (HI) becomes the main goal, and it is directly related to the prediction effect of remaining useful life. Usually, many kinds of features from original vibration signal can be chosen as the HI of bearing, however, inappropriate HI hardly reveal the tendency of bearing degradation and affect seriously the accuracy of prediction. Therefore, it is more conducive to prediction by extracting appropriate features and constructing efficient HI. In the existing literatures, time-domain features including root mean square, kurtosis and crest factor, or frequency-domain features containing frequency centre and mean square frequency, are the used commonly [82, 101]. Recently, time-frequency domain feature is attracted more attentions [88, 237], which mainly utilizes the techniques of information entropy, wavelet packet energy, EMD energy, and so on.

Among the three kinds of features mentioned above, variance from time-domain is simple and easy to calculate and has obvious physical significance, which is chosen as the feature in this section. Considering the fact that noise interference and insufficient degradation trend of the initial variance, further feature processing is essential. As a complete data analysis method, the complete ensemble empirical mode decomposition with adaptive noise (CEEMDAN) reduces the modal aliasing effect by adding adaptive white noise, and it can be used to reconstruct the variance feature. Actually, noise signals are generally distributed at high frequencies and it is discarded to obtain the original degradation trend of bearing. The specific degradation trend reconstruction and denoising process can be found in Algorithm 6.4, a cumulative function (CF) is used to construct a more trendy HI for subsequent remaining useful life prediction.

6.3.3 Remaining Useful Life Prediction of Bearings

Combining the TSP detected in Algorithm 6.3 with the well-chosen HI from Algorithm 6.4, we strive to achieve the most important goal of this article and to predict the remaining useful life of bearing in this section.

Continuous operation makes the bearings begin to wear out. Meanwhile, the extracted HI will also increase. In [238], the double exponential model is used to model the established bearing vibration signal HI and achieve acceptable results. Based on the analysis of the cumulative feature data of the bearings, it is found that the double exponential model can also track the degradation trend well in this section

$$y(k) = a \cdot exp(b \cdot k) + c \cdot exp(d \cdot k) \tag{6.24}$$

where $y(k)$ is bearing's cumulative feature, k is the time index, and a, b, c, and d are the parameters associated with the model. By using MATLAB curve fitting toolbox, model parameters a, b, c, and d can be determined.

In addition, some models based on the failure mechanism, such as Weibull model, Paris, etc., are also commonly used for bearing life prediction. The establishment of the Weibull model is based on the basic failure rate function under the Weibull distribution, and its form is as follows

$$\lambda(k) = (\beta/\eta)\,(k/\eta)^{\beta-1} \tag{6.25}$$

where $\beta > 0$ and $\eta > 0$ are the shape parameter and scale parameter, respectively. On the other hand, by assuming the failure growth following Paris's law, a fatigue

Algorithm 6.4 Var-CEEMDAN-CF

Input:

$\{S_i(k)\}_{i=1,\ldots,n,k=1,\ldots,K}$ represent the raw vibration signals of *i-th* sampling sequences at time *k*, n, K are the total number of sampling and sample length, respectively. $w_j(j=1,\ldots,J)$ is the Gaussian noise with standard deviation ε_j; $E_p(\cdot)$ $(p=0,1,\ldots,\ell)$ denotes the *p-th* EMD mode.

1 Calculate the variance of $S_i(k)$ with its mean value $\bar{S}_i(k)$ by

$$V(t)=\frac{1}{n}\sum_{i=1}^{n}(S_i(k)-\overline{S}(k))^2$$

2 Set $p=0, r_0(k)=V(k)$.

3 Add noise ε_k to $r_t(k)$, then construct a new signal $x_p^j(k)=x(k)+\varepsilon_p w_j(k)$.

4 Decompose the new signal x_p^j by EMD to obtain the mean and residual

$$\overline{I}_{p+1}(k)=\frac{1}{J}\sum_{j=1}^{J}I_{p+1}^{j}(k)$$

$$r_{p+1}(k)=r_p(k)-\overline{I}_{p+1}(k)$$

5 **if** $r_{p+1}(k)$ does not satisfy the EMD's decomposition condition **then**

6 **go to** step 7

7 **else** let $p=p+1$, **go to** step 2

8 **end if**

9 Remove high frequency noise

$$\bar{V}(k)=V(k)-\sum_{p=1}^{\lambda}\overline{I}_p(k).$$

10 Calculate the cumulative function by

$$C_F(k)=\frac{\sum_{i=1}^{k}\bar{V}(i)}{|\sum_{i=1}^{k}\bar{V}(i)|^{1/2}}$$

Output:

The constructed cumulative function $C_F(k)$.

crack growth model can be established, its simple form is

$$a_k=exp\left(b\cdot\triangle N\right)\cdot a_{k-1} \tag{6.26}$$

where a_k is the final crack length, which can usually be replaced with the extracted features, b is a parameter to be estimated, which is related to the material of the research object, and $\triangle N$ is the number of the increased load cycles, usually regarded as the measured cycle. In this section, the above two models are used to compare with the double exponential model used in the proposed method.

Before starting the remaining useful life prediction, a fault threshold should be determined first, which also has a great influence on the prediction accuracy.

Common methods such as the average or minimum value of the last point of the training set are not applicable to the cumulative features of this article. Based on the cumulative nature of CF, the slope is selected as the fault threshold. The final testing threshold was computed as the average of the slope at the last point of the training set, given by

$$\overline{Th} = \frac{1}{M}\sum_{i=1}^{M} th_i \tag{6.27}$$

where M is the number of training sets under the current operating conditions, and th_i is the threshold obtained from the ith bearing in the training set.

On the basis of Kalman filtering, the first-order Taylor series is obtained from the non-linear part, which is approximately converted into linear. Consider a dynamic system described by the following state space model:

$$X(k+1) = \Phi X(k) + W(k) \tag{6.28}$$

$$Y(k) = HX(k) + V(k) \tag{6.29}$$

where k is discrete time, the state of the system at time k is $X(k) \in R^n$; $Y(k) \in R^m$ is the observation signal of the corresponding state; $W(k) \sim N(0, Q)$ is the input white noise, and $V(k) \sim N(0, Q)$ is the observation noise. Equation (6.28) is the equation of state and Equation (6.29) is the observation equation. Φ is the state transition matrix, H is the observation matrix, and the extended Kalman filtering algorithm is shown in Algorithm 6.5.

6.3.4 Illustrative Examples

In this section, experimental data sets from IEEE-PHM-2012-Challenge [239] and XJTU-SY [240] are used to verify the effectiveness of the proposed remaining useful life prediction method. Meanwhile, multiple evaluation indicators are used to compare with other literature methods to highlight the advantages of this method.

IEEE-PHM-2012-Challenge data set is obtained from the PRONOSTIA test platform, which is developed by the FEMTO Technology Group. The experimental setup is shown in Figure 6.22. The experimental system consists of three main parts: rotating subsystem, load subsystem, and data acquisition system. Under different working conditions, the system gives different motor speeds and radial loads to the bearings, which can accelerate the degradation test of the bearings in a few hours. Two accelerometers are installed on bearings by DAQ system to measure horizontal and vertical vibration signals. The vibration signal was measured after every 10 s and at a sampling frequency of 25.6 kHz.

Three different operating condition information and the time to failure (TTF) values of the bearing are shown in Table 6.9. According to the relevant literature [241, 242], the vertical vibration signal provides less useful information on bearing degradation than the horizontal vibration signal, so in this experiment only the horizontal vibration signal data of all bearings were simulated and verified.

Before starting the degradation point detection, extract simple variance features from the original signal, and observe the overall trend of the brake. Next, use the linear Kalman filter to obtain the filtered features curve and the corresponding error curve. We can regard the former as the health factor of the bearing, and the latter as the failure factor. Then calculate the relevant error and determine the starting point of degradation according to the preset threshold. Finally, the degradation start point is used as the *TSP* for subsequent *RUL* prediction.

When the bearing begins to degenerate, because of the occurrence of the fault, usually a certain fault characteristic frequency will appear in the envelope

---

Algorithm 6.5 EKF

Input:

The constructed cumulative feature $CF(k), k = 1, \dots, K$; Initialize the initial state $X(0), Y(0), P0$.

1 Set $i = K + 1, Y(k) = CF(k)$

2 **while** $k < K$, **do**

3 State prediction

$$X(k|k-1) = \Phi X(k)$$

4 Observation prediction

$$Y(k|k-1) = HX(k|k-1)$$

5 Covariance prediction

$$P(k|k-1) = \Phi P(k-1)\Phi^T + Q$$

6 Calculate Kalman gain

$$Kg(k) = P(k|k-1)H^T[H\ P(k|k-1)H^T + R]^{-1}$$

7 State update

$$X(k) = X(k|k-1) + Kg(k)\ [Y(k) - Y(k|k-1)]$$

8 Covariance update

$$P(k) = [I_n - Kg(k)\ H]\ P(k|k-1)$$

9 **end while**

10 Substitute the model parameter $X(K)$ and time i estimated by the filter into the degraded model to predict the observed value $Z(i)$

11 **if** $Z(i) \geq \overline{Th}$ **then**

12 **go to** step 14

13 **else** $i = i + 1$, **go to** step 10

14 **end if**

Output:

The RUL value $r = i - K - 1$.

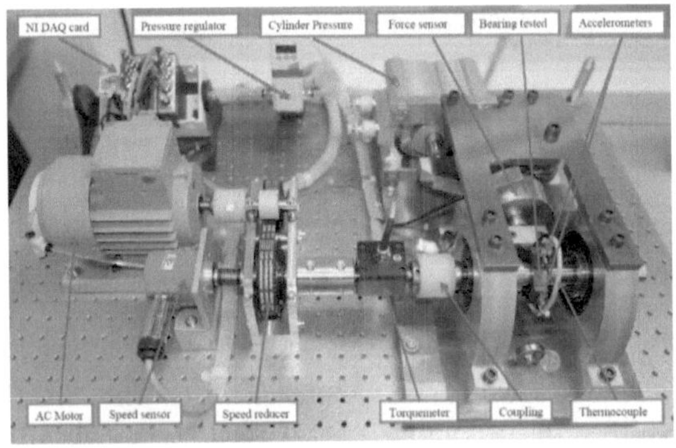

Figure 6.22: Overview of PRONOSTIA platform.

spectrum. Take the bearing in working condition 1 as an example, $f_{IRF} = 221.7$ Hz, $f_{ORF} = 168.3$ Hz, and $f_{BF} = 215.3$ Hz, which are the characteristic frequencies associated with the inner race fault, outer race fault, and ball fault, respectively [238]. Based on this, in order to further prove the accuracy of the proposed method for detecting the degenerate point, an envelope spectrum analysis is performed at the degenerate point. Taking bearings 1-3 as an example, vibration signal of the whole lifetime, vibration signal at TSP, Degradation Point Detection, and envelope spectrum of vibration signal at TSP are shown in Figure 6.23.

Table 6.10 shows the TSPs of four bearings under operating conditions 1. The method we used to determine the TSP is compared with Li et al.'s method [243] and Ahmad et al.'s approach [244]. In most cases, the method of determining TSP in this article can detect anomalies earlier.

After determining the TSP, start building HI and prepare for remaining useful life prediction. As described, calculate the variance features, then perform noise

Table 6.9: Observed TTF values for bearings (in seconds).

	Condition 1	Condition 2	Condition 3
Load (N)	4000	4200	5000
Speed (rpm)	1800	1650	1500
Training dataset	B1-1 (28030)	B2-1 (9110)	B3-1 (5150)
	B1-2 (8710)	B2-2 (7970)	B3-2 (16370)
Testing dataset	B1-3 (23750)	B2-3 (19550)	B3-3 (4340)
	B1-4 (14280)	B2-4 (7510)	
	B1-5 (24630)	B2-5 (23110)	
	B1-6 (24480)	B2-6 (7010)	
	B1-7 (22590)	B2-7 (2300)	

Table 6.10: Time to start prediction (TSPs).

	B1-1	B1-3	B1-4	B1-5
Li et al's method	14630 s	16440 s	10910 s	22130 s
Ahmad et al's approach	12970 s	13510 s	11600 s	24210s
Proposed method	14500 s	11630 s	10890 s	12850 s

reduction processing, and finally get the cumulative features as HI from the cumulative function. The effect of the cumulative function processing is illustrated in Figure 6.24. Compared with the variance feature, the corresponding CF is smoother and more suitable for modelling in the remaining useful life prediction stage.

The constructed HI should have a good correlation with the ageing state of the bearing, because the correlation will affect the final prediction. The trendability parameter can measure the degree of correlation between the constructed HI and time. Its range is between [0,1], the closer to 1 indicates the stronger the trendability. The trendability is defined as [245]

$$Trendability(X,T) = \frac{|N\sum X(t_N)t_N - \sum X(t_N)\sum t_N|}{\sqrt{[N\sum X^2(t_N) - (\sum X(t_N))^2][N\sum t_N^2 - (\sum t_N)^2]}} \tag{6.30}$$

where $X(t_N)$ is the degradation feature value at time t_N and N is total number of data points. In Table 6.11, the trend comparison of the variance feature and its cumulative feature under bearing condition 1 is given.

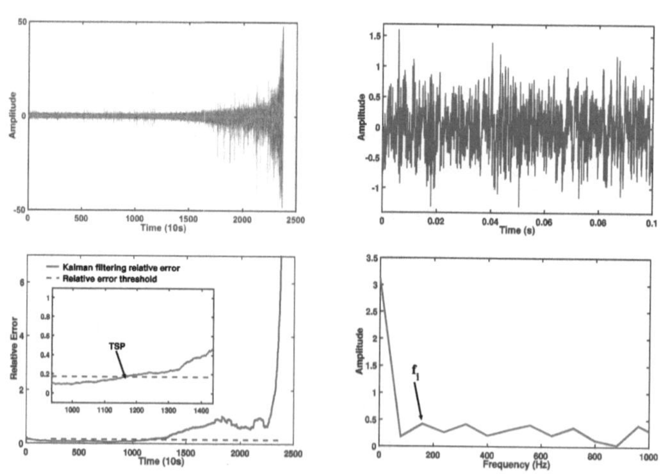

Figure 6.23: Bearing 1–3: (a) Vibration signal of the whole lifetime; (b) Vibration signal at TSP; (c) Degradation Point Detection; and (d) Envelope spectrum of vibration signal at TSP.

Table 6.11: Features trendability value comparison.

	B1-1	B1-2	B1-3	B1-4	B1-5	B1-6	B1-7
Var	0.4554	0.3682	0.3167	0.5913	0.1443	0.2047	0.2245
CF	0.9492	0.9556	0.8642	0.79	0.9824	0.987	0.9878

It can be seen from Table 6.11 that through the cumulative function, the trendability of all bearing data has been significantly improved. Because the 1–4 variance feature is very stable in the early stage and suddenly rises in the later period, the corresponding CF has a significant turning point, so the trend is not relatively obvious.

Once available TSP and suitable HI are obtained, remaining useful life prediction will start. For the HI of all bearings (training set and test set), the data before TSP is ignored. HI data from TSP to TTF in the training set is used for model fitting, through MATLAB curve fitting toolbox. Then, the initial parameters of the double exponential model under the corresponding working conditions can be determined, and at the same time, the slope threshold is calculated.

For the test set, from TSP to the end of the data is used as EKF input. Through this data sequence, EKF continuously corrects and updates the model parameters to fit the input data until the last point of the sequence. Without new data input, EKF iteratively calculates the predicted HI value at each subsequent point according to the last updated model parameters. When the slope of the predicted HI reaches

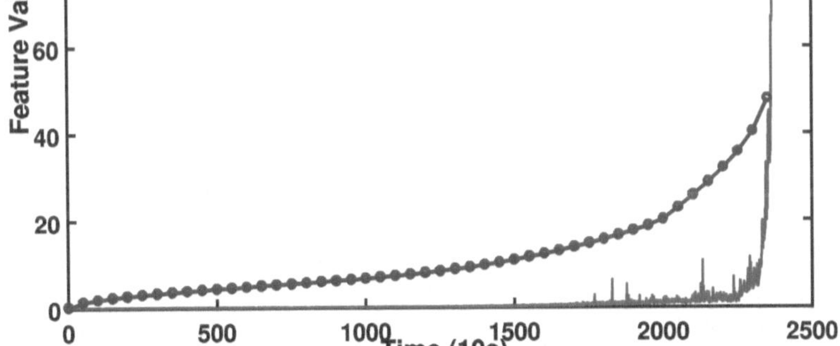

Figure 6.24: The variance and cumulative features of bearing 1–3.

Table 6.12: RUL prediction error comparison of PHM2012 dataset.

		B1-3	B1-4	B1-5	B1-6	B1-7	B2-3	B2-4	B2-5	B2-6	B2-7	B3-3
MEAN	Paris+EKF	732.68	259.4	415.5	414.5	1038	991.5	212.5	630.5	185	77	147
	MEAN Weibull+EKF	206.53	241.4	159.86	329.02	506.53	991.5	159.38	575.05	130.58	56.68	124.48
	D-exp + PF	259.93	256.96	415.5	**129.44**	1038	603.5	54.05	566.11	74.85	77	**59.45**
	D-exp + EKF	**4.7**	**128**	**90**	292.13	**404**	**157.77**	**21.05**	**349.62**	**69.58**	**54.19**	147
RMSE	Paris+EKF	738.4	259.76	440.62	442.44	1050.56	1000.98	216.64	657.17	187.75	77.74	151.64
	MEAN Weibull+EKF	236.44	257.8	177.75	351.37	580.88	1000.98	179.39	626.35	147.79	60.8	132.96
	D-exp + PF	289.45	262.32	440.62	**153.37**	1050.56	688.19	59.95	621.63	**81.67**	77.74	**73.28**
	D-exp + EKF	**5.43**	**139.91**	**101.57**	346.97	**461.08**	**194.74**	**24.37**	**412.76**	116.81	65.56	151.64

the threshold, the current time point is the predicted time to failure $\widehat{TTF}$, and the predicted remaining useful life of each point is obtained by calculating the interval between the current time t and $\widehat{TTF}$:

$$\widehat{RUL}(t) = \widehat{TTF} - t \tag{6.31}$$

In this section, the prediction results of the proposed method, the Paris model method [218] and the Weibull model method [246] are also compared. Figure 6.25 shows the remaining useful life prediction results of test set, and Figure 6.26 shows the remaining useful life prediction results of PF and EKF comparison. In these remaining useful life prediction results graphs, we also show a bandwidth around the true remaining useful life ($\pm$ 20% of the true remaining useful life value) to see how well the estimated remaining useful lifes fall within this band [241]. It should be noted that when the predicted remaining useful life exceeds 40% of the actual remaining useful life of the current point, remaining useful life is defined as 40% of the current point.

In order to compare the prediction results more accurately, the overall mean and root mean square of the remaining useful life prediction error are calculated, as shown in Table 6.12. Combining Figure 6.25, 6.27, and Table 6.12, the proposed method performs better in the overall remaining useful life prediction error of most bearings than the two commonly used physical models. This means that the double exponential model is more suitable for the constructed HI.

On the other hand, compared with the particle filtering method, the proposed method is not effective on some bearings, such as B1-6, B2-6, and B3-3. This may be due to the strong nonlinear characteristics of these bearings, which makes the performance obtained by PF better than the proposed method, but for most trending bearings, PF is not as good as EKF.

Besides, the performance of the proposed method is compared with [244], and the corresponding results are listed in Table 6.13. It can be seen from Table 6.13 that the overall robustness of the method in this section is better. At the same time, it is compared with some of the machine learning methods listed in [247], and the results are shown in Table 6.13. The Score index in Table 6.13 is based on the score function in the IEEE PHM 2012 challenge [239], but only the 5 test set bearings in Condition 1 are calculated. And the calculation formula of MAE and NRMSE indicators are as follows

$$MAE = \frac{1}{5}\sum_{i=1}^{5} |ActRUL_i - \widehat{RUL}_i| \tag{6.32}$$

Table 6.13: Comparison of RUL prediction errors with some machine learning methods.

Testing dataset	Current times(s)	Actual RUL(s)	MSCNN RUL(s)	RNN RUL(s)	SOM RUL(s)	SVR RUL(s)	Proposed Method RUL(s)
B1-3	18010	5730	4731	3250	5790	5970	5650
B1-4	11380	2900	2590	1100	410	1200	960
B1-5	23010	1610	3996	1980	6080	5040	1970
B1-6	23010	1460	1744	1150	1180	1230	1920
B1-7	15010	7570	6090	6220	8110	9120	7450
Score			0.3624	0.2798	0.3605	0.2657	**0.4123**
MAE			1091.8	1262	1568	1430	**590**
NRMSE			0.3514	0.5522	0.5342	0.4107	**0.2528**

$$NRMSE = \frac{\sqrt{\frac{1}{5}\sum_{i=1}^{5}(ActRUL_i - \widehat{RUL}_i)^2}}{\frac{1}{5}\sum_{i=1}^{5}\widehat{RUL}_i} \tag{6.33}$$

where $ActRUL_i$ and $\widehat{RUL}_i$ mean the actual RUL and the estimated RUL for the *ith* testing dataset. The results in Table 6.13 also show that the single-point prediction effect of the proposed method is better than other machine learning methods.

Another dataset named XJTU-SY is used to further verify the validity of the proposed method. Figure 6.28 shows the sampling platform of the data set. There are also two accelerometers used to collect vibration data, the sampling frequency is set to 25.6 kHz, a total of 32768 data points are recorded for each sampling, and the sampling period is equal to 1 min [240]. Bearing data as shown in Table 6.14.

In this section, in order to simplify the experiment, only the data of condition 1 is simulated and verified, and XJB1-1 was selected as the training set. For the remaining bearings under condition 1, the first 70% of data of each bearing is taken as the known

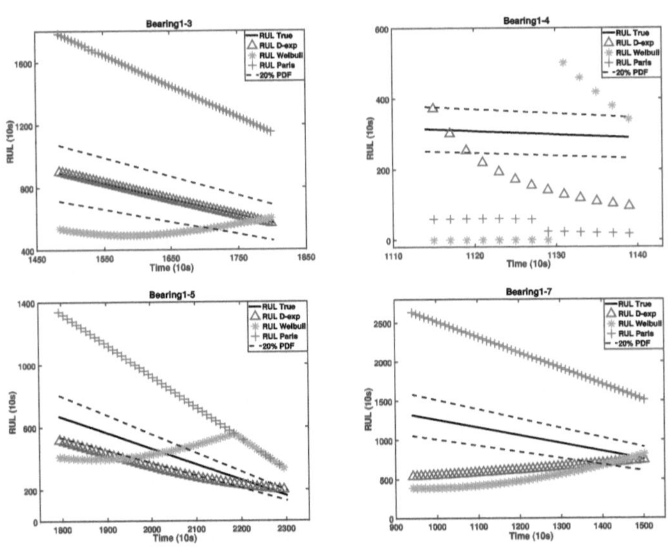

Figure 6.25: RUL prediction results comparison of different models based on EKF with bearing B1.

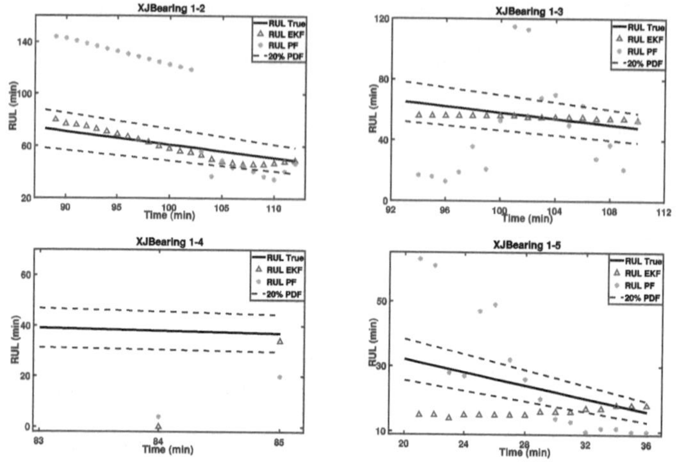

Figure 6.26: RUL prediction results comparison of EKF and PF based on XJTU-SY bearing datasets.

data, and the last 30% of the data is used to verify the prediction accuracy. Similarily, after determining TSP and building HI, RUL prediction is performed. The prediction results of XJTU-SY bearing datasets are shown in Figure 6.26. The method and PF are also compared in Figure 6.26, the performance results are shown in Table 6.15,

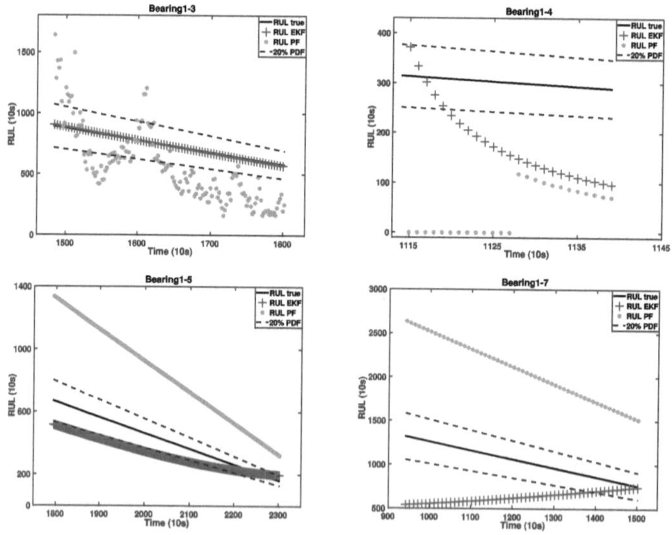

Figure 6.27: RUL prediction results comparison of EKF and PF based on double exponential model with bearing B1.

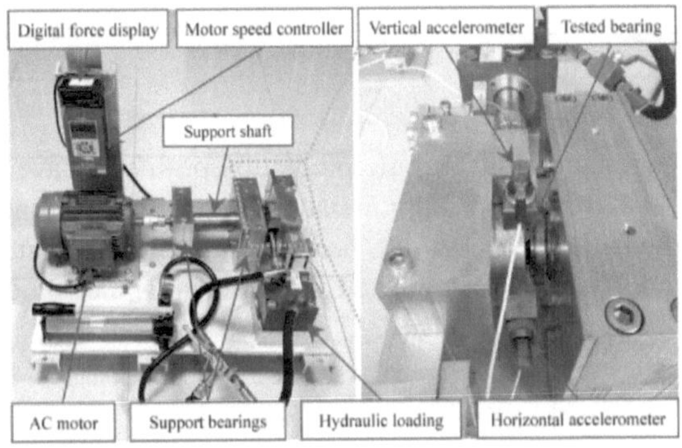

Figure 6.28: Bearing data sampling platform of XJTU-SY.

in which most of the results of the proposed method are better than PF. This proves that the method is still reliable in the XJTU-SY bearing data set.

Table 6.14: XJTU-SY bearing datasets.

	Load (N)	Speed (rpm)	Bearing dataset
Condition 1	12k	2100	XJB1-1 XJB1-2 XJB1-3 XJB1-4 XJB1-5
Condition 2	11k	2250	XJB2-1 XJB2-2 XJB2-3 XJB2-4 XJB2-5
Condition 3	10k	2400	XJB3-1 XJB3-2 XJB3-3 XJB3-4 XJB3-5

Table 6.15: RUL prediction error comparison of XJTU-SY bearing datasets.

		B1-2	B1-3	B1-4	B1-5
MEAN	D-exp + PF	43.04	28.18	25.5	10.88
	D-exp + EKF	**4.54**	**3.53**	**20.5**	**7.94**
RMSE	D-exp + PF	51.38	33.7	**26.88**	14.59
	D-exp + EKF	**5.18**	**4.16**	26.95	**9.52**

6.4 Conclusion

In this chapter, the remaining useful life prediction problems have been investigated by using the hybrid algorithm for lithium batteries. According to the characteristic of NASA data set, the battery capacity has been extracted as the degradation index and the state space model has been established and three integrated algorithms have been put forward.

7

Remaining Useful Life Prediction of Industrial Components with Machine Learning Methods

The process of data-driven remaining useful life (RUL) estimation usually consists of the health indicator (HI) construction and remaining useful life prediction, where the health indicator involves fault feature extraction and fault feature selection. In terms of feature extraction, some classical features will be extracted, including root-mean-square, kurtosis in time domain, or fast Fourier transform in frequency domain, and wavelet packet decomposition in time-frequency domain [248]. Note that, the limitation of those features should be emphasized that a lot of noises and fluctuations are included, which affect the prediction accuracy. To overcome the above disadvantages, the empirical mode decomposition (EMD) and its improved version [249] have been introduced to extract the trend features. Hong et al. [250] have proposed the wavelet packet-EMD, in which the last 4 intrinsic mode functions (IMFs) have been used to track the degradation process. Niu et al. [251] have employed the decomposed IMFs by complementary ensemble empirical mode decomposition (CEEMD) to build the support vector regression (SVR) model, and predict the remaining useful life with another optimized support vector regression by grey wolf optimizer. After the fault features are extracted, the next move is to select the suitable health indicator [252]. Compared to the traditional principal component analysis with contribution rate, the fusion of correlation and monotonicity may be a more suitable choice with clear physical meaning [253]. It is worth mentioning that the health indicator varies widely because of the complex external working environment, which causes great difficulties in the accurate remaining useful life prediction. Fortunately, the deep learning-based approaches are capable of handling such concerns caused by the health indicator varying. For example, Guo et al. [254] have adopted the recurrent neural network to establish the health indicator (which is called the RNN-HI) for remaining useful life prediction. Ren et al. [255] have proposed the multi-scale dense gate recurrent unit network to capture the sequence features and establish the health indicator of rolling element bearings. Zhao et al. [82] have introduced the reconstruction procedure to build the trend features, and developed the long-short-term memory (LSTM) to build the health indicator and predict the remaining useful life. Inspired by the above neural-network-based model, the gated recurrent unit (GRU) network [256] has been introduced to establish the health indicator which characterizes the degradation state and prediction the remaining useful life.

The rolling element bearings (REBs) are considered as the most common and critical mechanical components in rotating machinery, that have been widely used

DOI: 10.1201/9781003330998-7

in modern industries. Note that, due to the complexity and uncertainty of operating environment, rolling element bearings are easily damaged during the life of machines [257]. Once the rolling element bearings fails because they are fatigued, worn, corroded, etc., it may cause unexpected downtime of the machinery, or even lead to catastrophic damage, all of which directly affect the operational reliability of the entire mechanical equipment. Consequently, the timely detection of degradation anomaly point in the early stages of bearing failure and the accurately prediction of remaining useful life can effectively improve the reliability and operational safety of the rotating machinery [4], and reduce maintenance costs. In the past decades, the effective performance degradation assessment of rolling element bearings remains a challenge in academia or industry [258].

Since the lithium-ion batteries (LIBs) possess obvious advantages of high voltage, high energy density, low self-discharge rate, long cycle life, and high safety performance, lithium-ion batteries have become the main power sources in aircraft, electric vehicles, mobile phones, laptops and even aerospace. With the increase in the charge and discharge times, lithium-ion batteries will inevitably degrade and even fail. If the effective action cannot be taken before the battery failure, the equipment with lithium-ion batteries will not operate healthily, and the casualties are very likely to happen seriously.

Inspired by the above discussion, it is of great significance to handle the remaining useful life prediction problem in industrial systems. Section 7.1 is concerned with remaining useful Life prediction combining degenerate point detection and degenerate trend estimation. In Section 7.2, the degradation assessment of bearings with trend-reconstruct-based features selection and gated recurrent unit network is investigated. Section 7.3 develops the remaining useful life prediction of lithium-ion batteries with optimal input sequence selection and error compensation. Section 7.4 gives our conclusions.

7.1 Remaining Useful Life Prediction with WPT and Optimized SVR

For the purpose of achieving the remaining useful life of bearings, two-stage strategy combining degenerate point detection and degenerate trend estimation are presented in this section, where wavelet modulus maximum (WMM) is used to confirm the degenerate point, wavelet packet transform is adopted to extract a set of features, and genetic algorithm-optimized support vector machine is proposed to achieve remaining useful life prediction.

7.1.1 Degenerate Point Detection

The continuous wavelet transform of original signal $x(t)$ is defined as

$$W_x(\mu,\tau) = \frac{1}{\sqrt{\mu}} \int_{-\infty}^{+\infty} x(t)\psi^*(\frac{t-\tau}{\mu})dt \tag{7.1}$$

where $\psi^*(t)$ is the complex conjugate of the wavelet function, t is time or space, and μ and τ are the scale factor and time shifting of signal $x(t)$, respectively. Specifically, large scale factor μ means big wavelets and coarse features of $x(t)$, while small μ correspond to small wavelets and fine details.

Considering the continuous function $\psi(t)$, the wavelet transform (7.1) can be rewritten as

$$\begin{aligned} W_x(\mu,\tau) &= \frac{1}{\sqrt{\mu}}\int_{-\infty}^{+\infty} x(t)\frac{d}{d\tau}\psi^*(\frac{t-\tau}{\mu})dt \\ &= \frac{d}{d\tau}\{\frac{1}{\sqrt{\mu}}\int_{-\infty}^{+\infty} x(t)\psi^*(\frac{t-\tau}{\mu})dt\} \end{aligned} \tag{7.2}$$

$W_x(\mu,\tau)$ is proportional to the first order derivative of $x(t)$, and then the mutation points of $x(t)$ are directly related to the modulus maximum of $W_x(\mu,\tau)$. Utilizing the following assumptions:

$$\psi(t) = (-1)^n\theta^n(t) \tag{7.3}$$

where $\theta(t) = \lambda exp(-\frac{-t^2}{2\beta^2})$, $x(t) = x_0(t)[\frac{1}{\sigma\sqrt{2\pi}}exp(-\frac{-t^2}{2\sigma^2})]$, $x_0(t)$ is a uniform function with Lpschitz exponent α, and α satisfies $\alpha \leq n$ at time t. Then, the modulus of $W_x(\mu,\tau)$ satisfies the following inequality:

$$|W_x[\mu,\tau)| \leq A\mu^{\alpha+\frac{1}{2}}(1+\frac{\sigma^2}{(\beta\mu)^2}]^{-\frac{(n-\alpha)}{2}} \tag{7.4}$$

where A is a constant, n is the vanishing moments of selected wavelet basis.

Actually, if there exists a wavelet maximal point sequence $(\mu_p,\tau_p)_{p\in R^+}$ converges to s in a fine scale, that is $\lim_{p\to+\infty}\mu_p = 0$ and $\lim_{p\to+\infty}\tau_p = s$, then $x(t)$ is singular at s. In other words, all singularities of signal $x(t)$ can be detected by tracking the maximum modulus of wavelet transform $W_x(\mu,\tau)$ in fine scales.

When the working state of the industrial equipment changes in a short period of time, there will be a small mutation of the monitoring signal waveform. Especially, the mutant point can be amplified with the wavelet transform and the singularity of signal can be detected by using support vector regression [259]. Similarly, the accurate location of the degradation point can be realized.

The characteristic of wavelet has a great influence on the detection result of degenerate point. If the choice of wavelet function is improper, it is impossible to detect efficiently the degenerate point. Usually, regularity, number of vanishing moments, filter length and symmetry are selected as the evaluation criteria of wavelet choice [260]. Specifically, regularity ensures that there is no redundancy in the decomposition process and no information overlapping in the translation of wavelet basis function, symmetry determines that the original signal phase can be kept unchanged by a filter. Besides, large number of vanishing moments make the detection ability of mutation point strong, and long filter length is conducive to local analysis of signals in frequency domain.

It is worth mentioning that not only the wavelet function itself that plays an important role in degenerate point detection but also the combination of wavelet decomposition level and reconstruction order has a crucial influence on the detection result. Fig. 7.1 is a schematic diagram of wavelet basis function selection. Actually, the wavelet decomposition level N_1 and reconstruction order N_2 should be associated with the length of signal $x(t)$. Synchronously, the following condition should be satisfied:

$$N_1 \geq log_2[L_{x(t)}],\ N_2 = floor\{log_2[L_{x(t)}]\} \tag{7.5}$$

where N_1 and N_2 are integers, $L_{x(t)}$ represents the length of the one-dimensional signal $x(t)$, and $floor\{*\}$ is the floor function.

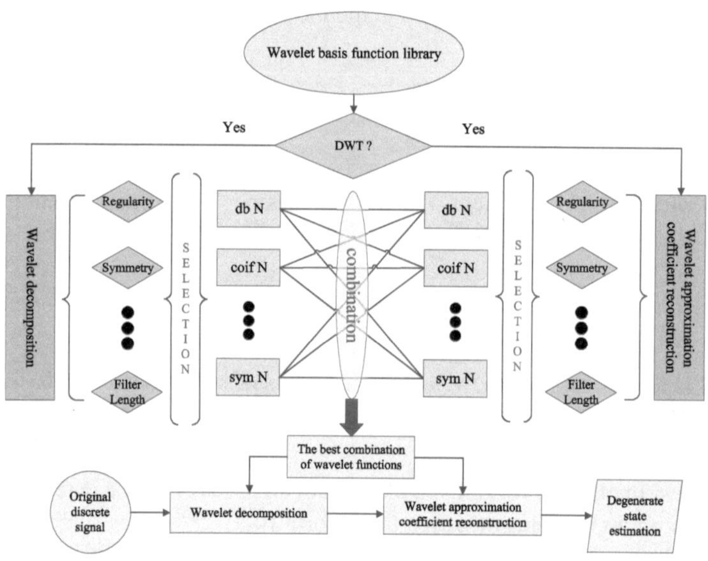

Figure 7.1: Wavelet basis function selection process.

In conclusion, symmetry wavelets with high vanishing moments, long filter length and good regularity should be selected. According to the length of the input signal, the appropriate decomposition level and reconstruction order can be identified, the main properties of several representative wavelets is shown in Table 7.1.

7.1.2 Remaining Useful Life Prediction of Turbine Engines

After the degenerate point of industrial equipment is determined with wavelet transform, the remaining process is degenerate trend estimation which includes two following parts: 1) feature extraction; 2) remaining useful life prediction. Fig. 7.2 shows the flowchart of the proposed two-stages prediction method.

Table 7.1: The main properties of commonly wavelets.

Wavelet function	Haar	Daubechies	Biorthogonal	Coilfets	Symlets	Daubechies
Example	haar	db2	bior3.3	coif4	sym8	db10
Regularity	Yes	Yes	No	Yes	Yes	Yes
Vanishing moments	1	2	2	8	8	10
Filter length	1	3	7	15	23	19
Symmetry	Yes	Near	No	Near	Near	Near

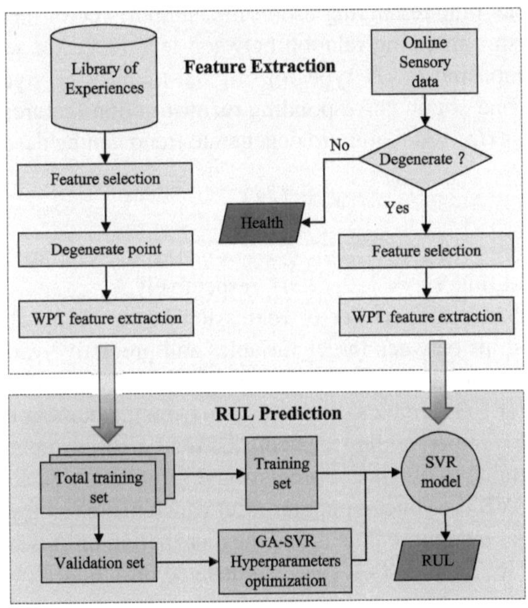

Figure 7.2: The flowchart of proposed two-stages approach.

A. Wavelet Packet Transform Feature Extraction

Wavelet transform is a suitable choice to denoise the signal and extract useful features [261], among which the popular orthogonal wavelet transform only does further decomposition on the low-frequency part of signal, while the high-frequency part is no longer applicable. In contrast, the wavelet packet transform (WPT) can provide more refined decomposition of the high-frequency part, and this decomposition is neither redundant nor negligent [262]. For the monitoring signal collected from industrial equipment, most of them are non-stationary mechanical vibration signal which contains large amount of medium and high frequency information. Therefore, wavelet packet transform is more suitable to analyze the time-frequency localization and draw the signal feature.

Undergoing the process of decomposition and reconstruction with wavelet packet transform, the acquired signal is suitable as the feature which contains the main information on both medium and high frequency, and the noisy information is removed in the same time.

B. Genetic Algorithm Optimized Support Vector Regression

In order to predict the remaining useful life, support vector regression [263] can be used to establish a mapping relation between feature vector set and degenerate trend. After decomposing the N types of original monitoring signals with wavelet packet transform, one obtain corresponding reconstruction features. Then, the mapping relation from extracted feature to degenerate trend can be described as:

$$\mathbf{y} \doteq f(\mathbf{x}) \tag{7.6}$$

where $\mathbf{x} = (x_1, x_2, \cdots, x_n), \mathbf{y} = (y_1, y_2, \cdots, y_n)$, and $x_i,\ y_i$ are the feature values and output signal at time $i(i = 1, 2, ..., n)$, respectively.

On the other hand, support vector regression has a strong ability to establish nonlinear relationships between target variables and predictive variables. Generally, support vector regression falls into two categories, namely, v-SVR and ϵ-SVR. Specially, the parameter v determines the proportion from the number of support vectors to the total sample number, while ϵ controls the total error of the model. Obviously, ϵ-SVR is more suitable to obtain the prediction results with high accuracy.

The goal of ϵ-SVR is to find a function $f(\mathbf{x})$ which makes all training samples deviate from the target y no more than ϵ, and the distribution of these deviations should be as flat as possible. Then, ϵ-SVR can be translated into the following optimization problem:

$$\min_{\mathbf{w},b}\{\frac{1}{2}\mathbf{w}^T\mathbf{w} + C\frac{1}{n}\sum_{i=1}^{n}\max(|y_i - f(x_i)| - \epsilon, 0)\} \tag{7.7}$$

where C has a strong tradeoff control capability between the flatness of function $f(*)$ and the allowable error ϵ. In order to increase the generalization and flexibility of formula (7.7), the relaxation variables ξ_i and ξ_i^* are introduced and the optimization problem (7.7) can be written in the following form:

$$\begin{aligned} &\min_{\mathbf{w},b,\xi_i,\xi_i^*}\{\frac{1}{2}\mathbf{w}^T\mathbf{w} + C\sum_{i=1}^{n}(\xi_i + \xi_i^*)\} \\ &subject\ to \begin{cases} f(x_i) - y_i \le \epsilon + \xi_i \\ y_i - f(x_i) \le \epsilon + \xi_i^* \\ \xi_i, \xi_i^* \ge 0, \quad i = 1, \cdots, n \end{cases} \end{aligned} \tag{7.8}$$

To solve the optimization problem (7.8) and increase the capability of nonlinear mapping between $\mathbf{x}$ and $\mathbf{y}$, by introducing kernel function $K(x_i, x_j)$, (7.8) can be translated into the following dual problem:

$$\min\{\frac{1}{2}\sum_{i,j=1}^{n}(a_i^* - a_i)(a_j^* - a_j)K(x_i, x_j) + \epsilon\sum_{i=1}^{n}(a_i^* + a_i)...$$

$$-\sum_{i}^{n} y_i(a_i^* - a_i)\}$$

$$subject\ to\ \sum_{i}^{n}(a_i^* - a_i) = 0,\ 0 \leq a_i, a_i^* \leq C \tag{7.9}$$

Different kernel functions for support vector regression will lead to different regression effects, in which gaussian radial basis function kernel has been proved to have better performance than other expression by experiment [264]. Therefore, the gaussian radial basis kernel function with $K(x_i, x_j) = exp(-\frac{\|x_i - x_j\|^2}{\gamma^2})$ is chosen as kernel function in this section. From what has been discussed above, we can draw a conclusion that the output of support vector regression process can be used to predict efficiently the remaining useful life of machinery equipment.

In data regression field, the strong generalization ability of support vector regression has been proved by lots of experimental results [265]. However, this ability depends largely on the selection of support vector regression for penalty coefficient C and kernel parameter γ. In terms of the parameter selection algorithm, the traditional exhaustive search algorithm is based on gradient descent principle and time consuming, it may get stuck into local optimum due to its non-convexity of generalization bounds. Actually, genetic algorithm (GA) overcomes above shortcoming and accomplishes a global optimization with strong parallel mode space searching ability [266].

In this section, by employing genetic algorithm to optimize support vector regression model, the best combination of penalty factors C and radial basis function kernel parameter γ can be achieved. As a byproduct, a good generalization ability for prediction is owned. The overall process of proposed GA-SVR algorithm is shown in Algorithm 7.1.

7.1.3 Illustrative Examples

In this section, the turbofan engine degradation data set is adopted to verify the validity of proposed two-stage prognosis method, which is available at the NASA prognostics data repository [267]. The remaining useful life prediction process of turbofan engine degradation can be described as follows:

A. Turbofan Engines Datasets

The data set is generated based on the simulation model of Commercial Modular Aero-Propulsion System Simulation (C-MAPSS) test-bed developed by NASA

Algorithm 7.1 GA-SVR model

1: Obtain the verification data set: a tenth of the training set randomly selected as the verification set. Denote MAXGEN as T, the population size as N, the crossover probability as P, the upper and lower limits of parameter pairs as (C, γ).
2: Produce randomly the first generation population $(C, \gamma)_t^N, N = 1$.
3: Set $i = 1$.
4: Take parameter pairs $(C, \gamma)_t^i$ as hyper-parameters.
5: Do 5-folded cross-validation and store its error.
6: Let $i = i + 1$.
7: **if** $i \leq N$ **then**
8:     **return** *step* 4.
9:     **else go to** *step* 11.
10: **end if**
11: Save $(C, \gamma)_t^N$ corresponding to min error.
12: Binary encoding of $(C, \gamma)_t^N$.
13: Selection, crossover, and mutation of codes.
14: $t = t + 1$.
15: Decode new codes to generate $(C, \gamma)_t^N$.
16: **if** $t \leq T$ **then**
17:     **return** *step* 3.
18:     **else go to** *step* 20.
19: **end if**
20: Ascertain the optimal parameter (C, γ) with $(C, \gamma) = \arg\min(error)$ for $(C, \gamma)_t^N$.
21: Utilize the obtained optimal parameter (C, γ) to build the optimal model.

[268]. For the details of various modules and their connections in the simulation, readers are referred to [267]. Specially, the target set FD001 is chosen to train and test the prognostics model.

B. Degradation Points Determination

With the increase in service time for turbofan engine, it runs healthily at the beginning phase and falls into the degradation stage at some time point. Therefore, it is of great significance to precisely determine the degradation point and then estimate the degradation trend. On the other hand, in order to comprehensively monitor the operation of the turbine engine, we need to obtain the signals of different monitoring objects. Actually, the difference in scale between different kind of signals may be very large, and the original data needs to be normalized so that effective analysis can be carried out smoothly. In at the same time, it is necessary to ignore some poor monotonicity and high volatility data to achieve more accurate degradation point.

For the train data set train_FD001, there are five different features in the first training unit, the normalized distribution curve can be obtained as Fig. 7.3. According to the selection criteria of wavelet in Table 7.1 and Fig. 7.1, $sym7$ and $db1$ are chosen as the wavelet basis functions, and the decomposition number of layers is 9. By employing wavelet transform to five different features in Fig. 7.3, respectively, the reconstruction curve is shown in Fig. 7.4 with the approximate coefficient of seven layer wavelet. During the whole period of the first training unit, a significant mutation occurs at 130th cycle, which can be identified as the degradation points of the turbine engines.

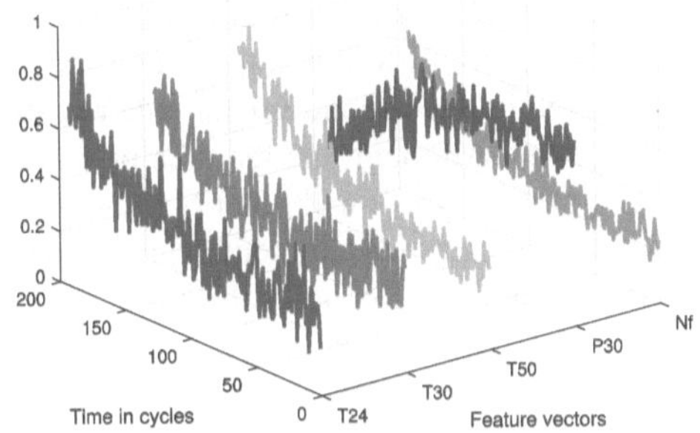

Figure 7.3: Different characteristics of the original signal distribution.

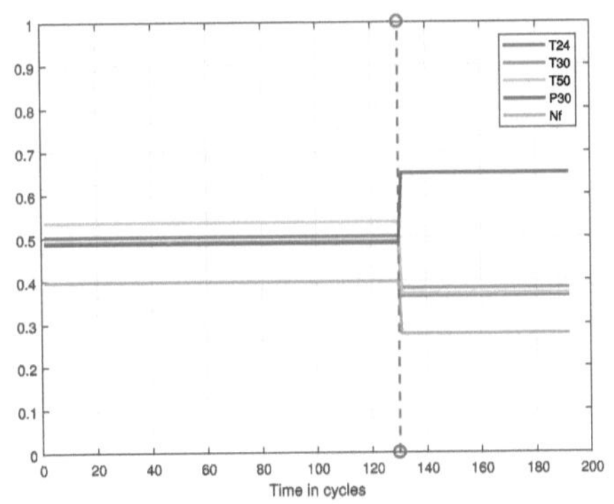

Figure 7.4: Wavelet transform approximation coefficient.

C. Wavelet Packet Transform Feature Extraction

Once the degradation is detected, the prediction unit is triggered to forecast the remaining useful life of equipment. Due to the signal collected by the sensor is always accompanied by obvious noise and big fluctuation, feature extraction is needed firstly. Based on the analysis of Section III-A, the data set is processed by using wavelet packet transform with Algorithm I, and the reconstructed signal can be extracted. It is observed that seven data sets among twenty-one data sets of sensor data are not changed over time, and it is not helpful for training the support vector regression model, therefore, these seven data sets are discarded and the remaining fourteen data sets are chosen as the features.

It is shown from Fig. 7.5 and the analysis of Section III-A that normalization is needed, and wavelet packet transform is chosen to dispose the normalized data. According to Table I, the optimal wavelet basis is set as $coif5$ and the decomposition layer number is chosen as 7. Fig. 7.6 is the reconstructed signal curve of 5 different features in the first training unit of train_FD001 after wavelet packet transform. Similarly, Fig. 7.7 shows corresponding life distribution and reconstruction features of 100 training data set collected by the second sensor, where different colour are corresponding to different engines. Although the features processed by wavelet packet transform satisfy strong monotony and small volatility, the wide range of life distribution which varies from 128 to 362 measurement cycles, it will greatly increase the difficulty of accurate prediction of remaining useful life.

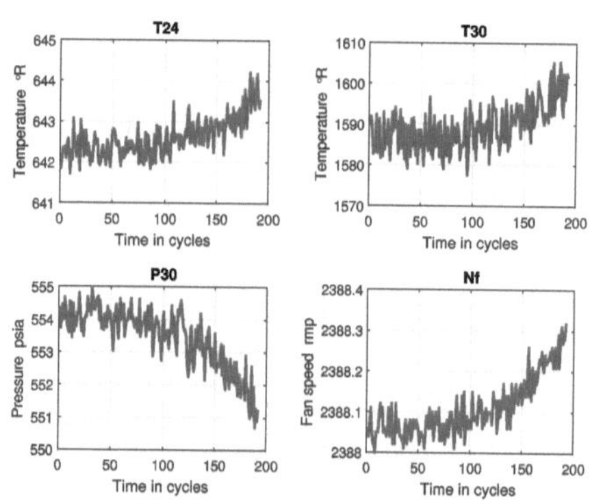

Figure 7.5: Partial sensor recording signals.

D. Evaluation Metrics

In order to compare and evaluate the performance of the presented method, the following two evaluation criteria are adopted:

1) Score: The score S is defined as the difference between the predicted remaining useful life $\tilde{R}_t$ and its true value R_t, and their specific expression can be written as

$$S = \sum_{t=1}^{N} s_t \tag{7.10}$$

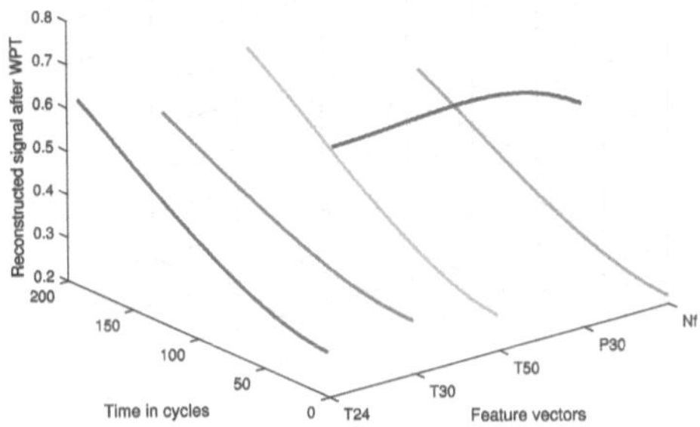

Figure 7.6: Reconstructed signal after WPT.

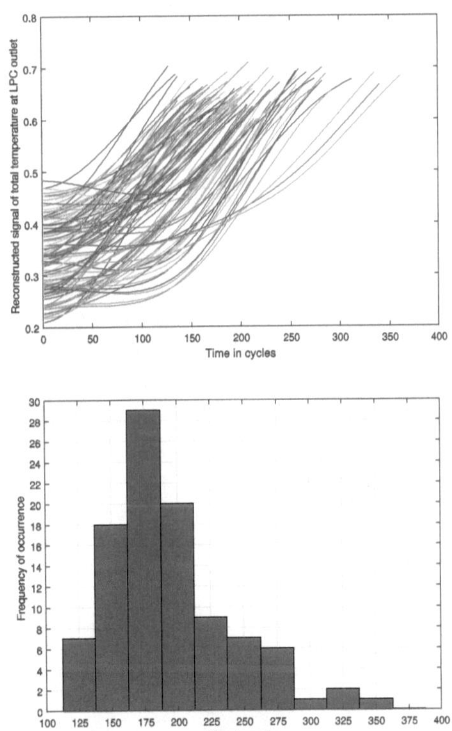

Figure 7.7: (a) Second sensor reconstructed signal of all units and (b) Life span distribution of all units.

where s_t satisfies

$$s_t = \begin{cases} e^{-(R_t - \tilde{R}_t)/10} - 1 \ \ if \ -(R_t - \tilde{R}_t) \leq 0 \\ e^{(R_t - \tilde{R}_t)/13} - 1 \ \ if \ -(R_t - \tilde{R}_t) > 0 \end{cases}$$

For the obtained prediction results, lower S means better forecast effect.

2) Performance: This metric represents the percentage of correct prediction by defining $\Delta = R_t - \tilde{R}_t$. The prediction result is regard as correct if its error falls within the interval $[-10, 13]$ for turbofan engine. In terms of industrial engineering practice, the harm of error caused by advanced prediction is far less than that caused by delayed prediction, then negative values are worse.

For the matrix defined above, *Score* represents the overall error and *Performance* reflects the accuracy of the algorithm. By introducing these two matrices, the advantages and disadvantages of different methods for remaining useful life prediction can be compared effectively.

E. GA-RUL Prediction

The end point of useful life is the time point after the degradation of equipment, therefore, the detection of degenerate point is of vital importance to the prediction performance. It can be seen from Fig. 7.4 that the degenerate point for data set "FD001" can be identified as 130 time cycles. As shown in Fig. 7.7, the reconstruction curve with wavelet packet transform is basically stable in the health state, and obvious monotonic with increase occurs in the degradation stage. After the degenerate point is confirmed, the main goal is to predict the useful life of the turbofan engine.

In order to improve the accuracy of remaining useful life prediction, support vector regression with intelligent optimization algorithm are adopted. By using genetic algorithm based hyper-parameters optimization described in Section III-C, the GA-SVR model can be established and the remaining useful life prediction of some engines for test_FD001 is shown in Fig. 7.8. Especially, to evaluate the prediction effect of proposed GA-SVR, remaining useful life of 100 engine data sets for test_FD001 are disposed and their prediction results are shown in Fig. 7.9. It can be observed from Fig. 7.9 that the prediction of the proposed method for most engines is accurate.

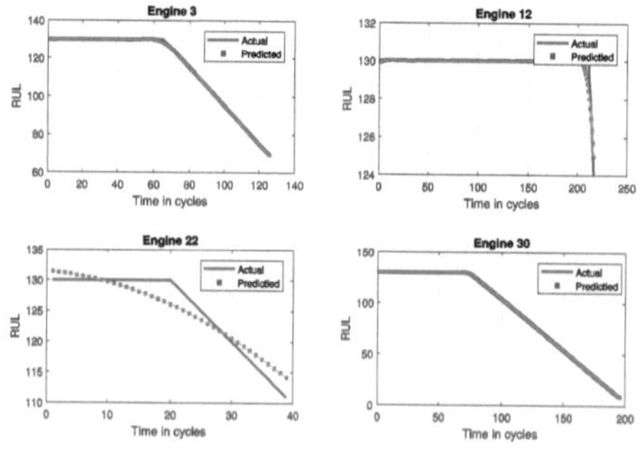

Figure 7.8: Actual versus predicted RULs of full cycle.

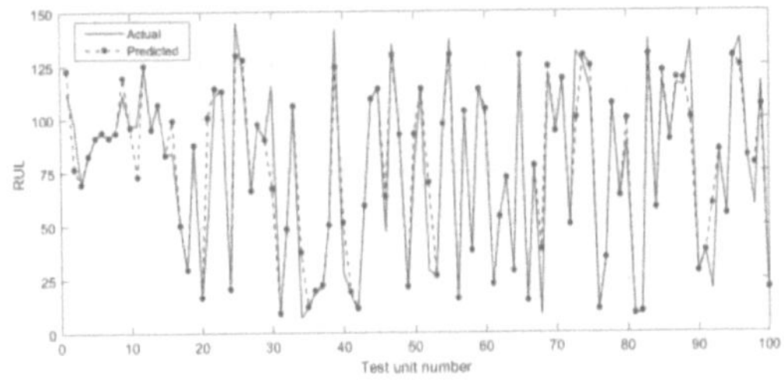

Figure 7.9: RUL prediction results of test_FD001.

Table 7.2 shows the comparison results for different optimization algorithms such as artificial bee colony algorithm (ABC), cuckoo search algorithm (CS), firefly algorithm (FA), particle swarm optimization algorithm (PSO), and grid search (GS). It can be founded that both the performance 81% and score 320.21 are all the best by using the GA-SVR method. For turbofan engine degradation data set, the composition of the data set "FD003" and "FD001" is similar. Therefore, the proposed method is efficient to the data set "FD003", the prediction results are shown in Fig. 7.10 with performance 87% and score 540.95.

Table 7.2: Comparison of different optimization algorithms.

Models	Bestc	Bestg	Score	Performance
GS	256	256	1133.96	85
ABC	1023.27	1021.35	812.56	81
CS	24.16	35.51	8284.04	62
FA	626.51	954.91	737.45	80
PSO	100	900.59	677.16	80
GA	99.97	337.36	**320.21**	**81**

The prediction results show that the presented method in this section have obvious advantages. The main reason lies in the following several aspects: i) support vector regression is used to detect degenerate points, which can avoid the data redundancy caused by useless information. ii) The features extracted by wavelet packet transform are characterized by strict monotony and small fluctuations, which are beneficial to support vector regression modelling. iii) The utilization of genetic optimization algorithm is helpful to get the best support vector regression model and more accurate prediction.

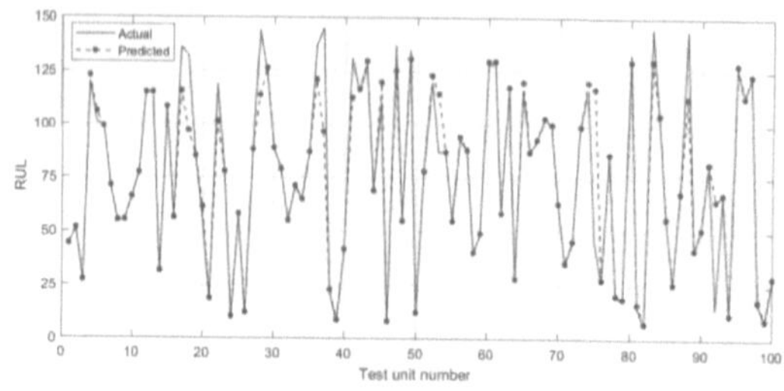

Figure 7.10: RUL prediction results of test_FD003.

7.2 Remaining Useful Life Prediction with Complete Ensemble EMD and GRU

The data-driven remaining useful life methodology is based on the bearing vibration signals. Although the vibration signals contain specific information about fault conditions, it is still difficult to detect and track the weak signals at an early stage. The objective of this subsection is to propose a trend-reconstruction-based feature selection and gated recurrent unit network, so that an effective HI can be established and the remaining useful life can be predicted accurately. The overall architecture of the proposed approach can be summarized in Fig. 7.11.

The main contributions are outlined as follows: 1) the degradation trend is reconstructed with the basic-characteristics-based complete ensemble empirical mode decomposition with adaptive noise (BC-CEEMDAN) method; 2) the features are selected by fusing the correlation and monotonicity from BC-CEEMDAN, and the HI is established by inputting the fused features into the gated recurrent unit network; and 3) the degradation point can be detected adaptively with the developed $\mu + 3\sigma$ technique, and the remaining useful life can be predicted effectively with the proposed BC-CEEMDAN-GRU method.

7.2.1 Health Indicator Construction

In this section, we first extract some basic features of time domain, frequency domain and time-frequency domain. Then, the BC-CEEMDAN is provided to reconstruct the degraded trend based on the basic features. In the end, a fusion method is introduced to choose the proper features by synthesizing the correlation and monotonicity.

A. Classical Feature Extraction

This subsection focuses on how to extract the representative features from the original vibration signals by using the following three kinds of signal processing techniques.

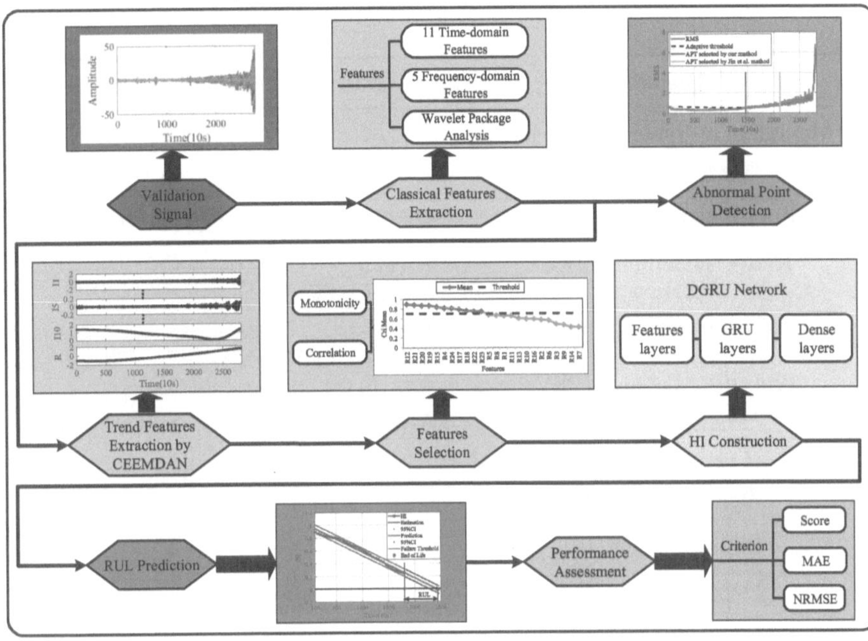

Figure 7.11: Integrated architecture for abnormal point detection and RUL prediction of REBs.

Time Domain: Denote $\{S_i(t)\}_{i=1,\cdots,n}$ as the ith sample sequences at time t for the original vibration signals $S(t)$, n is the number of sample points. $\{S_i(\mathrm{t})\}_{i=1:n}$ means $\{S_1(\mathrm{t}), ..., S_n(\mathrm{t})\}$. Eleven statistical features of time domain are summarized in Table 7.3, in which the detailed calculation and explanation of those features are included.

Frequency domain: Frequency domain features describe effectively the change of the spectrum information for rolling element bearings. Therefore, it can be chosen to monitor the health state of rolling element bearings. As a special frequency-domain feature, the fast Fourier transform (FFT) is a popular tool that can transform time series into the corresponding frequency coefficients. Once the vibration signals are processed by FFT, the complete spectrum signal f_{12} (0–12.8 kHZ) and four sub-spectrum signals f_{13} (0–3.2 kHZ), f_{14} (3.2–6.4 kHZ), f_{15} (6.4–9.6 kHZ), and f_{16} (9.6–12.8 kHZ) will be selected as the extracted features. Based on the obtained frequency domain features, the related-similarity feature [254] is used to describe the similarity measurement of data sequences between the current and initial times. The detailed calculation process is given as follows:

$$F_r(t) = \frac{\left|\sum_{m=1}^{M}\left(f_{\mathrm{r}}^{m}(0) - \overline{f_r(0)}\right)\left(f_r^{m}(t) - \overline{f_r(t)}\right)\right|}{\sqrt{\left[\sum_{m=1}^{M}\left(f_{\mathrm{r}}^{m}(0) - \overline{f_r(0)}\right)^2\right]\left[\sum_{m=1}^{M}\left(f_r^{m}(t) - \overline{f_r(t)}\right)^2\right]}} \tag{7.11}$$

Table 7.3: The statistical features in time domain.

Features	Computational formula
Maximum absolute value (MAV)	$F_1(t) = max\{\lvert S_i(t)\rvert\}_{i=1:n}$
Mean value (MV)	$F_2(t) = \frac{1}{n}\sum_{i=1}^{n} S_i(t)$
Root-mean-square (RMS)	$F_3(t) = \sqrt{\frac{1}{n}\sum_{i=1}^{n} S_i^2(t)}$
Kurtosis coefficient (KC)	$F_4(t) = \frac{\frac{1}{n}\sum_{i=1}^{n}(S_i(t)-F_2(t))^4}{\left(\sqrt{\frac{1}{n-1}\sum_{i=1}^{n}(S_i(t)-F_2(t))^2}\right)^4}$
Skewness coefficient (SC)	$F_5(t) = \frac{\frac{1}{n}\sum_{i=1}^{n}(S_i(t)-F_2(t))^3}{\left(\sqrt{\frac{1}{n-1}\sum_{i=1}^{n}(S_i(t)-F_2(t))^2}\right)^3}$
Peak-to-peak value (PPV)	$F_6(t) = \max\{S_i(t)\}_{i=1:n}$
Variance value (VV)	$F_7(t) = \frac{1}{n-1}\sum_{i=1}^{n}(S_i(t) - F_2(t))^2$
Crest factor (CF)	$F_8(t) = \frac{F_1(t)}{F_3(t)}$
Wave factor (WF)	$F_9(t) = \frac{F_3(t)}{\frac{1}{n}\sum_{i=1}^{n}\lvert S_i(t)\rvert}$
Impulse factor (IF)	$F_{10}(t) = \frac{F_1(t)}{\frac{1}{n}\sum_{i=1}^{n}\lvert S_i(t)\rvert}$
Margin factor (MF)	$F_{11}(t) = \frac{F_1(t)}{\left(\frac{1}{n}\sum_{i=1}^{n}\sqrt{\lvert S_i(t)\rvert}\right)^2}$

where M represents the length of the sequence $\{f_r(t)\}_{r=12,\cdots,16}$, $f_r(0)$ and $f_r(t)$ are the signal sequence f_r at initial time and current time, and $\overline{f_r(0)}$ and $\overline{f_r(t)}$ are the mean value $\{f_r^m(0)\}_{m=1:M}$ and $\{f_r^m(t)\}_{m=1:M}$, respectively. Therefore, the frequency domain features $\{F_r(t)\}_{r=12,\cdots,16}$ can be achieved.

Time-Frequency Domain: Usually, the wavelet transform (WT) can be characterized by the local characteristics of the signal in the time-frequency domain. Nevertheless, the high-frequency part of signals cannot be decomposed by WT. In order to deal with this problem, the wavelet package decomposition (WPD) is introduced to decompose both the high-frequency part and the low-frequency part of the signals [248]. Especially, the energy of wavelet packet is commonly used to identify the failure mode of rolling element bearings. Recur to the WPD with "harr" mother wavelet and three-level decomposition, eight energy ratio features $\{F_r(t)\}_{r=17,\cdots,24}$ are chosen as the time-frequency features.

By employing the above three kinds of efficient feature extraction methods, we can obtain the basic characteristics as $\{F_r(t)\}_{r=1,\cdots,24}$ which comprehensively capture the information from the vibration signal of bearing.

B. Trend Feature Reconstruction with BC-CEEMDAN

In [82], the empirical mode decomposition has been adopted to reconstruct the curve by considering that the degraded trends are inconspicuous and non-monotonic. By decomposing the signals into a finite number of IMFs and a smooth trend component, the technique of empirical mode decomposition series servers as a suitable choice to process the nonlinear and non-stationary data. Actually, mode mixing is a common phenomenon in the empirical mode decomposition technique that may bring challenges on data decomposition. Then, the EEMD has been proposed by adding the independent white noises to the original data and decomposing the noise-added data with empirical mode decomposition. Nevertheless, the EEMD method is time-consuming by calculating the decomposed average value, and the reconstruction errors are consequent due to the added white noises. To address such concerns in the EEMD method, the CEEMDAN method has been further proposed by adding the adaptive white noises to the residual components of each decomposition scale, and then averaging them [269]. The CEEMDAN method makes the decomposition results more thorough and effective when addressing the above problems.

In this section, the above 24 basic characteristics based CEEMDAN (BC-CEEMDAN) method is supplied. Compared with the conventional features in time domain, frequency domain and time-frequency domain, the BC-CEEMDAN exhibits two distinguished traits: 1) 24 features are selected by employing the vibration signal of rolling element bearings, which can effectively extract the characteristic information of bearing degradation in various aspects; and 2) the obtained residual with the BC-CEEMDAN method eliminates efficiently the fluctuation in the selected features, and then can track accurately the degradation process.

C. Degraded Feature Selection

After achieving the extraction of trend features, the next goal is to select proper features by discarding irrelevant and redundant characteristics so that sufficient degradation information can be obtained. Actually, a good degradation trait should have good monotony and a high correlation about the degradation process, which will improve the accuracy of the remaining useful life prediction and increase the computational efficiency. In our study, the monotonicity [253] and correlation [270] are selected as the sensitive features whose metrics are listed as follows:

$$\begin{cases} \text{Mon} = \dfrac{\left|\sum_{t=1}^{T}\delta(R(t+1)-R(t)) - \sum_{t=1}^{T}\delta(R(t)-R(t+1))\right|}{T-1} \\ \text{Cor} = \dfrac{\left|\sum_{t=1}^{T}\left(R(t)-\overline{R}\right)\left(L(t)-\overline{L}\right)\right|}{\sqrt{\sum_{t=1}^{T}\left(R(t)-\overline{R}\right)^2\sum_{t=1}^{T}\left(L(t)-\overline{L}\right)^2}} \end{cases} \tag{7.12}$$

where $\delta(\bullet)$ is the unit step function, $\{R(t)\}_{t=1:T}$ and $\{L(t)\}_{t=1:T}$ are the feature sequences at tth observation sample, $\overline{R}$ and $\overline{L}$ are their average values, respectively.

It can be seen from (7.12) that the above two metrics belong to $[0, 1]$, and they are positively correlated with the performance of the chosen characteristics. Therefore, a linear combination of Mon and Cor can be selected as the fused feature which can be described by

$$\begin{cases} \max \text{Cri} = \xi_1 \text{Mon} + \xi_2 \text{Cor} \\ \sum_{i=1}^{2}\xi_i = 1, \xi_i > 0 \end{cases} \tag{7.13}$$

where ξ_i is the weight coefficient. The larger Cri means that the selected features have a better reflection on the degradation process of rolling element bearings.

So far, we have completed the process of feature extraction and selection, where the corresponding steps include the BC-CEEMDAN and fused Cri (BC-CEEMDAN-FC). Based on the obtained efficient features, the degradation point detection and remaining useful life prediction will be carried out in the following part.

7.2.2 Remaining Useful Life Prediction of Bearings

Usually, the rolling element bearings are monitored routinely in the healthy stage. Once the abnormal/degradation point occurs, the rolling element bearings will enter the sub-health stage and intensive monitoring will be triggered. Different from the traditional remaining useful life prediction of rolling element bearings [82, 248, 254, 255, 269], we will investigate the joint detection of abnormal points and remaining useful life, which is more in line with industrial practice.

A. Abnormal Point Detection

Degradation point detection has attracted a lot of research attention [4]. Li et al. [243] have proposed an adaptive abnormal point selection approach based on the 3σ interval, where the kurtosis is adopted as the monitoring indexes. Jin et al. [238] have studied the degenerate points by using the Box–Cox transformation and Gauss distribution. Actually, the RMS reflects the increase of vibration energy with the development of degradation [243], thus being a suitable indicator to detect the abnormal point time (APT).

In order to detect adaptively the abnormal/degraded time, the 3σ criterion is introduced, which means that about 99.73% of data are within $(\mu - 3\sigma, \mu + 3\sigma)$ for a data set with Gauss distribution. The RMS value is assumed to be a random variable satisfying approximate normal distribution, whose mean and standard deviation are σ^2 and μ, respectively. Simultaneously, the abnormal/degraded threshold is defined by the upper bound $(\mu + 3\sigma)$, and the rolling element bearings are considered to degenerate when multiple consecutive values are outside the threshold.

B. Remaining Useful Life Prediction

Once the abnormal points are detected, we will establish the HI and predict the remaining useful life of rolling element bearings. It is well known that deep learning is the major method in data-driven remaining useful life prognosis, in which the recurrent neural network (RNN) has attracted special attention [254]. One problem in the RNN is that the gradient explosion or gradient disappearing may cause the optimization process of the training process falls into the local optimum. As a variant of RNN, the long short-term memory (LSTM) has been proposed [271] to avoid the long-term dependency problem, and has been employed in the remaining useful life prediction of a turbine engine. Nevertheless, the LSTM model has a complex form and needs a long time to train, which limits its applications. To overcome these shortcomings, the gated recurrent unit, as an improved strategy with a more simplified architecture, has been proposed by Cho [256].

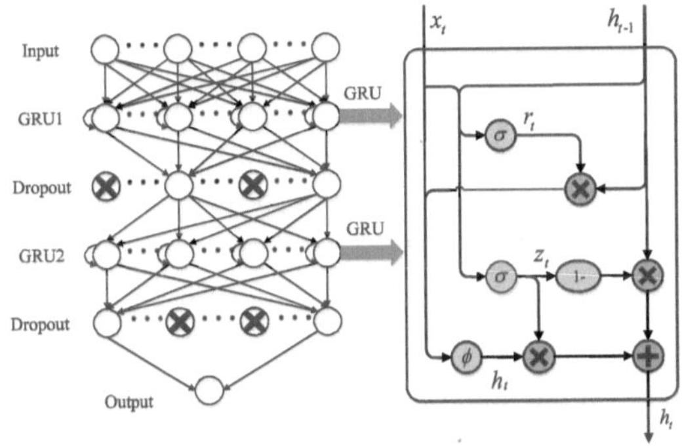

Figure 7.12: The flowchart of GRU.

In this subsection, the gated recurrent unit network is introduced to construct the health index (GRU-HI) and forecast the remaining useful life. Concretely, the proposed GRU-HI consists of one input layer, two gated recurrent unit layers, two dropout layers and an output layer. The structure of GRU-HI can be found in Fig. 7.12. The gated recurrent unit comprises reset gate r_t and update gate z_t at time t. The update gate refers to the degree to update the hidden state information. The reset gate determines how the current input information is combined with the previous hidden state. The operation of the gated recurrent unit hidden layer is defined as follows:

$$\begin{aligned} r_t &= \sigma\left(w_{xr}x_t + w_{hr}h_{t-1} + b_r\right) \\ z_t &= \sigma\left(w_{xz}x_t + w_{hz}h_{t-1} + b_z\right) \\ c_t &= \phi[(w_{xc}x_t) + w_{hc}(r_t \otimes h_{t-1})] \\ h_t &= (1 - z_t) \otimes h_{t-1} + z_t \otimes c_t \end{aligned} \tag{7.14}$$

where x_t is the input sequence, h_t is the output of the hidden layer, c_t is denoted as the candidate state, w_{xr}, w_{xz}, w_{xc} are the weight matrices from the input layer to reset gate, update gate and candidate state, respectively, w_{hr}, w_{hz}, w_{hc} represent the weight matrices of the cycle connections, $\otimes$ indicates an element-wise multiplication, and $\sigma(\bullet)$ and $\phi(\bullet)$ are the sigmoid function and the tanh function, respectively. The output of gated recurrent unit layer can be described as follows:

$$y_t^l = max\{0, h_t\} \tag{7.15}$$

To avoid model over-fitting, the dropout layer is alternated with the gated recurrent unit layer. During training, each neural unit is randomly omitted with a probability. In our study, the probability with the rate of 0.2 is used in the proposed network. Two gated recurrent unit and dropout layers are used to learn the input signal to obtain the learned features, which are then mapped into health index with a nonlinear mapping operation. In order to ensure that the output value of the network is within $[0, 1]$, the activation function of the output layer is the sigmoid function.

After the GRU-HI model is established, the training and test process will be triggered. In the training process, the length of training time series is T; the length of window is τ; the input sample is $x_{t+\tau} = [R(t), R(t+1), \cdots, R(t+\tau-1)]$; the corresponding label value is $y_{t+\tau}$; and the number of sample is $N = T - \tau$, where $R(t) \in \mathbb{R}^{r\times 1}$ and $y_t \in [0, 1]$ are the selected r features and remaining useful life percentage at time t, respectively. It is assumed that the end of life time is T_{end} and the current time is t. $y_{t+\tau}$ can be written by

$$y_{t+\tau} = \frac{T_{end} - (t+\tau)}{T_{end}} \tag{7.16}$$

In order to make output value $\hat{y}_t$ as close as possible to its corresponding label y_t, the root mean square error (RMSE) is chosen as the loss function, which can be defined by

$$\text{loss} = \sqrt{\frac{1}{N}\sum_{t=1}^{N}[y_{t+\tau} - \hat{y}_{t+\tau}]^2} \tag{7.17}$$

Moreover, the Adam optimization algorithm is used to update parameters. In the testing process, the selected features are directly put into the trained model.

C. Assessment Metric

To quantitatively evaluate the prediction effect of the proposed method, the following three evaluation criteria are introduced:

1) Score function: It is mainly used to describe the average of all test sets scores which can be calculated by

$$\text{Score} = \frac{1}{N_b}\sum_{b=1}^{N_b} A_b \tag{7.18}$$

where

$$A_b \triangleq \begin{cases} \exp\left[-ln(0.5)\cdot(Er_b/5)\right], & if\ Er_b \le 0 \\ \exp\left[+ln(0.5)\cdot(Er_b/20)\right], & if\ Er_b > 0 \end{cases}$$

$$Er_b \triangleq \frac{ActRUL_b - \widehat{RUL}_b}{ActRUL_b} \times 100\%,$$

and Er_b is the predicted percentage error of the bth test bearing, N_b is the total number of tested bearings.

2) Mean absolute error (MAE): It is the average absolute error between the true and predicted values, which is usually used to measure how close the predictions and the actual results are.

$$\text{MAE} = \frac{1}{N_b}\sum_{b=1}^{N_b}\left|ActRUL_b - \widehat{RUL}_b\right| \tag{7.19}$$

3) Normalized root mean square error (NRMSE): NRMSE can be described as

$$\text{NRMSE} = \frac{\sqrt{\frac{1}{N_b}\sum_{b=1}^{N_b}\left(ActRUL_b - \widehat{RUL}_b\right)^2}}{\left(\frac{1}{N_b}\sum_{b=1}^{N_b}\widehat{RUL}_b\right)} \tag{7.20}$$

Among the above matrices, $ActRUL_b$ and $\widehat{RUL}_b$ represent the actual remaining useful life and predicted remaining useful life of the bth test set, respectively. The score coefficient value belongs to [0,1], and if it is closer to 1, then a better prediction accuracy can be obtained. On the contrary, in terms of NRMSE and MAE, if the two indicators are more approximate to 0, then the accuracy of remaining useful life prediction is higher.

7.2.3 Illustrative Examples

In order to verify the effectiveness of the proposed remaining useful life prediction method, experimental data sets from IEEE-PHM-2012-Challenge [239] are adopted. At the same time, in terms of the comparison of some popular methods and the existing results, the advantages of the proposed techniques are emphasized.

The used data set is obtained through the PRONOSTIA platform, and the corresponding details about platforms and experiments can be found in [239]. According to the relevant research results [241,242], the useful information used to track bearing degradation provided by vertical vibration signals is less than horizontal vibration signals. Therefore, only horizontal signals in this part are adopted.

Three different operating conditions are selected from the experiment, in which the radial load and velocity are variable. Specially, there are totally 17 data sets, in which 6 run-to-failure data sets are used as training sets and the other 11 data sets are chosen as testing sets. By considering the similarity of degradation [238,243], only the data sets of operating condition 1 in this experiment are employed to verify the proposed methods. The run-to-failure bearing 1-1 and bearing 1-2 are used as the training set, and the remaining bearings are used as the testing set to predict the remaining life of the bearing in this operating condition.

With the help of the APT detection algorithm with the given initial time point $t_0 = 100$, the abnormal/degraded points can be detected and the corresponding experiment results are shown in Fig. 7.13. In some cases, anomaly conditions can be also observed in the run-in and useful life periods, but such an anomaly observation could lead to a false alarm. It's observed from Fig. 7.13 that some bearing degradation is steep and the others are moderative. Especially, for the sharp degeneration such as bearings 1–5 and 1–7, it is very difficult to predict the remaining useful life accurately. Compared with the existing results [272], the anomaly point can be detected later, and it is located in the front of the failure point. After the APT is detected, we will start preparing the maintenance plan so that the influence of wrong forecast results can be avoided.

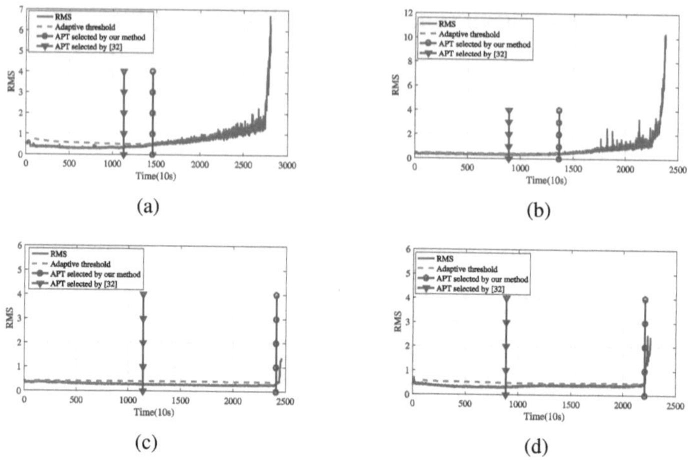

Figure 7.13: Abnormal point time detection: (a) Bearing 1–1; (b) Bearing 1–3; (c) Bearing 1–5; and (d) Bearing 1–7.

As described, 11 time domain features, 5 frequency domain features, and 8 time-frequency domain features are extracted from the original vibration signal. To obtain obvious tendency from the extracted 24 features $(R_1, R_2, \cdots, R_{24})$, the BC-CEEMDAN is performed to reconstruct the trend features.

To eliminate the redundant features and improve the remaining useful life prediction performance, the fused features can be acquired by using a linear combination of monotonicity and correlation. By denoting $\xi_1 = 0.6$, $\xi_2 = 0.4$ and the threshold as 0.7, the distribution of Cri (7.13) for all bearings under condition 1 is depicted in Fig. 7.14. In order to select the best feature subset, the Cri value of these extracted features is calculated, and their distribution is shown in Fig. 7.15. By choosing the fused features which are greater than the threshold, we obtain the feature sets including time domain features (R_4), frequency domain features $(R_{12}$ and $R_{15})$, and time-frequency domain features $(R_{17}, R_{18}, R_{19}, R_{20}, R_{21}, R_{22}, R_{23}$, and $R_{24})$. Once the above features have been selected, the most important of all is to construct the health index. In this section, we utilize these selected features as the inputs of gated recurrent unit neural network to obtain the fused health indicator.

Before the selected features enter into the gated recurrent unit network, they need to be normalized with the following min–max normalization method:

$$R_r^{nor}(t) = \frac{R_r(t) - R_r^{min}}{R_r^{max} - R_r^{min}} \tag{7.21}$$

where $R_r(t)$ denotes the tth measurement sequence of the rth feature, R_r^{min} and R_r^{max} are their minimum and maximum values, respectively.

If the degradation information is not too large, features extraction and selection are necessary for deep learning to establish an efficient health index, whose effect is shown in Fig. 7.16. In the training process, we set the length of window as $\tau = 100$.

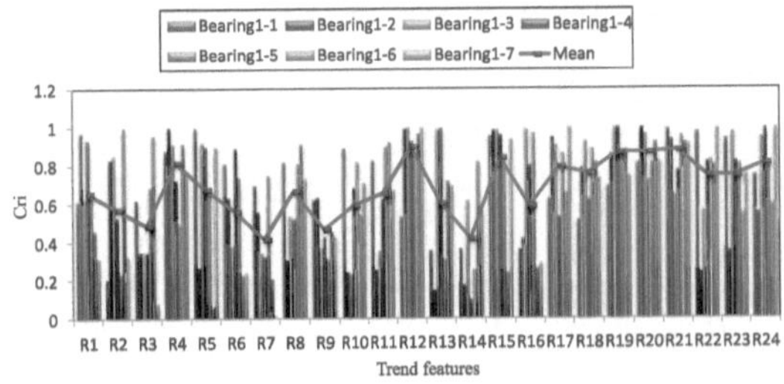

Figure 7.14: Cri of the extracted features in condition 1.

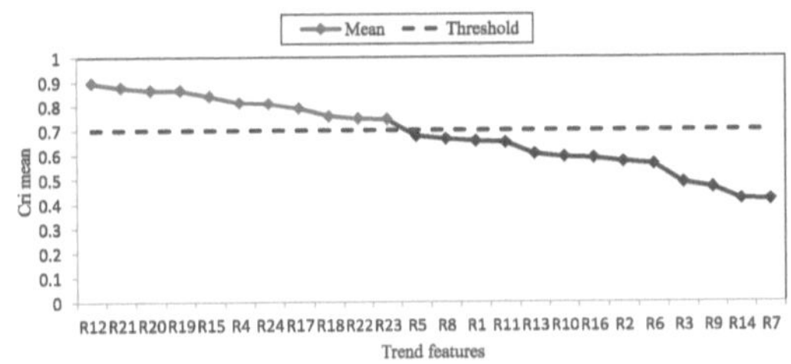

Figure 7.15: Feature selection with mean of Cri in condition 1.

In Fig. 7.16, the obtained health index is exhibited by inputting directly the selected 24 features into the gated recurrent unit network (named M1), from which we can see that the obvious fluctuations are found and the degradation trend is not obvious. In contrast, by preprocessing the basic features with BC-CEEMDAN-FC-GRU including current reconstruction and feature selection and feed in gated recurrent unit, it can be found that the values of the monotonic and time-correlated properties of health index are almost close to 1, and the degradation curves are given in Fig. 7.16.

After the available health index with BC-CEEMDAN-FC-GRU is constructed, the next task is to predict the remaining useful life. By observing the degradation trend of Fig. 7.16, it is found that the polynomial regression model tracks the tendency of rolling element bearings better. Therefore, a polynomial regression model based on a PF algorithm is constructed. The model can be described as the following formula (7.22):

$$H_I(t) = a_t t^3 + b_t t^2 + c_t t + d_t \tag{7.22}$$

where a_t, b_t, c_t, and d_t are the parameters to be determined, and t is the sample time.

The regression model (7.22) can be ascertained by employing the output data of BC-CEEMDAN-FC-GRU. The particle filtering algorithm is used to optimize the model parameters, and the health index is predicted until the failure threshold is breached. Actually, if the value of health index is equal to 0, then the bearing degradation reaches to failure threshold, and the corresponding time is the end of the life

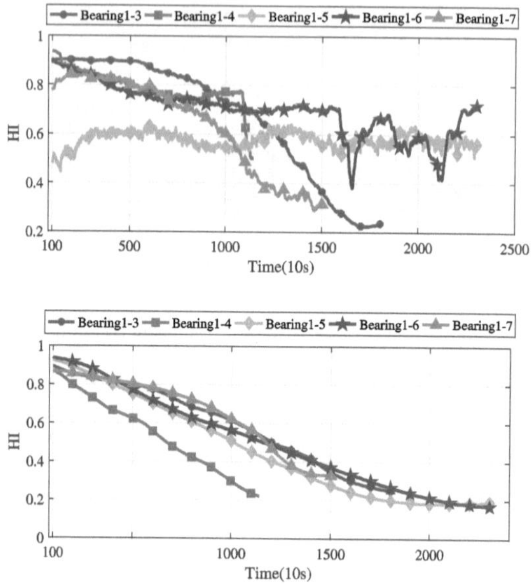

Figure 7.16: HIs of the test bearing: (a) The HI with M1 and (b) The HI with BC-CEEMDAN-FC and GRU.

(EOL). Fig. 7.17 shows remaining useful life of each particle at the current moment when it reaches the threshold and the median value and 95% confidence interval (CI) of bearing 1–4 and 1–7 are $3020s$ / $[2460s, 4600s]$ and $6330s$ / $[5760s, 804s]$, respectively. It can be observed from Fig. 7.17 that the BC-CEEMDAN-FC-GRU based health index has complement decreasing trend with little fluctuations. Meanwhile, the predicted remaining useful life with regression model (7.22) is very close to the real remaining useful life values.

As a representative data-driven method, machine learning has received more and more attention in the prognosis of remaining useful life for rolling element bearings. In [247], the time-frequency features have been extracted, where the multi-scale convolutional neural network has been employed to estimate the remaining useful life of bearings (named M2). In [254], some novel features have been proposed to construct the RNN-HI model (named M3). In [250], a wavelet packet-empirical mode decomposition method has been proposed to extract features, where the remaining useful life has been assessed by the self-organization mapping model (named M4). In [273], the remaining useful life estimation has been carried out with ϵ-support vector regression (named M5). To validate the advantages of the proposed approaches, the representative techniques of M1–M5 are applied. It can be seen that the proposed BC-CEEMDAN-FC-GRU method achieves the highest score with the lowest error. Obviously, the proposed method can efficiently improve the accuracy of remaining useful life and precisely track the degradation process of rolling element bearings.

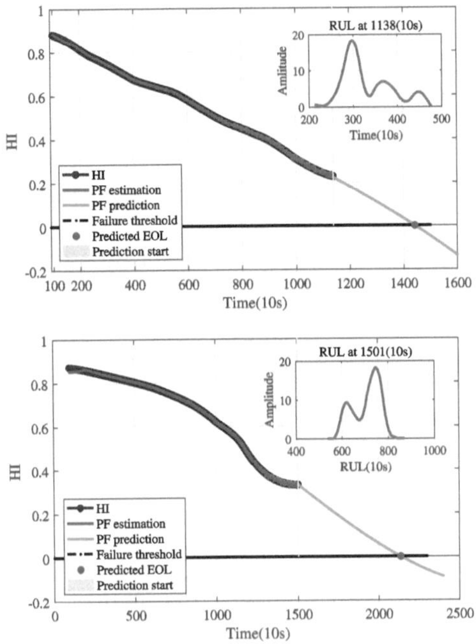

Figure 7.17: RUL prediction: (a) Bearing 1–3 and (b) Bearing 1–7.

7.3 Remaining Useful Life Prediction with PSR and Error Compensation

The capacity and voltage of the lithium-ion batteries, which can be used to quantify the aging precess of lithium-ion batteries and be utilized to determine the failure when that value transcends some usual standard, are typical HIs in battery health management [3,85,104,274]. In the following part of this chapter, the classical capacity and discharge voltage differences of equal time intervals are chosen as HIs. The ensemble empirical mode decomposition (EEMD) is adopted as the feature enhancement technique to reduce the influence. For the purpose of determining the maximum embedding dimension and delay time, the phase space reconstruction (PSR) method is first employed, and then, the optimal input sequence pattern of support vector regression can be obtained. The final remaining useful life prediction can be achieved with Error Compensation (EC). The framework of the proposed method is shown in Fig. 7.18.

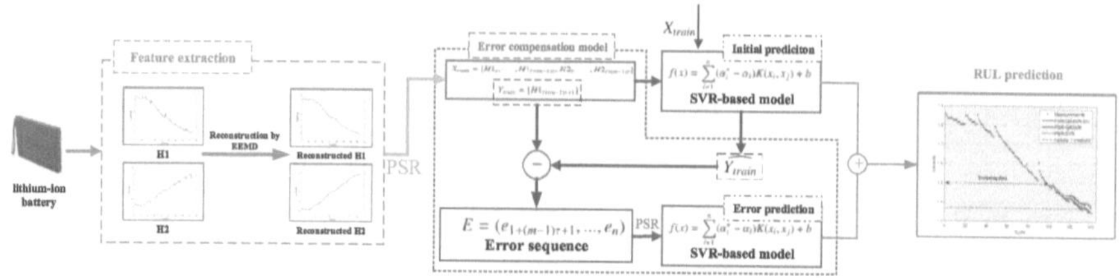

Figure 7.18: Flowchart of proposed approach.

7.3.1 Health Indicator Construction

The capacity is calculated by integrating discharge current over time and its definition derived from [215] is described as:

$$H1 = \int_{t_1}^{t_2} I dt \tag{7.23}$$

where t_1 and t_2 are the start and the end time of a discharge cycle, respectively, and I is the discharge current.

For lithium-ion batteries, voltage is one of the physical signals which is easy to measure, thus the exploration of voltage-related remaining useful life prediction will effectively avoid the error transfer in the conversion process from voltage to capacity, and it helps to obtain more accurate prediction results [275]. Especially, the increase of the charge and discharge times, will lead to persistent age of the lithium-ion batteries and continuous reduction of the capacity. As discussed in [276], for each cycle in the discharge phase, the evolution law of voltage difference will effectively reflect the degradation process of lithium-ion batteries within the same time interval. Therefore, discharging voltage difference (DVD) of the equal time interval for some cells can be chosen as the health indicator with the following form:

$$H_{2,i} = U_{i,t_2} - U_{i,t_1},\ i = 1, 2, 3..., n \tag{7.24}$$

where i means ith cycles, n represents the total number of cycles, t_1, t_2 are the sample points (times), and U_{i,t_1}, U_{i,t_2} are the corresponding discharging voltages. Based on (7.24), the HIs of discharging voltage difference are expressed as:

$$H2 = \{H_{2,1}, H_{2,2}, \cdots, H_{2,n}\} \tag{7.25}$$

In the case of the lithium-ion batteries data set from NASA Ames Prognostics Center of Excellence (PCE) [229], the evolution trend $H1, H2$ for batteries 05, 06, 07, and 18 can be reflected in Figure 7.22. It can be found that the capacity and the discharging voltage difference have different trends with the aging of lithium-ion batteries. Therefore, the information hidden in the battery degradation data can be better revealed by combining the two HIs.

Different from traditional filtering methods such as distributed filtering [277], recursive filtering [36], H_∞ filtering [278, 279] and Kalman filter [280], the empirical mode decomposition technique has an obvious advantage in reconstructing the degrade trends so that the influence of fluctuations caused by capacity regeneration and discharging voltage difference can be eliminated. Specifically, the main trend can be captured by decomposing the signals into a finite number of IMFs and a smooth trend component. Thus, the empirical mode decomposition technique is a suitable choice to process nonlinear and non-stationary data. As the development of empirical mode decomposition, ensemble empirical mode decomposition is a novel analysis method to deal with the unstable signals and mode aliasing phenomenon in the empirical mode decomposition process [88, 281], and the steps of data reconstruction are listed as follows:

Step 1. A Gaussian white noise sequence $v_z(k)$ is added to the raw data $\tilde{x}(k)$ and produce $x_z(k) = \tilde{x}(k) + v_z(k), z = 1, 2, 3...M.$, where $x_z(k)$ is the obtained sequences with added noise in ith trial.

Step 2. Each noised sequence $x_z(k)$ is decomposed into some IMFs component $c_{z,j}(k)$ and residual $r_z(k)$, that is $x_z(k) = \sum_{j=1}^{L} c_{z,j}(k) + r_z(k)$. The subscript j represents the jth IMFs component of $x_z(k)$, L is the maximum number of IMFs in each trial after empirical mode decomposition.

Step 3. Empirical mode decomposition will continue until the Mth trial is completed, and the empirical mode decomposition results are stored in each trial.

Step 4. After all trials are completed, the stored IMFs and residuals are used to calculate the ensemble means denoted $\bar{c}_j(k) = \sum_{z=1}^{M} c_{z,j}(k)/M, \bar{r}(k) = \sum_{z=1}^{M} r_z(k)/M$. where $\bar{c}_j(k)$ is ensemble means of $j - th$ IMF, and $\bar{r}(k)$ is ensemble means of residual.

Step 5. The result is represented by $\tilde{x}(k) = \sum_{j=1}^{L} \bar{c}_j(k) + \bar{r}(k)$, it is the ensemble means of each IMF component and residual.

Step 6. By discard the IMFs with the highest amplitude $\bar{c}_1(k)$, the raw data can be reconstructed as $x(k) = \sum_{j=2}^{L} \bar{c}_j(k) + \bar{r}(k)$.

7.3.2 Remaining Useful Life Prediction of Lithium-Ion Batteries

Based on the reconstructed data of HIs, the remaining useful life prediction can be carried out with phase space reconstruction, and the support vector regression and the error compensation are proposed.

Considering the advantage of the phase space reconstruction algorithm over identifying complex nonlinear data by transforming a one-dimensional time series into a high dimensional phase space, in this section, phase space reconstruction is used to establish the phase space of the nonlinear degradation process for lithium-ion batteries. In this algorithm, embedding dimension m and delay time τ are the two key parameters, and it has been suggested that delay time τ and embedding dimension

m are relevant through delay time window $t_w = (m-1)\tau$ [282]. For this reason, we use the C–C method based on the embedded window method in 1999 by Kim et al [282] to estimate delay time τ and embedded dimension m in this section.

Set the time series of reconstructed data as $\{x(i)\}$ $(i = 1, ..., N)$ and choose the following delay-coordinate means as the phase space reconstruction method.

$$X(i) = [x(i), x(i+\tau), \cdots, x(i+(m-1)\tau)] \tag{7.26}$$

where $i = 1, 2, ...M$, m is embedding dimensions, τ is delay times, $M = N - (m-1)\tau$ is the number of points in the phase space, and N is the length of time series. The reconstruction phase space can be described as

$$X = \begin{bmatrix} X_1 \\ X_2 \\ \vdots \\ X_M \end{bmatrix} = \begin{bmatrix} x(1) & x(1+\tau) & \cdots & x(1+(m-1)\tau) \\ x(2) & x(2+\tau) & \cdots & x(2+(m-1)\tau) \\ \vdots & \vdots & \cdots & \vdots \\ x(M) & x(M+\tau) & \cdots & x(M+(m-1)\tau) \end{bmatrix}$$

It can be seen from the previous experiments that the value of m and τ will affect the accuracy of forecast results of battery remaining useful life. Fortunately, C–C algorithm [282] is adopted to construct the statistics by using the correlation integral, then τ and $\bar{\tau} = (m-1)\tau$ can be obtained by employing the relation between the statistics and the time delay τ. At the same time, the embedded dimension m can be ascertained. More specifically, the correlation integral for the embedded time series is defined as

$$C(m, N, r, \tau) = \frac{2}{M(M-1)} \sum_{1 \le i < j \le M} \delta(r - \|X_i - X_j\|), r > 0 \tag{7.27}$$

where M is the number of phase point, r is the radius of neighborhood, and $\| * \|$ denotes the Euclidean distance between two points in a phase space, $\delta(x) = \begin{cases} 0, & x < 0 \\ 1, & x \ge 0 \end{cases}$.

If the time series data are divided into T disjoint sub-sequences, we have the following:

$$R(m, N, r, \tau) = \frac{1}{T} \sum_{s=1}^{T} \left[C_s(m, \frac{N}{T}, r, \tau) - C_s^m(m, \frac{N}{T}, r, \tau)\right] \tag{7.28}$$

Set $N \to \infty$, $R(m, N, r, \tau)$ can be written as

$$R(m, r, \tau) = \frac{1}{T} \sum_{s=1}^{T} \left[C_s(m, r, \tau) - C_s^m(m, r, \tau)\right] \tag{7.29}$$

Calculating the average of $R(m, r, t)$, we have

$$\bar{R}(m, r, \tau) = \frac{1}{MJ} \sum_{m=1}^{M} \sum_{j=1}^{J} R(m, r_j, \tau) \tag{7.30}$$

where J is the number of r. Select the minimum and maximum values of the radius r to define the difference:

$$\triangle R(m,\tau) = \max\{R(m,r_j,\tau)\} - \min\{R(m,r_j,\tau)\} \tag{7.31}$$

Similar to [282], the embedding dimensions are set as $m = 2,3,4,5$, the radius of neighborhood are chosen as set $r_j = i\sigma/2,\ i = 1,2,3,4$, where σ is the standard deviation of the time series, then the parameters can be calculated with the following equations:

$$\bar{R}_1(\tau) = \frac{1}{16}\sum_{m=2}^{5}\sum_{j=1}^{4} R(m,r_j,\tau) \tag{7.32}$$

$$\triangle\bar{R}(\tau) = \frac{1}{4}\sum_{m=2}^{5} \triangle R(m,\tau) \tag{7.33}$$

$$R_{cor}(\tau) = |\bar{R}_1(\tau)| + \triangle\bar{R}(\tau) \tag{7.34}$$

Since $\bar{R}_1(\tau)$ and $\triangle\bar{R}(\tau)$ reflect the autocorrelation properties of delay sequences, the optimal delay time τ_o can be obtained by seeking the value corresponding to the first zero crossing of $\bar{R}_1(\tau)$ or the first local minimum of $\triangle\bar{R}(\tau)$. With the help of $\bar{R}_1(\tau)$ and $\triangle\bar{R}(\tau)$, the best embedded window τ_ω can be found by searching the global minimum of $R_{cor}(\tau)$. Finally, the optimal embedding dimension m can be derived by using $\tau_w = (m-1)\tau_o$.

After the optimal pattern of input sequences are ascertained, next step is remaining useful life prediction. The support vector regression is used to effectively establish the nonlinear map relationship between the reconstructed data and remaining useful life. Especially, the support vector regression in this section is not only used to predict the remaining useful life of lithium-ion batteries but also employed to forecast the prediction error.

Consider a set of training samples: $\{(x_1,y_1),(x_2,y_3),\cdots,(x_N,y_N)\}$, where $x_i \in R(i = 1,2,\cdots,N)$ is the input samples and $y_i \in R$ is the target output. The main goal of the support vector regression is to establish a mapping relationship $f(x)$ based on the training data set, so that $f(x)$ is as close to the expected output y_i as possible. Similar to [63, 104], the support vector regression function is defined as:

$$f(x_i) = \omega^T x_i + b \tag{7.35}$$

where $f(x_i)$ represents the output value, x_i is the input value, and the coefficient $\omega \in R^n, b \in R$ is adjustable. Then the mapping (7.35) can be further transformed to the following optimization problem:

$$\min_{\omega,b,\xi_i,\xi_i^*} \{\frac{1}{2}\omega^T\omega + C\sum_{i=1}^{n}(\xi_i + \xi_i^*)\} \tag{7.36}$$

$$s.t. \begin{cases} f(x_i) - y_i \le \varepsilon + \xi_i\ , \\ y_i - f(x_i) \le \varepsilon + \xi_i^*\ , \\ \xi_i, \xi_i^* \ge 0,\ i = 1,2,3,...,n\ . \end{cases} \tag{7.37}$$

where $C > 0$ is a preset penalty coefficient, $\varepsilon > 0$ is the deviation between the trained output and the actual output, and ξ_i, ξ_i^* are slack variables. By introducing Lagrange multiplier α_i, α_i^*, we have the following Lagrangian function:

$$\begin{aligned} &L(\omega, b, \alpha_i, \alpha_i^*, \xi_i, \xi_i^*, \mu_i, \mu_i^*) \\ &= \frac{1}{2}\omega^T\omega + C\sum_{i=1}^{n}(\xi_i + \hat{\xi}_i) - \sum_{i=1}^{n}\mu_i\xi_i - \sum_{i=1}^{n}\mu_i^*\xi_i^* \\ &+ \sum_{i=1}^{n}\alpha_i(f(x_i) - y_i - \varepsilon - \xi_i) + \sum_{i=1}^{n}\alpha_i^*(f(x_i) - y_i - \varepsilon - \xi_i^*) \end{aligned} \tag{7.38}$$

Then, take the partial derivative and set it to zero, and we obtain the following support vector regression dual problem:

$$\begin{aligned} \max_{\alpha,\hat{\alpha}}\{&\sum_{i=1}^{n} y_i(\alpha_i^* - \alpha_i) - \varepsilon(\alpha_i^* - \alpha_i) \\ &- \frac{1}{2}\sum_{i=1}^{n}\sum_{j=1}^{n}(\alpha_i^* - \alpha_i)(\alpha_j^* - \alpha_j)x_i^T x_j\} \end{aligned} \tag{7.39}$$

$$s.t. \begin{cases} \sum_{i=1}^{n}(\alpha_i^* - \alpha_i) = 0\,, \\ 0 \le \alpha_i^*, \alpha_i \le C\,. \end{cases} \tag{7.40}$$

Similar to [227], the optimal solution with parameters ω, α_i, α_i^* can be written as:

$$f(x) = \sum_{i=1}^{n}(\alpha_i^* - \alpha_i)x_i^T x + b \tag{7.41}$$

Considering the nonlinear mapping features between x and y of lithium-ion batteries degradation, (7.41) can be translated into following equation with kernel function $K(x_i, x_j)$

$$f(x) = \sum_{i=1}^{n}(\alpha_i^* - \alpha_i)K(x_i, x_j) + b \tag{7.42}$$

Different kernel functions have different prediction effects, and it is found that the gaussian radial basis function kernel function $K(x_i, x_j) = exp(-\frac{\|x_i, x_j\|^2}{2\sigma^2})$ has a better performance in predicting the remaining useful life of lithium-ion batteries. Therefore, in this section, radial basis function is chosen as the kernel function to predict the remaining useful life.

In order to achieve more effective prognosis, the genetic algorithm is introduced to optimize the parameters of support vector regression. As an adaptive global optimization search method, genetic algorithm has some significant advantages such as strong global search ability, fast search speed, and high efficiency [193]. Accordingly, genetic algorithm is selected to optimize the parameters of support vector regression for remaining useful life prediction.

In order to fully exploit the hidden information of the battery data, We utilize capacity and discharging voltage difference as the HIs. Firstly, the embedding dimensions m_1, m_2 and delay times τ_1, τ_2 are calculated by the C–C method. Next, select $m = \max(m_1, m_2)$, $\tau = \max(\tau_1, \tau_2)$, and multi-variable phase space reconstruction of time series can be performed.

Denote the input $X_{train}^{(l)}$ and output $Y_{train}^{(l)}$ as

$$\begin{aligned} X_{train}^{(l)} &= [H1_l, \cdots, H1_{l+(m-1)\tau}, H2_l, \cdots, H2_{l+(m-1)\tau}] \\ Y_{train}^{(l)} &= H1_{l+(m-1)\tau+1}, \quad l = 1, \cdots, n-(m-1)\tau - 1 \end{aligned}$$

Then the training samples $(X_{train}^{(l)}, Y_{train}^{(l)})$ can be utilized to build the support vector regression prediction model with follows form:

$$Y_{train}^{(l)} = f(X_{train}^{(l)}) = \sum_{l=1}^{n-(m-1)\tau-1} (\alpha_l^* - \alpha_l) K(X_l, X_{train}^{(l)}) + b \tag{7.43}$$

Based on the used m data $X^{(n-(m-1)\tau)} = [H1_{n-(m-1)\tau}, \cdots, H1_n, H2_{n-(m-1)\tau}, \cdots, H2_n]$, the prediction relation at the next time can be established:

$$\begin{aligned} \widehat{H1}_{n+1} &= f(X^{(n-(m-1)\tau)}) \\ &= \sum_{l=1}^{n-(m-1)\tau-1} (\alpha_l^* - \alpha_l) K(X_l, X^{(n-(m-1)\tau)}) + b \end{aligned} \tag{7.44}$$

The degradation amplitude of lithium-ion batteries is relatively gentle, and few significant mutations occur [85, 104, 193]. Therefore, we forecast the error which is produced by support vector regression prediction in the training phase, and take the prediction of error as the compensation part. By combining error compensation and normal prediction, we achieve more accurate prediction results. The corresponding flowchart can be found in Fig. 7.18 and specific process can be shown through the following steps:

Step 1. **Initial prediction**: Initial prediction support vector regression model is established for preliminary prediction, and the training samples are fed to the trained SVR-based model to get the prediction of the training label.

$$\begin{aligned} \widehat{Y}_{train} &= f(X_{train}) \\ &= \sum_{l=1}^{n-(m-1)\tau-1} (\alpha_l^* - \alpha_l) K(X_l, X_{train}) + b \end{aligned} \tag{7.45}$$

Step 2. **Generation of forecast error**: Forecast errors is obtained by subtracting the known value from the preliminary prediction of training label:

$$E = [e_{1+(m-1)\tau}, \cdots, e_n] = Y_{train} - \widehat{Y}_{train} \tag{7.46}$$

Step 3. Error prediction: Embedding dimensions m_{er} and delay τ_{er} of error sequence can be achieved by C–C method, and the input $E_{train}^{(P)}$ and output $Lab_{train}^{(p)}$ of the error compensation model is defined as:

$$\begin{aligned} E_{train}^{(P)} &= [e_p, \cdots, e_{p+(m_{er}-1)\tau_{er}}] & (7.47) \\ Lab_{train}^{(p)} &= e_{p+(m_{er}-1)\tau_{er}+1} & (7.48) \\ p &= 1+(m_{er}-1)\tau+1, \cdots, n-(m_{er}-1)\tau_{er}-1 \end{aligned}$$

Thus, training samples $(E_{train}^{(P)}, Lab_{train}^{(p)})$ is utilized to build error compensation model based on support vector regression that can be expressed as:

$$Lab_{train}^{(p)} = f(E_{train}^{(P)}) = \sum_{p=1+(m-1)\tau+1}^{n-(m_{er}-1)\tau_{er}-1} (\alpha_p^* - \alpha_p)K(E_p, E_{train}^{(p)}) + b \tag{7.49}$$

Relying on the past m_{er} set of error $E^{(n-(m_{er}-1)\tau_{er})} = [e_{n-(m_{er}-1)\tau_{er}}, \cdots, e_n]$, the prediction of error at next time can be established as:

$$\begin{aligned} \widehat{E}_{n+1} &= f(E^{(n-(m_{er}-1)\tau_{er})}) \\ &= \sum_{p=1+(m-1)\tau+1}^{n-(m_{er}-1)\tau_{er}-1} (\alpha_p^* - \alpha_p)K(E_p, E^{(n-(m_{er}-1)\tau_{er})}) + b & (7.50) \end{aligned}$$

Step 4. Final remaining useful life with error compensation: The final prediction is obtained through summarizing the initial prediction and prediction of error sequence, and it is realized by the following formula:

$$\widehat{H1}_{final} = \widehat{H1} + \widehat{E} \tag{7.51}$$

Based on the above description, we can obtain the complete remaining useful life prediction process based on the optimal input sequence and the error compensation. The validity of the above method will be verified by using the lithium-ion batteries data set in the following section.

7.3.3 Illustrative Examples

A. LIB Data Description

The LIB degradation data used in this section was derived from the Prognostics Center of Excellence (PCE) [229]. The battery runs through three different operational profiles (charge, discharge, and impedance) at room temperature, and it is first charged in 1.5 A constant current mode until the voltage reaches 4.2 V. Then, it continues to charge in constant-voltage mode until the charging current drops to 20 mA. During the discharge phase, the battery is discharged at a constant current of 2 A until the battery voltage drops to 2.7 V, 2.5 V, and 2.2 V, respectively. In general, the failure threshold of the battery capacity is defined as about 70–80% of the rated capacity. In this simulation, in order to facilitate the analysis and compare with other approach, the failure thresholds of batteries 05, 06, 07, and 18 are defined as 74%, 65%, 78%, and 74% of the rated health indicator, respectively.

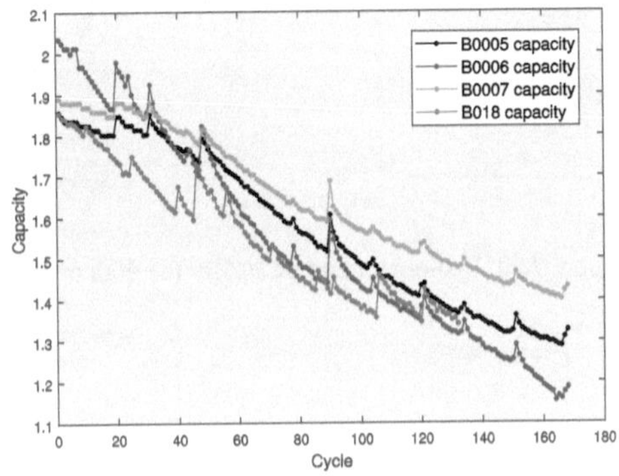

Figure 7.19: Evolution trend of $H1$.

B. Feature Extraction

As stated above, the capacity and discharging voltage difference are chosen as H1 and H2, respectively. In particular, reasonable time interval selection for discharging voltage difference affects the effect of remaining useful life prediction. Because capacity is a widely adopted health indicator, we ascertain the time interval by comparing the Pearson correlation between discharging voltage difference (H2) and capacity (H1). Especially, the HIs decay curves of capacity and discharging voltage difference for batteries 05, 06, 07, and 18 are shown in Figs. 7.19 and 7.20.

Affected by various interferences (i.e. measurement error, electromagnetic interference, and complex chemical reactions inside the battery), the degraded curve of lithium-ion batteries demonstrates some phenomenons including capacity regeneration and fluctuation, which will significantly reduce the prediction accuracy of remaining useful life. In order to reduce the impact of the above factors, this section adopts ensemble empirical mode decomposition to separate the frequency components in the original series, and the decomposition results based on battery 05 are shown in Fig. 7.21. The decomposition results reveal that the original series is decomposed into 7 components and are denoted as $IMF_1, IMF_2, IMF_3, IMF_4, IMF_5, IMF_6$, and residual, respectively. Considering the fact that each component represents different natural oscillatory modes in the original series, this section abandons the IMF_1 considering the highest vibration frequency. The reconstructed health indicator can be obtained by formula (7.52), and the corresponding curves can be found in Fig. 7.22:

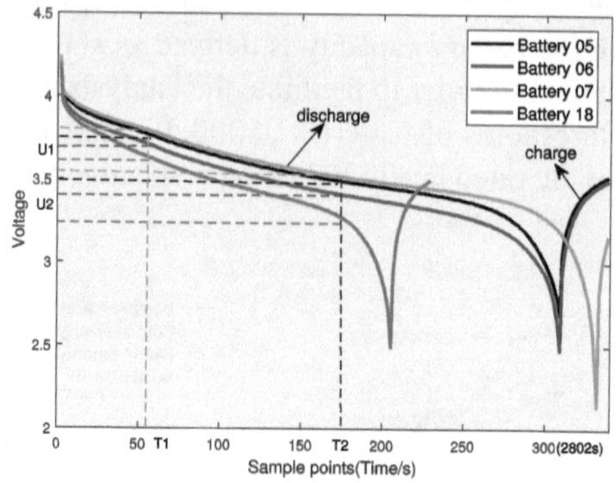

Figure 7.20: Evolution trend of $H2$ for the 90th cycles.

$$HI_r = \sum_{i=2}^{6} IMF_i + r_{re} \tag{7.52}$$

where IMF_i is the decomposed component, r_{re} is the decomposed residual, and HI_r is the reconstructed health indicator.

C. Evaluation Criteria

Similar to [85, 104, 193], three classic evaluation criteria are taken to evaluate the prediction performance:

(1) Root mean square error (RMSE)

$$\text{RMSE} = \sqrt{\frac{1}{N}\sum_{k}^{N}(y_k - f(x_k))} \tag{7.53}$$

(2) Mean absolute percent error (MAPE)

$$\text{MAPE} = \frac{1}{N}\sum_{k}^{N}\left|\frac{y_k - f(x_k)}{y_k}\right| \tag{7.54}$$

(3) Accuracy error (AE)

$$\text{AE} = |RUL_{prediction} - RUL_{ture}| \tag{7.55}$$

where N is the data length, y_k is the real data at kth cycle, and $f(x_k)$ is the predicted data at kth cycle. The smaller values of the three criteria represent the higher prediction accuracy.

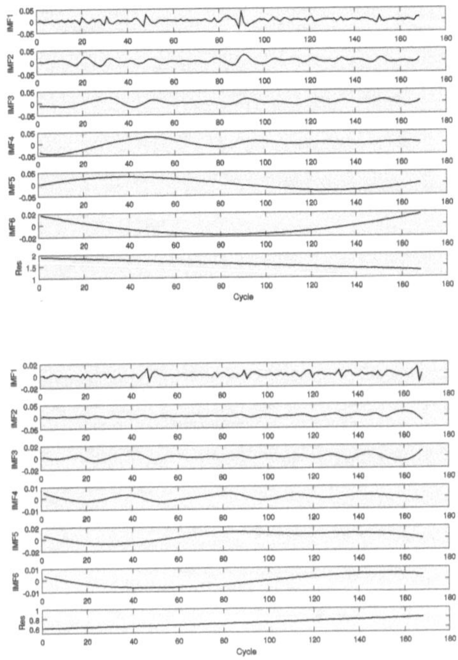

Figure 7.21: (a) Decomposed series of original H1 and (b) Decomposed series of original H2.

D. Simulation Result Analysis

To verify the advantages of the proposed phase space reconstruction with C–C method, several sliding windows and delay examples are presented. In this experiment, 1-60 cycles are used as the training set and 61-100 cycles as the validation set for batteries 05, 06, 07, and 1-40 cycles are used as the training set and 61-80 cycles as validation set for battery 18. To be specific, embedding dimensions $m = 3$ and delay times $\tau = 6$ can be obtained by using the optimal input sequence selection pattern with C–C method. Conclusively, PSR-SVR can not only greatly improves the prediction accuracy, but also reduces the dependence on prior knowledge or attempts when defining the training set.

Batteries 05, 06, 07, and 18 are used to verify the prediction performance of the hybrid PSR-GASVR-EC method. For the selected predictive performance criteria RMSE, MAPE, and AE, it can be seen from Fig. 7.23 and Tables 7.4, 7.5 that the hybrid PSR-GASVR-EC method owned higher precision than PSR-GASVR, PSR-SVR, and other classic methods such as combination Gaussian process functional regression with a linear mean function [233], multiscale Gaussian process regression with squared exponential (SE-MGPR) [234], rest time-based prognostic framework (RTPF) [235], and the support vector regression with particle swarm optimization

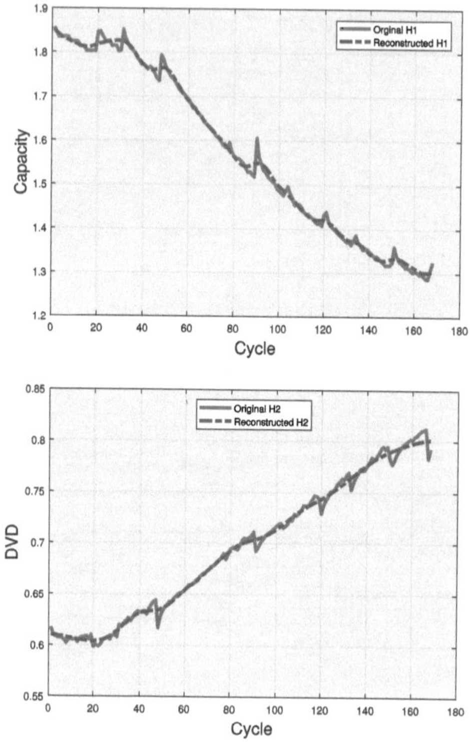

Figure 7.22: (a) Reconstructed H1 and (b) Reconstructed H2.

(PSO-SVR) [236]. Based on a widespread fact, the predictive performance of machine learning depends on an abundance of high-quality data. In this section, different training sample lengths, including 1-80 cycles, 1-90 cycles, and 1-100 cycles for batteries 05, 06, 07, and 1-60 cycles, 1-70 cycles, and 1-80 cycles for battery 18 are adopted, the experimental results verify the fact that more sufficient data will lead to higher prediction accuracy.

Table 7.4: Comparison of prediction performance with three different algorithms.

Method	Battery 05			Battery 06			Battery 07			Battery 08		
	RMSE	MAPE (%)	AE	RMSE	MAPE (%)	AE	RMSE	MAPE (%)	AE	RMSE	MAPE (%)	AE
PSR-SVR	0.1402	1.5424	17	0.1908	0.0634	9	0.1361	1.3924	18	0.1	1.7513	6
PSR-GASVR	0.1166	1.2967	7	0.0634	0.827	4	0.0927	0.9532	7	0.0765	1.161	3
PSR-GASVR-EC	0.0191	0.2126	0	0.0081	0.0956	1	0.0101	1.048	1	0.0051	0.2722	0

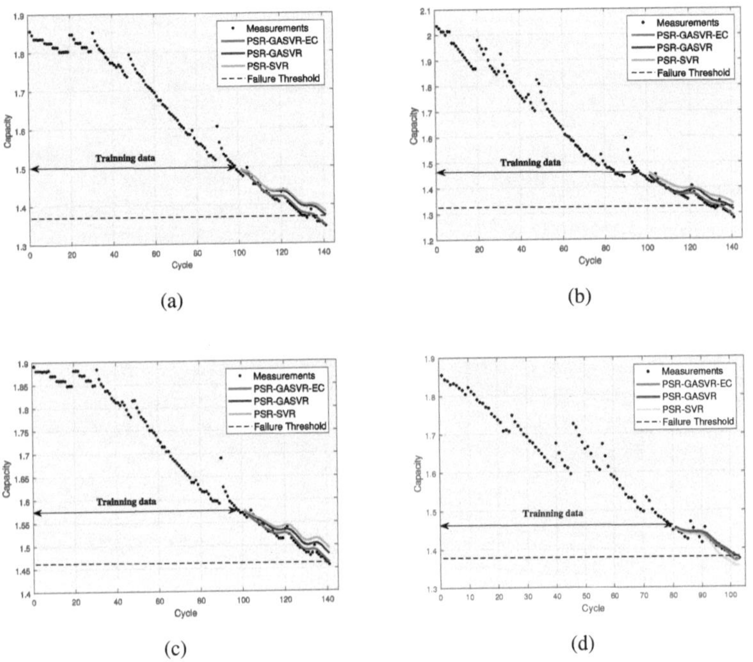

Figure 7.23: (a) Prediction for battery 05, (b) Prediction for battery 06, (c) Prediction for battery 07, and (d) Prediction for battery 18.

Table 7.5: Comparison of prediction performance with the existed results.

Method	Battery 05		Battery 06		Battery 07	
	RMSE	MAPE (%)	RMSE	MAPE (%)	RMSE	MAPE (%)
CLGPFR [233]	1.36	1.6	6.86	10.2	1.73	1.7
SE-MGPR [234]	1.2	1.38	2.11	2.93	1.07	1.02
RTPF [235]	0.68	0.76	0.93	1.25	0.43	0.44
PSO-SVR [236]	0.75	0.82	0.93	1.25	0.97	1.02

7.4 Conclusion

In this chapter, remaining useful life prognostics methods that rely on degenerate point detection and degenerate trend estimation are proposed. After the degenerate point of industrial equipment is determined, the remaining process is degenerate trend estimation which includes two following parts: 1) health indicator construction; 2) remaining useful life prediction. NASA data set of turbofan engines, and PCE battery data are used for verify the effectiveness of those proposed approaches.

8

Conclusions and Future Topics

This book has been focused on filter-based fault diagnosis and remaining useful life prediction. On the one hand, from the perspective of system analysis, for networked multi-rate sampling systems subject to the influence of signal quantization, data missing, channel fading and event-triggered mechanism, some typical safety analysis topics have been investigated under the filter-based framework, that include state estimation, fault detection, fault location and fault estimation. In our obtained theoretical results, some interesting implicit relationships between network-induced factors and diagnostic performance have been constructed. On the other hand, with the fusion framework of data and model, the fault diagnosis and remaining useful life prediction of electromechanical equipment have been achieved by means of the filter technique, where the high performance accuracy have be obtained.

On the whole, the filter-based frameworks have been established in this book for studying fault diagnosis and remaining useful life prediction problems, which involve the latest system analysis theory and machine learning methods. It should be stressed that the established results can be regarded as the theoretical basis of the filter-based fault diagnosis and remaining useful life prediction. Some of the related topics for the future research are listed as follows:

1) Other network-induced complexities: With the development of networked systems, some new network-induced complexities appear constantly, which will affect fault diagnosis performance of state monitoring, fault detection, fault isolation and fault estimation. The latest network-induced complexities involve channel fadings, event-triggered mechanisms, cyber attacks, constraints of bit rate allocation protocol, intermittent measurement outliers and constraints of encoding–decoding mechanisms.

2) Fault-tolerant control: When the system degrades or a fault appears, it is of great significance to design fault-tolerant controllers to ensure the normal operation of industrial systems, that consists of an important branch of our future topics. In these topics, special attention should be paid to the sufficient use of system state information and conduct fault detection, fault isolation and fault estimation comprehensively, and then evaluate the influence of the aforementioned network-induced complexities, which is important to design an efficient fault-tolerant controller.

3) Privacy preserving: Machine learning has become the core technology of data-driven fault diagnosis and remaining useful life prediction. Various privacy

DOI: 10.1201/9781003330998-8

threats of machine learning and the corresponding defense mechanisms have attracted more and more attention from academia and industry. Our future research will focus on the privacy protection of typical machine learning methods, such as semi-supervised, unsupervised and reinforcement learning approaches.

4) Interpretability: The interactive mechanism between the expert experience of industrial production and machine learning models should be established in conjunction with the existing safety analysis methods. Interpretable fault diagnosis and remaining useful life prediction include explainable feature extraction, explainable machine learning architecture and explainable result evaluation, which deserve our more attention.

Bibliography

[1] S. X. Ding, Model-based Fault Diagnosis Techniques, Springer Berlin Heidelberg, 2008. doi:10.1007/978-3-540-76304-8.

[2] Y. Ju, X. Tian, H. Liu, L. Ma, Fault detection of networked dynamical systems: a survey of trends and techniques, International Journal of Systems Science 52 (16) (2021) 3390–3409. doi:10.1080/00207721.2021.1998722.

[3] L. Liao, F. Köttig, Review of hybrid prognostics approaches for remaining useful life prediction of engineered systems, and an application to battery life prediction, IEEE Transactions on Reliability 63 (1) (2014) 191–207. doi:10.1109/TR.2014.2299152.

[4] Y. Lei, N. Li, L. Guo, N. Li, T. Yan, J. Lin, Machinery health prognostics: A systematic review from data acquisition to RUL prediction, Mechanical Systems and Signal Processing 104 (2018) 799–834. doi:10.1016/j.ymssp.2017.11.016.

[5] D. Zhou, Y. Zhao, Z. Wang, X. He, M. Gao, Review on diagnosis techniques for intermittent faults in dynamic systems, IEEE Transactions on Industrial Electronics 67 (3) (2020) 2337–2347. doi:10.1109/TIE.2019.2907500.

[6] D. Cardoso, L. Ferreira, Application of predictive maintenance concepts using artificial intelligence tools, Applied Sciences 11 (1) (2020) 18. doi:10.3390/app11010018.

[7] X. Zhang, M. Polycarpou, T. Parisini, A robust detection and isolation scheme for abrupt and incipient faults in nonlinear systems, IEEE Transactions on Automatic Control 47 (4) (2002) 576–593. doi:10.1109/9.995036.

[8] I. Hwang, S. Kim, Y. Kim, C. E. Seah, A survey of fault detection, isolation, and reconfiguration methods, IEEE Transactions on Control Systems Technology 18 (3) (2010) 636–653. doi:10.1109/TCST.2009.2026285.

[9] L. Liu, L. Ma, J. Zhang, Y. Bo, Distributed non-fragile set-membership filtering for nonlinear systems under fading channels and bias injection attacks, International Journal of Systems Science 52 (6) (2021) 1192–1205. doi:10.1080/00207721.2021.1872118.

[10] J. Mao, Y. Sun, X. Yi, H. Liu, D. Ding, Recursive filtering of networked nonlinear systems: a survey, International Journal of Systems Science 52 (4) (2021) 1110–1128. doi:10.1080/00207721.2020.1868615.

[11] H. Geng, H. Liu, L. Ma, X. Yi, Multi-sensor filtering fusion meets censored measurements under a constrained network environment: advances, challenges and prospects, International Journal of Systems Science 52 (16) (2021) 3410–3436. doi:10.1080/00207721.2021.2005178.

[12] Y. Zhang, Z. Liu, H. Fang, H. Chen, $H\infty$ fault detection for nonlinear networked systems with multiple channels data transmission pattern, Information Sciences 221 (2013) 534–543. doi:10.1016/j.ins.2012.09.026.

[13] Y. Liu, Z. Wang, X. He, D. Zhou, A class of observer-based fault diagnosis schemes under closed-loop control: performance evaluation and improvement, IET Control Theory & Applications 11 (3) (2017) 135–141. doi:10.1049/iet-cta.2016.0504.

[14] M. Zhong, S. X. Ding, D. Zhou, A new scheme of fault detection for linear discrete time-varying systems, IEEE Transactions on Automatic Control 61 (9) (2016) 2597–2602. doi:10.1109/tac.2015.2497899.

[15] Q. Zhang, M. Basseville, Statistical detection and isolation of additive faults in linear time-varying systems, Automatica 50 (10) (2014) 2527–2538. doi:10.1016/j.automatica.2014.09.004.

[16] Y. Wan, W. Dong, H. Ye, Integrated trade-off design of fault detection system for linear discrete time-varying systems, IET Control Theory & Applications 7 (3) (2013) 455–463. doi:10.1049/iet-cta.2011.0727.

[17] L. Ma, Z. Wang, H.-K. Lam, N. Kyriakoulis, Distributed event-based set-membership filtering for a class of nonlinear systems with sensor saturations over sensor networks, IEEE Transactions on Cybernetics 47 (11) (2017) 3772–3783. doi:10.1109/tcyb.2016.2582081.

[18] Y. Zhang, Z. Wang, L. Ma, F. E. Alsaadi, Annulus-event-based fault detection, isolation and estimation for multirate time-varying systems: Applications to a three-tank system, Journal of Process Control 75 (2019) 48–58. doi:10.1016/j.jprocont.2018.12.005.

[19] Y. Zhang, Z. Wang, F. E. Alsaadi, Detection of intermittent faults for nonuniformly sampled multi-rate systems with dynamic quantisation and missing measurements, International Journal of Control 93 (4) (2018) 898–909. doi:10.1080/00207179.2018.1487083.

[20] X. Wan, T. Han, J. An, M. Wu, Hidden Markov model based fault detection for networked singularly perturbed systems, IEEE Transactions on Systems, Man, and Cybernetics: Systems 51 (10) (2021) 6445–6456. doi:10.1109/tsmc.2019.2961978.

[21] J. L. Wang, G.-H. Yang, J. Liu, An LMI approach to H_- index and mixed H_-/H_∞ fault detection observer design, Automatica 43 (9) (2007) 1656–1665. doi:https://doi.org/10.1016/j.automatica.2007.02.019.

[22] J. Xu, J. Wang, I. Izadi, T. Chen, Performance assessment and design for univariate alarm systems based on far, mar, and aad, IEEE Transactions on Automation Science and Engineering 9 (2) (2012) 296–307. doi:10.1109/TASE.2011.2176490.

[23] S. X. Ding, B. Shen, Z. Wang, M. Zhong, A fault detection scheme for linear discrete-time systems with an integrated online performance evaluation, International Journal of Control 87 (12) (2014) 2511–2521. doi:10.1080/00207179.2014.930183.

[24] Y. Q. Wang, H. Ye, G. Z. Wang, Fault detection of ncs based on eigendecomposition, adaptive evaluation and adaptive threshold, International Journal of Control 80 (12) (2007) 1903–1911. doi:10.1080/00207170701474167.

[25] Z. Yong, H. Fang, Y. Zheng, X. Li, Torus-event-based fault diagnosis for stochastic multirate time-varying systems with constrained fault, IEEE Transactions on Cybernetics 50 (6) (2020) 2803–2813. doi:10.1109/TCYB.2019.2895238.

[26] R. N. Clark, Instrument fault detection, IEEE Transactions on Aerospace and Electronic Systems AES-14 (3) (1978) 456–465. doi:10.1109/taes.1978.308607.

[27] P. M. Frank, Fault diagnosis in dynamic systems using analytical and knowledge-based redundancy: A survey and some new results, Automatica 26 (3) (1990) 459–474. doi:https://doi.org/10.1016/0005-1098(90)90018-D.

[28] X. Zhang, T. Parisini, M. Polycarpou, Sensor bias fault isolation in a class of nonlinear systems, IEEE Transactions on Automatic Control 50 (3) (2005) 370–376. doi:10.1109/TAC.2005.843875.

[29] X. He, Z. Wang, Y. Liu, D. H. Zhou, Least-squares fault detection and diagnosis for networked sensing systems using a direct state estimation approach, IEEE Transactions on Industrial Informatics 9 (3) (2013) 1670–1679. doi:10.1109/TII.2013.2251891.

[30] H. Dong, Z. Wang, S. X. Ding, H. Gao, Finite-horizon estimation of randomly occurring faults for a class of nonlinear time-varying systems, Automatica 50 (12) (2014) 3182–3189. doi:https://doi.org/10.1016/j.automatica.2014.10.026.

[31] J. Hu, Z. Wang, H. Gao, Joint state and fault estimation for time-varying nonlinear systems with randomly occurring faults and sensor saturations, Automatica 97 (2018) 150–160. doi:https://doi.org/10.1016/j.automatica.2018.07.027.

[32] K. Zhang, B. Jiang, P. Shi, Fault estimation observer design for discrete-time takagi–sugeno fuzzy systems based on piecewise lyapunov functions, IEEE Transactions on Fuzzy Systems 20 (1) (2012) 192–200. doi:10.1109/TFUZZ.2011.2168961.

[33] B. Ding, H. Fang, International Journal of Robust and Nonlinear Control 28 (4) (2017) 1165–1181. doi:10.1002/rnc.3925.
URL https://doi.org/10.1002%2Frnc.3925

[34] R. Yan, X. He, Z. Wang, D. H. Zhou, Detection, isolation and diagnosability analysis of intermittent faults in stochastic systems, International Journal of Control 91 (2) (2017) 480–494. doi:10.1080/00207179.2017.1286039.

[35] M. Zhong, T. Xue, S. X. Ding, A survey on model-based fault diagnosis for linear discrete time-varying systems, Neurocomputing 306 (2018) 51–60. doi:https://doi.org/10.1016/j.neucom.2018.04.037.

[36] L. Zou, Z. Wang, Q.-L. Han, D. Zhou, Recursive filtering for time-varying systems with random access protocol, IEEE Transactions on Automatic Control (2018) 1–1doi:10.1109/tac.2018.2833154.

[37] M. Gao, S. Yang, L. Sheng, D. Zhou, Fault diagnosis for time-varying systems with multiplicative noises over sensor networks subject to round-robin protocol, Neurocomputing 346 (2019) 65–72. doi:https://doi.org/10.1016/j.neucom.2018.08.087.

[38] H. Wang, G.-H. Yang, D. Ye, Fault detection and isolation for affine fuzzy systems with sensor faults, IEEE Transactions on Fuzzy Systems 24 (5) (2016) 1058–1071. doi:10.1109/tfuzz.2015.2501414.

[39] L. Li, M. Chadli, S. X. Ding, J. Qiu, Y. Yang, Diagnostic observer design for t–s fuzzy systems: Application to real-time-weighted fault-detection approach, IEEE Transactions on Fuzzy Systems 26 (2) (2018) 805–816. doi:10.1109/TFUZZ.2017.2690627.

[40] Y. Zhang, Z. Liu, B. Wang, Robust fault detection for nonlinear networked systems with stochastic interval delay characteristics, Isa Transactions 50 (4) (2011) 521–528. doi:10.1016/j.isatra.2011.07.001.

[41] S. Wang, X. Tian, H. Fang, Event-based state and fault estimation for nonlinear systems with logarithmic quantization and missing measurements, Journal of the Franklin Institute 356 (7) (2019) 4076–4096. doi:10.1016/j.jfranklin.2018.11.044.

[42] Y. Yuan, X. Tang, W. Zhou, W. Pan, X. Li, H.-T. Zhang, H. Ding, J. Goncalves, Data driven discovery of cyber physical systems, Nature Communications 10 (1). doi:10.1038/s41467-019-12490-1.

[43] S. Yin, S. X. Ding, A. Haghani, H. Hao, P. Zhang, A comparison study of basic data-driven fault diagnosis and process monitoring methods on the benchmark Tennessee Eastman process, Journal of Process Control 22 (9) (2012) 1567–1581. doi:10.1016/j.jprocont.2012.06.009.

[44] W. Yu, C. Zhao, Online fault diagnosis in industrial processes using multimodel exponential discriminant analysis algorithm, IEEE Transactions on Control Systems Technology 27 (3) (2019) 1317–1325. doi:10.1109/TCST.2017.2789188.

[45] J. Huang, X. Yan, Dynamic process fault detection and diagnosis based on dynamic principal component analysis, dynamic independent component analysis and bayesian inference, Chemometrics and Intelligent Laboratory Systems 148 (2015) 115–127. doi:10.1016/j.chemolab.2015.09.010.

[46] K. Peng, K. Zhang, B. You, J. Dong, Z. Wang, A quality-based nonlinear fault diagnosis framework focusing on industrial multimode batch processes, IEEE Transactions on Industrial Electronics 63 (4) (2016) 2615–2624. doi:10.1109/TIE.2016.2520906.

[47] L. Zhou, Y. Wang, Z. Ge, Z. Song, Multirate factor analysis models for fault detection in multirate processes, IEEE Transactions on Industrial Informatics 15 (7) (2019) 4076–4085. doi:10.1109/TII.2018.2889750.

[48] Y. Zheng, S. Mao, S. Liu, D. S.-H. Wong, Y.-W. Wang, Normalized relative RBC-based minimum risk bayesian decision approach for fault diagnosis of industrial process, IEEE Transactions on Industrial Electronics 63 (12) (2016) 7723–7732. doi:10.1109/tie.2016.2591902.

[49] Y. Yuan, G. Ma, C. Cheng, B. Zhou, H. Zhao, H.-T. Zhang, H. Ding, A general end-to-end diagnosis framework for manufacturing systems, National Science Review 7 (2) (2019) 418–429. doi:10.1093/nsr/nwz190.

[50] C. Li, Z. Liu, Y. Zhang, L. Chai, B. Xu, Diagnosis and location of the open-circuit fault in modular multilevel converters: An improved machine learning method, Neurocomputing 331 (2019) 58–66. doi:10.1016/j.neucom.2018.09.041.

[51] Y. Zhao, T. Li, X. Zhang, C. Zhang, Artificial intelligence-based fault detection and diagnosis methods for building energy systems: Advantages, challenges and the future, Renewable and Sustainable Energy Reviews 109 (2019) 85–101. doi:10.1016/j.rser.2019.04.021.

[52] M. M. M. Islam, J.-M. Kim, Reliable multiple combined fault diagnosis of bearings using heterogeneous feature models and multiclass support vector machines, Reliability Engineering & System Safety 184 (2018) 55–66. doi:10.1016/j.ress.2018.02.012.

[53] F. Lu, J. Wu, J. Huang, X. Qiu, Restricted-Boltzmann-based extreme learning machine for gas path fault diagnosis of turbofan engine, IEEE Transactions on Industrial Informatics 16 (2) (2020) 959–968. doi:10.1109/TII.2019.2921032.

[54] H. Darong, K. Lanyan, C. Xiaoyan, Z. Ling, M. Bo, Fault diagnosis for the motor drive system of urban transit based on improved

hidden Markov model, Microelectronics Reliability 82 (2018) 179–189. doi:https://doi.org/10.1016/j.microrel.2018.01.017.

[55] Y. Qin, X. Wang, J. Zou, The optimized deep belief networks with improved logistic sigmoid units and their application in fault diagnosis for planetary gearboxes of wind turbines, IEEE Transactions on Industrial Electronics 66 (5) (2019) 3814–3824. doi:10.1109/TIE.2018.2856205.

[56] L. Ke, Y. Zhang, B. Yang, Z. Luo, Z. Liu, Fault diagnosis with synchrosqueezing transform and optimized deep convolutional neural network: An application in modular multilevel converters, Neurocomputing 430 (7) (2020) 24–33. doi:10.1016/j.neucom.2020.11.037.

[57] K. Yan, J. Su, J. Huang, Y. Mo, Chiller fault diagnosis based on vae-enabled generative adversarial networks, IEEE Transactions on Automation Science and Engineering 19 (1) (2022) 387–395. doi:10.1109/TASE.2020.3035620.

[58] S. Hao, F.-X. Ge, Y. Li, J. Jiang, Multisensor bearing fault diagnosis based on one-dimensional convolutional long short-term memory networks, Measurement 159 (2020) 107802. doi:https://doi.org/10.1016/j.measurement.2020.107802.

[59] P. Peng, W. Zhang, Y. Zhang, Y. Xu, H. Wang, H. Zhang, Cost sensitive active learning using bidirectional gated recurrent neural networks for imbalanced fault diagnosis, Neurocomputing 407 (2020) 232–245. doi:https://doi.org/10.1016/j.neucom.2020.04.075.

[60] B. Yang, Y. Lei, F. Jia, N. Li, Z. Du, A polynomial kernel induced distance metric to improve deep transfer learning for fault diagnosis of machines, IEEE Transactions on Industrial Electronics 67 (11) (2020) 9747–9757. doi:10.1109/TIE.2019.2953010.

[61] L. Xie, D. Pi, X. Zhang, J. Chen, Y. Luo, W. Yu, Graph neural network approach for anomaly detection, Measurement 180 (1) (2021) 109546. doi:10.1016/j.measurement.2021.109546.

[62] Q. Shi, H. Zhang, Fault diagnosis of an autonomous vehicle with an improved svm algorithm subject to unbalanced datasets, IEEE Transactions on Industrial Electronics 68 (7) (2021) 6248–6256. doi:10.1109/TIE.2020.2994868.

[63] Z. Xue, Y. Zhang, C. Cheng, G. Ma, Remaining useful life prediction of lithium-ion batteries with adaptive unscented kalman filter and optimized support vector regression, Neurocomputing 376 (2020) 95–102. doi:10.1016/j.neucom.2019.09.074.

[64] M. Yan, X. Wang, B. Wang, M. Chang, I. Muhammad, Bearing remaining useful life prediction using support vector machine and hybrid degradation tracking model, ISA Transactions 98 (2020) 471–482. doi:10.1016/j.isatra.2019.08.058.

[65] C. He, T. Wu, R. Gu, Z. Jin, R. Ma, H. Qu, Rolling bearing fault diagnosis based on composite multiscale permutation entropy and reverse cognitive fruit fly optimization algorithm – extreme learning machine, Measurement 173 (2021) 108636. doi:https://doi.org/10.1016/j.measurement.2020.108636.

[66] K. Javed, R. Gouriveau, N. Zerhouni, A new multivariate approach for prognostics based on extreme learning machine and fuzzy clustering, IEEE Transactions on Cybernetics 45 (12) (2015) 2626–2639. doi:10.1109/tcyb.2014.2378056.

[67] Z. Pan, Z. Meng, Z. Chen, W. Gao, Y. Shi, A two-stage method based on extreme learning machine for predicting the remaining useful life of rolling-element bearings, Mechanical Systems and Signal Processing 144 (2020) 106899. doi:10.1016/j.ymssp.2020.106899.

[68] J. Liu, Q. Li, W. Chen, T. Cao, A discrete hidden Markov model fault diagnosis strategy based on k-means clustering dedicated to pem fuel cell systems of tramways, International Journal of Hydrogen Energy 43 (27) (2018) 12428–12441. doi:https://doi.org/10.1016/j.ijhydene.2018.04.163.

[69] W. Li, T. Liu, Time varying and condition adaptive hidden Markov model for tool wear state estimation and remaining useful life prediction in micro-milling, Mechanical Systems and Signal Processing 131 (2019) 689–702. doi:10.1016/j.ymssp.2019.06.021.

[70] T. Liu, K. Zhu, A switching hidden semi-Markov model for degradation process and its application to time-varying tool wear monitoring, IEEE Transactions on Industrial Informatics 17 (4) (2021) 2621–2631. doi:10.1109/tii.2020.3004445.

[71] Y. Zhang, J. Ji, B. Ma, Fault diagnosis of reciprocating compressor using a novel ensemble empirical mode decomposition-convolutional deep belief network, Measurement 156 (2020) 107619. doi:https://doi.org/10.1016/j.measurement.2020.107619.

[72] S.-B. Zhu, Z.-L. Li, S.-M. Zhang, Ying-Yu, H.-F. Zhang, Deep belief network-based internal valve leakage rate prediction approach, Measurement 133 (2019) 182–192. doi:10.1016/j.measurement.2018.10.020.

[73] X. Kong, C. Li, F. Zheng, C. Wang, Improved deep belief network for short-term load forecasting considering demand-side management, IEEE Transactions on Power Systems 35 (2) (2020) 1531–1538. doi:10.1109/tpwrs.2019.2943972.

[74] H. Wang, J. Xu, R. Yan, R. X. Gao, A new intelligent bearing fault diagnosis method using sdp representation and se-cnn, IEEE Transactions on Instrumentation and Measurement 69 (5) (2020) 2377–2389. doi:10.1109/tim.2019.2956332.

[75] B. Yang, R. Liu, E. Zio, Remaining useful life prediction based on a double-convolutional neural network architecture, IEEE Transactions on Industrial Electronics 66 (12) (2019) 9521–9530. doi:10.1109/tie.2019.2924605.

[76] C. Cheng, G. Ma, Y. Zhang, M. Sun, F. Teng, H. Ding, Y. Yuan, A deep learning-based remaining useful life prediction approach for bearings, IEEE/ASME Transactions on Mechatronics 25 (3) (2020) 1243–1254. doi:10.1109/tmech.2020.2971503.

[77] Z. Li, T. Zheng, Y. Wang, Z. Cao, Z. Guo, H. Fu, A novel method for imbalanced fault diagnosis of rotating machinery based on generative adversarial networks, IEEE Transactions on Instrumentation and Measurement 70 (2021) 1–17. doi:10.1109/TIM.2020.3009343.

[78] X. Zhang, Y. Qin, C. Yuen, L. Jayasinghe, X. Liu, Time-series regeneration with convolutional recurrent generative adversarial network for remaining useful life estimation, IEEE Transactions on Industrial Informatics 17 (10) (2021) 6820–6831. doi:10.1109/tii.2020.3046036.

[79] Y. Tang, Y. Huang, Prognostics with variational auto-encoder by generative adversarial learning (aug 2020). doi:10.36227/techrxiv.12751172.

[80] F. Li, Y. Chen, J. Wang, X. Zhou, B. Tang, A reinforcement learning unit matching recurrent neural network for the state trend prediction of rolling bearings, Measurement 145 (2019) 191–203. doi:https://doi.org/10.1016/j.measurement.2019.05.093.

[81] Z. An, S. Li, J. Wang, X. Jiang, A novel bearing intelligent fault diagnosis framework under time-varying working conditions using recurrent neural network, ISA Transactions 100 (2020) 155–170. doi:https://doi.org/10.1016/j.isatra.2019.11.010.

[82] S. Zhao, Y. Zhang, S. Wang, B. Zhou, C. Cheng, A recurrent neural network approach for remaining useful life prediction utilizing a novel trend features construction method, Measurement 146 (2019) 279–288. doi:10.1016/j.measurement.2019.06.004.

[83] S. Wu, Y. Jiang, H. Luo, S. Yin, Remaining useful life prediction for ion etching machine cooling system using deep recurrent neural network-based approaches, Control Engineering Practice 109 (2021) 104748. doi:10.1016/j.conengprac.2021.104748.

[84] J. Lei, C. Liu, D. Jiang, Fault diagnosis of wind turbine based on long short-term memory networks, Renewable Energy 133 (2019) 422–432. doi:10.1016/j.renene.2018.10.031.

[85] G. Ma, Y. Zhang, C. Cheng, B. Zhou, P. Hu, Y. Yuan, Remaining useful life prediction of lithium-ion batteries based on false nearest neighbors and a hybrid neural network, Applied Energy 253 (2019) 113626. doi:10.1016/j.apenergy.2019.113626.

[86] Y. Cheng, J. Wu, H. Zhu, S. W. Or, X. Shao, Remaining useful life prognosis based on ensemble long short-term memory neural network, IEEE Transactions on Instrumentation and Measurement 70 (2021) 1–12. doi:10.1109/tim.2020.3031113.

[87] Y. Tao, X. Wang, R.-V. Sánchez, S. Yang, Y. Bai, Spur gear fault diagnosis using a multilayer gated recurrent unit approach with vibration signal, IEEE Access 7 (2019) 56880–56889. doi:10.1109/ACCESS.2019.2914181.

[88] L. Xiao, Z. Liu, Y. Zhang, Y. Zheng, C. Cheng, Degradation assessment of bearings with trend-reconstruct-based features selection and gated recurrent unit network, Measurement 165 (2020) 108064. doi:10.1016/j.measurement.2020.108064.

[89] Z. Que, X. Jin, Z. Xu, Remaining useful life prediction for bearings based on a gated recurrent unit, IEEE Transactions on Instrumentation and Measurement 70 (2021) 1–11. doi:10.1109/tim.2021.3054025.

[90] C. Chen, F. Shen, J. Xu, R. Yan, Domain adaptation-based transfer learning for gear fault diagnosis under varying working conditions, IEEE Transactions on Instrumentation and Measurement 70 (2021) 1–10. doi:10.1109/TIM.2020.3011584.

[91] W. Mao, J. He, M. J. Zuo, Predicting remaining useful life of rolling bearings based on deep feature representation and transfer learning, IEEE Transactions on Instrumentation and Measurement 69 (4) (2020) 1594–1608. doi:10.1109/tim.2019.2917735.

[92] Y. Che, Z. Deng, X. Lin, L. Hu, X. Hu, Predictive battery health management with transfer learning and online model correction, IEEE Transactions on Vehicular Technology 70 (2) (2021) 1269–1277. doi:10.1109/tvt.2021.3055811.

[93] K. Guo, Y. Hu, Z. Qian, H. Liu, K. Zhang, Y. Sun, J. Gao, B. Yin, Optimized graph convolution recurrent neural network for traffic prediction, IEEE Transactions on Intelligent Transportation Systems 22 (2) (2021) 1138–1149. doi:10.1109/tits.2019.2963722.

[94] X. Chen, Y. Bin, C. Gao, N. Sang, H. Tang, Relevant region prediction for crowd counting, Neurocomputing 407 (2020) 399–408. doi:10.1016/j.neucom.2020.04.117.

[95] M. Kordestani, M. Saif, M. E. Orchard, R. Razavi-Far, K. Khorasani, Failure prognosis and applications—a survey of recent literature, IEEE Transactions on Reliability 70 (2) (2021) 728–748. doi:10.1109/tr.2019.2930195.

[96] Z. Zhang, X. Si, C. Hu, Y. Lei, Degradation data analysis and remaining useful life estimation: A review on wiener-process-based methods, European Journal of Operational Research 271 (3) (2018) 775–796. doi:10.1016/j.ejor.2018.02.033.

[97] J. Yu, Local and nonlocal preserving projection for bearing defect classification and performance assessment, IEEE Transactions on Industrial Electronics 59 (5) (2012) 2363–2376. doi:10.1109/tie.2011.2167893.

[98] L. Chen, Y. Zhang, Y. Zheng, X. Li, X. Zheng, Remaining useful life prediction of lithium-ion battery with optimal input sequence selection and error compensation, Neurocomputing 414 (2020) 245–254. doi:https://doi.org/10.1016/j.neucom.2020.07.081.

[99] X. Yang, Y. Zheng, Y. Zhang, D. S.-H. Wong, W. Yang, Bearing remaining useful life prediction based on regression shapalet and graph neural network, IEEE Transactions on Instrumentation and Measurement 71 (2022) 1–12. doi:10.1109/tim.2022.3151169.

[100] L. Cui, X. Wang, H. Wang, J. Ma, Research on remaining useful life prediction of rolling element bearings based on time-varying kalman filter, IEEE Transactions on Instrumentation and Measurement 69 (6) (2020) 2858–2867. doi:10.1109/tim.2019.2924509.

[101] X. Jin, Z. Que, Y. Sun, Y. Guo, W. Qiao, A data-driven approach for bearing fault prognostics (sep 2018). doi:10.1109/ias.2018.8544586.

[102] P. Shrivastava, T. K. Soon, M. Y. I. B. Idris, S. Mekhilef, Overview of model-based online state-of-charge estimation using kalman filter family for lithium-ion batteries, Renewable and Sustainable Energy Reviews 113 (2019) 109233. doi:10.1016/j.rser.2019.06.040.

[103] X. Zheng, H. Fang, An integrated unscented kalman filter and relevance vector regression approach for lithium-ion battery remaining useful life and short-term capacity prediction, Reliability Engineering & System Safety 144 (2015) 74–82. doi:10.1016/j.ress.2015.07.013.

[104] Y. Chang, H. Fang, Y. Zhang, A new hybrid method for the prediction of the remaining useful life of a lithium-ion battery, Applied Energy 206 (2017) 1564–1578. doi:10.1016/j.apenergy.2017.09.106.

[105] M. E. Orchard, G. J. Vachtsevanos, A particle-filtering approach for on-line fault diagnosis and failure prognosis, Transactions of the Institute of Measurement and Control 31 (3-4) (2009) 221–246. doi:10.1177/0142331208092026.

[106] Y. Chang, H. Fang, A hybrid prognostic method for system degradation based on particle filter and relevance vector machine, Reliability Engineering & System Safety 186 (2019) 51–63. doi:10.1016/j.ress.2019.02.011.

[107] Q. Liu, Z. Wang, X. He, Stochastic Control and Filtering over Constrained Communication Networks, Springer International Publishing, 2019. doi:10.1007/978-3-030-00157-5.

[108] D. Ding, Z. Wang, G. Wei, Performance Analysis and Synthesis for Discrete-Time Stochastic Systems with Network-Enhanced Complexities, CRC Press, 2018. doi:10.1201/9780429465901.

[109] B. Shen, Z. Wang, H. Shu, $H\infty$ filtering with randomly occurring sensor saturations and missing measurements, in: Nonlinear Stochastic Systems with Incomplete Information, Springer London, 2013, pp. 103–117. doi:10.1007/978-1-4471-4914-9_5.

[110] S. Marano, V. Matta, P. Willett, Quantizer precision for distributed estimation in a large sensor network, IEEE Transactions on Signal Processing 54 (10) (2006) 4073–4078. doi:10.1109/tsp.2006.879259.

[111] H. Ishii, T. Başar, Remote control of LTI systems over networks with state quantization, Systems & Control Letters 54 (1) (2005) 15–31. doi:10.1016/j.sysconle.2004.06.002.

[112] D. Ding, Z. Wang, B. Shen, H. Dong, Envelope-constrained H_∞ filtering with fading measurements and randomly occurring nonlinearities: the finite horizon case, Automatica 55 (2015) 37–45. doi:10.1016/j.automatica.2015.02.024.

[113] N. Elia, Remote stabilization over fading channels, Systems & Control Letters 54 (3) (2005) 237–249. doi:10.1016/j.sysconle.2004.08.009.

[114] M. K. Simon, M. S. Alouini, Digital communication over fading channels, second edition Edition, Wiley, 2005.

[115] A. S. Leong, S. Dey, J. S. Evans, Asymptotics and power allocation for state estimation over fading channels, IEEE Transactions on Aerospace and Electronic Systems 47 (1) (2011) 611–633. doi:10.1109/TAES.2011.5705695.

[116] X. Wang, M. D. Lemmon, Event-triggering in distributed networked control systems, IEEE Transactions on Automatic Control 56 (3) (2011) 586–601. doi:10.1109/TAC.2010.2057951.

[117] H. Gao, H. Dong, Z. Wang, F. Han, An event-triggering approach to recursive filtering for complex networks with state saturations and random coupling strengths, IEEE Transactions on Neural Networks and Learning Systems 31 (10) (2020) 4279–4289. doi:10.1109/TNNLS.2019.2953649.

[118] S. Hu, D. Yue, Q.-L. Han, X. Xie, X. Chen, C. Dou, Observer-based event-triggered control for networked linear systems subject to denial-of-service attacks, IEEE Transactions on Cybernetics 50 (5) (2020) 1952–1964. doi:10.1109/TCYB.2019.2903817.

[119] J. Hu, J. Liang, D. Chen, D. Ji, J. Du, A recursive approach to non-fragile filtering for networked systems with stochastic uncertainties and incomplete measurements, Journal of the Franklin Institute 352 (5) (2015) 1946–1962. doi:https://doi.org/10.1016/j.jfranklin.2015.02.002.

[120] Y. Hung, F. Yang, Robust $H\infty$ filtering with error variance constraints for discrete time-varying systems with uncertainty, Automatica 39 (7) (2003) 1185–1194. doi:10.1016/s0005-1098(03)00117-1.

[121] A. Subramanian, A. Sayed, Multiobjective filter design for uncertain stochastic time-delay systems, IEEE Transactions on Automatic Control 49 (1) (2004) 149–154. doi:10.1109/tac.2003.821422.

[122] L. Ma, Z. Wang, J. Hu, Q.-L. Han, Probability-guaranteed envelope-constrained filtering for nonlinear systems subject to measurement outliers, IEEE Transactions on Automatic Control 66 (7) (2021) 3274–3281. doi:10.1109/tac.2020.3016767.

[123] J. Kullaa, Detection, identification, and quantification of sensor fault in a sensor network, Mechanical Systems and Signal Processing 40 (1) (2013) 208–221. doi:10.1016/j.ymssp.2013.05.007.

[124] G. Heredia, A. Ollero, M. Bejar, R. Mahtani, Sensor and actuator fault detection in small autonomous helicopters, Mechatronics 18 (2) (2008) 90–99. doi:10.1016/j.mechatronics.2007.09.007.

[125] Z. Wang, F. Yang, D. Ho, X. Liu, Robust H_∞ control for networked systems with random packet losses, IEEE Transactions on Systems, Man and Cybernetics, Part B (Cybernetics) 37 (4) (2007) 916–924. doi:10.1109/tsmcb.2007.896412.

[126] M. Fu, L. Xie, The sector bound approach to quantized feedback control, IEEE Transactions on Automatic Control 50 (11) (2005) 1698–1711. doi:10.1109/tac.2005.858689.

[127] W.-A. Zhang, G. Feng, L. Yu, Multi-rate distributed fusion estimation for sensor networks with packet losses, Automatica 48 (9) (2012) 2016–2028. doi:10.1016/j.automatica.2012.06.027.

[128] Z. Wang, D. Ho, X. Liu, Variance-constrained filtering for uncertain stochastic systems with missing measurements, IEEE Transactions on Automatic Control 48 (7) (2003) 1254–1258. doi:10.1109/tac.2003.814272.

[129] S. Boyd, L. E. Ghaoui, E. Feron, V. Balakrishnan, Linear Matrix Inequalities in System and Control Theory, Society for Industrial and Applied Mathematics, 1994. doi:10.1137/1.9781611970777.

[130] Y. Liang, T. Chen, Q. Pan, Multi-rate stochastic h_∞ filtering for networked multi-sensor fusion, Automatica 46 (2) (2010) 437–444. doi:10.1016/j.automatica.2009.11.019.

[131] Y. Zhang, Z. Wang, L. Ma, Variance-constrained state estimation for networked multi-rate systems with measurement quantization and probabilistic sensor failures, International Journal of Robust and Nonlinear Control 26 (16) (2016) 3507–3523. doi:10.1002/rnc.3520.

[132] F. Amato, M. Ariola, Finite-time control of discrete-time linear systems, IEEE Transactions on Automatic Control 50 (5) (2005) 724–729. doi:10.1109/TAC.2005.847042.

[133] A. Liu, L. Yu, D. Zhang, W. an Zhang, Finite-time $H\infty$ control for discrete-time genetic regulatory networks with random delays and partly unknown transition probabilities, Journal of the Franklin Institute 350 (7) (2013) 1944–1961. doi:10.1016/j.jfranklin.2013.05.016.

[134] Z. Zhang, Z. Zhang, H. Zhang, P. Shi, H. Reza Karimi, Finite-time H_∞ filtering for t–s fuzzy discrete-time systems with time-varying delay and norm-bounded uncertainties, IEEE Transactions on Fuzzy Systems 23 (6) (2015) 2427–2434. doi:10.1109/TFUZZ.2015.2394380.

[135] J. Baillieul, P. J. Antsaklis, Control and communication challenges in networked real-time systems, Proceedings of the IEEE 95 (1) (2007) 9–28. doi:10.1109/jproc.2006.887290.

[136] D. Ding, Z. Wang, J. Lam, B. Shen, Finite-horizon H_∞ control for discrete time-varying systems with randomly occurring nonlinearities and fading measurements, IEEE Transactions on Automatic Control 60 (9) (2015) 2488–2493. doi:10.1109/TAC.2014.2380671.

[137] H. Fang, H. Ye, M. Zhong, Fault diagnosis of networked control systems, Annual Reviews in Control 31 (1) (2007) 55–68. doi:https://doi.org/10.1016/j.arcontrol.2007.01.001.

[138] Y. Long, G.-H. Yang, Fault detection for a class of networked control systems with finite-frequency servo inputs and random packet dropouts, IET Control Theory & Applications 6 (15) (2012) 2397–2408. doi:10.1049/iet-cta.2012.0333.

[139] J. Feng, S. Wang, Q. Zhao, Closed-loop design of fault detection for networked non-linear systems with mixed delays and packet losses, Iet Control Theory & Applications 7 (6) (2013) 858–868. doi:10.1049/iet-cta.2012.0987.

[140] Y. Zheng, H. Fang, H. O. Wang, Takagi-sugeno fuzzy-model-based fault detection for networked control systems with Markov delays, IEEE Transactions on Systems Man & Cybernetics Part B Cybernetics A Publication of the IEEE Systems Man & Cybernetics Society 36 (4) (2006) 924–929. doi:10.1109/tsmcb.2005.861879.

[141] Z. Mao, B. Jiang, P. Shi, Protocol and fault detection design for nonlinear networked control systems, IEEE Transactions on Circuits and Systems II: Express Briefs 56 (3) (2009) 255–259. doi:10.1109/tcsii.2008.2011600.

[142] G. Mustafa, T. Chen, filtering for nonuniformly sampled systems: A Markovian jump systems approach, Systems & Control Letters 60 (10) (2011) 871–876. doi:10.1016/j.sysconle.2011.07.005.

[143] Y. Zhang, Z. Wang, L. Zou, H. Fang, Event-based finite-time filtering for multirate systems with fading measurements, IEEE Transactions on Aerospace and Electronic Systems 53 (3) (2017) 1431–1441. doi:10.1109/taes.2017.2671498.

[144] Y. Zhang, Z. Wang, L. Zou, Z. Liu, Fault detection filter design for networked multi-rate systems with fading measurements and randomly occurring faults, IET Control Theory & Applications 10 (5) (2016) 573–581. doi:10.1049/iet-cta.2015.0582.

[145] M. S. Fadali, Observer-based robust fault detection of multirate linear system using a lift reformulation, Computers & Electrical Engineering 29 (1) (2003) 235–243. doi:10.1016/s0045-7906(01)00008-8.

[146] M. Zhong, H. Ye, S. X. Ding, G. Wang, Observer-based fast rate fault detection for a class of multirate sampled-data systems, IEEE Transactions on Automatic Control 52 (3) (2007) 520–525. doi:10.1109/tac.2006.890488.

[147] A. S. Leong, S. Dey, G. N. Nair, P. Sharma, Power allocation for outage minimization in state estimation over fading channels, IEEE Transactions on Signal Processing 59 (7) (2011) 3382–3397. doi:10.1109/TSP.2011.2135350.

[148] J. J. Gertler, Fault Detection and Diagnosis in Engineering Systems, CRC Press, 2017. doi:10.1201/9780203756126.
URL `https://doi.org/10.1201%2F9780203756126`

[149] X. He, Z. Wang, Y. D. Ji, D. H. Zhou, Robust fault detection for networked systems with distributed sensors, IEEE Transactions on Aerospace and Electronic Systems 47 (1) (2011) 166–177. doi:10.1109/TAES.2011.5705667.

[150] C. de Souza, Robust stability and stabilization of uncertain discrete-time Markovian jump linear systems, IEEE Transactions on Automatic Control 51 (5) (2006) 836–841. doi:10.1109/tac.2006.875012.

[151] L. Zhang, E.-K. Boukas, $H\infty$control for discrete-time Markovian jump linear systems with partly unknown transition probabilities, International Journal of Robust and Nonlinear Control 19 (8) (2009) 868–883. doi:10.1002/rnc.1355.

[152] Z. Mao, B. Jiang, P. Shi, $H\infty$ fault detection filter design for networked control systems modelled by discrete Markovian jump systems, IET Control Theory & Applications 1 (5) (2007) 1336–1343. doi:10.1049/iet-cta:20060431.

[153] J. Xiong, J. Lam, Stabilization of linear systems over networks with bounded packet loss, Automatica 43 (1) (2007) 80–87. doi:10.1016/j.automatica.2006.07.017.

[154] J. Sheng, T. Chen, S. L. Shah, Generalized predictive control for non-uniformly sampled systems, Journal of Process Control 12 (8) (2002) 875–885. doi:10.1016/s0959-1524(02)00009-4.

[155] W.-A. Zhang, S. Liu, L. Yu, Fusion estimation for sensor networks with nonuniform estimation rates, IEEE Transactions on Circuits and Systems I: Regular Papers 61 (5) (2014) 1485–1498. doi:10.1109/tcsi.2013.2285693.

[156] D. Liberzon, Hybrid feedback stabilization of systems with quantized signals, Automatica 39 (9) (2003) 1543–1554. doi:10.1016/s0005-1098(03)00151-1.

[157] J. Chen, R. J. Patton, Robust model-based fault diagnosis for dynamic systems, International Journal of Robust and Nonlinear Control 11 (14) (2001) 1400–1401. doi:10.1002/rnc.615.

[158] L. Chen, R. J. Patton, A mixed $H/H\infty$ LPV approach to adaptive fault compensation for a nonlinear UAV, IFAC Proceedings Volumes 45 (20) (2012) 830–835. doi:10.3182/20120829-3-mx-2028.00235.

[159] L. Chen, R. Patton, A time-domain LPV $H_{-}/H\infty$ fault detection filter, IFAC-PapersOnLine 50 (1) (2017) 8600–8605. doi:10.1016/j.ifacol.2017.08.1427.

[160] Y. Liu, Z. Wang, X. He, D. H. Zhou, Event-triggered filtering and fault estimation for nonlinear systems with stochastic sensor saturations, International Journal of Control 90 (5) (2016) 1052–1062. doi:10.1080/00207179.2016.1199916.

[161] K. Zhang, B. Jiang, P. Shi, J. Xu, Analysis and design of robust hinfty fault estimation observer with finite-frequency specifications for discrete-time fuzzy systems, IEEE Transactions on Cybernetics 45 (7) (2015) 1225–1235. doi:10.1109/tcyb.2014.2347697.

[162] H. Dong, Z. Wang, S. X. Ding, H. Gao, On h-infinity estimation of randomly occurring faults for a class of nonlinear time-varying systems with fading channels, IEEE Transactions on Automatic Control 61 (2) (2016) 479–484. doi:10.1109/tac.2015.2437526.

[163] K. Liu, E. Fridman, K. H. Johansson, Y. Xia, Quantized control under round-robin communication protocol, IEEE Transactions on Industrial Electronics 63 (7) (2016) 4461–4471. doi:10.1109/tie.2016.2539259.

[164] L. Qian, Z. Gajic, Variance minimization stochastic power control in CDMA systems (2002). doi:10.1109/icc.2002.997151.

[165] V. Pappala, I. Erlich, K. Rohrig, J. Dobschinski, A stochastic model for the optimal operation of a wind-thermal power system, IEEE Transactions on Power Systems 24 (2) (2009) 940–950. doi:10.1109/tpwrs.2009.2016504.

[166] Z. Mao, Y. Zhan, G. Tao, B. Jiang, X.-G. Yan, Sensor fault detection for rail vehicle suspension systems with disturbances and stochastic noises, IEEE Transactions on Vehicular Technology 66 (6) (2017) 4691–4705. doi:10.1109/tvt.2016.2628054.

[167] L. Sheng, Z. Wang, W. Wang, F. E. Alsaadi, Output-feedback control for nonlinear stochastic systems with successive packet dropouts and uniform quantization effects, IEEE Transactions on Systems, Man, and Cybernetics: Systems (2016) 1–11doi:10.1109/tsmc.2016.2563393.

[168] H. Dong, Z. Wang, B. Shen, D. Ding, Variance-constrained $H\infty$ control for a class of nonlinear stochastic discrete time-varying systems: The event-triggered design, Automatica 72 (2016) 28–36. doi:10.1016/j.automatica.2016.05.012.

[169] T. Chen, B. A. Francis, $H\infty$-optimal SD control, in: Optimal Sampled-Data Control Systems, Springer London, 1995, pp. 309–347. doi:10.1007/978-1-4471-3037-6_13.

[170] Z. Wang, H. Dong, B. Shen, H. Gao, Finite-horizon H_∞ filtering with missing measurements and quantization effects, IEEE Transactions on Automatic Control 58 (7) (2013) 1707–1718. doi:10.1109/tac.2013.2241492.

[171] L. Zou, Z. Wang, H. Gao, Set-membership filtering for time-varying systems with mixed time-delays under round-robin and weighted try-once-discard protocols, Automatica 74 (2016) 341–348. doi:10.1016/j.automatica.2016.07.025.

[172] H. Tan, B. Shen, Y. Liu, A. Alsaedi, B. Ahmad, Event-triggered multi-rate fusion estimation for uncertain system with stochastic nonlinearities and colored measurement noises, Information Fusion 36 (2017) 313–320. doi:10.1016/j.inffus.2016.12.003.

[173] L. Zhang, H. Gao, O. Kaynak, Network-induced constraints in networked control systems—a survey, IEEE Transactions on Industrial Informatics 9 (1) (2013) 403–416. doi:10.1109/tii.2012.2219540.

[174] Z.-W. Liu, Z.-H. Guan, X. Shen, G. Feng, Consensus of multi-agent networks with aperiodic sampled communication via impulsive algorithms using position-only measurements, IEEE Transactions on Automatic Control 57 (10) (2012) 2639–2643. doi:10.1109/TAC.2012.2214451.

[175] X.-H. Chang, Y.-M. Wang, Peak-to-peak filtering for networked nonlinear dc motor systems with quantization, IEEE Transactions on Industrial Informatics 14 (12) (2018) 5378–5388. doi:10.1109/tii.2018.2805707.

[176] X.-J. Li, G.-H. Yang, Fault detection for t–s fuzzy systems with unknown membership functions, IEEE Transactions on Fuzzy Systems 22 (1) (2014) 139–152. doi:10.1109/TFUZZ.2013.2249519.

[177] X.-J. Li, G.-H. Yang, Fault detection in finite frequency domain for Takagi-Sugeno fuzzy systems with sensor faults, IEEE Transactions on Cybernetics 44 (8) (2014) 1446–1458. doi:10.1109/TCYB.2013.2286209.

[178] Y. Gu, X.-J. Li, Fault detection for sector-bounded non-linear systems with servo inputs and sensor stuck faults, Journal of Control and Decision 6 (3) (2018) 147–165. doi:10.1080/23307706.2018.1439778.

[179] A. Xu, Q. Zhang, Residual generation for fault diagnosis in linear time-varying systems, IEEE Transactions on Automatic Control 49 (5) (2004) 767–772. doi:10.1109/tac.2004.825983.

[180] S. Hu, D. Yue, X. Xie, Y. Ma, X. Yin, Stabilization of neural-network-based control systems via event-triggered control with nonperiodic sampled data, IEEE Transactions on Neural Networks and Learning Systems 29 (3) (2018) 573–585. doi:10.1109/TNNLS.2016.2636875.

[181] P. Zhang, S. X. Ding, G. Z. Wang, D. H. Zhou, Fault detection for multi-rate sampled-data systems with time delays, International Journal of Control 75 (18) (2002) 1457–1471. doi:10.1080/0020717021000031475.

[182] I. Izadi, Q. Zhao, T. Chen, An optimal scheme for fast rate fault detection based on multirate sampled data, Journal of Process Control 15 (3) (2005) 307–319. doi:10.1016/j.jprocont.2004.06.008.

[183] G. H. B. Foo, X. Zhang, D. M. Vilathgamuwa, A sensor fault detection and isolation method in interior permanent-magnet synchronous motor drives based on an extended kalman filter, IEEE Transactions on Industrial Electronics 60 (8) (2013) 3485–3495. doi:10.1109/TIE.2013.2244537.

[184] S. Debnath, J. Qin, B. Bahrani, M. Saeedifard, P. Barbosa, Operation, control, and applications of the modular multilevel converter: A review, IEEE Transactions on Power Electronics 30 (1) (2015) 37–53. doi:10.1109/tpel.2014.2309937.

[185] U. N. Gnanarathna, A. M. Gole, R. P. Jayasinghe, Efficient modeling of modular multilevel HVDC converters (MMC) on electromagnetic transient simulation programs, IEEE Transactions on Power Delivery 26 (1) (2011) 316–324. doi:10.1109/TPWRD.2010.2060737.

[186] M. Saeedifard, R. Iravani, J. Pou, A space vector modulation strategy for a back-to-back five-level HVDC converter system, IEEE Transactions on Industrial Electronics 56 (2) (2009) 452–466. doi:10.1109/tie.2008.2008360.

[187] F. Deng, Z. Chen, M. R. Khan, R. Zhu, Fault detection and localization method for modular multilevel converters, IEEE Transactions on Power Electronics 30 (5) (2015) 2721–2732. doi:10.1109/TPEL.2014.2348194.

[188] S. Shao, P. W. Wheeler, J. C. Clare, A. J. Watson, Fault detection for modular multilevel converters based on sliding mode observer, IEEE Transactions on Power Electronics 28 (11) (2013) 4867–4872. doi:10.1109/tpel.2013.2242093.

[189] Q. Yang, J. Qin, M. Saeedifard, Analysis, detection, and location of open-switch submodule failures in a modular multilevel converter, IEEE Transactions on Power Delivery 31 (1) (2016) 155–164. doi:10.1109/tpwrd.2015.2477476.

[190] L. Ke, Z. Liu, Y. Zhang, Diagnosis and location of open-circuit fault in modular multilevel converters based on high-order harmonic analysis, Tehniki vjesnik 27 (3) (2018) 898–905.

[191] H. Dong, N. Hou, Z. Wang, Fault estimation for complex networks with randomly varying topologies and stochastic inner couplings, Automatica 112 (2020) 108734. doi:https://doi.org/10.1016/j.automatica.2019.108734.

[192] M. Gao, W. Zhang, L. Sheng, D. Zhou, Distributed fault estimation for delayed complex networks with round-robin protocol based on unknown input observer, Journal of the Franklin Institute 357 (13) (2020) 8678–8702. doi:10.1016/j.jfranklin.2020.04.012.

[193] C. Cheng, J. Ding, Y. Zhang, A koopman operator approach for machinery health monitoring and prediction with noisy and low-dimensional industrial time series, Neurocomputing 406 (2020) 204–214. doi:10.1016/j.neucom.2020.04.005.

[194] Y. Zhang, H. Hu, Z. Liu, M. Zhao, L. Cheng, Concurrent fault diagnosis of modular multilevel converter with kalman filter and optimized support vector machine, Systems Science & Control Engineering 7 (3) (2019) 43–53. doi:10.1080/21642583.2019.1650840.

[195] N. Zeng, H. Zhang, B. Song, W. Liu, Y. Li, A. M. Dobaie, Facial expression recognition via learning deep sparse autoencoders, Neurocomputing 273 (2018) 643–649. doi:10.1016/j.neucom.2017.08.043.

[196] D. Zhao, T. Wang, F. Chu, Deep convolutional neural network based planet bearing fault classification, Computers in Industry 107 (2019) 59–66. doi:10.1016/j.compind.2019.02.001.

[197] L. Zou, Z. Wang, J. Hu, D. Zhou, Moving horizon estimation with unknown inputs under dynamic quantization effects, IEEE Transactions on Automatic Control 65 (12) (2020) 5368–5375. doi:10.1109/tac.2020.2968975.

[198] B. Shen, Z. Wang, D. Wang, H. Liu, Distributed state-saturated recursive filtering over sensor networks under round-robin protocol, IEEE Transactions on Cybernetics 50 (8) (2020) 3605–3615. doi:10.1109/tcyb.2019.2932460.

[199] S. Wang, Z. Wang, H. Dong, F. E. Alsaadi, Recursive state estimation for linear systems with lossy measurements under time-correlated multiplicative noises, Journal of the Franklin Institute 357 (3) (2020) 1887–1908. doi:10.1016/j.jfranklin.2019.11.031.

[200] L. Xu, M. Cao, B. Song, J. Zhang, Y. Liu, F. E. Alsaadi, Open-circuit fault diagnosis of power rectifier using sparse autoencoder based deep neural network, Neurocomputing 311 (2018) 1–10. doi:10.1016/j.neucom.2018.05.040.

[201] Q. Tu, Z. Xu, L. Xu, Reduced switching-frequency modulation and circulating current suppression for modular multilevel converters, IEEE Transactions on Power Delivery 26 (3) (2011) 2009–2017. doi:10.1109/tpwrd.2011.2115258.

[202] B. Li, S. Shi, B. Wang, G. Wang, W. Wang, D. Xu, Fault diagnosis and tolerant control of single igbt open-circuit failure in modular multilevel converters, IEEE Transactions on Power Electronics 31 (4) (2016) 3165–3176. doi:10.1109/tpel.2015.2454534.

[203] T. G. Kolda and B. W. Bader, Tensor decompositions and applications, SIAM Review 51 (3) (2009) 455–500.

[204] B. Cyganek, B. Krawczyk, M. Woźniak, Multidimensional data classification with chordal distance based kernel and support vector machines, Engineering Applications of Artificial Intelligence 46 (NOV.PT.A) (2015) 10–22. doi:10.1016/j.engappai.2015.08.001.

[205] M. Signoretto, L. D. Lathauwer, J. A. Suykens, A kernel-based framework to tensorial data analysis, Neural Networks 24 (8) (2011) 861–874. doi:10.1016/j.neunet.2011.05.011.

[206] J. Gosme, C. Richard, Beyond standard classes of generalized joint signal representations of arbitrary variables: Mercer kernel-based representations, IEEE Signal Processing Letters 12 (1) (2005) 25–28. doi:10.1109/LSP.2004.838212.

[207] S. S. Keerthi, C.-J. Lin, Asymptotic behaviors of support vector machines with gaussian kernel, Neural Computation 15 (7) (2003) 1667–1689. doi:10.1162/089976603321891855.

[208] W. Liu, Z. Wang, X. Liu, N. Zeng, Y. Liu, F. E. Alsaadi, A survey of deep neural network architectures and their applications, Neurocomputing 234 (2017) 11–26. doi:10.1016/j.neucom.2016.12.038.

[209] R. Yang, P. V. Er, Z. Wang, K. K. Tan, An RBF neural network approach towards precision motion system with selective sensor fusion, Neurocomputing 199 (2016) 31–39. doi:10.1016/j.neucom.2016.01.093.

[210] B. Scrosati, J. Garche, Lithium batteries: Status, prospects and future, Journal of Power Sources 195 (9) (2010) 2419–2430. doi:10.1016/j.jpowsour.2009.11.048.

[211] L. Ungurean, G. Cârstoiu, M. V. Micea, V. Groza, Battery state of health estimation: a structured review of models, methods and commercial devices, International Journal of Energy Research 41 (2) (2016) 151–181. doi:10.1002/er.3598.

[212] M. H. Lipu, M. Hannan, A. Hussain, M. Hoque, P. J. Ker, M. Saad, A. Ayob, A review of state of health and remaining useful life estimation methods for lithium-ion battery in electric vehicles: Challenges and recommendations, Journal of Cleaner Production 205 (2018) 115–133. doi:https://doi.org/10.1016/j.jclepro.2018.09.065.

[213] L. Wu, X. Fu, Y. Guan, Review of the remaining useful life prognostics of vehicle lithium-ion batteries using data-driven methodologies, Applied Sciences 6 (6) (2016) 166. doi:10.3390/app6060166.

[214] A. Nuhic, T. Terzimehic, T. Soczka-Guth, M. Buchholz, K. Dietmayer, Health diagnosis and remaining useful life prognostics of lithium-ion batteries using data-driven methods, Journal of Power Sources 239 (2013) 680–688. doi:10.1016/j.jpowsour.2012.11.146.

[215] M. A. Patil, P. Tagade, K. S. Hariharan, S. M. Kolake, T. Song, T. Yeo, S. Doo, A novel multistage support vector machine based approach for li ion battery remaining useful life estimation, Applied Energy 159 (2015) 285–297. doi:10.1016/j.apenergy.2015.08.119.

[216] X. Li, X. Shu, J. Shen, R. Xiao, W. Yan, Z. Chen, An on-board remaining useful life estimation algorithm for lithium-ion batteries of electric vehicles, Energies 10 (5) (2017) 691–706. doi:10.3390/en10050691.

[217] Q. Zhao, X. Qin, H. Zhao, W. Feng, A novel prediction method based on the support vector regression for the remaining useful life of lithium-ion batteries, Microelectronics Reliability 85 (2018) 99–108. doi:10.1016/j.microrel.2018.04.007.

[218] L. Liao, Discovering prognostic features using genetic programming in remaining useful life prediction, IEEE Transactions on Industrial Electronics 61 (5) (2014) 2464–2472. doi:10.1109/TIE.2013.2270212.

[219] A. Doucet, S. Godsill, C. Andrieu, On sequential monte carlo sampling methods for bayesian filtering, Statistics and Computing 10 (3) (2000) 197–208.

[220] W. Li, Z. Wang, Y. Yuan, L. Guo, Two-stage particle filtering for non-gaussian state estimation with fading measurements, Automatica 115 (2020) 108882. doi:https://doi.org/10.1016/j.automatica.2020.108882.

[221] M. Jouin, R. Gouriveau, D. Hissel, M.-C. Péra, N. Zerhouni, Particle filter-based prognostics: Review, discussion and perspectives, Mechanical Systems and Signal Processing 72-73 (2016) 2–31. doi:https://doi.org/10.1016/j.ymssp.2015.11.008.

[222] R. Jiao, K. Peng, J. Dong, Remaining useful life prediction of lithium-ion batteries based on conditional variational autoencoders-particle filter, IEEE Transactions on Instrumentation and Measurement 69 (11) (2020) 8831–8843. doi:10.1109/TIM.2020.2996004.

[223] Y. Qian, R. Yan, Remaining useful life prediction of rolling bearings using an enhanced particle filter, IEEE Transactions on Instrumentation and Measurement 64 (10) (2015) 2696–2707. doi:10.1109/tim.2015.2427891.

[224] S. Yin, xiangping zhu, Intelligent particle filter and its application on fault detection of nonlinear system, IEEE Transactions on Industrial Electronics (2015) 1–1doi:10.1109/tie.2015.2399396.

[225] Y. Xing, E. W. Ma, K.-L. Tsui, M. Pecht, An ensemble model for predicting the remaining useful performance of lithium-ion batteries, Microelectronics Reliability 53 (6) (2013) 811–820. doi:10.1016/j.microrel.2012.12.003.

[226] E. A. Wan, R. V. D. Merwe, The unscented kalman filter for nonlinear estimation (2020). doi:10.1109/asspcc.2000.882463.

[227] R. Khelif, B. Chebel-Morello, S. Malinowski, E. Laajili, F. Fnaiech, N. Zerhouni, Direct remaining useful life estimation based on support vector regression, IEEE Transactions on Industrial Electronics 64 (3) (2017) 2276–2285. doi:10.1109/TIE.2016.2623260.

[228] I. Younas, F. Kamrani, M. Bashir, J. Schubert, Efficient genetic algorithms for optimal assignment of tasks to teams of agents, Neurocomputing 314 (2018) 409–428. doi:https://doi.org/10.1016/j.neucom.2018.07.008.

[229] B. Saha, K. Goebel, Battery data set, `http://ti.arc.nasa.gov/project/prognostic-data-repository` (2007).

[230] E. H. Houssein, M. R. Saad, F. A. Hashim, H. Shaban, M. Hassaballah, Lévy flight distribution: A new metaheuristic algorithm for solving engineering optimization problems, Engineering Applications of Artificial Intelligence 94 (2020) 103731. doi:https://doi.org/10.1016/j.engappai.2020.103731.

[231] R. N. Mantegna, Fast, accurate algorithm for numerical simulation of lévy stable stochastic processes, Physical Review E 49 (5) (1994) 4677–4683. doi:10.1103/physreve.49.4677.

[232] X. S. Yang, S. Deb, Engineering optimisation by cuckoo search, International Journal of Mathematical Modelling and Numerical Optimisation 1 (4) (2010) 330. doi:10.1504/ijmmno.2010.035430.

[233] D. Liu, J. Pang, J. Zhou, Y. Peng, M. Pecht, Prognostics for state of health estimation of lithium-ion batteries based on combination gaussian process functional regression, Microelectronics Reliability 53 (6) (2013) 832–839. doi:10.1016/j.microrel.2013.03.010.

[234] Y.-J. He, J.-N. Shen, J.-F. Shen, Z.-F. Ma, State of health estimation of lithium-ion batteries: a multiscale gaussian process regression modeling approach, AIChE J 61 (2015) 1589–1600. doi:10.1002/aic.14760.

[235] T. Qin, S. Zeng, J. Guo, Z. Skaf, A rest time-based prognostic framework for state of health estimation of lithium-ion batteries with regeneration phenomena, Energies 9 (11) (2016) 896. doi:10.3390/en9110896.

[236] T. Qin, S. Zeng, J. Guo, Robust prognostics for state of health estimation of lithium-ion batteries based on an improved PSO–SVR model, Microelectronics Reliability 55 (9-10) (2015) 1280–1284. doi:10.1016/j.microrel.2015.06.133.

[237] G. Qiu, Y. Gu, J. Chen, Selective health indicator for bearings ensemble remaining useful life prediction with genetic algorithm and weibull proportional hazards model, Measurement 150 (2020) 107097. doi:10.1016/j.measurement.2019.107097.

[238] X. Jin, Y. Sun, Z. Que, Y. Wang, T. W. S. Chow, Anomaly detection and fault prognosis for bearings, IEEE TRANSACTIONS ON INSTRUMENTATION AND MEASUREMENT 65 (9) (2016) 2046–2054. doi:10.1109/tim.2016.2570398.

[239] P. Nectoux, R. Gouriveau, K. Medjaher, E. Ramasso, C. Varnier, Pronostia: An experimental platform for bearings accelerated degradation tests, in: IEEE International Conference on Prognostics and Health Management, 2012, pp. 1–8.

[240] B. Wang, Y. Lei, N. Li, N. Li, A hybrid prognostics approach for estimating remaining useful life of rolling element bearings, IEEE Transactions on Reliability 69 (1) (2020) 401–412. doi:10.1109/tr.2018.2882682.

[241] R. K. Singleton, E. G. Strangas, S. Aviyente, Extended kalman filtering for remaining-useful-life estimation of bearings, IEEE Transactions on Industrial Electronics 62 (3) (2015) 1781–1790. doi:10.1109/tie.2014.2336616.

[242] X. Li, W. Zhang, Q. Ding, Deep learning-based remaining useful life estimation of bearings using multi-scale feature extraction, Reliability Engineering & System Safety 182 (2019) 208–218. doi:https://doi.org/10.1016/j.ress.2018.11.011.

[243] N. Li, Y. Lei, J. Lin, S. X. Ding, An improved exponential model for predicting remaining useful life of rolling element bearings, IEEE Transactions on Industrial Electronics 62 (12) (2015) 7762–7773. doi:10.1109/TIE.2015.2455055.

[244] W. Ahmad, S. A. Khan, M. M. M. Islam, J.-M. Kim, A reliable technique for remaining useful life estimation of rolling element bearings using dynamic regression models, Reliability Engineering & System Safety 184 (2019) 67–76. doi:10.1016/j.ress.2018.02.003.

[245] B. Zhang, L. Zhang, J. Xu, Degradation feature selection for remaining useful life prediction of rolling element bearings, Quality and Reliability Engineering International 32 (2) (2015) 547–554. doi:10.1002/qre.1771.

[246] J. B. Ali, B. Chebel-Morello, L. Saidi, S. Malinowski, F. Fnaiech, Accurate bearing remaining useful life prediction based on weibull distribution and artificial neural network, Mechanical Systems & Signal Processing 56-57 (May) (2015) 150–172. doi:10.1016/j.ymssp.2014.10.014.

[247] J. Zhu, N. Chen, W. Peng, Estimation of bearing remaining useful life based on multiscale convolutional neural network, IEEE Transactions on Industrial Electronics 66 (4) (2019) 3208–3216. doi:10.1109/tie.2018.2844856.

[248] F. Yang, M. S. Habibullah, T. Zhang, Z. Xu, P. Lim, S. Nadarajan, Health index-based prognostics for remaining useful life predictions in electrical machines, IEEE Transactions on Industrial Electronics 63 (4) (2016) 2633–2644. doi:10.1109/tie.2016.2515054.

[249] Z. Qian, Y. Pei, H. Zareipour, N. Chen, A review and discussion of decomposition-based hybrid models for wind energy forecasting applications, Applied Energy 235 (2019) 939–953. doi:10.1016/j.apenergy.2018.10.080.

[250] S. Hong, Z. Zhou, E. Zio, K. Hong, Condition assessment for the performance degradation of bearing based on a combinatorial feature extraction method, Digital Signal Processing 27 (2014) 159–166. doi:10.1016/j.dsp.2013.12.010.

[251] M. Niu, Y. Wang, S. Sun, Y. Li, A novel hybrid decomposition-and-ensemble model based on CEEMD and GWO for short-term PM2.5 concentration forecasting, Atmospheric Environment 134 (2016) 168–180. doi:10.1016/j.atmosenv.2016.03.056.

[252] J. Chen, D. Zhou, C. Lyu, C. Lu, A novel health indicator for PEMFC state of health estimation and remaining useful life prediction, International Journal of Hydrogen Energy 42 (31) (2017) 20230–20238. doi:10.1016/j.ijhydene.2017.05.241.

[253] K. Javed, R. Gouriveau, N. Zerhouni, P. Nectoux, Enabling health monitoring approach based on vibration data for accurate prognostics, IEEE Transactions on Industrial Electronics 62 (1) (2015) 647–656. doi:10.1109/TIE.2014.2327917.

[254] L. Guo, N. Li, F. Jia, Y. Lei, J. Lin, A recurrent neural network based health indicator for remaining useful life prediction of bearings, Neurocomputing 240 (2017) 98–109. doi:https://doi.org/10.1016/j.neucom.2017.02.045.

[255] L. Ren, X. Cheng, X. Wang, J. Cui, L. Zhang, Multi-scale dense gate recurrent unit networks for bearing remaining useful life prediction, Future Generation Computer Systems 94 (2019) 601–609. doi:10.1016/j.future.2018.12.009.

[256] R. Zhao, D. Wang, R. Yan, K. Mao, F. Shen, J. Wang, Machine health monitoring using local feature-based gated recurrent unit networks, IEEE Transactions on Industrial Electronics 65 (2) (2018) 1539–1548. doi:10.1109/tie.2017.2733438.

[257] M. Cerrada, R.-V. Sánchez, C. Li, F. Pacheco, D. Cabrera, J. V. de Oliveira, R. E. Vásquez, A review on data-driven fault severity assessment in rolling bearings, Mechanical Systems and Signal Processing 99 (Jan.15) (2018) 169–196. doi:10.1016/j.ymssp.2017.06.012.

[258] D. Wang, K.-L. Tsui, Q. Miao, Prognostics and health management: A review of vibration based bearing and gear health indicators, IEEE Access 6 (2018) 665–676. doi:10.1109/access.2017.2774261.

[259] Y. guo Chen, Z. ping Tian, Z. qing Miao, Detection of singularities in the pressure fluctuations of circulating fluidized beds based on wavelet modulus maximum method, Chemical Engineering Science 59 (17) (2004) 3569–3575. doi:10.1016/j.ces.2004.06.001.

[260] O. Poisson, P. Rioual, M. Meunier, Detection and measurement of power quality disturbances using wavelet transform. doi:10.1109/ichqp.1998.760196.

[261] J. Kim, Discrete wavelet transform-based feature extraction of experimental voltage signal for li-ion cell consistency, IEEE Transactions on Vehicular Technology 65 (3) (2016) 1150–1161. doi:10.1109/TVT.2015.2414936.

[262] M. Chafii, J. Palicot, R. Gribonval, F. Bader, Adaptive wavelet packet modulation, IEEE Transactions on Communications 66 (2018) 2947–2957. doi:10.1109/tcomm.2018.2809586.

[263] A. J. Smola, B. Schölkopf, A tutorial on support vector regression, Statistics and Computing 14 (3) (2004) 199–222. doi:10.1023/b:stco.0000035301.49549.88.

[264] Y. Li, T. Zhao, P. Wang, H. B. Gooi, L. Wu, Y. Liu, J. Ye, Optimal operation of multimicrogrids via cooperative energy and reserve scheduling, IEEE Transactions on Industrial Informatics 14 (8) (2018) 3459–3468. doi:10.1109/TII.2018.2792441.

[265] M. Fernández-Delgado, M. Sirsat, E. Cernadas, S. Alawadi, S. Barro, M. Febrero-Bande, An extensive experimental survey of regression methods, Neural Networks 111 (2019) 11–34. doi:https://doi.org/10.1016/j.neunet.2018.12.010.

[266] N. Srinivas, K. Deb, Muiltiobjective optimization using nondominated sorting in genetic algorithms, Evolutionary Computation 2 (3) (1994) 221–248. doi:10.1162/evco.1994.2.3.221.

[267] A. Saxena, K. Goebel, D. Simon, N. Eklund, Damage propagation modeling for aircraft engine run-to-failure simulation, in: 2008 International Conference on Prognostics and Health Management, IEEE, 2008. doi:10.1109/phm.2008.4711414.

[268] D. K. Frederick, J. A. Decastro, J. S. Litt, User's guide for the commercial modular aero-propulsion system simulation (c-mapss), NASA Technical Manuscript.

[269] J. Wu, C. Wu, S. Cao, S. W. Or, C. Deng, X. Shao, Degradation data-driven time-to-failure prognostics approach for rolling element bearings in electrical machines, IEEE Transactions on Industrial Electronics 66 (1) (2019) 529–539. doi:10.1109/tie.2018.2811366.

[270] K. Javed, R. Gouriveau, N. Zerhouni, P. Nectoux, A feature extraction procedure based on trigonometric functions and cumulative descriptors to enhance prognostics modeling, in: 2013 IEEE Conference on Prognostics and Health Management (PHM), 2013, pp. 1–7. doi:10.1109/ICPHM.2013.6621413.

[271] S. Hochreiter, J. Schmidhuber, Long short-term memory, Neural Computation 9 (8) (1997) 1735–1780. doi:10.1162/neco.1997.9.8.1735.

[272] X. Jin, Y. Sun, J. Shan, Y. Wang, Z. Xu, Health monitoring and fault detection using wavelet packet technique and multivariate process control method, in: 2014 Prognostics and System Health Management Conference (PHM-2014 Hunan), 2014, pp. 257–260. doi:10.1109/PHM.2014.6988174.

[273] T. H. Loutas, D. Roulias, G. Georgoulas, Remaining useful life estimation in rolling bearings utilizing data-driven probabilistic e-support vectors regression, IEEE Transactions on Reliability 62 (4) (2013) 821–832. doi:10.1109/tr.2013.2285318.

[274] Y. Zhou, M. Huang, Y. Chen, Y. Tao, A novel health indicator for online lithium-ion batteries remaining useful life prediction, Journal of Power Sources 321 (2016) 1–10. doi:10.1016/j.jpowsour.2016.04.119.

[275] F.-K. Wang, T. Mamo, A hybrid model based on support vector regression and differential evolution for remaining useful lifetime prediction of lithium-ion batteries, Journal of Power Sources 401 (2018) 49–54. doi:10.1016/j.jpowsour.2018.08.073.

[276] D. Liu, J. Zhou, H. Liao, Y. Peng, X. Peng, A health indicator extraction and optimization framework for lithium-ion battery degradation modeling and prognostics, IEEE Transactions on Systems, Man, and Cybernetics: Systems 45 (6) (2015) 915–928. doi:10.1109/TSMC.2015.2389757.

[277] L. Ma, Z. Wang, Y. Liu, F. E. Alsaadi, Distributed filtering for nonlinear time-delay systems over sensor networks subject to multiplicative link noises and switching topology, International Journal of Robust and Nonlinear Control 29 (10) (2019) 2941–2959. doi:10.1002/rnc.4535.

[278] Y. Shen, Z. Wang, B. Shen, F. E. Alsaadi, F. E. Alsaadi, Fusion estimation for multi-rate linear repetitive processes under weighted try-once-discard protocol, Information Fusion 55 (2020) 281–291. doi:10.1016/j.inffus.2019.08.013.

[279] Y. Chen, Z. Wang, L. Wang, W. Sheng, Mixed H_2/H_∞ state estimation for discrete-time switched complex networks with random coupling strengths through redundant channels, IEEE Transactions on Neural Networks and Learning Systems 31 (10) (2020) 4130–4142. doi:10.1109/TNNLS.2019.2952249.

[280] S. Liu, Z. Wang, Y. Chen, G. Wei, Protocol-based unscented kalman filtering in the presence of stochastic uncertainties, IEEE Transactions on Automatic Control 65 (3) (2020) 1303–1309. doi:10.1109/tac.2019.2929817.

[281] Z. Wu, N. E. Huang, Ensemble empirical mode decomposition: a noise-assisted data analysis method, Advances in Adaptive Data Analysis 01 (01) (2009) 1–41. doi:10.1142/s1793536909000047.

[282] H. Kim, R. Eykholt, J. Salas, Nonlinear dynamics, delay times, and embedding windows, Physica D: Nonlinear Phenomena 127 (1-2) (1999) 48–60. doi:10.1016/s0167-2789(98)00240-1.

Index

Convolutional neural network, 140

Data-driven fault diagnosis, 6
Degradation point detection, 186, 212

Empirical mode decomposition, 183, 210
Event-based relay, 29
Event-triggered communication schemes, 13
Exponential mean-square stability, 19
Extended Kalman filtering, 185

Fading measurements, 30, 35, 52, 55
Fault Diagnosis, 1
Fault diagnosis and location, 151
Fault isolation, 95, 110
Filter-based Fault Diagnosis, 1
Finite-time filtering, 35

Gated recurrent unit network, 213
Genetic algorithm, 157, 201

Levy flight, 166
Long short-term memory network, 169

Measurement quantization, 14

Networked multi-rate systems, 18
Nonuniformly sampled multi-rate systems, 65

Particle filter, 165
Phase space reconstruction, 221

Randomly occurring faults, 52, 55
Recursive matrix inequalities, 89
Remaining Useful Life Prediction, 7

Sensor degradation, 103
Sensor failures, 14
Set-membership estimation, 104
Stochastically finite-time bounded, 35
Support vector regression, 157, 201
Synchrosqueezing transform, 139

Unscented Kalman filter, 155

Variance constraints, 21

Wavelet packet transform, 199